Charles D. Miller • Vern E. Heeren • John Hornsby
Christopher Heeren
Margaret L. Morrow
for the chapter on Graph Theory

Mathematical Ideas

Fourth Custom Edition for Luzerne County Community College

Taken from:
Mathematical Ideas, Thirteenth Edition
by Charles D. Miller, Vern E. Heeren, John Hornsby, and Christopher Heeren

Pearson Learning Solutions, 330 Hudson Street, New York, New York 10013
A Pearson Education Company
www.pearsoned.com

Printed in the United States of America

1 2 3 4 5 6 7 8 9 10 V092 17 16 15

000200010271974690

SL

PEARSON   ISBN 10: 1-323-13346-1
ISBN 13: 978-1-323-13346-0

CONTENTS

v

11 Probability 583

12 Statistics 643

NOTE: Trigonometry module and Metrics module available in MyMathLab or online at www.pearsonhighered.com/mathstatsresources.

PREFACE

After twelve editions and over four decades, *Mathematical Ideas* continues to be one of the most popular textbooks in liberal arts mathematics education. We are proud to present the thirteenth edition of a text that offers non-physical science students a practical coverage that connects mathematics to the world around them. It is a flexible text that has evolved alongside changing trends but remains steadfast to its original objectives.

Mathematical Ideas is written with a variety of students in mind. It is well suited for several courses, including those geared toward the aforementioned liberal arts audience and survey courses in mathematics or finite mathematics. Students taking these courses will pursue careers in nursing and healthcare, the construction trades, communications, hospitality, information technology, criminal justice, retail management and sales, computer programming, political science, school administration, and a myriad of other careers. Accordingly, we have chosen to increase our focus on showcasing how the math in this course will be relevant in this wide array of career options.

- Chapter openers now address how the chapter topics can be applied within the context of work and future careers.

- We made sure to retain the hundreds of examples and exercises from the previous edition that pertain to these interests.

- Every chapter also contains the brand new *When Will I Ever Use This?* features that help students connect mathematics to the workplace.

Interesting and mathematically pertinent movie and television applications and references are still interspersed throughout the chapters.

Ample topics are included for a two-term course, yet the variety of topics and flexibility of sequence makes the text suitable for shorter courses as well. Our main objectives continue to be comprehensive coverage, appropriate organization, clear exposition, an abundance of examples, and well-planned exercise sets with numerous applications.

New to This Edition

- New chapter openers connect the mathematics of the chapter to a particular career area, or in some cases, to an everyday life situation that will be important to people in virtually any career.

- *When Will I Ever Use This*? features in each chapter also connect chapter topics to career or workplace situations and answer that age-old question.

- Career applications have taken on greater prominence.

- Every section of every chapter now begins with a list of clear learning objectives for the student.

- An extensive summary at the end of each chapter includes the following components.

  - A list of *Key Terms* for each section of the chapter

  - *New Symbols,* with definitions, to clarify newly introduced symbols

  - *Test Your Word Power* questions that allow students to test their knowledge of new vocabulary

  - A *Quick Review* that gives a brief summary of concepts covered in the chapter, along with examples illustrating those concepts

- All exercise sets have once again been updated, with over 1000 new or modified exercises, many with a new emphasis on career applications.

- Since Intermediate Algebra is often a prerequisite for the liberal arts course, the algebra chapters have been streamlined to focus in on key concepts, many of which will aid in comprehension of other chapters' content.

- The presentation has been made more uniform whenever clarity for the reader could be served.

- The general style has been freshened, with more pedagogical use of color, new photos and art, and opening of the exposition.

- NEW! An Integrated Review MyMathLab course option provides embedded review of select developmental topics in a Ready to Go format with assignments pre-assigned. This course solution can be used in a co-requisite course model, or simply to help under-prepared students master prerequisite skills and concepts.

- Expanded online resources

  ○ **NEW! Interactive, conceptual videos** with assignable MML questions walk students through a concept and then ask them to answer a question within the video. If students answer correctly, the concept is summarized. If students select one of the two incorrect answers, the video continues focusing on why students probably selected that answer and works to correct that line of thinking and explain the concept. Then students get another chance to answer a question to prove mastery.

  ○ **NEW! Learning Catalytics**   This student engagement, assessment and classroom intelligence system gives instructors real-time feedback on student learning.

  ○ **NEW! "When Will I Ever Use This?" videos** bring the ideas in the feature to life in a fun, memorable way.

  ○ **NEW!** An **Integrated Review MyMathLab** course option provides an embedded review of selected developmental topics. Assignments are pre-assigned in this course, which includes a Skills Check quiz on skills that students will need in order to learn effectively at the chapter level. Students who demonstrate mastery can move on to the *Mathematical Ideas* content, while students who need additional review can polish up their skills by using the videos supplied and can benefit from the practice they gain from the Integrated Review Worksheets. This course solution can be used either in a co-requisite course model, or simply to help underprepared students master prerequisite skills and concepts.

  ○ The Trigonometry and Metrics content that was previously in the text is now found in the MyMathLab course, including the assignable MML questions.

  ○ Extensions previously in the text are now found in the MyMathLab course, along with any assignable MML questions.

Overview of Chapters

- **Chapter 1 (The Art of Problem Solving)** introduces the student to inductive reasoning, pattern recognition, and problem-solving techniques. We continue to provide exercises based on the monthly Calendar from *Mathematics Teacher* and have added new ones throughout this edition. The new chapter opener recounts the solving of the Rubik's cube by a college professor. The *When Will I Ever Use This?* feature (p. 31) shows how estimation techniques may be used by a group home employee charged with holiday grocery shopping.

- **Chapter 2 (The Basic Concepts of Set Theory)** includes updated examples and exercises on surveys. The chapter opener and the *When Will I Ever Use This?* feature (p. 73) address the future job outlook for the nursing profession and the allocation of work crews in the building trade, respectively.

- **Chapter 3 (Introduction to Logic)** introduces the fundamental concepts of inductive and deductive logic. The chapter opener connects logic with fantasy literature, and new exercises further illustrate this relationship. A new *For Further Thought* (p. 99) and new exercises address logic gates in computers. One *When Will I Ever Use This?* feature (p. 108) connects circuit logic to the design and installation of home monitoring systems. Another (p. 128) shows a pediatric nurse applying a logical flowchart and truth tables to a child's vaccination protocol.

- **Chapter 4 (Numeration Systems)** covers historical numeration systems, including Egyptian, Roman, Chinese, Babylonian, Mayan, Greek, and Hindu-Arabic systems. A connection between base conversions in positional numeration systems and computer network design is suggested in the new chapter opener and illustrated in the *When Will I Ever Use This?* feature (p. 168), a new example, and new exercises.

- **Chapter 5 (Number Theory)** presents an introduction to the prime and composite numbers, the Fibonacci sequence, and a cross section of related historical developments, including the fairly new topic of "prime number splicing." The largest currently known prime numbers of various categories are identified, and recent progress on Goldbach's conjecture and the twin prime conjecture are noted. The chapter opener and one *When Will I Ever Use This?* feature (p. 189) apply cryptography and modular arithmetic to criminal justice, relating to cyber security. Another *When Will I Ever Use This?* feature (p. 205) shows how a nurse may use the concept of least common denominator in determining proper drug dosage.

- **Chapter 6 (The Real Numbers and Their Representations)** introduces some of the basic concepts of real numbers, their various forms of representation, and operations of arithmetic with them. The chapter opener and *When Will I Ever Use This?* feature (p. 273) connect percents and basic algebraic procedures to pricing, markup and discount, student grading, and market share analysis, as needed by a retail manager, a teacher, a salesperson, a fashion merchandiser, and a business owner.

- **Chapter 7 (The Basic Concepts of Algebra)** can be used to present the basics of algebra (linear and quadratic equations, applications, exponents, polynomials, and factoring) to students for the first time, or as a review of previous courses. The chapter opener connects proportions to an automobile owner's determination of fuel mileage, and the *When Will I Ever Use This?* feature (p. 330) relates inequalities to a test-taker's computation of the score needed to maintain a certain grade point average.

- **Chapter 8 (Graphs, Functions, and Systems of Equations and Inequalities)** is the second of our two algebra chapters. It continues with graphs, equations, and applications of linear, quadratic, exponential, and logarithmic functions and models, along with systems of equations. The chapter opener shows how an automobile owner can use a linear graph to relate price per gallon, amount purchased, and total cost. The *When Will I Ever Use This?* feature (p. 416) connects logarithms with the interpretation of earthquake reporting in the news.

- **Chapter 9 (Geometry)** covers elementary plane geometry, transformational geometry, basic geometric constructions, non-Euclidean geometry, and chaos and fractals. Section 9.7 now includes projective geometry. At reviewer request, the discussion of networks (the Königsberg Bridge problem) has been moved to Chapter 14 (Graph Theory). The chapter opener and one *When Will I Ever Use This?* feature (p. 497) connect geometric volume formulas to a video game programmer's job of designing the visual field of a game screen. A second *When Will I Ever Use This?* feature (p. 470) relates right triangle geometry to a forester's determining of safe tree-felling parameters.

- **Chapter 10 (Counting Methods)** focuses on elementary counting techniques, in preparation for the probability chapter. The chapter opener relates how a restaurateur used counting methods to help design the sales counter signage in a new restaurant. The *When Will I Ever Use This?* feature (p. 534) describes an entrepreneur's use of probability and sports statistics in designing a game and in building a successful company based on it.

- **Chapter 11 (Probability)** covers the basics of probability, odds, and expected value. The chapter opener relates to the professions of weather forecaster, actuary, baseball manager, and corporate manager, applying probability, statistics, and expected value to interpreting forecasts, determining insurance rates, selecting optimum strategies, and making business decisions. One *When Will I Ever Use This?* feature (p. 586) shows how a tree diagram helps a decision maker provide equal chances of winning to three players in a game of chance. A second such feature (p. 606) shows how knowledge of probability can help a television game show contestant determine the best winning strategy.

- **Chapter 12 (Statistics)** is an introduction to statistics that focuses on the measures of central tendency, dispersion, and position and discusses the normal distribution and its applications. The chapter opener and two *When Will I Ever Use This?* features (pp. 656, 661) connect probability and graph construction and interpretation to how a psychological therapist may motivate and carry out treatment for alcohol and tobacco addiction.

- **Chapter 13 (Personal Financial Management)** provides the student with the basics of the mathematics of finance as applied to inflation, consumer debt, and house buying. We also include a section on investing, with emphasis on stocks, bonds, and mutual funds. Tables, examples, and exercises have been updated to reflect current interest rates and investment returns. New margin notes feature smart apps for financial calculations. Additions in response to reviewer requests include a *When Will I Ever Use This?* feature (p. 741) connecting several topics of the chapter to how a financial planner can provide comparisons between renting and buying a house, and exercises comparing different mortgage options. Another *When Will I Ever Use This?* feature (p. 732) explores the cost-effectiveness of solar energy, using chapter topics essential for a solar energy salesperson. The chapter opener connects the time value of money to how a financial planner can help clients make wise financial decisions.

- **Chapter 14 (Graph Theory)** covers the basic concepts of graph theory and its applications. The chapter opener shows how a writer can apply graph theory to the analysis of poetic rhyme. One *When Will I Ever Use This?* feature (p. 800) connects graph theory to how a postal or delivery service manager could determine the most efficient delivery routes. Another (p. 818) tells of a unique use by an entrepreneur who developed a business based on finding time-efficient ways to navigate theme parks.

- **Chapter 15 (Voting and Apportionment)** deals with issues in voting methods and apportionment of representation, topics that have become increasingly popular in liberal arts mathematics courses. The Adams method of apportionment, as well as the Huntington-Hill method (currently used in United States presidential elections) are now included in the main body of the text. To illustrate the important work of a political consultant, the chapter opener connects different methods of analyzing votes. One *When Will I Ever Use This?* feature (p. 859) relates voting methods to the functioning of governing boards. Another (p. 874) gives an example of how understanding apportionment methods can help in the work of a school administrator.

Course Outline Considerations

Chapters in the text are, in most cases, independent and may be covered in the order chosen by the instructor. The few exceptions are as follows:

- Chapter 6 contains some material dependent on the ideas found in Chapter 5.
- Chapter 6 should be covered before Chapter 7 if student background so dictates.
- Chapters 7 and 8 form an algebraic "package" and should be covered in sequential order.
- A thorough coverage of Chapter 11 depends on knowledge of Chapter 10 material, although probability can be covered without teaching extensive counting methods by avoiding the more difficult exercises.

Features of the Thirteenth Edition

NEW! Chapter Openers   In keeping with the career theme, chapter openers address a situation related to a particular career. All are new to this edition. Some openers include a problem that the reader is asked to solve. We hope that you find these chapter openers useful and practical.

ENHANCED! Varied Exercise Sets   We continue to present a variety of exercises that integrate drill, conceptual, and applied problems, and there are over 1000 new or modified exercises in this edition. The text contains a wealth of exercises to provide students with opportunities to practice, apply, connect, and extend the mathematical skills they are learning. We have updated the exercises that focus on real-life data and have retained their titles for easy identification. Several chapters are enriched with new applications, particularly Chapters 6, 7, 8, 11, 12, and 13. We continue to use graphs, tables, and charts when appropriate. Many of the graphs use a style similar to that seen by students in today's print and electronic media.

UPDATED! Emphasis on Real Data in the Form of Graphs, Charts, and Tables   We continue to use up-to-date information from magazines, newspapers, and the Internet to create real applications that are relevant and meaningful.

Problem-Solving Strategies   Special paragraphs labeled "Problem-Solving Strategy" relate the discussion of problem-solving strategies to techniques that have been presented earlier.

For Further Thought   These entries encourage students to share their reasoning processes among themselves to gain a deeper understanding of key mathematical concepts.

ENHANCED! Margin Notes   This popular feature is a hallmark of this text and has been retained and updated where appropriate. These notes are interspersed throughout the text and are drawn from various sources, such as lives of mathematicians, historical vignettes, anecdotes on mathematics textbooks of the past, newspaper and magazine articles, and current research in mathematics.

Optional Graphing Technology   We continue to provide sample graphing calculator screens to show how technology can be used to support results found analytically. It is not essential, however, that a student have a graphing calculator to study from this text. *The technology component is optional.*

NEW! Chapter Summaries   Extensive summaries at the end of each chapter include Key Terms, New Symbols with definitions, Test Your Word Power vocabulary checks, and a Quick Review that provides a brief summary of concepts (with examples) covered in the chapter.

Chapter Tests   Each chapter concludes with a chapter test so that students can check their mastery of the material.

Resources For Success

MyMathLab® Online Course (access code required)

MyMathLab delivers **proven results** in helping individual students succeed. It provides **engaging experiences** that personalize, stimulate, and measure learning for each student. And it comes from an **experienced partner** with educational expertise and an eye on the future. **MyMathLab** helps prepare students and gets them thinking more conceptually and visually through the following features:

Personalized Homework ▶

Attaching a personalized homework assignment to a quiz or test enables students to focus on the topics they did not master and those with which they need additional practice, providing an individualized experience.

Homework   modify

Current Score   78.57% (11 points out of 14)
This homework will **not** affect your Study Plan score.
Number of times you can complete each question: unlimited

⚠ Changes made here WILL affect your score. Go to Results to practice without affecting your score.

➤ You received automatic credit (11 pts) for topics you mastered on sample pretest.
  You only need to work on questions that are links below.

[Show All] [Show What I Need to Do]

Questions: 14	Scored: 11	Correct: 11	Partial Credit: 0	Incorrect: 0
✓ Question 1 (1/1)	✓ Question 2 (1/1)	✓ Question 3 (1/1)		
✓ Question 4 (1/1)	Question 5 (0/1)	Question 6 (0/1)		
Question 7 (0/1)	✓ Question 8 (1/1)	✓ Question 9 (1/1)		
✓ Question 10 (1/1)	✓ Question 11 (1/1)	✓ Question 12 (1/1)		
✓ Question 13 (1/1)	✓ Question 14 (1/1)			

Study Plan

Study Plan

You have earned 1 of 592 mastery points (MP).
Practice these objectives and then take a Quiz Me to prove mastery and earn more points.

What to work on next

1.1 Solving Problems by Inductive Reasoning

📌 Identify reasoning as deductive or inductive.   [Practice] [Quiz Me]   0 of 1 MP

➊ 📌 More Objectives to practice and master   View all chapters

6.1 Real Numbers, Order, and Absolute Value

📌 Understand number classifications.   [Practice] [Quiz Me]   0 of 1 MP

📌 Use integers to represent numbers from real life.   [Practice] [Quiz Me]   0 of 1 MP

📌 Graph numbers on a number line.   [Practice] [Quiz Me]   0 of 1 MP

1.1 Solving Problems by Inductive Reasoning

📌 Use inductive reasoning to predict the next term.   [Practice] [Quiz Me]   0 of 1 MP

◀ Adaptive Study Plan

The Study Plan makes studying more efficient and effective for every student. Performance and activity are assessed continually in real time. Then data and analytics are used to provide personalized content-reinforcing concepts that target each student's strengths and weaknesses.

PEARSON   ALWAYS LEARNING

YOUR TURN

Find the *y*-intercept:

a) $\left(0, -\frac{13}{2}\right)$

b) $(0, 13)$

c) $(0, 2)$

1. Finding the y-intercept
 ○ a)
 ○ b)
 ○ c) $2x - 13$
                              [Close]

◀ Interactive Concept Videos

NEW! Conceptual videos require students' input as they walk students through a concept and pause to ask them questions. Correct answers are followed by reinforcement of the concept. Incorrect answers are followed by a video that explains the correct answer, as well as addressing the misconception that may have led to that particular mistake. Instructors can also assign trackable exercises that correspond to the videos to check student understanding.

Learning Catalytics ▶

NEW! Integrated into **MyMathLab**, the Learning Catalytics feature uses students' devices in the classroom for an engagement, assessment, and classroom intelligence system that gives instructors real-time feedback on student learning.

●●●○○ AT&T ⬙   10:19 AM   ◁ ⚡ 72% ▮
              ⬙ learningcatalytics.com
[Refresh]   Session 3339:   [Logout]

sketch question

This is a graph of $f(x) = \ln x$. Sketch a graph of the derivative $f'(x)$.

Your response:

Skills for Success Modules are integrated within the **MyMathLab** course to help students succeed in college courses and prepare for future professions.

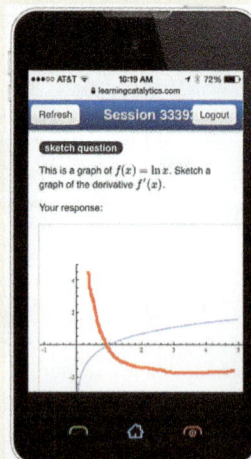

Instructor Resources

Additional resources can be downloaded from www.pearsonhighered.com or hardcopy resources can be ordered from your sales representative.

Integrated Review Ready to Go MyMathLab® Course

The Ready to Go **MyMathLab** course option makes it even easier to get started, including author-chosen preassigned homework, integrated review of prerequisite topics, and more.

TestGen®

TestGen® (www.pearsoned.com/testgen) enables instructors to build, edit, print, and administer tests using a computerized bank of questions developed to cover all the objectives of the text.

PowerPoint® Lecture Slides

Fully editable slides correlated with the textbook are available.

Annotated Instructor's Edition

When possible, answers are on the page with the exercises. Longer answers are in the back of the book.

Instructor's Resource and Solutions Manual

This manual includes fully worked solutions to all text exercises, as well as the Collaborative Investigations that were formerly in the text.

Instructor's Testing Manual

This manual includes tests with answer keys for each chapter of the text.

Student Resources

Additional resources are available to support student success.

Updated Video Program

Available in **MyMathLab,** video lectures cover every section in the text and have been updated for this edition where necessary. New interactive concept videos and new *"When Will I Ever Use This?"* videos complete the video package, reinforcing students' conceptual understanding, while also engaging them with the math in context.

Student Solutions Manual

This manual provides detailed worked-out solutions to odd-numbered exercises.

Integrated Review Worksheets

Intended to be used with the Integrated Review **MyMathLab** course, these worksheets give students an opportunity to review and practice prerequisite topics from developmental math that are needed for each chapter in *Mathematical Ideas.*

ACKNOWLEDGMENTS

We wish to thank the following reviewers for their helpful comments and suggestions for this and previous editions of the text. (Reviewers of the thirteenth edition are noted with an asterisk.)

H. Achepohl, *College of DuPage*

Shahrokh Ahmadi, *Northern Virginia Community College*

Richard Andrews, *Florida A&M University*

Cindy Anfinson, *Palomar College*

Erika Asano, *University of South Florida, St. Petersburg*

Elaine Barber, *Germanna Community College*

Anna Baumgartner, *Carthage College*

James E. Beamer, *Northeastern State University*

*Brad Beauchamp, *Vernon College*

Elliot Benjamin, *Unity College*

Jaime Bestard, *Barry University*

Joyce Blair, *Belmont University*

Gus Brar, *Delaware County Community College*

Roger L. Brown, *Davenport College*

Douglas Burke, *Malcolm X College*

John Busovicki, *Indiana University of Pennsylvania*

Ann Cascarelle, *St. Petersburg Junior College*

Kenneth Chapman, *St. Petersburg Junior College*

Gordon M. Clarke, *University of the Incarnate Word*

M. Marsha Cupitt, *Durham Technical Community College*

James Curry, *American River College*

Rosemary Danaher, *Sacred Heart University*

Ken Davis, *Mesa State College*

Nancy Davis, *Brunswick Community College*

George DeRise, *Thomas Nelson Community College*

Catherine Dermott, *Hudson Valley Community College*

Greg Dietrich, *Florida Community College at Jacksonville*

Vincent Dimiceli, *Oral Roberts University*

*Qiang Dotzel, *University of Missouri, St. Louis*

Diana C. Dwan, *Yavapai College*

Laura Dyer, *Belleville Area College*

Jan Eardley, *Barat College*

Joe Eitel, *Folsom College*

Azin Enshai, *American River College*

Gayle Farmer, *Northeastern State University*

Michael Farndale, *Waldorf College*

Gordon Feathers, *Passaic County Community College*

Thomas Flohr, *New River Community College*

Bill Fulton, *Black Hawk College—East*

Anne Gardner, *Wenatchee Valley College*

Justin M. Gash, *Franklin College*

Donald Goral, *Northern Virginia Community College*

Glen Granzow, *Idaho State University*

Larry Green, *Lake Tahoe Community College*

Arthur D. Grissinger, *Lock Haven University*

Don Hancock, *Pepperdine University*

Denis Hanson, *University of Regina*

Marilyn Hasty, *Southern Illinois University*

Shelby L. Hawthorne, *Thomas Nelson Community College*

Jeff Heiking, *St. Petersburg Junior College*

Laura Hillerbrand, *Broward Community College*

Corinne Irwin, *University of Texas at Austin*

*Neha Jain, *Northern Virginia Community College*

Jacqueline Jensen, *Sam Houston State University*

Emanuel Jinich, *Endicott College*

Frank Juric, *Brevard Community College–Palm Bay*

Karla Karstens, *University of Vermont*

Najam Khaja, *Centennial College*

Hilary Kight, *Wesleyan College*

Barbara J. Kniepkamp, *Southern Illinois University at Edwardsville*

Suda Kunyosying, *Shepherd College*

Yu-Ju Kuo, *Indiana University of Pennsylvania*

Stephane Lafortune, *College of Charleston*

Pam Lamb, *J. Sargeant Reynolds Community College*

John Lattanzio, *Indiana University of Pennsylvania*

John W. Legge, *Pikeville College*

Dawn Locklear, *Crown College*

Bin Lu, *California State University, Sacramento*

Leo Lusk, *Gulf Coast Community College*

Sherrie Lutsch, *Northwest Indian College*

Rhonda Macleod, *Florida State University*

Andrew Markoe, *Rider University*

Darlene Marnich, *Point Park College*

Victoria Martinez, *Okaloosa Walton Community College*

Chris Mason, *Community College of Vermont*

Mark Maxwell, *Maryville University*

Carol McCarron, *Harrisburg Area Community College*

Delois McCormick, *Germanna Community College*

Daisy McCoy, *Lyndon State College*

Cynthia McGinnis, *Okaloosa Walton Community College*

Vena McGrath, *Davenport College*

Robert Moyer, *Fort Valley State University*

Shai Neumann, *Brevard Community College*

Barbara Nienstedt, *Gloucester County College*

Chaitanya Nigam, *Gateway Community-Technical College*

Vladimir Nikiforov, *University of Memphis*

Vicky Ohlson, *Trenholm State Technical College*

Jean Okumura, *Windward Community College*

Stan Perrine, *Charleston Southern University*

*Mary Anne Petruska, *Pensacola State College*
 Bob Phillips, *Mesabi Range Community College*
 Kathy Pinchback, *University of Memphis*
*Fatima Prioleau, *Borough of Manhattan Community College*
 Priscilla Putman, *New Jersey City University*
 Scott C. Radtke, *Davenport College*
 Doraiswamy Ramachandran, *California State University, Sacramento*
 John Reily, *Montclair State University*
 Beth Reynolds, *Mater Dei College*
 Shirley I. Robertson, *High Point University*
 Andrew M. Rockett, *CW Post Campus of Long Island University*
 Kathleen Rodak, *St. Mary's College of Ave Maria University*
 Cynthia Roemer, *Union County College*
 Lisa Rombes, *Washtenaw Community College*
 Abby Roscum, *Marshalltown Community College*
*Catherine A. Sausville, *George Mason University*
 D. Schraeder, *McLennan Community College*
 Wilfred Schulte, *Cosumnes River College*
 Melinda Schulteis, *Concordia University*
 Gary D. Shaffer, *Allegany College of Maryland*
 Doug Shaw, *University of North Iowa*

Jane Sinibaldi, *York College of Pennsylvania*
Nancy Skocik, *California University of Pennsylvania*
Larry Smith, *Peninsula College*
Marguerite Smith, *Merced College*
Charlene D. Snow, *Lower Columbia College*
H. Jeannette Stephens, *Whatcom Community College*
Suzanne J. Stock, *Oakton Community College*
Dawn M. Strickland, *Winthrop University*
Dian Thom, *McKendree College*
Claude C. Thompson, *Hollins University*
Mark Tom, *College of the Sequoias*
Ida Umphers, *University of Arkansas at Little Rock*
Karen Villarreal, *University of New Orleans*
Dr. Karen Walters, *Northern Virginia Community College*
Wayne Wanamaker, *Central Florida Community College*
David Wasilewski, *Luzerne County Community College*
William Watkins, *California State University, Northridge*
Alice Williamson, *Sussex County Community College*
Susan Williford, *Columbia State Community College*
Tom Witten, *Southwest Virginia Community College*
Fred Worth, *Henderson State University*
Rob Wylie, *Carl Albert State College*
Henry Wyzinski, *Indiana University Northwest*

A project of this magnitude cannot be accomplished without the help of many other dedicated individuals. Marnie Greenhut served as acquisitions editor for this edition. Carol Merrigan provided excellent production supervision. Anne Kelly, Greg Tobin, Sherry Berg, Alicia Frankel, and Patty Bergin of Pearson gave us their unwavering support.

Terry McGinnis gave her usual excellent behind-the-scenes guidance. Thanks go to Dr. Margaret L. Morrow of Plattsburgh State University and Dr. Jill Van Newenhizen of Lake Forest College, who wrote the material on graph theory and voting/apportionment, respectively. Paul Lorczak, Renato Mirollo, Beverly Fusfield, Jack Hornsby, and Perian Herring did an outstanding job of accuracy- and answer-checking. And finally, we thank our loyal users over these many editions for making this book one of the most successful in its market.

Vern E. Heeren
John Hornsby
Christopher Heeren

ABOUT THE AUTHORS

Vern Heeren grew up in the Sacramento Valley of California. After earning a Bachelor of Arts degree in mathematics, with a minor in physics, at Occidental College, and completing his Master of Arts degree in mathematics at the University of California, Davis, he began a 38-year teaching career at American River College, teaching math and a little physics. He coauthored *Mathematical Ideas* in 1968 with office mate Charles Miller, and he has enjoyed researching and revising it over the years. It has been a joy for him to complete the thirteenth edition, along with long-time coauthor John Hornsby, and now also with son Christopher.

These days, besides pursuing his mathematical interests, Vern enjoys spending time with his wife Carole and their family, exploring the wonders of nature near their home in central Oregon.

John Hornsby joined the author team of Margaret Lial, Charles Miller, and Vern Heeren in 1988. In 1990, the sixth edition of *Mathematical Ideas* became the first of nearly 150 titles he has coauthored for Scott Foresman, HarperCollins, Addison-Wesley, and Pearson in the years that have followed. His books cover the areas of developmental and college algebra, precalculus, trigonometry, and mathematics for the liberal arts. He is a native and resident of New Roads, Louisiana.

Christopher Heeren is a native of Sacramento, California. While studying engineering in college, he had an opportunity to teach a math class at a local high school, and this sparked both a passion for teaching and a change of major. He received a Bachelor of Arts degree and a Master of Arts degree, both in mathematics, from California State University, Sacramento. Chris has taught mathematics at the middle school, high school, and college levels, and he currently teaches at American River College in Sacramento. He has a continuing interest in using technology to bring mathematics to life. When not writing, teaching, or preparing to teach, Chris enjoys spending time with his lovely wife Heather and their three children (and two dogs and a guinea pig).

The Art of Problem Solving

1

Professor Terry Krieger, of Rochester (Minnesota) Community College, shares his thoughts about why he decided to become a mathematics teacher. He is an expert at the Rubik's Cube. Here, he explains how he mastered this classic problem.

*From a very young age I always enjoyed solving problems, especially problems involving numbers and patterns. There is something inherently beautiful in the process of discovering mathematical truth. Mathematics may be the only discipline in which different people, using wildly varied but logically sound methods, will arrive at the **same** correct result—not just once, but every time! It is this aspect of mathematics that led me to my career as an educator. As a mathematics instructor, I get to be part of, and sometimes guide, the discovery process.*

I received a Rubik's Cube as a gift my junior year of high school. I was fascinated by it. I devoted the better part of three months to solving it for the first time, sometimes working 3 or 4 hours per day on it.

There was a lot of trial and error involved. I devised a process that allowed me to move only a small number of pieces at a time while keeping other pieces in their places. Most of my moves affect only three or four of the 26 unique pieces of the puzzle. What sets my solution apart from those found in many books is that I hold the cube in a consistent position and work from the top to the bottom. Most book solutions work upward from the bottom.

My first breakthrough came when I realized that getting a single color on one face of the cube was not helpful if the colors along the edges of that face were placed improperly. In other words, it does no good to make the top of the cube all white if one of the edges along the white top shows green, yellow, and blue. It needs to be all green, for example.

I worked on the solution so much that I started seeing cube moves in my sleep. In fact, I figured out the moves for one of my most frustrating sticking points while sleeping. I just woke up knowing how to do it.

The eight corners of the cube represented a particularly difficult challenge for me. Finding a consistent method for placing the corners appropriately took many, many hours. To this day, the amount of time that it takes for me to solve a scrambled cube depends largely on the amount of time that it takes for me to place the corners.

When I first honed my technique, I was able to consistently solve the cube in 2 to 3 minutes. My average time is now about 65 seconds. My fastest time is 42 seconds.

Since figuring out how to solve the cube, I have experimented with other possible color patterns that can be formed. The most complicated one I have created leaves the cube with three different color stripes on all six faces. I have never met another person who can accomplish this arrangement.

1.1   SOLVING PROBLEMS BY INDUCTIVE REASONING

OBJECTIVES

1 Be able to distinguish between inductive and deductive reasoning.

2 Understand that in some cases, inductive reasoning may not lead to valid conclusions.

Characteristics of Inductive and Deductive Reasoning

The development of mathematics can be traced to the Egyptian and Babylonian cultures (3000 B.C.–A.D. 260) as a necessity for counting and problem solving. To solve a problem, a cookbook-like recipe was given, and it was followed repeatedly to solve similar problems. By observing that a specific method worked for a certain type of problem, the Babylonians and the Egyptians concluded that the same method would work for any similar type of problem. Such a conclusion is called a *conjecture*. A **conjecture** is an educated guess based on repeated observations of a particular process or pattern.

The method of reasoning just described is called *inductive reasoning*.

INDUCTIVE REASONING

Inductive reasoning is characterized by drawing a general conclusion (making a conjecture) from repeated observations of specific examples. The conjecture may or may not be true.

In testing a conjecture obtained by inductive reasoning, it takes only one example that does not work to prove the conjecture false. Such an example is called a **counterexample.**

Inductive reasoning provides a powerful method of drawing conclusions, but there is no assurance that the observed conjecture will always be true. For this reason, mathematicians are reluctant to accept a conjecture as an absolute truth until it is formally proved using methods of *deductive reasoning.* Deductive reasoning characterized the development and approach of Greek mathematics, as seen in the works of Euclid, Pythagoras, Archimedes, and others. During the classical Greek period (600 B.C.–A.D. 450), general concepts were applied to specific problems, resulting in a structured, logical development of mathematics.

DEDUCTIVE REASONING

Deductive reasoning is characterized by applying general principles to specific examples.

We now look at examples of these two types of reasoning. In this chapter, we often refer to the **natural,** or **counting, numbers.**

$$1, 2, 3, \ldots \quad \text{Natural (counting) numbers}$$

Ellipsis points

The three dots (*ellipsis points*) indicate that the numbers continue indefinitely in the pattern that has been established. The most probable rule for continuing this pattern is "Add 1 to the previous number," and this is indeed the rule that we follow.

Now consider the following list of natural numbers:

$$2, 9, 16, 23, 30.$$

What is the next number of this list? What is the pattern? After studying the numbers, we might see that $2 + 7 = 9$, and $9 + 7 = 16$. Do we add 16 and 7 to get 23? Do we add 23 and 7 to get 30? Yes. It seems that any number in the given list can be found by adding 7 to the preceding number, so the next number in the list would be $30 + 7 = 37$.

We set out to find the "next number" by reasoning from observation of the numbers in the list. We may have jumped from these observations to the general statement that any number in the list is 7 more than the preceding number. This is an example of inductive reasoning.

By using inductive reasoning, we concluded that 37 was the next number. Suppose the person making up the list has another answer in mind. The list of numbers

$$2, 9, 16, 23, 30$$

actually gives the dates of Mondays in June if June 1 falls on a Sunday. The next Monday after June 30 is July 7. With this pattern, the list continues as

$$2, 9, 16, 23, 30, 7, 14, 21, 28, \ldots.$$

See the calendar in **Figure 1.** The correct answer would then be 7. The process used to obtain the rule "add 7" in the preceding list reveals a main flaw of inductive reasoning. *We can never be sure that what is true in a specific case will be true in general. Inductive reasoning does not guarantee a true result, but it does provide a means of making a conjecture.*

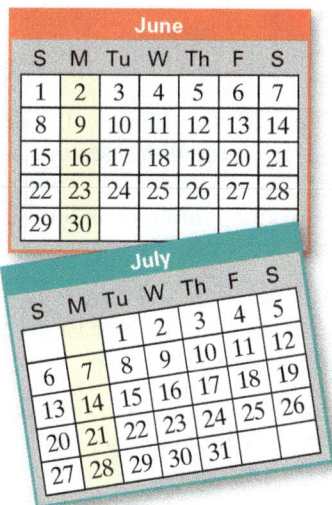

Figure 1

We now review some basic notation. Throughout this book, we use *exponents* to represent repeated multiplication.

Base → $4^3 = 4 \cdot 4 \cdot 4 = 64$ 4 is used as a factor 3 times.

↑
Exponent

EXPONENTIAL EXPRESSION

If a is a number and n is a counting number $(1, 2, 3, \ldots)$, then the exponential expression a^n is defined as follows.

$$a^n = \underbrace{a \cdot a \cdot a \cdot \ldots \cdot a}_{n \text{ factors of } a}$$

The number a is the **base** and n is the **exponent.**

With deductive reasoning, we use general statements and apply them to specific situations. For example, a basic rule for converting feet to inches is to multiply the number of feet by 12 in order to obtain the equivalent number of inches. This can be expressed as a formula.

Number of inches $= 12 \times$ number of feet

This general rule can be applied to any specific case. For example, the number of inches in 3 feet is $12 \times 3 = 36$ inches.

Reasoning through a problem usually requires certain *premises*. A **premise** can be an assumption, law, rule, widely held idea, or observation. Then reason inductively or deductively from the premises to obtain a **conclusion.** The premises and conclusion make up a **logical argument.**

EXAMPLE 1 Identifying Premises and Conclusions

Identify each premise and the conclusion in each of the following arguments. Then tell whether each argument is an example of inductive or deductive reasoning.

(a) Our house is made of brick. Both of my next-door neighbors have brick houses. Therefore, all houses in our neighborhood are made of brick.

(b) All keyboards have the symbol @. I have a keyboard. My keyboard has the symbol @.

(c) Today is Tuesday. Tomorrow will be Wednesday.

Solution

(a) The premises are "Our house is made of brick" and "Both of my next-door neighbors have brick houses." The conclusion is "Therefore, all houses in our neighborhood are made of brick." Because the reasoning goes from specific examples to a general statement, the argument is an example of inductive reasoning (although it may very well be faulty).

(b) Here the premises are "All keyboards have the symbol @" and "I have a keyboard." The conclusion is "My keyboard has the symbol @." This reasoning goes from general to specific, so deductive reasoning was used.

(c) There is only one premise here, "Today is Tuesday." The conclusion is "Tomorrow will be Wednesday." The fact that Wednesday immediately follows Tuesday is being used, even though this fact is not explicitly stated. Because the conclusion comes from general facts that apply to this special case, deductive reasoning was used. ∎

While inductive reasoning may, at times, lead to false conclusions, in many cases it does provide correct results if we look for the most *probable* answer.

The Fibonacci Sequence

1, 1, 2, 3, 5, 8, 13, 21, ...

In the 2003 movie *A Wrinkle in Time*, young Charles Wallace, played by David Dorfman, is challenged to identify a particular sequence of numbers. He correctly identifies it as the **Fibonacci sequence.**

EXAMPLE 2 Predicting the Next Number in a Sequence

Use inductive reasoning to determine the *probable* next number in each list below.

(a) 5, 9, 13, 17, 21, 25, 29   **(b)** 1, 1, 2, 3, 5, 8, 13, 21   **(c)** 2, 4, 8, 16, 32

Solution

(a) Each number in the list is obtained by adding 4 to the previous number. The probable next number is $29 + 4 = 33$. (This is an example of an *arithmetic sequence*.)

(b) Beginning with the third number in the list, 2, each number is obtained by adding the two previous numbers in the list. That is,

$$1 + 1 = 2, \quad 1 + 2 = 3, \quad 2 + 3 = 5,$$

and so on. The probable next number in the list is $13 + 21 = 34$. (These are the first few terms of the *Fibonacci sequence*.)

(c) It appears here that to obtain each number after the first, we must double the previous number. Therefore, the probable next number is $32 \times 2 = 64$. (This is an example of a *geometric sequence*.) ■

EXAMPLE 3 Predicting the Product of Two Numbers

Consider the list of equations. Predict the next multiplication fact in the list.

$$37 \times \ 3 = 111$$
$$37 \times \ 6 = 222$$
$$37 \times \ 9 = 333$$
$$37 \times 12 = 444$$

Solution

The left side of each equation has two factors, the first 37 and the second a multiple of 3, beginning with 3. Each product (answer) consists of three digits, all the same, beginning with 111 for 37×3. Thus, the next multiplication fact would be

$$37 \times 15 = 555, \quad \text{which is indeed true.} \quad ■$$

Pitfalls of Inductive Reasoning

There are pitfalls associated with inductive reasoning. A classic example involves the maximum number of regions formed when chords are constructed in a circle. When two points on a circle are joined with a line segment, a *chord* is formed.

Locate a single point on a circle. Because no chords are formed, a single interior region is formed. See **Figure 2(a)** on the next page. Locate two points and draw a chord. Two interior regions are formed, as shown in **Figure 2(b).** Continue this pattern. Locate three points, and draw all possible chords. Four interior regions are formed, as shown in **Figure 2(c).** Four points yield 8 regions and five points yield 16 regions. See **Figures 2(d) and 2(e).**

The results of the preceding observations are summarized in **Table 1.** The pattern formed in the column headed "Number of Regions" is the same one we saw in **Example 2(c),** where we predicted that the next number would be 64. It seems here that for each additional point on the circle, the number of regions doubles.

Table 1

Number of Points	Number of Regions
1	1
2	2
3	4
4	8
5	16

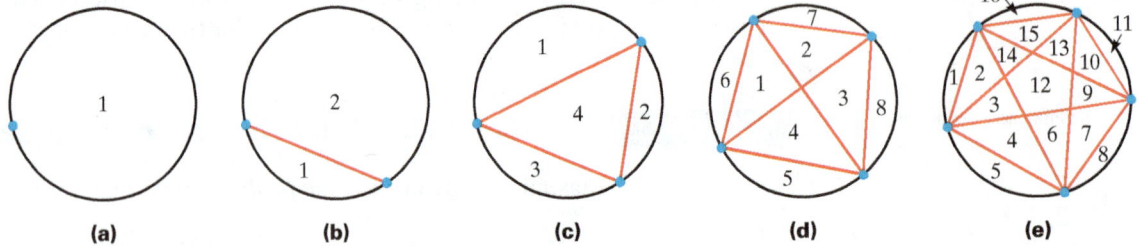

Figure 2

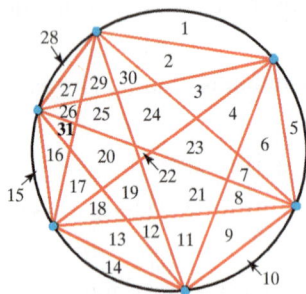

Figure 3

A reasonable inductive conjecture would be that for six points, 32 regions would be formed. But as **Figure 3** indicates, there are *only 31 regions*. The pattern of doubling ends when the sixth point is considered. Adding a seventh point would yield 57 regions. The numbers obtained here are

$$1, 2, 4, 8, 16, 31, 57.$$

For *n* points on the circle, the number of regions is given by the formula

$$\frac{n^4 - 6n^3 + 23n^2 - 18n + 24}{24}. \text{ *}$$

1.1 EXERCISES

In Exercises 1–16, determine whether the reasoning is an example of deductive or inductive reasoning.

1. The next number in the pattern 2, 4, 6, 8, 10 is 12.

2. My dog barked and woke me up at 1:02 a.m., 2:03 a.m., and 3:04 a.m. So he will bark again and wake me up at 4:05 a.m.

3. To find the perimeter *P* of a square with side of length *s*, I can use the formula $P = 4s$. So the perimeter of a square with side of length 7 inches is $4 \times 7 = 28$ inches.

4. A company charges a 10% re-stocking fee for returning an item. So when I return a radio that cost $150, I will only get $135 back.

5. If the mechanic says that it will take seven days to repair your SUV, then it will actually take ten days. The mechanic says, "I figure it'll take exactly one week to fix it, ma'am." Then you can expect it to be ready ten days from now.

6. If you take your medicine, you'll feel a lot better. You take your medicine. Therefore, you'll feel a lot better.

7. It has rained every day for the past six days, and it is raining today as well. So it will also rain tomorrow.

8. Carrie's first five children were boys. If she has another baby, it will be a boy.

9. The 2000 movie *Cast Away* stars Tom Hanks as the only human survivor of a plane crash, stranded on a tropical island. He approximates his distance from where the plane lost radio contact to be 400 miles (a radius), and uses the formula for the area of a circle,

$$\text{Area} = \pi \, (\text{radius})^2$$

to determine that a search party would have to cover an area of over 500,000 square miles to look for him and his "pal" Wilson.

*For more information on this and other similar patterns, see "Counting Pizza Pieces and Other Combinatorial Problems," by Eugene Maier, in the January 1988 issue of *Mathematics Teacher*, pp. 22–26.

10. If the same number is subtracted from both sides of a true equation, the new equation is also true. I know that $9 + 18 = 27$. Therefore, $(9 + 18) - 13 = 27 - 13$.

11. If you build it, they will come. You build it. Therefore, they will come.

12. All men are mortal. Socrates is a man. Therefore, Socrates is mortal.

13. It is a fact that every student who ever attended Delgado University was accepted into graduate school. Because I am attending Delgado, I can expect to be accepted to graduate school, too.

14. For the past 126 years, a rare plant has bloomed in Columbia each summer, alternating between yellow and green flowers. Last summer, it bloomed with green flowers, so this summer it will bloom with yellow flowers.

15. In the sequence 5, 10, 15, 20, 25, ..., the most probable next number is 30.

16. (This anecdote is adapted from a story by Howard Eves in *In Mathematical Circles*.) A scientist had a group of 100 fleas, and one by one he would tell each flea "Jump," and the flea would jump. Then with the same fleas, he yanked off their hind legs and repeated "Jump," but the fleas would not jump. He concluded that when a flea has its hind legs yanked off, it cannot hear.

17. Discuss the differences between inductive and deductive reasoning. Give an example of each.

18. Give an example of faulty inductive reasoning.

Determine the most probable next term in each of the following lists of numbers.

19. 6, 9, 12, 15, 18

20. 13, 18, 23, 28, 33

21. 3, 12, 48, 192, 768

22. 32, 16, 8, 4, 2

23. 3, 6, 9, 15, 24, 39

24. $\dfrac{1}{3}, \dfrac{3}{5}, \dfrac{5}{7}, \dfrac{7}{9}, \dfrac{9}{11}$

25. $\dfrac{1}{2}, \dfrac{3}{4}, \dfrac{5}{6}, \dfrac{7}{8}, \dfrac{9}{10}$

26. 1, 4, 9, 16, 25

27. 1, 8, 27, 64, 125

28. 2, 6, 12, 20, 30, 42

29. 4, 7, 12, 19, 28, 39

30. 27, 21, 16, 12, 9

31. 5, 3, 5, 5, 3, 5, 5, 5, 3, 5, 5, 5, 5, 3, 5, 5, 5, 5

32. 8, 2, 8, 2, 2, 8, 2, 2, 2, 8, 2, 2, 2, 2, 8, 2, 2, 2, 2

33. Construct a list of numbers similar to those in **Exercise 19** such that the most probable next number in the list is 60.

34. Construct a list of numbers similar to those in **Exercise 30** such that the most probable next number in the list is 8.

Use the list of equations and inductive reasoning to predict the next equation, and then verify your conjecture.

35.
$$(9 \times 9) + 7 = 88$$
$$(98 \times 9) + 6 = 888$$
$$(987 \times 9) + 5 = 8888$$
$$(9876 \times 9) + 4 = 88{,}888$$

36.
$$(1 \times 9) + 2 = 11$$
$$(12 \times 9) + 3 = 111$$
$$(123 \times 9) + 4 = 1111$$
$$(1234 \times 9) + 5 = 11{,}111$$

37.
$$3367 \times 3 = 10{,}101$$
$$3367 \times 6 = 20{,}202$$
$$3367 \times 9 = 30{,}303$$
$$3367 \times 12 = 40{,}404$$

38.
$$15873 \times 7 = 111{,}111$$
$$15873 \times 14 = 222{,}222$$
$$15873 \times 21 = 333{,}333$$
$$15873 \times 28 = 444{,}444$$

39.
$$34 \times 34 = 1156$$
$$334 \times 334 = 111{,}556$$
$$3334 \times 3334 = 11{,}115{,}556$$

40.
$$11 \times 11 = 121$$
$$111 \times 111 = 12{,}321$$
$$1111 \times 1111 = 1{,}234{,}321$$

41.
$$3 = \frac{3(2)}{2}$$
$$3 + 6 = \frac{6(3)}{2}$$
$$3 + 6 + 9 = \frac{9(4)}{2}$$
$$3 + 6 + 9 + 12 = \frac{12(5)}{2}$$

42.
$$2 = 4 - 2$$
$$2 + 4 = 8 - 2$$
$$2 + 4 + 8 = 16 - 2$$
$$2 + 4 + 8 + 16 = 32 - 2$$

43.
$$5(6) = 6(6 - 1)$$
$$5(6) + 5(36) = 6(36 - 1)$$
$$5(6) + 5(36) + 5(216) = 6(216 - 1)$$
$$5(6) + 5(36) + 5(216) + 5(1296) = 6(1296 - 1)$$

44.
$$3 = \frac{3(3-1)}{2}$$

$$3 + 9 = \frac{3(9-1)}{2}$$

$$3 + 9 + 27 = \frac{3(27-1)}{2}$$

$$3 + 9 + 27 + 81 = \frac{3(81-1)}{2}$$

45.
$$\frac{1}{2} = 1 - \frac{1}{2}$$

$$\frac{1}{2} + \frac{1}{4} = 1 - \frac{1}{4}$$

$$\frac{1}{2} + \frac{1}{4} + \frac{1}{8} = 1 - \frac{1}{8}$$

$$\frac{1}{2} + \frac{1}{4} + \frac{1}{8} + \frac{1}{16} = 1 - \frac{1}{16}$$

46.
$$\frac{1}{1 \cdot 2} = \frac{1}{2}$$

$$\frac{1}{1 \cdot 2} + \frac{1}{2 \cdot 3} = \frac{2}{3}$$

$$\frac{1}{1 \cdot 2} + \frac{1}{2 \cdot 3} + \frac{1}{3 \cdot 4} = \frac{3}{4}$$

$$\frac{1}{1 \cdot 2} + \frac{1}{2 \cdot 3} + \frac{1}{3 \cdot 4} + \frac{1}{4 \cdot 5} = \frac{4}{5}$$

Legend has it that the great mathematician Carl Friedrich Gauss (1777–1855) at a very young age was told by his teacher to find the sum of the first 100 *counting numbers. While his classmates toiled at the problem, Carl simply wrote down a single number and handed the correct answer in to his teacher. The young Carl explained that he observed that there were* 50 *pairs of numbers that each added up to* 101. *(See below.) So the sum of all the numbers must be* 50 × 101 = 5050.

$$1 + 2 + 3 + \cdots + 98 + 99 + 100$$

50 sums of 101 = 50 × 101 = 5050

Use the method of Gauss to find each sum.

47. $1 + 2 + 3 + \cdots + 200$ **48.** $1 + 2 + 3 + \cdots + 400$

49. $1 + 2 + 3 + \cdots + 800$ **50.** $1 + 2 + 3 + \cdots + 2000$

51. Modify the procedure of Gauss to find the sum $1 + 2 + 3 + \cdots + 175$.

52. Explain in your own words how the procedure of Gauss can be modified to find the sum $1 + 2 + 3 + \cdots + n$, where n is an odd natural number. (When an odd natural number is divided by 2, it leaves a remainder of 1.)

53. Modify the procedure of Gauss to find the sum $2 + 4 + 6 + \cdots + 100$.

54. Use the result of **Exercise 53** to find the sum $4 + 8 + 12 + \cdots + 200$.

55. What is the most probable next number in this list?

$$12, 1, 1, 1, 2, 1, 3$$

(*Hint:* Think about a clock with chimes.)

56. What is the next term in this list?

$$O, T, T, F, F, S, S, E, N, T$$

(*Hint:* Think about words and their relationship to numbers.)

57. Choose any three-digit number with all different digits, and follow these steps.

(a) Reverse the digits, and subtract the smaller from the larger. Record your result.

(b) Choose another three-digit number and repeat this process. Do this as many times as it takes for you to see a pattern in the different results you obtain. (*Hint:* What is the middle digit? What is the sum of the first and third digits?)

(c) Write an explanation of this pattern.

58. Choose any number, and follow these steps.

(a) Multiply by 2. (b) Add 6.

(c) Divide by 2. (d) Subtract the number you started with.

(e) Record your result.

Repeat the process, except in Step (b), add 8. Record your final result. Repeat the process once more, except in Step (b), add 10. Record your final result.

(f) Observe what you have done. Then use inductive reasoning to explain how to predict the final result.

59. Complete the following.

142,857 × 1 = _____	142,857 × 2 = _____
142,857 × 3 = _____	142,857 × 4 = _____
142,857 × 5 = _____	142,857 × 6 = _____

What pattern exists in the successive answers? Now multiply 142,857 by 7 to obtain an interesting result.

60. Refer to **Figures 2(b)–(e)** and **Figure 3.** Instead of counting interior regions of the circle, count the chords formed. Use inductive reasoning to predict the number of chords that would be formed if seven points were used.

1.2	**AN APPLICATION OF INDUCTIVE REASONING: NUMBER PATTERNS**

OBJECTIVES

1 Be able to recognize arithmetic and geometric sequences.

2 Be able to apply the method of successive differences to predict the next term in a sequence.

3 Be able to recognize number patterns.

4 Be able to use sum formulas.

5 Be able to recognize triangular, square, and pentagonal numbers.

Number Sequences

An ordered list of numbers such as

$$3, 9, 15, 21, 27, \ldots$$

is called a *sequence.* A **number sequence** is a list of numbers having a first number, a second number, a third number, and so on, called the **terms** of the sequence.

The sequence that begins

$$5, 9, 13, 17, 21, \ldots$$

is an *arithmetic sequence,* or *arithmetic progression.* In an **arithmetic sequence,** each term after the first is obtained by adding the same number, called the **common difference,** to the preceding term. To find the common difference, choose any term after the first and subtract from it the preceding term. If we choose $9 - 5$ (the second term minus the first term), for example, we see that the common difference is 4. To find the term following 21, we add 4 to get $21 + 4 = 25$.

The sequence that begins

$$2, 4, 8, 16, 32, \ldots$$

is a *geometric sequence,* or *geometric progression.* In a **geometric sequence,** each term after the first is obtained by multiplying the preceding term by the same number, called the **common ratio.** To find the common ratio, choose any term after the first and divide it by the preceding term. If we choose $\frac{4}{2}$ (the second term divided by the first term), for example, we see that the common ratio is 2. To find the term following 32, we multiply by 2 to get $32 \cdot 2 = 64$.

EXAMPLE 1	**Identifying Arithmetic and Geometric Sequences**

For each sequence, determine if it is an *arithmetic sequence,* a *geometric sequence,* or *neither.* If it is either arithmetic or geometric, give the next term in the sequence.

(a) $5, 10, 15, 20, 25, \ldots$　　**(b)** $3, 12, 48, 192, 768, \ldots$　　**(c)** $1, 4, 9, 16, 25, \ldots$

Solution

(a) If we choose *any* term after the first term, and subtract the preceding term, we find that the common difference is 5.

$$10 - 5 = 5 \qquad 15 - 10 = 5 \qquad 20 - 15 = 5 \qquad 25 - 20 = 5$$

Therefore, this is an arithmetic sequence. The next term in the sequence is

$$25 + 5 = 30.$$

(b) If any term after the first is multiplied by 4, the following term is obtained.

$$\frac{12}{3} = 4 \qquad \frac{48}{12} = 4 \qquad \frac{192}{48} = 4 \qquad \frac{768}{192} = 4$$

Therefore, this is a geometric sequence. The next term in the sequence is

$$768 \cdot 4 = 3072.$$

(c) Although there is a pattern here (the terms are the squares of the first five counting numbers), there is neither a common difference nor a common ratio. This is neither an arithmetic nor a geometric sequence. ∎

Successive Differences

Some sequences may present more difficulty than our earlier examples when making a conjecture about the next term. Often the **method of successive differences** may be applied in such cases. Consider the sequence

$$2, 6, 22, 56, 114, \ldots.$$

Because the next term is not obvious, subtract the first term from the second term, the second from the third, the third from the fourth, and so on.

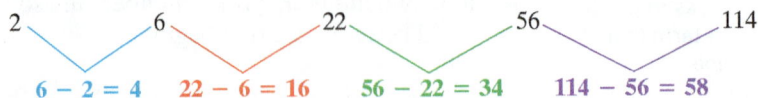

2   6   22   56   114

$6 - 2 = 4$ $22 - 6 = 16$ $56 - 22 = 34$ $114 - 56 = 58$

Now repeat the process with the sequence 4, 16, 34, 58 and continue repeating until the difference is a constant value, as shown in line (4).

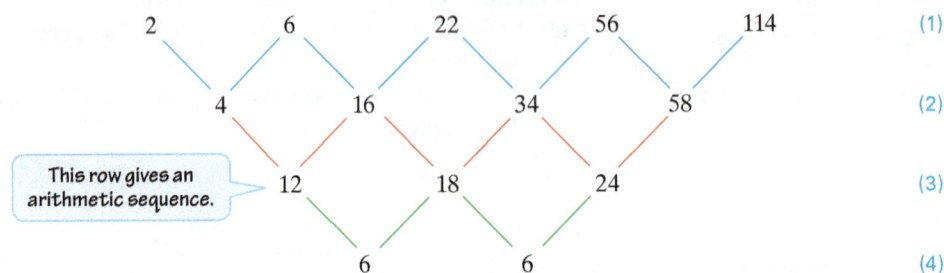

2	6	22	56	114	(1)
	4	16	34	58	(2)
		12	18	24	(3)
			6	6	(4)

> This row gives an arithmetic sequence.

Once a line of constant values is obtained, simply work "backward" by adding until the desired term of the given sequence is obtained. Thus, for this pattern to continue, another 6 should appear in line (4), meaning that the next term in line (3) would have to be $24 + 6 = 30$. The next term in line (2) would be $58 + 30 = 88$. Finally, the next term in the given sequence would be $114 + 88 = \mathbf{202}$.

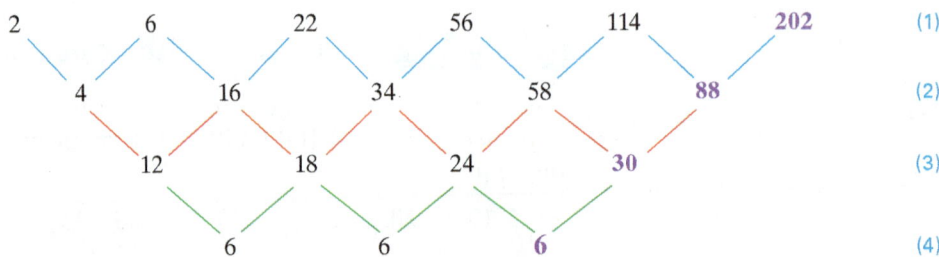

2	6	22	56	114	**202**	(1)
	4	16	34	58	88	(2)
		12	18	24	30	(3)
			6	6	6	(4)

EXAMPLE 2 Using Successive Differences

Use successive differences to determine the next number in each sequence.

(a) 14, 22, 32, 44, . . .   **(b)** 5, 15, 37, 77, 141, . . .

Solution

(a) Subtract a term from the one that follows it, and continue until a pattern is observed.

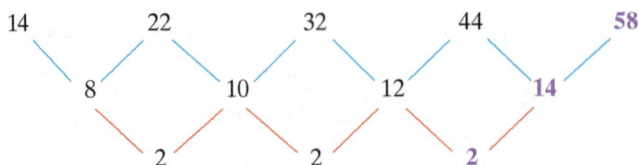

14	22	32	44	**58**
	8	10	12	**14**
		2	2	**2**

Once the row of 2s was obtained and extended, we were able to obtain

$$12 + 2 = 14, \quad \text{and} \quad 44 + 14 = 58 \quad \text{as shown above.}$$

The next number in the sequence is **58**.

(b) Proceed as before to obtain the following diagram.

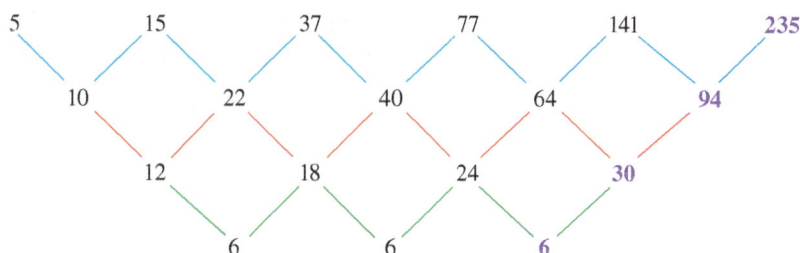

The numbers in the "diagonal" at the far right were obtained by adding: $24 + 6 = 30$, $64 + 30 = 94$, and $141 + 94 = 235$. The next number in the sequence is **235**. ∎

The method of successive differences does not always work. For example, try it on the Fibonacci sequence in **Example 2(b)** of **Section 1.1** and see what happens.

Number Patterns and Sum Formulas

Observe the following number pattern.

$$1 = 1^2$$

$$1 + 3 = 2^2$$

$$1 + 3 + 5 = 3^2$$

$$1 + 3 + 5 + 7 = 4^2$$

$$1 + 3 + 5 + 7 + 9 = 5^2$$

In each case, the left side of the equation is the indicated sum of consecutive odd counting numbers beginning with 1, and the right side is the square of the number of terms on the left side. Inductive reasoning would suggest that the next line in this pattern is as follows.

$$1 + 3 + 5 + 7 + 9 + 11 = 6^2$$

Evaluating each side shows that each side simplifies to 36.

We cannot conclude that this pattern will continue indefinitely, because observation of a finite number of examples does *not* guarantee that the pattern will continue. However, mathematicians have proved that this pattern does indeed continue indefinitely, using a method of proof called **mathematical induction.** (See any standard college algebra text.)

Any even counting number may be written in the form $2k$, where k is a counting number. It follows that the kth odd counting number is written $2k - 1$. For example, the **third** odd counting number, 5, can be written

$$2(3) - 1.$$

Using these ideas, we can write the result obtained above as follows.

SUM OF THE FIRST *n* ODD COUNTING NUMBERS

If *n* is any counting number, then the following is true.

$$1 + 3 + 5 + \cdots + (2n - 1) = n^2$$

EXAMPLE 3 **Predicting the Next Equation in a List**

In each of the following, several equations are given illustrating a suspected number pattern. Determine what the next equation would be, and verify that it is indeed a true statement.

(a)
$$1^2 = 1^3$$
$$(1 + 2)^2 = 1^3 + 2^3$$
$$(1 + 2 + 3)^2 = 1^3 + 2^3 + 3^3$$
$$(1 + 2 + 3 + 4)^2 = 1^3 + 2^3 + 3^3 + 4^3$$

(b)
$$1 = 1^3$$
$$3 + 5 = 2^3$$
$$7 + 9 + 11 = 3^3$$
$$13 + 15 + 17 + 19 = 4^3$$

(c)
$$1 = \frac{1 \cdot 2}{2}$$
$$1 + 2 = \frac{2 \cdot 3}{2}$$
$$1 + 2 + 3 = \frac{3 \cdot 4}{2}$$
$$1 + 2 + 3 + 4 = \frac{4 \cdot 5}{2}$$

(d) $12{,}345{,}679 \times 9 = 111{,}111{,}111$
$12{,}345{,}679 \times 18 = 222{,}222{,}222$
$12{,}345{,}679 \times 27 = 333{,}333{,}333$
$12{,}345{,}679 \times 36 = 444{,}444{,}444$
$12{,}345{,}679 \times 45 = 555{,}555{,}555$

> *Notice that there is no 8 here.*

Solution

(a) The left side of each equation is the square of the sum of the first n counting numbers, and the right side is the sum of their cubes. The next equation in the pattern would be

$$(1 + 2 + 3 + 4 + 5)^2 = 1^3 + 2^3 + 3^3 + 4^3 + 5^3.$$

Each side simplifies to 225, so the pattern is true for this equation.

(b) The left sides of the equations contain the sum of odd counting numbers, starting with the first (1) in the first equation, the second and third (3 and 5) in the second equation, the fourth, fifth, and sixth (7, 9, and 11) in the third equation, and so on. Each right side contains the cube (third power) of the number of terms on the left side. Following this pattern, the next equation would be

$$21 + 23 + 25 + 27 + 29 = 5^3,$$

which can be verified by computation.

(c) The left side of each equation gives the indicated sum of the first n counting numbers, and the right side is always of the form

$$\frac{n(n + 1)}{2}.$$

For the pattern to continue, the next equation would be

$$1 + 2 + 3 + 4 + 5 = \frac{5 \cdot 6}{2}.$$

Because each side simplifies to 15, the pattern is true for this equation.

(d) In each case, the first factor on the left is 12,345,679 and the second factor is a multiple of 9 (that is, 9, 18, 27, 36, 45). The right side consists of a nine-digit number, all digits of which are the same (that is, 1, 2, 3, 4, 5). For the pattern to continue, the next equation would be as follows.

$$12{,}345{,}679 \times 54 = 666{,}666{,}666$$

Verify that this is a true statement.   ∎

The patterns established in **Examples 3(a) and 3(c)** can be written as follows.

SPECIAL SUM FORMULAS

For any counting number n, the following are true.

$$(1 + 2 + 3 + \cdots + n)^2 = 1^3 + 2^3 + 3^3 + \cdots + n^3$$

$$1 + 2 + 3 + \cdots + n = \frac{n(n + 1)}{2}$$

We can provide a general deductive argument showing how the second equation is obtained.

Let S represent the sum $1 + 2 + 3 + \cdots + n$. This sum can also be written as $S = n + (n - 1) + (n - 2) + \cdots + 1$. Write these two equations as follows.

$$S = 1 \qquad + 2 \qquad + 3 \qquad + \cdots + n$$

$$\underline{S = n \qquad + (n - 1) + (n - 2) + \cdots + 1}$$

$$2S = (n + 1) + (n + 1) + (n + 1) + \cdots + (n + 1) \qquad \text{Add the corresponding sides.}$$

$$2S = n(n + 1) \qquad \text{There are } n \text{ terms of } n + 1.$$

$$S = \frac{n(n + 1)}{2} \qquad \text{Divide both sides by 2.}$$

Figurate Numbers

Pythagoras and his Pythagorean brotherhood studied numbers of geometric arrangements of points, such as **triangular numbers, square numbers,** and **pentagonal numbers. Figure 4** illustrates the first few of each of these types of numbers.

The **figurate numbers** possess numerous interesting patterns. For example, every square number greater than 1 is the sum of two consecutive triangular numbers. ($9 = 3 + 6$, $25 = 10 + 15$, and so on.)

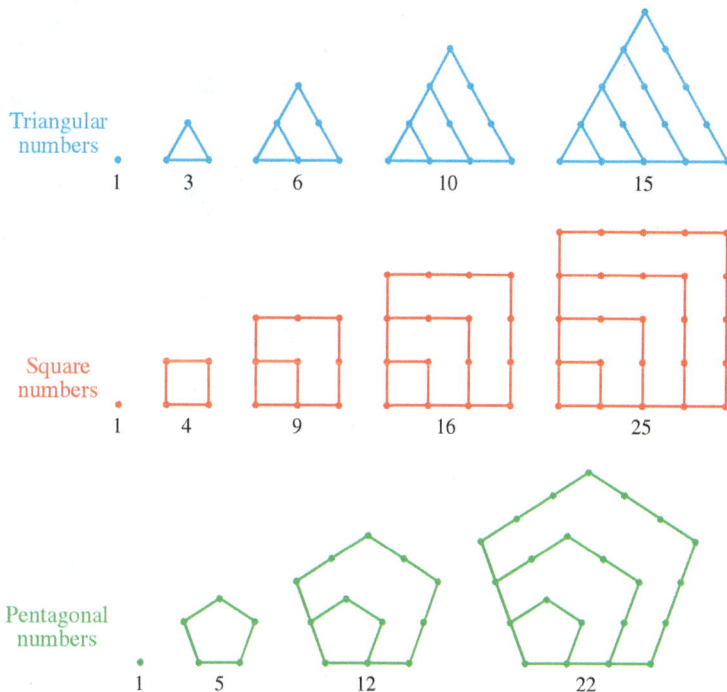

Triangular numbers
. 1   3   6   10   15

Square numbers
. 1   4   9   16   25

Pentagonal numbers
. 1   5   12   22

Figure 4

Every pentagonal number can be represented as the sum of a square number and a triangular number. (For example, $5 = 4 + 1$ and $12 = 9 + 3$.) Many other such relationships exist.

In the expression T_n, n is called a **subscript.** T_n is read **"T sub n,"** and it represents the triangular number in the nth position in the sequence. For example,

$$T_1 = 1, \quad T_2 = 3, \quad T_3 = 6, \quad \text{and} \quad T_4 = 10.$$

S_n and P_n represent the nth square and pentagonal numbers, respectively.

FORMULAS FOR TRIANGULAR, SQUARE, AND PENTAGONAL NUMBERS

For any natural number n, the following are true.

The nth triangular number is given by $\quad T_n = \dfrac{n(n + 1)}{2}$.

The nth square number is given by $\quad\quad S_n = n^2$.

The nth pentagonal number is given by $\quad P_n = \dfrac{n(3n - 1)}{2}$.

EXAMPLE 4 **Using the Formulas for Figurate Numbers**

Use the formulas to find each of the following.

(a) seventh triangular number   **(b)** twelfth square number

(c) sixth pentagonal number

Solution

(a) $T_7 = \dfrac{n(n + 1)}{2} = \dfrac{7(7 + 1)}{2} = \dfrac{7(8)}{2} = \dfrac{56}{2} = 28$ Formula for a triangular number, with $n = 7$

(b) $S_{12} = n^2 = 12^2 = 144$ Formula for a square number, with $n = 12$

$12^2 = 12 \cdot 12$

Inside the brackets, multiply first and then subtract.

(c) $P_6 = \dfrac{n(3n - 1)}{2} = \dfrac{6[3(6) - 1]}{2} = \dfrac{6(18 - 1)}{2} = \dfrac{6(17)}{2} = 51$ ■

EXAMPLE 5 **Illustrating a Figurate Number Relationship**

Show that the sixth pentagonal number is equal to the sum of 6 and 3 times the fifth triangular number.

Solution

From **Example 4(c),** $P_6 = 51$. The fifth triangular number is 15. Thus,

$$51 = 6 + 3(15) = 6 + 45 = 51.$$ ■

The general relationship examined in **Example 5** can be written as follows.

$$P_n = n + 3 \cdot T_{n-1} \quad (n \geq 2)$$

EXAMPLE 6   **Predicting the Value of a Pentagonal Number**

The first five pentagonal numbers are 1, 5, 12, 22, 35. Use the method of successive differences to predict the sixth pentagonal number.

Solution

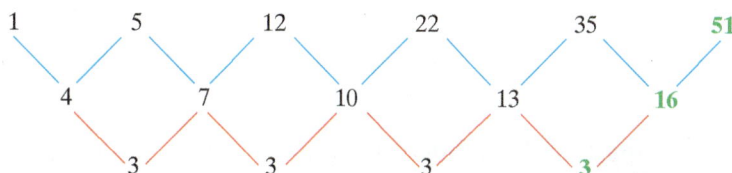

After the second line of successive differences, we work backward to find that the sixth pentagonal number is **51**, which was also found in **Example 4(c).**   ∎

FOR FURTHER THOUGHT

Kaprekar Constants

Take any four-digit number whose digits are all different. Arrange the digits in decreasing order, and then arrange them in increasing order. Now subtract. Repeat the process, called the *Kaprekar routine,* until the same result appears.

For example, suppose that we choose a number whose digits are 1, 5, 7, and 9, such as 1579.

9751	8721	7443	9963	6642	7641
−1579	−1278	−3447	−3699	−2466	−1467
8172	7443	3996	6264	4176	**6174**

Note that we have obtained the number 6174, and the process will lead to 6174 again. The number 6174 is called a **Kaprekar constant**. This number 6174 will always be generated eventually if this process is applied to such a four-digit number.

For Group or Individual Investigation

1. Apply the process of Kaprekar to a four-digit number of your choice, in which the digits are all different. How many steps did it take for you to arrive at 6174?

2. See **Exercises 77 and 78** in the exercise set that follows.

1.2   EXERCISES

For each sequence, determine if it is an arithmetic *sequence, a* geometric *sequence, or* neither. *If it is either arithmetic or geometric, give the next term in the sequence.*

1. 6, 16, 26, 36, 46, . . .

2. 8, 16, 24, 32, 40, . . .

3. 5, 15, 45, 135, 405, . . .

4. 2, 12, 72, 432, 2592, . . .

5. 1, 8, 27, 81, 243, . . .

6. 2, 8, 18, 32, 50, . . .

7. 256, 128, 64, 32, 16, . . .

8. 4096, 1024, 256, 64, 16, . . .

9. 1, 3, 4, 7, 11, . . .

10. 0, 1, 1, 2, 3, . . .

11. 12, 14, 16, 18, 20, . . .

12. 10, 50, 90, 130, 170, . . .

Use the method of successive differences to determine the next number in each sequence.

13. 1, 4, 11, 22, 37, 56, . . .

14. 3, 14, 31, 54, 83, 118, . . .

15. 6, 20, 50, 102, 182, 296, . . .

16. 1, 11, 35, 79, 149, 251, . . .

17. 0, 12, 72, 240, 600, 1260, 2352, . . .

18. 2, 57, 220, 575, 1230, 2317, . . .

19. 5, 34, 243, 1022, 3121, 7770, 16799, . . .

20. 3, 19, 165, 771, 2503, 6483, 14409, . . .

21. Refer to **Figures 2 and 3** in **Section 1.1.** The method of successive differences can be applied to the sequence of interior regions,

$$1, 2, 4, 8, 16, 31,$$

to find the number of regions determined by seven points on the circle. What is the next term in this sequence? How many regions would be determined by eight points? Verify this using the formula given at the end of that section.

22. The 1952 film *Hans Christian Andersen* stars Danny Kaye as the Danish writer of fairy tales. In a scene outside a schoolhouse window, Kaye sings a song to an inchworm. *Inchworm* was written for the film by the composer Frank Loesser and has been recorded by many artists, including Paul McCartney and Kenny Loggins. It was once featured on an episode of *The Muppets* and sung by Charles Aznavour.

As Kaye sings the song, the children in the school room are heard chanting addition facts: $2 + 2 = 4$, $4 + 4 = 8$, $8 + 8 = 16$, and so on.

(a) Use patterns to state the next addition fact (as heard in the movie).

(b) If the children were to extend their facts to the next four in the pattern, what would those facts be?

In Exercises 23–32, several equations are given illustrating a suspected number pattern. Determine what the next equation would be, and verify that it is indeed a true statement.

23. $(1 \times 9) - 1 = 8$
$(21 \times 9) - 1 = 188$
$(321 \times 9) - 1 = 2888$

24. $(1 \times 8) + 1 = 9$
$(12 \times 8) + 2 = 98$
$(123 \times 8) + 3 = 987$

25. $999,999 \times 2 = 1,999,998$
$999,999 \times 3 = 2,999,997$

26. $101 \times 101 = 10,201$
$10,101 \times 10,101 = 102,030,201$

27. $3^2 - 1^2 = 2^3$
$6^2 - 3^2 = 3^3$
$10^2 - 6^2 = 4^3$
$15^2 - 10^2 = 5^3$

28. $1 = 1^2$
$1 + 2 + 1 = 2^2$
$1 + 2 + 3 + 2 + 1 = 3^2$
$1 + 2 + 3 + 4 + 3 + 2 + 1 = 4^2$

29. $2^2 - 1^2 = 2 + 1$
$3^2 - 2^2 = 3 + 2$
$4^2 - 3^2 = 4 + 3$

30. $1^2 + 1 = 2^2 - 2$
$2^2 + 2 = 3^2 - 3$
$3^2 + 3 = 4^2 - 4$

31. $1 = 1 \times 1$
$1 + 5 = 2 \times 3$
$1 + 5 + 9 = 3 \times 5$

32. $1 + 2 = 3$
$4 + 5 + 6 = 7 + 8$
$9 + 10 + 11 + 12 = 13 + 14 + 15$

Use the formula $S = \frac{n(n+1)}{2}$ to find each sum.

33. $1 + 2 + 3 + \cdots + 300$

34. $1 + 2 + 3 + \cdots + 500$

35. $1 + 2 + 3 + \cdots + 675$

36. $1 + 2 + 3 + \cdots + 825$

Use the formula $S = n^2$ to find each sum. (Hint: To find n, add 1 to the last term and divide by 2.)

37. $1 + 3 + 5 + \cdots + 101$

38. $1 + 3 + 5 + \cdots + 49$

39. $1 + 3 + 5 + \cdots + 999$

40. $1 + 3 + 5 + \cdots + 301$

41. Use the formula for finding the sum

$$1 + 2 + 3 + \cdots + n$$

to discover a formula for finding the sum

$$2 + 4 + 6 + \cdots + 2n.$$

42. State in your own words the following formula discussed in this section.

$$(1 + 2 + 3 + \cdots + n)^2 = 1^3 + 2^3 + 3^3 + \cdots + n^3$$

43. Explain how the following diagram geometrically illustrates the formula $1 + 3 + 5 + 7 + 9 = 5^2$.

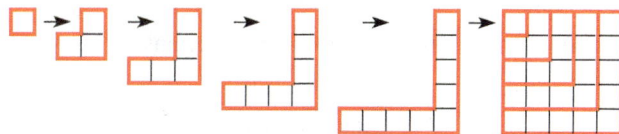

44. Explain how the following diagram geometrically illustrates the formula $1 + 2 + 3 + 4 = \frac{4 \times 5}{2}$.

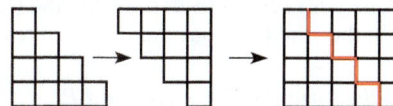

45. Use patterns to complete the table below.

Figurate Number	1st	2nd	3rd	4th	5th	6th	7th	8th
Triangular	1	3	6	10	15	21		
Square	1	4	9	16	25			
Pentagonal	1	5	12	22				
Hexagonal	1	6	15					
Heptagonal	1	7						
Octagonal	1							

46. The first five triangular, square, and pentagonal numbers can be obtained using sums of terms of sequences as shown below.

Triangular	Square	Pentagonal
$1 = 1$	$1 = 1$	$1 = 1$
$3 = 1 + 2$	$4 = 1 + 3$	$5 = 1 + 4$
$6 = 1 + 2 + 3$	$9 = 1 + 3 + 5$	$12 = 1 + 4 + 7$
$10 = 1 + 2 + 3 + 4$	$16 = 1 + 3 + 5 + 7$	$22 = 1 + 4 + 7 + 10$
$15 = 1 + 2 + 3 + 4 + 5$	$25 = 1 + 3 + 5 + 7 + 9$	$35 = 1 + 4 + 7 + 10 + 13$

Notice the successive differences of the added terms on the right sides of the equations. The next type of figurate number is the **hexagonal** number. (A hexagon has six sides.) Use the patterns above to predict the first five hexagonal numbers.

47. Eight times any triangular number, plus 1, is a square number. Show that this is true for the first four triangular numbers.

48. Divide the first triangular number by 3 and record the remainder. Divide the second triangular number by 3 and record the remainder. Repeat this procedure several more times. Do you notice a pattern?

49. Repeat **Exercise 48,** but instead use square numbers and divide by 4. What pattern is determined?

50. Exercises 48 and 49 are specific cases of the following: When the numbers in the sequence of n-agonal numbers are divided by n, the sequence of remainders obtained is a repeating sequence. Verify this for $n = 5$ and $n = 6$.

51. Every square number can be written as the sum of two triangular numbers. For example, $16 = 6 + 10$. This can be represented geometrically by dividing a square array of dots with a line as shown.

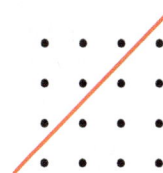

The triangular arrangement above the line represents 6, the one below the line represents 10, and the whole arrangement represents 16. Show how the square numbers 25 and 36 may likewise be geometrically represented as the sum of two triangular numbers.

52. A fraction is in *lowest terms* if the greatest common factor of its numerator and its denominator is 1. For example, $\frac{3}{8}$ is in lowest terms, but $\frac{4}{12}$ is not.

(a) For $n = 2$ to $n = 8$, form the fractions

$$\frac{n\text{th square number}}{(n+1)\text{st square number}}.$$

(b) Repeat part (a) with triangular numbers.

(c) Use inductive reasoning to make a conjecture based on your results from parts (a) and (b), observing whether the fractions are in lowest terms.

In addition to the formulas for T_n, S_n, and P_n, the following formulas are true for **hexagonal** *numbers* (H), **heptagonal** *numbers* (Hp), *and* **octagonal** *numbers* (O):

$$H_n = \frac{n(4n-2)}{2}, \quad Hp_n = \frac{n(5n-3)}{2}, \quad O_n = \frac{n(6n-4)}{2}.$$

Use these formulas to find each of the following.

53. the sixteenth square number

54. the eleventh triangular number

55. the ninth pentagonal number

56. the seventh hexagonal number

57. the tenth heptagonal number

58. the twelfth octagonal number

59. Observe the formulas given for H_n, Hp_n, and O_n, and use patterns and inductive reasoning to predict the formula for N_n, the nth **nonagonal** number. (A nonagon has nine sides.) Then use the fact that the sixth nonagonal number is 111 to further confirm your conjecture.

60. Use the result of **Exercise 59** to find the tenth nonagonal number.

Use inductive reasoning to answer each question.

61. If you add two consecutive triangular numbers, what kind of figurate number do you get?

62. If you add the squares of two consecutive triangular numbers, what kind of figurate number do you get?

63. Square a triangular number. Square the next triangular number. Subtract the smaller result from the larger. What kind of number do you get?

64. Choose a value of n greater than or equal to 2. Find T_{n-1}, multiply it by 3, and add n. What kind of figurate number do you get?

In an arithmetic sequence, the nth term a_n is given by the formula

$$a_n = a_1 + (n-1)d,$$

where a_1 is the first term and d is the common difference. Similarly, in a geometric sequence, the nth term is given by

$$a_n = a_1 \cdot r^{n-1}.$$

Here r is the common ratio. In Exercises 65–76, use these formulas to determine the indicated term in the given sequence.

65. The eleventh term of $2, 6, 10, 14, \ldots$

66. The sixteenth term of $5, 15, 25, 35, \ldots$

67. The 21st term of $19, 39, 59, 79, \ldots$

68. The 36th term of $8, 38, 68, 98, \ldots$

69. The 101st term of $\frac{1}{2}, 1, \frac{3}{2}, 2, \ldots$

70. The 151st term of $0.75, 1.50, 2.25, 3.00, \ldots$

71. The eleventh term of $2, 4, 8, 16, \ldots$

72. The ninth term of $1, 4, 16, 64, \ldots$

73. The 12th term of $1, \frac{1}{2}, \frac{1}{4}, \frac{1}{8}, \ldots$

74. The 10th term of $1, \frac{1}{3}, \frac{1}{9}, \frac{1}{27}, \ldots$

75. The 8th term of $40, 10, \frac{5}{2}, \frac{5}{8}, \ldots$

76. The 9th term of $10, 2, \frac{2}{5}, \frac{2}{25}, \ldots$

77. In the *For Further Thought* investigation of this section, we saw that the number 6174 is a Kaprekar constant. Use the procedure described there, starting with a three-digit number of your choice whose digits are all different. You should arrive at a particular three-digit number that has the same property described for 6174. What is this three-digit number?

78. Applying the Kaprekar routine to a five-digit number does not reach a single repeating result but instead reaches one of the following ten numbers and then cycles repeatedly through a subset of these ten numbers.

> 53955, 59994, 61974, 62964, 63954,
>
> 71973, 74943, 75933, 82962, 83952

(a) Start with the number 45986 and determine which one of the ten numbers above is reached first.

(b) Start with a five-digit number of your own, and determine which one of the ten numbers is eventually reached first.

*The mathematical array of numbers known as **Pascal's triangle** consists of rows of numbers, each of which contains one more entry than the previous row. The first six rows are shown here.*

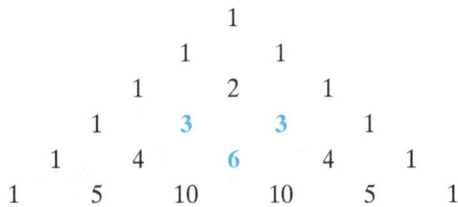

```
              1
           1     1
        1     2     1
     1     3     3     1
  1     4     6     4     1
1     5    10    10     5     1
```

Refer to this array to answer Exercises 79–82.

79. Each row begins and ends with a 1. Discover a method whereby the other entries in a row can be determined from the entries in the row immediately above it. (*Hint:* See the entries in color above.) Find the next three rows of the triangle, and prepare a copy of the first nine rows for later reference.

80. Find the sum of the entries in each of the first eight rows. What is the pattern that emerges? Predict the sum of the entries in the ninth row, and confirm your prediction.

81. The first six rows of the triangle are arranged "flush left" here.

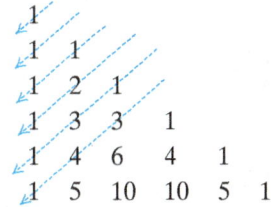

```
1
1   1
1   2   1
1   3   3   1
1   4   6   4   1
1   5  10  10   5   1
```

Add along the blue diagonal lines. Write these sums in order from left to right. What sequence is this?

82. Find the values of the first four powers of the number 11, starting with 11^0, which by definition is equal to 1. Predict what the next power of 11 will equal by observing the rows of Pascal's triangle. Confirm your prediction by actual computation.

1.3 STRATEGIES FOR PROBLEM SOLVING

OBJECTIVES

1. Know George Polya's four-step method of problem solving.
2. Be able to apply various strategies for solving problems.

A General Problem-Solving Method

In the first two sections of this chapter we stressed the importance of pattern recognition and the use of inductive reasoning in solving problems. Probably the most famous study of problem-solving techniques was developed by George Polya (1888–1985), among whose many publications was the modern classic *How to Solve It.* In this book, Polya proposed a four-step method for problem solving.

POLYA'S FOUR-STEP METHOD FOR PROBLEM SOLVING

Step 1 **Understand the problem.** You cannot solve a problem if you do not understand what you are asked to find. The problem must be read and analyzed carefully. You may need to read it several times. After you have done so, ask yourself, *"What must I find?"*

Step 2 **Devise a plan.** There are many ways to attack a problem. Decide what plan is appropriate for the particular problem you are solving.

Step 3 **Carry out the plan.** Once you know how to approach the problem, carry out your plan. You may run into "dead ends" and unforeseen roadblocks, but be persistent.

Step 4 **Look back and check.** Check your answer to see that it is reasonable. *Does it satisfy the conditions of the problem? Have you answered all the questions the problem asks? Can you solve the problem a different way and come up with the same answer?*

In Step 2 of Polya's problem-solving method, we are told to devise a plan. Here are some strategies that may prove useful.

George Polya, author of the classic *How to Solve It*, died at the age of 97 on September 7, 1985. A native of Budapest, Hungary, he was once asked why there were so many good mathematicians to come out of Hungary at the turn of the century. He theorized that it was because mathematics is the cheapest science. It does not require any expensive equipment, only pencil and paper.

Polya authored or coauthored more than 250 papers in many languages, wrote a number of books, and was a brilliant lecturer and teacher. Yet, interestingly enough, he never learned to drive a car.

Problem-Solving Strategies

- Make a table or a chart.
- Look for a pattern.
- Solve a similar, simpler problem.
- Draw a sketch.
- Use inductive reasoning.
- Write an equation and solve it.

- If a formula applies, use it.
- Work backward.
- Guess and check.
- Use trial and error.
- Use common sense.
- Look for a "catch" if an answer seems too obvious or impossible.

Using a Table or Chart

EXAMPLE 1   Solving Fibonacci's Rabbit Problem

A man put a pair of rabbits in a cage. During the first month the rabbits produced no offspring but each month thereafter produced one new pair of rabbits. If each new pair thus produced reproduces in the same manner, how many pairs of rabbits will there be at the end of 1 year? (This problem is a famous one in the history of mathematics and first appeared in *Liber Abaci*, a book written by the Italian mathematician Leonardo Pisano (also known as Fibonacci) in the year 1202.)

Solution

Step 1   Understand the problem. We can reword the problem as follows:

> *How many pairs of rabbits will the man have at the end of one year if he starts with one pair, and they reproduce this way: During the first month of life, each pair produces no new rabbits, but each month thereafter each pair produces one new pair?*

Step 2   Devise a plan. Because there is a definite pattern to how the rabbits will reproduce, we can construct **Table 2.**

Fibonacci (1170–1250) discovered the sequence named after him in a problem on rabbits. Fibonacci ("son of Bonaccio") is one of several names for Leonardo of Pisa. His father managed a warehouse in present-day Bougie (or Bejaia), in Algeria. Thus it was that Leonardo Pisano studied with a Moorish teacher and learned the "Indian" numbers that the Moors and other Moslems brought with them in their westward drive.

Fibonacci wrote books on algebra, geometry, and trigonometry.

Table 2

Month	Number of Pairs at Start	Number of New Pairs Produced	Number of Pairs at End of Month
1st			
2nd			
3rd			
4th			
5th			
6th			
7th			
8th			
9th			
10th			
11th			
12th			

The answer will go here.

Step 3 **Carry out the plan.** At the start of the first month, there is only one pair of rabbits. No new pairs are produced during the first month, so there is $1 + 0 = 1$ pair present at the end of the first month. This pattern continues. In **Table 3,** we add the number in the first column of numbers to the number in the second column to get the number in the third.

Table 3

Month	Number of Pairs at Start	+	Number of New Pairs Produced	=	Number of Pairs at End of Month	
1st	1		0		1	$1 + 0 = 1$
2nd	1		1		2	$1 + 1 = 2$
3rd	2		1		3	$2 + 1 = 3$
4th	3		2		5	•
5th	5		3		8	•
6th	8		5		13	•
7th	13		8		21	•
8th	21		13		34	•
9th	34		21		55	•
10th	55		34		89	•
11th	89		55		144	•
12th	144		89		233	$144 + 89 = 233$

The answer is the final entry.

There will be **233** pairs of rabbits at the end of one year.

Step 4 **Look back and check.** Go back and make sure that we have interpreted the problem correctly. Double-check the arithmetic. We have answered the question posed by the problem, so the problem is solved. ∎

Working Backward

EXAMPLE 2 **Determining a Wager at the Track**

Ronnie goes to the racetrack with his buddies on a weekly basis. One week he tripled his money, but then lost $12. He took his money back the next week, doubled it, but then lost $40. The following week he tried again, taking his money back with him. He quadrupled it, and then played well enough to take that much home, a total of $224. How much did he start with the first week?

Solution

This problem asks us to find Ronnie's starting amount. Since we know his final amount, the method of working backward can be applied.

Because his final amount was $224 and this represents four times the amount he started with on the third week, we *divide* $224 by 4 to find that he started the third week with $56. Before he lost $40 the second week, he had this $56 plus the $40 he lost, giving him $96.

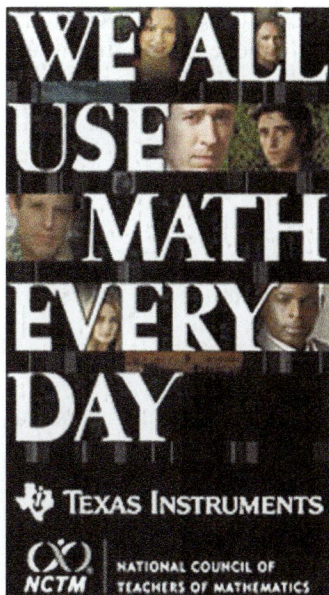

On January 23, 2005, the CBS television network presented the first episode of *NUMB3RS*, a show focusing on how mathematics is used in solving crimes. David Krumholtz plays Charlie Eppes, a brilliant mathematician who assists his FBI agent brother (Rob Morrow).

In the first-season episode "Sabotage" (2/25/2005), one of the agents admits that she was not a good math student, and Charlie uses the **Fibonacci sequence** and its relationship to nature to enlighten her.

The sequence shown in color in **Table 3** is the Fibonacci sequence, mentioned in **Example 2(b)** of **Section 1.1.**

Augustus De Morgan was an English mathematician and philosopher, who served as professor at the University of London. He wrote numerous books, one of which was *A Budget of Paradoxes*. His work in set theory and logic led to laws that bear his name and are covered in other chapters.

The $96 represented double what he started with, so he started with $96 *divided by* 2, or $48, the second week. Repeating this process once more for the first week, before his $12 loss he had

$$\$48 + \$12 = \$60,$$

which represents triple what he started with. Therefore, he started with

$$\$60 \div 3 = \$20. \quad \text{Answer}$$

To check, observe the following equations that depict winnings and losses.

First week: $(3 \times \$20) - \$12 = \$60 - \$12 = \$48$
Second week: $(2 \times \$48) - \$40 = \$96 - \$40 = \$56$
Third week: $(4 \times \$56) = \$224$ His final amount   ■

Using Trial and Error

Recall that $5^2 = 5 \cdot 5 = 25$. That is, 5 squared is 25. Thus, 25 is called a **perfect square,** a term that we use in **Example 3.**

$$1, \quad 4, \quad 9, \quad 16, \quad 25, \quad 36, \quad \text{and so on} \quad \text{Perfect squares}$$

EXAMPLE 3 Finding Augustus De Morgan's Birth Year

The mathematician Augustus De Morgan lived in the nineteenth century. He made the following statement: "I was x years old in the year x^2." In what year was he born?

Solution

We must find the year of De Morgan's birth. The problem tells us that he lived in the nineteenth century, which is another way of saying that he lived during the 1800s. One year of his life was a perfect square, so we must find a number between 1800 and 1900 that is a perfect square. Use trial and error.

$$42^2 = 42 \cdot 42 = 1764$$
$$43^2 = 43 \cdot 43 = 1849 \quad \longleftarrow \boxed{\text{1849 is between 1800 and 1900.}}$$
$$44^2 = 44 \cdot 44 = 1936$$

The only natural number whose square is between 1800 and 1900 is 43, because $43^2 = 1849$. Therefore, De Morgan was 43 years old in 1849. The final step in solving the problem is to subtract 43 from 1849 to find the year of his birth.

$$1849 - 43 = 1806 \quad \longleftarrow \boxed{\text{He was born in 1806.}}$$

To check this answer, look up De Morgan's birth date in a book dealing with mathematics history, such as *An Introduction to the History of Mathematics,* Sixth Edition, by Howard W. Eves.   ■

Guessing and Checking

As mentioned above, $5^2 = 25$. The inverse procedure for squaring a number is called taking the **square root.** We indicate the positive square root using a **radical symbol** $\sqrt{}$. Thus, $\sqrt{25} = 5$. Also,

$$\sqrt{4} = 2, \quad \sqrt{9} = 3, \quad \sqrt{16} = 4, \quad \text{and so on.} \quad \text{Square roots}$$

The next problem deals with a square root and dates back to Hindu mathematics, circa 850.

EXAMPLE 4 Finding the Number of Camels

One-fourth of a herd of camels was seen in the forest. Twice the square root of that herd had gone to the mountain slopes, and 3 times 5 camels remained on the river-bank. What is the numerical measure of that herd of camels?

Solution

The numerical measure of a herd of camels must be a counting number. Because the problem mentions "one-fourth of a herd" and "the square root of that herd," the number of camels must be both a multiple of 4 and a perfect square, so that only whole numbers are used. The least counting number that satisfies both conditions is 4.

We write an equation where x represents the numerical measure of the herd, and then substitute 4 for x to see if it is a solution.

One-fourth of the herd	+	Twice the square root of that herd	+	3 times 5 camels	=	The numerical measure of the herd.
$\frac{1}{4}x$	+	$2\sqrt{x}$	+	$3 \cdot 5$	=	x

$$\frac{1}{4}(4) + 2\sqrt{4} + 3 \cdot 5 \stackrel{?}{=} 4 \quad \text{Let } x = 4.$$

$$1 + 4 + 15 \stackrel{?}{=} 4 \quad \sqrt{4} = 2$$

$$20 \neq 4$$

Because 4 is not the solution, try **16**, the next perfect square that is a multiple of 4.

$$\frac{1}{4}(16) + 2\sqrt{16} + 3 \cdot 5 \stackrel{?}{=} 16 \quad \text{Let } x = 16.$$

$$4 + 8 + 15 \stackrel{?}{=} 16 \quad \sqrt{16} = 4$$

$$27 \neq 16$$

Because 16 is not a solution, try **36**.

$$\frac{1}{4}(36) + 2\sqrt{36} + 3 \cdot 5 \stackrel{?}{=} 36 \quad \text{Let } x = 36.$$

$$9 + 12 + 15 \stackrel{?}{=} 36 \quad \sqrt{36} = 6$$

$$36 = 36$$

Thus, 36 is the numerical measure of the herd.

Check: "One-fourth of 36, plus twice the square root of 36, plus 3 times 5" gives 9 plus 12 plus 15, which equals 36. ∎

Considering a Similar, Simpler Problem

EXAMPLE 5 Finding the Units Digit of a Power

The digit farthest to the right in a counting number is called the *ones* or *units* digit, because it tells how many ones are contained in the number when grouping by tens is considered. What is the ones (or units) digit in 2^{4000}?

Solution

Recall that 2^{4000} means that 2 is used as a factor 4000 times.

$$2^{4000} = \underbrace{2 \times 2 \times 2 \times \cdots \times 2}_{\text{4000 factors}}$$

Mathematics to Die For In the 1995 movie *Die Hard: With a Vengeance,* detective John McClane (Bruce Willis) is tormented by the villain Simon Gruber (Jeremy Irons), who has planted a bomb in a park. Gruber presents McClane and his partner Zeus Carver (Samuel L. Jackson) with a riddle for disarming it, but there is a time limit. The riddle requires that they get exactly 4 gallons of water using 3-gallon and 5-gallon jugs having no markers. They are able to solve it and defuse the bomb. Can you do it? (See the next page for the answer.)

Solution to the Jugs-of-Water Riddle
This is one way to do it: With both jugs empty, fill the 3-gallon jug and pour its contents into the 5-gallon jug. Then fill the 3-gallon jug again, and pour it into the 5-gallon jug until the latter is filled. There is now $(3 + 3) - 5 = 1$ gallon in the 3-gallon jug. Empty the 5-gallon jug, and pour the 1 gallon of water from the 3-gallon jug into the 5-gallon jug. Finally, fill the 3-gallon jug and pour all of it into the 5-gallon jug, resulting in $1 + 3 = 4$ gallons in the 5-gallon jug.

(*Note:* There is another way to solve this problem. See if you can discover the alternative solution.)

To answer the question, we examine some smaller powers of 2 and then look for a pattern. We start with the exponent 1 and look at the first twelve powers of 2.

$$2^1 = 2 \qquad 2^5 = 32 \qquad 2^9 = 512$$
$$2^2 = 4 \qquad 2^6 = 64 \qquad 2^{10} = 1024$$
$$2^3 = 8 \qquad 2^7 = 128 \qquad 2^{11} = 2048$$
$$2^4 = 16 \qquad 2^8 = 256 \qquad 2^{12} = 4096$$

Notice that in any one of the four rows above, the ones digit is the same all the way across the row. The final row, which contains the exponents 4, 8, and 12, has the ones digit 6. Each of these exponents is divisible by 4, and because 4000 is divisible by 4, we can use inductive reasoning to predict that the units digit in 2^{4000} is 6.

(*Note:* The units digit for any other power can be found if we divide the exponent by 4 and consider the remainder. Then compare the result to the list of powers above. For example, to find the units digit of 2^{543}, divide 543 by 4 to get a quotient of 135 and a remainder of 3. The units digit is the same as that of 2^3, which is 8.) ■

Drawing a Sketch

EXAMPLE 6 Connecting the Dots

Figure 5

An array of nine dots is arranged in a 3×3 square, as shown in **Figure 5.** Is it possible to join the dots with exactly four straight line segments if you are not allowed to pick up your pencil from the paper and may not trace over a segment that has already been drawn? If so, show how.

Solution

Figure 6 shows three attempts. In each case, something is wrong. In the first sketch, one dot is not joined. In the second, the figure cannot be drawn without picking up your pencil from the paper or tracing over a line that has already been drawn. In the third figure, all dots have been joined, but you have used five line segments as well as retraced over the figure.

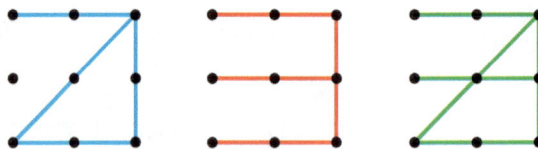

Figure 6

The conditions of the problem can be satisfied, as shown in **Figure 7.** We "went outside of the box," which was not prohibited by the conditions of the problem. This is an example of creative thinking—we used a strategy that often is not considered at first. ■

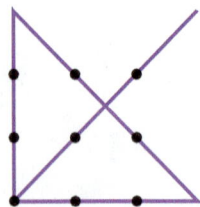

Figure 7

Using Common Sense

Problem-Solving Strategies

Some problems involve a "catch." They seem too easy or perhaps impossible at first because we tend to overlook an obvious situation. Look carefully at the use of language in such problems. And, of course, never forget to use common sense.

EXAMPLE 7 Determining Coin Denominations

Two currently minted United States coins together have a total value of $1.05. One is not a dollar. What are the two coins?

Solution

Our initial reaction might be, "The only way to have two such coins with a total of $1.05 is to have a nickel and a dollar, but the problem says that one of them is not a dollar." This statement is indeed true. What we must realize here is that the one that is not a dollar is the nickel, and the *other* coin is a dollar! So the two coins are a dollar and a nickel. ■

1.3 EXERCISES

One of the most popular features in the journal Mathematics Teacher, *published by the National Council of Teachers of Mathematics, is the monthly calendar. It provides an interesting, unusual, or challenging problem for each day of the month. Some of these exercises, and others later in this text, are chosen from these calendars (the day, month, and year of publication of each problem are indicated). The authors want to thank the many contributors for permission to use these problems.*

Use the various problem-solving strategies to solve each problem. In many cases there is more than one possible approach, so be creative.

1. *Broken Elevator* A man enters a building on the first floor and runs up to the third floor in 20 seconds. At this rate, how many seconds would it take for the man to run from the first floor up to the sixth floor? (October 3, 2010)

2. *Saving Her Dollars* Every day Sally saved a penny, a dime, and a quarter. What is the least number of days required for her to save an amount equal to an integral (counting) number of dollars? (January 11, 2012)

3. *Do You Have a Match?* Move 4 of the matches in the figure to create exactly 3 equilateral triangles. (An *equilateral triangle* has all three sides the same length.) (February 20, 2011)

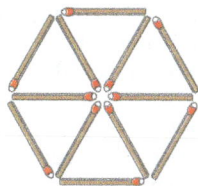

4. *Sudoku* Sudoku is an $n \times n$ puzzle that requires the solver to fill in all the squares using the integers 1 through n. Each row, column, and subrectangle contains exactly one of each number. Complete the $n \times n$ puzzle. (June 25, 2012)

5. *Break This Code* Each letter of the alphabet is assigned an integer, starting with A = 0, B = 1, and so on. The numbers repeat after every seven letters, so that G = 6, H = 0, and I = 1, continuing on to Z. What two-letter word is represented by the digits 16? (October 3, 2011)

6. *A Real Problem* We are given the following sequence:

PROBLEMSOLVINGPROBLEMSOLVINGPROB . . .

If the pattern continues, what letter will be in the 2012th position? (November 1, 2012)

7. *How Old Is Mommy?* A mother has two children whose ages differ by 5 years. The sum of the squares of their ages is 97. The square of the mother's age can be found by writing the squares of the children's ages one after the other as a four-digit number. How old is the mother? (March 9, 2012)

8. *An Alarming Situation* You have three alarms in your room. Your cell phone alarm is set to ring every 30 minutes, your computer alarm is set to ring every 20 minutes, and your clock alarm is set to ring every 45 minutes. If all three alarms go off simultaneously at 12:34 p.m., when is the next time that they will go off simultaneously? (November 2, 2012)

9. *Laundry Day* Every Monday evening, a mathematics teacher stops by the dry cleaners, drops off the shirts that he wore for the week, and picks up his previous week's load. If he wears a clean shirt every day, including Saturday and Sunday, what is the minimum number of shirts that he can own? (April 1, 2012)

10. *Pick an Envelope* Three envelopes contain a total of six bills. One envelope contains two $10 bills, one contains two $20 bills, and the third contains one $10 and one $20 bill. A label on each envelope indicates the sum of money in one of the other envelopes. It is possible to select one envelope, see one bill in that envelope, and then state the contents of all of the envelopes. Which envelope should you choose? (May 1, 2012)

11. *Class Members* A classroom contains an equal number of boys and girls. If 8 girls leave, twice as many boys as girls remain. What was the original number of students present? (May 24, 2008)

12. *Give Me a Digit* Given a two-digit number, make a three-digit number by putting a 6 as the rightmost digit. Then add 6 to the resulting three-digit number and remove the rightmost digit to obtain another two-digit number. If the result is 76, what is the original two-digit number? (October 18, 2009)

13. *Missing Digit* Look for a pattern and find the missing digit x.

3	2	4	8
7	2	1	3
8	4	x	5
4	3	6	9

(February 14, 2009)

14. *Abundancy* An integer $n > 1$ is **abundant** if the sum of its proper divisors (positive integer divisors smaller than n) is greater than n. Find the smallest abundant integer. (November 27, 2009)

15. *Cross-Country Competition* The schools in an athletic conference compete in a cross-country meet to which each school sends three participants. Erin, Katelyn, and Iliana are the three representatives from one school. Erin finished the race in the middle position; Katelyn finished after Erin, in the 19th position; and Iliana finished 28th. How many schools took part in the race? (May 27, 2008)

16. *Gone Fishing* Four friends go fishing one day and bring home a total of 11 fish. If each person caught at least 1 fish, then which of the following *must* be true?

A. One person caught exactly 2 fish.

B. One person caught exactly 3 fish.

C. One person caught fewer than 3 fish.

D. One person caught more than 3 fish.

E. Two people each caught more than 1 fish.

(May 24, 2008)

17. *Cutting a Square in Half* In how many ways can a single straight line cut a square in half? (October 2, 2008)

18. *You Lie!* Max, Sam, and Brett were playing basketball. One of them broke a window, and the other two saw him break it. Max said, "I am innocent." Sam said, "Max and I are both innocent." Brett said, "Max and Sam are both innocent." If only one of them is telling the truth, who broke the window? (September 21, 2008)

19. *Bookworm Snack* A 26-volume encyclopedia (one for each letter) is placed on a bookshelf in alphabetical order from left to right. Each volume is 2 inches thick, including the front and back covers. Each cover is $\frac{1}{4}$ inch thick. A bookworm eats straight through the encyclopedia, beginning inside the front cover of volume A and ending after eating through the back cover of volume Z. How many inches of book did the bookworm eat? (November 12, 2008)

20. *Pick a Card, Any Card* Three face cards from an ordinary deck of playing cards lie facedown in a horizontal row and are arranged such that immediately to the right of a king is a queen or two queens, immediately to the left of a queen is a queen or two queens, immediately to the left of a heart is a spade or two spades, and immediately to the right of a spade is a spade or two spades. Name the three cards in order. (April 23, 2008)

21. *Catwoman's Cats* If you ask Batman's nemesis, Catwoman, how many cats she has, she answers with a riddle: "Five-sixths of my cats plus seven." How many cats does Catwoman have? (April 20, 2003)

22. *Pencil Collection* Bob gave four-fifths of his pencils to Barbara, then he gave two-thirds of the remaining pencils to Bonnie. If he ended up with ten pencils for himself, with how many did he start? (October 12, 2003)

23. *Adding Gasoline* The gasoline gauge on a van initially read $\frac{1}{8}$ full. When 15 gallons were added to the tank, the gauge read $\frac{3}{4}$ full. How many more gallons are needed to fill the tank? (November 25, 2004)

24. *Gasoline Tank Capacity* When 6 gallons of gasoline are put into a car's tank, the indicator goes from $\frac{1}{4}$ of a tank to $\frac{5}{8}$. What is the total capacity of the gasoline tank? (February 21, 2004)

25. *Number Pattern* What is the relationship between the rows of numbers?

18,	38,	24,	46,	42
8,	24,	8,	24,	8

(May 26, 2005)

26. *Locking Boxes* You and I each have one lock and a corresponding key. I want to mail you a box with a ring in it, but any box that is not locked will be emptied before it reaches its recipient. How can I safely send you the ring? (Note that you and I each have keys to our own lock but not to the other lock.) (May 4, 2004)

27. *Number in a Sequence* In the sequence 16, 80, 48, 64, A, B, C, D, each term beyond the second term is the arithmetic mean (average) of the two previous terms. What is the value of D? (April 26, 2004)

28. *Unknown Number* Cindy was asked by her teacher to subtract 3 from a certain number and then divide the result by 9. Instead, she subtracted 9 and then divided the result by 3, giving an answer of 43. What would her answer have been if she had worked the problem correctly? (September 3, 2004)

29. *Unfolding and Folding a Box* An unfolded box is shown below.

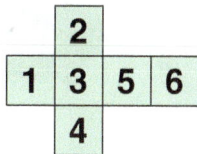

Which figure shows the box folded up? (November 7, 2001)

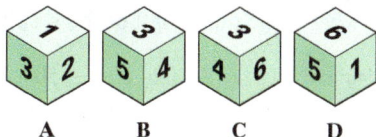

30. *Vertical Symmetry in States' Names* (If a vertical line is drawn through the center of a figure, and the left and right sides are reflections of each other across this line, the figure is said to have vertical symmetry.) When spelled with all capital letters, each letter in HAWAII has vertical symmetry. Find the name of a state whose letters all have vertical and horizontal symmetry. (September 11, 2001)

31. *Labeling Boxes* You are working in a store that has been very careless with the stock. Three boxes of socks are each incorrectly labeled. The labels say *red socks*, *green socks*, and *red and green socks*. How can you relabel the boxes correctly by taking only one sock out of one box, without looking inside the boxes? (October 22, 2001)

32. *Mr. Green's Age* At his birthday party, Mr. Green would not directly tell how old he was. He said, "If you add the year of my birth to this year, subtract the year of my tenth birthday and the year of my fiftieth birthday, and then add my present age, the result is eighty." How old was Mr. Green? (December 14, 1997)

33. *Sum of Hidden Dots on Dice* Three dice with faces numbered 1 through 6 are stacked as shown. Seven of the eighteen faces are visible, leaving eleven faces hidden on the back, on the bottom, and between dice. The total number of dots not visible in this view is _____.

A. 21
B. 22
C. 31
D. 41
E. 53

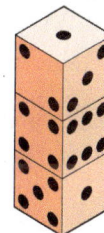

(September 17, 2001)

34. *Age of the Bus Driver* Today is your first day driving a city bus. When you leave downtown, you have twenty-three passengers. At the first stop, three people exit and five people get on the bus. At the second stop, eleven people exit and eight people get on the bus. At the third stop, five people exit and ten people get on. How old is the bus driver? (April 1, 2002)

35. *Matching Triangles and Squares* How can you connect each square with the triangle that has the same number? Lines cannot cross, enter a square or triangle, or go outside the diagram. (October 15, 1999)

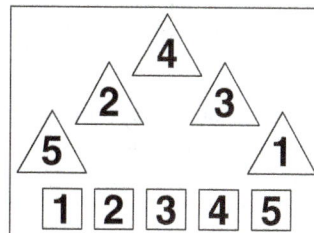

36. Forming Perfect Square Sums How must one place the integers from 1 to 15 in each of the spaces below in such a way that no number is repeated and the sum of the numbers in any two consecutive spaces is a perfect square? (November 11, 2001)

37. Difference Triangle Balls numbered 1 through 6 are arranged in a **difference triangle.** Note that in any row, the difference between the larger and the smaller of two successive balls is the number of the ball that appears below them. Arrange balls numbered 1 through 10 in a difference triangle. (May 6, 1998)

38. Clock Face By drawing two straight lines, divide the face of a clock into three regions such that the numbers in the regions have the same total. (October 28, 1998)

39. Alphametric If a, b, and c are digits for which

$$\begin{array}{r} 7\ \ a\ \ 2 \\ -4\ \ 8\ \ b \\ \hline c\ \ 7\ \ 3, \end{array}$$

then $a + b + c =$ _____.

A. 14 **B.** 15 **C.** 16 **D.** 17 **E.** 18

(September 22, 1999)

40. Perfect Square Only one of these numbers is a perfect square. Which one is it? (October 8, 1997)

329476   389372   964328

326047   724203

41. Sleeping on the Way to Grandma's House While traveling to his grandmother's for Christmas, George fell asleep halfway through the journey. When he awoke, he still had to travel half the distance that he had traveled while sleeping. For what part of the entire journey had he been asleep? (December 25, 1998)

42. Buckets of Water You have brought two unmarked buckets to a stream. The buckets hold 7 gallons and 3 gallons of water, respectively. How can you obtain exactly 5 gallons of water to take home? (October 19, 1997)

43. Counting Puzzle (Rectangles) How many rectangles are in the figure? (March 27, 1997)

44. Digit Puzzle Place each of the digits 1, 2, 3, 4, 5, 6, 7, and 8 in separate boxes so that boxes that share common corners do not contain successive digits. (November 29, 1997)

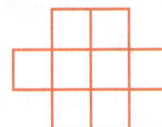

45. Palindromic Number (*Note*: A **palindromic number** is a number whose digits read the same left to right as right to left. For example, 383, 12321, and 9876789 are palindromic.) The odometer of the family car read 15951 when the driver noticed that the number was palindromic. "Curious," said the driver to herself. "It will be a long time before that happens again." But 2 hours later, the odometer showed a new palindromic number. (*Author's note:* Assume it was the next possible one.) How fast was the car driving in those 2 hours? (December 26, 1998)

46. How Much Is That Doggie in the Window? A man wishes to sell a puppy for $11. A customer who wants to buy it has only foreign currency. The exchange rate for the foreign currency is as follows: 11 round coins = $15, 11 square coins = $16, 11 triangular coins = $17. How many of each coin should the customer pay? (April 20, 2008)

47. Final Digits of a Power of 7 What are the final two digits of 7^{1997}? (November 29, 1997)

48. Units Digit of a Power of 3 If you raise 3 to the 324th power, what is the units digit of the result?

49. Summing the Digits When $10^{50} - 50$ is expressed as a single whole number, what is the sum of its digits? (April 7, 2008)

50. *Frog Climbing up a Well*   A frog is at the bottom of a 20-foot well. Each day it crawls up 4 feet, but each night it slips back 3 feet. After how many days will the frog reach the top of the well?

51. *Units Digit of a Power of 7*   What is the units digit in 7^{491}?

52. *Money Spent at a Bazaar*   Christine bought a book for $10 and then spent half her remaining money on a train ticket. She then bought lunch for $4 and spent half her remaining money at a bazaar. She left the bazaar with $8. How much money did she start with?

53. *Going Postal*   Joanie wants to mail a package that requires $1.53 in postage. If she has only 5-cent and 8-cent stamps, what is the smallest number of stamps she could use that would total exactly $1.53? (August 20, 2008)

54. *Counting Puzzle (Squares)*   How many squares are in the figure?

55. *Matching Socks*   A drawer contains 20 black socks and 20 white socks. If the light is off and you reach into the drawer to get your socks, what is the minimum number of socks you must pull out in order to be sure that you have a matching pair?

56. *Counting Puzzle (Triangles)*   How many triangles are in the figure?

57. *Perfect Number*   A **perfect number** is a counting number that is equal to the sum of all its counting number divisors except itself. For example, 28 is a perfect number because its divisors other than itself are 1, 2, 4, 7, and 14, and $1 + 2 + 4 + 7 + 14 = 28$. What is the least perfect number?

58. *Naming Children*   Becky's mother has three daughters. She named her first daughter Penny and her second daughter Nichole. What did she name her third daughter?

59. *Growth of a Lily Pad*   A lily pad grows so that each day it doubles its size. On the twentieth day of its life, it completely covers a pond. On what day was the pond half covered?

60. *Interesting Property of a Sentence*   Comment on an interesting property of this sentence: "A man, a plan, a canal, Panama." (*Hint:* See **Exercise 45.**)

61. *High School Graduation Year of Author*   One of the authors of this book graduated from high school in the year that satisfies these conditions: (1) The sum of the digits is 23; (2) The hundreds digit is 3 more than the tens digit; (3) No digit is an 8. In what year did he graduate?

62. *Where in the World Is Matt Lauer?*   Matt Lauer is one of the hosts of the *Today* show on the NBC television network. From time to time he travels the world and is in a new location each day of the week, which is unknown even to his co-hosts back in the studio in New York. On one day, he decided to give them a riddle as a hint to where he would be on the following day. Here's the riddle:

> This country is an ANAGRAM of a SYNONYM of a HOMOPHONE of an EVEN PRIME NUMBER.

In what country was Matt going to be the following day? (If you are unfamiliar with some of the terms in capital letters, look up their definitions.)

63. *Adam and Eve's Assets*   Eve said to Adam, "If you give me one dollar, then we will have the same amount of money." Adam then replied, "Eve, if you give me one dollar, I will have double the amount of money you are left with." How much does each have?

64. *Missing Digits Puzzle*   In the addition problem below, some digits are missing, as indicated by the blanks. If the problem is done correctly, what is the sum of the missing digits?

$$
\begin{array}{r}
\underline{}\ 3\ 5 \\
8\ \underline{}\ 6 \\
+\ 1\ 4\ \underline{} \\
\hline
\underline{}\ 4\ 0\ 8
\end{array}
$$

65. *Missing Digits Puzzle*   Fill in the blanks so that the multiplication problem below uses all digits 0, 1, 2, 3, ..., 9 exactly once and is worked correctly.

$$
\begin{array}{r}
\underline{}\ 0\ 2 \\
\times\ \ 3\ \underline{} \\
\hline
\underline{}\ 5,\ \underline{}\ \underline{}\ \underline{}
\end{array}
$$

66. Magic Square A **magic square** is a square array of numbers that has the property that the sum of the numbers in any row, column, or diagonal is the same. Fill in the square below so that it becomes a magic square, and all digits 1, 2, 3, . . . , 9 are used exactly once.

6		8
	5	
		4

67. Magic Square Refer to **Exercise 66.** Complete the magic square below so that all counting numbers 1, 2, 3, . . . , 16 are used exactly once, and the sum in each row, column, or diagonal is 34.

6			9
	15		14
11		10	
16		13	

68. Decimal Digit What is the 100th digit in the decimal representation for $\frac{1}{7}$?

69. Pitches in a Baseball Game What is the minimum number of pitches that a baseball player who pitches a complete game can make in a regulation 9-inning baseball game?

70. Weighing Coins You have eight coins. Seven are genuine and one is a fake, which weighs a little less than the other seven. You have a balance scale, which you may use only three times. Tell how to locate the bad coin in three weighings. (Then show how to detect the bad coin in only *two* weighings.)

71. Geometry Puzzle When the diagram shown is folded to form a cube, what letter is opposite the face marked Z?

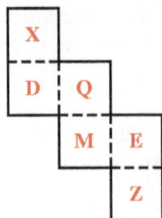

X		
D	Q	
	M	E
		Z

72. Geometry Puzzle Draw the following figure without picking up your pencil from the paper and without tracing over a line you have already drawn.

73. Geometry Puzzle Repeat **Exercise 72** for this figure.

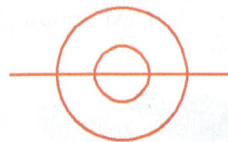

74. Books on a Shelf Volumes 1 and 2 of *The Complete Works of Wally Smart* are standing in numerical order from left to right on your bookshelf. Volume 1 has 450 pages and Volume 2 has 475 pages. Excluding the covers, how many pages are between page 1 of Volume 1 and page 475 of Volume 2?

75. Paying for a Mint Brian has an unlimited number of cents (pennies), nickels, and dimes. In how many different ways can he pay 15¢ for a chocolate mint? (For example, one way is 1 dime and 5 pennies.)

76. Teenager's Age A teenager's age increased by 2 gives a perfect square. Her age decreased by 10 gives the square root of that perfect square. She is 5 years older than her brother. How old is her brother?

77. Area and Perimeter Triangle *ABC* has sides 10, 24, and 26 cm long. A rectangle that has an area equal to that of the triangle is 3 cm wide. Find the perimeter of the rectangle. (November 13, 2008)

78. Making Change In how many different ways can you make change for a half dollar using currently minted U.S. coins, if cents (pennies) are not allowed?

79. Ages James, Dan, Jessica, and Cathy form a pair of married couples. Their ages are 36, 31, 30, and 29. Jessica is married to the oldest person in the group. James is older than Jessica but younger than Cathy. Who is married to whom, and what are their ages?

80. Final Digit What is the last digit of $49{,}327^{1783}$? (April 11, 2009)

81. Geometry Puzzle What is the maximum number of small squares in which we may place crosses (✕) and not have any row, column, or diagonal completely filled with crosses? Illustrate your answer.

82. Making Change Webster has some pennies, dimes, and quarters in his pocket. When Josefa asks him for change for a dollar, Webster discovers that he cannot make the change exactly. What is the largest possible total value of the coins in his pocket? (October 5, 2009)

WHEN Will I Ever USE This ?

Suppose that you are an employee of a group home, and you have been assigned to provide the necessary items for a Thanksgiving meal for ten of the residents of the home, who will be preparing the meal for themselves. You decide that the dinner will be a traditional one: turkey, stuffing, cranberry sauce, yams, green bean casserole, rolls, and iced tea. You must go to the supermarket for these items and have decided to use cash to pay for them. Of course, you do not want to approach the checkout counter and not have enough for the purchase. You also know that a sales tax of 8.75% will be added to the total. You take your allotment of $80 in cash with you.

In a situation like this, it is not difficult to get a fairly accurate idea of what the total will be by mentally rounding each item (up or down) to the nearest dollar and keeping a running total along the way as you place items in the shopping cart. Here is an example of this procedure.

Item	Actual Cost	Estimate
18-lb turkey	$26.82	$30
12-pack of dinner rolls	2.29	2
15-oz container of margarine	2.79	3
40-oz can of yams	3.34	3
28-oz can of cranberry sauce	3.97	4
26-oz can of cream of mushroom soup	2.99	3
14-oz bag of herb stuffing	3.59	4
1-lb bag of pecans	8.79	9
50-oz can of green beans	2.59	3
28-oz can of onion flakes	2.19	2
22-bag pack of tea bags	4.29	4

Before reaching the checkout counter, look at the items in the basket (move from the top to the bottom in the third column) and add them, rounding off whenever doing so helps to simplify the computation. Here is one of many ways this can be done. (Remember, this is not an *exact* computation, so thought processes will vary.)

"30 plus 2 gives 32; I notice that the next three items total 10, so 32 plus 10 equals 42, which I will round down to 40; plus 3 gives 43; plus 4 gives 47, which I will round up to 50; I will round 9 to 10 and add 50 to 10 to give 60; the final three items total 9, which I will round down to 8 (because I rounded up just before) and add to 60 to get 68; 68 is about 70."

Before tax is added, the items total about $70. The sales tax is 8.75%, so round this to 10%. Taking 10% of a number is simple: Just move the decimal point one unit to the left. In this case, the tax will be about $7. So add 7 to 70 to get an approximate total of $77. It looks like $80 will cover the total cost.

Now, as the cashier rings up the purchase, the screen shows that the actual total cost is $63.65, the sales tax is $5.57, and the grand total is $69.22. The estimate is a bit high in this example (because most of the roundoffs were "upward"), but this is what "being in the ballpark" means: The estimate is close enough for our purpose.

Happy Thanksgiving.

1.4 NUMERACY IN TODAY'S WORLD

OBJECTIVES

1 Be able to use a calculator for routine mathematical operations.

2 Be able to use estimation techniques.

3 Be able to interpret information by reading circle, bar, and line graphs.

4 Be able to use writing skills to convey information about mathematics.

The familiar term *literacy* applies to language in the same way that the term **numeracy** applies to mathematics. It is virtually impossible to function in the world today without understanding fundamental number concepts. The basic ideas of calculating, estimating, interpreting data from graphs, and conveying mathematics via language and writing are among the skills required to be "numerate."

Calculation

The search for easier ways to calculate and compute has culminated in the development of hand-held calculators and computers. For the general population, a calculator that performs the operations of arithmetic and a few other functions is sufficient. These are known as **four-function calculators.** Students who take higher mathematics courses (engineers, for example) usually need the added power of **scientific calculators. Graphing calculators,** which actually plot graphs on small screens, are also available. *Always refer to your owner's manual if you need assistance in performing an operation with your calculator. If you need further help, ask your instructor or another student who is using the same model.*

Today's smartphones routinely include a calculator application (app). For example, Apple's iPhone has an app that serves as a four-function calculator when the phone is held vertically, but it becomes a scientific calculator when held horizontally. Furthermore, graphing calculator apps are available at little or no cost. Although it is not necessary to have a graphing calculator to study the material presented in this text, we occasionally include graphing calculator screens to support results obtained or to provide supplemental information.*

The screens that follow illustrate some common entries and operations.

Screen A	
3+9	12
7−2	5
4*5	20

$3 + 9 = 12$
$7 - 2 = 5$
$4 \times 5 = 20$

A

Screen B	
24/20	1.2
Ans▶Frac	$\frac{6}{5}$
5−(8−7)	4

$\frac{24}{20} = 1.2$
$1.2 = \frac{6}{5}$
$5 - (8 - 7) = 4$

B

Screen C	
7^2	49
5^3	125
$\sqrt{81}$	9

$7^2 = 49$
$5^3 = 125$
$\sqrt{81} = 9$

C

Shown here is an example of a **calculator app** from a smartphone. Since the introduction of hand-held calculators in the early 1970s, the methods of everyday arithmetic have been drastically altered. One of the first consumer models available was the Texas Instruments SR-10, which sold for nearly $150 in 1973. It could perform the four operations of arithmetic and take square roots, but it could do very little more.

Screen A illustrates how two numbers can be added, subtracted, or multiplied. Screen B shows how two numbers can be divided, how the decimal quotient (stored in the memory cell Ans) can be converted into a fraction, and how parentheses can be used in a computation. Screen C shows how a number can be squared, how it can be cubed, and how its square root can be taken.

Screen D	
$\sqrt[3]{27}$	3
$\sqrt[4]{16}$	2
5^{-1}	.2

$\sqrt[3]{27} = 3$
$\sqrt[4]{16} = 2$
$5^{-1} \left(\text{or } \frac{1}{5} \right) = .2$

D

Screen E	
π	3.141592654
5!	120
6265804*8980591	5.627062301E13

$\pi \approx 3.141592654$
5! (or $1 \times 2 \times 3 \times 4 \times 5$) = 120
6,265,804 $\times$ 8,980,591 $\approx 5.627062301 \times 10^{13}$
$\approx$ indicates "is approximately equal to"

E

*Because it is one of the most popular graphing calculators, we include screens similar to those generated by the TI-83 Plus and TI-84 Plus from Texas Instruments.

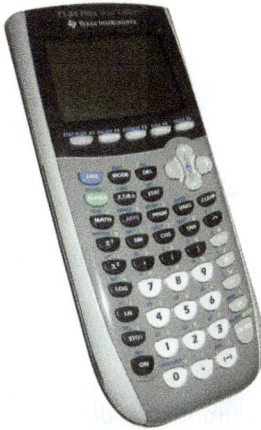

The popular **TI-84 Plus** graphing calculator is shown here.

Screen D shows how other roots (cube root and fourth root) can be found, and how the reciprocal of a number can be found using −1 as an exponent. Screen E shows how π can be accessed with its own special key, how a *factorial* (as indicated by !) can be found, and how a result might be displayed in *scientific notation*. (The "E13" following 5.627062301 means that this number is multiplied by 10^{13}. This answer is still only an approximation, because the product $6,265,804 \times 8,980,591$ contains more digits than the calculator can display.)

In **Section 1.3** we presented a list of problem-solving strategies. As Terry Krieger points out in the chapter opener, "mathematics may be the only discipline in which different people, using wildly varied but logically sound methods, will arrive at the same correct result" with respect to solving problems. Sometimes more than one strategy can be used in a particular situation. In **Example 1,** we present a type of problem that has been around for thousands of years in various forms, and is solved by observing a pattern within a table. One ancient form of this problem deals with doubling a kernel of corn for each square on a checkerbord. (See http://mathforum.org/sanders/geometry/GP11Fable.html.) It illustrates an example of exponential growth, covered in more detail in a later chapter.

EXAMPLE 1 Calculating a Sum That Involves a Pattern

Following her success in *I Love Lucy,* Lucille Ball starred in *The Lucy Show,* which aired for six seasons on CBS in the 1960s. She worked for Mr. Mooney (Gale Gordon), who was very careful with his money.

In the September 26, 1966, show "Lucy, the Bean Queen," Lucy learned a lesson about saving money. Mr. Mooney had refused to lend her $1500 to buy furniture, because he claimed she did not know the value of money. He explained to her that if she were to save so that she would have one penny on Day 1, two pennies on Day 2, four pennies on Day 3, and so on, she would have more than enough money to buy her furniture after only nineteen days. Use a calculator to verify this fact.

Solution

A calculator will help in constructing **Table 4.**

Table 4

Day Number	Accumulated Savings on That Day	Day Number	Accumulated Savings on That Day
1	$0.01	11	$10.24
2	0.02	12	20.48
3	0.04	13	40.96
4	0.08	14	81.92
5	0.16	15	163.84
6	0.32	16	327.68
7	0.64	17	655.36
8	1.28	18	1310.72
9	2.56	19	2621.44
10	5.12		

So, indeed, Lucy will have accumulated $2621.44 on day 19, which is enough to buy the furniture.)

Estimation

Although calculators can make life easier when it comes to computations, many times we need only estimate an answer to a problem, and in these cases, using a calculator may not be necessary or appropriate.

EXAMPLE 2 Estimating an Appropriate Number of Birdhouses

A birdhouse for swallows can accommodate up to 8 nests. How many birdhouses would be necessary to accommodate 58 nests?

Solution

If we divide 58 by 8 either by hand or with a calculator, we get 7.25. Can this possibly be the desired number? Of course not, because we cannot consider fractions of birdhouses. Do we need 7 or 8 birdhouses?

To provide nesting space for the nests left over after the 7 birdhouses (as indicated by the decimal fraction), we should plan to use 8 birdhouses. In this problem, we must round our answer *up* to the next counting number. ∎

EXAMPLE 3 Approximating Average Number of Yards per Carry

In 2013, Fred Jackson of the Buffalo Bills carried the football a total of 206 times for 890 yards (*Source:* www.nfl.com). Approximate his average number of yards per carry that year.

Solution

Because we are told only to find Jackson's approximate average, we can say that he carried about 200 times for about 900 yards, and his average was therefore about $\frac{900}{200} = 4.5$ yards per carry. (A calculator shows that his average to the nearest tenth was 4.3 yards per carry. Verify this.) ∎

Interpretation of Graphs

In a **circle graph,** or **pie chart,** a circle is used to indicate the total of all the data categories represented. The circle is divided into sectors, or wedges (like pieces of a pie), whose sizes show the relative magnitudes of the categories. The sum of all the fractional parts must be 1 (for one whole circle).

EXAMPLE 4 Interpreting Information in a Circle Graph

In a recent month there were about 2100 million (2.1 billion) Internet users worldwide. The circle graph in **Figure 8** shows the approximate shares of these users living in various regions of the world.

(a) Which region had the largest share of Internet users? What was that share?

(b) Estimate the number of Internet users in North America.

(c) How many actual Internet users were there in North America?

Worldwide Internet Users by Region

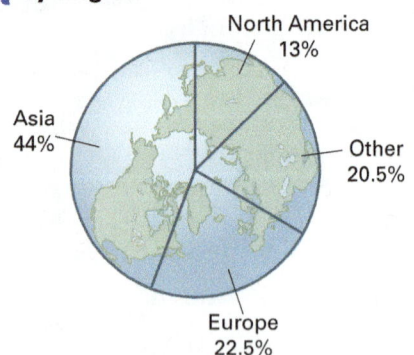

Asia 44%, North America 13%, Other 20.5%, Europe 22.5%

Source: www.internetworldstats.com

Figure 8

Solution

(a) In the circle graph, the sector for Asia is the largest, so Asia had the largest share of Internet users, 44%.

(b) A share of 13% can be rounded down to 10%. Then find 10% of 2100 million by finding $\frac{1}{10}$ of 2100, or 210. There were *about* 210 million users in North America. (This estimate is low, because we rounded *down*.)

(c) To find the actual number of users, find 13% of 2100 million. We do this by multiplying 0.13×2100 million.

$$\underbrace{0.13}_{13\%} \quad \times \quad \underbrace{2100 \text{ million}}_{\text{total}} \quad = \quad \underbrace{273 \text{ million}}_{\substack{\text{Actual number} \\ \text{of users in} \\ \text{North America}}}$$

A **bar graph** is used to show comparisons. It consists of a series of bars (or simulations of bars) arranged either vertically or horizontally. In a bar graph, values from two categories are paired with each other (for example, years with dollar amounts).

EXAMPLE 5   Interpreting Information in a Bar Graph

The bar graph in **Figure 9** shows annual per-capita spending on health care in the United States for the years 2004 through 2009.

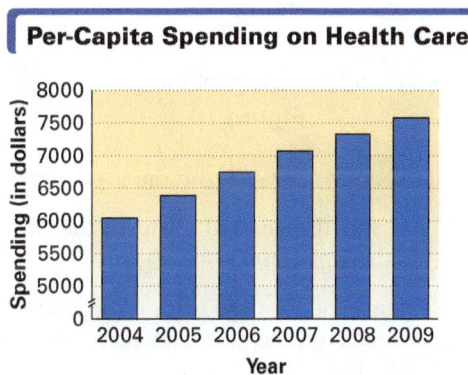

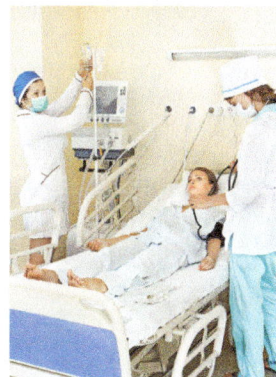

Per-Capita Spending on Health Care

Source: U.S. Centers for Medicare and Medicaid Services.

Figure 9

(a) In what years was per-capita health care spending greater than $7000?

(b) Estimate per-capita health care spending in 2004 and 2008.

(c) Describe the change in per-capita spending as the years progressed.

Solution

(a) Locate 7000 on the vertical axis and follow the line across to the right. Three years—2007, 2008, and 2009—have bars that extend above the line for 7000, so per-capita health care spending was greater than $7000 in those years.

(b) Locate the top of the bar for 2004 and move horizontally across to the vertical scale to see that it is about 6000. Per-capita health care spending for 2004 was about $6000.

Similarly, follow the top of the bar for 2008 across to the vertical scale to see that it lies a little more than halfway between 7000 and 7500, so per-capita health care spending in 2008 was about $7300.

(c) As the years progressed, per-capita spending on health care increased steadily, from about $6000 in 2004 to about $7500 in 2009. ∎

A **line graph** is used to show changes or trends in data over time. To form a line graph, we connect a series of points representing data with line segments.

EXAMPLE 6 Interpreting Information in a Line Graph

The line graph in **Figure 10** shows average prices of a gallon of regular unleaded gasoline in the United States for the years 2004 through 2011.

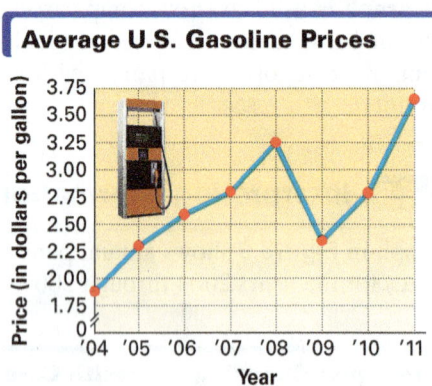

Average U.S. Gasoline Prices

Source: U.S. Department of Energy.

Figure 10

(a) Between which years did the average price of a gallon of gasoline decrease?

(b) What was the general trend in the average price of a gallon of gasoline from 2004 through 2008?

(c) Estimate the average price of a gallon of gasoline in 2004 and in 2008. About how much did the price increase between 2004 and 2008?

Solution

(a) The line between 2008 and 2009 falls, so the average price of a gallon of gasoline decreased from 2008 to 2009.

(b) The line graph rises from 2004 to 2008, so the average price of a gallon of gasoline increased over those years.

(c) Move up from 2004 on the horizontal scale to the point plotted for 2004. This point is about halfway between the lines on the vertical scale for $1.75 and $2.00. Halfway between $1.75 and $2.00 would be about $1.88. Therefore, a gallon of gasoline cost about $1.88 in 2004.

Similarly, locate the point plotted for 2008. Moving across to the vertical scale, the graph indicates that the price for a gallon of gasoline in 2008 was about $3.25.

Between 2004 and 2008, the average price of a gallon of gasoline increased by about

$$\$3.25 - \$1.88 = \$1.37.$$
∎

Communicating Mathematics through Language Skills

Research has indicated that the ability to express mathematical observations in writing can serve as a positive force in one's continued development as a mathematics student. The implementation of writing in the mathematics class can use several approaches.

One way of using writing in mathematics is to keep a **journal** in which you spend a few minutes explaining what happened in class that day. The journal entries may be general or specific, depending on the topic covered, the degree to which you understand the topic, your interest level at the time, and so on. Journal entries are usually written in informal language.

Although journal entries are for the most part informal writings in which the student's thoughts are allowed to roam freely, entries in **learning logs** are typically more formal. An instructor may pose a specific question for a student to answer in a learning log. In this text, we intersperse in each exercise set exercises that require written answers that are appropriate for answering in a learning log.

Mathematical writing takes many forms. One of the most famous author/mathematicians was **Charles Dodgson** (1832–1898), who used the pen name **Lewis Carroll.**

Dodgson was a mathematics lecturer at Oxford University in England. Queen Victoria told Dodgson how much she enjoyed *Alice's Adventures in Wonderland* and how much she wanted to read his next book; he is said to have sent her *Symbolic Logic,* his most famous mathematical work.

The *Alice* books made Carroll famous. Late in life, however, Dodgson shunned attention and denied that he and Carroll were the same person, even though he gave away hundreds of signed copies to children and children's hospitals.

EXAMPLE 7   Writing an Answer to a Conceptual Exercise

Exercise 17 of **Section 1.1** reads as follows.

> ***Discuss the differences between inductive and deductive reasoning. Give an example of each.***

Write a short paragraph to answer this exercise.

Solution

Here is one possible response.

> Deductive reasoning occurs when you go from general ideas to specific ones. For example, I know that I can multiply both sides of $\frac{1}{2}x = 6$ by 2 to get $x = 12$, because I can multiply both sides of any equation by whatever I want (except 0). Inductive reasoning goes the other way. If I have a general conclusion from specific observations, that's inductive reasoning. Example – in the numbers 4, 8, 12, 16, and so on, I can conclude that the next number is 20, since I always add 4 to get the next number.

The motto "Publish or perish" has long been around, implying that a scholar in pursuit of an academic position must publish in a journal in his or her field. There are numerous such journals in mathematics research and/or mathematics education. The National Council of Teachers of Mathematics publishes *Teaching Children Mathematics, Mathematics Teaching in the Middle School, Mathematics Teacher, Journal for Research in Mathematics Education, Mathematics Teacher Educator,* and *Student Explorations in Mathematics.* Refer to the Web site www.nctm.org to access these journals, or refer to print copies in your local library.

Writing a report on a journal article can help you understand what mathematicians do and what ideas mathematics teachers use to convey concepts to their students. Many professors in mathematics survey courses require short term papers of their students. In doing such research, students can become aware of the plethora of books and articles on mathematics and mathematicians, many written specifically for the layperson.

A list of important mathematicians, philosophers, and scientists follows.

Abel, N.	Cardano, G.	Gauss, C.	Noether, E.
Agnesi, M. G.	Copernicus, N.	Hilbert, D.	Pascal, B.
Agnesi, M. T.	De Morgan, A.	Kepler, J.	Plato
Al-Khowârizmi	Descartes, R.	Kronecker, L.	Polya, G.
Apollonius	Euler, L.	Lagrange, J.	Pythagoras
Archimedes	Fermat, P.	Leibniz, G.	Ramanujan, S.
Aristotle	Fibonacci	L'Hôspital, G.	Riemann, G.
Babbage, C.	(Leonardo	Lobachevsky, N.	Russell, B.
Bernoulli, Jakob	of Pisa)	Mandelbrot, B.	Somerville, M.
Bernoulli,	Galileo (Galileo	Napier, J.	Tartaglia, N.
Johann	Galilei)	Nash, J.	Whitehead, A.
Cantor, G.	Galois, E.	Newton, I.	Wiles, A.

The following topics in the history and development of mathematics can also be used for term papers.

Babylonian mathematics

Egyptian mathematics

The origin of zero

Plimpton 322

The Rhind papyrus

Origins of the Pythagorean theorem

The regular (Platonic) solids

The Pythagorean brotherhood

The Golden Ratio (Golden Section)

The three famous construction problems of the Greeks

The history of the approximations of π

Euclid and his *Elements*

Early Chinese mathematics

Early Hindu mathematics

Origin of the word *algebra*

Magic squares

Figurate numbers

The Fibonacci sequence

The Cardano/Tartaglia controversy

Historical methods of computation (logarithms, the abacus, Napier's rods, the slide rule, etc.)

Pascal's triangle

The origins of probability theory

Women in mathematics

Mathematical paradoxes

Unsolved problems in mathematics

The four-color theorem

The proof of Fermat's Last Theorem

The search for large primes

Fractal geometry

The co-inventors of calculus

The role of the computer in the study of mathematics

Mathematics and music

Police mathematics

The origins of complex numbers

Goldbach's conjecture

The use of the Internet in mathematics education

The development of graphing calculators

Mathematics education reform movement

Multicultural mathematics

The Riemann Hypothesis

1.4 EXERCISES

Perform the indicated operations, and give as many digits in your answer as shown on your calculator display. (The number of displayed digits may vary depending on the model used.)

1. $39.7 + (8.2 - 4.1)$

2. $2.8 \times (3.2 - 1.1)$

3. $\sqrt{5.56440921}$

4. $\sqrt{37.38711025}$

5. $\sqrt[3]{418.508992}$

6. $\sqrt[3]{700.227072}$

7. 2.67^2

8. 3.49^3

9. 5.76^5

10. 1.48^6

11. $\dfrac{14.32 - 8.1}{2 \times 3.11}$

12. $\dfrac{12.3 + 18.276}{3 \times 1.04}$

13. $\sqrt[5]{1.35}$

14. $\sqrt[6]{3.21}$

15. $\dfrac{\pi}{\sqrt{2}}$

16. $\dfrac{2\pi}{\sqrt{3}}$

17. $\sqrt[4]{\dfrac{2143}{22}}$

18. $\dfrac{12{,}345{,}679 \times 72}{\sqrt[3]{27}}$

19. $\dfrac{\sqrt{2}}{\sqrt[3]{6}}$

20. $\dfrac{\sqrt[3]{12}}{\sqrt{3}}$

21. Choose any number consisting of five digits. Multiply it by 9 on your calculator. Now add the digits in the answer. If the sum is more than 9, add the digits of this sum, and repeat until the sum is less than 10. Your answer will always be 9. Repeat the exercise with a number consisting of six digits. Does the same result hold?

22. Use your calculator to *square* the following two-digit numbers ending in 5: 15, 25, 35, 45, 55, 65, 75, 85. Write down your results, and examine the pattern that develops. Then use inductive reasoning to predict the value of 95^2. Write an explanation of how you can mentally square a two-digit number ending in 5.

Perform each calculation and observe the answers. Then fill in the blank with the appropriate response.

23. $\boxed{\dfrac{-3}{-8}}$; $\boxed{\dfrac{-5}{-4}}$; $\boxed{\dfrac{-2.7}{-4.3}}$

Dividing a negative number by another negative number gives a _____ product.
(negative/positive)

24. $\boxed{5 * -4}$; $\boxed{-3 * 8}$; $\boxed{2.7 * -4.3}$

Multiplying a negative number by a positive number gives a _____ product.
(negative/positive)

25. $\boxed{5.6^0}$; $\boxed{\pi^0}$; $\boxed{2^0}$; $\boxed{120^0}$

Raising a nonzero number to the power 0 gives a result of _____.

26. $\boxed{1^2}$; $\boxed{1^3}$; $\boxed{1^{-3}}$; $\boxed{1^0}$

Raising 1 to any power gives a result of _____.

27. $\boxed{\dfrac{1}{7}}$; $\boxed{\dfrac{1}{-4}}$; $\boxed{\dfrac{1}{3}}$; $\boxed{\dfrac{1}{-8}}$

The sign of the reciprocal of a number is _____ the sign of the number.
(the same as/different from)

28. $\boxed{5 \div 0}$; $\boxed{9 \div 0}$; $\boxed{0 \div 0}$

Dividing a number by 0 gives a(n) _____ on a calculator.

29. $\boxed{0 \div 8}$; $\boxed{0 \div -2}$; $\boxed{0 \div \pi}$

Zero divided by a nonzero number gives a quotient of _____.

30. $\boxed{\sqrt{-3}}$; $\boxed{\sqrt{-4}}$; $\boxed{\sqrt{-10}}$

Taking the square root of a negative number gives a(n) _____ on a calculator.

31. $\boxed{-3 * -4 * -5}$; $\boxed{-3 * -4 * -5 * -6 * -7}$; $\boxed{-3 * -4 * -5 * -6 * -7 * -8 * -9}$

Multiplying an *odd* number of negative numbers gives a _____ product.
(positive/negative)

32. $\boxed{-3 * -4}$; $\boxed{-3 * -4 * -5 * -6}$; $\boxed{-3 * -4 * -5 * -6 * -7 * -8}$

Multiplying an *even* number of negative numbers gives a _____ product.
(positive/negative)

33. Find the decimal representation of $\frac{1}{6}$ on your calculator. Following the decimal point will be a 1 and a string of 6s. The final digit will be a 7 if your calculator *rounds off* or a 6 if it *truncates*. Which kind of calculator do you have?

34. Choose any three-digit number and enter the digits into a calculator. Then enter them again to get a six-digit number. Divide this six-digit number by 7. Divide the result by 13. Divide the result by 11. What is interesting about your answer? Explain why this happens.

35. Choose any digit except 0. Multiply it by 429. Now multiply the result by 259. What is interesting about your answer? Explain why this happens.

36. Refer to **Example 1.** If Lucy continues to double her savings amount each day, on what day will she become a millionaire?

Give an appropriate counting number answer to each question in Exercises 37–40. (Find the least counting number that will work.)

37. Pages to Store Trading Cards A plastic page designed to hold trading cards will hold up to 9 cards. How many pages will be needed to store 563 cards?

38. Drawers for DVDs A sliding drawer designed to hold DVD cases has 20 compartments. If Chris wants to house his collection of 408 Disney DVDs, how many such drawers will he need?

39. Containers for African Violets A gardener wants to fertilize 800 African violets. Each container of fertilizer will supply up to 60 plants. How many containers will she need to do the job?

40. Fifth-Grade Teachers Needed False River Academy has 155 fifth-grade students. The principal has decided that each fifth-grade teacher should have a maximum of 24 students. How many fifth-grade teachers does he need?

In Exercises 41–46, use estimation to determine the choice closest to the correct answer.

41. Price per Acre of Land To build a "millennium clock" on Mount Washington in Nevada that would tick once each year, chime once each century, and last at least 10,000 years, the nonprofit Long Now Foundation purchased 80 acres of land for $140,000. Which one of the following is the closest estimate to the price per acre?

   A. $1000 **B.** $2000 **C.** $4000 **D.** $11,200

42. Time of a Round Trip The distance from Seattle, Washington, to Springfield, Missouri, is 2009 miles. About how many hours would a round trip from Seattle to Springfield and back take a bus that averages 50 miles per hour for the entire trip?

   A. 60 **B.** 70 **C.** 80 **D.** 90

43. People per Square Mile Baton Rouge, LA has a population of 230,058 and covers 76.9 square miles. About how many people per square mile live in Baton Rouge?

   A. 3000 **B.** 300 **C.** 30 **D.** 30,000

44. Revolutions of Mercury The planet Mercury takes 88.0 Earth days to revolve around the sun once. Pluto takes 90,824.2 days to do the same. When Pluto has revolved around the sun once, about how many times will Mercury have revolved around the sun?

   A. 100,000 **B.** 10,000 **C.** 1000 **D.** 100

45. Reception Average In 2013, A. J. Green of the Cincinnati Bengals caught 98 passes for 1426 yards. His approximate number of yards gained per catch was _____.

   A. $\frac{1}{14}$ **B.** 0.07 **C.** 139,748 **D.** 14

46. Area of the Sistine Chapel The Sistine Chapel in Vatican City measures 40.5 meters by 13.5 meters.

Which is the closest approximation to its area?

   A. 110 meters   **B.** 55 meters
   C. 110 square meters **D.** 600 square meters

Foreign-Born Americans *Approximately 37.5 million people living in the United States in a recent year were born in other countries. The circle graph gives the share from each region of birth for these people. Use the graph to answer each question in Exercises 47–50.*

U.S. Foreign-Born Population by Region of Birth

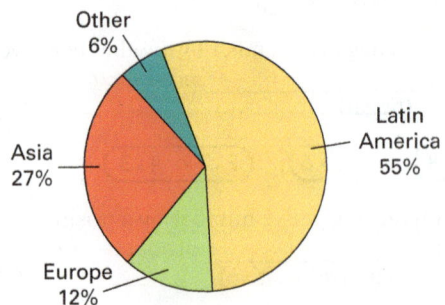

Other 6%, Latin America 55%, Asia 27%, Europe 12%

Source: U.S. Census Bureau.

47. What share was from other regions?

48. What share was from Latin America or Asia?

49. How many people (in millions) were born in Europe?

50. How many more people (in millions) were born in Latin America than in Asia?

Milk Production   *The bar graph shows total U.S. milk production (in billions of pounds) for the years 2004 through 2010. Use the bar graph to work Exercises 51–54.*

U.S. Milk Production

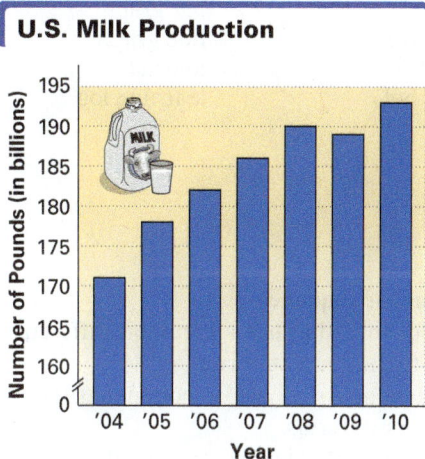

Source: U.S. Department of Agriculture.

51. In what years was U.S. milk production greater than 185 billion pounds?

52. In what years was U.S. milk production about the same?

53. Estimate U.S. milk production in 2004 and 2010.

54. Describe the change from 2004 to 2010.

U.S. Car Imports   *The line graph shows the number of new and used passenger cars (in millions) imported into the United States over the years 2005 through 2010. Use the line graph to work Exercises 55–58.*

Passenger Cars Imported into the U.S.

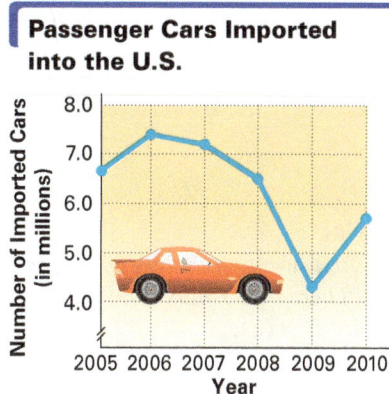

Source: U.S. Census Bureau.

55. Over which two consecutive years did the number of imported cars increase the most? About how much was this increase?

56. Estimate the number of cars imported during 2007, 2008, and 2009.

57. Describe the trend in car imports from 2006 to 2009.

58. During which year(s) were fewer than 6 millions cars imported into the United States?

Use effective writing skills to address the following.

59. *Mathematics Web Sites*   The following Web sites provide a fascinating list of mathematics-related topics. Investigate, choose a topic that interests you, and report on it according to the guidelines provided by your instructor.

www.mathworld.wolfram.com

world.std.com/~reinhold/mathmovies.html

www.maths.surrey.ac.uk/hosted-sites/R.Knott/

http://dir.yahoo.com/Science/Mathematics/

www.cut-the-knot.org

www.ics.uci.edu/~eppstein/recmath.html

www.coolmathguy.com

ptri1.tripod.com

mathforum.org

rosettacode.org

plus.maths.org

60. *The Simpsons*   The longest-running animated television series is *The Simpsons,* having begun in 1989. The Web site www.simpsonsmath.com explores the occurrence of mathematics in the episodes on a season-by-season basis. Watch several episodes and elaborate on the mathematics found in them.

61. *Donald in Mathmagic Land*   One of the most popular mathematical films of all time is *Donald in Mathmagic Land,* a 1959 Disney short that is available on DVD. Spend an entertaining half-hour watching this film, and write a report on it according to the guidelines provided by your instructor.

62. *Mathematics in Hollywood*   A theme of mathematics-related scenes in movies and television is found throughout this text. Prepare a report on one or more such scenes, and determine whether the mathematics involved is correct or incorrect. If correct, show why. If incorrect, find the correct answer.

CHAPTER 1 SUMMARY

KEY TERMS

1.1

conjecture
inductive reasoning
counterexample
deductive reasoning
natural (counting) numbers
base
exponent
premise
conclusion
logical argument

1.2

number sequence
terms of a sequence
arithmetic sequence
common difference
geometric sequence
common ratio
method of successive
    differences
mathematical
    induction

triangular, square, and
    pentagonal numbers
figurate number
subscript
Kaprekar constant

1.3

perfect square
square root
radical symbol

1.4

numeracy
four-function calculator
scientific calculator
graphing calculator
circle graph (pie chart)
bar graph
line graph
journal
learning log

TEST YOUR WORD POWER

See how well you have learned the vocabulary in this chapter.

1. A **conjecture** is
   A. a statement that has been proved to be true.
   B. an educated guess based on repeated observations.
   C. an example that shows that a general statement is false.
   D. an example of deductive reasoning.

2. An example of a **natural number** is
   A. 0.   B. $\frac{1}{2}$.   C. -1.   D. 1.

3. An **arithmetic sequence** is
   A. a sequence that has a common difference between any two successive terms.
   B. a sequence that has a common sum of any two successive terms.
   C. a sequence that has a common ratio between any two successive terms.
   D. a sequence that can begin 1, 1, 2, 3, 5. . . .

4. A **geometric sequence** is
   A. a sequence that has a common difference between any two successive terms.
   B. a sequence that has a common sum of any two successive terms.
   C. a sequence that has a common ratio between any two successive terms.
   D. A sequence that can begin 1, 1, 2, 3, 5, ...

5. The symbol T_n, which uses the **subscript** n, is read
   A. "T to the nth power."   B. "T times n."
   C. "T of n."   D. "T sub n."

ANSWERS
1. B   2. D   3. A   4. C   5. D

QUICK REVIEW

Concepts	Examples
1.1 Solving Problems by Inductive Reasoning	
Inductive Reasoning Inductive reasoning is characterized by drawing a general conclusion (making a conjecture) from repeated observations of specific examples. The conjecture may or may not be true.	Consider the following: *When I square the first twenty numbers ending in 5, the result always ends in 25. Therefore, I make the conjecture that this happens in the twenty-first case.*
A general conclusion from inductive reasoning can be shown to be false by providing a single counterexample.	This is an example of inductive reasoning because a general conclusion follows from repeated observations.

Concepts	Examples

Deductive Reasoning
Deductive reasoning is characterized by applying general principles to specific examples.

Consider the following:
The formula for finding the perimeter P of a rectangle with length L and width W is P = 2L + 2W. Therefore, the perimeter P is
$$2(5) + 2(3) = 16.$$

W = 3

L = 5

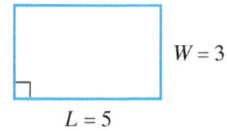

This is an example of deductive reasoning because a specific conclusion follows from a mathematical formula that is true in general.

1.2 An Application of Inductive Reasoning: Number Patterns

Sequences
A number sequence is a list of numbers having a first number, a second number, a third number, and so on, which are called the terms of the sequence.

Arithmetic Sequence
In an arithmetic sequence, each term after the first is obtained by adding the same number, called the common difference.

The arithmetic sequence that begins
$$2, 4, 6, 8$$
has common difference $4 - 2 = 2$, and the next term in the sequence is $8 + 2 = 10$.

Geometric Sequence
In a geometric sequence, each term after the first is obtained by multiplying by the same number, called the common ratio.

The geometric sequence that begins
$$4, 20, 100, 500$$
has common ratio $\frac{20}{4} = 5$, and the next term in the sequence is $500 \times 5 = 2500$.

Method of Successive Differences
The next term in a sequence can sometimes be found by computing successive differences between terms until a pattern can be established.

The sequence that begins
$$7, 15, 25, 37$$
has the following successive differences.

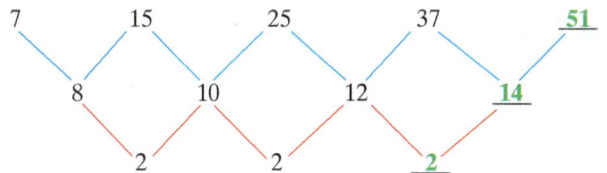

7   15   25   37   **51**
   8   10   12   **14**
      2   2   **2**

The next term in the sequence is $37 + 14 = 51$.

Figurate Numbers
Figurate numbers, such as triangular, square, and pentagonal numbers, can be represented by geometric arrangements of points.

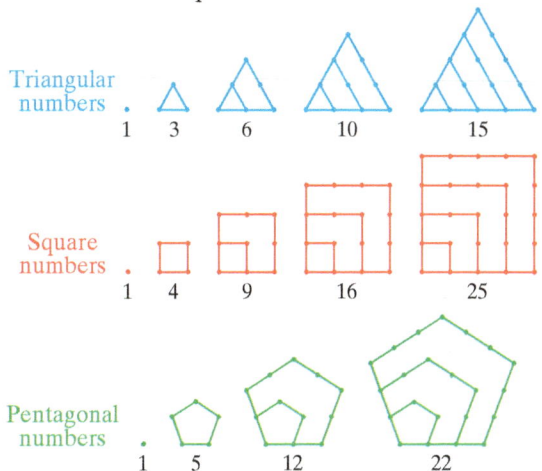

Triangular numbers: 1 3 6 10 15

Square numbers: 1 4 9 16 25

Pentagonal numbers: 1 5 12 22

Concepts | **Examples**

1.3 Strategies for Problem Solving

Polya's Four-Step Method for Problem Solving

Step 1 Understand the problem.

Step 2 Devise a plan.

Step 3 Carry out the plan.

Step 4 Look back and check.

Problem-Solving Strategies

- Make a table or a chart.
- Look for a pattern.
- Solve a similar, simpler problem.
- Draw a sketch.
- Use inductive reasoning.
- Write an equation and solve it.
- If a formula applies, use it.
- Work backward.
- Guess and check.
- Use trial and error.
- Use common sense.
- Look for a "catch" if an answer seems too obvious or impossible.

What is the ones, or units, digit in 7^{350}?

Solution
We can observe a pattern in the table of simpler powers of 7. (Use a calculator.)

$7^1 = 7$ $7^5 = 16,807$ $7^9 = 40,353,607$

$7^2 = 49$ $7^6 = 117,649$ ...

$7^3 = 343$ $7^7 = 823,543$ ...

$7^4 = 2401$ $7^8 = 5,764,801$ ...

The ones digit appears in a pattern of four digits over and over: 7, 9, 3, 1, 7, 9, 3, 1, ... If the exponent is divided by 4, the remainder helps predict the ones digit. If we divide the exponent 350 by 4, the quotient is 87 and the remainder is 2, just as it is in the second row above for 7^2 and 7^6, where the units digit is 9. So the units digit in 7^{350} is 9.

How many ways are there to make change equivalent to one dollar using only nickels, dimes, and quarters? You do not need at least one coin of each denomination. (December 26, 2013)

Solution (Verify each of the following by trial and error.)

If we start with 4 quarters, there is 1 way to make change for a dollar.

If we start with 3 quarters, there are 3 ways.

If we start with 2 quarters, there are 6 ways. Thus, there are $1 + 3 + 6 + 8 + 11 = 29$ ways in all.

If we start with 1 quarter, there are 8 ways.

If we start with 0 quarters, there are 11 ways.

1.4 Numeracy in Today's World

There are a variety of types of calculators available; four-function, scientific, and graphing calculators are some of them.

In practical applications, it is often convenient to simply approximate to get an idea of an answer to a problem.

Circle graphs (pie charts), bar graphs, and line graphs are used in today's media to illustrate data in a compact way.

Percents of Students Who Return for Second Year (2-Year Public Institutions)

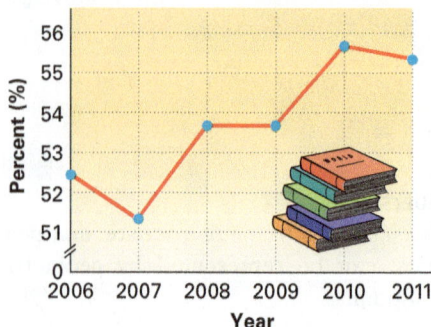

Source: ACT.

Over which two consecutive years did the percent stay the same? Estimate this percent.

Solution
The graph is horizontal between 2008 and 2009, at about 53.7%.

CHAPTER 1	TEST

In Exercises 1 and 2, decide whether the reasoning involved is an example of inductive or deductive reasoning.

1. Michelle is a sales representative for a publishing company. For the past 16 years, she has exceeded her annual sales goal, primarily by selling mathematics textbooks. Therefore, she will also exceed her annual sales goal this year.

2. For all natural numbers n, n^2 is also a natural number. 176 is a natural number. Therefore, 176^2 is a natural number.

3. *Counting Puzzle (Rectangles)* How many rectangles of any size are in the figure shown? (September 10, 2001)

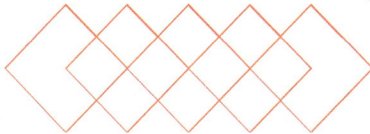

4. Use the list of equations and inductive reasoning to predict the next equation, and then verify your conjecture.

$$65,359,477,124,183 \times 17 = 1,111,111,111,111,111$$
$$65,359,477,124,183 \times 34 = 2,222,222,222,222,222$$
$$65,359,477,124,183 \times 51 = 3,333,333,333,333,333$$

5. Use the method of successive differences to find the next term in the sequence

$$3, 11, 31, 69, 131, 223, \dots.$$

6. Find the sum $1 + 2 + 3 + \cdots + 250$.

7. Consider the following equations, where the left side of each is an octagonal number.

$$1 = 1$$
$$8 = 1 + 7$$
$$21 = 1 + 7 + 13$$
$$40 = 1 + 7 + 13 + 19$$

Use the pattern established on the right sides to predict the next octagonal number. What is the next equation in the list?

8. Use the result of **Exercise 7** and the method of successive differences to find the first eight octagonal numbers. Then divide each by 4 and record the remainder. What is the pattern obtained?

9. Describe the pattern used to obtain the terms of the Fibonacci sequence below. What is the next term?

$$1, 1, 2, 3, 5, 8, 13, 21, \dots.$$

Use problem-solving strategies to solve each problem, taken from the date indicated in the monthly calendar of Mathematics Teacher.

10. *Building a Fraction* Each of the four digits 2, 4, 6, and 9 is placed in one of the boxes to form a fraction. The numerator and the denominator are both two-digit whole numbers. What is the smallest value of all the common fractions that can be formed? Express your answer as a common fraction. (November 17, 2004)

11. *Units Digit of a Power of 9* What is the units digit (ones digit) in the decimal representation of 9^{1997}? (January 27, 1997)

12. *Counting Puzzle (Triangles)* How many triangles are in this figure? (January 6, 2000)

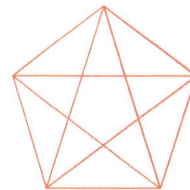

13. *Make Them Equal* Consider the following:

$$1\ 2\ 3\ 4\ 5\ 6\ 7\ 8\ 9\ 0 = 100.$$

Leaving all the numerals in the order given, insert addition and subtraction signs into the expression to make the equation true. (March 23, 2008)

14. *Shrinkage* Dr. Small is 36 inches tall, and Ms. Tall is 96 inches tall. If Dr. Small shrinks 2 inches per year and Ms. Tall grows $\frac{2}{3}$ of an inch per year, how tall will Ms. Tall be when Dr. Small disappears altogether? (November 2, 2007)

15. *Units Digit of a Sum* Find the units digit (ones digit) of the decimal numeral representing the number

$$11^{11} + 14^{14} + 16^{16}. \text{ (February 14, 1994)}$$

16. Based on your knowledge of elementary arithmetic, describe the pattern that can be observed when the following operations are performed.

$$9 \times 1, \quad 9 \times 2, \quad 9 \times 3, \dots, \quad 9 \times 9$$

(*Hint:* Add the digits in the answers. What do you notice?)

Use your calculator to evaluate each of the following. Give as many decimal places as the calculator displays.

17. $\sqrt{98.16}$

18. 3.25^3

19. *Basketball Scoring Results* During the 2012–13 NCAA women's basketball season, Brittney Griner of Baylor made 148 of her 208 free throw attempts. This means that for every 20 attempts, she made approximately _____ of them.

A. 10   **B.** 14   **C.** 8   **D.** 11

20. *Unemployment Rate*   The line graph shows the overall unemployment rate in the U.S. civilian labor force for the years 2003 through 2010.

(a) Between which pairs of consecutive years did the unemployment rate decrease?

(b) What was the general trend in the unemployment rate between 2007 and 2010?

(c) Estimate the overall unemployment rate in 2008 and 2009. About how much did the unemployment rate increase between 2008 and 2009?

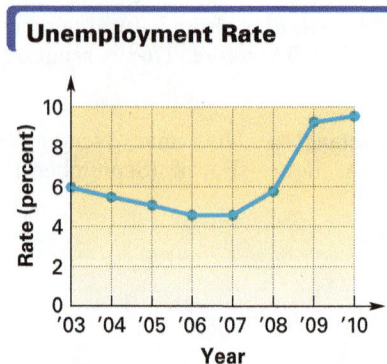

Unemployment Rate

Source: Bureau of Labor Statistics.

The Basic Concepts of Set Theory

2

For many reasons, the job outlook in the United States is good—and improving—for all types of nursing careers, as well as other health-related categories. Students may be considering groups of different training programs; lists of certificate or degree objectives; an array of employment opportunities; the pros and cons of different opportunities relative to upward mobility, stress level, and flexibility; and many others. The properties of these arrays, groups, and collections can be better understood by treating them all as *sets* (the mathematical term) and applying the methods presented in this chapter. (See, for example, the opening discussion of **Section 2.4** on page 71.)

Refer to the table on the next page. To the nearest tenth of a percent, what percentage increase is predicted for number of jobs over the decade 2012–2022 for RNs? For LPNs? (See page 50 for the answer.)

	2012 Median Salary	Number of Jobs in 2012	Predicted Increase in Number of Jobs, 2012–2022
Registered Nurses, RNs	$65,470	2,711,500	526,800
Licensed Practical (or Vocational) Nurses, LPNs (or LVNs)	$41,540	738,400	182,900

Source of data: www.bls.gov

2.1 SYMBOLS AND TERMINOLOGY

OBJECTIVES

1 Use three methods to designate sets.

2 Understand important categories of numbers, and determine cardinal numbers of sets.

3 Distinguish between finite and infinite sets.

4 Determine whether two sets are equal.

The basic ideas of set theory were developed by the German mathematician **Georg Cantor** (1845–1918) in about 1875. Cantor created a new field of theory and at the same time continued the long debate over infinity that began in ancient times. He developed counting by one-to-one correspondence to determine how many objects are contained in a set. Infinite sets differ from finite sets by not obeying the familiar law that the whole is greater than any of its parts.

Designating Sets

A **set** is a collection of objects. The objects belonging to the set are called the **elements,** or **members,** of the set. Sets are designated using the following three methods: (1) *word description,* (2) the *listing method,* and (3) *set-builder notation.*

The set of even counting numbers less than 10   Word description

$\{2, 4, 6, 8\}$ Listing method

$\{x \mid x$ is an even counting number less than 10$\}$ Set-builder notation

The set-builder notation above is read "the set of all x such that x is an even counting number less than 10." Set-builder notation uses the algebraic idea of a *variable.* (Any symbol would do, but just as in other algebraic applications, the letter x is a common choice.)

Variable representing an element in general ↓

$\{x \mid x$ is an even counting number less than 10$\}$

↑ Criteria by which an element qualifies for membership in the set

Sets are commonly given names (usually capital letters), such as E for the set of all letters of the English alphabet.

$E = \{a, b, c, d, e, f, g, h, i, j, k, l, m, n, o, p, q, r, s, t, u, v, w, x, y, z\}$

The listing notation can often be shortened by establishing the pattern of elements included and using ellipsis points to indicate a continuation of the pattern.

$E = \{a, b, c, d, \ldots, x, y, z\},$ or $E = \{a, b, c, d, e, \ldots, z\}$

The set containing no elements is called the **empty set,** or **null set.** The symbol $\varnothing$ is used to denote the empty set, so $\varnothing$ and $\{\ \}$ have the same meaning. We do *not* denote the empty set with the symbol $\{\varnothing\}$ because this notation represents a set with one element (that element being the empty set).

EXAMPLE 1 Listing Elements of Sets

Give a complete listing of all the elements of each set.

(a) the set of counting numbers between eight and thirteen

(b) $\{5, 6, 7, \ldots, 13\}$

(c) $\{x \mid x \text{ is a counting number between 4 and 5}\}$

Solution

(a) This set can be denoted $\{9, 10, 11, 12\}$. (Notice that the word *between* excludes the endpoint values.)

(b) This set begins with the element 5, then 6, then 7, and so on, with each element obtained by adding 1 to the previous element in the list. This pattern stops at 13, so a complete listing is

$$\{5, 6, 7, 8, 9, 10, 11, 12, 13\}.$$

(c) There are no counting numbers between 4 and 5, so this is the empty set: $\{ \ \}$, or $\varnothing$.

For a set to be useful, it must be *well defined*. For example, the preceding set E of the letters of the English alphabet is well defined. Given the letter q, we know that q is an element of E. Given the Greek letter θ (theta), we know that it is not an element of set E.

However, given the set C of all good singers, and a particular singer, Adilah, it may not be possible to say whether

Adilah is an element of C or   Adilah is *not* an element of C.

The problem is the word "good"; how good is good? Because we cannot necessarily decide whether a given singer belongs to set C, set C is not well defined.

The fact that the letter q is an element of set E is denoted by using the symbol $\in$.

$$q \in E \qquad \text{This is read "q is an element of set } E.\text{"}$$

The letter θ is not an element of E. To show this, $\in$ with a slash mark is used.

$$\theta \notin E \qquad \text{This is read "}\theta\text{ is not an element of set } E.\text{"}$$

Many other mathematical symbols also have their meanings negated by use of a **slash mark.** The most common example, $\neq$, means "does not equal" or "is not equal to."

EXAMPLE 2 Applying the Symbol $\in$

Decide whether each statement is *true* or *false*.

(a) $4 \in \{1, 2, 5, 8, 13\}$ **(b)** $0 \in \{0, 1, 2, 3\}$ **(c)** $\frac{1}{5} \notin \left\{\frac{1}{3}, \frac{1}{4}, \frac{1}{6}\right\}$

Solution

(a) Because 4 is *not* an element of the set $\{1, 2, 5, 8, 13\}$, the statement is *false*.

(b) Because 0 is indeed an element of the set $\{0, 1, 2, 3\}$, the statement is *true*.

(c) This statement says that $\frac{1}{5}$ is not an element of the set $\left\{\frac{1}{3}, \frac{1}{4}, \frac{1}{6}\right\}$, which is *true*. ∎

Sets of Numbers and Cardinality

Important categories of numbers are summarized below.

> **SETS OF NUMBERS**
>
> **Natural numbers (or counting numbers)** $\{1, 2, 3, 4, \ldots\}$
> **Whole numbers** $\{0, 1, 2, 3, 4, \ldots\}$
> **Integers** $\{\ldots, -3, -2, -1, 0, 1, 2, 3, \ldots\}$
> **Rational numbers** $\left\{\frac{p}{q} \mid p \text{ and } q \text{ are integers, and } q \neq 0\right\}$
> (*Examples:* $\frac{3}{5}, -\frac{7}{9}, 5, 0$. Any rational number may be written as a terminating decimal number, such as 0.25, or a repeating decimal number, such as 0.666 . . . .)
> **Real numbers** $\{x \mid x \text{ is a number that can be expressed as a decimal}\}$
> **Irrational numbers** $\{x \mid x \text{ is a real number and } x \text{ cannot be expressed as a quotient of integers}\}$
> (*Examples:* $\sqrt{2}$, $\sqrt[3]{4}$, π. Decimal representations of irrational numbers are neither terminating nor repeating.)

The number of elements in a set is called the **cardinal number,** or **cardinality,** of the set. The symbol

$$n(A), \quad \text{which is read "} n \text{ of } A \text{,"}$$

represents the cardinal number of set A. If elements are repeated in a set listing, they should not be counted more than once when determining the cardinal number of the set.

EXAMPLE 3 Finding Cardinal Numbers

Find the cardinal number of each set.

(a) $K = \{3, 9, 27, 81\}$ **(b)** $M = \{0\}$ **(c)** $B = \{1, 1, 2, 3, 2\}$

(d) $R = \{7, 8, \ldots, 15, 16\}$ **(e)** $\varnothing$

Solution

(a) Set K contains four elements, so the cardinal number of set K is 4, and $n(K) = 4$.

(b) Set M contains only one element, 0, so $n(M) = 1$.

(c) Do not count repeated elements more than once. Set B has only three *distinct* elements, so $n(B) = 3$.

(d) Although only four elements are listed, the ellipsis points indicate that there are other elements in the set. Counting them all, we find that there are ten elements, so $n(R) = 10$.

(e) The empty set, $\varnothing$, contains no elements, so $n(\varnothing) = 0$. ∎

Finite and Infinite Sets

If the cardinal number of a set is a particular whole number (0 or a counting number), as in all parts of **Example 3,** we call that set a **finite set.** Given enough time, we could finish counting all the elements of any finite set and arrive at its cardinal number.

Some sets, however, are so large that we could never finish the counting process. The counting numbers themselves are such a set. Whenever a set is so large that its cardinal number is not found among the whole numbers, we call that set an **infinite set.**

A close-up of a camera lens shows the **infinity symbol,** ∞, defined in this case as any distance greater than 1000 times the focal length of a lens.

The sign was invented by the mathematician John Wallis in 1655. Wallis used $1/\infty$ to represent an infinitely small quantity.

EXAMPLE 4 Designating an Infinite Set

Designate all odd counting numbers by the three common methods of set notation.

Solution

The set of all odd counting numbers   Word description

$\{1, 3, 5, 7, 9, \dots\}$ Listing method

$\{x \mid x \text{ is an odd counting number}\}$ Set-builder notation ■

Equality of Sets

> **SET EQUALITY**
>
> Set A is **equal** to set B provided the following two conditions are met:
>
> **1.** Every element of A is an element of B, and
>
> **2.** Every element of B is an element of A.

Two sets are equal if they contain exactly the same elements, regardless of order.

$$\{a, b, c, d\} = \{a, c, d, b\}$$ Both sets contain exactly the same elements.

Repetition of elements in a set listing does not add new elements.

$$\{1, 0, 1, 5, 3, 3\} = \{0, 1, 3, 5\}$$ Both sets contain exactly the same elements.

EXAMPLE 5 Determining Whether Two Sets Are Equal

Are $\{-4, 3, 2, 5\}$ and $\{-4, 0, 3, 2, 5\}$ equal sets?

Solution

Every element of the first set is an element of the second. However, 0 is an element of the second and not of the first. The sets do not contain exactly the same elements.

$$\{-4, 3, 2, 5\} \neq \{-4, 0, 3, 2, 5\}$$ The sets are not equal. ■

Two sets are **equivalent** if they have the *same number* of elements. (See **Exercises 87–90.**) Georg Cantor extended the idea of equivalence to infinite sets, used one-to-one correspondence to establish equivalence, and showed that, surprisingly, the natural numbers, the whole numbers, the integers, and the rational numbers are all equivalent. The elements of any one of these sets will match up, one-to-one, with those of any other, with no elements left over in either set. All these sets have cardinal number $\aleph_0$ (which is read **aleph null**).

However, the irrational numbers and the real numbers, though equivalent to one another, are of a higher infinite order than the sets mentioned above. Their cardinal number is denoted **c** (representing the **continuum** of points on a line).

EXAMPLE 6 Determining Whether Two Sets Are Equal

Decide whether each statement is *true* or *false*.

(a) $\{3\} = \{x \mid x \text{ is a counting number between 1 and 5}\}$

(b) $\{x \mid x \text{ is a negative whole number}\} = \{y \mid y \text{ is a number that is both rational and irrational}\}$

(c) $\{(0, 0), (1, 1), (2, 4)\} = \{(x, y) \mid x \text{ is a natural number less than 3, and } y = x^2\}$

Solution

(a) The set on the right contains *all* counting numbers between 1 and 5, namely 2, 3, and 4, while the set on the left contains *only* the number 3. Because the sets do not contain exactly the same elements, they are not equal. The statement is *false*.

(b) No whole numbers are negative, so the set on the left is $\varnothing$. By definition, if a number is rational, it cannot be irrational, so the set on the right is also $\varnothing$. Because each set is the empty set, the sets are equal. The statement is *true*.

(c) The first listed ordered pair in the set on the left has x-value 0, which is not a natural number. Therefore, the ordered pair (0, 0) is not an element of the set on the right, even though the relationship $y = x^2$ is true for (0, 0). Thus the sets are not equal. The statement is *false*. ■

2.1 EXERCISES

Match each set in Group I with the appropriate description in Group II.

I

1. $\{1, 3, 5, 7, 9\}$

2. $\{x \mid x$ is an even integer greater than 4 and less than 6$\}$

3. $\{\ldots, -4, -3, -2, -1\}$

4. $\{\ldots, -5, -3, -1, 1, 3, 5, \ldots\}$

5. $\{2, 4, 8, 16, 32\}$

6. $\{\ldots, -4, -2, 0, 2, 4, \ldots\}$

7. $\{2, 4, 6, 8, 10\}$

8. $\{2, 4, 6, 8\}$

II

A. the set of all even integers

B. the set of the five least positive integer powers of 2

C. the set of even positive integers less than 10

D. the set of all odd integers

E. the set of all negative integers

F. the set of odd positive integers less than 10

G. $\varnothing$

H. the set of the five least positive integer multiples of 2

List all the elements of each set. Use set notation and the listing method to describe the set.

9. the set of all counting numbers less than or equal to 6

10. the set of all whole numbers greater than 8 and less than 18

11. the set of all whole numbers not greater than 4

12. the set of all natural numbers between 4 and 14

13. $\{6, 7, 8, \ldots, 14\}$

14. $\{3, 6, 9, 12, \ldots, 30\}$

15. $\{2, 4, 8, \ldots, 256\}$

16. $\{90, 87, 84, \ldots, 69\}$

17. $\{x \mid x$ is an even whole number less than 11$\}$

18. $\{x \mid x$ is an odd integer between -8 and 7$\}$

Denote each set by the listing method. There may be more than one correct answer.

19. the set of all multiples of 20 that are greater than 200

20. $\{x \mid x$ is a negative multiple of 6$\}$

21. the set of U.S. Great Lakes

22. the set of U.S. presidents who served after Richard Nixon and before Barack Obama

23. $\{x \mid x$ is the reciprocal of a natural number$\}$

24. $\{x \mid x$ is a positive integer power of 4$\}$

25. $\{(x, y) \mid x$ and y are whole numbers and $x^2 + y^2 = 25\}$

26. $\{(x, y) \mid x$ and y are integers, and $x^2 = 9y^2 + 16\}$

Denote each set by set-builder notation, using x as the variable. There may be more than one correct answer.

27. the set of all rational numbers

28. the set of all even natural numbers

29. $\{1, 3, 5, \ldots, 75\}$

30. $\{35, 40, 45, \ldots, 95\}$

Give a word description for each set. There may be more than one correct answer.

31. $\{-9, -8, -7, \ldots, 7, 8, 9\}$

32. $\left\{\dfrac{1}{2}, \dfrac{2}{3}, \dfrac{3}{4}, \ldots\right\}$

33. $\{$Alabama, Alaska, Arizona, $\ldots$, Wisconsin, Wyoming$\}$

34. $\{$Alaska, California, Hawaii, Oregon, Washington$\}$

Identify each set as finite *or* infinite.

35. $\{2, 4, 6, \ldots, 932\}$

36. $\{6, 12, 18\}$

37. $\left\{1, \dfrac{1}{2}, \dfrac{1}{3}, \dfrac{1}{4}, \ldots\right\}$

38. $\{3, 6, 9, \ldots\}$

39. $\{x \mid x$ is a natural number greater than 50$\}$

40. $\{x \mid x$ is a natural number less than 50$\}$

41. $\{x \mid x$ is a rational number$\}$

42. $\{x \mid x$ is a rational number between 0 and 1$\}$

Find n(A) for each set.

43. $A = \{0, 1, 2, 3, 4, 5, 6, 7\}$

44. $A = \{-3, -1, 1, 3, 5, 7, 9\}$

45. $A = \{2, 4, 6, \ldots, 1000\}$

46. $A = \{0, 1, 2, 3, \ldots, 2000\}$

47. $A = \{a, b, c, \ldots, z\}$

48. $A = \{x \mid x$ is a vowel in the English alphabet$\}$

49. $A =$ the set of integers between -20 and 20

50. $A =$ the set of sanctioned U.S. senate seats

51. $A = \left\{\dfrac{1}{3}, \dfrac{2}{4}, \dfrac{3}{5}, \dfrac{4}{6}, \ldots, \dfrac{27}{29}, \dfrac{28}{30}\right\}$

52. $A = \left\{\dfrac{1}{2}, -\dfrac{1}{2}, \dfrac{1}{3}, -\dfrac{1}{3}, \ldots, \dfrac{1}{10}, -\dfrac{1}{10}\right\}$

53. Although x is a consonant, why can we write

"x is a vowel in the English alphabet"

in **Exercise 48?**

54. Explain how **Exercise 51** can be answered without actually listing and then counting all the elements.

Identify each set as well defined *or* not well defined.

55. $\{x \mid x$ is a real number$\}$

56. $\{x \mid x$ is a good athlete$\}$

57. $\{x \mid x$ is a difficult course$\}$

58. $\{x \mid x$ is a counting number less than 2$\}$

Fill each blank with either $\in$ *or* $\notin$ *to make each statement true.*

59. 3 ___ $\{2, 4, 5, 7\}$

60. -4 ___ $\{4, 7, 8, 12\}$

61. 8 ___ $\{3, 8, 12, 18\}$

62. 0 ___ $\{-2, 0, 5, 9\}$

63. 8 ___ $\{10 - 2, 10\}$

64. $\{6\}$ ___ $\{5 + 1, 6 + 1\}$

65. Is the statement $\{0\} = \varnothing$ true, or is it false?

66. The statement

$$3 \in \{9 - 6, 8 - 6, 7 - 6\}$$

is true even though the *symbol* 3 does not appear in the set. Explain.

Write true *or* false *for each statement.*

67. $3 \in \{2, 5, 6, 8\}$

68. $m \in \{l, m, n, o, p\}$

69. $c \in \{c, d, a, b\}$

70. $2 \in \{-2, 5, 8, 9\}$

71. $\{k, c, r, a\} = \{k, c, a, r\}$

72. $\{e, h, a, n\} = \{a, h, e, n\}$

73. $\{5, 8, 9\} = \{5, 8, 9, 0\}$

74. $\{3, 7\} = \{3, 7, 0\}$

75. $\{4\} \in \{\{3\}, \{4\}, \{5\}\}$

76. $4 \in \{\{3\}, \{4\}, \{5\}\}$

77. $\{x \mid x$ is a natural number less than 3$\} = \{1, 2\}$

78. $\{x \mid x$ is a natural number greater than 10$\}$
$= \{11, 12, 13, \ldots\}$

Write true *or* false *for each statement in Exercises 79–84.*

$$\text{Let} \quad A = \{2, 4, 6, 8, 10, 12\}, \quad B = \{2, 4, 8, 10\},$$
$$\text{and} \quad C = \{4, 10, 12\}.$$

79. $4 \in A$

80. $10 \in B$

81. $4 \notin C$

82. $10 \notin A$

83. Every element of C is also an element of A.

84. Every element of C is also an element of B.

85. The human mind likes to create collections. Why do you suppose this is so? In your explanation, use one or more particular "collections," mathematical or otherwise.

86. Explain the difference between a well-defined set and a set that is not well defined. Give examples, and use terms introduced in this section.

Two sets are **equal** *if they contain identical elements. Two sets are* **equivalent** *if they contain the same number of elements (but not necessarily the same elements). For each condition, give an example or explain why it is impossible.*

87. two sets that are neither equal nor equivalent

88. two sets that are equal but not equivalent

89. two sets that are equivalent but not equal

90. two sets that are both equal and equivalent

91. *Hiring Nurses* A medical organization plans to hire three nurses from the pool of applicants shown.

Name	Certification
Bernice	RN
Heather	RN
Marcy	LVN
Natalie	LVN
Susan	RN

Show all possible sets of hires that would include

(a) two RNs and one LVN.

(b) one RN and two LVNs.

(c) no LVNs.

92. *Burning Calories* Candice Cotton likes cotton candy, each serving of which contains 220 calories. To burn off unwanted calories, Candice participates in her favorite activities, shown in the next column, in increments of 1 hour and never repeats a given activity on a given day.

Activity	Symbol	Calories Burned per Hour
Volleyball	v	160
Golf	g	260
Canoeing	c	340
Swimming	s	410
Running	r	680

(a) On Monday, Candice has time for no more than two hours of activities. List all possible sets of activities that would burn off at least the number of calories obtained from three cotton candies.

(b) Assume that Candice can afford up to three hours of time for activities on Wednesday. List all sets of activities that would burn off at least the number of calories in five cotton candies.

(c) Candice can spend up to four hours in activities on Saturday. List all sets of activities that would burn off at least the number of calories in seven cotton candies.

2.2	**VENN DIAGRAMS AND SUBSETS**

OBJECTIVES

1 Use Venn diagrams to depict set relationships.

2 Determine the complement of a set within a universal set.

3 Determine whether one set is a subset of another.

4 Understand the distinction between a subset and a proper subset.

5 Determine the number of subsets of a given set.

Venn Diagrams

In most discussions, there is either a stated or an implied **universe of discourse.** The universe of discourse includes all things under discussion at a given time. For example, if the topic of interest is what courses to offer at a vocational school, the universe of discourse might be all students at the school, or the board of trustees of the school, or the members of a local overseer board, or the members of a state regulatory agency, or perhaps all these groups of people.

In set theory, the universe of discourse is called the **universal set,** typically designated by the letter U. The universal set might change from one discussion to another.

Also in set theory, we commonly use **Venn diagrams,** developed by the logician John Venn (1834–1923). In these diagrams, the universal set is represented by a rectangle, and other sets of interest within the universal set are depicted by circular regions (sometimes ovals or other shapes). See **Figure 1.**

Complement of a Set

The colored region inside U and outside the circle in **Figure 1** is labeled A' (read "*A* prime"). This set, called the *complement* of A, contains all elements that are contained in U but are not contained in A.

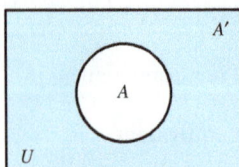

The entire region bounded by the rectangle represents the universal set U, and the portion bounded by the circle represents set A.

Figure 1

> **THE COMPLEMENT OF A SET**
>
> For any set A within a universal set U, the **complement** of A, written A', is the set of elements of U that are not elements of A. That is,
>
> $$A' = \{x \mid x \in U \text{ and } x \notin A\}.$$

EXAMPLE 1 Finding Complements

Find each set.

Let $U = \{a, b, c, d, e, f, g, h\}$, $M = \{a, b, e, f\}$, and $N = \{b, d, e, g, h\}$.

(a) M' **(b)** N'

Solution

(a) Set M' contains all the elements of set U that are *not* in set M. Because set M contains a, b, e, and f, these elements will be disqualified from belonging to set M'.

$$M' = \{c, d, g, h\}$$

(b) Set N' contains all the elements of U that are not in set N, so $N' = \{a, c, f\}$. ■

Consider the complement of the universal set, U'. The set U' is found by selecting all the elements of U that do not belong to U. There are no such elements, so there can be no elements in set U'. This means that for any universal set U,

$$U' = \varnothing.$$

Now consider the complement of the empty set, $\varnothing'$. The set $\varnothing'$ includes all elements of U that do *not* belong to $\varnothing$. All elements of U qualify, because none of them belongs to $\varnothing$. Therefore, for any universal set U,

$$\varnothing' = U.$$

Subsets of a Set

Suppose that we are given the universal set $U = \{1, 2, 3, 4, 5\}$, while $A = \{1, 2, 3\}$. Every element of set A is also an element of set U. Because of this, set A is called a *subset* of set U, written

$$A \subseteq U.$$

("A is not a subset of set U" would be written $A \nsubseteq U$.)

A Venn diagram showing that set M is a subset of set N is shown in **Figure 2**.

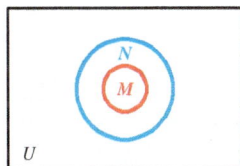

Figure 2

SUBSET OF A SET

Set A is a **subset** of set B if every element of A is also an element of B. This is written $A \subseteq B$.

EXAMPLE 2 Determining If One Set Is a Subset of Another

Write $\subseteq$ or $\nsubseteq$ in each blank to make a true statement.

(a) $\{3, 4, 5, 6\}$ _____ $\{3, 4, 5, 6, 8\}$ **(b)** $\{1, 2, 6\}$ _____ $\{2, 4, 6, 8\}$

(c) $\{5, 6, 7, 8\}$ _____ $\{6, 5, 8, 7\}$

Solution

(a) Because every element of $\{3, 4, 5, 6\}$ is also an element of $\{3, 4, 5, 6, 8\}$, the first set is a subset of the second, so $\subseteq$ goes in the blank.

$$\{3, 4, 5, 6\} \subseteq \{3, 4, 5, 6, 8\}$$

(b) $\{1, 2, 6\} \nsubseteq \{2, 4, 6, 8\}$ 1 does not belong to $\{2, 4, 6, 8\}$.

(c) $\{5, 6, 7, 8\} \subseteq \{6, 5, 8, 7\}$ ■

As **Example 2(c)** suggests, every set is a subset of itself.

$$B \subseteq B, \quad \text{for any set } B.$$

SET EQUALITY (ALTERNATIVE DEFINITION)

Suppose A and B are sets. Then $A = B$ if $A \subseteq B$ and $B \subseteq A$ are both true.

Proper Subsets

Suppose that we are given the following sets.

$$B = \{5, 6, 7, 8\} \quad \text{and} \quad A = \{6, 7\}$$

A is a subset of B, but A is not all of B. There is at least one element in B that is not in A. (Actually, in this case there are two such elements, 5 and 8.) In this situation, A is called a *proper subset* of B, written $A \subset B$.

Notice the similarity of the subset symbols, $\subset$ and $\subseteq$, to the inequality symbols from algebra, $<$ and $\leq$.

PROPER SUBSET OF A SET

Set A is a **proper subset** of set B if $A \subseteq B$ and $A \neq B$. This is written $A \subset B$.

EXAMPLE 3 Determining Subsets and Proper Subsets

Decide whether $\subset$, $\subseteq$, or both could be placed in each blank to make a true statement.

(a) $\{5, 6, 7\}$ _____ $\{5, 6, 7, 8\}$ (b) $\{a, b, c\}$ _____ $\{a, b, c\}$

Solution

(a) Every element of $\{5, 6, 7\}$ is contained in $\{5, 6, 7, 8\}$, so $\subseteq$ could be placed in the blank. Also, the element 8 belongs to $\{5, 6, 7, 8\}$ but not to $\{5, 6, 7\}$, making $\{5, 6, 7\}$ a proper subset of $\{5, 6, 7, 8\}$. Thus $\subset$ could also be placed in the blank.

(b) The set $\{a, b, c\}$ is a subset of $\{a, b, c\}$. Because the two sets are equal, $\{a, b, c\}$ is not a proper subset of $\{a, b, c\}$. Only $\subseteq$ may be placed in the blank. ■

Set A is a subset of set B if every element of set A is also an element of set B. Alternatively, we say that set A is a subset of set B if there are no elements of A that are not also elements of B. Thus, the empty set is a subset of any set.

$$\varnothing \subseteq B, \quad \text{for any set } B.$$

One-to-one correspondence was employed by Georg Cantor to establish many controversial facts about infinite sets. For example, the correspondence

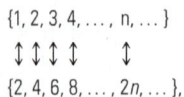

$$\{1, 2, 3, 4, \ldots, n, \ldots\}$$
$$\updownarrow \updownarrow \updownarrow \updownarrow \quad \updownarrow$$
$$\{2, 4, 6, 8, \ldots, 2n, \ldots\},$$

which can be continued indefinitely without leaving any elements in either set unpaired, shows that the counting numbers and the even counting numbers are equivalent (have the same number of elements), even though logic may seem to say that the first set has twice as many elements as the second.

This is true because it is not possible to find any element of $\varnothing$ that is not also in B. (There are no elements in $\varnothing$.) The empty set $\varnothing$ is a proper subset of every set except itself.

$$\varnothing \subset B \quad \text{if } B \text{ is any set other than } \varnothing.$$

Every set (except $\varnothing$) has at least two subsets, $\varnothing$ and the set itself.

EXAMPLE 4 Listing All Subsets of a Set

Find all possible subsets of each set.

(a) $\{7, 8\}$ (b) $\{a, b, c\}$

Solution

(a) By trial and error, the set $\{7, 8\}$ has four subsets: $\varnothing, \{7\}, \{8\}, \{7, 8\}$.

(b) Here, trial and error leads to eight subsets for $\{a, b, c\}$:

$$\varnothing, \{a\}, \{b\}, \{c\}, \{a, b\}, \{a, c\}, \{b, c\}, \{a, b, c\}.$$ ■

Counting Subsets

In **Example 4,** the subsets of $\{7, 8\}$ and the subsets of $\{a, b, c\}$ were found by trial and error. An alternative method involves drawing a **tree diagram,** a systematic way of listing all the subsets of a given set. See **Figure 3.**

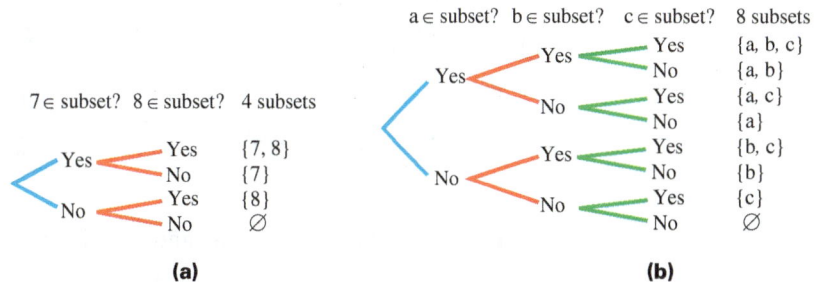

Figure 3

In **Example 4,** we determined the number of subsets of a given set by making a list of all such subsets and then counting them. The tree diagram method also produced a list of all possible subsets. To obtain a formula for finding the number of subsets, we use inductive reasoning. That is, we observe particular cases to try to discover a general pattern.

Begin with the set containing the least number of elements possible—the empty set. This set, $\varnothing$, has only one subset, $\varnothing$ itself. Next, a set with one element has only two subsets, itself and $\varnothing$. These facts, together with those obtained in **Example 4** for sets with two and three elements, are summarized here.

Number of elements	0	1	2	3
Number of subsets	1	2	4	8

This chart suggests that as the number of elements of the set increases by one, the number of subsets doubles. If so, then the number of subsets in each case might be a power of 2. Since every number in the second row of the chart is indeed a power of 2, add this information to the chart.

Number of elements	0	1	2	3
Number of subsets	$1 = 2^0$	$2 = 2^1$	$4 = 2^2$	$8 = 2^3$

This chart shows that the number of elements in each case is the same as the exponent on the base 2. Inductive reasoning gives the following generalization.

NUMBER OF SUBSETS

The number of subsets of a set with n elements is 2^n.

Because the value 2^n includes the set itself, we must subtract 1 from this value to obtain the number of proper subsets of a set containing n elements.

NUMBER OF PROPER SUBSETS

The number of proper subsets of a set with n elements is $2^n - 1$.

Powers of 2

$2^0 = 1$
$2^1 = 2$
$2^2 = 2 \cdot 2 = 4$
$2^3 = 2 \cdot 2 \cdot 2 = 8$
$2^4 = 2 \cdot 2 \cdot 2 \cdot 2 = 16$
$2^5 = 32$
$2^6 = 64$
$2^7 = 128$
$2^8 = 256$
$2^9 = 512$
$2^{10} = 1024$
$2^{11} = 2048$
$2^{12} = 4096$
$2^{15} = 32,768$
$2^{20} = 1,048,576$
$2^{25} = 33,554,432$
$2^{30} = 1,073,741,824$

As shown in **Chapter 1,** although inductive reasoning is a good way of *discovering* principles or arriving at a *conjecture*, it does not provide a proof that the conjecture is true in general. The two formulas above are true, by observation, for $n = 0, 1, 2,$ and 3. (For a general proof, see **Exercise 63** at the end of this section.)

EXAMPLE 5   **Finding Numbers of Subsets and Proper Subsets**

Find the number of subsets and the number of proper subsets of each set.

(a) $\{3, 4, 5, 6, 7\}$ **(b)** $\{1, 2, 3, 4, 5, 9, 12, 14\}$

Solution

(a) This set has 5 elements and $2^5 = 2 \cdot 2 \cdot 2 \cdot 2 \cdot 2 = 32$ subsets. Of these,

$$2^5 - 1 = 32 - 1 = 31 \text{ are proper subsets.}$$

(b) This set has 8 elements. There are $2^8 = 256$ subsets and 255 proper subsets. ∎

2.2   EXERCISES

Match each set or sets in Column I with the appropriate description in Column II.

I	II
1. $\{p\}, \{q\}, \{p, q\}, \varnothing$	**A.** the proper subsets of $\{p, q\}$
2. $\{p\}, \{q\}, \varnothing$	**B.** the complement of $\{c, d\}$, if $U = \{a, b, c, d\}$
3. $\{a, b\}$	**C.** the complement of U
4. $\varnothing$	**D.** the subsets of $\{p, q\}$

Insert $\subseteq$ or $\nsubseteq$ in each blank to obtain a true statement.

5. $\{-2, 0, 2\}$ ____ $\{-2, -1, 1, 2\}$

6. $\{M, W, F\}$ ____ $\{S, M, T, W, Th\}$

7. $\{2, 5\}$ ____ $\{0, 1, 5, 3, 7, 2\}$

8. $\{a, n, d\}$ ____ $\{r, a, n, d, y\}$

9. $\varnothing$ ____ $\{a, b, c, d, e\}$

10. $\varnothing$ ____ $\varnothing$

11. $\{-5, 2, 9\}$ ____ $\{x \mid x \text{ is an odd integer}\}$

12. $\left\{1, 2, \dfrac{9}{3}\right\}$ ____ the set of rational numbers

Decide whether $\subset$, $\subseteq$, both, or neither can be placed in each blank to make the statement true.

13. $\{P, Q, R\}$ ____ $\{P, Q, R, S\}$

14. $\{\text{red, blue, yellow}\}$ ____ $\{\text{yellow, blue, red}\}$

15. $\{9, 1, 7, 3, 5\}$ ____ $\{1, 3, 5, 7, 9\}$

16. $\{S, M, T, W, Th\}$ ____ $\{W, E, E, K\}$

17. $\varnothing$ ____ $\{0\}$

18. $\varnothing$ ____ $\varnothing$

19. $\{0, 1, 2, 3\}$ ____ $\{1, 2, 3, 4\}$

20. $\left\{\dfrac{5}{6}, \dfrac{9}{8}\right\}$ ____ $\left\{\dfrac{6}{5}, \dfrac{8}{9}\right\}$

For Exercises 21–36, tell whether each statement is true *or* false. *U is the universal set.*

Let $U = \{a, b, c, d, e, f, g\}$, $A = \{a, e\}$,

$B = \{a, b, e, f, g\}$, $C = \{b, f, g\}$,   and $D = \{d, e\}$.

21. $A \subset U$

22. $C \nsubseteq U$

23. $D \subseteq B$

24. $D \nsubseteq A$

25. $A \subset B$

26. $B \subseteq C$

27. $\varnothing \nsubseteq A$

28. $\varnothing \subseteq D$

29. $D \nsubseteq B$

30. $A \nsubseteq B$

31. There are exactly 6 subsets of C.

32. There are exactly 31 subsets of B.

33. There are exactly 3 proper subsets of A.

34. There are exactly 4 subsets of D.

35. The Venn diagram below correctly represents the relationship among sets A, D, and U.

36. The Venn diagram below correctly represents the relationship among sets B, C, and U.

For Exercises 37–40, find **(a)** *the number of subsets and* **(b)** *the number of proper subsets of each set.*

37. $\{a, b, c, d, e, f\}$

38. the set of days of the week

39. $\{x \mid x$ is an odd integer between -4 and $6\}$

40. $\{x \mid x$ is an even whole number less than $4\}$

For Exercises 41–44, let $U = \{1, 2, 3, 4, 5, 6, 7, 8, 9, 10\}$ *and find the complement of each set.*

41. U **42.** $\varnothing$

43. $\{1, 2, 3, 4, 6, 8\}$ **44.** $\{2, 5, 9, 10\}$

Vacationing in California *Terry is planning a trip with her two sons to California. In weighing her options concerning whether to fly or drive from their home in Iowa, she has listed the following considerations.*

Fly to California	Drive to California
Higher cost	Lower cost
Educational	Educational
More time to see the sights in California	Less time to see the sights in California
Cannot visit friends along the way	Can visit friends along the way

Refer to the table for Exercises 45–50.

45. Find the smallest universal set U that contains all listed considerations of both options.

Let F represent the set of considerations of the flying option and let D represent the set of considerations of the driving option. Use the universal set from **Exercise 45.**

46. Give the set F'.   **47.** Give the set D'.

Find the set of elements common to both sets in Exercises 48–50.

48. F and D **49.** F' and D'

50. F and D'

Meeting in the Conference Room *Amie, Bruce, Corey, Dwayne, and Eric, members of an architectural firm, plan to meet in the company conference room to discuss the project coordinator's plans for their next project. Denoting these five people by A, B, C, D, and E, list all the possible sets of this group in which the given number of them can gather.*

51. five people   **52.** four people

53. three people   **54.** two people

55. one person   **56.** no people

57. Find the total number of ways that members of this group can gather. (*Hint:* Find the total number of sets in your answers to **Exercises 51–56.**)

58. How does your answer in **Exercise 57** compare with the number of subsets of a set of five elements? Interpret the answer to **Exercise 57** in terms of subsets.

59. Selecting a Club Delegation The twenty-five members of the mathematics club must send a delegation to a meeting for student groups at their school. The delegation can include as many members of the club as desired, but at least one member must attend. How many different delegations are possible? (*Mathematics Teacher* calendar problem)

60. In **Exercise 59**, suppose ten of the club members say they do not want to be part of the delegation. Now how many delegations are possible?

61. Selecting Bills Suppose you have the bills shown here.

(a) How many sums of money can you make using nonempty subsets of these bills?

(b) Repeat part (a) without the condition "nonempty."

62. *Selecting Coins* The photo shows a group of obsolete U.S. coins, consisting of one each of the penny, nickel, dime, quarter, and half dollar. Repeat **Exercise 61,** replacing "bill(s)" with "coin(s)."

63. In discovering the expression (2^n) for finding the number of subsets of a set with n elements, we observed that for the first few values of n, increasing the number of elements by one doubles the number of subsets.

Here, you can prove the formula in general by showing that the same is true for any value of n. Assume set A has n elements and s subsets. Now add one additional element, say e, to the set A. (We now have a new set, say B, with $n + 1$ elements.) Divide the subsets of B into those that do not contain e and those that do.

(a) How many subsets of B do not contain e? (*Hint:* Each of these is a subset of the original set A.)

(b) How many subsets of B do contain e? (*Hint:* Each of these would be a subset of the original set A, with the additional element e included.)

(c) What is the total number of subsets of B?

(d) What do you conclude?

64. Explain why $\varnothing$ is both a subset and an element of $\{\varnothing\}$.

2.3	SET OPERATIONS

OBJECTIVES

1 Determine intersections of sets.

2 Determine unions of sets.

3 Determine the difference of two sets.

4 Understand ordered pairs and their uses.

5 Determine Cartesian products of sets.

6 Analyze sets and set operations with Venn diagrams.

7 Apply De Morgan's laws for sets.

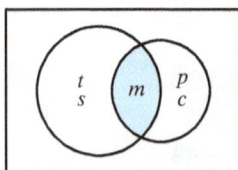

Figure 4

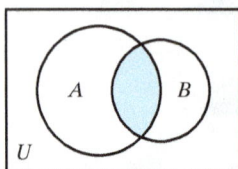

$A \cap B$

Figure 5

Intersection of Sets

Two candidates, Aimee and Darien, are running for a seat on the city council. A voter deciding for whom she should vote recalled the campaign promises, each given a code letter, made by the candidates.

Honest Aimee	**Determined Darien**
Spend less money, *m*	**Spend less money, *m***
Emphasize traffic law enforcement, *t*	Crack down on crooked politicians, *p*
Increase service to suburban areas, *s*	Increase service to the city, *c*

The only promise common to both candidates is promise m, to spend less money. Suppose we take each candidate's promises to be a set. The promises of Aimee give the set $\{m, t, s\}$, while the promises of Darien give $\{m, p, c\}$. The common element m belongs to the *intersection* of the two sets, as shown in color in the Venn diagram in **Figure 4.**

$$\{m, t, s\} \cap \{m, p, c\} = \{m\} \qquad \cap \text{ represents set intersection.}$$

The intersection of two sets is itself a set.

> **INTERSECTION OF SETS**
>
> The **intersection** of sets A and B, written $A \cap B$, is the set of elements common to both A and B.
>
> $$A \cap B = \{x \mid x \in A \text{ and } x \in B\}$$

Form the intersection of sets A and B by taking all the elements included in both sets, as shown in color in **Figure 5.**

White light can be viewed as the intersection of the three primary colors.

A   B

U

Disjoint sets

Figure 6

t s *m* *p c*

Figure 7

A   B

U

A ∪ B

Figure 8

EXAMPLE 1 Finding Intersections

Find each intersection.

(a) $\{3, 4, 5, 6, 7\} \cap \{4, 6, 8, 10\}$ **(b)** $\{9, 14, 25, 30\} \cap \{10, 17, 19, 38, 52\}$

(c) $\{5, 9, 11\} \cap \varnothing$

Solution

(a) The elements common to both sets are 4 and 6.

$$\{3, 4, 5, 6, 7\} \cap \{4, 6, 8, 10\} = \{4, 6\}$$

(b) These two sets have no elements in common.

$$\{9, 14, 25, 30\} \cap \{10, 17, 19, 38, 52\} = \varnothing$$

(c) There are no elements in $\varnothing$, so there can be no elements belonging to both $\{5, 9, 11\}$ and $\varnothing$.

$$\{5, 9, 11\} \cap \varnothing = \varnothing$$

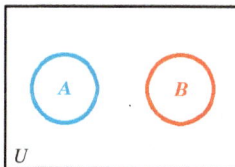

Examples 1(b) and 1(c) show two sets that have no elements in common. Sets with no elements in common are called **disjoint sets.** (See **Figure 6.**) A set of dogs and a set of cats would be disjoint sets.

Sets A and B are disjoint if $A \cap B = \varnothing$.

Union of Sets

Referring again to the lists of campaign promises, suppose a pollster wants to summarize the types of promises made by the candidates. The pollster would need to study *all* the promises made by *either* candidate, or the set

$$\{m, t, s, p, c\}.$$

This set is the *union* of the sets of promises, as shown in color in the Venn diagram in **Figure 7.**

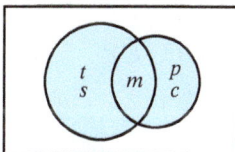

Be careful not to confuse this symbol with the universal set U.

$$\{m, t, s\} \cup \{m, p, c\} = \{m, t, s, p, c\}$$ ∪ denotes set union.

Again, the union of two sets is a set.

UNION OF SETS

The **union** of sets A and B, written $A \cup B$, is the set of all elements belonging to either A or B.

$$A \cup B = \{x \mid x \in A \text{ or } x \in B\}$$

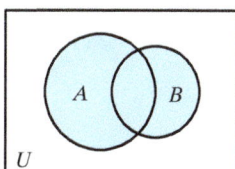

Form the union of sets A and B by first taking every element of set A and then also including every element of set B that is not already listed. See Figure 8.

EXAMPLE 2 **Finding Unions**

Find each union.

(a) $\{2, 4, 6\} \cup \{4, 6, 8, 10, 12\}$ **(b)** $\{a, b, d, f, g, h\} \cup \{c, f, g, h, k\}$

(c) $\{3, 4, 5\} \cup \varnothing$

Solution

(a) Start by listing all the elements from the first set, 2, 4, and 6. Then list all the elements from the second set that are not in the first set, 8, 10, and 12. The union is made up of *all* these elements.

$$\{2, 4, 6\} \cup \{4, 6, 8, 10, 12\} = \{2, 4, 6, 8, 10, 12\}$$

(b) $\{a, b, d, f, g, h\} \cup \{c, f, g, h, k\} = \{a, b, c, d, f, g, h, k\}$

(c) Because there are no elements in $\varnothing$, the union of $\{3, 4, 5\}$ and $\varnothing$ contains only the elements 3, 4, and 5.

$$\{3, 4, 5\} \cup \varnothing = \{3, 4, 5\}$$ ∎

Recall from the previous section that A' represents the *complement* of set A. *Set A' is formed by taking every element of the universal set U that is not in set A.*

EXAMPLE 3 **Finding Intersections and Unions of Complements**

Find each set. Let

$$U = \{1, 2, 3, 4, 5, 6, 9\}, \quad A = \{1, 2, 3, 4\}, \quad B = \{2, 4, 6\}, \quad \text{and} \quad C = \{1, 3, 6, 9\}.$$

(a) $A' \cap B$ **(b)** $B' \cup C'$ **(c)** $A \cap (B \cup C')$ **(d)** $(B \cup C)'$

Solution

(a) First identify the elements of set A', the elements of U that are not in set A.

$$A' = \{5, 6, 9\}$$

Now, find $A' \cap B$, the set of elements belonging both to A' and to B.

$$A' \cap B = \{5, 6, 9\} \cap \{2, 4, 6\} = \{6\}$$

(b) $B' \cup C' = \{1, 3, 5, 9\} \cup \{2, 4, 5\} = \{1, 2, 3, 4, 5, 9\}$

(c) First find the set inside the parentheses.

$$B \cup C' = \{2, 4, 6\} \cup \{2, 4, 5\} = \{2, 4, 5, 6\}$$

Now, find the intersection of this set with A.

$$A \cap (B \cup C') = A \cap \{2, 4, 5, 6\}$$
$$= \{1, 2, 3, 4\} \cap \{2, 4, 5, 6\}$$
$$= \{2, 4\}$$

(d) $B \cup C = \{2, 4, 6\} \cup \{1, 3, 6, 9\} = \{1, 2, 3, 4, 6, 9\}$,   so

$$(B \cup C)' = \{5\}.$$ ∎

Comparing **Examples 3(b) and 3(d),** we see that, interestingly, $(B \cup C)'$ is not the same as $B' \cup C'$. This fact will be investigated further later in this section.

Comparing Properties

The arithmetic operations of addition and multiplication, when applied to numbers, have some familiar properties. If a, b, and c are *real numbers,* then the **commutative property of addition** says that the order of the numbers being added makes no difference:

$$a + b = b + a.$$

(Is there a **commutative property of multiplication?**) The **associative property of addition** says that when three numbers are added, the grouping used makes no difference:

$$(a + b) + c = a + (b + c).$$

(Is there an **associative property of multiplication?**) The number 0 is called the **identity element for addition** since adding it to any number does not change that number:

$$a + 0 = a.$$

(What is the **identity element for multiplication?**) Finally, the **distributive property of multiplication over addition** says that

$$a(b + c) = ab + ac.$$

(Is there a distributive property of addition over multiplication?)

For Group or Individual Investigation

Now consider the operations of union and intersection, applied to sets. By recalling definitions, trying examples, or using Venn diagrams, answer the following questions.

1. Is set union commutative?   Set intersection?
2. Is set union associative?   Set intersection?
3. Is there an identity element for set union? If so, what is it? How about set intersection?
4. Is set intersection distributive over set union? Is set union distributive over set intersection?

EXAMPLE 4 **Describing Sets in Words**

Describe each set in words.

(a) $A \cap (B \cup C')$ (b) $(A' \cup C') \cap B'$

Solution

(a) This set might be described as "the set of all elements that are in A, and also are in B or not in C."

(b) One possibility is "the set of all elements that are not in A or not in C, and also are not in B."

Difference of Sets

Suppose that $A = \{1, 2, 3, \ldots, 10\}$ and $B = \{2, 4, 6, 8, 10\}$. If the elements of B are excluded (or taken away) from A, the set $C = \{1, 3, 5, 7, 9\}$ is obtained. C is called the *difference* of sets A and B.

DIFFERENCE OF SETS

The **difference** of sets A and B, written $A - B$, is the set of all elements belonging to set A and not to set B.

$$A - B = \{x \mid x \in A \text{ and } x \notin B\}$$

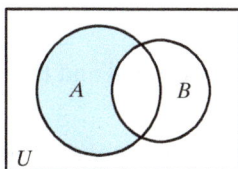

$A - B$

Figure 9

Assume a universal set U containing both A and B. Then because $x \notin B$ has the same meaning as $x \in B'$, the set difference $A - B$ can also be described as

$$\{x \mid x \in A \text{ and } x \in B'\}, \quad \text{or} \quad A \cap B'.$$

Figure 9 illustrates the idea of set difference. The region in color represents $A - B$.

EXAMPLE 5 Finding Set Differences

Find each set.

$$\text{Let}\quad U = \{1, 2, 3, 4, 5, 6, 7\},\quad A = \{1, 2, 3, 4, 5, 6\},$$
$$B = \{2, 3, 6\},\quad \text{and}\quad C = \{3, 5, 7\}.$$

(a) $A - B$ **(b)** $B - A$ **(c)** $(A - B) \cup C'$

Solution

(a) Begin with set A and exclude any elements found also in set B.

$$A - B = \{1, 2, 3, 4, 5, 6\} - \{2, 3, 6\} = \{1, 4, 5\}$$

(b) To be in $B - A$, an element must be in set B and not in set A. But all elements of B are also in A. Thus, $B - A = \varnothing$.

(c) From part (a), $A - B = \{1, 4, 5\}$. Also, $C' = \{1, 2, 4, 6\}$.

$$(A - B) \cup C' = \{1, 2, 4, 5, 6\}$$ ∎

The results in **Examples 5(a) and 5(b)** illustrate that, in general,

$$A - B \neq B - A.$$

Ordered Pairs

When writing a set that contains several elements, the order in which the elements appear is not relevant. For example,

$$\{1, 5\} = \{5, 1\}.$$

However, there are many instances in mathematics where, when two objects are paired, the order in which the objects are written is important. This leads to the idea of *ordered pair*. When writing ordered pairs, use parentheses rather than braces, which are reserved for writing sets.

ORDERED PAIRS

In the **ordered pair** (a, b), a is called the **first component** and b is called the **second component.** In general, $(a, b) \neq (b, a)$.

Two ordered pairs (a, b) and (c, d) are **equal** provided that their first components are equal and their second components are equal.

$$(a, b) = (c, d)\quad \text{if and only if}\quad a = c\quad \text{and}\quad b = d.$$

EXAMPLE 6 Determining Equality of Sets and of Ordered Pairs

Decide whether each statement is *true* or *false*.

(a) $(3, 4) = (5 - 2, 1 + 3)$ **(b)** $\{3, 4\} \neq \{4, 3\}$ **(c)** $(7, 4) = (4, 7)$

Solution

(a) Because $3 = 5 - 2$ and $4 = 1 + 3$, the first components are equal and the second components are equal. The statement is *true*.

(b) Because these are sets and not ordered pairs, the order in which the elements are listed is not important. Because these sets are equal, the statement is *false*.

(c) The ordered pairs $(7, 4)$ and $(4, 7)$ are not equal because their corresponding components are not equal. The statement is *false*. ∎

Cartesian Product of Sets

A set may contain ordered pairs as elements. If A and B are sets, then each element of A can be paired with each element of B, and the results can be written as ordered pairs. The set of all such ordered pairs is called the *Cartesian product* of A and B, which is written $A \times B$ and read **"A cross B."** The name comes from that of the French mathematician René Descartes, profiled in **Chapter 8.**

CARTESIAN PRODUCT OF SETS

The **Cartesian product** of sets A and B is defined as follows.

$$A \times B = \{(a, b) \mid a \in A \text{ and } b \in B\}$$

EXAMPLE 7 Finding Cartesian Products

Let $A = \{1, 5, 9\}$ and $B = \{6, 7\}$. Find each set.

(a) $A \times B$ **(b)** $B \times A$

Solution

(a) Pair each element of A with each element of B. Write the results as ordered pairs, with the element of A written first and the element of B written second. Write as a set.

$$A \times B = \{(1, 6), (1, 7), (5, 6), (5, 7), (9, 6), (9, 7)\}$$

(b) Because B is listed first, this set will consist of ordered pairs that have their components interchanged when compared to those in part (a).

$$B \times A = \{(6, 1), (7, 1), (6, 5), (7, 5), (6, 9), (7, 9)\}$$

The order in which the ordered pairs themselves are listed is not important. For example, another way to write $B \times A$ in **Example 7(b)** would be

$$\{(6, 1), (6, 5), (6, 9), (7, 1), (7, 5), (7, 9)\}.$$

From **Example 7** it can be seen that, in general,

$$A \times B \neq B \times A,$$

because they do not contain exactly the same ordered pairs. However, each set contains the same *number* of elements, six. Furthermore, $n(A) = 3$, $n(B) = 2$, and $n(A \times B) = n(B \times A) = 6$. Because $3 \cdot 2 = 6$, one might conclude that the cardinal number of the Cartesian product of two sets is equal to the product of the cardinal numbers of the sets. In general, this conclusion is correct.

CARDINAL NUMBER OF A CARTESIAN PRODUCT

If $n(A) = a$ and $n(B) = b$, then the following is true.

$$n(A \times B) = n(B \times A) = n(A) \cdot n(B) = n(B) \cdot n(A) = ab = ba$$

EXAMPLE 8 **Finding Cardinal Numbers of Cartesian Products**

Find $n(A \times B)$ and $n(B \times A)$ from the given information.

(a) $A = \{a, b, c, d, e, f, g\}$ and $B = \{2, 4, 6\}$ **(b)** $n(A) = 24$ and $n(B) = 5$

Solution

(a) Because $n(A) = 7$ and $n(B) = 3$, $n(A \times B)$ and $n(B \times A)$ both equal $7 \cdot 3$, or 21.

(b) $n(A \times B) = n(B \times A) = 24 \cdot 5 = 5 \cdot 24 = 120$ ∎

An **operation** is a rule or procedure by which one or more objects are used to obtain another object. The most common operations on sets are summarized in the following box.

SET OPERATIONS

Let A and B be any sets within a universal set U.

The **complement** of A, written A', is

$$A' = \{x \mid x \in U \text{ and } x \notin A\}.$$

The **intersection** of A and B is

$$A \cap B = \{x \mid x \in A \text{ and } x \in B\}.$$

The **union** of A and B is

$$A \cup B = \{x \mid x \in A \text{ or } x \in B\}.$$

The **difference** of A and B is

$$A - B = \{x \mid x \in A \text{ and } x \notin B\}.$$

The **Cartesian product** of A and B is

$$A \times B = \{(x, y) \mid x \in A \text{ and } y \in B\}.$$

More on Venn Diagrams

It is often helpful to use numbers in Venn diagrams, as in **Figures 10, 11, and 12,** depending on whether the discussion involves one, two, or three (distinct) sets, respectively. In each case, the numbers are neither elements nor cardinal numbers, but simply arbitrary labels for the various regions within the diagram.

In **Figure 11,** region 3 includes the elements belonging to both A and B, while region 4 includes those elements (if any) belonging to B but not to A. How would you describe region 7 in **Figure 12?**

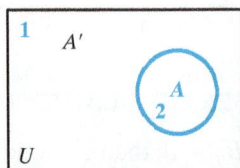

Figure 10

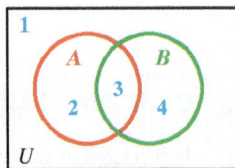

Figure 11

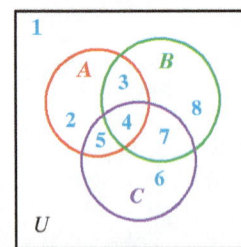

Figure 12

EXAMPLE 9 Shading Venn Diagrams to Represent Sets

Draw a Venn diagram similar to **Figure 11,** and shade the region or regions representing each set.

(a) $A' \cap B$ **(b)** $A' \cup B'$

Solution

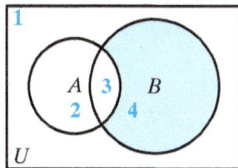

Figure 13

(a) See **Figure 11.** Set A' contains all the elements outside of set A—in other words, the elements in regions 1 and 4. Set B contains the elements in regions 3 and 4. The intersection of sets A' and B is made up of the elements in the region common to (1 and 4) and (3 and 4), which is region 4. Thus, $A' \cap B$ is represented by region 4, shown in color in **Figure 13.** This region can also be described as $B - A$.

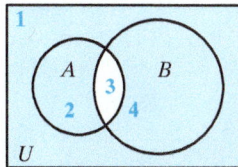

Figure 14

(b) Again, set A' is represented by regions 1 and 4, and B' is made up of regions 1 and 2. The union of A' and B', the set $A' \cup B'$, is made up of the elements belonging to the union of (1 and 4) with (1 and 2)—that is, regions 1, 2, and 4, shown in color in **Figure 14.** ■

EXAMPLE 10 Locating Elements in a Venn Diagram

Place the elements of the sets in their proper locations in a Venn diagram.

Let $U = \{q, r, s, t, u, v, w, x, y, z\}$, $A = \{r, s, t, u, v\}$,   and $B = \{t, v, x\}$.

Solution

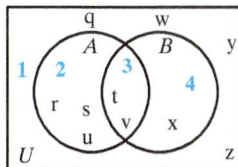

Figure 15

Because $A \cap B = \{t, v\}$, elements t and v are placed in region 3 in **Figure 15.** The remaining elements of A, that is, r, s, and u, go in region 2. The figure shows the proper placement of all other elements. ■

EXAMPLE 11 Shading a Set in a Venn Diagram

Shade the set $(A' \cap B') \cap C$ in a Venn diagram similar to the one in **Figure 12.**

Solution

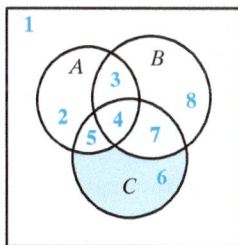

$(A' \cap B') \cap C$

Figure 16

Work first inside the parentheses. Set A' is made up of the regions outside set A, or regions 1, 6, 7, and 8. Set B' is made up of regions 1, 2, 5, and 6. The intersection of these sets is given by the overlap of regions 1, 6, 7, 8 and 1, 2, 5, 6, or regions 1 and 6.

For the final Venn diagram, find the intersection of regions 1 and 6 with set C. Set C is made up of regions 4, 5, 6, and 7. The overlap of regions 1, 6 and 4, 5, 6, 7 is region 6, the region shown in color in **Figure 16.** ■

EXAMPLE 12 Verifying a Statement Using a Venn Diagram

Suppose $A, B \subseteq U$. Is the statement $(A \cap B)' = A' \cup B'$ true for every choice of sets A and B?

Solution

Use the regions labeled in **Figure 11.** Set $A \cap B$ is made up of region 3, so $(A \cap B)'$ is made up of regions 1, 2, and 4. These regions are shown in color in **Figure 17(a)** on the next page.

To identify set $A' \cup B'$, proceed as in **Example 9(b).** The result, shown in **Figure 14,** is repeated in **Figure 17(b)** on the next page.

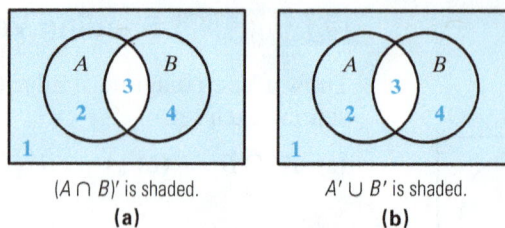

$(A \cap B)'$ is shaded.
(a)

$A' \cup B'$ is shaded.
(b)

Figure 17

The fact that the same regions are in color in both Venn diagrams suggests that

$$(A \cap B)' = A' \cup B'.$$ ∎

De Morgan's Laws

The result of **Example 12** can be stated in words.

> *The complement of the intersection of two sets is equal to the union of the complements of the two sets.*

Interchanging the words "intersection" and "union" produces another true statement.

> *The complement of the union of two sets is equal to the intersection of the complements of the two sets.*

Both of these "laws" were established by the British logician Augustus De Morgan (1806–1871), profiled on **page 22**. They are stated in set symbols as follows.

DE MORGAN'S LAWS FOR SETS

For any sets A and B, where $A, B \subseteq U$,

$$(A \cap B)' = A' \cup B' \quad \text{and} \quad (A \cup B)' = A' \cap B'.$$

The Venn diagrams in **Figure 17** strongly suggest the truth of the first of De Morgan's laws. They provide a *conjecture*. Actual proofs of De Morgan's laws would require methods used in more advanced courses in set theory.

EXAMPLE 13 **Describing Venn Diagram Regions Using Symbols**

For the Venn diagrams, write several symbolic descriptions of the region in color, using $A, B, C, \cap, \cup, -$, and $'$ as necessary.

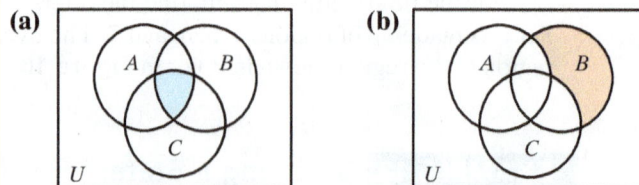

(a)

(b)

Solution

(a) The region in color can be described as belonging to all three sets, A and B and C. Therefore, the region corresponds to

$$(A \cap B) \cap C, \quad \text{or} \quad A \cap (B \cap C), \quad \text{or} \quad A \cap B \cap C.$$

(b) The region in color is in set B and is not in A and is not in C. Because it is not in A, it is in A', and similarly it is in C'. The region can be described as

$$B \cap A' \cap C', \quad \text{or} \quad B - (A \cup C), \quad \text{or} \quad B \cap (A \cup C)'.$$ ∎

2.3   EXERCISES

Match each term in Group I with the appropriate description from A–F in Group II. Assume that A and B are sets.

I

1. the intersection of A and B

2. the union of A and B

3. the difference of A and B

4. the complement of A

5. the Cartesian product of A and B

6. the difference of B and A

II

A. the set of elements in A that are not in B

B. the set of elements common to both A and B

C. the set of elements in the universal set that are not in A

D. the set of elements in B that are not in A

E. the set of ordered pairs such that each first element is from A and each second element is from B, with every element of A paired with every element of B

F. the set of elements that are in A or in B or in both A and B

Perform the indicated operations, and designate each answer using the listing method.

Let $U = \{a, b, c, d, e, f, g\}$, $X = \{a, c, e, g\}$,
$Y = \{a, b, c\}$, and $Z = \{b, c, d, e, f\}$.

7. $X \cap Y$ 8. $X \cup Y$

9. $Y \cup Z$ 10. $Y \cap Z$

11. X' 12. Y'

13. $X' \cap Y'$ 14. $X' \cap Z$

15. $X \cup (Y \cap Z)$ 16. $Y \cap (X \cup Z)$

17. $X - Y$ 18. $Y - X$

19. $(Z \cup X')' \cap Y$ 20. $(Y \cap X')' \cup Z'$

21. $X \cap (X - Y)$ 22. $Y \cup (Y - X)$

23. $X' - Y$ 24. $Y' - (X \cap Z)$

Describe each set in words.

25. $A \cup (B' \cap C')$ 26. $(A \cap B') \cup (B \cap A')$

27. $(C - B) \cup A$ 28. $(A' \cap B') \cup C'$

Adverse Effects of Tobacco and Alcohol   *The table lists some common adverse effects of prolonged tobacco and alcohol use.*

Tobacco	Alcohol
Emphysema, e	Liver damage, l
Heart damage, h	Brain damage, b
Cancer, c	Heart damage, h

In Exercises 29–32, let T be the set of listed effects of tobacco and A be the set of listed effects of alcohol. Find each set.

29. the smallest possible universal set U that includes all the effects listed

30. $T \cap A$ 31. $T \cup A$ 32. $T \cap A'$

An accountant is sorting tax returns in her files that require attention in the next week.

Describe in words each set in Exercises 33–36.

Let U = the set of all tax returns in the file,
   A = the set of all tax returns with itemized deductions,
   B = the set of all tax returns showing business income,
   C = the set of all tax returns filed in 2014,
   D = the set of all tax returns selected for audit.

33. $C - A$ 34. $D \cup A'$

35. $(A \cup B) - D$ 36. $(C \cap A) \cap B'$

For Exercises 37–40, assume that A and B represent any two sets. Identify each statement as either always true *or* not always true.

37. $(A \cap B) \subseteq A$

38. $A \subseteq (A \cap B)$

39. $n(A \cup B) = n(A) + n(B)$

40. $n(A \cup B) = n(A) + n(B) - n(A \cap B)$

For Exercises 41–44, use your results in parts (a) and (b) to answer part (c).

Let $U = \{1, 2, 3, 4, 5\}$, $X = \{1, 3, 5\}$, $Y = \{1, 2, 3\}$,
and $Z = \{3, 4, 5\}$.

41. (a) Find $X \cup Y$. **(b)** Find $Y \cup X$.

(c) State a conjecture.

42. (a) Find $X \cap Y$. **(b)** Find $Y \cap X$.

(c) State a conjecture.

43. (a) Find $X \cup (Y \cup Z)$. **(b)** Find $(X \cup Y) \cup Z$.

(c) State a conjecture.

44. (a) Find $X \cap (Y \cap Z)$. **(b)** Find $(X \cap Y) \cap Z$.

(c) State a conjecture.

Decide whether each statement is true *or* false.

45. $(3, 2) = (5 - 2, 1 + 1)$

46. $(2, 13) = (13, 2)$

47. $\{6, 3\} = \{3, 6\}$

48. $\{(5, 9), (4, 8), (4, 2)\} = \{(4, 8), (5, 9), (2, 4)\}$

Find $A \times B$ and $B \times A$, for A and B defined as follows.

49. $A = \{d, o, g\}$, $B = \{p, i, g\}$

50. $A = \{3, 6, 9, 12\}$, $B = \{6, 8\}$

Use the given information to find $n(A \times B)$ and $n(B \times A)$.

51. $n(A) = 35$ and $n(B) = 6$

52. $n(A) = 13$ and $n(B) = 5$

Find the cardinal number specified.

53. If $n(A \times B) = 72$ and $n(A) = 12$, find $n(B)$.

54. If $n(A \times B) = 300$ and $n(B) = 30$, find $n(A)$.

Place the elements of these sets in the proper locations in the given Venn diagram.

55. Let $U = \{a, b, c, d, e, f, g\}$,
$A = \{b, d, f, g\}$,
$B = \{a, b, d, e, g\}$.

56. Let $U = \{5, 6, 7, 8, 9, 10, 11, 12, 13\}$,
$M = \{5, 8, 10, 11\}$,
$N = \{5, 6, 7, 9, 10\}$.

Use a Venn diagram similar to the one shown below to shade each set.

57. $A' \cup B$ **58.** $A' \cap B'$

59. $B \cap A'$ **60.** $A \cup B$

61. $B' \cap B$ **62.** $A' \cup A$

63. $B' \cup (A' \cap B')$ **64.** $(A - B) \cup (B - A)$

In Exercises 65 and 66, place the elements of the sets in the proper locations in a Venn diagram similar to the one shown below.

65. Let $U = \{m, n, o, p, q, r, s, t, u, v, w\}$,
$A = \{m, n, p, q, r, t\}$,
$B = \{m, o, p, q, s, u\}$,
$C = \{m, o, p, r, s, t, u, v\}$.

66. Let $U = \{1, 2, 3, 4, 5, 6, 7, 8, 9\}$,
$A = \{1, 3, 5, 7\}$,
$B = \{1, 3, 4, 6, 8\}$,
$C = \{1, 4, 5, 6, 7, 9\}$.

Use a Venn diagram to shade each set.

67. $(A \cap B) \cap C$ **68.** $(A' \cap B) \cap C$

69. $(A' \cap B') \cap C$ **70.** $(A \cap C') \cap B$

71. $(A \cap B') \cap C'$ **72.** $(A \cap B)' \cup C$

Write a description of each shaded area. Use the symbols A, B, C, ∩, ∪, −, and '. Different answers are possible.

73.

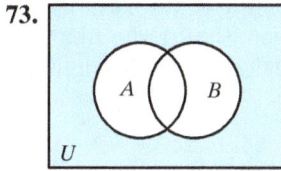

74.

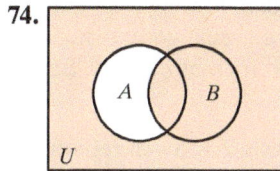

75.

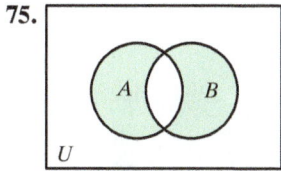

76.

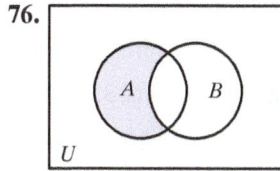

77.

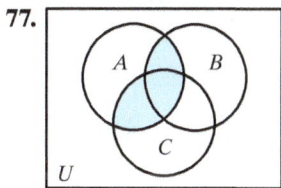

78.

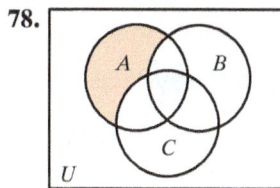

Suppose A and B are sets. Describe the conditions under which each statement would be true.

79. $A = A - B$

80. $A = B - A$

81. $A = A - \varnothing$

82. $A \cap \varnothing = \varnothing$

83. $A \cup B = A$

84. $A \cap B = B$

For Exercises 85–88, draw two Venn diagrams to decide whether the statement is always true *or* not always true.

85. $(A \cap B) \subseteq A$

86. $(A \cup B) \subseteq A$

87. If $A \subseteq B$, then $A \cup B = A$.

88. If $A \subseteq B$, then $A \cap B = A$.

89. Explain why, if A and B are sets, it is not necessarily true that $n(A - B) = n(A) - n(B)$. Give a counterexample.

90. The five set operations listed on **page 66** are applied to subsets of U (that is, to A and/or B). Is the result always a subset of U also? Explain why or why not.

2.4 SURVEYS AND CARDINAL NUMBERS

OBJECTIVES

1 Analyze survey results.
2 Apply the cardinal number formula.
3 Interpret information from tables.

Surveys

As suggested in the chapter opener, the techniques of set theory can be applied to many different groups of people (or objects). The problems addressed sometimes require analyzing known information about certain subsets to obtain cardinal numbers of other subsets. In this section, we apply three problem-solving strategies to such problems:

Venn diagrams, cardinal number formulas, and tables.

The "known information" is quite often (although not always) obtained by conducting a survey.

Suppose a group of attendees at an educational seminar, all of whom desire to become registered nurses (RNs), are asked their preferences among the three traditional ways to become an RN, and the following information is produced.

23 would consider pursuing the Bachelor of Science in Nursing (BSN).

16 would consider pursuing the Associate Degree in Nursing (ADN).

7 would consider a Diploma program (Diploma).

10 would consider both the BSN and the ADN.

5 would consider both the BSN and the Diploma.

3 would consider both the ADN and the Diploma.

2 would consider all three options.

4 are looking for an option other than these three.

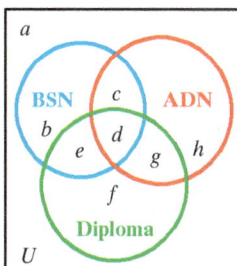

Figure 18

To determine the total number of attendees surveyed, we cannot just add the eight numbers above because there is some overlap. For example, in **Figure 18,** the 23 who like the BSN option should not be positioned in region b but, rather, should be distributed among regions b, c, d, and e in a way that is consistent with all the data. (Region b actually contains those who like the BSN but neither of the other two options.)

Because, at the start, we do not know how to distribute the 23 who like the BSN, we look first for some more manageable data. The smallest total listed, the 2 who like all three options, can be placed in region d (the intersection of the three sets). The 4 who like none of the three must go into region a. Then, the 10 who like the BSN and the ADN must go into regions c and d. Because region d already contains 2 attendees, we must place

$$10 - 2 = 8 \quad \text{in region } c.$$

Because 5 like BSN and Diploma (regions d and e), we place

$$5 - 2 = 3 \quad \text{in region } e.$$

Now that regions c, d, and e contain 8, 2, and 3, respectively, we must place

$$23 - 8 - 2 - 3 = 10 \quad \text{in region } b.$$

By similar reasoning, all regions are assigned their correct numbers. See **Figure 19.**

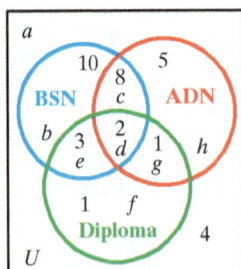

Figure 19

EXAMPLE 1 Analyzing a Survey

Using the survey data on personal preferences for nursing preparation, as summarized in **Figure 19,** answer each question.

(a) How many persons like the BSN option only?

(b) How many persons like exactly two of these three options?

(c) How many persons were surveyed?

Solution

(a) A person who likes BSN only does not like ADN and does not like Diploma. These persons are inside the regions for BSN and outside the regions for ADN and Diploma. Region b is the appropriate region in **Figure 19,** and we see that ten persons like BSN only.

(b) The persons in regions c, e, and g like exactly two of the three options. The total number of such persons is

$$8 + 3 + 1 = 12.$$

(c) Each person surveyed has been placed in exactly one region of **Figure 19,** so the total number surveyed is the sum of the numbers in all eight regions:

$$4 + 10 + 8 + 2 + 3 + 1 + 1 + 5 = 34. \qquad ■$$

Cardinal Number Formula

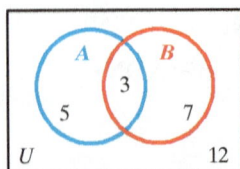

Figure 20

If the numbers shown in **Figure 20** are the cardinal numbers of the individual regions, then

$$n(A) = 5 + 3 = 8, \quad n(B) = 3 + 7 = 10, \quad n(A \cap B) = 3,$$

and

$$n(A \cup B) = 5 + 3 + 7 = 15.$$

Notice that $n(A \cup B) = n(A) + n(B) - n(A \cap B)$ because $15 = 8 + 10 - 3$. This relationship is true for any two sets A and B.

CARDINAL NUMBER FORMULA

For any two sets A and B, the following is true.

$$n(A \cup B) = n(A) + n(B) - n(A \cap B)$$

The cardinal number formula can be rearranged to find any one of its four terms when the others are known.

> **EXAMPLE 2** Applying the Cardinal Number Formula
>
> Find $n(A)$ if $n(A \cup B) = 22$, $n(A \cap B) = 8$, and $n(B) = 12$.
>
> **Solution**
>
> We solve the cardinal number formula for $n(A)$.
>
> $$n(A) = n(A \cup B) - n(B) + n(A \cap B)$$
> $$= 22 - 12 + 8$$
> $$= 18$$

Sometimes, even when information is presented as in **Example 2,** it is more convenient to fit that information into a Venn diagram as in **Example 1.**

WHEN Will I Ever USE This ?

Suppose you run a small construction company, building a few "spec" homes at a time. This week's work will require the following jobs:

Hanging drywall (D), Installing roofing (R), Doing electrical work (E).

You will assign 9 workers, with job skills as described here.

6 of the 9 can do D	5 can do both D and E
4 can do R	4 can do both R and E
7 can do E	4 can do all three

1. Construct a Venn diagram to decide how many of your workers have

   (a) exactly two of the three skills

   (b) none of the three skills

   (c) no more than one of the three skills.

2. Which skill is common to the greatest number of workers, and how many possess that skill?

Answers: 1. (a) 1 (b) 1 (c) 4 2. Electrical work; 7

EXAMPLE 3 Analyzing Data in a Report

Scott, who leads a group of software engineers who investigate illegal activities on social networking sites, reported the following information.

T = the set of group members following patterns on Twitter

F = the set of group members following patterns on Facebook

L = the set of group members following patterns on LinkedIn

$$n(T) = 13 \quad n(T \cap F) = 9 \quad n(T \cap F \cap L) = 5$$
$$n(F) = 16 \quad n(F \cap L) = 10 \quad n(T' \cap F' \cap L') = 3$$
$$n(L) = 13 \quad n(T \cap L) = 6$$

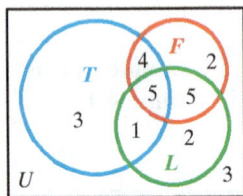

How many engineers are in Scott's group?

Solution

The data supplied by Scott are reflected in **Figure 21.** The sum of the numbers in the diagram gives the total number of engineers in the group.

$$3 + 3 + 1 + 2 + 5 + 5 + 4 + 2 = 25$$

Figure 21

Tables

Sometimes information appears in a table rather than a Venn diagram, but the basic ideas of union and intersection still apply.

EXAMPLE 4 Analyzing Data in a Table

Melanie, the officer in charge of the cafeteria on a military base, wanted to know if the beverage that enlisted men and women preferred with lunch depended on their ages. On a given day, Melanie categorized her lunch patrons according to age and preferred beverage, recording the results in a table.

		Beverage			
		Cola (C)	Iced Tea (I)	Sweet Tea (S)	Totals
Age	18–25 (Y)	45	10	35	90
	26–33 (M)	20	25	30	75
	Over 33 (O)	5	30	20	55
	Totals	70	65	85	220

Using the letters in the table, find the number of people in each set.

(a) $Y \cap C$ **(b)** $O' \cup I$

Solution

(a) The set Y includes all personnel represented across the top row of the table (90 in all), while C includes the 70 down the left column. The intersection of these two sets is just the upper left entry, 45 people.

(b) The set O' excludes the bottom row, so it includes the first and second rows. The set I includes the middle column only. The union of the two sets represents

$$45 + 10 + 35 + 20 + 25 + 30 + 30 = 195 \text{ people.}$$

2.4 EXERCISES

Use the numerals representing cardinalities in the Venn diagrams to give the cardinality of each set specified.

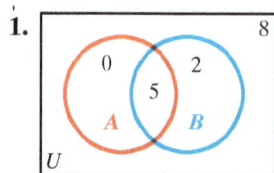

1.

(a) $A \cap B$ (b) $A \cup B$
(c) $A \cap B'$ (d) $A' \cap B$
(e) $A' \cap B'$

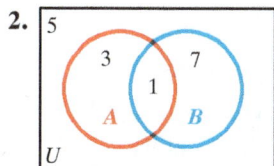

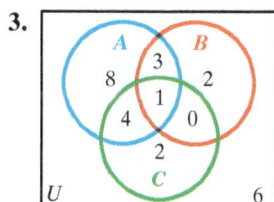

2.

(a) $A \cap B$ (b) $A \cup B$
(c) $A \cap B'$ (d) $A' \cap B$
(e) $A' \cap B'$

3.

(a) $A \cap B \cap C$ (b) $A \cap B \cap C'$
(c) $A \cap B' \cap C$ (d) $A' \cap B \cap C$
(e) $A' \cap B' \cap C$ (f) $A \cap B' \cap C'$
(g) $A' \cap B \cap C'$ (h) $A' \cap B' \cap C'$

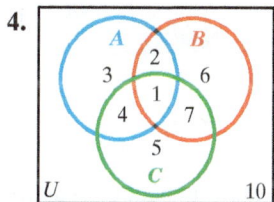

4.

(a) $A \cap B \cap C$ (b) $A \cap B \cap C'$
(c) $A \cap B' \cap C$ (d) $A' \cap B \cap C$
(e) $A' \cap B' \cap C$ (f) $A \cap B' \cap C'$
(g) $A' \cap B \cap C'$ (h) $A' \cap B' \cap C'$

In Exercises 5–10, make use of an appropriate formula.

5. Find the value of $n(A \cup B)$ if $n(A) = 12, n(B) = 14,$ and $n(A \cap B) = 5.$

6. Find the value of $n(A \cup B)$ if $n(A) = 16, n(B) = 28,$ and $n(A \cap B) = 5.$

7. Find the value of $n(A \cap B)$ if $n(A) = 20, n(B) = 12,$ and $n(A \cup B) = 25.$

8. Find the value of $n(A \cap B)$ if $n(A) = 20, n(B) = 24,$ and $n(A \cup B) = 30.$

9. Find the value of $n(A)$ if $n(B) = 35, n(A \cap B) = 15,$ and $n(A \cup B) = 55.$

10. Find the value of $n(B)$ if $n(A) = 20, n(A \cap B) = 6,$ and $n(A \cup B) = 30.$

Draw a Venn diagram and use the given information to fill in the number of elements in each region.

11. $n(A) = 19, \ n(B) = 13, \ n(A \cup B) = 25, \ n(A') = 11$

12. $n(U) = 43, \ n(A) = 25, \ n(A \cap B) = 5, \ n(B') = 30$

13. $n(A') = 25, \ n(B) = 28, \ n(A' \cup B') = 40,$ $n(A \cap B) = 10$

14. $n(A \cup B) = 15, \ n(A \cap B) = 8, \ n(A) = 13,$ $n(A' \cup B') = 11$

15. $n(A) = 57, \ n(A \cap B) = 35, \ n(A \cup B) = 81,$ $n(A \cap B \cap C) = 15, \ n(A \cap C) = 21, n(B \cap C) = 25,$ $n(C) = 49, \ n(B') = 52$

16. $n(A) = 24, \ n(B) = 24, \ n(C) = 26, \ n(A \cap B) = 10,$ $n(B \cap C) = 8, \ n(A \cap C) = 15, \ n(A \cap B \cap C) = 6,$ $n(U) = 50$

17. $n(A) = 15, \ n(A \cap B \cap C) = 5, \ n(A \cap C) = 13,$ $n(A \cap B') = 9, n(B \cap C) = 8, n(A' \cap B' \cap C') = 21,$ $n(B \cap C') = 3, \ n(B \cup C) = 32$

18. $n(A \cap B) = 21, \ n(A \cap B \cap C) = 6, \ n(A \cap C) = 26,$ $n(B \cap C) = 7, \ n(A \cap C') = 20, \ n(B \cap C') = 25,$ $n(C) = 40, \ n(A' \cap B' \cap C') = 2$

Use Venn diagrams to work each problem.

19. ***Writing and Producing Music*** Joe Long worked on 9 music projects last year.

Joe Long, Bob Gaudio, Tommy DeVito, and Frankie Valli
The Four Seasons

He wrote and produced 3 projects.

He wrote a total of 5 projects.

He produced a total of 7 projects.

(a) How many projects did he write but not produce?

(b) How many projects did he produce but not write?

20. *Compact Disc Collection* Gitti is a fan of the music of Paul Simon and Art Garfunkel. In her collection of 25 compact discs, she has the following:

> 5 on which both Simon and Garfunkel sing
>
> 7 on which Simon sings
>
> 8 on which Garfunkel sings
>
> 15 on which neither Simon nor Garfunkel sings.

(a) How many of her compact discs feature only Paul Simon?

(b) How many of her compact discs feature only Art Garfunkel?

(c) How many feature at least one of these two artists?

(d) How many feature at most one of these two artists?

21. *Student Response to Classical Composers* The 65 students in a classical music lecture class were polled, with the following results:

> 37 like Wolfgang Amadeus Mozart
>
> 36 like Ludwig van Beethoven
>
> 31 like Franz Joseph Haydn
>
> 14 like Mozart and Beethoven
>
> 21 like Mozart and Haydn
>
> 14 like Beethoven and Haydn
>
> 8 like all three composers.

How many of these students like:

(a) exactly two of these composers?

(b) exactly one of these composers?

(c) none of these composers?

(d) Mozart, but neither Beethoven nor Haydn?

(e) Haydn and exactly one of the other two?

(f) no more than two of these composers?

22. *Financial Aid for Students* At a southern university, half of the 48 mathematics majors were receiving federal financial aid.

> 5 had Pell Grants
>
> 14 participated in the College Work Study Program
>
> 4 had TOPS scholarships
>
> 2 had TOPS scholarships and participated in Work Study.
>
> Those with Pell Grants had no other federal aid.

How many of the 48 math majors had:

(a) no federal aid?

(b) more than one of these three forms of aid?

(c) federal aid other than these three forms?

(d) a TOPS scholarship or Work Study?

(e) exactly one of these three forms of aid?

(f) no more than one of these three forms of aid?

23. *Animated Movies* A middle school counselor, attempting to correlate school performance with leisure interests, found the following information for a group of students:

> 34 had seen *Despicable Me*
>
> 29 had seen *Epic*
>
> 26 had seen *Turbo*
>
> 16 had seen *Despicable Me* and *Epic*
>
> 12 had seen *Despicable Me* and *Turbo*
>
> 10 had seen *Epic* and *Turbo*
>
> 4 had seen all three of these films
>
> 5 had seen none of the three films.

(a) How many students had seen *Turbo* only?

(b) How many had seen exactly two of the films?

(c) How many students were surveyed?

24. *Non-Mainline Religious Beliefs* 140 U.S. adults were surveyed.

Let A = the set of respondents who believe in astrology,

R = the set of respondents who believe in reincarnation,

Y = the set of respondents who believe in the spirituality of yoga.

The survey revealed the following information:

$$n(A) = 35 \qquad n(R \cap Y) = 8$$
$$n(R) = 36 \qquad n(A \cap Y) = 10$$
$$n(Y) = 32 \qquad n(A \cap R \cap Y) = 6$$
$$n(A \cap R) = 19$$

How many of the respondents believe in:

(a) astrology but not reincarnation?

(b) at least one of these three things?

(c) reincarnation but neither of the others?

(d) exactly two of these three things?

(e) none of the three?

25. Survey on Attitudes toward Religion Researchers interviewed a number of people and recorded the following data. Of all the respondents:

240 think Hollywood is unfriendly toward religion

160 think the media are unfriendly toward religion

181 think scientists are unfriendly toward religion

145 think both Hollywood and the media are unfriendly toward religion

122 think both scientists and the media are unfriendly toward religion

80 think exactly two of these groups are unfriendly toward religion

110 think all three groups are unfriendly toward religion

219 think none of these three groups is unfriendly toward religion.

How many respondents:

(a) were surveyed?

(b) think exactly one of these three groups is unfriendly toward religion?

26. Student Goals Sofia, who sells college textbooks, interviewed first-year students on a community college campus to find out the main goals of today's students.

Let W = the set of those who want to be wealthy,

F = the set of those who want to raise a family,

E = the set of those who want to become experts in their fields.

Sofia's findings are summarized here.

$$n(W) = 160 \qquad n(E \cap F) = 90$$
$$n(F) = 140 \qquad n(W \cap F \cap E) = 80$$
$$n(E) = 130 \qquad n(E') = 95$$
$$n(W \cap F) = 95 \qquad n[(W \cup F \cup E)'] = 10$$

Find the total number of students interviewed.

27. Hospital Patient Symptoms A survey was conducted among 75 patients admitted to a hospital cardiac unit during a two-week period.

Let B = the set of patients with high blood pressure,

C = the set of patients with high cholesterol levels,

S = the set of patients who smoke cigarettes.

The survey produced the following data.

$$n(B) = 47 \qquad n(B \cap S) = 33$$
$$n(C) = 46 \qquad n(B \cap C) = 31$$
$$n(S) = 52 \qquad n(B \cap C \cap S) = 21$$
$$n[(B \cap C) \cup (B \cap S) \cup (C \cap S)] = 51$$

Find the number of these patients who:

(a) had either high blood pressure or high cholesterol levels, but not both

(b) had fewer than two of the indications listed

(c) were smokers but had neither high blood pressure nor high cholesterol levels

(d) did not have exactly two of the indications listed.

28. Song Themes It was once said that country-western songs emphasize three basic themes: love, prison, and trucks. A survey of the local country-western radio station produced the following data.

12 songs about a truck driver who is in love while in prison

13 about a prisoner in love

28 about a person in love

18 about a truck driver in love

3 about a truck driver in prison who is not in love

2 about people in prison who are not in love and do not drive trucks

8 about people who are out of prison, are not in love, and do not drive trucks

16 about truck drivers who are not in prison

(a) How many songs were surveyed?

Find the number of songs about:

(b) truck drivers

(c) prisoners

(d) truck drivers in prison

(e) people not in prison

(f) people not in love.

29. Use the figure to find the numbers of the regions belonging to each set.

(a) $A \cap B \cap C \cap D$

(b) $A \cup B \cup C \cup D$

(c) $(A \cap B) \cup (C \cap D)$

(d) $(A' \cap B') \cap (C \cup D)$

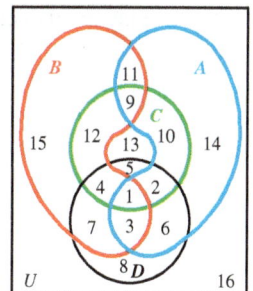

30. *Sports Viewing* A survey of 130 TV viewers was taken.

   52 watch football

   56 watch basketball

   62 watch tennis

   60 watch golf

   21 watch football and basketball

   19 watch football and tennis

   22 watch basketball and tennis

   27 watch football and golf

   30 watch basketball and golf

   21 watch tennis and golf

   3 watch football, basketball, and tennis

   15 watch football, basketball, and golf

   10 watch football, tennis, and golf

   10 watch basketball, tennis, and golf

   3 watch all four of these sports

   5 don't watch any of these four sports

Use a Venn diagram to answer these questions.

(a) How many of these viewers watch football, basketball, and tennis, but not golf?

(b) How many watch exactly one of these four sports?

(c) How many watch exactly two of these four sports?

Solve each problem.

31. *Basketball Positions* Sakda runs a basketball program in California. On the first day of the season, 60 young women showed up and were categorized by age level and by preferred basketball position, as shown in the following table.

	Position			
	Guard (G)	Forward (F)	Center (N)	Totals
Junior High (J)	9	6	4	19
Age Senior High (S)	12	5	9	26
College (C)	5	8	2	15
Totals	26	19	15	60

Using the set labels (letters) in the table, find the number of players in each of the following sets.

(a) $J \cap G$ **(b)** $S \cap N$

(c) $N \cup (S \cap F)$ **(d)** $S' \cap (G \cup N)$

(e) $(S \cap N') \cup (C \cap G')$ **(f)** $N' \cap (S' \cap C')$

32. *Army Housing* A study of U.S. Army housing trends categorized personnel as commissioned officers (C), warrant officers (W), or enlisted (E), and categorized their living facilities as on-base (B), rented off-base (R), or owned off-base (O). One survey yielded the following data.

		Facilities			
		B	R	O	Totals
	C	12	29	54	95
Personnel	W	4	5	6	15
	E	374	71	285	730
	Totals	390	105	345	840

Find the number of personnel in each of the following sets.

(a) $W \cap O$

(b) $C \cup B$

(c) $R' \cup W'$

(d) $(C \cup W) \cap (B \cup R)$

(e) $(C \cap B) \cup (E \cap O)$

(f) $B \cap (W \cup R)'$

33. Could the information of **Example 4** have been presented in a Venn diagram similar to those in **Examples 1 and 3**? If so, construct such a diagram. Otherwise, explain the essential difference of **Example 4.**

34. Explain how a cardinal number formula can be derived for the case where *three* sets occur. Specifically, give a formula relating $n(A \cup B \cup C)$ to

$$n(A), \ n(B), \ n(C), \ n(A \cap B), \ n(A \cap C),$$
$$n(B \cap C), \ \text{and} \ n(A \cap B \cap C).$$

Illustrate with a Venn diagram.

CHAPTER 2   SUMMARY

KEY TERMS

2.1

set
elements
members
empty (null) set
natural (counting) numbers
whole numbers
integers
rational numbers
real numbers
irrational numbers
cardinal number
  (cardinality)
finite set
infinite set

2.2

universal set
Venn diagram
complement (of a set)
subset (of a set)
proper subset (of a set)
tree diagram

2.3

intersection (of sets)
disjoint sets
union (of sets)
difference (of sets)
ordered pairs
Cartesian product (of sets)
operation (on sets)

NEW SYMBOLS

$\varnothing$	empty set (or null set)	$\not\subseteq$	is not a subset of
$\in$	is an element of	$\subset$	is a proper subset of
$\notin$	is not an element of	$\cap$	intersection (of sets)
$n(A)$	cardinal number of set A	$\cup$	union (of sets)
$\aleph_0$	aleph null	$-$	difference (of sets)
U	universal set	(a, b)	ordered pair
A'	complement of set A	$\times$	Cartesian product
$\subseteq$	is a subset of		

TEST YOUR WORD POWER

See how well you have learned the vocabulary in this chapter.

1. In an **ordered pair,**
A. the components are always numbers of the same type.
B. the first component must be less than the second.
C. the components can be any kinds of objects.
D. the first component is a subset of the second component.

2. The **complement** of a set
A. contains only some, but not all, of that set's elements.
B. contains the same number of elements as the given set.
C. always contains fewer elements than the given set.
D. cannot be determined until a universal set is given.

3. Any **subset** of set A
A. must have fewer elements than A.
B. has fewer elements than A only if it is a proper subset.
C. is an element of A.
D. contains all the elements of A, plus at least one additional element.

4. Examples of **operations on sets** are
A. union, intersection, and subset.
B. Cartesian product, difference, and proper subset.
C. complement, supplement, and intersection.
D. complement, union, and Cartesian product.

5. The **set difference $A - B$** must have cardinality
A. $n(A) - n(B)$.
B. less than or equal to $n(A)$.
C. greater than or equal to $n(B)$.
D. less than $n(A)$.

6. The **cardinal number formula** says that
A. $n(A \cup B) = n(A) + n(B) - n(A \cap B)$.
B. $n(A \cup B) = n(A) + n(B) + n(A \cap B)$.
C. $n(A \cup B) = n(A) + n(B)$.
D. $n(A \cup B) = n(A) \cdot n(B)$.

ANSWERS

1. C **2.** D **3.** B **4.** D **5.** B **6.** A

QUICK REVIEW

Concepts	Examples

2.1 Symbols and Terminology

Designating Sets

Sets are designated using the following methods:

(1) word descriptions,

(2) the listing method,

(3) set-builder notation.

The set of odd counting numbers less than 7

$\{1, 3, 5\}$ — All are equal.

$\{x \mid x \text{ is an odd counting number and } x < 7\}$

The **cardinal number** of a set is the number of elements it contains.

If $A = \{10, 20, 30, ..., 80\}$, then $n(A) = 8$.

Equal sets have exactly the same elements.

$\{a, e, i, o, u\} = \{i, o, u, a, e\}, \quad \{q, r, s, t\} \neq \{q, p, s, t\}$

2.2 Venn Diagrams and Subsets

Sets are normally discussed within the context of a designated **universal set, U.**

Let $U = \{x \mid x \text{ is a whole number}\}$

and

$A = \{x \mid x \text{ is an even whole number}\}$.

The **complement** of a set A contains all elements in U that are not in A.

Then $A' = \{x \mid x \text{ is an odd whole number}\}$.

Set B is a **subset** of set A if every element of B is also an element of A.

Let $U = \{2, 3, 5, 7\}$, $A = \{3, 5, 7\}$, and $B = \{5\}$.
Then $A' = \{2\}$ and $B \subseteq A$.

Set B is a **proper subset** of A if $B \subseteq A$ and $B \neq A$

For the sets A and B given above, $B \subset A$.

If a set has cardinal number n, it has 2^n subsets and $2^n - 1$ proper subsets.

If $D = \{x, y, z\}$, then $n(D) = 3$.
D has $2^3 = 8$ subsets and $2^3 - 1 = 8 - 1 = 7$ proper subsets.

2.3 Set Operations

Intersection of sets

$$A \cap B = \{x \mid x \in A \text{ and } x \in B\}$$

$$\{1, 2, 7\} \cap \{5, 7, 9, 11\} = \{7\}$$

Union of sets

$$A \cup B = \{x \mid x \in A \text{ or } x \in B\}$$

$$\{20, 40, 60\} \cup \{40, 60, 80\} = \{20, 40, 60, 80\}$$

Difference of sets

$$A - B = \{x \mid x \in A \text{ and } x \notin B\}$$

$$\{5, 6, 7, 8\} - \{1, 3, 5, 7\} = \{6, 8\}$$

Cartesian product of sets

$$A \times B = \{(a, b) \mid a \in A \text{ and } b \in B\}$$

$\{1, 2\} \times \{30, 40, 50\}$
$= \{(1, 30), (1, 40), (1, 50), (2, 30), (2, 40), (2, 50)\}$

Cardinal number of a Cartesian product

If $n(A) = a$ and $n(B) = b$, then

$$n(A \times B) = n(A) \cdot n(B) = ab.$$

If $A = \{2, 4, 6, 8\}$ and $B = \{3, 7\}$, then

$$n(A \times B) = n(A) \cdot n(B) = 4 \cdot 2 = 8.$$

Verify by listing the elements of $A \times B$.

$\{(2, 3), (2, 7), (4, 3), (4, 7), (6, 3), (6, 7), (8, 3), (8, 7)\}$

It, indeed, has 8 elements.

Concepts	*Examples*

Numbering the regions in a Venn diagram facilitates identification of various sets and relationships among them.

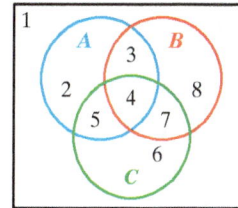

A consists of regions 2, 3, 4, and 5.

$A \cap B$ consists of regions 3 and 4.

$B \cup C$ consists of regions 3, 4, 5, 6, 7, and 8.

$(A \cup B) \cap C'$ consists of regions 2, 3, and 8.

$A \cap B \cap C$ consists of region 4.

$(A \cup B \cup C)'$ consists of region 1.

$A - C$ consists of regions 2 and 3.

De Morgan's Laws

For any sets A and B, where $A, B \subseteq U$, the following are true.

$$(A \cap B)' = A' \cup B'$$

Let $U = \{1, 2, 3, 4, \ldots, 9\}$, $A = \{2, 4, 6, 8\}$, and $B = \{4, 5, 6, 7\}$.

Then $A \cap B = \{4, 6\}$,

so $(A \cap B)' = \{1, 2, 3, 5, 7, 8, 9\}$.

Also $A' = \{1, 3, 5, 7, 9\}$ and $B' = \{1, 2, 3, 8, 9\}$,   — Same

so $A' \cup B' = \{1, 2, 3, 5, 7, 8, 9\}$.

Using the same sets A and B as above,

$$A \cup B = \{2, 4, 5, 6, 7, 8\},$$

$$(A \cup B)' = A' \cap B'$$

so $(A \cup B)' = \{1, 3, 9\}$.

Also, $A' = \{1, 3, 5, 7, 9\}$ and $B' = \{1, 2, 3, 8, 9\}$,   — Same

so $A' \cap B' = \{1, 3, 9\}$.

2.4   Surveys and Cardinal Numbers

Cardinal Number Formula

For any two sets A and B, the following is true.

$$n(A \cup B) = n(A) + n(B) - n(A \cap B)$$

Suppose $n(A) = 5, n(B) = 12$, and $n(A \cup B) = 10$. Then, to find $n(A \cap B)$, first solve the formula for that term.

$$n(A \cap B) = n(A) + n(B) - n(A \cup B)$$
$$= 5 + 12 - 10 = 7$$

Enter known facts in a Venn diagram to find desired facts.

Given $n(A) = 12, n(B) = 27, n(A \cup B) = 32$, and $n(U) = 50$, find $n(A - B)$ and $n[(A \cap B)']$.

$$n(A \cap B) = n(A) + n(B) - n(A \cup B)$$
$$= 12 + 27 - 32$$
$$= 7$$

Then the cardinalities of all regions can be entered in the diagram, starting with 7 in the center.

Now observe that $n(A - B) = 5$ and $n[(A \cap B)'] = 18 + 5 + 20 = 43$.

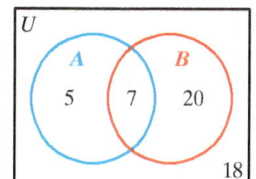

CHAPTER 2   **TEST**

In Exercises 1–14, let

$$U = \{a, b, c, d, e, f, g, h\}, \quad A = \{a, b, c, d\},$$
$$B = \{b, e, a, d\}, \quad and \quad C = \{e, a\}.$$

Find each set.

1. $A \cup C$

2. $B \cap A$

3. B'

4. $A - (B \cap C')$

Identify each statement as true *or* false.

5. $e \in A$

6. $C \subseteq B$

7. $B \subset (A \cup C)$

8. $c \notin C$

9. $n[(A \cup B) - C] = 4$

10. $\varnothing \not\subset C$

11. $A \cap B'$ is equivalent to $B \cap A'$

12. $(A \cup B)' = A' \cap B'$

Find each of the following.

13. $n(B \times A)$

14. the number of proper subsets of B

Give a word description for each set.

15. $\{-3, -1, 1, 3, 5, 7, 9\}$

16. $\{Sun, Mon, Tue, \dots, Sat\}$

Express each set in set-builder notation.

17. $\{-1, -2, -3, -4, \dots\}$

18. $\{24, 32, 40, 48, \dots, 88\}$

Place $\subset$, $\subseteq$, *both,* or *neither in each blank to make a true statement.*

19. $\varnothing$ _____ $\{x \mid x$ is a counting number between 20 and 21$\}$

20. $\{3, 5, 7\}$ _____ $\{4, 5, 6, 7, 8, 9, 10\}$

Shade each set in an appropriate Venn diagram.

21. $X \cup Y'$

22. $X' \cap Y'$

23. $(X \cup Y) - Z$

24. $[(X \cap Y) \cup (X \cap Z)] - (Y \cap Z)$

25. State De Morgan's laws for sets in words rather than symbols.

Facts about Inventions The table lists ten inventions, together with other pertinent data.

Invention	Date	Inventor	Nation
Adding machine	1642	Pascal	France
Baking powder	1843	Bird	England
Electric razor	1917	Schick	U.S.
Fiber optics	1955	Kapany	England
Geiger counter	1913	Geiger	Germany
Pendulum clock	1657	Huygens	Holland
Radar	1940	Watson-Watt	Scotland
Telegraph	1837	Morse	U.S.
Thermometer	1593	Galileo	Italy
Zipper	1891	Judson	U.S.

Let $U =$ the set of all ten inventions,

$A =$ the set of items invented in the United States,

and $T =$ the set of items invented in the twentieth century.

List the elements of each set.

26. $A \cap T$ **27.** $(A \cup T)'$ **28.** $T' - A'$

29. The numerals in the Venn diagram indicate the number of elements in each particular subset. Determine the number of elements in each set.

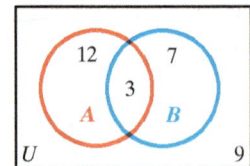

(a) $A \cup B$ **(b)** $A \cap B'$

(c) $(A \cap B)'$

30. *Financial Aid to College Students*   Three sources of financial aid are government grants, private scholarships, and the colleges themselves. Susan, a financial aid director of a private college, surveyed the records of 100 sophomores and found the following:

49 receive government grants

55 receive private scholarships

43 receive aid from the college

23 receive government grants and private scholarships

18 receive government grants and aid from the college

28 receive private scholarships and aid from the college

8 receive help from all three sources.

How many of the students in the survey:

(a) have government grants only?

(b) have scholarships but not government grants?

(c) receive financial aid from only one of these sources?

(d) receive aid from exactly two of these sources?

(e) receive no financial aid from any of these sources?

(f) receive no aid from the college or the government?

Introduction to Logic

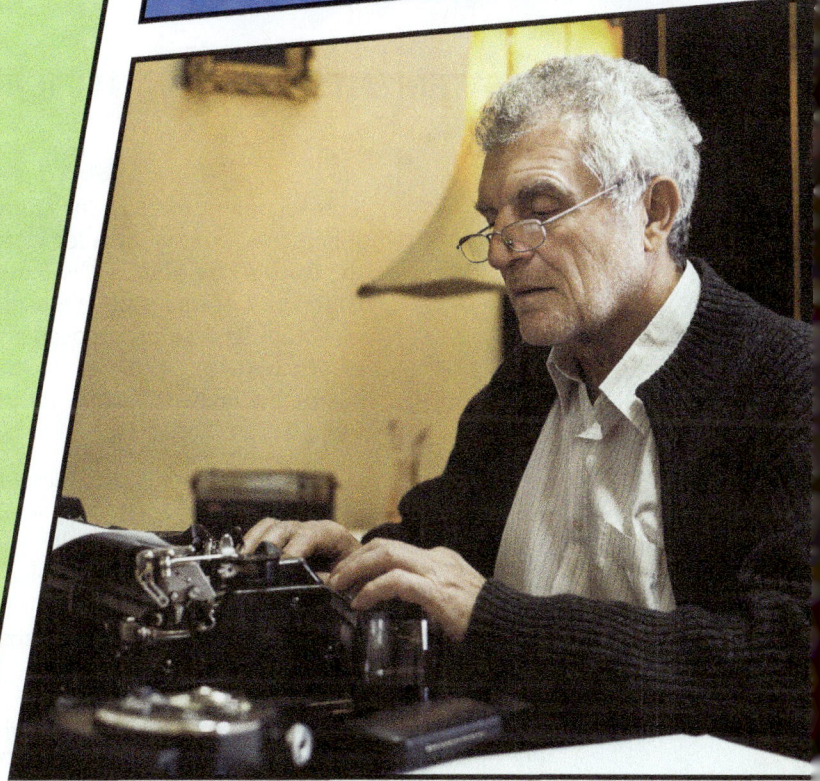

To capture the imagination of their readers, authors of fantasy literature (such as J. R. R. Tolkien, J. K. Rowling, and Lewis Carroll) have long used magical imagery, including riddles. Such a riddle confronts a hero named Humphrey, who finds himself trapped in a magical maze with only two exits.

One door to freedom opens wide,
The other naught but dungeon hides.
Twin watchers, one by truth is bound,
The other never truth will sound.

The watcher here shall speak when he
Spies a man who would be free.
Name him Lying Troll or Truthful,
Rightly named he shall be useful.

If he moves, then be not late,
Choose a door and seal thy fate.

As Humphrey approaches the troll, it speaks:

> *"If truly Truthful Troll I be,*
> *then go thou east, and be thou free."*

Logic, the topic of this chapter, is useful to authors in the creation of such puzzles—and also to readers in solving them.

3.1 STATEMENTS AND QUANTIFIERS

OBJECTIVES

1 Distinguish between statements and non-statements.

2 Compose negations of statements.

3 Translate between words and symbols.

4 Interpret statements with quantifiers and form their negations.

5 Find truth values of statements involving quantifiers and number sets.

Statements

This section introduces the study of **symbolic logic,** which uses letters to represent statements, and symbols for words such as *and, or, not.* Logic is used to determine the **truth value** (that is, the truth or falsity) of statements with multiple parts. The truth value of such statements depends on their components.

Many kinds of sentences occur in ordinary language, including factual statements, opinions, commands, and questions. Symbolic logic discusses only statements that involve facts. A **statement** is a declarative sentence that is either true or false, but not both simultaneously.

> Electronic mail provides a means of communication. } Statements
> $12 + 6 = 13$ } Each is either true or false.

Access the file.
Did the Seahawks win the Super Bowl?
Dustin Pedroia is a better baseball player than Miguel Cabrera.
This sentence is false.

} Not statements
Each cannot be identified as being either true or false.

Of the sentences that are not statements, the first is a command, and the second is a question. The third is an opinion. "This sentence is false" is a paradox: If we assume it is true, then it is false, and if we assume it is false, then it is true.

A **compound statement** may be formed by combining two or more statements. The statements making up a compound statement are called **component statements.** Various **logical connectives,** or simply **connectives,** such as *and, or, not,* and *if . . . then,* can be used in forming compound statements. (Although a statement such as "Today is not Tuesday" does not consist of two component statements, for convenience it is considered compound, because its truth value is determined by noting the truth value of a different statement, "Today is Tuesday.")

EXAMPLE 1 Deciding Whether a Statement Is Compound

Decide whether each statement is compound. If so, identify the connective.

(a) Lord Byron wrote sonnets, and the poem exhibits iambic pentameter.

(b) You can pay me now, or you can pay me later.

(c) If it's on the Internet, then it must be true.

(d) My pistol was made by Smith and Wesson.

Solution

(a) This statement is compound, because it is made up of the component statements "Lord Byron wrote sonnets" and "the poem exhibits iambic pentameter." The connective is *and*.

(b) The connective here is *or*. The statement is compound.

Gottfried Leibniz (1646–1716) was a wide-ranging philosopher and a universalist who tried to patch up Catholic–Protestant conflicts. He promoted cultural exchange between Europe and the East. Chinese ideograms led him to search for a universal symbolism. He was an early inventor of **symbolic logic.**

(c) The connective here is *if . . . then*, discussed in more detail in **Section 3.3.** The statement is compound.

(d) Although the word "and" is used in this statement, it is not used as a *logical* connective. It is part of the name of the manufacturer. The statement is not compound. ∎

Negations

The sentence "Anthony Mansella has a red truck" is a statement. The **negation** of this statement is "Anthony Mansella does not have a red truck." ***The negation of a true statement is false, and the negation of a false statement is true.***

EXAMPLE 2 Forming Negations

Form the negation of each statement.

(a) That city has a mayor.   (b) The moon is not a planet.

Solution

(a) To negate this statement, we introduce *not* into the sentence: "That city does not have a mayor."

(b) The negation is "The moon is a planet." ∎

One way to detect incorrect negations is to check truth values. ***A negation must have the opposite truth value from the original statement.***

The next example uses some of the inequality symbols in **Table 1.** In the case of an inequality involving a variable, the negation must have the opposite truth value for *any* replacement of the variable.

```
TEST  LOGIC
1: =
2: ≠
3: >
4: ≥
5: <
6: ≤
```

The TEST menu of the TI-83/84 Plus calculator allows the user to test the truth value of statements involving =, ≠, >, ≥, <, and ≤. If a statement is true, it returns a 1. If a statement is false, it returns a 0.

Table 1

Symbolism	Meaning	Examples	
$a < b$	a is less than b	$4 < 9$	$\frac{1}{2} < \frac{3}{4}$
$a > b$	a is greater than b	$6 > 2$	$-5 > -11$
$a \leq b$	a is less than or equal to b	$8 \leq 10$	$3 \leq 3$
$a \geq b$	a is greater than or equal to b	$-2 \geq -3$	$-5 \geq -5$

```
4<9
          1
4>9
          0
```

$4 < 9$ is true, as indicated by the 1.
$4 > 9$ is false, as indicated by the 0.

EXAMPLE 3 Negating Inequalities

Give a negation of each inequality. Do *not* use a slash symbol.

(a) $x < 9$ (b) $7x + 11y \geq 77$

Solution

(a) The negation of "x is less than 9" is "x is *not* less than 9." Because we cannot use "not," which would require writing $x \not< 9$, phrase the negation as "x is greater than or equal to 9," or

$$x \geq 9.$$

(b) The negation, with no slash, is

$$7x + 11y < 77.$$ ∎

Symbols

The study of logic uses symbols. Statements are represented with letters, such as p, q, or r. Several symbols for connectives are shown in **Table 2**.

Table 2

Connective	Symbol	Type of Statement
and	$\wedge$	Conjunction
or	$\vee$	Disjunction
not	$\sim$	Negation

The symbol $\sim$ represents the connective *not*. If p represents the statement "Barack Obama was president in 2014," then $\sim p$ represents "Barack Obama was *not* president in 2014."

EXAMPLE 4 Translating from Symbols to Words

Let p represent the statement "Nursing informatics is a growing field," and let q represent "Critical care will always be in demand." Translate each statement from symbols to words.

(a) $p \vee q$ **(b)** $\sim p \wedge q$ **(c)** $\sim(p \vee q)$ **(d)** $\sim(p \wedge q)$

Solution

(a) From the table, $\vee$ symbolizes *or*. Thus, $p \vee q$ represents

Nursing informatics is a growing field, or critical care will always be in demand.

(b) Nursing informatics is not a growing field and critical care will always be in demand.

(c) It is not the case that nursing informatics is a growing field or critical care will always be in demand. (This is usually translated as **"Neither p nor q."**)

(d) It is not the case that nursing informatics is a growing field and critical care will always be in demand. ∎

Quantifiers

Quantifiers are used to indicate *how many* cases of a particular situation exist. The words ***all, each, every,*** and ***no(ne)*** are **universal quantifiers,** while words and phrases such as ***some, there exists,*** and ***(for) at least one*** are **existential quantifiers.**

Be careful when forming the negation of a statement involving quantifiers. A statement and its negation must have opposite truth values in all possible cases. Consider this statement.

All girls in the group are named Mary.

Many people would write the negation of this statement as "No girls in the group are named Mary" or "All girls in the group are not named Mary." But neither of these is correct. To see why, look at the three groups below.

Group I:	Mary Jane Payne, Mary Meyer, Mary O'Hara
Group II:	Mary Johnson, Lisa Pollak, Margaret Watson
Group III:	Donna Garbarino, Paula Story, Rhonda Alessi, Kim Falgout

These groups contain all possibilities that need to be considered. In Group I, *all* girls are named Mary. In Group II, *some* girls are named Mary (and some are not). In Group III, *no* girls are named Mary.

Consider **Table 3**. Keep in mind that "some" means "at least one (and possibly all)."

Table 3 Truth Value as Applied to:

	Group I	Group II	Group III
(1) All girls in the group are named Mary. **(Given)**	T	F	F
(2) No girls in the group are named Mary. **(Possible negation)**	F	F	T
(3) All girls in the group are not named Mary. **(Possible negation)**	F	F	T
(4) Some girls in the group are not named Mary. **(Possible negation)**	F	T	T

Negation

The negation of the given statement (1) must have opposite truth values in *all* cases. It can be seen that statements (2) and (3) do not satisfy this condition (for Group II), but statement (4) does. It may be concluded that the correct negation for "All girls in the group are named Mary" is "Some girls in the group are not named Mary." Other ways of stating the negation include the following.

Not all girls in the group are named Mary.

It is not the case that all girls in the group are named Mary.

At least one girl in the group is not named Mary.

Table 4 shows how to find the negation of a statement involving quantifiers.

Table 4 Negations of Quantified Statements

Statement	Negation
All do.	Some do not. (Equivalently: Not all do.)
Some do.	None do. (Equivalently: All do not.)

The negation of the negation of a statement is simply the statement itself. For instance, the negations of the statements in the Negation column are simply the corresponding original statements in the Statement column.

EXAMPLE 5 **Forming Negations of Quantified Statements**

Form the negation of each statement.

(a) Some cats have fleas.   **(b)** Some cats do not have fleas.

(c) No cats have fleas.

Solution

(a) Because *some* means "at least one," the statement "Some cats have fleas" is really the same as "At least one cat has fleas." The negation of this is

"No cat has fleas."

(b) The statement "Some cats do not have fleas" claims that at least one cat, somewhere, does not have fleas. The negation of this is

"All cats have fleas."

(c) The negation is "Some cats have fleas." ⎯⎯ Avoid the incorrect answer "All cats have fleas."

Quantifiers and Number Sets

Earlier we introduced sets of numbers.

SETS OF NUMBERS
Natural numbers (or counting numbers) $\{1, 2, 3, 4, \ldots\}$
Whole numbers $\{0, 1, 2, 3, 4, \ldots\}$
Integers $\{\ldots, -3, -2, -1, 0, 1, 2, 3, \ldots\}$
Rational numbers $\left\{\frac{p}{q} \,\middle|\, p \text{ and } q \text{ are integers, and } q \neq 0\right\}$
Real numbers $\{x \,|\, x \text{ is a number that can be written as a decimal}\}$
Irrational numbers $\{x \,|\, x \text{ is a real number and } x \text{ cannot be written as a quotient of integers}\}$

EXAMPLE 6 **Deciding Whether Quantified Statements Are True or False**

Decide whether each statement involving a quantifier is *true* or *false*.

(a) There exists a whole number that is not a natural number.

(b) Every integer is a natural number.

(c) Every natural number is a rational number.

(d) There exists an irrational number that is not real.

Solution

(a) Because there is such a whole number (it is 0), this statement is true.

(b) This statement is false, because we can find at least one integer that is not a natural number. For example, -1 is an integer but is not a natural number.

(c) Because every natural number can be written as a fraction with denominator 1, this statement is true.

(d) In order to be an irrational number, a number must first be real. Because we cannot give an irrational number that is not real, this statement is false. (Had we been able to find at least one, the statement would have been true.) ∎

3.1 EXERCISES

Decide whether each is a statement or is not a statement.

1. February 2, 2009, was a Monday.

2. The ZIP code for Oscar, Louisiana, is 70762.

3. Listen, my children, and you shall hear of the midnight ride of Paul Revere.

4. Did you yield to oncoming traffic?

5. $5 + 9 \neq 14$ and $4 - 1 = 12$

6. $5 + 9 \neq 12$ or $4 - 2 = 5$

7. Some numbers are positive.

8. Grover Cleveland was president of the United States in 1885 and 1897.

9. Accidents are the main cause of deaths of children under the age of 7.

10. It is projected that in the United States between 2010 and 2020, there will be over 500,000 job openings per year for elementary school teachers, with median annual salaries of about $51,000.

11. Where are you going tomorrow?

12. Behave yourself and sit down.

13. Kevin "Catfish" McCarthy once took a prolonged continuous shower for 340 hours, 40 minutes.

14. One gallon of milk weighs more than 3 pounds.

Decide whether each statement is compound.

15. I read the *Detroit Free Press,* and I read the *Sacramento Bee.*

16. My brother got married in Copenhagen.

17. Tomorrow is Saturday.

18. Jing is younger than 18 years of age, and so is her friend Shu-fen.

19. Jay's wife loves Ben and Jerry's ice cream.

20. The sign on the back of the car read "Canada or bust!"

21. If Lorri sells her quota, then Michelle will be happy.

22. If Bobby is a politician, then Mitch is a crook.

Write a negation for each statement.

23. Her aunt's name is Hermione.

24. No rain fell in southern California today.

25. Some books are longer than this book.

26. All students present will get another chance.

27. No computer repairman can play blackjack.

28. Some people have all the luck.

29. Everybody loves somebody sometime.

30. Everyone needs a friend.

31. The trash needs to be collected.

32. Every architect who wants a job can find one.

Give a negation of each inequality. Do not use a slash symbol.

33. $x > 12$

34. $x < -6$

35. $x \geq 5$

36. $x \leq 19$

37. Try to negate the sentence "The exact number of words in this sentence is ten" and see what happens. Explain the problem that arises.

38. Explain why the negation of "$x > 5$" is not "$x < 5$."

Let p represent the statement "She has green eyes" and let q represent the statement "He is 60 years old." Translate each symbolic compound statement into words.

39. $\sim p$

40. $\sim q$

41. $p \wedge q$

42. $p \vee q$

43. $\sim p \vee q$

44. $p \wedge \sim q$

45. $\sim p \vee \sim q$

46. $\sim p \wedge \sim q$

47. $\sim (\sim p \wedge q)$

48. $\sim (p \vee \sim q)$

Let p represent the statement "Tyler collects DVDs" and let q represent the statement "Josh is an art major." Convert each compound statement into symbols.

49. Tyler collects DVDs and Josh is not an art major.

50. Tyler does not collect DVDs or Josh is not an art major.

51. Tyler does not collect DVDs or Josh is an art major.

52. Josh is an art major and Tyler does not collect DVDs.

53. Neither Tyler collects DVDs nor Josh is an art major.

54. Either Josh is an art major or Tyler collects DVDs, and it is not the case that both Josh is an art major and Tyler collects DVDs.

55. Incorrect use of quantifiers often is heard in everyday language. Suppose you hear that a local electronics chain is having a 40% off sale, and the radio advertisement states "All items are not available in all stores." Do you think that, literally translated, the ad really means what it says? What do you think is really meant? Explain your answer.

56. Repeat **Exercise 55** for the following: "All people don't have the time to devote to maintaining their vehicles properly."

Refer to the groups of art labeled A, B, and C, and identify by letter the group or groups that satisfy the given statements involving quantifiers in Exercises 57–64.

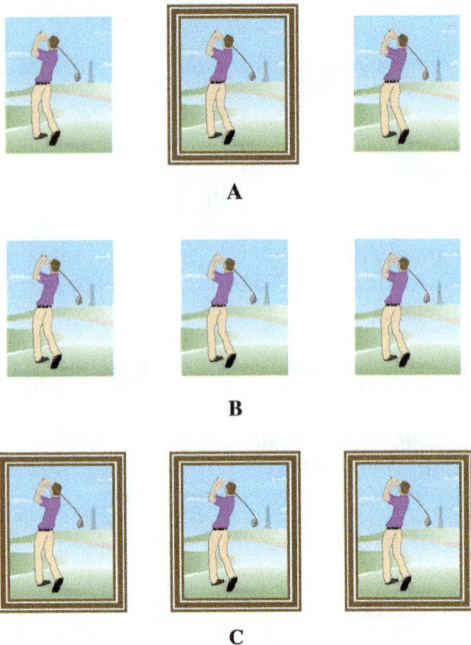

A

B

C

57. All pictures have frames.

58. No picture has a frame.

59. At least one picture does not have a frame.

60. Not every picture has a frame.

61. At least one picture has a frame.

62. No picture does not have a frame.

63. All pictures do not have frames.

64. Not every picture does not have a frame.

Decide whether each statement in Exercises 65–74 involving a quantifier is true *or* false.

65. Every whole number is an integer.

66. Every integer is a whole number.

67. There exists a natural number that is not an integer.

68. There exists an integer that is not a natural number.

69. All rational numbers are real numbers.

70. All irrational numbers are real numbers.

71. Some rational numbers are not integers.

72. Some whole numbers are not rational numbers.

73. Each whole number is a positive number.

74. Each rational number is a positive number.

75. Explain the difference between the statements "All students did not pass the test" and "Not all students passed the test."

76. The statement "For some real number x, $x^2 \geq 0$" is true. However, your friend does not understand why, because he claims that $x^2 \geq 0$ is true for *all* real numbers x (and not *some*). How would you explain his misconception to him?

77. Write the following statement using "every": There is no one here who has not made mistakes before.

78. Only one of these statements is true. Which one is it?

   A. For some real number x, $x \not< 0$.

   B. For all real numbers x, $x^3 > 0$.

   C. For all real numbers x less than 0, x^2 is also less than 0.

   D. For some real number x, $x^2 < 0$.

Symbolic logic also uses symbols for quantifiers. The symbol for the existential quantifier is

$$\exists \quad (a \text{ rotated } E),$$

and the symbol for the universal quantifier is

$$\forall \quad (an \text{ inverted } A).$$

The statement "For some x, p is true" can be symbolized

$$(\exists x)(p).$$

The statement "For all x, p is true" can be symbolized

$$(\forall x)(p).$$

The negation of $(\exists x)(p)$ is

$$(\forall x)(\sim p),$$

and the negation of $(\forall x)(p)$ is

$$(\exists x)(\sim p).$$

79. Refer to **Example 5.** If we let c represent "cat" and f represent "The cat has fleas," then the statement "Some cats have fleas" can be represented by $(\exists c)(f)$. Use symbols to express the negation of this statement.

80. Use symbols to express the statements for parts (b) and (c) of **Example 5** and their negations. Verify that the symbolic expressions translate to the negations found in the text.

3.2 TRUTH TABLES AND EQUIVALENT STATEMENTS

OBJECTIVES

1 Find the truth value of a conjunction.

2 Find the truth value of a disjunction.

3 Find the truth values for compound mathematical statements.

4 Construct truth tables for compound statements.

5 Understand and determine equivalence of statements.

6 Use De Morgan's laws to find negations of compound statements.

Conjunctions

Truth values of component statements are used to find truth values of compound statements. To begin, we must decide on truth values of the **conjunction** *p and q,* symbolized $p \wedge q$. Here, the connective *and* implies the idea of "both." The following statement is true, because each component statement is true.

Monday immediately follows Sunday, and March immediately follows February.

True

On the other hand, the following statement is false, even though part of the statement (Monday immediately follows Sunday) is true.

Monday immediately follows Sunday, and March immediately follows January.

False

For the conjunction $p \wedge q$ to be true, both p and q must be true. This result is summarized by a table, called a **truth table,** which shows truth values of $p \wedge q$ for all four possible combinations of truth values for the component statements *p* and *q*.

TRUTH TABLE FOR THE CONJUNCTION *p and q*

p and q

p	*q*	$p \wedge q$
T	T	T
T	F	F
F	T	F
F	F	F

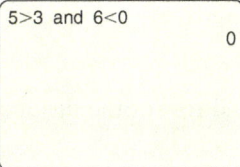

5>3 and 6<0
                    0

The calculator returns a 0 for

5 > 3 *and* 6 < 0,

indicating that the statement is false.

EXAMPLE 1 Finding the Truth Value of a Conjunction

Let *p* represent "5 > 3" and let *q* represent "6 < 0." Find the truth value of $p \wedge q$.

Solution

Here *p* is true and *q* is false. The second row of the conjunction truth table shows that $p \wedge q$ is false in this case. ■

In some cases, the logical connective *but* is used in compound statements.

He wants to go to the mountains but she wants to go to the beach.

Here, *but* is used in place of *and* to give a different emphasis to the statement. We consider this statement as we would consider the conjunction using the word *and.* The truth table for the conjunction, given above, would apply.

Disjunctions

In ordinary language, the word *or* can be ambiguous. The expression "this or that" can mean either "this or that or both," or "this or that but not both." For example, consider the following statement.

I will paint the wall or I will paint the ceiling.

This statement probably means: "I will paint the wall or I will paint the ceiling or I will paint both."

On the other hand, consider the following statement.

I will wear my glasses or my contact lenses.

It probably means "I will wear my glasses, or I will wear my contacts, but I will not wear both."

The symbol ∨ represents the first *or* described. That is,

_p_ ∨ _q_   means   "_p_ or _q_ or both."   Disjunction

With this meaning of *or*, _p_ ∨ _q_ is called the **inclusive disjunction,** or just the **disjunction** of _p_ and _q_. In everyday language, the disjunction implies the idea of "either." For example, consider the following disjunction.

I have a quarter or I have a dime.

It is true whenever I have either a quarter, a dime, or both. The only way this disjunction could be false would be if I had neither coin. *The disjunction _p_ ∨ _q_ is false only if both component statements are false.*

TRUTH TABLE FOR THE DISJUNCTION   _p_ or _q_

_p_ or _q_

_p_	_q_	_p_ ∨ _q_
T	T	T
T	F	T
F	T	T
F	F	F

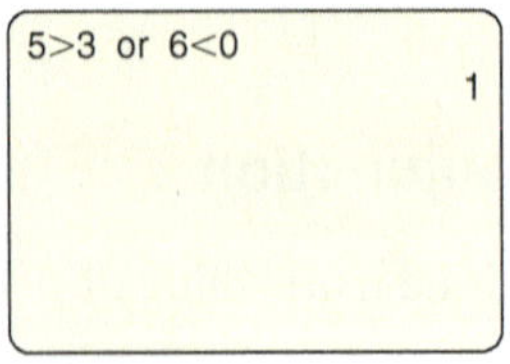

EXAMPLE 2 **Finding the Truth Value of a Disjunction**

Let _p_ represent "5 > 3" and let _q_ represent "6 < 0." Find the truth value of _p_ ∨ _q_.

Solution

Here, as in **Example 1,** _p_ is true and _q_ is false. The second row of the disjunction truth table shows that _p_ ∨ _q_ is true.

The symbol ≥ is read **"is greater than or equal to,"** while ≤ is read **"is less than or equal to."** If _a_ and _b_ are real numbers, then $a \leq b$ is true if $a < b$ or $a = b$. See **Table 5.**

5>3 or 6<0

1

The calculator returns a 1 for

5 > 3 *or* 6 < 0,

indicating that the statement is true.

Table 5

Statement	Reason It Is True
$8 \geq 8$	$8 = 8$
$3 \geq 1$	$3 > 1$
$-5 \leq -3$	$-5 < -3$
$-4 \leq -4$	$-4 = -4$

Negations

The **negation** of a statement _p_, symbolized **~_p_,** must have the opposite truth value from the statement _p_ itself. This leads to the truth table for the negation.

TRUTH TABLE FOR THE NEGATION   not _p_

not _p_

_p_	~_p_
T	F
F	T

EXAMPLE 3 **Finding the Truth Value of a Compound Statement**

Suppose p is false, q is true, and r is false. What is the truth value of the compound statement $\sim p \wedge (q \vee \sim r)$?

Solution

Here parentheses are used to group q and $\sim r$ together. Work first inside the parentheses. Because r is false, $\sim r$ will be true. Because $\sim r$ is true and q is true, the truth value of $q \vee \sim r$ is T, as shown in the first row of the *or* truth table.

Because p is false, $\sim p$ is true, and the truth value of $\sim p \wedge (q \vee \sim r)$ is found in the top row of the *and* truth table. The statement

$$\sim p \wedge (q \vee \sim r) \quad \text{is true.}$$

We can use a shortcut symbolic method that involves replacing the statements with their truth values, letting T represent a true statement and F represent a false statement.

$$\sim p \wedge (q \vee \sim r)$$
$$\sim F \wedge (T \vee \sim F) \quad \boxed{\text{Work within parentheses first.}}$$
$$T \wedge (T \vee T) \qquad \sim F \text{ gives T.}$$
$$T \wedge T \qquad\qquad T \vee T \text{ gives T.}$$

The compound statement is true. $\rightarrow$ **T** $T \wedge T$ gives T.   ∎

Mathematical Statements

We can use truth tables to determine the truth values of compound mathematical statements.

EXAMPLE 4 **Deciding Whether Compound Mathematical Statements Are True or False**

Let p represent the statement $3 > 2$, q represent $5 < 4$, and r represent $3 < 8$. Decide whether each statement is *true* or *false*.

(a) $\sim p \wedge \sim q$ **(b)** $\sim (p \wedge q)$ **(c)** $(\sim p \wedge r) \vee (\sim q \wedge \sim p)$

Solution

(a) Because p is true, $\sim p$ is false. By the *and* truth table, if one part of an "and" statement is false, the entire statement is false.

$$\sim p \wedge \sim q \quad \text{is false.}$$

(b) For $\sim (p \wedge q)$, first work within the parentheses. Because p is true and q is false, $p \wedge q$ is false by the *and* truth table. Next, apply the negation. The negation of a false statement is true.

$$\sim (p \wedge q) \quad \text{is true.}$$

(c) Here p is true, q is false, and r is true. This makes $\sim p$ false and $\sim q$ true. By the *and* truth table, $\sim p \wedge r$ is false, and $\sim q \wedge \sim p$ is also false. By the *or* truth table,

$$(\sim p \wedge r) \vee (\sim q \wedge \sim p) \quad \text{is false.}$$
$$\quad \downarrow \qquad\qquad \downarrow$$
$$\quad F \quad \vee \quad F$$

(Alternatively, see **Example 8(b)**.)   ∎

---

not(3>2) and not (5<4)   0

not((3>2) and (5<4))   1

Example 4(a) explains why

$$\sim (3 > 2) \wedge \sim (5 < 4)$$

is false. The calculator returns a 0. For a true statement such as

$$\sim [(3 > 2) \wedge (5 < 4)]$$

from **Example 4(b)**, it returns a 1.

When a quantifier is used with a conjunction or a disjunction, we must be careful in determining the truth value, as shown in the following example.

George Boole (1815–1864) grew up in poverty. His father, a London tradesman, gave him his first mathematics lessons and taught him to make optical instruments. Boole was largely self-educated. At 16 he worked in an elementary school and by age 20 had opened his own school. He studied mathematics in his spare time. He died of lung disease at age 49.

Boole's ideas have been used in the design of computers, telephone systems, and search engines.

EXAMPLE 5 **Deciding Whether Quantified Mathematical Statements Are True or False**

Decide whether each statement is *true* or *false*.

(a) For some real number x, $x < 5$ and $x > 2$.

(b) For every real number x, $x > 0$ or $x < 1$.

(c) For all real numbers x, $x^2 > 0$.

Solution

(a) Because "some" is an *existential* quantifier, we need only find one real number x that makes both component statements true, and $x = 3$ is such a number. The statement is true by the *and* truth table.

(b) No matter which real number might be tried as a replacement for x, at least one of the two statements

$$x > 0, \quad x < 1$$

will be true. Because an "or" statement is true if one or both component statements are true, the entire statement as given is true.

(c) Because "for all" is a *universal* quantifier, we need only find one case in which the inequality is false to make the entire statement false. Can we find a real number whose square is not positive (that is, not greater than 0)? Yes, we can— 0 is such a number. In fact 0 is the *only* real number whose square is not positive. This statement is false.   ■

FOR FURTHER THOUGHT

Whose Picture Am I Looking At?

Raymond Smullyan is one of today's foremost writers of logic puzzles. This professor of mathematics and philosophy is now retired from Indiana University and has written several books on recreational logic, including *What Is the Name of This Book?*, *The Lady or the Tiger?*, and *The Gödelian Puzzle Book.* The first of these includes the following puzzle, which has been around for many years.

For Group or Individual Investigation

A man was looking at a portrait. Someone asked him, "Whose picture are you looking at?" He replied: "Brothers and sisters, I have none, but this man's father is my father's son." ("This man's father" means, of course, the father of the man in the picture.)

Whose picture was the man looking at? (The answer is on **page 96.**)

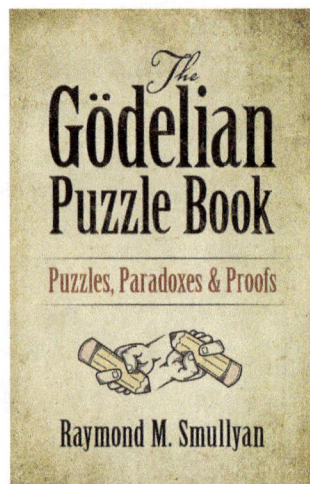

The **Gödelian** Puzzle Book

Puzzles, Paradoxes & Proofs

Raymond M. Smullyan

p	q	Compound Statement
T	T	
T	F	
F	T	
F	F	

Truth Tables

In the preceding examples, the truth value for a given statement was found by going back to the basic truth tables. It is generally easier to first create a complete truth table for the given statement itself. Then final truth values can be read directly from this table.

In this book we use the standard format shown in the margin for listing the possible truth values in compound statements involving two component statements.

EXAMPLE 6   Constructing a Truth Table

Consider the statement $(\sim p \wedge q) \vee \sim q$.

(a) Construct a truth table.

(b) Suppose both p and q are true. Find the truth value of the compound statement.

Solution

(a) As shown below, begin by listing all possible combinations of truth values for p and q. Then list the truth values of $\sim p$, which are the opposite of those of p.

Use the "$\sim p$" and "q" columns, along with the *and* truth table, to find the truth values of $\sim p \wedge q$. List them in a separate column.

Next include a column for $\sim q$.

p	q	$\sim p$	$\sim p \wedge q$	$\sim q$
T	T	F	F	F
T	F	F	F	T
F	T	T	T	F
F	F	T	F	T

Finally, make a column for the entire compound statement. To find the truth values, use *or* to combine $\sim p \wedge q$ with $\sim q$ and refer to the *or* truth table.

p	q	$\sim p$	$\sim p \wedge q$	$\sim q$	$(\sim p \wedge q) \vee \sim q$
T	T	F	F	F	F
T	F	F	F	T	T
F	T	T	T	F	T
F	F	T	F	T	T

(b) Look in the first row of the final truth table above, where both p and q have truth value T. Read across the row to find that the compound statement is false. ∎

EXAMPLE 7   Constructing a Truth Table

Construct the truth table for $p \wedge (\sim p \vee \sim q)$.

Solution

p	q	$\sim p$	$\sim q$	$\sim p \vee \sim q$	$p \wedge (\sim p \vee \sim q)$
T	T	F	F	F	F
T	F	F	T	T	T
F	T	T	F	T	F
F	F	T	T	T	F

∎

If a compound statement involves three component statements p, q, and r, we will use the following standard format in setting up the truth table.

p	q	r	Compound Statement
T	T	T	
T	T	F	
T	F	T	
T	F	F	
F	T	T	
F	T	F	
F	F	T	
F	F	F	

Emilie, Marquise du Châtelet
(1706–1749) participated in the scientific activity of the generation after Newton and Leibniz. Educated in science, music, and literature, she was studying mathematics at the time (1733) she began a long intellectual relationship with the philosopher **François Voltaire** (1694–1778). She and Voltaire competed independently in 1738 for a prize offered by the French Academy on the subject of fire. Although du Châtelet did not win, her dissertation was published by the academy in 1744.

EXAMPLE 8 **Constructing a Truth Table**

Consider the statement $(\sim p \wedge r) \vee (\sim q \wedge \sim p)$.

(a) Construct a truth table.

(b) Suppose p is true, q is false, and r is true. Find the truth value of this statement.

Solution

(a) There are three component statements: p, q, and r. The truth table thus requires eight rows to list all possible combinations of truth values of p, q, and r. The final truth table can be found in much the same way as the ones earlier.

p	q	r	$\sim p$	$\sim p \wedge r$	$\sim q$	$\sim q \wedge \sim p$	$(\sim p \wedge r) \vee (\sim q \wedge \sim p)$
T	T	T	F	F	F	F	F
T	T	F	F	F	F	F	F
T	F	T	F	F	T	F	F
T	F	F	F	F	T	F	F
F	T	T	T	T	F	F	T
F	T	F	T	F	F	F	F
F	F	T	T	T	T	T	T
F	F	F	T	F	T	T	T

(b) By the third row of the truth table in part (a), the compound statement is false. (This is an alternative method for working part (c) of **Example 4**.) ∎

Problem-Solving Strategy

One strategy for problem solving is to notice a pattern and then use inductive reasoning. This strategy is applied in the next example.

EXAMPLE 9 **Using Inductive Reasoning**

Suppose that n is a counting number, and a logical statement is composed of n component statements. How many rows will appear in the truth table for the compound statement?

Solution

We examine some of the earlier truth tables in this section. The truth table for the negation has one statement and two rows. The truth tables for the conjunction and the disjunction have two component statements, and each has four rows. The truth table in **Example 8(a)** has three component statements and eight rows.

Summarizing these in **Table 6** (seen in the margin) reveals a pattern encountered earlier. Inductive reasoning leads us to the conjecture that if a logical statement is composed of n component statements, it will have 2^n rows. This can be proved using more advanced concepts. ∎

Table 6

Number of Statements	Number of Rows
1	$2 = 2^1$
2	$4 = 2^2$
3	$8 = 2^3$

The result of **Example 9** is reminiscent of the formula for the number of subsets of a set having n elements.

NUMBER OF ROWS IN A TRUTH TABLE

A logical statement having n component statements will have 2^n rows in its truth table.

Alternative Method for Constructing Truth Tables

After making a reasonable number of truth tables, some people prefer the shortcut method shown in **Example 10,** which repeats **Examples 6 and 8.**

EXAMPLE 10 Constructing Truth Tables

Construct the truth table for each compound statement.

(a) $(\sim p \wedge q) \vee \sim q$ **(b)** $(\sim p \wedge r) \vee (\sim q \wedge \sim p)$

Solution

(a) Start by inserting truth values for $\sim p$ and for q. Then use the *and* truth table to obtain the truth values for $\sim p \wedge q$.

p	q	($\sim p$	$\wedge$	q)	$\vee$	$\sim q$
T	T	F		T		
T	F	F		F		
F	T	T		T		
F	F	T		F		

p	q	($\sim p$	$\wedge$	q)	$\vee$	$\sim q$
T	T	F	F	T		
T	F	F	F	F		
F	T	T	T	T		
F	F	T	F	F		

Now disregard the two preliminary columns of truth values for $\sim p$ and for q, and insert truth values for $\sim q$. Finally, use the *or* truth table.

p	q	($\sim p \wedge q$)	$\vee$	$\sim q$
T	T	F		F
T	F	F		T
F	T	T		F
F	F	F		T

p	q	($\sim p \wedge q$)	$\vee$	$\sim q$
T	T	F	F	F
T	F	F	T	T
F	T	T	T	F
F	F	F	T	T

These steps can be summarized as follows.

p	q	($\sim p$	$\wedge$	q)	$\vee$	$\sim q$
T	T	F	F	T	F	F
T	F	F	F	F	T	T
F	T	T	T	T	T	F
F	F	T	F	F	T	T
		①	②	①	④	③

> The circled numbers indicate the order in which the various columns of the truth table were found.

(b) Work as follows.

p	q	r	($\sim p$	$\wedge$	r)	$\vee$	($\sim q$	$\wedge$	$\sim p$)
T	T	T	F	F	T	F	F	F	F
T	T	F	F	F	F	F	F	F	F
T	F	T	F	F	T	F	T	F	F
T	F	F	F	F	F	F	T	F	F
F	T	T	T	T	T	T	F	F	T
F	T	F	T	F	F	F	F	F	T
F	F	T	T	T	T	T	T	T	T
F	F	F	T	F	F	T	T	T	T
			①	②	①	⑤	③	④	③

> The circled numbers indicate the order.

Equivalent Statements and De Morgan's Laws

Two statements are **equivalent** if they have the same truth value in *every* possible situation. The columns of the two truth tables that were the last to be completed will be the same for equivalent statements.

Ada Lovelace (1815–1852) was born Augusta Ada Byron. Her talents as a mathematician and logician led to her work with **Charles Babbage** (1791–1871) on his Analytical Engine, the first programmable mechanical computer. Lovelace's notes on this machine include what is regarded by many as the first computer program and reveal her visionary belief that computing machines would have applications beyond numerical calculations.

EXAMPLE 11 Deciding Whether Two Statements Are Equivalent

Are the following two statements equivalent?

$$\sim p \wedge \sim q \quad \text{and} \quad \sim(p \vee q)$$

Solution

Construct a truth table for each statement.

p	q	$\sim p \wedge \sim q$	p	q	$\sim(p \vee q)$
T	T	F	T	T	F
T	F	F	T	F	F
F	T	F	F	T	F
F	F	T	F	F	T

Because the truth values are the same in all cases, as shown in the columns in color, the statements $\sim p \wedge \sim q$ and $\sim(p \vee q)$ are equivalent.

$$\sim p \wedge \sim q \equiv \sim(p \vee q) \quad \text{The symbol} \equiv \text{denotes equivalence.} \qquad \blacksquare$$

In the same way, the statements $\sim p \vee \sim q$ and $\sim(p \wedge q)$ are equivalent. We call these equivalences *De Morgan's laws*.

Compare **De Morgan's Laws** for logical statements with the set theoretical version on **page 68**.

DE MORGAN'S LAWS FOR LOGICAL STATEMENTS

For any statements p and q, the following equivalences are valid.

$$\sim(p \vee q) \equiv \sim p \wedge \sim q \quad \text{and} \quad \sim(p \wedge q) \equiv \sim p \vee \sim q$$

EXAMPLE 12 Applying De Morgan's Laws

Find a negation of each statement by applying De Morgan's laws.

(a) I got an A or I got a B.   **(b)** She won't try and he will succeed.

(c) $\sim p \vee (q \wedge \sim p)$

Solution

(a) If p represents "I got an A" and q represents "I got a B," then the compound statement is symbolized $p \vee q$. The negation of $p \vee q$ is $\sim(p \vee q)$. By one of De Morgan's laws, this is equivalent to $\sim p \wedge \sim q$, or, in words,

I didn't get an A and I didn't get a B.

This negation is reasonable—the original statement says that I got either an A or a B. The negation says that I didn't get *either* grade.

(b) From De Morgan's laws, $\sim(p \wedge q) \equiv \sim p \vee \sim q$, so the negation becomes

She will try or he won't succeed.

(c) Negate both component statements and change $\vee$ to $\wedge$.

$$\sim[\sim p \vee (q \wedge \sim p)] \equiv p \wedge \sim(q \wedge \sim p)$$
$$\equiv p \wedge (\sim q \vee \sim(\sim p)) \quad \text{Apply De Morgan's law again.}$$
$$\equiv p \wedge (\sim q \vee p)$$

A truth table will show that the statements

$$\sim p \vee (q \wedge \sim p) \quad \text{and} \quad p \wedge (\sim q \vee p) \quad \text{are negations of each other.} \qquad \blacksquare$$

FOR FURTHER THOUGHT

The Logic Behind Computers

Computer designers use *logic gates* at the foundation of digital circuits. Logic gates treat the binary digits 0 and 1 as logical values, and give a single output of 1 or 0 based on the input value(s), which also have a value of 1 or 0. Since 1 and 0 are the only digits in the binary number system, that system is the perfect bridge between logic and numerical computation (as Leibniz realized).

There are three basic logic gates: the AND gate, the OR gate, and the NOT gate, whose symbols and truth tables follow.

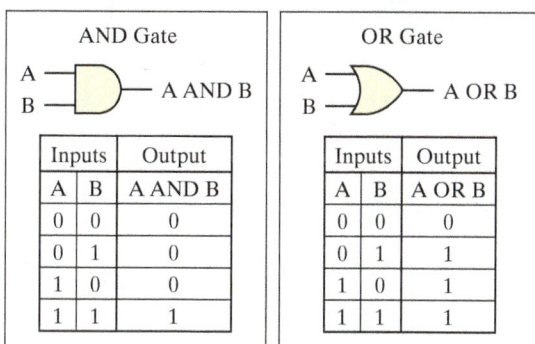

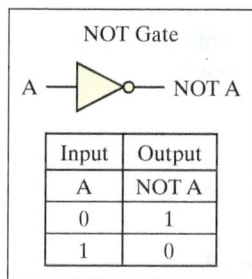

AND Gate

A ─┐
)── A AND B
B ─┘

Inputs		Output
A	B	A AND B
0	0	0
0	1	0
1	0	0
1	1	1

OR Gate

A ─┐
)── A OR B
B ─┘

Inputs		Output
A	B	A OR B
0	0	0
0	1	1
1	0	1
1	1	1

NOT Gate

A ──▷○── NOT A

Input	Output
A	NOT A
0	1
1	0

Other logic gates include NAND, NOR (the negations of AND and OR, respectively), XOR (exclusive OR—see **Exercise 77** in this section), and XNOR (the negation of XOR). Their symbols are shown below.

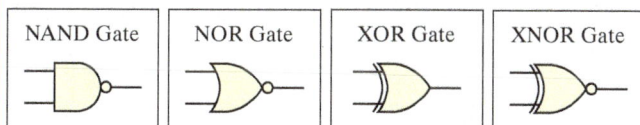

NAND Gate	NOR Gate	XOR Gate	XNOR Gate

Early logic gates were constructed with electrical relays and vacuum tubes, but the development of semiconductors and the transistor have made possible the placement of millions of logic gates in integrated circuits, microchips that perform billions of logical calculations per second. More recently, logic gates have even been fashioned from biological organisms and DNA.

There are some excellent Android Apps (such as Logic Simulator, free in the Android Play Store) for simulating logic circuits. The simulator shown allows the user to add logic gates, switches (connected to inputs), and light bulbs (connected to outputs). Red highlighting indicates where current is flowing. These four screenshots show the truth values for the XOR logic gate.

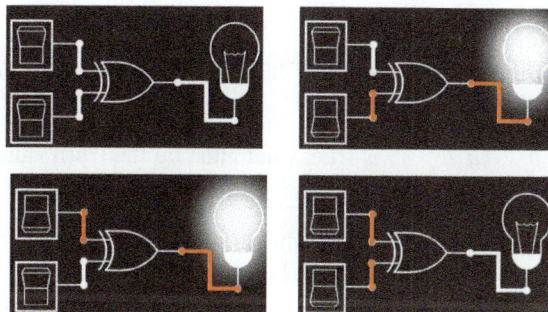

For Group or Individual Investigation

1. Create the truth table for A XOR B using the accompanying screenshots. (Remember that the red nodes are "hot," which is indicated by a 1 or a T in the truth table.)

2. Charles Peirce (1839–1914) showed that all of the logic gates we have mentioned could be constructed using only NAND gates. Construct a truth table for the logic circuit below to show that it is equivalent to the XOR gate. *Hint:* The output of the circuit shown is

   [A NAND (A NAND B)] NAND [(A NAND B) NAND B],

   which is equivalent to

   $\sim(\sim(A \wedge \sim(A \wedge B)) \wedge \sim(\sim(A \wedge B) \wedge B))$.

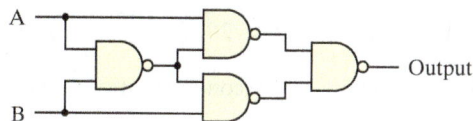

3. Download and install an app to simulate logic gates, and use it to verify your truth table from **Exercise 2.** One image of the simulated circuit is shown below.

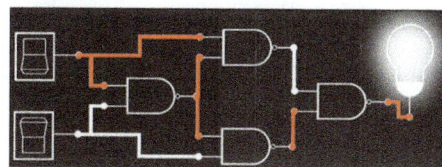

4. Use the simulator to build circuits to verify De Morgan's laws for logic.

Use the concepts introduced in this section to answer Exercises 1–8.

1. If q is false, what must be the truth value of the statement $(p \wedge \sim q) \wedge q$?

2. If q is true, what must be the truth value of the statement $q \vee (q \wedge \sim p)$?

3. If the statement $p \wedge q$ is true, and p is true, then q must be _____.

4. If the statement $p \vee q$ is false, and p is false, then q must be _____.

5. If $p \vee (q \wedge \sim q)$ is true, what must be the truth value of p?

6. If $p \wedge \sim (q \vee r)$ is true, what must be the truth value of the component statements?

7. If $\sim (p \vee q)$ is true, what must be the truth values of the component statements?

8. If $\sim (p \wedge q)$ is false, what must be the truth values of the component statements?

Let p represent a false statement and let q represent a true statement. Find the truth value of the given compound statement.

9. $\sim p$ **10.** $\sim q$

11. $p \vee q$ **12.** $p \wedge q$

13. $p \vee \sim q$ **14.** $\sim p \wedge q$

15. $\sim p \vee \sim q$ **16.** $p \wedge \sim q$

17. $\sim (p \wedge \sim q)$ **18.** $\sim (\sim p \vee \sim q)$

19. $\sim [\sim p \wedge (\sim q \vee p)]$ **20.** $\sim [(\sim p \wedge \sim q) \vee \sim q]$

21. Is the statement $6 \geq 2$ a conjunction or a disjunction? Why?

22. Why is the statement $8 \geq 3$ true? Why is $5 \geq 5$ true?

Let p represent a true statement, and let q and r represent false statements. Find the truth value of the given compound statement.

23. $(p \wedge r) \vee \sim q$ **24.** $(q \vee \sim r) \wedge p$

25. $p \wedge (q \vee r)$ **26.** $(\sim p \wedge q) \vee \sim r$

27. $\sim (p \wedge q) \wedge (r \vee \sim q)$ **28.** $(\sim r \wedge \sim q) \vee (\sim r \wedge q)$

29. $\sim [(\sim p \wedge q) \vee r]$ **30.** $\sim [r \vee (\sim q \wedge \sim p)]$

31. $\sim [\sim q \vee (r \wedge \sim p)]$ **32.** $\sim (p \vee q) \wedge \sim (p \wedge q)$

Let p represent the statement $16 < 8$, let q represent the statement $5 \not> 4$, and let r represent the statement $17 \leq 17$. Find the truth value of the given compound statement.

33. $p \wedge r$ **34.** $p \vee \sim q$

35. $\sim q \vee \sim r$ **36.** $\sim p \wedge \sim r$

37. $(p \wedge q) \vee r$ **38.** $\sim p \vee (\sim r \vee \sim q)$

39. $(\sim r \wedge q) \vee \sim p$ **40.** $\sim (p \vee \sim q) \vee \sim r$

Give the number of rows in the truth table for each compound statement.

41. $p \vee \sim r$

42. $p \wedge (r \wedge \sim s)$

43. $(\sim p \wedge q) \vee (\sim r \vee \sim s) \wedge r$

44. $[(p \vee q) \wedge (r \wedge s)] \wedge (t \vee \sim p)$

45. $[(\sim p \wedge \sim q) \wedge (\sim r \wedge s \wedge \sim t)] \wedge (\sim u \vee \sim v)$

46. $[(\sim p \wedge \sim q) \vee (\sim r \vee \sim s)]$
$\qquad \vee [(\sim m \wedge \sim n) \wedge (u \wedge \sim v)]$

47. If the truth table for a certain compound statement has 64 rows, how many distinct component statements does it have?

48. Is it possible for the truth table of a compound statement to have exactly 54 rows? Why or why not?

Construct a truth table for each compound statement.

49. $\sim p \wedge q$ **50.** $\sim p \vee \sim q$

51. $\sim (p \wedge q)$ **52.** $p \vee \sim q$

53. $(q \vee \sim p) \vee \sim q$

54. $(p \wedge \sim q) \wedge p$

55. $(p \vee \sim q) \wedge (p \wedge q)$

56. $(\sim p \wedge \sim q) \vee (\sim p \vee q)$

57. $(\sim p \wedge q) \wedge r$

58. $r \vee (p \wedge \sim q)$

59. $(\sim p \wedge \sim q) \vee (\sim r \vee \sim p)$

60. $(\sim r \vee \sim p) \wedge (\sim p \vee \sim q)$

61. $\sim (\sim p \wedge \sim q) \vee (\sim r \vee \sim s)$

62. $(\sim r \vee s) \wedge (\sim p \wedge q)$

Use one of De Morgan's laws to write the negation of each statement.

63. You can pay me now or you can pay me later.

64. I am not going or she is going.

65. It is summer and there is no snow.

66. $\frac{1}{2}$ is a positive number and -9 is less than zero.

67. I said yes but she said no.

68. Dan tried to sell the software, but he was unable to do so.

69. $6 - 1 = 5$ and $9 + 13 \neq 7$

70. $8 < 10$ or $5 \neq 2$

71. Prancer or Vixen will lead Santa's reindeer sleigh next Christmas.

72. The lawyer and the client appeared in court.

Identify each statement as true *or* false.

73. For every real number x, $x < 14$ or $x > 6$.

74. For every real number x, $x > 9$ or $x < 9$.

75. There exists an integer n such that $n > 0$ and $n < 0$.

76. For some integer n, $n \geq 3$ and $n \leq 3$.

77. Complete the truth table for *exclusive disjunction*. The symbol $\underline{\vee}$ represents "one or the other is true, but not both."

p	q	$p \underline{\vee} q$
T	T	
T	F	
F	T	
F	F	

Exclusive disjunction

78. Attorneys sometimes use the phrase "and/or." This phrase corresponds to which usage of the word *or:* inclusive or exclusive disjunction?

Decide whether each compound statement is true *or* false. *Remember that* $\underline{\vee}$ *is the exclusive disjunction of* **Exercise 77.**

79. $3 + 1 = 4 \underline{\vee} 2 + 5 = 7$

80. $3 + 1 = 4 \underline{\vee} 2 + 5 = 10$

81. $3 + 1 = 6 \underline{\vee} 2 + 5 = 7$

82. $3 + 1 = 12 \underline{\vee} 2 + 5 = 10$

83. In his book *The Lady or the Tiger and Other Logic Puzzles*, Raymond Smullyan proposes the following problem. It is taken from the classic Frank Stockton short story, in which a prisoner must make a choice between two doors: behind one is a beautiful lady, and behind the other is a hungry tiger.

What if each door has a sign, and the man knows that only one sign is true?

The sign on Door 1 reads:

IN THIS ROOM THERE IS A LADY AND IN THE OTHER ROOM THERE IS A TIGER.

The sign on Door 2 reads:

IN ONE OF THESE ROOMS THERE IS A LADY AND IN ONE OF THESE ROOMS THERE IS A TIGER.

With this information, the man is able to choose the correct door. Can you?

84. (a) Build truth tables for

$$p \vee (q \wedge r) \quad \text{and} \quad (p \vee q) \wedge (p \vee r).$$

Decide whether it can be said that "OR distributes over AND." Explain.

(b) Build truth tables for

$$p \wedge (q \vee r) \quad \text{and} \quad (p \wedge q) \vee (p \wedge r).$$

Decide whether it can be said that "AND distributes over OR." Explain.

(c) Describe how the logical equivalences developed in parts (a) and (b) are related to the set-theoretical equations

$$X \cup (Y \cap Z) = (X \cup Y) \cap (X \cup Z)$$

and

$$X \cap (Y \cup Z) = (X \cap Y) \cup (X \cap Z).$$

85. De Morgan's law

$$\sim(p \vee q) \equiv \sim p \wedge \sim q$$

can be stated verbally, "The negation of a disjunction is equivalent to the conjunction of the negations." Give a similar verbal statement of

$$\sim(p \wedge q) \equiv \sim p \vee \sim q.$$

3.3 THE CONDITIONAL AND CIRCUITS

Conditionals

"If truly Truthful Troll I be,
then go thou east, and be thou free."

This of course is the statement uttered by the troll on the second page of this chapter. A more modern paraphrase would be, "*If* I am the troll who always tells the truth, *then* the door to the east is the one that leads to freedom."

The troll's utterance is an example of a conditional statement. A **conditional** statement is a compound statement that uses the connective *if . . . then.*

> *If* I read for too long, *then* I get tired.
>
> *If* looks could kill, *then* I would be dead.
>
> *If* he doesn't get back soon, *then* you should go look for him.

Conditional statements

In each of these conditional statements, the component coming after the word *if* gives a condition (but not necessarily the only condition) under which the statement coming after *then* will be true. For example, "If it is over 90°, then I'll go to the mountains" tells one possible condition under which I will go to the mountains—if the temperature is over 90°.

The conditional is written with an arrow and symbolized as follows.

$$p \rightarrow q \qquad \text{If } p, \text{ then } q.$$

We read $p \rightarrow q$ as "*p* **implies** *q*" or "**If** *p*, **then** *q*." In the conditional $p \rightarrow q$, the statement *p* is the **antecedent,** while *q* is the **consequent.**

The conditional connective may not always be explicitly stated. That is, it may be "hidden" in an everyday expression. For example, consider the following statement.

Quitters never win.

It can be written in *if . . . then* form as

If you're a quitter, *then* you will never win.

As another example, consider this statement.

It is difficult to study when you are distracted.

It can be written

If you are distracted, *then* it is difficult to study.

In the quotation "If you aim at nothing, you will hit it every time," the word "then" is not stated but understood from the context of the statement. "You aim at nothing" is the antecedent, and "you will hit it every time" is the consequent.

The conditional truth table is more difficult to define than the tables in the previous section. To see how to define the conditional truth table, imagine you have bought a used car (with financing from the car dealer), and the used-car salesman says,

If you fail to make your payment on time, *then* your car will be taken.

Let *p* represent "You fail to make your payment on time," and let *q* represent "Your car will be taken." There are four combinations of truth values for the two component statements.

In his April 21, 1989, five-star review of *Field of Dreams,* the *Chicago Sun-Times* movie critic Roger Ebert gave an explanation of why the movie has become an American classic.

There is a speech in this movie about baseball that is so simple and true that it is heartbreaking. And the whole attitude toward the players reflects that attitude. Why do they come back from the great beyond and play in this cornfield? Not to make any kind of vast, earthshattering statement, but simply to hit a few and field a few, and remind us of a good and innocent time.

The photo above was taken in 2007 in Dyersville, Iowa, at the actual scene of the filming. The carving "Ray Loves Annie" in the bleacher seats can be seen in a quick shot during the movie. It has weathered over time.

What famous **conditional statement** inspired Ray to build a baseball field in his cornfield?

As we consider these four possibilities, it is helpful to ask,

"Did the salesman lie?"

If so, then the conditional statement is considered false. Otherwise, the conditional statement is considered true.

Possibility	Failed to Pay on Time?	Car Taken?	
1	Yes	Yes	p is T, q is T.
2	Yes	No	p is T, q is F.
3	No	Yes	p is F, q is T.
4	No	No	p is F, q is F.

The four possibilities are as follows.

1. In the first case, assume you failed to make your payment on time, and your car *was* taken (p is T, q is T). The salesman told the truth, so place T in the first row of the truth table. (We do not claim that your car was taken *because* you failed to pay on time. It may be gone for a completely different reason.)

2. In the second case, assume that you failed to make your payment on time, and your car was *not* taken (p is T, q is F). The salesman lied (gave a false statement), so place an F in the second row of the truth table.

3. In the third case, assume that you paid on time, but your car was taken anyway (p is F, q is T). The salesman did *not* lie. He only said what would happen if you were late on payments, not what would happen if you paid on time. (The crime here is stealing, not lying.) Since we cannot say that the salesman lied, place a T in the third row of the truth table.

4. Finally, assume that you made timely payment, and your car was not taken (p is F, q is F). This certainly does not contradict the salesman's statement, so place a T in the last row of the truth table.

The completed truth table for the conditional is defined as follows.

TRUTH TABLE FOR THE CONDITIONAL   If p, then q

If p, then q

p	q	$p \rightarrow q$
T	T	T
T	F	F
F	T	T
F	F	T

The use of the conditional connective in no way implies a cause-and-effect relationship. Any two statements may have an arrow placed between them to create a compound statement. Consider this example.

If I pass mathematics, then the sun will rise the next day.

It is true, because the consequent is true. (See the special characteristics following **Example 1** on the next page.) There is, however, no cause-and-effect connection between my passing mathematics and the rising of the sun. The sun will rise no matter what grade I get.

```
PROGRAM:PARITY
: Prompt N
: If N/2 = int (N/2)
:Then
: Disp "EVEN"
:
:
```

```
PROGRAM:PARITY
:
: Else
: Disp "ODD"
: End
:
```

```
prgmPARITY
N = ?7
ODD
                   Done
```

```
prgmPARITY
N = ?6
EVEN
                   Done
```

Conditional statements are useful in writing programs. The short program in the first two screens determines whether an integer is even. Notice the lines that begin with *If* and *Then*.

EXAMPLE 1 Finding the Truth Value of a Conditional

Given that p, q, and r are all false, find the truth value of the following statement.

$$(p \rightarrow \sim q) \rightarrow (\sim r \rightarrow q)$$

Solution

Using the shortcut method explained in **Example 3** of the previous section, we can replace p, q, and r with F (since each is false) and proceed as before, using the negation and conditional truth tables as necessary.

$$
\begin{array}{rcll}
(p \rightarrow \sim q) & \rightarrow & (\sim r \rightarrow q) & \\
(F \rightarrow \sim F) & \rightarrow & (\sim F \rightarrow F) & \\
(F \rightarrow T) & \rightarrow & (T \rightarrow F) & \text{Use the negation truth table.} \\
T & \rightarrow & F & \text{Use the conditional truth table.} \\
& \mathbf{F} & &
\end{array}
$$

The statement $(p \rightarrow \sim q) \rightarrow (\sim r \rightarrow q)$ is false when p, q, and r are all false. ∎

SPECIAL CHARACTERISTICS OF CONDITIONAL STATEMENTS

1. $p \rightarrow q$ is false only when the antecedent is *true* and the consequent is *false*.

2. If the antecedent is *false*, then $p \rightarrow q$ is automatically *true*.

3. If the consequent is *true*, then $p \rightarrow q$ is automatically *true*.

EXAMPLE 2 Determining Whether Conditionals Are True or False

Write *true* or *false* for each statement. Here T represents a true statement, and F represents a false statement.

(a) $T \rightarrow (7 = 3)$ **(b)** $(8 < 2) \rightarrow F$ **(c)** $(4 \neq 3 + 1) \rightarrow T$

Solution

(a) Because the antecedent is true, while the consequent, $7 = 3$, is false, the given statement is false by the first point mentioned above.

(b) The antecedent is false, so the given statement is true by the second observation.

(c) The consequent is true, making the statement true by the third characteristic of conditional statements. ∎

EXAMPLE 3 Constructing Truth Tables

Construct a truth table for each statement.

(a) $(\sim p \rightarrow \sim q) \rightarrow (\sim p \wedge q)$ **(b)** $(p \rightarrow q) \rightarrow (\sim p \vee q)$

Solution

(a) Insert the truth values of $\sim p$ and $\sim q$. Find the truth values of $\sim p \rightarrow \sim q$.

p	q	$\sim p$	$\sim q$	$\sim p \rightarrow \sim q$
T	T	F	F	T
T	F	F	T	T
F	T	T	F	F
F	F	T	T	T

Next use $\sim p$ and q to find the truth values of $\sim p \wedge q$.

p	q	$\sim p$	$\sim q$	$\sim p \rightarrow \sim q$	$\sim p \wedge q$
T	T	F	F	T	F
T	F	F	T	T	F
F	T	T	F	F	T
F	F	T	T	T	F

Now find the truth values of $(\sim p \rightarrow \sim q) \rightarrow (\sim p \wedge q)$.

p	q	$\sim p$	$\sim q$	$\sim p \rightarrow \sim q$	$\sim p \wedge q$	$(\sim p \rightarrow \sim q) \rightarrow (\sim p \wedge q)$
T	T	F	F	T	F	F
T	F	F	T	T	F	F
F	T	T	F	F	T	T
F	F	T	T	T	F	F

(b) For $(p \rightarrow q) \rightarrow (\sim p \vee q)$, go through steps similar to the ones above.

p	q	$p \rightarrow q$	$\sim p$	$\sim p \vee q$	$(p \rightarrow q) \rightarrow (\sim p \vee q)$
T	T	T	F	T	T
T	F	F	F	F	T
F	T	T	T	T	T
F	F	T	T	T	T

As the truth table in **Example 3(b)** shows, the statement

$$(p \rightarrow q) \rightarrow (\sim p \vee q)$$

is always true, no matter what the truth values of the components. Such a statement is called a **tautology.** Several other examples of tautologies (as can be checked by forming truth tables) are

$$p \vee \sim p, \quad p \rightarrow p, \quad \text{and} \quad (\sim p \vee \sim q) \rightarrow \sim (p \wedge q). \quad \text{Tautologies}$$

The truth tables in **Example 3** also could have been found by the alternative method shown in **Section 3.2.**

Writing a Conditional as a Disjunction

p	q	$p \rightarrow q$
T	T	T
T	F	F
F	T	T
F	F	T

Recall that the truth table for the conditional (repeated in the margin) shows that $p \rightarrow q$ is false only when p is true and q is false. But we also know that the disjunction is false for only one combination of component truth values, and it is easy to see that $\sim p \vee q$ will be false only when p is true and q is false, as the following truth table indicates.

p	q	$\sim p$	$\vee$	q
T	T	F	T	T
T	F	F	F	F
F	T	T	T	T
F	F	T	T	F

Thus we see that the disjunction $\sim p \vee q$ is equivalent to the conditional $p \rightarrow q$.

WRITING A CONDITIONAL AS A DISJUNCTION

$p \rightarrow q$ is equivalent to $\sim p \vee q.$

We now know that

$$p \rightarrow q \equiv \,\sim p \vee q,$$

so the negation of the conditional is

$$\sim(p \rightarrow q) \equiv \,\sim(\sim p \vee q).$$

Applying De Morgan's law to the right side of the above equivalence gives the negation as a conjunction.

> **NEGATION OF $p \rightarrow q$**
>
> The negation of $p \rightarrow q$ is $p \wedge \sim q.$

EXAMPLE 4 Determining Negations

Determine the negation of each statement.

(a) If I'm hungry, I will eat.   **(b)** All dogs have fleas.

Solution

(a) If p represents "I'm hungry" and q represents "I will eat," then the given statement can be symbolized by $p \rightarrow q$. The negation of $p \rightarrow q$, as shown earlier, is $p \wedge \sim q$, so the negation of the statement is

> I'm hungry and I will not eat.

(b) First, we must restate the given statement in *if . . . then* form.

> If it is a dog, then it has fleas.

Based on our earlier discussion, the negation is

> It is a dog and it does not have fleas. ■

> Do not try to negate a conditional with another conditional.

As seen in **Example 4,** the negation of a conditional statement is written as a conjunction.

EXAMPLE 5 Determining Statements Equivalent to Conditionals

Write each conditional as an equivalent statement without using *if . . . then.*

(a) If the Indians win the pennant, then Johnny will go to the World Series.

(b) If it's Borden's, it's got to be good.

Solution

(a) Because the conditional $p \rightarrow q$ is equivalent to $\sim p \vee q$, let p represent "The Indians win the pennant" and q represent "Johnny will go to the World Series." Restate the conditional as

> The Indians do not win the pennant or Johnny will go to the World Series.

(b) If p represents "it's Borden's" and if q represents "it's got to be good," the conditional may be restated as

> It's not Borden's or it's got to be good. ■

Figure 1

Figure 2

Series circuit

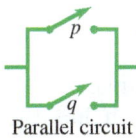
Parallel circuit

Figure 3

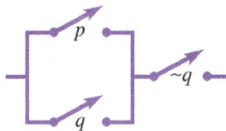

Figure 4

Switch Position	Current	Truth Value
Closed	flows	T
Open	stops	F

Circuits

One of the first nonmathematical applications of symbolic logic was seen in the master's thesis of Claude Shannon in 1937. Shannon showed how logic could be used to design electrical circuits. His work was immediately used by computer designers. Then in the developmental stage, computers could be simplified and built for less money using the ideas of Shannon.

To see how Shannon's ideas work, look at the electrical switch shown in **Figure 1.** We assume that current will flow through this switch when it is closed and not when it is open.

Figure 2 shows two switches connected in *series.* In such a circuit, current will flow only when both switches are closed. Note how closely a series circuit corresponds to the conjunction $p \wedge q$. We know that $p \wedge q$ is true only when both p and q are true.

A circuit corresponding to the disjunction $p \vee q$ can be found by drawing a *parallel* circuit, as in **Figure 3.** Here, current flows if either p *or* q is closed or if both p *and* q are closed.

The circuit in **Figure 4** corresponds to the statement $(p \vee q) \wedge {\sim}q$, which is a compound statement involving both a conjunction and a disjunction.

Simplifying an electrical circuit depends on the idea of equivalent statements from **Section 3.2.** Recall that two statements are equivalent if they have the same truth table final column. The symbol $\equiv$ is used to indicate that the two statements are equivalent. Some equivalent statements are shown in the following box.

EQUIVALENT STATEMENTS USED TO SIMPLIFY CIRCUITS

$$p \vee (q \wedge r) \equiv (p \vee q) \wedge (p \vee r) \qquad p \vee p \equiv p$$
$$p \wedge (q \vee r) \equiv (p \wedge q) \vee (p \wedge r) \qquad p \wedge p \equiv p$$
$$p \rightarrow q \equiv {\sim}q \rightarrow {\sim}p \qquad {\sim}(p \wedge q) \equiv {\sim}p \vee {\sim}q$$
$$p \rightarrow q \equiv {\sim}p \vee q \qquad {\sim}(p \vee q) \equiv {\sim}p \wedge {\sim}q$$

If T represents any true statement and F represents any false statement, then

$$p \vee T \equiv T \qquad p \vee {\sim}p \equiv T$$
$$p \wedge F \equiv F \qquad p \wedge {\sim}p \equiv F.$$

Circuits can be used as models of compound statements, with a closed switch corresponding to T (current flowing) and an open switch corresponding to F (current not flowing).

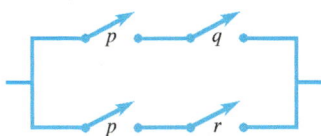

Figure 5

EXAMPLE 6 Simplifying a Circuit

Simplify the circuit of **Figure 5.**

Solution

At the top of **Figure 5,** p and q are connected in series, and at the bottom, p and r are connected in series. These are interpreted as the compound statements $p \wedge q$ and $p \wedge r$, respectively. These two conjunctions are connected in parallel, as indicated by the figure treated as a whole.

Write the disjunction of the two conjunctions.

$$(p \wedge q) \vee (p \wedge r)$$

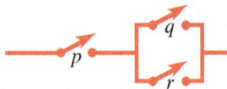

Figure 6

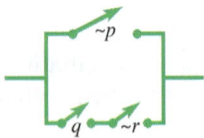

Figure 7

(Think of the two switches labeled "*p*" as being controlled by the same lever.) By one of the pairs of equivalent statements in the preceding box,

$$(p \wedge q) \vee (p \wedge r) \equiv p \wedge (q \vee r),$$

which has the circuit of **Figure 6.** This circuit is logically equivalent to the one in **Figure 5,** and yet it contains only three switches instead of four—which might well lead to a large savings in manufacturing costs. ∎

EXAMPLE 7 **Drawing a Circuit for a Conditional Statement**

Draw a circuit for $p \rightarrow (q \wedge \sim r)$.

Solution

From the list of equivalent statements in the box, $p \rightarrow q$ is equivalent to $\sim p \vee q$. This equivalence gives $p \rightarrow (q \wedge \sim r) \equiv \sim p \vee (q \wedge \sim r)$, which has the circuit diagram in **Figure 7.** ∎

WHEN **Will I** **Ever USE** **This** **?**

Suppose you are a home-monitoring and control system designer. A homeowner wants the capability of turning his air-conditioning system on or off at home via his smartphone (in case he forgets before leaving for vacation). You will need to install a control module for the AC unit that receives signals from your customer's smartphone.

Draw a circuit that will allow the AC unit to turn on when both a master switch and the control module are activated, and when either (or both) of two thermostats is (or are) triggered. Then write a logical statement for the circuit.

Since the master switch (*m*) and the control module (*c*) both need to be activated, they must be connected in series, followed by two thermostats T_1 and T_2 in parallel (because only one needs to be triggered). The circuit is shown below, and the corresponding logical statement is $m \wedge c \wedge (T_1 \vee T_2)$.

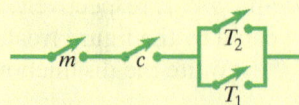

3.3 **EXERCISES**

Rewrite each statement using the if . . . then *connective. Rearrange the wording or add words as necessary.*

1. You can do it if you just believe.

2. It must be bad for you if it's sweet.

3. Every even integer divisible by 5 is divisible by 10.

4. No perfect square integers have units digit 2, 3, 7, or 8.

5. No grizzly bears live in California.

6. No guinea pigs get lonely.

7. Surfers can't stay away from the beach.

8. Running Bear loves Little White Dove.

Decide whether each statement is true *or* false.

9. If the antecedent of a conditional statement is false, the conditional statement is true.

10. If the consequent of a conditional statement is true, the conditional statement is true.

11. If q is true, then $(p \wedge (q \rightarrow r)) \rightarrow q$ is true.

12. If p is true, then $\sim p \rightarrow (q \vee r)$ is true.

13. The negation of "If pigs fly, I'll believe it" is "If pigs don't fly, I won't believe it."

14. The statements "If it flies, then it's a bird" and "It does not fly or it's a bird" are logically equivalent.

15. Given that $\sim p$ is true and q is false, the conditional $p \rightarrow q$ is true.

16. Given that $\sim p$ is false and q is false, the conditional $p \rightarrow q$ is true.

17. Explain why the statement "If $3 > 5$, then $4 < 6$" is true.

18. In a few sentences, explain how to determine the truth value of a conditional statement.

Tell whether each conditional is true (T) *or* false (F).

19. $T \rightarrow (7 < 3)$

20. $F \rightarrow (4 \neq 8)$

21. $F \rightarrow (5 \neq 5)$

22. $(8 \geq 8) \rightarrow F$

23. $(5^2 \neq 25) \rightarrow (8 - 8 = 16)$

24. $(5 = 12 - 7) \rightarrow (9 > 0)$

Let s represent "She sings for a living," *let p represent* "he fixes cars," *and let m represent* "they collect classics." *Express each compound statement in words.*

25. $\sim m \rightarrow p$

26. $p \rightarrow \sim m$

27. $s \rightarrow (m \wedge p)$

28. $(s \wedge p) \rightarrow m$

29. $\sim p \rightarrow (\sim m \vee s)$

30. $(\sim s \vee \sim m) \rightarrow \sim p$

Let b represent "I take my ball," *let s represent* "it is sunny," *and let p represent* "the park is open." *Write each compound statement in symbols.*

31. If I take my ball, then the park is open.

32. If I do not take my ball, then it is not sunny.

33. The park is open, and if it is sunny then I do not take my ball.

34. I take my ball, or if the park is open then it is sunny.

35. It is sunny if the park is open.

36. I'll take my ball if it is not sunny.

Find the truth value of each statement. Assume that p and r are false, and q is true.

37. $\sim r \to q$ **38.** $q \to p$

39. $p \to q$ **40.** $\sim r \to p$

41. $\sim p \to (q \wedge r)$ **42.** $(\sim r \vee p) \to p$

43. $\sim q \to (p \wedge r)$

44. $(\sim p \wedge \sim q) \to (p \wedge \sim r)$

45. $(p \to \sim q) \to (\sim p \wedge \sim r)$

46. $[(p \to \sim q) \wedge (p \to r)] \to r$

47. Explain why we know that

$$[r \vee (p \vee s)] \to [(p \vee q) \vee \sim p]$$

is true, even if we are not given the truth values of p, q, r, and s.

48. Construct a true statement involving a conditional, a conjunction, a disjunction, and a negation (not necessarily in that order) that consists of component statements p, q, and r, with all of these component statements false.

Construct a truth table for each statement. Identify any tautologies.

49. $\sim q \to p$

50. $(\sim q \to \sim p) \to \sim q$

51. $(\sim p \to q) \to p$

52. $(p \wedge q) \to (p \vee q)$

53. $(p \vee q) \to (q \vee p)$

54. $(\sim p \to \sim q) \to (p \wedge q)$

55. $[(r \vee p) \wedge \sim q] \to p$

56. $[(r \wedge p) \wedge (p \wedge q)] \to p$

57. $(\sim r \to s) \vee (p \to \sim q)$

58. $(\sim p \wedge \sim q) \to (s \to r)$

59. What is the minimum number of Fs that must appear in the final column of a truth table for us to be assured that the statement is not a tautology?

60. If all truth values in the final column of a truth table are F, how can we easily transform the statement into a tautology?

Write the negation of each statement. Remember that the negation of $p \to q$ is $p \wedge \sim q$.

61. If that is an authentic Coach bag, I'll be surprised.

62. If Muley Jones hits that note, he will shatter glass.

63. If the bullfighter doesn't get going, he's going to get gored.

64. If you don't say "I do," then you'll regret it for the rest of your life.

65. "If you want to be happy for the rest of your life, never make a pretty woman your wife." *Jimmy Soul*

66. "If I had a hammer, I'd hammer in the morning." *Lee Hayes and Pete Seeger*

Write each statement as an equivalent statement that does not use the if ... then *connective. Remember that*

$$p \to q \quad \text{is equivalent to} \quad \sim p \vee q.$$

67. If you give your plants tender, loving care, they flourish.

68. If you scratch my back, I'll scratch yours.

69. If she doesn't, he will.

70. If I say "black," she says "white."

71. All residents of Pensacola are residents of Florida.

72. All women were once girls.

Use truth tables to decide which of the pairs of statements are equivalent.

73. $p \to q$; $\sim p \vee q$ **74.** $\sim (p \to q)$; $p \wedge \sim q$

75. $p \to q$; $\sim q \to \sim p$ **76.** $p \to q$; $q \to p$

77. $p \to \sim q$; $\sim p \vee \sim q$ **78.** $\sim p \wedge q$; $\sim p \to q$

79. $q \to \sim p$; $p \to \sim q$

80. $\sim (p \lor q) \to r$; $(p \lor q) \lor r$

Write a logical statement representing each of the following circuits. Simplify each circuit when possible.

81.

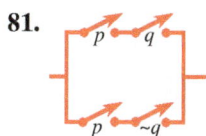

82.

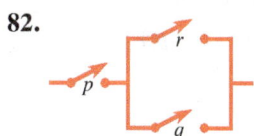

83.

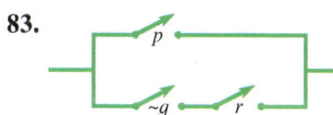

84.

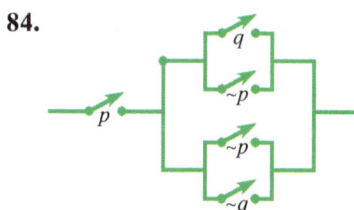

85.

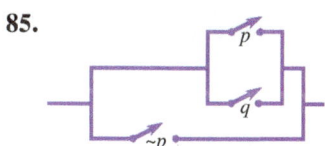

86.

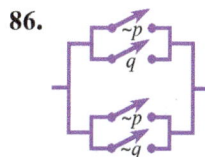

Draw circuits representing the following statements as they are given. Simplify if possible.

87. $p \land (q \lor \sim p)$ **88.** $(\sim p \land \sim q) \land \sim r$

89. $(p \lor q) \land (\sim p \land \sim q)$

90. $(\sim q \land \sim p) \lor (\sim p \lor q)$

91. $[(p \lor q) \land r] \land \sim p$

92. $[(\sim p \land \sim r) \lor \sim q] \land (\sim p \land r)$

93. $\sim q \to (\sim p \to q)$ **94.** $\sim p \to (\sim p \lor \sim q)$

95. Refer to **Figures 5 and 6** in **Example 6.** Suppose the cost of the use of one switch for an hour is $0.06. By using the circuit in **Figure 6** rather than the circuit in **Figure 5,** what is the savings for a year of 365 days, assuming that the circuit is in continuous use?

96. Explain why the circuit shown will always have exactly one open switch. What does this circuit simplify to?

97. Refer to the "For Further Thought" at the end of **Section 3.2.** Verify that the logic circuit shown below is equivalent to the conditional statement A → B.

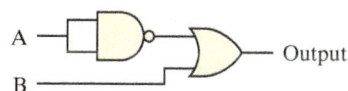

| 3.4 | THE CONDITIONAL AND RELATED STATEMENTS |

OBJECTIVES

1 Determine the converse, inverse, and contrapositive of a conditional statement.

2 Translate conditional statements into alternative forms.

3 Understand the structure of the biconditional.

4 Summarize the truth tables of compound statements.

Converse, Inverse, and Contrapositive

Many mathematical properties and theorems are stated in *if . . . then* form. Any conditional statement $p \to q$ is made up of an antecedent p and a consequent q. If they are interchanged, negated, or both, a new conditional statement is formed. Suppose that we begin with a conditional statement.

If you stay, then I go.   Conditional statement

By interchanging the antecedent ("you stay") and the consequent ("I go"), we obtain a new conditional statement.

If I go, then you stay.   Converse

This new conditional is called the **converse** of the given conditional statement.

By negating both the antecedent and the consequent, we obtain the **inverse** of the given conditional statement.

If you do not stay, then I do not go.   Inverse

Alfred North Whitehead (1861–1947) and Bertrand Russell worked together on *Principia Mathematica*. During that time, Whitehead was teaching mathematics at Cambridge University and had written *Universal Algebra*. In 1910 he went to the University of London, exploring not only the philosophical basis of science but also the "aims of education" (as he called one of his books). It was as a philosopher that he was invited to Harvard University in 1924. Whitehead died at the age of 86 in Cambridge, Massachusetts.

If the antecedent and the consequent are both interchanged *and* negated, the **contrapositive** of the given conditional statement is formed.

> If I do not go, then you do not stay.   Contrapositive

These three related statements for the conditional $p \rightarrow q$ are summarized below.

RELATED CONDITIONAL STATEMENTS		
Conditional Statement	$p \rightarrow q$	(If p, then q.)
Converse	$q \rightarrow p$	(If q, then p.)
Inverse	$\sim p \rightarrow \sim q$	(If not p, then not q.)
Contrapositive	$\sim q \rightarrow \sim p$	(If not q, then not p.)

Notice that the inverse is the contrapositive of the converse.

EXAMPLE 1 **Determining Related Conditional Statements**

Determine each of the following, given the conditional statement

> If I am running, then I am moving.

(a) the converse   **(b)** the inverse   **(c)** the contrapositive

Solution

(a) Let p represent "I am running" and q represent "I am moving." Then the given statement may be written $p \rightarrow q$. The converse, $q \rightarrow p$, is

> If I am moving, then I am running.

The converse is not necessarily true, even though the given statement is true.

(b) The inverse of $p \rightarrow q$ is $\sim p \rightarrow \sim q$. Thus the inverse is

> If I am not running, then I am not moving.

Again, this is not necessarily true.

(c) The contrapositive, $\sim q \rightarrow \sim p$, is

> If I am not moving, then I am not running.

The contrapositive, like the given conditional statement, is true. ∎

Example 1 shows that the converse and inverse of a true statement need not be true. They *can* be true, but they need not be. The relationships between the related conditionals are shown in the truth table that follows.

		Conditional	Converse	Inverse	Contrapositive
p	q	$p \rightarrow q$	$q \rightarrow p$	$\sim p \rightarrow \sim q$	$\sim q \rightarrow \sim p$
T	T	T	T	T	T
T	F	F	T	T	F
F	T	T	F	F	T
F	F	T	T	T	T

Equivalent (Conditional ↔ Contrapositive); Equivalent (Converse ↔ Inverse)

As this truth table shows,

1. *A conditional statement and its contrapositive always have the same truth value,* making it possible to replace any statement with its contrapositive without affecting the logical meaning.

2. *The converse and inverse always have the same truth value.*

Bertrand Russell (1872–1970) was a student of Whitehead's before they wrote the *Principia*. Like his teacher, Russell turned toward philosophy. His works include a critique of Leibniz, analyses of mind and of matter, and a history of Western thought.

Russell became a public figure because of his involvement in social issues. Deeply aware of human loneliness, he was "passionately desirous of finding ways of diminishing this tragic isolation." During World War I he was an antiwar crusader, and he was imprisoned briefly. Again in the 1960s he championed peace. He wrote many books on social issues, winning the Nobel Prize for Literature in 1950.

> **EQUIVALENCES**
>
> A conditional statement and its contrapositive are equivalent. Also, the converse and the inverse are equivalent.

EXAMPLE 2 **Determining Related Conditional Statements**

For the conditional statement $\sim p \to q$, write each of the following.

(a) the converse   **(b)** the inverse   **(c)** the contrapositive

Solution

(a) The converse of $\sim p \to q$ is $q \to \sim p$.

(b) The inverse is $\sim (\sim p) \to \sim q$, which simplifies to $p \to \sim q$.

(c) The contrapositive is $\sim q \to \sim (\sim p)$, which simplifies to $\sim q \to p$. ∎

Alternative Forms of "If *p*, then *q*"

The conditional statement "If *p*, then *q*" can be stated in several other ways in English. Consider this statement.

> If you take Tylenol, then you will find relief from your symptoms.

It can also be written as follows.

> Taking Tylenol is *sufficient* for relieving your symptoms.

According to this statement, taking Tylenol is enough to relieve your symptoms. Taking other medications or using other treatment techniques *might* also result in symptom relief, but at least we *know* that taking Tylenol will. Thus $p \to q$ can be written "*p* is sufficient for *q*." Knowing that *p* has occurred is sufficient to guarantee that *q* will also occur.

On the other hand, consider this statement, which has a different structure.

> Fresh ingredients are necessary for making a good pizza.   (*)

This statement claims that fresh ingredients are one condition for making a good pizza. But there may be other conditions (such as a working oven, for example). The statement labeled (*) could be written as

> If you want good pizza, then you need fresh ingredients.

As this example suggests, $p \to q$ is the same as "*q* is necessary for *p*." In other words, if *q* doesn't happen, then neither will *p*. Notice how this idea is closely related to the idea of equivalence between a conditional statement and its contrapositive.

COMMON TRANSLATIONS OF $p \rightarrow q$

The conditional $p \rightarrow q$ can be translated in any of the following ways, none of which depends on the truth or falsity of $p \rightarrow q$.

If p, then q.	p is sufficient for q.
If p, q.	q is necessary for p.
p implies q.	All p are q.
p only if q.	q if p.

Kurt Gödel (1906–1978) is widely regarded as the most influential mathematical logician of the twentieth century. He proved by his Incompleteness Theorem that the search for a set of axioms from which all mathematical truths could be proved was futile. In particular, "the vast structure of the *Principia Mathematica* of Whitehead and Russell was inadequate for deciding all mathematical questions."

After the death of his friend **Albert Einstein** (1879–1955), Gödel developed paranoia, and his life ended tragically when, convinced he was being poisoned, he refused to eat, essentially starving himself to death.

Example: If you live in Alamogordo, then you live in New Mexico.   Statement

> You live in New Mexico if you live in Alamogordo.
> You live in Alamogordo only if you live in New Mexico.
> Living in New Mexico is necessary for living in Alamogordo.
> Living in Alamogordo is sufficient for living in New Mexico.
> All residents of Alamogordo are residents of New Mexico.
> Being a resident of Alamogordo implies residency in New Mexico.

Common translations

EXAMPLE 3 **Rewording Conditional Statements**

Rewrite each statement in the form "If p, then q."

(a) You'll get sick if you eat that.

(b) Go to the doctor only if your temperature exceeds 101°F.

(c) Everyone at the game had a great time.

Solution

(a) If you eat that, then you'll get sick.

(b) If you go to the doctor, then your temperature exceeds 101°F.

(c) If you were at the game, then you had a great time.   ∎

EXAMPLE 4 **Translating from Words to Symbols**

Let p represent "A triangle is equilateral," and let q represent "A triangle has three sides of equal length." Write each of the following in symbols.

(a) A triangle is equilateral if it has three sides of equal length.

(b) A triangle is equilateral only if it has three sides of equal length.

Solution

(a) $q \rightarrow p$ **(b)** $p \rightarrow q$ ∎

Biconditionals

The compound statement **p if and only if q** (often abbreviated **p iff q**) is called a **biconditional**. It is symbolized $p \leftrightarrow q$ and is interpreted as the conjunction of the two conditionals $p \rightarrow q$ and $q \rightarrow p$.

Using symbols, the conjunction of the conditionals $p \rightarrow q$ and $q \rightarrow p$ is written $(q \rightarrow p) \land (p \rightarrow q)$ so that, by definition,

$$p \leftrightarrow q \equiv (q \rightarrow p) \land (p \rightarrow q). \quad \text{Biconditional}$$

The truth table for the biconditional $p \leftrightarrow q$ can be determined using this definition.

TRUTH TABLE FOR THE BICONDITIONAL p if and only if q

p if and only if *q*

p	q	$p \leftrightarrow q$
T	T	T
T	F	F
F	T	F
F	F	T

A biconditional is true when both component statements have the same truth value. It is false when they have different truth values.

EXAMPLE 5 **Determining Whether Biconditionals Are True or False**

Determine whether each biconditional statement is *true* or *false*.

(a) $6 + 8 = 14$ if and only if $11 + 5 = 16$

(b) $6 = 5$ if and only if $12 \neq 12$

(c) Mars is a moon if and only if Jupiter is a planet.

Solution

(a) Both $6 + 8 = 14$ and $11 + 5 = 16$ are true. By the truth table for the biconditional, this biconditional is true.

(b) Both component statements are false, so by the last line of the truth table for the biconditional, this biconditional statement is true.

(c) Because the first component Mars is a moon is false, and the second is true, this biconditional statement is false. ∎

Summary of Truth Tables

Truth tables have been derived for several important types of compound statements.

SUMMARY OF BASIC TRUTH TABLES

1. $\sim p$, the **negation** of p, has truth value opposite that of p.

2. $p \land q$, the **conjunction,** is true only when both p and q are true.

3. $p \lor q$, the **disjunction,** is false only when both p and q are false.

4. $p \rightarrow q$, the **conditional,** is false only when p is true and q is false.

5. $p \leftrightarrow q$, the **biconditional,** is true only when both p and q have the same truth value.

3.4 EXERCISES

*For each given conditional statement (or statement that can be written as a conditional), write (**a**) the converse, (**b**) the inverse, and (**c**) the contrapositive in* if . . . then *form. In some of the exercises, it may be helpful to first restate the given statement in* if . . . then *form.*

1. If beauty were a minute, then you would be an hour.

2. If you lead, then I will follow.

3. If it ain't broke, don't fix it.

4. If I had a nickel for each time that happened, I would be rich.

5. Walking in front of a moving car is dangerous to your health.

6. Milk contains calcium.

7. Birds of a feather flock together.

8. A rolling stone gathers no moss.

9. If you build it, he will come.

10. Where there's smoke, there's fire.

11. $p \to {\sim}q$

12. ${\sim}p \to q$

13. ${\sim}p \to {\sim}q$

14. ${\sim}q \to {\sim}p$

15. $p \to (q \lor r)$ (*Hint:* Use one of De Morgan's laws as necessary.)

16. $(r \lor {\sim}q) \to p$ (*Hint:* Use one of De Morgan's laws as necessary.)

17. Discuss the equivalences that exist among a given conditional statement, its converse, its inverse, and its contrapositive.

18. State the contrapositive of "If the square of a natural number is odd, then the natural number is odd." The two statements must have the same truth value. Use several examples and inductive reasoning to decide whether both are true or both are false.

Write each statement in the form "if p, then q."

19. If the Kings go to the playoffs, pigs will fly.

20. If I score 90% or higher on my test, I'll go to a movie.

21. Legs of 3 and 4 imply a hypotenuse of 5.

22. "This is a leap year" implies that next year is not.

23. All whole numbers are rational numbers.

24. No irrational numbers are rational.

25. Doing logic puzzles is sufficient for driving me crazy.

26. Being in Kalamazoo is sufficient for being in Michigan.

27. Two coats of paint are necessary to cover the graffiti.

28. Being an environmentalist is necessary for being elected.

29. Employment will improve only if the economy recovers.

30. The economy will recover only if employment improves.

31. No whole numbers are not integers.

32. No integers are irrational numbers.

33. The Phillies will win the pennant when their pitching improves.

34. The grass will be greener when we're on the other side.

35. A rectangle is a parallelogram with perpendicular adjacent sides.

36. A square is a rectangle with two adjacent sides equal.

37. A triangle with two perpendicular sides is a right triangle.

38. A parallelogram is a four-sided figure with opposite sides parallel.

39. The square of a three-digit number whose units digit is 5 will end in 25.

40. An integer whose units digit is 0 or 5 is divisible by 5.

41. One of the following statements is not equivalent to all the others. Which one is it?

    A. *r* only if *s*.   **B.** *r* implies *s*.

    C. If *r*, then *s*.   **D.** *r* is necessary for *s*.

42. Many students have difficulty interpreting *necessary* and *sufficient*. Use the statement "Being in Vancouver is sufficient for being in North America" to explain why "*p* is sufficient for *q*" translates as "if *p*, then *q*."

43. Use the statement "To be an integer, it is necessary that a number be rational" to explain why "*p* is necessary for *q*" translates as "if *q*, then *p*."

44. Explain why the statement "A week has eight days if and only if October has forty days" is true.

Identify each statement as true *or* false.

45. $6 = 9 - 3$ if and only if $8 + 2 = 10$.

46. $3 + 1 \neq 7$ if and only if $8 \neq 8$.

47. $8 + 7 \neq 15$ if and only if $3 \times 5 \neq 8$.

48. $6 \times 2 = 18$ if and only if $9 + 7 \neq 16$.

49. George H. W. Bush was president if and only if George W. Bush was not president.

50. McDonald's sells Whoppers if and only if Apple manufactures Ipods.

Two statements that can both be true about the same object are **consistent.** *For example, "It is green" and "It weighs 60 pounds" are consistent statements. Statements that cannot both be true about the same object are called* **contrary.** *"It is a Nissan" and "It is a Mazda" are contrary. In Exercises 51–55, label each pair of statements as either* contrary *or* consistent.

51. Michael Jackson is alive. Michael Jackson is dead.

52. That book is nonfiction. That same book costs more than $150.

53. This number is a whole number. This same number is irrational.

54. This number is positive. This same number is a natural number.

55. This number is an integer. This same number is a rational number.

56. Refer to the "For Further Thought" on **page 99** at the end of **Section 3.2.** Verify that the logic circuit at the top of the next column (consisting of only NOR gates) is equivalent to the biconditional $A \leftrightarrow B$. Build a truth table for the statement $\sim(p \veebar q)$ (the negation of the Exclusive OR statement). What is the relationship between this and the biconditional?

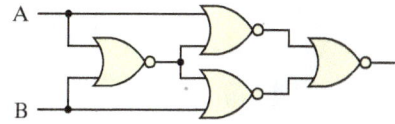

Exercises 57 and 58 refer to the Chapter Opener on **page 83.** *Humphrey the hero deduced from the riddle that there are twin trolls who take shifts guarding the magical doors. One of them tells only truths, the other only lies. When the troll standing guard sees Humphrey, he says,*

"If truly Truthful Troll I be, then go thou east and be thou free."

Humphrey needs to decide which troll he is addressing, call him by name, and tell him which door he would like opened.

There are two things Humphrey must get right: the name of the troll and the proper door. If he misidentifies the troll, no door will be opened. If he correctly names the troll and picks the wrong door, he will be confined to the dungeon behind it.

57. Because the troll either always lies or always tells the truth, Humphrey knows that

(**1**) if the troll is Truthful Troll, then the conditional statement he uttered is true, and

(**2**) if the conditional statement he uttered is true, then he is Truthful Troll.

Let p represent "the troll is Truthful Troll" and let q represent "the door to the east leads to freedom." Express the statements in (1) and (2) in symbolic form.

58. The conjunction of the answers from **Exercise 57** is a biconditional that must be true.

(**a**) Build a truth table for this biconditional.

(**b**) Use the fact that it *must* be true to solve Humphrey's riddle.

3.5 ANALYZING ARGUMENTS WITH EULER DIAGRAMS

OBJECTIVES

1 Define logical arguments.

2 Use Euler diagrams to analyze arguments with universal quantifiers.

3 Use Euler diagrams to analyze arguments with existential quantifiers.

Logical Arguments

With inductive reasoning we observe patterns to solve problems. Now we study how deductive reasoning may be used to determine whether logical arguments are valid or invalid.

A logical argument is made up of **premises** (assumptions, laws, rules, widely held ideas, or observations) and a **conclusion.** Recall that *deductive* reasoning involves drawing specific conclusions from given general premises. When reasoning from the premises of an argument to obtain a conclusion, we want the argument to be valid.

Leonhard Euler (1707–1783) won the Academy prize and edged out du Châtelet and Voltaire. That was a minor achievement, as was the invention of "Euler circles" (which antedated Venn diagrams). Euler was the most prolific mathematician of his generation despite blindness that forced him to dictate from memory.

> **VALID AND INVALID ARGUMENTS**
>
> An argument is **valid** if the fact that all the premises are true forces the conclusion to be true. An argument that is not valid is **invalid.** It is called a **fallacy.**

"Valid" and "true" do not have the same meaning—an argument can be valid even though the conclusion is false (see Example 4), *or invalid even though the conclusion is true* (see Example 6).

Arguments with Universal Quantifiers

Several techniques can be used to check whether an argument is valid. One such technique is based on **Euler diagrams.**

Leonhard Euler (pronounced "Oiler") was one of the greatest mathematicians who ever lived. He is immortalized in mathematics history with the important irrational number *e*, named in his honor. This number appears throughout mathematics and is discussed in **Chapters 6 and 8.**

EXAMPLE 1 Using an Euler Diagram to Determine Validity

Is the following argument valid?

> No accidents happen on purpose.
> Spilling the beans was an accident.
> _____
> The beans were not spilled on purpose.

Solution

To begin, draw regions to represent the first premise. Because no accidents happen on purpose, the region for "accidents" goes outside the region for "things that happen on purpose," as shown in **Figure 8.**

The second premise, "Spilling the beans was an accident," suggests that "spilling the beans" belongs in the region representing "accidents." Let *x* represent "spilling the beans." **Figure 9** shows that "spilling the beans" is not in the region for "things that happen on purpose." If both premises are true, the conclusion that the beans were not spilled on purpose is also true. The argument is valid. ∎

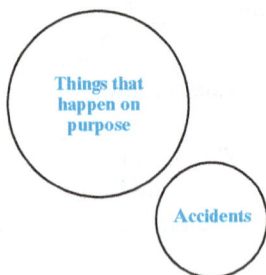

Figure 8

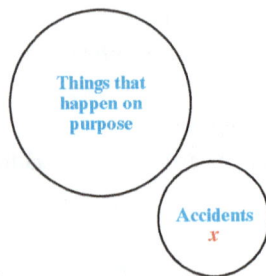

Figure 9

EXAMPLE 2 Using an Euler Diagram to Determine Validity

Is the following argument valid?

> All rainy days are cloudy.
> Today is not cloudy.
> _____
> Today is not rainy.

Solution

In **Figure 10,** the region for "rainy days" is drawn entirely inside the region for "cloudy days." Since "Today is *not* cloudy," place an *x* for "today" *outside* the region for "cloudy days." See **Figure 11.** Placing the *x* outside the region for "cloudy days" forces it also to be outside the region for "rainy days." Thus, if the two premises are true, then it is also true that today is not rainy. The argument is valid. ∎

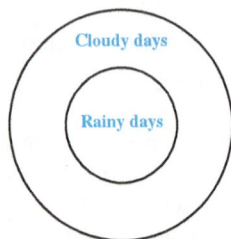

Figure 10

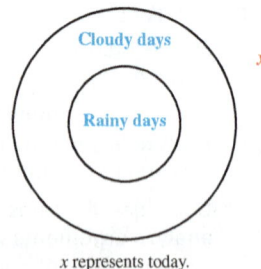

x represents today.

Figure 11

EXAMPLE 3 Using an Euler Diagram to Determine Validity

Is the following argument valid?

> All magnolia trees have green leaves.
> That plant has green leaves.
> _____
> That plant is a magnolia tree.

Solution

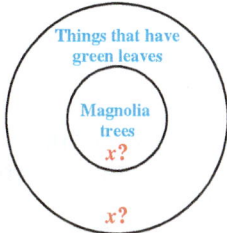

Figure 12

The region for "magnolia trees" goes entirely inside the region for "things that have green leaves." See **Figure 12.** The x that represents "that plant" must go inside the region for "things that have green leaves," but it can go either inside or outside the region for "magnolia trees." Even if the premises are true, we are not forced to accept the conclusion as true. This argument is invalid. It is a fallacy. ■

EXAMPLE 4 Using an Euler Diagram to Determine Validity

Is the following argument valid?

> All expensive things are desirable.
> All desirable things make you feel good.
> All things that make you feel good make you live longer.
> _____
> All expensive things make you live longer.

Solution

A diagram for the argument is given in **Figure 13.** If each premise is true, then the conclusion must be true because the region for "expensive things" lies completely within the region for "things that make you live longer." Thus, the argument is valid. (This argument is an example of the fact that a *valid* argument need *not* have a true conclusion.)

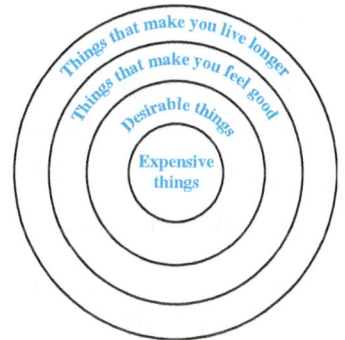

Figure 13 ■

Arguments with Existential Quantifiers

EXAMPLE 5 Using an Euler Diagram to Determine Validity

Is the following argument valid?

> Many students drive Hondas.
> I am a student.
> _____
> I drive a Honda.

Solution

Figure 14

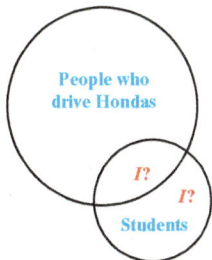

Figure 15

The first premise is sketched in **Figure 14,** where many (but not necessarily *all*) students drive Hondas. There are two possibilities for *I*, as shown in **Figure 15.** One possibility is that *I* drive a Honda. The other is that *I* don't. Since the truth of the premises does not force the conclusion to be true, the argument is invalid. ■

EXAMPLE 6   **Using an Euler Diagram to Determine Validity**

Is the following argument valid?

All fish swim.

All whales swim.

A whale is not a fish.

Solution

The premises lead to two possibilities. **Figure 16** shows the set of fish and the set of whales as intersecting, while **Figure 17** does not. Both diagrams are valid interpretations of the given premises, but only one supports the conclusion.

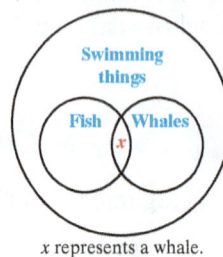

x represents a whale.

Figure 16

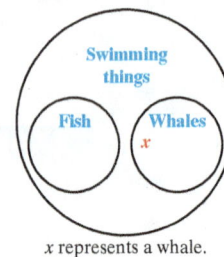

x represents a whale.

Figure 17

Because the truth of the premises does not force the conclusion to be true, the argument is invalid. Even though we know the conclusion to be true, this knowledge is not deduced from the premises.   ∎

FOR FURTHER THOUGHT

Common Fallacies

Discussions in everyday conversation, politics, and advertising provide a nearly endless stream of examples of **fallacies**—arguments exhibiting illogical reasoning. There are many general forms of fallacies, and we now present descriptions and examples of some of the more common ones. (Much of this list is adapted from the document Reader Mission Critical, located on the San Jose State University Web site, www.sjsu.edu)

1. **Circular Reasoning** (also called *Begging the Question*) The person making the argument assumes to be true what he is trying to prove.

   Husband: What makes you say that this dress makes you look fat?

   Wife: Because it does.

   Here the wife makes her case by stating what she wants to prove.

2. **False Dilemma** (also called the *Either-Or Fallacy*, or the *Black and White Fallacy*) Presenting two options with the assumption that they are contradictions (that is, the truth of one implies the falsity of the other) when in fact they are not, is the basis of a common fallacy.

Politician: America: Love it or leave it.

This argument implies only two choices. It is possible that someone may love America and yet leave, while someone else may not love America and yet stay.

3. **Loaded Question and Complex Claims** This fallacy involves one person asking a question or making a statement that is constructed in such a way as to obtain an answer in which the responder agrees to something with which he does not actually agree.

   Teenager Beth to her father: I hope you enjoyed embarrassing me in front of my friends.

   If Beth gets the expected response "No, I didn't enjoy it," the answer allows Beth to interpret that while her father didn't enjoy it, he did indeed embarrass her.

4. **Post Hoc Reasoning** An argument that is based on the false belief that if event A preceded event B, then A must have caused B is called *post hoc reasoning*.

   Johnny: I wore my Hawaiian shirt while watching all three playoff games, and my team won all three games. So I am going to wear that shirt every time I watch them.

The fact that Johnny put the same shirt on before each game has nothing to do with the outcomes of the games.

5. **Red Herring** (also called *Smoke Screen,* or *Wild Goose Chase*) This fallacy involves introducing an irrelevant topic to divert attention away from the original topic, allowing the person making the argument to seemingly prevail.

(The following script is from a political advertisement during the 2008 presidential campaign, intended to establish that John McCain lacked understanding of the economy.)

Maybe you're struggling just to pay the mortgage on your home. But recently, John McCain said, "The fundamentals of our economy are strong." Hmm. Then again, that same day, when asked how many houses he owns, McCain lost track. He couldn't remember. Well, it's seven. Seven houses. And here's one house America can't afford to let John McCain move into (showing a picture of the White House).

The advertisement shifted the focus to the number of houses McCain owned, which had nothing to do with the state of the economy, or the ability of "average" citizens to make their mortgage payments.

6. **Shifting the Burden of Proof** A person making a claim usually is required to support that claim. In this fallacy, if the claim is difficult to support, that person turns the burden of proof of that claim over to someone else.

Employee: You accuse me of embezzling money? That's ridiculous.

Employer: Well, until you can prove otherwise, you will just have to accept it as true.

If money has been disappearing, it is up to the employer to prove that this employee is guilty. The burden of proof is on the employer, but he is insinuating that the employee must prove that he is not the one taking the money.

7. **Straw Man** This fallacy involves creating a false image (like a scarecrow, or straw man) of someone else's position in an argument.

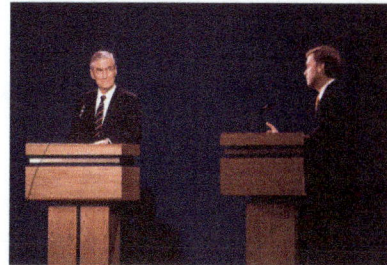

Dan Quayle: I have as much experience in the Congress as Jack Kennedy did when he sought the presidency.

Lloyd Bentsen: Senator, I served with Jack Kennedy. I knew Jack Kennedy. Jack Kennedy was a friend of mine. And Senator, you're no Jack Kennedy.

Dan Quayle: That was really uncalled for, Senator.

Lloyd Bentsen: You're the one that was making the comparison, Senator.

While this was the defining moment of the 1988 vice-presidential debate, Bentsen expertly used the straw man fallacy. Quayle did not compare himself or his accomplishments to those of Kennedy, but merely stated that he had spent as much time in Congress as Kennedy had when the latter ran for president.

For Group or Individual Investigation

Use the Internet to investigate the following additional logical fallacies.

Appeal to Authority	Appeal to Common Belief	Common Practice
Two Wrongs		Wishful Thinking
Appeal to Fear	Indirect Consequences	Appeal to Pity
Appeal to Prejudice	Appeal to Loyalty	Appeal to Vanity
Guilt by Association	Appeal to Spite	Hasty Generalization
	Slippery Slope	

3.5   EXERCISES

Decide whether each argument is valid or invalid.

1. All amusement parks have thrill rides.
   <u>*Universal Orlando* is an amusement park.</u>
   Universal Orlando has thrill rides.

2. All disc jockeys play music.
   <u>Calvin is a disc jockey.</u>
   Calvin plays music.

3. All celebrities have problems.
That man has problems.
_____
That man is a celebrity.

4. All Southerners speak with an accent.
Nick speaks with an accent.
_____
Nick is a Southerner.

5. All dogs love to bury bones.
Puddles does not love to bury bones.
_____
Puddles is not a dog.

6. All vice presidents use cell phones.
Bob does not use a cell phone.
_____
Bob is not a vice president.

7. All residents of Colorado know how to breathe thin air.
Julie knows how to breathe thin air.
_____
Julie lives in Colorado.

8. All drivers must have a photo I.D.
Kay has a photo I.D.
_____
Kay is a driver.

9. Some dinosaurs were plant eaters.
Danny was a plant eater.
_____
Danny was a dinosaur.

10. Some philosophers are absent minded.
Nicole is a philosopher.
_____
Nicole is absent minded.

11. Many nurses belong to unions.
Heather is a nurse.
_____
Heather belongs to a union.

12. Some trucks have sound systems.
Some trucks have gun racks.
_____
Some trucks with sound systems have gun racks.

13. Refer to **Example 3.** If the second premise and the conclusion were interchanged, would the argument then be valid?

14. Refer to **Example 4.** Give a different conclusion from the one given there so that the argument is still valid.

Construct a valid argument based on the Euler diagram shown.

15.

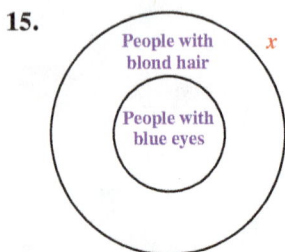

_x_ represents Erin.

16.

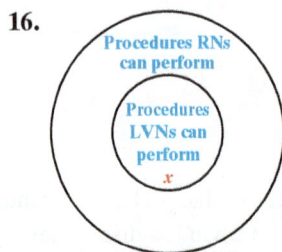

_x_ represents vaccinations.

As mentioned in the text, an argument can have a true conclusion yet be invalid. In these exercises, each argument has a true conclusion. Identify each argument as valid _or_ invalid.

17. All birds fly.
All planes fly.
_____
A bird is not a plane.

18. All actors have cars.
All cars use gas.
_____
All actors have gas.

19. All chickens have beaks.
All hens are chickens.
_____
All hens have beaks.

20. All chickens have beaks.
All birds have beaks.
_____
All chickens are birds.

21. Amarillo is northeast of El Paso.
Amarillo is northeast of Deming.
_____
El Paso is northeast of Deming.

22. Beaverton is north of Salem.
Salem is north of Lebanon.
_____
Beaverton is north of Lebanon.

23. No whole numbers are negative.
-3 is negative.
_____
-3 is not a whole number.

24. A scalene triangle has a longest side.
A scalene triangle has a largest angle.
_____
The largest angle in a scalene triangle is opposite the longest side.

In Exercises 25–30, the premises marked A, B, _and_ C _are followed by several possible conclusions. Take each conclusion in turn, and check whether the resulting argument is_ valid _or_ invalid.

A. _All people who drive contribute to air pollution._

B. _All people who contribute to air pollution make life a little worse._

C. _Some people who live in a suburb make life a little worse._

25. Some people who live in a suburb contribute to air pollution.

26. Some people who live in a suburb drive.

27. Suburban residents never drive.

28. Some people who contribute to air pollution live in a suburb.

29. Some people who make life a little worse live in a suburb.

30. All people who drive make life a little worse.

3.6 ANALYZING ARGUMENTS WITH TRUTH TABLES

OBJECTIVES

1 Use truth tables to determine validity of arguments.

2 Recognize valid and invalid argument forms.

3 Be familiar with the arguments of Lewis Carroll.

Using Truth Tables to Determine Validity

In **Section 3.5** we used Euler diagrams to test the validity of arguments. While Euler diagrams often work well for simple arguments, difficulties can develop with more complex ones, because Euler diagrams must show every possible case. In complex arguments, it is hard to be sure that all cases have been considered.

In deciding whether to use Euler diagrams to test the validity of an argument, look for quantifiers such as "all," "some," or "no." These words often indicate arguments best tested by Euler diagrams. If these words are absent, it may be better to use truth tables to test the validity of an argument.

EXAMPLE 1 Using a Truth Table to Determine Validity

Determine whether the argument is *valid* or *invalid*.

> If there is a problem, then I must fix it.
> There is a problem.
> _____
> I must fix it.

Solution

To test the validity of this argument, we begin by assigning the letters p and q to represent these statements.

p represents "There is a problem."

q represents "I must fix it."

Now we write the two premises and the conclusion in symbols.

$$\text{Premise 1:} \quad p \rightarrow q$$
$$\text{Premise 2:} \quad p$$
$$\overline{\text{Conclusion:} \quad q}$$

To decide if this argument is valid, we must determine whether the conjunction of both premises implies the conclusion for all possible combinations of truth values for p and q. Therefore, write the conjunction of the premises as the antecedent of a conditional statement, and write the conclusion as the consequent.

$$[(p \rightarrow q) \quad \wedge \quad p] \quad \rightarrow \quad q$$

premise and premise implies conclusion

Finally, construct the truth table for this conditional statement, as shown below.

p	q	$p \rightarrow q$	$(p \rightarrow q) \wedge p$	$[(p \rightarrow q) \wedge p] \rightarrow q$
T	T	T	T	T
T	F	F	F	T
F	T	T	F	T
F	F	T	F	T

Because the final column, shown in color, indicates that the conditional statement that represents the argument is true for all possible truth values of p and q, the statement is a tautology. Thus, the argument is valid. ∎

In the 2007 Spanish film **La Habitacion de Fermat (Fermat's Room),** four mathematicians are invited to dinner, only to discover that the room in which they are meeting is designed to eventually crush them as walls creep in closer and closer. The only way for them to delay the inevitable is to answer enigmas, questions, puzzles, problems, and riddles that they are receiving on a cell phone.

One of the enigmas deals with a hermetically sealed room that contains a single light bulb. There are three switches outside the room, all of them are off, and only one of these switches controls the bulb. You are allowed to flip any or all of the switches as many times as you wish before you enter the room, but once you enter, you cannot return to the switches outside. How can you determine which one controls the bulb? (The answer is on **page 124.**)

Answer to the Light Bulb question on page 123.

Label the switches 1, 2, and 3. Turn switch 1 on and leave it on for several minutes. Then turn switch 1 off, turn switch 2 on, and then immediately enter the room. If the bulb is on, then you know that switch 2 controls it. If the bulb is off, touch it to see if it is still warm. If it is, then switch 1 controls it. If the bulb is not warm, then switch 3 controls it.

The pattern of the argument in **Example 1**

$$p \to q$$
$$\dfrac{p}{q}$$

is called **modus ponens,** or the *law of detachment.*

To test the validity of an argument using a truth table, follow the steps in the box.

TESTING THE VALIDITY OF AN ARGUMENT WITH A TRUTH TABLE

Step 1 Assign a letter to represent each component statement in the argument.

Step 2 Express each premise and the conclusion symbolically.

Step 3 Form the symbolic statement of the entire argument by writing the *conjunction* of *all* the premises as the antecedent of a conditional statement, and the conclusion of the argument as the consequent.

Step 4 Complete the truth table for the conditional statement formed in Step 3. If it is a tautology, then the argument is valid; otherwise, it is invalid.

EXAMPLE 2 Using a Truth Table to Determine Validity

Determine whether the argument is *valid* or *invalid.*

If my check arrives in time, I'll register for fall semester.
I've registered for fall semester.
My check arrived in time.

Solution

Let p represent "My check arrives (arrived) in time." Let q represent "I'll register (I've registered) for fall semester." The argument can be written as follows.

$$p \to q$$
$$\dfrac{q}{p}$$

To test for validity, construct a truth table for the statement $[(p \to q) \land q] \to p$.

p	q	$p \to q$	$(p \to q) \land q$	$[(p \to q) \land q] \to p$
T	T	T	T	T
T	F	F	F	T
F	T	T	T	F
F	F	T	F	T

The final column of the truth table contains an F. The argument is invalid. ■

If a conditional and its converse were logically equivalent, then an argument of the type found in **Example 2** would be valid. Because a conditional and its converse are *not* equivalent, the argument is an example of what is sometimes called the **fallacy of the converse.**

EXAMPLE 3 Using a Truth Table to Determine Validity

Determine whether the argument is *valid* or *invalid.*

If I can avoid sweets, I can avoid the dentist.
I can't avoid the dentist.
I can't avoid sweets.

Solution

If p represents "I can avoid sweets" and q represents "I can avoid the dentist," the argument is written as follows.

$$p \rightarrow q$$
$$\frac{\sim q}{\sim p}$$

The symbolic statement of the entire argument is as follows.

$$[(p \rightarrow q) \land \sim q] \rightarrow \sim p$$

The truth table for this argument indicates a tautology, and the argument is valid.

p	q	$p \rightarrow q$	$\sim q$	$(p \rightarrow q) \land \sim q$	$\sim p$	$[(p \rightarrow q) \land \sim q] \rightarrow \sim p$
T	T	T	F	F	F	T
T	F	F	T	F	F	T
F	T	T	F	F	T	T
F	F	T	T	T	T	T

The pattern of reasoning of this example is called **modus tollens,** or the *law of contraposition,* or *indirect reasoning.* ∎

With reasoning similar to that used to name the fallacy of the converse, the fallacy

$$p \rightarrow q$$
$$\frac{\sim p}{\sim q}$$

Concluding $\sim q$ from $\sim p$ wrongly assumes $\sim p \rightarrow \sim q$, the *inverse* of the given premise $p \rightarrow q$.

is called the **fallacy of the inverse.** An example of such a fallacy is "If it rains, I get wet. It doesn't rain. Therefore, I don't get wet."

EXAMPLE 4   Using a Truth Table to Determine Validity

Determine whether the argument is *valid* or *invalid.*

> I'll buy a car or I'll take a vacation.
> I won't buy a car.
> ────────────────────────
> I'll take a vacation.

Solution

If p represents "I'll buy a car" and q represents "I'll take a vacation," the argument is symbolized as follows.

$$p \lor q$$
$$\frac{\sim p}{q}$$

We must set up a truth table for the statement $[(p \lor q) \land \sim p] \rightarrow q$.

p	q	$p \lor q$	$\sim p$	$(p \lor q) \land \sim p$	$[(p \lor q) \land \sim p] \rightarrow q$
T	T	T	F	F	T
T	F	T	F	F	T
F	T	T	T	T	T
F	F	F	T	F	T

The statement is a tautology and the argument is valid. Any argument of this form is valid by the law of **disjunctive syllogism.** ∎

EXAMPLE 5 **Using a Truth Table to Determine Validity**

Determine whether the argument is *valid* or *invalid*.

> If it squeaks, then I use WD-40.
> If I use WD-40, then I must go to the hardware store.
> ──────────────
> If it squeaks, then I must go to the hardware store.

Solution

Let p represent "It squeaks," let q represent "I use WD-40," and let r represent "I must go to the hardware store." The argument takes on the following form.

$$p \rightarrow q$$
$$q \rightarrow r$$
$$\overline{p \rightarrow r}$$

Make a truth table for this statement, which requires eight rows.

$$[(p \rightarrow q) \wedge (q \rightarrow r)] \rightarrow (p \rightarrow r)$$

p	q	r	$p \rightarrow q$	$q \rightarrow r$	$p \rightarrow r$	$(p \rightarrow q) \wedge (q \rightarrow r)$	$[(p \rightarrow q) \wedge (q \rightarrow r)] \rightarrow (p \rightarrow r)$
T	T	T	T	T	T	T	T
T	T	F	T	F	F	F	T
T	F	T	F	T	T	F	T
T	F	F	F	T	F	F	T
F	T	T	T	T	T	T	T
F	T	F	T	F	T	F	T
F	F	T	T	T	T	T	T
F	F	F	T	T	T	T	T

This argument is valid because the final statement is a tautology. This pattern of argument is called **reasoning by transitivity**, or the *law of hypothetical syllogism.* ■

Valid and Invalid Argument Forms

A summary of the valid forms of argument presented so far follows.

VALID ARGUMENT FORMS

Modus Ponens	Modus Tollens	Disjunctive Syllogism	Reasoning by Transitivity
$p \rightarrow q$	$p \rightarrow q$	$p \vee q$	$p \rightarrow q$
p	$\sim q$	$\sim p$	$q \rightarrow r$
q	$\sim p$	q	$p \rightarrow r$

The following is a summary of invalid forms (or fallacies).

INVALID ARGUMENT FORMS (FALLACIES)

Fallacy of the Converse	Fallacy of the Inverse
$p \rightarrow q$	$p \rightarrow q$
q	$\sim p$
p	$\sim q$

In a scene near the beginning of the 1974 film *Monty Python and the Holy Grail,* an amazing application of **poor logic** leads to the apparent demise of a supposed witch. Some peasants have forced a young woman to wear a nose made of wood. The convoluted argument they make is this: Witches and wood are both burned, and because witches are made of wood, and wood floats, and ducks also float, if she weighs the same as a duck, then she is made of wood and, therefore, is a witch!

Setting the Table Correctly If an argument has the form

$$p_1$$
$$p_2$$
$$\vdots$$
$$\underline{p_n,}$$
$$c$$

then **Step 3** in the testing process calls for the statement

$$(p_1 \land p_2 \land \ldots \land p_n) \to c.$$

EXAMPLE 6 Using a Truth Table to Determine Validity

Determine whether the argument is *valid* or *invalid*.

> If Eddie goes to town, then Mabel stays at home.
> If Mabel does not stay at home, then Rita will cook.
> Rita will not cook. Therefore, Eddie does not go to town.

Solution

In an argument written in this manner, the premises are given first, and the conclusion is the statement that follows the word "Therefore." Let p represent "Eddie goes to town," let q represent "Mabel stays at home," and let r represent "Rita will cook." Then the argument is symbolized as follows.

$$p \to q$$
$$\sim q \to r$$
$$\underline{\sim r}$$
$$\sim p$$

When an argument contains more than two premises, it is necessary to determine the truth values of the conjunction of *all* of them.

> **If at least one premise in a conjunction of several premises is false, then the entire conjunction is false.**

To test validity, set up a truth table for this statement.

$$[(p \to q) \land (\sim q \to r) \land \sim r] \to \sim p$$

p	q	r	$p \to q$	$\sim q$	$\sim q \to r$	$\sim r$	$(p \to q) \land (\sim q \to r) \land \sim r$	$\sim p$	$[(p \to q) \land (\sim q \to r) \land \sim r] \to \sim p$
T	T	T	T	F	T	F	F	F	T
T	T	F	T	F	T	T	T	F	F
T	F	T	F	T	T	F	F	F	T
T	F	F	F	T	F	T	F	F	T
F	T	T	T	F	T	F	F	T	T
F	T	F	T	F	T	T	T	T	T
F	F	T	T	T	T	F	F	T	T
F	F	F	T	T	F	T	F	T	T

Because the final column does not contain all Ts, the statement is not a tautology. The argument is invalid. ∎

Arguments of Lewis Carroll

Consider the following verse, which has been around for many years.

> *For want of a nail, the shoe was lost.*
> *For want of a shoe, the horse was lost.*
> *For want of a horse, the rider was lost.*
> *For want of a rider, the battle was lost.*
> *For want of a battle, the war was lost.*
> *Therefore, for want of a nail, the war was lost.*

Each line of the verse may be written as an *if . . . then* statement. For example, the first line may be restated as "If a nail is lost, then the shoe is lost." The conclusion, "For want of a nail, the war was lost," follows from the premises, because repeated use of the law of transitivity applies. Arguments such as the one used by Lewis Carroll in the next example often take a similar form.

Alice's Adventures in Wonderland is the most famous work of **Charles Dodgson** (1832–1898), better known as **Lewis Carroll,** who was a mathematician and logician. He popularized recreational mathematics with this story and its sequel, *Through the Looking-Glass.* More than a century later, Raymond Smullyan continues this genre in his book *Alice in Puzzle-land* and many others.

EXAMPLE 7 Supplying a Conclusion to Ensure Validity

Supply a conclusion that yields a valid argument for the following premises.

> Babies are illogical.
> Nobody is despised who can manage a crocodile.
> Illogical persons are despised.

Solution

First, write each premise in the form *if . . . then. . . .*

> If you are a baby, then you are illogical.
> If you can manage a crocodile, then you are not despised.
> If you are illogical, then you are despised.

Let p represent "you are a baby," let q represent "you are logical," let r represent "you can manage a crocodile," and let s represent "you are despised." The statements can be written symbolically.

$$p \rightarrow \sim q$$
$$r \rightarrow \sim s$$
$$\sim q \rightarrow s$$

Begin with any letter that appears only once. Here p appears only once. Using the contrapositive of $r \rightarrow \sim s$, which is $s \rightarrow \sim r$, rearrange the statements as follows.

$$p \rightarrow \sim q$$
$$\sim q \rightarrow s$$
$$s \rightarrow \sim r$$

From the three statements, repeated use of reasoning by transitivity gives the conclusion

$$p \rightarrow \sim r, \text{ which leads to a valid argument.}$$

In words, the conclusion is "If you are a baby, then you cannot manage a crocodile," or, as Lewis Carroll would have written it, "Babies cannot manage crocodiles." ◼

WHEN Will I Ever USE This ?

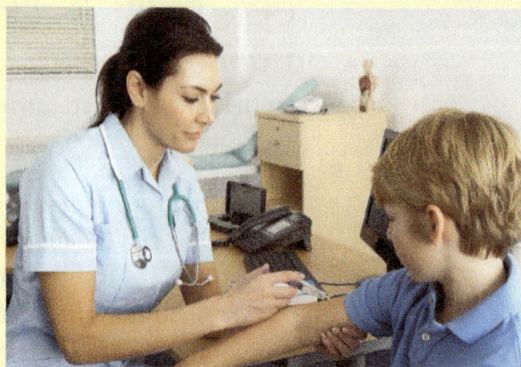

Suppose you are a pediatric nurse administering flu vaccination for a 6-year-old patient. The flowchart on the next page is a dosing algorithm provided by the CDC for children 6 months to 8 years of age.

Influenza Dosing Algorithm

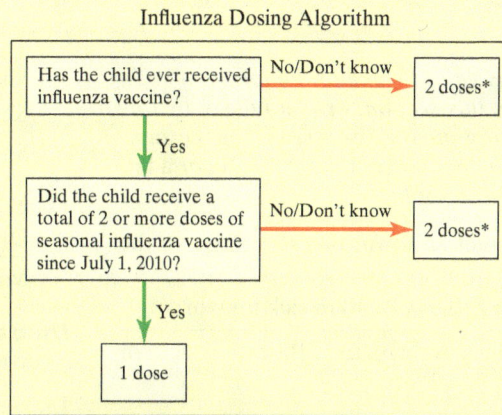

* Doses should be administered at least 4 weeks apart.
Source: cdc.gov

The patient was vaccinated for the first time last year, so you place the patient on a 2-dose regimen. Use a truth table to determine the validity of your action.

We let

p be "The child has received flu vaccine in the past,"

q be "The child has received 2 or more doses of flu vaccine since July 1, 2010,"

and r be "The child needs 2 doses this season."

Then the algorithm and your treatment decision can be expressed symbolically by the following argument.

$$(p \wedge q) \leftrightarrow \sim r$$
$$\underline{\sim q}$$
$$r$$

Make a truth table for the argument. Since there are three component statements, we require eight rows.

p	q	r	[((p	$\wedge$	q)	$\leftrightarrow$	$\sim r$)	$\wedge$	$\sim q$]	$\rightarrow$	r
T	T	T								T	T
T	T	F	T	T	T	T	T	F	F	T	F
T	F	T								T	T
T	F	F	T	F	F	F	T	F	T	T	F
F	T	T								T	T
F	T	F	F	F	T	F	T	F	F	T	F
F	F	T								T	T
F	F	F	F	F	F	F	T	F	T	T	F
			①	②	①	④	③	⑥	⑤	⑧	⑦

The truth table shows that your argument is a tautology, so your action was valid according to the CDC.

Note that in the odd rows of the table, we used the fact that the consequent r was true to conclude that the conditional was true, thus avoiding the effort required to "build" those rows piece by piece. In the even rows, we used the shortcut method introduced in **Example 10** of **Section 3.2**.

3.6   EXERCISES

Each argument either is valid by one of the forms of valid arguments discussed in this section, or is a fallacy by one of the forms of invalid arguments discussed. (See the summary boxes.) Decide whether the argument is valid *or a fallacy, and give the form that applies.*

1. If Rascal Flatts comes to town, then I will go to the concert.
   If I go to the concert, then I'll call in sick for work.
   ___
   If Rascal Flatts comes to town, then I'll call in sick for work.

2. If you use binoculars, then you get a glimpse of the bald eagle.
   If you get a glimpse of the bald eagle, then you'll be amazed.
   ___
   If you use binoculars, then you'll be amazed.

3. If Marina works hard enough, she will get a promotion.
   Marina works hard enough.
   ___
   She will get a promotion.

4. If Isaiah's ankle heals on time, he'll play this season.
   His ankle heals on time.
   ___
   He'll play this season.

5. If he doesn't have to get up at 3:00 A.M., he's ecstatic.
   He's ecstatic.
   ___
   He doesn't have to get up at 3:00 A.M.

6. "A mathematician is a device for turning coffee into theorems." (quote from Paul Erdos)
   You turn coffee into theorems.
   ___
   You are a mathematician.

7. If Clayton pitches, the Dodgers win.
   The Dodgers do not win.
   ___
   Clayton does not pitch.

8. If Josh plays, the opponent gets shut out.
   The opponent does not get shut out.
   ___
   Josh does not play.

9. "If you're going through hell, keep going." (quote from Winston Churchill)
   You're not going through hell.
   ___
   Don't keep going.

10. "If you can't get rid of the skeleton in your closet, you'd best teach it to dance." (quote from George Bernard Shaw)
    You can get rid of the skeleton in your closet.
    ___
    You'd best not teach it to dance.

11. She uses e-commerce or she pays by credit card.
    She does not pay by credit card.
    ___
    She uses e-commerce.

12. Mia kicks or Drew passes.
    Drew does not pass.
    ___
    Mia kicks.

Use a truth table to determine whether the argument is valid *or* invalid.

13. $p \lor q$
    p
    ___
    $\sim q$

14. $p \land \sim q$
    p
    ___
    $\sim q$

15. $\sim p \rightarrow \sim q$
    q
    ___
    p

16. $p \lor \sim q$
    p
    ___
    $\sim q$

17. $p \rightarrow q$
    $q \rightarrow p$
    ___
    $p \land q$

18. $\sim p \rightarrow q$
    p
    ___
    $\sim q$

19. $p \rightarrow \sim q$
    q
    ___
    $\sim p$

20. $p \rightarrow \sim q$
    $\sim p$
    ___
    $\sim q$

21. $(p \rightarrow q) \land (q \rightarrow p)$
    p
    ___
    $p \lor q$

22. $(p \land q) \lor (p \lor q)$
    q
    ___
    p

23. $(\sim p \lor q) \land (\sim p \rightarrow q)$
    p
    ___
    $\sim q$

24. $(r \land p) \rightarrow (r \lor q)$
    $q \land p$
    ___
    $r \lor p$

25. $(\sim p \land r) \rightarrow (p \lor q)$
    $\sim r \rightarrow p$
    ___
    $q \rightarrow r$

26. $(p \rightarrow \sim q) \lor (q \rightarrow \sim r)$
    $p \lor \sim r$
    ___
    $r \rightarrow p$

27. Earlier we showed how to analyze arguments using Euler diagrams. Refer to **Example 5** in this section, restate each premise and the conclusion using a quantifier, and then draw an Euler diagram to illustrate the relationship.

28. Explain in a few sentences how to determine the statement for which a truth table will be constructed so that the arguments that follow in **Exercises 29–38** can be analyzed for validity.

Determine whether each argument is valid *or* invalid.

29. Joey loves to watch movies. If Terry likes to jog, then Joey does not love to watch movies. If Terry does not like to jog, then Carrie drives a school bus. Therefore, Carrie drives a school bus.

30. If Hurricane Gustave hit that grove of trees, then the trees are devastated. People plant trees when disasters strike and the trees are not devastated. Therefore, if people plant trees when disasters strike, then Hurricane Gustave did not hit that grove of trees.

31. If Yoda is my favorite *Star Wars* character, then I hate Darth Vader. I hate Luke Skywalker or Darth Vader. I don't hate Luke Skywalker. Therefore, Yoda is not my favorite character.

32. Carrie Underwood sings or Joe Jonas is not a teen idol. If Joe Jonas is not a teen idol, then Jennifer Hudson does not win a Grammy. Jennifer Hudson wins a Grammy. Therefore, Carrie Underwood does not sing.

33. The Cowboys will make the playoffs if and only if Troy comes back to play. Jerry doesn't coach the Cowboys or Troy comes back to play. Jerry does coach the Cowboys. Therefore, the Cowboys will not be in the playoffs.

34. If I've got you under my skin, then you are deep in the heart of me. If you are deep in the heart of me, then you are not really a part of me. You are deep in the heart of me or you are really a part of me. Therefore, if I've got you under my skin, then you are really a part of me.

35. If Dr. Hardy is a department chairman, then he lives in Atlanta. He lives in Atlanta and his first name is Larry. Therefore, if his first name is not Larry, then he is not a department chairman.

36. If I were your woman and you were my man, then I'd never stop loving you. I've stopped loving you. Therefore, I am not your woman or you are not my man.

37. All men are created equal. All people who are created equal are women. Therefore, all men are women.

38. All men are mortal. Socrates is a man. Therefore, Socrates is mortal.

39. A recent DirecTV commercial had the following script: "When the cable company keeps you on hold, you feel trapped. When you feel trapped, you need to feel free. When you need to feel free, you try hang-gliding. When you try hang-gliding, you crash into things. When you crash into things, the grid goes down. When the grid goes down, crime goes up, and when crime goes up, your dad gets punched over a can of soup . . . "

(a) Use reasoning by transitivity and all the component statements to draw a valid conclusion.

(b) If we added the line, "Your dad does not get punched over a can of soup," what valid conclusion could be drawn?

40. Molly made the following observation: "If I want to determine whether an argument leading to the statement

$$[(p \rightarrow q) \land \sim q] \rightarrow \sim p$$

is valid, I only need to consider the lines of the truth table that lead to T for the column that is headed $(p \rightarrow q) \land \sim q$." Molly was very perceptive. Can you explain why her observation was correct?

In the arguments used by Lewis Carroll, it is helpful to restate a premise in if . . . then *form in order to more easily identify a valid conclusion. The following premises come from Lewis Carroll. Write each premise in* if . . . then *form.*

41. All my poultry are ducks.

42. None of your sons can do logic.

43. Guinea pigs are hopelessly ignorant of music.

44. No teetotalers are pawnbrokers.

45. No teachable kitten has green eyes.

46. Opium-eaters have no self-command.

47. I have not filed any of them that I can read.

48. All of them written on blue paper are filed.

Exercises 49–54 involve premises from Lewis Carroll. Write each premise in symbols, and then, in the final part, give a conclusion that yields a valid argument.

49. Let *p* be "it is a duck," *q* be "it is my poultry," *r* be "one is an officer," and *s* be "one is willing to waltz."

(a) No ducks are willing to waltz.

(b) No officers ever decline to waltz.

(c) All my poultry are ducks.

(d) Give a conclusion that yields a valid argument.

50. Let p be "one is able to do logic," q be "one is fit to serve on a jury," r be "one is sane," and s be "he is your son."

   (a) Everyone who is sane can do logic.

   (b) No lunatics are fit to serve on a jury.

   (c) None of your sons can do logic.

   (d) Give a conclusion that yields a valid argument.

51. Let p be "one is honest," q be "one is a pawnbroker," r be "one is a promise-breaker," s be "one is trustworthy," t be "one is very communicative," and u be "one is a wine-drinker."

   (a) Promise-breakers are untrustworthy.

   (b) Wine-drinkers are very communicative.

   (c) A person who keeps a promise is honest.

   (d) No teetotalers are pawnbrokers. (*Hint:* Assume "teetotaler" is the opposite of "wine-drinker.")

   (e) One can always trust a very communicative person.

   (f) Give a conclusion that yields a valid argument

52. Let p be "it is a guinea pig," q be "it is hopelessly ignorant of music," r be "it keeps silent while the *Moonlight Sonata* is being played," and s be "it appreciates Beethoven."

   (a) Nobody who really appreciates Beethoven fails to keep silent while the *Moonlight Sonata* is being played.

   (b) Guinea pigs are hopelessly ignorant of music.

   (c) No one who is hopelessly ignorant of music ever keeps silent while the *Moonlight Sonata* is being played.

   (d) Give a conclusion that yields a valid argument.

53. Let p be "it begins with 'Dear Sir'," q be "it is crossed," r be "it is dated," s be "it is filed," t be "it is in black ink," u be "it is in the third person," v be "I can read it," w be "it is on blue paper," x be "it is on one sheet," and y be "it is written by Brown."

   (a) All the dated letters are written on blue paper.

   (b) None of them are in black ink, except those that are written in the third person.

   (c) I have not filed any of them that I can read.

   (d) None of them that are written on one sheet are undated.

   (e) All of them that are not crossed are in black ink.

   (f) All of them written by Brown begin with "Dear Sir."

   (g) All of them written on blue paper are filed.

   (h) None of them written on more than one sheet are crossed.

   (i) None of them that begin with "Dear Sir" are written in the third person.

   (j) Give a conclusion that yields a valid argument.

54. Let p be "he is going to a party," q be "he brushes his hair," r be "he has self-command," s be "he looks fascinating," t be "he is an opium-eater," u be "he is tidy," and v be "he wears white kid gloves."

   (a) No one who is going to a party ever fails to brush his hair.

   (b) No one looks fascinating if he is untidy.

   (c) Opium-eaters have no self-command.

   (d) Everyone who has brushed his hair looks fascinating.

   (e) No one wears white kid gloves unless he is going to a party. (*Hint:* "a unless b" $\equiv \sim b \rightarrow a$.)

   (f) A man is always untidy if he has no self-command.

   (g) Give a conclusion that yields a valid argument.

CHAPTER 3 · SUMMARY

KEY TERMS

3.1

symbolic logic
truth value
statement
compound
   statement
component
   statements
connectives
negation
quantifiers

3.2

conjunction
truth table
disjunction
equivalent statements

3.3

conditional statement
antecedent
consequent
tautology

3.4

converse
inverse
contrapositive
biconditional

3.5

argument
premises
conclusion
valid

fallacy
Euler diagram

3.6

modus ponens
modus tollens
disjunctive syllogism
fallacy of the converse
fallacy of the inverse
reasoning by
   transitivity

NEW SYMBOLS

∨ disjunction
∧ conjunction
~ negation

→ implication
↔ biconditional
≡ equivalence

TEST YOUR WORD POWER

See how well you have learned the vocabulary in this chapter.

1. A **statement** is
   A. a sentence that asks a question, the answer to which may be true or false.
   B. a directive giving specific instructions.
   C. a sentence declaring something that is either true or false, but not both at the same time.
   D. a paradoxical sentence with no truth value.

2. A **disjunction** (inclusive) is
   A. a compound statement that is true only if both of its component statements are true.
   B. a compound statement that is true if one or both of its component statements is/are true.
   C. a compound statement that is false if either of its component statements is false.
   D. a compound statement that is true if exactly one of its component statements is true.

3. A **conditional** statement is
   A. a statement that may be true or false, depending on some condition.
   B. an idea that can be stated only under certain conditions.
   C. a statement using the connective *if . . . then.*
   D. a statement the antecedent of which is implied by the consequent.

4. The **inverse** of a conditional statement is
   A. the result when the antecedent and consequent are negated.
   B. the result when the antecedent and consequent are interchanged.
   C. the result when the antecedent and consequent are interchanged and negated.
   D. logically equivalent to the conditional.

5. A **fallacy** is
   A. an argument with a false conclusion.
   B. an argument whose conclusion is not supported by the premises.
   C. a valid argument.
   D. an argument containing at least one false premise.

6. **Fallacy of the converse** is
   A. the reason the converse of a conditional is not equivalent to the conditional.
   B. an invalid argument form that assumes the converse of a premise.
   C. an invalid argument form that denies the converse of a premise.
   D. an invalid argument form that assumes the inverse of a premise.

ANSWERS
1. C 2. B 3. C 4. A 5. B 6. B

QUICK REVIEW

Concepts	Examples

3.1 Statements and Quantifiers

A **statement** is a declarative sentence that is either true or false (not both simultaneously).

A **compound statement** is made up of two or more **component statements** joined by **connectives** (*not, and, or, if . . . then*).

Quantifiers indicate how many members in a group being considered exhibit a particular property or characteristic. Universal quantifiers indicate *all* members, and existential quantifiers indicate *at least one* member.

The **negation** of a statement has the opposite truth value of that statement in all cases.

Consider the following statement.

   "If it rains this month, then we'll have a green spring."

It is a compound statement made up of the following two component statements joined by the connective *if . . . then.*

   "It rains this month"   and   "We'll have a green spring,"

The statement

   "All five of those birds can fly"

contains a universal quantifier. Its negation is

   "At least one of those five birds cannot fly,"

which contains an existential quantifier.

Concepts	Examples

3.2 Truth Tables and Equivalent Statements

p	q	$p \wedge q$	$p \vee q$
T	T	T	T
T	F	F	T
F	T	F	T
F	F	F	F

Given two component statements p and q, their **conjunction** is symbolized $p \wedge q$ and is true only when both component statements are true.

Their **disjunction**, symbolized $p \vee q$, is false only when both component statements are false.

If p represents "$7 < 10$" and q represents "$4 < 3$," then the second row of the truth table above shows that $p \wedge q$ is false and $p \vee q$ is true.

The truth value of a compound statement is found by substituting T or F for each component statement, and then working from inside parentheses out, determining truth values for larger parts of the overall statement, until the entire statement has been evaluated.
  When this process is carried out for all possible combinations of truth values for the component statements, a **truth table** results.

The truth table for the statement $\sim p \vee (q \wedge p)$ is shown below, with circled numbers indicating the order in which columns were determined.

p	q	$\sim p$	$\vee$	$(q$	$\wedge$	$p)$
T	T	F	T	T	T	T
T	F	F	F	F	F	T
F	T	T	T	T	F	F
F	F	T	T	F	F	F
		①	③	①	②	①

Equivalent statements have the same truth value for all combinations of truth values for the component statements. To determine whether two statements are equivalent, construct truth tables for both and see if the final truth values agree in all rows.

The statements $\sim(p \wedge q)$ and $\sim p \vee \sim q$ are equivalent, as shown in the table.

p	q	$\sim$	$(p$	$\wedge$	$q)$	$\sim p$	$\vee$	$\sim q$
T	T	F	T	T	T	F	F	F
T	F	T	T	F	F	F	T	T
F	T	T	F	F	T	T	T	F
F	F	T	F	F	F	T	T	T

De Morgan's laws can be used to quickly find negations of disjunctions and conjunctions.

$$\sim(p \vee q) \equiv \sim p \wedge \sim q$$
$$\sim(p \wedge q) \equiv \sim p \vee \sim q$$

To find the negation of the statement "I love chess and I had breakfast," let p represent "I love chess" and let q represent "I had breakfast." Then the above statement becomes $p \wedge q$. Its negation $\sim(p \wedge q) \equiv \sim p \vee \sim q$ translates to

  "I don't love chess or I didn't have breakfast."

3.3 The Conditional and Circuits

A **conditional statement** uses the *if . . . then* connective and is symbolized $p \to q$, where p is the **antecedent** and q is the **consequent.**

If p represents "You are mighty" and q represents "I am flighty," then the conditional statement $p \to q$ is expressed as

  "*If* you are mighty, *then* I am flighty."

The conditional is false if the antecedent is true and if the consequent is false. Otherwise, the conditional is true. This is because q is only *required* to be true on the *condition* that p is true, but q may "voluntarily" be true even if p is false. That is, p is sufficient for q, but not necessary.

The statement $(6 < 1) \to (3 = 7)$ is true because the antecedent is false.

The statement "If you are reading this book, then it is the year 1937" is false, because the antecedent is true and the consequent is false.

Concepts	Examples

Concepts

The conditional $p \rightarrow q$ is equivalent to the disjunction $\sim p \vee q$, and its negation is $p \wedge \sim q$, as shown by a comparison of their truth tables.

Examples

p	q	$p \rightarrow q$	$\sim p \vee q$	$p \wedge \sim q$
T	T	T T T	F T T	T F F
T	F	T F F	F F F	T T T
F	T	F T T	T T T	F F F
F	F	F T F	T T F	F F T

The statement "All mice love cheese" can be stated, "*If* it's a mouse, *then* it loves cheese." This is equivalent to saying, "It's not a mouse or it loves cheese." The negation of this statement is "It's a mouse and it does not love cheese."

Electrical circuits are analogous to logical statements, with *parallel* circuits corresponding to disjunctions, and *series* circuits corresponding to conjunctions. Each switch (modeled by an arrow) represents a component statement. When a switch is closed, it allows current to pass through. The circuit represents a true statement when current flows from one end of the circuit to the other.

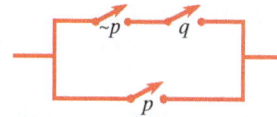

This corresponds to the logical statement $p \vee (\sim p \wedge q)$. Current will flow from one end to the other if p is true *or* if p is false and q is true. This statement is equivalent to

$$(p \vee \sim p) \wedge (p \vee q), \quad \text{which simplifies to} \quad p \vee q.$$

3.4 The Conditional and Related Statements

Given a conditional statement $p \rightarrow q$, its **converse, inverse,** and **contrapositive** are defined as follows.

Converse: $q \rightarrow p$

Inverse: $\sim p \rightarrow \sim q$

Contrapositive: $\sim q \rightarrow \sim p$

The conditional $p \rightarrow q$ can be translated in many ways:

If p, then q.	p is sufficient for q.
If p, q.	q is necessary for p.
p implies q.	All p are q.
p only if q.	q if p.

For the **biconditional** statement "p if and only if q,"

$$p \leftrightarrow q \equiv (p \rightarrow q) \wedge (q \rightarrow p).$$

It is true only when p and q have the same truth value.

Consider the statement "If it's a pie, it tastes good."

Converse: "If it tastes good, it's a pie.

Inverse: "If it's not a pie, it doesn't taste good."

Contrapositive: "If it doesn't taste good, it's not a pie."

Statement	If . . . then form
You'll be sorry if I go.	If I go, then you'll be sorry.
Today is Tuesday only if yesterday was Monday.	If today is Tuesday, then yesterday was Monday.
All nurses wear comfortable shoes.	If you are a nurse, then you wear comfortable shoes.

The statement "$5 < 9$ if and only if $3 > 7$" is false because the component statements have opposite truth values. The first is true, while the second is false.

Summary of Basic Truth Tables

1. $\sim p$, the **negation** of p, has truth value opposite that of p.
2. $p \wedge q$, the **conjunction,** is true only when both p and q are true.
3. $p \vee q$, the **disjunction,** is false only when both p and q are false.
4. $p \rightarrow q$, the **conditional,** is false only when p is true and q is false.
5. $p \leftrightarrow q$, the **biconditional,** is true only when both p and q have the same truth value.

p	$\sim p$
T	F
F	T

p	q	$p \wedge q$	$p \vee q$	$p \rightarrow q$	$p \leftrightarrow q$
T	T	T T T	T T T	T T T	T T T
T	F	T F F	T T F	T F F	T F F
F	T	F F T	F T T	F T T	F F T
F	F	F F F	F F F	F T F	F T F

Concepts	Examples

3.5 Analyzing Arguments with Euler Diagrams

A logical **argument** consists of premises and a conclusion. An argument is considered **valid** if the truth of the premises forces the conclusion to be true. Otherwise, it is **invalid.**

Euler diagrams can be used to determine whether an argument is valid or invalid.

To draw a Euler diagram, follow these steps:

1. Use the first premise to draw regions. (Arguments with multiple premises may involve multiple regions.)

2. Place an x in the diagram to represent the subject of the argument.

Consider this argument. Notice the universal quantifier "all."

All dogs are animals.

Dotty is a dog.
_____

Dotty is an animal.

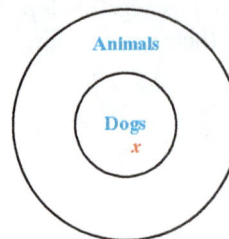

x represents Dotty.

We see from the Euler diagram that the truth of the premises forces the conclusion, that Dotty is an animal, to be true. Thus the argument is valid.

Consider the following argument. Notice the existential quantifier "some."

Some animals are warmblooded.

Albie is an animal.
_____

Albie is warmblooded.

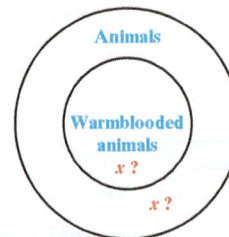

x represents Albie.

We see from the Euler diagram that the location of x is uncertain, so the truth of the premises does not force the conclusion to be true. Thus, the argument is invalid.

3.6 Analyzing Arguments with Truth Tables

An argument with premises $p_1, p_2, \ldots, p_n$ and conclusion c can be tested for validity by constructing a truth table for the statement

$$(p_1 \wedge p_2 \wedge \ldots \wedge p_n) \rightarrow c.$$

If all rows yield T for this statement (that is, if it is a tautology), then the argument is valid. Otherwise, the argument is invalid.

Valid Argument Forms

Modus Ponens	Modus Tollens	Disjunctive Syllogism	Reasoning by Transitivity
$p \rightarrow q$	$p \rightarrow q$	$p \vee q$	$p \rightarrow q$
p	$\sim q$	$\sim p$	$q \rightarrow r$
q	$\sim p$	q	$p \rightarrow r$

Consider this argument. "If I eat ice cream, I regret it later. I didn't regret it later. Therefore, I didn't eat ice cream."

Let p represent "I eat ice cream" and q represent "I regret it later." Then the argument can be expressed as the compound statement

$$[(p \rightarrow q) \wedge \sim q] \rightarrow \sim p.$$

Test its validity using a truth table, which shows that the statement is a tautology. The argument is valid (by **modus tollens**).

p	q	[(p	$\rightarrow$	q)	$\wedge$	$\sim q$]	$\rightarrow$	$\sim p$
T	T				F	F	T	F
T	F	T	F	F	F	T	T	F
F	T						T	T
F	F						T	T

Concepts	Examples

Invalid Argument Forms

Fallacy of the Converse	Fallacy of the Inverse
$p \rightarrow q$	$p \rightarrow q$
q	$\sim p$
p	$\sim q$

Consider this argument.

> If I drink coffee, I get jittery.
> I didn't drink coffee.
> _____
> I don't get jittery.

Let p represent "I drink coffee" and q represent "I get jittery."

Test the argument, using a truth table for the statement

$$[(p \rightarrow q) \wedge \sim p] \rightarrow \sim q.$$

p	q	$[(p$	$\rightarrow$	$q)$	$\wedge$	$\sim p]$	$\rightarrow$	$\sim q$
T	T			T		F	F	T	F
T	F						T	T
F	T			T		T	T	F	F
F	F						T	T

This argument is invalid by **fallacy of the inverse.**

CHAPTER 3   TEST

Write a negation for each statement.

1. $6 - 3 = 3$

2. All men are created equal.

3. Some members of the class went on the field trip.

4. If I fall in love, it will be forever.

5. She applied and did not get a student loan.

Let p represent "You will love me" and let q represent "I will love you." Write each statement in symbols.

6. If you won't love me, then I will love you.

7. I will love you if you will love me.

8. I won't love you if and only if you won't love me.

Using the same statements as for Exercises 6–8, write each of the following in words.

9. $\sim p \wedge q$

10. $\sim (p \vee \sim q)$

In each of the following, assume that p is true and that q and r are false. Find the truth value of each statement.

11. $\sim q \wedge \sim r$

12. $r \vee (p \wedge \sim q)$

13. $r \rightarrow (s \vee r)$ (The truth value of the statement s is unknown.)

14. $p \leftrightarrow (p \rightarrow q)$

15. Explain in your own words why, if p is a statement, the biconditional $p \leftrightarrow \sim p$ must be false.

16. State the necessary conditions for each of the following.

  (a) a conditional statement to be false

  (b) a conjunction to be true

  (c) a disjunction to be false

  (d) a biconditional to be true

Construct a truth table for each of the following.

17. $p \wedge (\sim p \vee q)$

18. $\sim (p \wedge q) \rightarrow (\sim p \vee \sim q)$

Decide whether each statement is true *or* false.

19. Some negative integers are whole numbers.

20. All irrational numbers are real numbers.

Write each conditional statement in if . . . then *form.*

21. All integers are rational numbers.

22. Being a rhombus is sufficient for a polygon to be a quadrilateral.

23. Being divisible by 2 is necessary for a number to be divisible by 4.

24. She digs dinosaur bones only if she is a paleontologist.

*For each statement in Exercises 25 and 26, write (**a**) the converse, (**b**) the inverse, and (**c**) the contrapositive.*

25. If a picture paints a thousand words, the graph will help me understand it.

26. $\sim p \rightarrow (q \wedge r)$ (Use one of De Morgan's laws as necessary.)

27. Use an Euler diagram to determine whether the argument is *valid* or *invalid*.

All members of that athletic club save money.

Don is a member of that athletic club.
_____
Don saves money.

28. Match each argument in parts (a)–(d) in the next column with the law that justifies its validity, or the fallacy of which it is an example, in choices A–F.

A. Modus ponens

B. Modus tollens

C. Reasoning by transitivity

D. Disjunctive syllogism

E. Fallacy of the converse

F. Fallacy of the inverse

(a) If he eats liver, then he'll eat anything.

He eats liver.
_____
He'll eat anything.

(b) If you use your seat belt, you will be safer.

You don't use your seat belt.
_____
You won't be safer.

(c) If I hear *Mr. Bojangles,* I think of her.

If I think of her, I smile.
_____
If I hear *Mr. Bojangles,* I smile.

(d) She sings or she dances.

She does not sing.
_____
She dances.

Use a truth table to determine whether each argument is valid or invalid.

29. If I write a check, it will bounce. If the bank guarantees it, then it does not bounce. The bank guarantees it. Therefore, I don't write a check.

30. $\sim p \rightarrow \sim q$

$\quad q \rightarrow p$
_____
$\quad p \vee q$

Number Theory

5

The modern criminal justice system, like most other areas of government, the military, commerce, and personal life, is highly dependent on computer technology and the mathematical theory and algorithms underlying it. Our money, safety, and identity are protected by public key **cryptography,** which includes the processes of **encrypting** (coding) and **decrypting** (decoding) crucial information. Cryptography involves two kinds of problems:

1. Given two prime numbers, find their product.
2. Given the product, find its two prime factors.

You will learn about prime numbers in **Section 5.1.**

Cryptography systems utilize extremely large prime numbers, which are the subject of **Section 5.2.** The success of encryption depends on the fact that, given today's state of computer hardware and software, and given large enough primes, **Problem 1** above can be done, but **Problem 2** *cannot* be done, even with the most powerful computers available.

OBJECTIVES
1 Identify prime and composite numbers.
2 Apply divisibility tests for natural numbers.
3 Apply the fundamental theorem of arithmetic.

Do not confuse $b\,|\,a$ with b/a. The expression $b\,|\,a$ denotes the **statement** "b divides a." For example, $3\,|\,12$ is a true statement, while $5\,|\,14$ is a false statement. On the other hand, b/a denotes the **operation** "b divided by a." For example, $28/4$ yields the result 7.

The ideas of **even** and **odd natural numbers** are based on the concept of divisibility. A natural number is even if it is divisible by 2 and odd if it is not. Every even number can be written in the form $2k$ (for some natural number k), while every odd number can be written in the form $2k - 1$. Here is another way to say the same thing: 2 divides every even number but fails to divide every odd number. (If a is even, then $2\,|\,a$, whereas if a is odd, then $2\nmid a$.)

For any natural number a, it is true that $a\,|\,a$ and also that $1\,|\,a$.

Primes, Composites, and Divisibility

The famous German mathematician Carl Friedrich Gauss once remarked, "Mathematics is the Queen of Science, and number theory is the Queen of Mathematics." **Number theory** is devoted to the study of the properties of the **natural numbers,** which are also called the **counting numbers** and the **positive integers.**

$$N = \{1, 2, 3, \dots \}$$

A key concept of number theory is the idea of *divisibility*. One counting number is *divisible* by another if dividing the first by the second leaves a remainder 0.

DIVISIBILITY

The natural number a is **divisible** by the natural number b if there exists a natural number k such that $a = bk$. If b divides a, then we write $b\,|\,a$.

Notice that if b divides a, then the quotient a/b or $\frac{a}{b}$ is a natural number. For example, 4 divides 20 because there exists a natural number k such that

$$20 = 4k.$$

The value of k here is 5, because

$$20 = 4 \cdot 5.$$

The natural number 20 is not divisible by 7, since there is no natural number k satisfying $20 = 7k$. Alternatively, "20 divided by 7 gives quotient 2 with remainder 6," and since there is a nonzero remainder, divisibility does not hold. We write $7\nmid 20$ to indicate that 7 does *not* divide 20.

If the natural number a is divisible by the natural number b, then b is a **factor** (or **divisor**) of a, and a is a **multiple** of b. For example, 5 is a factor of 30, and 30 is a multiple of 5. Also, 6 is a factor of 30, and 30 is a multiple of 6. The number 30 equals $6 \cdot 5$. This product $6 \cdot 5$ is called a **factorization** of 30. Other factorizations of 30 include

$$3 \cdot 10, \quad 2 \cdot 15, \quad 1 \cdot 30, \quad \text{and} \quad 2 \cdot 3 \cdot 5.$$

EXAMPLE 1 Checking Divisibility

Decide whether the first number is divisible by the second.

(a) 45; 9 **(b)** 60; 7 **(c)** 19; 19 **(d)** 26; 1

Solution

(a) Is there a natural number k that satisfies $45 = 9k$? The answer is yes, because $45 = 9 \cdot 5$, and 5 is a natural number. Therefore, 9 divides 45, written $9\,|\,45$.

(b) Because the quotient $60 \div 7$ is not a natural number, 60 is not divisible by 7, written $7\nmid 60$.

(c) The quotient $19 \div 19$ is the natural number 1, so 19 is divisible by 19. (*In fact, any natural number is divisible by itself.*)

(d) The quotient $26 \div 1$ is the natural number 26, so 26 is divisible by 1. (*In fact, any natural number is divisible by 1.*)

EXAMPLE 2 Finding Factors

Find all the natural number factors of each number.

(a) 36 **(b)** 50 **(c)** 11

Solution

(a) To find the factors of 36, try to divide 36 by 1, 2, 3, 4, 5, 6, and so on. This gives the following natural number factors of 36: 1, 2, 3, 4, 6, 9, 12, 18, and 36.

(b) The factors of 50 are 1, 2, 5, 10, 25, and 50.

(c) The only natural number factors of 11 are 11 and 1. ■

How to Use Up Lots of Chalk In 1903, the mathematician F. N. Cole presented before a meeting of the American Mathematical Society his discovery of a factorization of the number

$$2^{67} - 1.$$

He walked up to the chalkboard, raised 2 to the 67th power, and then subtracted 1. Then he moved over to another part of the board and multiplied out

$$193{,}707{,}721 \times 761{,}838{,}257{,}287.$$

The two calculations agreed, and Cole received a standing ovation for a presentation that did not include a single word.

PRIME AND COMPOSITE NUMBERS

A natural number greater than 1 that has only itself and 1 as factors is called a **prime number.** A natural number greater than 1 that is not prime is called **composite.**

Mathematicians agree that the natural number 1 is neither prime nor composite.

ALTERNATIVE DEFINITION OF A PRIME NUMBER

A **prime number** is a natural number that has *exactly* two different natural number factors (which clarifies that 1 is not a prime).

There is a systematic method for identifying prime numbers in a list of numbers: 2, 3, . . . , n. The method, known as the **Sieve of Eratosthenes,** is named after the Greek geographer, poet, astronomer, and mathematician (about 276–192 B.C.).

To construct such a sieve, list all the natural numbers from 2 through some given natural number n, such as 100. The number 2 is prime, but all other multiples of 2 (4, 6, 8, 10, and so on) are composite. Circle the prime 2, and cross out all other multiples of 2. The next number not crossed out and not circled is 3, the next prime. Circle the 3, and cross out all other multiples of 3 (6, 9, 12, 15, and so on) that are not already crossed out. Circle the next prime, 5, and cross out all other multiples of 5 not already crossed out. Continue this process for all primes less than or equal to the square root of the last number in the list. For this list, we may stop with 7, because the next prime, 11, is greater than the square root of 100, which is 10. At this stage, simply circle all remaining numbers that are not crossed out.

Table 1 shows the Sieve of Eratosthenes for 2, 3, 4, . . . , 100.

Table 1 Sieve of Eratosthenes

The 25 primes between 1 and 100 are circled.

In the October 1, 1994 issue of *Science News*, Ivars Peterson gave a fascinating account of the discovery of a 75-year-old **factoring machine** ("Cranking Out Primes: Tracking Down a Long-lost Factoring Machine"). In 1989, Jeffrey Shallit of the University of Waterloo in Ontario came across an article in an obscure 1920 French journal, in which the author, Eugene Olivier Carissan, reported his invention of the factoring apparatus. Shallit and two colleagues embarked on a search for the machine. They contacted all telephone subscribers in France named Carissan and received a reply from Eugene Carissan's daughter. The machine was still in existence and in working condition, stored in a drawer at an astronomical observatory in Floirac, near Bordeaux.

Peterson explains in the article how the apparatus works. Using the machine, Carissan took just ten minutes to prove that 708,158,977 is a prime number, and he was able to factor a 13-digit number. While this cannot compare to what technology can accomplish today, it was a significant achievement for Carissan's day.

EXAMPLE 3 **Identifying Prime and Composite Numbers**

Decide whether each number is prime or composite.

(a) 89   **(b)** 83,572   **(c)** 629

Solution

(a) Because 89 is circled in **Table 1,** it is prime. If 89 had a smaller prime factor, 89 would have been crossed out as a multiple of that factor.

(b) The number 83,572 is even, so it is divisible by 2. It is composite.

> *There is only one even prime, the number 2 itself.*

(c) For 629 to be composite, there must be a number other than 629 and 1 that divides into it with remainder 0. Start by trying 2, and then 3. Neither works. There is no need to try 4. (If 4 divides with remainder 0 into a number, then 2 will also.) Try 5. There is no need to try 6 or any succeeding even number. (Why?) Try 7. Try 11. (Why not try 9?) Try 13. Keep trying numbers until one works, or until a number is tried whose square exceeds the given number, 629. Try 17.

$$629 \div 17 = 37$$

The number 629 is composite, since

$$629 = 17 \cdot 37. \qquad \blacksquare$$

An aid in determining whether a natural number is divisible by another natural number is called a **divisibility test. Table 2** shows tests for divisibility by the natural numbers 2 through 12 (except for 7 and 11, which are covered in the exercises).

Table 2   Divisibility Tests for Natural Numbers

Divisible by	Test	Example
2	Number ends in 0, 2, 4, 6, or 8. (The last digit is even.)	9,489,994 ends in 4; it is divisible by 2.
3	Sum of the digits is divisible by 3.	897,432 is divisible by 3, since $8 + 9 + 7 + 4 + 3 + 2 = 33$ is divisible by 3.
4	Last two digits form a number divisible by 4.	7,693,432 is divisible by 4, since 32 is divisible by 4.
5	Number ends in 0 or 5.	890 and 7635 are divisible by 5.
6	Number is divisible by both 2 and 3.	27,342 is divisible by 6 since it is divisible by both 2 and 3.
8	Last three digits form a number divisible by 8.	1,437,816 is divisible by 8, since 816 is divisible by 8.
9	Sum of the digits is divisible by 9.	428,376,105 is divisible by 9 since the sum of the digits is 36, which is divisible by 9.
10	The last digit is 0.	897,463,940 is divisible by 10.
12	Number is divisible by both 4 and 3.	376,984,032 is divisible by 12.

The following program, written by Charles W. Gantner and provided courtesy of Texas Instruments, can be used on the TI-83 Plus calculator to list all primes less than or equal to a given natural number N.

```
PROGRAM: PRIMES
: Disp "INPUT N ≥ 2"
: Disp "TO GET"
: Disp "PRIMES ≤ N"
: Input N
: 2 → T
: Disp T
: 1 → A
: Lbl 1
: A + 2 → A
: 3 → B
: If A > N
: Stop
: Lbl 2
: If B ≤ √(A)
: Goto 3
: Disp A
: Pause
: Goto 1
: Lbl 3
: If A/B ≤ int (A/B)
: Goto 1
: B + 2 → B
: Goto 2
```

```
prgmPRIMES

INPUT N≥2
TO GET
PRIMES ≤ N
?6
```

```
TO GET
PRIMES ≤ N
?6
                    2
                    3
                    5
                 Done
```

The display indicates that the primes less than or equal to 6 are 2, 3, and 5.

EXAMPLE 4 Applying Divisibility Tests

In each case, decide whether the first number is divisible by the second.

(a) 2,984,094; 4   (b) 2,429,806,514; 9

Solution

(a) The last two digits form the number 94. Since 94 is not divisible by 4, the given number is not divisible by 4.

(b) The sum of the digits is

$$2 + 4 + 2 + 9 + 8 + 0 + 6 + 5 + 1 + 4 = 41,$$

which is not divisible by 9. The given number is, therefore, not divisible by 9. ∎

The Fundamental Theorem of Arithmetic

A *composite* number can be thought of as "composed" of smaller factors. For example, 42 is composite since $42 = 6 \cdot 7$. If the smaller factors are all primes, then we have a *prime factorization*. For example, $42 = 2 \cdot 3 \cdot 7$.

THE FUNDAMENTAL THEOREM OF ARITHMETIC

Every natural number can be expressed in one and only one way as a product of primes (if the order of the factors is disregarded). This unique product of primes is called the **prime factorization** of the natural number.

Because a prime natural number is not composed of smaller factors, its prime factorization is simply itself. For example, $17 = 17$.

EXAMPLE 5 Finding the Unique Prime Factorization of a Composite Number

Find the prime factorization of the number 1320.

Solution

We use a "factor tree." The factor tree can start with $1320 = 2 \cdot 660$, as shown below on the left. Then $660 = 2 \cdot 330$, and so on, until every branch of the tree ends with a prime. All the resulting prime factors are shown circled in the diagram.

Alternatively, the same factorization is obtained by repeated division by primes, as shown on the right. (In general, you would divide by the primes 2, 3, 5, 7, 11, and so on, each as many times as possible, until the answer is no longer composite.)

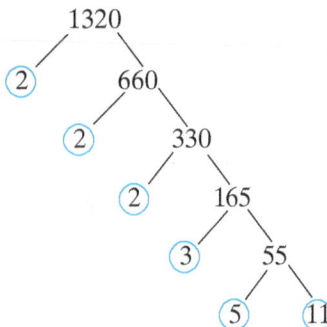

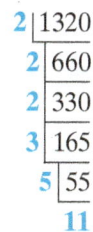

By either method, the prime factorization, in exponential form, is

$$1320 = 2^3 \cdot 3 \cdot 5 \cdot 11. \quad 2 \cdot 2 \cdot 2 = 2^3$$ ∎

FOR FURTHER THOUGHT

Prime Factor Splicing

Prime factor splicing (PFS) has received increasing attention in recent years. A good basic explanation can be found in the article "Home Primes and Their Families," in *Mathematics Teacher*, vol. 107, no. 8, April 2014. Like many topics in number theory, PFS concerns easily stated problems but quickly leads to difficult and unanswered questions. Here is the process:

- Begin with a composite natural number, say 15.
- Obtain its prime factorization: $15 = 3 \cdot 5$
- Form a new natural number by arranging all the prime factors, in ascending order: 35
- Repeat the steps until a prime number is produced, called the **home prime**. For 15, we can summarize the entire process as shown here.

$15 = 3 \cdot 5 \rightarrow 35 = 5 \cdot 7 \rightarrow 57 = 3 \cdot 19 \rightarrow 319 = 11 \cdot 29$

$\rightarrow 1129$ (a prime)

If we call the prime 1129 the "parent," then the composite 15 is its "child." If two or more children are found to have the same parent, call them "siblings." The Sieve of Eratosthenes (see **page 179**) provides plenty of composites to get started. We consider just a few arbitrarily chosen examples here. To do very much prime factor splicing, you would want to use some computer algebra system (CAS) for the factoring. In the case above, the child is 15, the parent is 1129, and the four arrows show that it took four iterations to get from child to parent.

For Group or Individual Investigation

Complete the table, and then answer the questions that follow it.

Child	Number of Iterations	Parent
4	_____	_____
6	_____	_____
9	_____	_____
10	_____	_____
12	_____	_____
22	_____	_____
25	_____	_____
33	_____	_____
46	_____	_____
55	_____	_____

Questions

- Is it possible for a child to have more than one parent?
- Can a parent have more than one child?
- It is possible for PFS to require at least _____ iterations.
- It is possible for a family to include at least _____ siblings.
- Show that if 49 and 77 both have parents, they must be the same.

Interesting Facts

- The child 80 requires 31 iterations and has a 48-digit prime parent.
- 49 and 77 have been "spliced" to at least 110 and 109 iterations, respectively, and no one knows whether a parent exists for them.

5.1   EXERCISES

Decide whether each statement is true *or* false.

1. If n is a natural number and $9\,|\,n$, then $3\,|\,n$.

2. If n is a natural number and $5\,|\,n$, then $10\,|\,n$.

3. Every natural number is divisible by 1.

4. There are no even prime numbers.

5. 1 is the least prime number.

6. Every natural number is both a factor and a multiple of itself.

7. If 16 divides a natural number, then 2, 4, and 8 must also divide that natural number.

8. The prime number 53 has exactly two natural number factors.

Find all natural number factors of each number.

9. 12

10. 20

11. 28

12. 172

Use divisibility tests to decide whether the given number is divisible by each number.

(a) 2   **(b)** 3   **(c)** 4   **(d)** 5   **(e)** 6   **(f)** 8
(g) 9   **(h)** 10   **(i)** 12

13. 321

14. 540

15. 36,360

16. 123,456,789

17. (a) In constructing the Sieve of Eratosthenes for 2 through 100, we said that any composite in that range had to be a multiple of some prime less than or equal to 7 (since the next prime, 11, is greater than the square root of 100). Explain.

(b) To extend the Sieve of Eratosthenes to 200, what is the largest prime whose multiples would have to be considered?

(c) Complete this statement: In seeking prime factors of a given number, we need only consider all primes up to and including the _____ of that number, since a prime factor greater than the _____ can occur only if there is at least one other prime factor less than the _____.

(d) Complete this statement: If no prime less than or equal to $\sqrt{n}$ divides n, then n is a _____ number.

18. (a) Continue the Sieve of Eratosthenes in **Table 1** from 101 to 200 and list the primes between 100 and 200. How many are there?

(b) From your list in part (a), verify that the numbers 197 and 199 are both prime.

19. In your list for **Exercise 18(a),** consider the six largest primes less than 200. Which pairs of these would have products that end in the digit 7?

20. By checking your pairs of primes from **Exercise 19,** give the prime factorization of 35,657.

21. List two primes that are consecutive natural numbers. Can there be any others?

22. Can there be three primes that are consecutive natural numbers? Explain.

23. For a natural number to be divisible by both 2 and 5, what must be true about its last digit?

24. Consider the divisibility tests for 2, 4, and 8 (all powers of 2). Use inductive reasoning to predict the divisibility test for 16. Then use the test to show that 456,882,320 is divisible by 16.

25. Redraw the factor tree of **Example 5,** assuming that you first observe that $1320 = 12 \cdot 110$, then that $12 = 3 \cdot 4$ and $110 = 10 \cdot 11$. Complete the process and give the resulting prime factorization.

26. Explain how your result in **Exercise 25** illustrates the fundamental theorem of arithmetic.

Find the prime factorization of each composite number.

27. 126

28. 825

29. 1183

30. 340

Here is a divisibility test for 7.

(a) *Double the last digit of the given number, and subtract this value from the given number with the last digit omitted.*

(b) *Repeat the process of part (a) as many times as necessary until it is clear whether the number obtained is divisible by 7.*

(c) *If the final number obtained is divisible by 7, then the given number also is divisible by 7. If the final number is not divisible by 7, then neither is the given number.*

Use this divisibility test to determine whether each number is divisible by 7.

31. 226,233

32. 548,184

33. 496,312

34. 368,597

Here is a divisibility test for 11.

(a) *Starting at the left of the given number, add together every other digit.*

(b) *Add together the remaining digits.*

(c) *Subtract the smaller of the two sums from the larger. (If they are the same, the difference is 0.)*

(d) *If the final number obtained is divisible by 11, then the given number also is divisible by 11. If the final number is not divisible by 11, then neither is the given number.*

Use this divisibility test to determine whether each number is divisible by 11.

35. 6,524,846

36. 108,410,313

37. 60,128,459,358

38. 29,630,419,088

39. Consider the divisibility test for the composite number 6, and make a conjecture for the divisibility test for the composite number 15.

40. Give two factorizations of the number 75 that are not prime factorizations.

Determine all possible digit replacements for x so that the first number is divisible by the second. For example, 37,58x is divisible by 2 if

$$x = 0, 2, 4, 6, or\ 8.$$

41. 398,87x; 2   **42.** 2,45x,765; 3

43. 64,537,84x; 4   **44.** 2,143,89x; 5

45. 985,23x; 6   **46.** 23, x54,470; 10

*There is a method to determine the **number of divisors** of a composite number. To do this, write the composite number in its prime factored form, using exponents. Add 1 to each exponent and multiply these numbers. Their product gives the number of divisors of the composite number. For example,*

$$24 = 2^3 \cdot 3 = 2^3 \cdot 3^1.$$

Now add 1 to each exponent:

$$3 + 1 = 4,\ 1 + 1 = 2.$$

Multiply 4 · 2 to get 8. There are 8 divisors of 24. (Because 24 is rather small, this can be verified easily. The divisors are 1, 2, 3, 4, 6, 8, 12, and 24—a total of eight, as predicted.)

Find the number of divisors of each composite number.

47. 105   **48.** 156

49. $5^8 \cdot 29^2$ **50.** $2^4 \cdot 7^2 \cdot 13^3$

Leap years occur when the year number is divisible by 4. An exception to this occurs when the year number is divisible by 100 (that is, it ends in two zeros). In such a case, the number must be divisible by 400 in order for the year to be a leap year. Determine which years are leap years.

51. 1556   **52.** 1990

53. 2200   **54.** 2400

55. Why is the following *not* a valid divisibility test for the number 8? "A number is divisible by 8 if it is divisible by both 4 and 2." Support your answer with an appropriate example.

56. Choose any three consecutive natural numbers, multiply them together, and divide the product by 6. Repeat this several times, using different choices of three consecutive numbers. Make a conjecture concerning the result.

57. Explain why the product of three consecutive natural numbers must be divisible by 6. Include examples in your explanation.

58. Choose any 6-digit number consisting of three digits followed by the same three digits in the same order (for example, 467,467). Divide by 13. Divide by 11. Divide by 7. What do you notice? Why do you think this happens?

*One of the authors has three sons who were born, from eldest to youngest, on August 30, August 31, and October 14. For most (but not all) of each year, their ages are spaced two years from the eldest to the middle and three years from the middle to the youngest. In 2011, their ages were three consecutive prime numbers for a period of exactly 44 days. This same situation had also occurred in exactly two previous years. Use this information for **Exercises 59–62.** (Hint: Consult the Sieve of Eratosthenes on **page 179** for a listing of all primes less than 100.)*

59. Which were the "two previous years" referred to above?

60. What were the years of birth of the three sons?

61. In what year will the same situation next occur?

62. What will be the ages of the three sons at that time?

5.2   LARGE PRIME NUMBERS

OBJECTIVES

1 Understand the infinitude of primes.

2 Investigate several categories of prime numbers.

3 Learn how large primes are identified.

The Infinitude of Primes

One important basic result about prime numbers was proved by Euclid around 300 B.C., namely that there are infinitely many primes. This means that no matter how large a prime we identify, there are always others even larger. Euclid's proof remains today as one of the most elegant proofs in all of mathematics. (An *elegant* mathematical proof is one that demonstrates the desired result in a most direct, concise manner. Mathematicians strive for elegance in their proofs.) It is called a **proof by contradiction.**

A statement can be proved by contradiction as follows: Assume that the negation of the statement is true, and use that assumption to produce some sort of contradiction, or absurdity. Logically, the fact that the negation of the original statement leads to a contradiction means that the original statement must be true.

To better understand a particular part of the proof that there are infinitely many primes, first examine the following argument.

Suppose that $M = 2 \cdot 3 \cdot 5 \cdot 7 + 1 = 211$. Now M is the product of the first four prime numbers, plus 1. If we divide 211 by each of the primes 2, 3, 5, and 7, the remainder is always 1.

$$
\begin{array}{cccc}
105 & 70 & 42 & 30 \\
2\overline{)211} & 3\overline{)211} & 5\overline{)211} & 7\overline{)211} \\
\underline{210} & \underline{210} & \underline{210} & \underline{210} \\
1 & 1 & 1 & 1
\end{array}
$$
All remainders are 1.

Thus 211 is not divisible by any of the primes 2, 3, 5, and 7.

Now we can present Euclid's proof that there are infinitely many primes. If *there is no largest prime number,* then there must be infinitely many primes.

EXAMPLE 1   Proving the Infinitude of Primes

Prove by contradiction that there are infinitely many primes.

Solution

Suppose there is a largest prime number, called P. Form the number M such that

$$M = p_1 \cdot p_2 \cdot p_3 \cdot \cdots \cdot P + 1,$$

where $p_1, p_2, p_3, \ldots, P$ represent all the primes less than or equal to P. Now the number M must be either prime or composite.

1. Suppose that M is prime.
    M is obviously larger than P, so if M is prime, it is larger than the assumed largest prime P. We have reached a *contradiction.*

2. Suppose that M is composite.
    If M is composite, it must have a prime factor. But none of $p_1, p_2, p_3, \ldots, P$ is a factor of M, because division by each will leave a remainder of 1. (Recall the above argument.) So if M has a prime factor, it must be greater than P. But this is a *contradiction,* because P is the assumed largest prime.

In either case 1 or case 2, we reach a contradiction. The whole argument was based on the assumption that a largest prime exists, but this leads to contradictions, so there must be no largest prime, or, equivalently, ***there are infinitely many primes.*** ■

We could never investigate all infinitely many primes directly. So, historically, people have observed properties of the smaller, familiar, ones and then tried to either "disprove" the property (usually by finding counterexamples) or prove it (usually by some deductive argument).

Here is one way to **partition** the natural numbers (divide them into disjoint subsets).

Set A: all natural numbers of the form $4k$

Set B: all natural numbers of the form $4k + 1$

Set C: all natural numbers of the form $4k + 2$

Set D: all natural numbers of the form $4k + 3$

In each case, k is some whole number, except that for set A, k cannot be 0. (Why?) Every natural number is now contained in exactly one of the sets A, B, C, or D. Any number of the form $4k + 4$ would be in subset A, because

$$4k + 4 = 4(k + 1).$$

Any number of the form $4k + 5$ would be in subset B, because

$$4k + 5 = 4(k + 1) + 1, \quad \text{and so on.}$$

The subset A consists of all the multiples of 4. Each of these is divisible by 4; hence, each is not a prime. Now consider C. Because

$$4k + 2 = 2(2k + 1),$$

each member is divisible by 2; hence, each is not a prime (except for 2 itself).

Under this partitioning, all primes other than 2 must be in either B or D.

EXAMPLE 2 Partitioning the Natural Numbers

List the first eight members of each of the infinite sets A, B, C, and D.

Solution

$A = \{4, 8, 12, 16, 20, 24, 28, 32, \dots\};$ $B = \{1, 5, 9, 13, 17, 21, 25, 29, \dots\}$

$C = \{2, 6, 10, 14, 18, 22, 26, 30, \dots\};$ $D = \{3, 7, 11, 15, 19, 23, 27, 31, \dots\}$ ■

As mentioned earlier, sets A and C contain no primes except 2. All other primes must lie in sets B and D. (In fact, there are infinitely many primes in each.)

EXAMPLE 3 Identifying Primes of the Forms $4k + 1$ and $4k + 3$

Identify all primes *specifically listed* in the following infinite sets of **Example 2.**

(a) set B **(b)** set D

Solution

(a) 5, 13, 17, 29   **(b)** 3, 7, 11, 19, 23, 31 ■

Pierre de Fermat (profiled on **page 201**) proved that every prime number of the form $4k + 1$ can be expressed as the sum of two squares.

EXAMPLE 4 Expressing Primes as Sums of Squares

Express each prime in the solution of **Example 3(a)** as a sum of two squares.

Solution

$5 = 1^2 + 2^2;$ $13 = 2^2 + 3^2;$ $17 = 1^2 + 4^2;$ $29 = 2^2 + 5^2$ ■

The Search for Large Primes

Identifying larger and larger prime numbers and factoring large composite numbers into their prime components is of great practical importance today, because it is the basis of modern **cryptography systems.** (See the chapter opener.)

No reasonable formula has ever been found that will consistently generate prime numbers, much less "generate all primes." The most useful attempt, named to honor the French monk Marin Mersenne (1588–1648), follows.

MERSENNE NUMBERS AND MERSENNE PRIMES

For $n = 1, 2, 3, \dots$, the **Mersenne numbers** are those generated by the formula

$$M_n = 2^n - 1.$$

(1) If n is composite, then M_n is also composite.

(2) If n is prime, then M_n may be either prime or composite.

The prime values of M_n are called the **Mersenne primes.** Large primes being verified currently are typically Mersenne primes.

Marin Mersenne (1588–1648), in his *Cogitata Physico-Mathematica* (1644), claimed that M_n was prime for $n = 2, 3, 5, 7, 13, 17, 19, 31, 67, 127,$ and 257, and composite for all other prime numbers n less than 257. Other mathematicians at the time knew that Mersenne could not have actually tested all these values, but no one else could prove or disprove them either. It was more then 300 years before all primes up to 257 were legitimately checked out, and Mersenne was finally revealed to have made five errors:

M_{61} is prime.
M_{67} is composite.
M_{89} is prime.
M_{107} is prime.
M_{257} is composite.

At one time, $2^{11,213} - 1$ was the largest known **Mersenne prime.** To honor its discovery, the Urbana, Illinois, post office used the cancellation picture above.

Long before Mersenne's time, there was general agreement on statement (1) in the box. (**Exercises 23–25** show how to find a factor of $2^n - 1$ whenever n is composite.) However, some early writers did not agree with statement (2), believing instead (incorrectly) that a prime n would always produce a prime M_n.

EXAMPLE 5 Finding Mersenne Numbers

Find each Mersenne number M_n for $n = 2, 3,$ and 5.

Solution

$$M_2 = 2^2 - 1 = 3 \qquad 2^2 = 2 \cdot 2 = 4$$

$$M_3 = 2^3 - 1 = 7 \qquad 2^3 = 2 \cdot 2 \cdot 2 = 8$$

$$M_5 = 2^5 - 1 = 31 \qquad 2^5 = 2 \cdot 2 \cdot 2 \cdot 2 \cdot 2 = 32$$

Note that all three values, 3, 7, and 31, are indeed primes. ∎

It turns out that $M_7 = 2^7 - 1 = 127$ is also a prime (see **Exercise 18(a)** of **Section 5.1**), but it was discovered in 1536 that

$$M_{11} = 2^{11} - 1 = 2047 \quad \text{is not prime (since it is } 23 \cdot 89\text{).}$$

*So prime values of n do **not** always produce prime M_n.* Which prime values of n *do* produce prime Mersenne numbers (the so-called **Mersenne primes**)? No way was ever found to identify, in general, which prime values of n result in Mersenne primes. It is a matter of checking out each prime n value individually—not an easy task given that the Mersenne numbers rapidly become very large.

The Mersenne prime search yielded results slowly. By about 1600, M_n had been verified as prime for all prime n up to 19 (except for 11, as mentioned above). The next one was M_{31}, verified by Euler sometime between 1752 and 1772.

In 1876, French mathematician Edouard Lucas used a clever test he had developed to show that M_{127} (a 39-digit number) is prime. In the 1930s Lucas's method was further simplified by D. H. Lehmer, and the testing of Mersenne numbers for primality has been done ever since with the Lucas-Lehmer test. In 1952 an early computer verified that M_{521}, M_{607}, M_{1279}, M_{2203}, and M_{2281} are primes.

Over the last half century, most new record-breaking primes have been identified by computer algorithms devised and implemented by mathematicians and programmers. In 1996, the **Great Internet Mersenne Prime Search (GIMPS)** was launched and now involves well over 900,000 CPUs worldwide. Of the 48 Mersenne primes presently known, the GIMPS program has discovered the fourteen largest ones. The latest one found (as of summer 2014),

$$M_{57,885,161} = 2^{57,885,161} - 1,$$

has 17,425,170 digits.

During the same general period that Mersenne was thinking about prime numbers, Pierre de Fermat (about 1601–1665) conjectured that the formula

$$2^{2^n} + 1$$

would always produce a prime, for any whole number value of n. **Table 3** on the next page shows how this formula generates the first four **Fermat numbers,** which are all primes. The fifth Fermat number (from $n = 4$) is likewise prime. Fermat had verified these first five by around 1630. But the sixth Fermat number (from $n = 5$) turns out to be 4,294,967,297, which is *not* prime. (See **Exercises 15 and 16.**)

Prime Does Pay Since August 23, 2008, GIMPS has offered participants (individuals or groups) these awards:

- $3000 GIMPS research award for a new Mersenne prime with fewer than 100,000,000 (decimal) digits
- $50,000 for the first prime discovered with at least 100,000,000 (decimal) digits

This second award is one-third of $150,000 offered by the Electronic Frontier Foundation (EFF). GIMPS plans to retain another $50,000 for expenses and the final $50,000 to fund past and/or future awards. EFF seeks, by its sponsorship, to "encourage the harmonious integration of Internet innovation into the whole of society."

Distributed computing is a way of achieving great computer power by having lots of individual machines do separate parts of the computation. One example is the **Great Internet Mersenne Prime Search (GIMPS),** described in this section. Another example is **SETI@home (Search for Extraterrestrial Intelligence),** which assigns the analysis of signal data from small patches of the "sky" to participants. (The movie *Contact* was fiction, but still the search goes on.)

A third example, **Folding@home,** based at Stanford University, investigates the folding of proteins in living organisms into complex shapes and how they interact with other biological molecules.

In July 2014, **Prime Grid,** another distributed computing project, making use of the Berkeley Open Infrastructure for Network Computing (BOINC), confirmed the largest Fermat number known to be composite,

$$F_{3329780},$$

by discovering its prime factor

$$193 \times 2^{3329782} + 1,$$

a prime with more than one million decimal digits.

Source: www.wikipedia.org

As of late 2014, no more Fermat primes had been found. All F_n were known to be composite for $5 \le n \le 32$, but only F_0 to F_{11} had been completely factored.

Table 3 The Generation of Fermat Numbers

n	2^n	2^{2^n}	$2^{2^n} + 1$
0	1	2	3
1	2	4	5
2	4	16	17
3	8	256	257

Of historical interest are a couple of polynomial formulas that produce primes. (A *polynomial* in a given variable involves adding or subtracting integer multiples of whole number powers of the variable. Discussed in **Section 7.6,** polynomials are among the most basic mathematical functions.) In 1732, Leonhard Euler offered the formula

$$n^2 - n + 41, \quad \text{Euler's formula}$$

which generates primes for n up to 40 and fails at $n = 41$. In 1879, E. B. Escott produced more primes with the formula

$$n^2 - 79n + 1601, \quad \text{Escott's formula}$$

which first fails at $n = 80$.

EXAMPLE 6 **Finding Numbers Using Euler's and Escott's Formulas**

Find the first five numbers produced by each of the polynomial formulas of Euler and Escott.

Solution
Table 4 shows the required numbers.

Table 4 A Few Polynomial-Generated Prime Numbers

n	Euler formula $n^2 - n + 41$	Escott formula $n^2 - 79n + 1601$
1	41	1523
2	43	1447
3	47	1373
4	53	1301
5	61	1231

All values found here are primes. (Use **Table 1** to verify the Euler values.) ■

Actually, it is not hard to prove that there can be no polynomial that will consistently generate primes. More complicated mathematical formulas exist for generating primes, but none produced so far can be practically applied in a reasonable amount of time, even using the fastest computers.

Imagine working for the U.S. Department of Justice as part of the effort to control cybercrime. Your work would surely involve Internet security. A recent ransomware scheme involved encrypting victims' computer files and then demanding payment for the decryption key to regain access.)

The chapter opener described the central role of multiplying large primes and factoring their products in public key cryptography. Another key mathematical component is **modular systems.** The partitioning of the natural numbers described on **page 185** is an example. In that case every natural number n was, in a sense, equated with one of just four numbers, 0, 1, 2, 3 — the remainders when n is divided by 4. In general, the divisor, any natural number greater than 1, is called the modulus (m), and the remainder, called the **residue,** is denoted n (mod m). For example,

$$849{,}657{,}221 \ (\text{mod } 215)$$

denotes the remainder when 849,657,221 is divided by 215. Using a calculator may produce

$$\frac{849{,}657{,}221}{215} = 3951894.051.$$

Subtract the integer part, 3951894. (We want the remainder only, represented by the fractional part.) Now the calculator may display

$$.0511628.$$

Multiply this result by the divisor, 215, to obtain 11. (Round if necessary — A slight error may result from the calculator's inherent limited number of significant digits.) The desired remainder, or residue, is 11. This establishes that

$$849{,}657{,}221 = 3{,}951{,}894 \cdot 215 + 11.$$

Now consider an even larger number. Try to find the remainder when 11^{14} is divided by 18. The number 11^{14} is so large that the calculator displays it in exponential notation:

$$11^{14} = 3.797498336\text{E}14$$

and

$$\frac{11^{14}}{18} = 2.109721298\text{E}13.$$

The calculator cannot display enough significant digits to reveal the integer portion of the result, so we cannot proceed as before. But we can apply a rule of exponents (see **Chapter 7**) and modular arithmetic as follows.

$$11^{14} \text{ (mod 18)} = 11^{7+7} \text{ (mod 18)} \qquad \text{\small 14 = 7 + 7}$$

$$= (11^7 \cdot 11^7) \text{ (mod 18)} \qquad \text{\small Rule of exponents}$$

$$= [11^7 \text{ (mod 18)}] \cdot [11^7 \text{ (mod 18)}] \qquad \text{\small Modular arithmetic}$$

Now

$$11^7 = 19{,}487{,}171 \qquad \text{\small Use a calculator.}$$

$$\frac{19{,}487{,}171}{18} = 1082620.6\overline{1}$$

and

$$.6\overline{1} \cdot 18 = 11.$$

Therefore,

$$11^7 \text{ (mod 18)} = 11$$

$$11^{14} \text{ (mod 18)} = 11 \cdot 11 = 121$$

$$\frac{121}{18} = 6.7\overline{2}$$

$$.7\overline{2} \cdot 18 = 13.$$

The desired remainder, or residue, is 13.

These examples give a hint of the mathematics involved. But in practice the numbers would be much, much larger, and high-powered computers would be used.

5.2   EXERCISES

In Exercises 1–6 decide whether each statement is true *or* false.

1. A proof by contradiction assumes the negation of a statement and proceeds until a contradiction is encountered.

2. Marin Mersenne gave the first proof that there are infinitely many prime numbers.

3. If n is prime, then $2^n - 1$ must be prime also.

4. There are infinitely many Fermat numbers.

5. There can be no polynomial formula that consistently generates prime numbers.

6. As of late 2014, only five Fermat primes had ever been found.

7. Find the next three primes, of the form $4k + 1$, *not* listed specifically in **Example 3(a),** and express them as sums of squares.

8. Recall the first few perfect squares: 1, 4, 9, 16, 25. Try writing the numbers of the form $4k + 3$ listed in **Example 3(b)** as sums of two squares. Then complete this statement: The primes tested, of the form $4k + 3$, _____ be expressed as the sum of two squares. (Fermat claimed, but did not prove, that *no* prime of the form $4k + 3$ was the sum of two squares. Euler proved it 100 years later.)

9. Explain how the expressions $4k$, $4k + 1$, $4k + 2$, and $4k + 3$ serve to "partition" the natural numbers.

10. Explain why **Example 4** does not prove that every prime of the form $4k + 1$ can be expressed as a sum of two squares.

11. Evaluate Euler's polynomial formula for each of the following, and determine whether each value is prime or composite. If composite, give the prime factorization.

    (a) $n = 41$

    (b) $n = 42$

    (c) $n = 43$

12. Consider your answers for **Exercise 11,** and choose the correct completion: For $n > 41$, Euler's formula produces a prime

    A. never.   **B.** sometimes.   **C.** always.

13. Evaluate Escott's polynomial formula for each of the following, and determine whether each value is prime or composite. If composite, give the prime factorization.

    (a) $n = 80$

    (b) $n = 81$

    (c) $n = 82$

14. Consider your answers for **Exercise 13,** and choose the correct completion: For $n > 80$, Escott's formula produces a prime

    A. never.   **B.** sometimes.   **C.** always.

15. (a) Evaluate the Fermat number F_4: $2^{2^n} + 1$ for $n = 4$.

    (b) In seeking possible prime factors of the Fermat number of part (a), what is the largest potential prime factor that one would have to try? (As stated in the text, F_4 is in fact prime.)

16. (a) Verify the value given in the text for the "sixth" Fermat number ($2^{2^5} + 1$).

    (b) Divide this Fermat number by 641 and express it in factored form. (Euler discovered this factorization in 1732, proving that the sixth Fermat number is not prime.)

17. The 48th Mersenne prime was identified in the text. Write a short report on when, how, and by whom it was found.

18. The margin note on Marin Mersenne on **page 186** cites a 1644 claim that was not totally resolved for some 300 years. Find out when, and by whom, Mersenne's five errors were demonstrated. (*Hint:* One was mentioned in the margin note on **page 179**.)

19. Why do you suppose it normally takes up to a few years to discover each new Mersenne prime?

20. In Euclid's proof that there is no largest prime, we formed a number M by taking the product of primes and adding 1. Observe the pattern below.

$M = 2 + 1 = 3$	(3 is prime)
$M = 2 \cdot 3 + 1 = 7$	(7 is prime)
$M = 2 \cdot 3 \cdot 5 + 1 = 31$	(31 is prime)
$M = 2 \cdot 3 \cdot 5 \cdot 7 + 1 = 211$	(211 is prime)
$M = 2 \cdot 3 \cdot 5 \cdot 7 \cdot 11 + 1 = 2311$	(2311 is prime)

    It may seem as though this pattern will always yield a prime number. Now evaluate

    $$M = 2 \cdot 3 \cdot 5 \cdot 7 \cdot 11 \cdot 13 + 1.$$

21. Is the final value of M computed in **Exercise 20** prime or composite? If it is composite, give its prime factorization.

22. Explain in your own words the proof by Euclid that there is no largest prime.

The text stated that the Mersenne number M_n is composite whenever n is composite. Exercises 23–26 develop one way in which you can always find a factor of such a Mersenne number.

23. For the composite number $n = 6$, find

    $$M_n = 2^n - 1.$$

24. Notice that $p = 3$ is a prime factor of $n = 6$. Find $2^p - 1$ for $p = 3$. Is $2^p - 1$ a factor of $2^n - 1$?

25. Complete this statement: If p is a prime factor of n, then _____ is a factor of the Mersenne number $2^n - 1$.

26. Find $M_n = 2^n - 1$ for $n = 10$.

27. Use the statement of **Exercise 25** to find two distinct factors of M_{10}.

28. Do you think this procedure will always produce *prime* factors of M_n for composite n? (*Hint:* Consider $n = 22$ and its prime factor $p = 11$, and recall the statement following **Example 5.**) Explain.

29. Explain why large prime numbers are important in modern cryptography systems.

30. Describe the difference between Mersenne *numbers* and Mersenne *primes*.

5.3 SELECTED TOPICS FROM NUMBER THEORY

OBJECTIVES

1 Understand and identify perfect numbers.

2 Understand and identify deficient and abundant numbers.

3 Understand amicable (friendly) numbers.

4 State and evaluate Goldbach's conjecture.

5 Understand and identify twin primes.

6 State and evaluate Fermat's Last Theorem.

The mathematician **Albert Wilansky,** when phoning his brother-in-law, Mr. Smith, noticed an interesting property concerning Smith's phone number (493–7775). The number 4,937,775 is composite, and its prime factorization is

$$3 \cdot 5 \cdot 5 \cdot 65{,}837.$$

When the digits of the phone number are added, the result, 42, is equal to the sum of the digits in the prime factors:

$$3 + 5 + 5 + 6 + 5 + 8 + 3 + 7 = 42.$$

Wilansky termed such a number a **Smith number.** In 1985 it was proved that there are infinitely many Smith numbers, but there still are many unanswered questions about them.

Perfect Numbers

In **Chapter 1,** we introduced figurate numbers, a topic investigated by the Pythagoreans, a group of Greek mathematicians and musicians who held their meetings in secret. In this section we examine some of the other special numbers that fascinated the Pythagoreans and are still studied by mathematicians today.

Divisors of a natural number were covered in **Section 5.1.** The **proper divisors** of a natural number include all divisors of the number except the number itself. For example, the proper divisors of 8 are 1, 2, and 4. (8 is *not* a proper divisor of 8.)

> **PERFECT NUMBERS**
>
> A natural number is said to be **perfect** if it is equal to the sum of its proper divisors.

Is 8 perfect? No, because $1 + 2 + 4 = 7$, and $7 \neq 8$. The least perfect number is 6, because the proper divisors of 6 are 1, 2, and 3, and

$$1 + 2 + 3 = 6. \quad \text{6 is perfect.}$$

EXAMPLE 1 Verifying a Perfect Number

Show that 28 is a perfect number.

Solution

The proper divisors of 28 are 1, 2, 4, 7, and 14. The sum of these is 28:

$$1 + 2 + 4 + 7 + 14 = 28.$$

By the definition, 28 is perfect. ∎

The numbers 6 and 28 are the two least perfect numbers. The next two are 496 and 8128. The pattern of these first four perfect numbers led early writers to conjecture that

1. The *n*th perfect number contains exactly *n* digits.
2. The even perfect numbers end in the digits 6 and 8, alternately.

> Conjectures **NOT NECESSARILY TRUE**

(**Exercises 39–41** will help you analyze these conjectures.)

There still are many unanswered questions about perfect numbers. Euclid showed that the following is true.

If $2^n - 1$ is prime, then $2^{n-1}(2^n - 1)$ is perfect, and conversely.

Because the prime values of $2^n - 1$ are the Mersenne primes (discussed in the previous section), this means that for every new Mersenne prime discovered, another perfect number is automatically revealed. (Hence, as of summer 2014, there were 48 known perfect numbers.) Also, it is known that the following is true.

All even perfect numbers must take the form $2^{n-1}(2^n - 1)$.

It is strongly suspected that no odd perfect numbers exist. (Any odd one would have at least eight different prime factors and would have at least 300 decimal digits.) Therefore, Euclid and the early Greeks most likely identified the form of all perfect numbers.

Deficient and Abundant Numbers

Earlier we saw that 8 is not perfect because it is not equal to the sum of its proper divisors ($8 \neq 7$). Next we define two alternative categories for natural numbers that are *not* perfect.

DEFICIENT AND ABUNDANT NUMBERS

A natural number is **deficient** if it is greater than the sum of its proper divisors. It is **abundant** if it is less than the sum of its proper divisors.

Based on this definition, a *deficient number* is one with proper divisors that add up to less than the number itself, while an *abundant number* is one with proper divisors that add up to more than the number itself. For example, because the proper divisors of 8 (1, 2, and 4) add up to 7, which is less than 8, the number 8 is deficient.

EXAMPLE 2 Identifying Deficient and Abundant Numbers

Decide whether each number is deficient or abundant.

(a) 12   **(b)** 10

Solution

(a) The proper divisors of 12 are 1, 2, 3, 4, and 6. The sum of these divisors is 16. Because $16 > 12$, the number 12 is abundant.

(b) The proper divisors of 10 are 1, 2, and 5. Since $1 + 2 + 5 = 8$, and $8 < 10$, the number 10 is deficient.   ∎

Amicable (Friendly) Numbers

Suppose that we add the proper divisors of 284.

$$1 + 2 + 4 + 71 + 142 = 220$$

Their sum is 220. Now, add the proper divisors of 220.

$$1 + 2 + 4 + 5 + 10 + 11 + 20 + 22 + 44 + 55 + 110 = 284$$

The sum of the proper divisors of 220 is 284, while the sum of the proper divisors of 284 is 220. Number pairs with this property are said to be *amicable*, or *friendly*.

An extension of the idea of amicable numbers results in **sociable numbers.** In a chain of sociable numbers, the sum of the proper divisors of each number is the next number in the chain, and the sum of the proper divisors of the last number in the chain is the first number. Here is a 5-link chain of sociable numbers:

12,496

14,288

15,472

14,536

14,264.

The number 14,316 starts a 28-link chain of sociable numbers.

AMICABLE OR FRIENDLY NUMBERS

The natural numbers a and b are **amicable, or friendly,** if the sum of the proper divisors of a is b, and the sum of the proper divisors of b is a.

The smallest pair of amicable numbers, 220 and 284, was known to the Pythagoreans, but it was not until 1636 that Fermat found the next pair, 17,296 and 18,416. Many more pairs were found over the next few decades, but it took a 16-year-old Italian boy named Nicolo Paganini to discover, in the year 1866, that the pair of amicable numbers 1184 and 1210 had been overlooked for centuries!

Today, powerful computers continually extend the lists of known amicable pairs. The last time we checked, nearly twelve million pairs were known. It still is unknown, however, whether there are infinitely many such pairs. No one has found an amicable pair without prime factors in common, but the possibility of such a pair has not been eliminated.

A Dull Number? The Indian mathematician **Srinivasa Ramanujan** (1887–1920) developed many ideas in number theory. His friend and collaborator on occasion was G. H. Hardy, also a number theorist and professor at Cambridge University in England.

A story has been told about Ramanujan that illustrates his genius. Hardy once mentioned to Ramanujan that he had just taken a taxicab with a rather dull number: 1729. Ramanujan countered by saying that this number isn't dull at all; it is the smallest natural number that can be expressed as the sum of two cubes in two different ways:

$$1^3 + 12^3 = 1729$$
and $$9^3 + 10^3 = 1729.$$

Show that 85 can be written as the sum of two *squares* in two ways.

Mathematics professor Gregory Larkin, played by Jeff Bridges, woos colleague Rose Morgan (Barbra Streisand) in the 1996 film *The Mirror Has Two Faces*. Larkin's research and book focus on the **twin prime conjecture,** which he correctly states in a dinner scene. He is amazed that his nonmathematician friend actually understands what he is talking about.

Goldbach's Conjecture

The mathematician Christian Goldbach (1690–1764) stated the following conjecture (guess), which is one of the most famous unsolved problems in mathematics. Mathematicians have tried to prove the conjecture but have not succeeded. However, the conjecture has been verified (as of late 2014) for numbers up to 4×10^{18}.

GOLDBACH'S CONJECTURE (NOT PROVED)

Every even number greater than 2 can be written as the sum of two prime numbers.

Examples: $8 = 5 + 3$
$10 = 5 + 5$ (or $10 = 7 + 3$)

EXAMPLE 3 **Expressing Numbers as Sums of Primes**

Write each even number as the sum of two primes.

(a) 18   **(b)** 60

Solution

(a) $18 = 5 + 13$. Another way of writing it is $7 + 11$. Notice that $1 + 17$ is *not* valid because by definition 1 is not a prime number.

(b) $60 = 7 + 53$. Can you find other ways? Why is $3 + 57$ not valid? ∎

Twin Primes

Prime numbers that differ by 2 are called **twin primes.** Some twin prime pairs are

3 and 5,   5 and 7,   11 and 13,   and so on.

Like Goldbach's conjecture, the following conjecture about twin primes has never been proved. Interestingly, however, on May 13, 2013, substantial progress was made toward proofs of *both* the twin prime conjecture and Goldbach's conjecture. Two mathematicians, one in New Hampshire and one in Paris, proved conditions that get us significantly closer to proofs of the two famous conjectures.

TWIN PRIME CONJECTURE (NOT PROVED)

There are infinitely many pairs of twin primes.

You may wish to verify that there are eight such pairs less than 100, using the Sieve of Eratosthenes in **Table 1.** As of summer 2014, the largest known twin primes were

$$3{,}756{,}801{,}695{,}685 \cdot 2^{666{,}669} \pm 1. \quad \text{Each contains 200,700 digits.}$$

Recall from **Section 5.2** that Euclid's proof of the infinitude of primes used numbers of the form

$$p_1 \cdot p_2 \cdot p_3 \cdots \cdot p_n + 1,$$

that is, the product of the first n primes, plus 1. It may seem that any such number must be prime, but that is not so. (See **Exercises 20 and 21** of **Section 5.2.**) However, this form often does produce primes (as does the same form with the plus replaced by a minus).

Sophie Germain (1776–1831) studied at the École Polytechnique in Paris in a day when female students were not admitted. A **Sophie Germain prime** is a prime p for which $2p + 1$ also is prime. Lately, large Sophie Germain primes have been discovered at the rate of one or more per year. As of late 2014, the largest one known was $18{,}543{,}637{,}900{,}515 \cdot 2^{666{,}667} - 1$, which has 200,701 digits.

(*Source:* www.primes.utm.edu)

The popular animated television series *The Simpsons* provides not only humor and social commentary but also lessons in mathematics. One episode depicted the equation

$$1782^{12} + 1841^{12} = 1922^{12},$$

which, according to **Fermat's Last Theorem,** cannot be true. Your calculator may indicate that the equation is true, but this is because it cannot accurately display powers of this size. Actually, 1782^{12} must be an *even* number because *an even number to any power is even.* Also, 1841^{12} must be *odd,* because *an odd number to any power is odd.* So the sum on the left must be *odd,* because

$$even + odd = odd.$$

Similarly, 1922^{12} must be *even.* So the equation states that an odd number equals an even number, which is impossible. (See www.simpsonsmath.com)

When *all* the primes up to p_n are included, the resulting numbers, if prime, are called **primorial primes.** They are denoted

$$p\# \pm 1.$$

For example, $5\# + 1 = 2 \cdot 3 \cdot 5 + 1 = 31$ is a primorial prime. (In late 2014, the largest known primorial prime was $1{,}098{,}133\# - 1$, a number with 476,311 digits.) The primorial primes are a popular place to look for twin primes.

EXAMPLE 4   Verifying Twin Primes

Verify that the primorial formula $p\# \pm 1$ produces twin prime pairs for both **(a)** $p = 3$ and **(b)** $p = 5$.

Solution

(a) $3\# \pm 1 = 2 \cdot 3 \pm 1 = 6 \pm 1 = 5$ and 7 Twin primes

> Multiply, then add and subtract.

(b) $5\# \pm 1 = 2 \cdot 3 \cdot 5 \pm 1 = 30 \pm 1 = 29$ and 31 Twin primes ∎

Fermat's Last Theorem

In any right triangle with shorter sides (legs) a and b, and longest side (hypotenuse) c, the equation $a^2 + b^2 = c^2$ will hold true. This is the famous Pythagorean theorem. For example,

$$3^2 + 4^2 = 5^2 \qquad a = 3,\ b = 4,\ c = 5$$
$$9 + 16 = 25 \qquad \text{Apply the exponents.}$$
$$25 = 25. \qquad \text{True}$$

It is known that there are infinitely many such triples (a, b, c) that satisfy the equation $a^2 + b^2 = c^2$. Is something similar true of the equation

$$a^n + b^n = c^n$$

for natural numbers $n \geq 3$? Pierre de Fermat, who is profiled in a margin note on **page 201,** thought that not only were there not infinitely many such triples, but that there were, in fact, none. He made the following claim in the 1600s.

> **FERMAT'S LAST THEOREM (PROVED IN THE 1990s)**
>
> For *any* natural number $n \geq 3$, there are *no* triples (a, b, c) that satisfy the equation
>
> $$a^n + b^n = c^n.$$

Fermat's assertion was the object of some 350 years of attempts by mathematicians to provide a suitable proof. While it was verified for many specific cases (Fermat himself proved it for $n = 3$), a proof of the general case could not be found until the Princeton mathematician Andrew Wiles announced a proof in the spring of 1993. Although some flaws were discovered in his argument, Wiles was able, by the fall of 1994, to repair the proof.

There were probably about 100 mathematicians around the world qualified to understand the Wiles proof. ***Today Fermat's Last Theorem finally is regarded by the mathematics community as officially proved.***

EXAMPLE 5   Applying a Theorem Proved by Fermat

One of the theorems legitimately proved by Fermat is as follows:

Every odd prime can be expressed as the difference of two squares in one and only one way.

Express each odd prime as the difference of two squares.

(a) 3   **(b)** 7

Solution

(a) $3 = 4 - 1 = 2^2 - 1^2$ **(b)** $7 = 16 - 9 = 4^2 - 3^2$ ■

FOR FURTHER THOUGHT

Curious and Interesting

One of the most remarkable books on number theory is *The Penguin Dictionary of Curious and Interesting Numbers* (1986) by David Wells. This book contains fascinating numbers and their properties, including the following.

- There are only three sets of three digits that form prime numbers in all possible arrangements: {1, 1, 3}, {1, 9, 9}, {3, 3, 7}.

- Find the sum of the cubes of the digits of 136:

$$1^3 + 3^3 + 6^3 = 244.$$

Repeat the process with the digits of 244:

$$2^3 + 4^3 + 4^3 = 136.$$

We're back to where we started.

- 635,318,657 is the least number that can be expressed as the sum of two fourth powers in two ways:

$$635{,}318{,}657 = 59^4 + 158^4 = 133^4 + 134^4.$$

- The number 24,678,050 has an interesting property:

$$24{,}678{,}050 = 2^8 + 4^8 + 6^8 + 7^8 + 8^8 + 0^8$$
$$+ 5^8 + 0^8.$$

- The number 54,748 has a similar interesting property:

$$54{,}748 = 5^5 + 4^5 + 7^5 + 4^5 + 8^5.$$

- The number 3435 has this property:

$$3435 = 3^3 + 4^4 + 3^3 + 5^5.$$

For anyone whose curiosity is piqued by such facts, the book mentioned above is for you!

For Group or Individual Investigation

Have each student in the class choose a three-digit number that is a multiple of 3. Add the cubes of the digits. Repeat the process until the same number is obtained over and over. Then, have the students compare their results. What is curious and interesting about this process?

5.3   EXERCISES

In Exercises 1–10 decide whether each statement is true *or* false.

1. Given a prime number, no matter how large, there is always another prime even larger.

2. The prime numbers 2 and 3 are twin primes.

3. The first and third perfect numbers both end in the digit 6, and the second and fourth perfect numbers both end in the digits 28.

4. For every Mersenne prime, there is a corresponding perfect number.

5. All prime numbers are deficient.

6. The equation $17 + 51 = 68$ verifies Goldbach's conjecture for the number 68.

7. Even perfect numbers are more plentiful than Mersenne primes.

8. The twin prime conjecture was proved in 2013.

9. The number $2^5(2^6 - 1)$ is perfect.

10. Every natural number greater than 1 must be one of the following: prime, abundant, or deficient.

11. The proper divisors of 496 are 1, 2, 4, 8, 16, 31, 62, 124, and 248. Use this information to verify that 496 is perfect.

12. The proper divisors of 8128 are 1, 2, 4, 8, 16, 32, 64, 127, 254, 508, 1016, 2032, and 4064. Use this information to verify that 8128 is perfect.

13. As mentioned in the text, when $2^n - 1$ is prime,

$$2^{n-1}(2^n - 1)$$

is perfect. By letting $n = 2, 3, 5$, and 7, we obtain the first four perfect numbers. Show that $2^n - 1$ is prime for $n = 13$, and then find the decimal digit representation for the fifth perfect number.

14. In the summer of 2014, the largest known prime number was $2^{57,885,161} - 1$. Use the formula in **Exercise 13** to write an expression for the perfect number generated by this prime number.

15. It has been proved that the reciprocals of *all* the positive divisors of a perfect number have a sum of 2. Verify this for the perfect number 28.

16. Consider the following equations.

$$6 = 1 + 2 + 3$$
$$28 = 1 + 2 + 3 + 4 + 5 + 6 + 7$$

Show that a similar equation is valid for the third perfect number, 496.

Determine whether each number is abundant *or* deficient.

17. 32   **18.** 60   **19.** 84   **20.** 75

21. There are four abundant numbers between 1 and 25. Find them. (*Hint:* They are all even, and no prime number is abundant.)

22. Explain why a prime number must be deficient.

23. There are no odd abundant numbers less than 945. If 945 has proper divisors 1, 3, 5, 7, 9, 15, 21, 27, 35, 45, 63, 105, 135, 189, and 315, determine whether it is abundant or deficient.

24. Explain in your own words the terms *perfect number, abundant number,* and *deficient number.*

25. The proper divisors of 1184 are 1, 2, 4, 8, 16, 32, 37, 74, 148, 296, and 592. The proper divisors of 1210 are 1, 2, 5, 10, 11, 22, 55, 110, 121, 242, and 605. Verify that 16-year-old Nicolo Paganini was correct about these two numbers in 1866. See the subsection "Amicable (Friendly) Numbers."

26. An Arabian mathematician of the ninth century stated the following: "If the three numbers

$$x = 3 \cdot 2^{n-1} - 1,$$
$$y = 3 \cdot 2^n - 1,$$

and $\quad z = 9 \cdot 2^{2n-1} - 1$

are all prime and $n \geq 2$, then $2^n xy$ and $2^n z$ are amicable numbers."

(a) Use $n = 2$, and show that the result is the least pair of amicable numbers, namely 220 and 284.

(b) Use $n = 4$ to obtain another pair of amicable numbers.

Write each even number as the sum of two primes. (There may be more than one way to do this.)

27. 12   **28.** 24   **29.** 40   **30.** 54

31. Joseph Louis Lagrange (1736–1813) conjectured that every odd natural number greater than 5 can be written as a sum $a + 2b$, where a and b are both primes.

(a) Verify this for the odd natural number 11.

(b) Verify that the odd natural number 17 can be written in this form in four different ways.

32. Another unproved conjecture in number theory states that every natural number multiple of 6 can be written as the difference of two primes. Verify this for 12 and 24.

Find one pair of twin primes between the two numbers given.

33. 45, 65   **34.** 65, 85

Pierre de Fermat provided proofs of many theorems in number theory. Exercises 35–38 investigate some of these theorems.

35. If p is prime and the natural numbers a and p have no common factor except 1, then $a^{p-1} - 1$ is divisible by p.

(a) Verify this for $p = 5$ and $a = 3$.

(b) Verify this for $p = 7$ and $a = 2$.

36. Every odd prime can be expressed as the difference of two squares in one and only one way.

(a) Find this one way for the prime number 5.

(b) Find this one way for the prime number 11.

37. There is only one solution in natural numbers for $a^2 + 2 = b^3$, and it is $a = 5, b = 3$. Verify this solution.

38. There are only two solutions in integers for $a^2 + 4 = b^3$. One solution is $a = 2, b = 2$. Find the other solution.

The first four perfect numbers were identified in the text: 6, 28, 496, and 8128. The next two are 33,550,336 and 8,589,869,056. Use this information about perfect numbers to work Exercises 39–41.

39. Verify that each of these six perfect numbers ends in either 6 or 28. (In fact, this is true of all even perfect numbers.)

40. Is conjecture (1) in the text (that the nth perfect number contains exactly n digits) true or false? Explain.

41. Is conjecture (2) in the text (that the even perfect numbers end in the digits 6 and 8, alternately) true or false? Explain.

*According to the Web site www.shyamsundergupta.com/ amicable.htm, a natural number is **happy** if the process of repeatedly summing the squares of its decimal digits finally ends in 1. For example, the least natural number (greater than 1) that is happy is 7, as shown here.*

$$7^2 = 49, \quad 4^2 + 9^2 = 97, \quad 9^2 + 7^2 = 130,$$

$$1^2 + 3^2 + 0^2 = 10, \quad 1^2 + 0^2 = 1.$$

*An amicable pair is a **happy amicable pair** if and only if both members of the pair are happy numbers. (The first 5000 amicable pairs include only 111 that are happy amicable pairs.) For each amicable pair, determine whether neither, one, or both of the members are happy, and whether the pair is a happy amicable pair.*

42. 220 and 284

43. 1184 and 1210

44. 10,572,550 and 10,854,650

45. 35,361,326 and 40,117,714

46. If the early Greeks knew the form of all even perfect numbers, namely $2^{n-1}(2^n - 1)$, then why did they not discover all the ones that are known today?

47. Explain why the primorial formula $p\# \pm 1$ does not result in a pair of twin primes for the prime value $p = 2$.

48. (a) What two numbers does the primorial formula produce for $p = 7$?

  (b) Which, if either, of these numbers is prime?

49. Choose the correct completion: The primorial formula produces twin primes

  A. never. **B.** sometimes. **C.** always.

*See the margin note (on **page 195**) defining a Sophie Germain prime, and complete this table.*

	p	$2p+1$	Is p a Sophie Germain prime?
50.	2	____	____
51.	3	____	____
52.	5	____	____
53.	7	____	____

Factorial primes *are of the form $n! \pm 1$ for natural numbers n. ($n!$ denotes "n factorial," the product of all natural numbers up to n, not just the primes as in the primorial primes. For example, $4! = 1 \cdot 2 \cdot 3 \cdot 4 = 24$.) As of late 2014, the largest verified factorial prime was $150,209! + 1$, which has 712,355 digits. Find the missing entries in the following table.*

n	$n!$	$n! - 1$	$n! + 1$	Is $n! - 1$ prime?	Is $n! + 1$ prime?
2	2	1	3	no	yes
54. 3	____	____	____	____	____
55. 4	____	____	____	____	____
56. 5	____	____	____	____	____

57. Explain why the factorial prime formula does not give twin primes for $n = 2$.

Based on the preceding table, complete each statement with one of the following:

A. *never,* **B.** *sometimes, or* **C.** *always.*

When applied to particular values of n, the factorial prime formula $n! \pm 1$ produces

58. no primes ____   **59.** exactly one prime ____

60. twin primes ____

*Because it does not equal the sum of its proper divisors, an abundant number is not perfect, but it is called **pseudoperfect** if it is equal to the sum of a subset of its proper divisors. For example, 12, an abundant number, has proper divisors 1, 2, 3, 4, and 6. $12 \neq 1 + 2 + 3 + 4 + 6$, but $12 = 2 + 4 + 6$. (Also, $12 = 1 + 2 + 3 + 6$.)*
  Show that each number is pseudoperfect.

61. 18   (Show it with two different sums.)

62. 24   (Show it with five different sums.)

*Abundant numbers are so commonly pseudoperfect that when we find one that isn't, we call it **weird**. There are no weird numbers less than 70. Do the following exercises to investigate 70.*

63. Show that 70 is abundant.

64. Show that 70 is *not* pseudoperfect and must therefore be the smallest weird number. (Among the first 10,000 counting numbers, there are only seven weird ones: 70, 863, 4030, 5830, 7192, 7912, and 9272.)

The following number is known as Belphegor's prime.

One nonillion, sixty-six quadrillion, six hundred trillion, one

Belphegor, who is referred to in many literary works, is one of the seven princes of hell, known for tempting men toward particular evils. Exercises 65–68 refer to Belphegor's prime.

65. Write out the decimal form of the number. (It is a "palindrome"—that is, it reads the same backward and forward. And in keeping with its name, it actually is a prime.)

66. What are the "middle three" digits? (The biblical book of Revelation calls this "the number of the beast.")

67. How many zeroes are on each side of the middle three digits?

68. With what digit does Belphegor's prime begin and end?

5.4 GREATEST COMMON FACTOR AND LEAST COMMON MULTIPLE

OBJECTIVES

1 Find the greatest common factor by several methods.

2 Find the least common multiple by several methods.

Greatest Common Factor

The *greatest common factor* is defined as follows.

GREATEST COMMON FACTOR

The **greatest common factor (GCF)** of a group of natural numbers is the largest natural number that is a factor of all the numbers in the group.

Examples: 18 is the GCF of 36 and 54, because 18 is the largest natural number that divides both 36 and 54.

1 is the GCF of 7 and 16.

The greatest common factor is often called the *greatest common divisor*. Greatest common factors can be found by using prime factorizations. To determine the GCF of 36 and 54, first write the prime factorization of each number.

$$36 = 2^2 \cdot 3^2$$
$$54 = 2^1 \cdot 3^3$$

The GCF is the product of the primes common to the factorizations, with each prime raised to the power indicated by the *least* exponent that it has in any factorization. Here, the prime 2 has 1 as the least exponent (in $54 = 2^1 \cdot 3^3$), while the prime 3 has 2 as the least exponent (in $36 = 2^2 \cdot 3^2$).

$$GCF = 2^1 \cdot 3^2 = 2 \cdot 9 = 18$$

We summarize as follows.

FINDING THE GREATEST COMMON FACTOR (PRIME FACTORS METHOD)

Step 1 Write the prime factorization of each number.

Step 2 Choose all primes common to *all* factorizations, with each prime raised to the *least* exponent that it has in any factorization.

Step 3 Form the product of all the numbers in Step 2. This product is the greatest common factor.

EXAMPLE 1 **Finding the Greatest Common Factor by the Prime Factors Method**

Find the greatest common factor of 360 and 2700.

Solution

Write the prime factorization of each number.

$$360 = 2^3 \cdot 3^2 \cdot 5 \qquad 2700 = 2^2 \cdot 3^3 \cdot 5^2$$

Find the primes common to both factorizations, with each prime having as its exponent the *least* exponent from either product.

Use the least exponents.

$$GCF = 2^2 \cdot 3^2 \cdot 5 = 180$$

The greatest common factor of 360 and 2700 is 180.

gcd(360,2700)
180

The calculator shows that the greatest common divisor (factor) of 360 and 2700 is 180. Compare with **Example 1.**

EXAMPLE 2 Finding the Greatest Common Factor by the Prime Factors Method

Find the greatest common factor of 720, 1000, and 1800.

Solution

Write the prime factorization for each number.

$$720 = 2^4 \cdot 3^2 \cdot 5$$
$$1000 = 2^3 \cdot 5^3$$
$$1800 = 2^3 \cdot 3^2 \cdot 5^2$$

Use the smallest exponent on each prime common to the factorizations.

$$GCF = 2^3 \cdot 5 = 40$$

(The prime 3 is not used in the greatest common factor because it does not appear in the prime factorization of 1000.) ∎

EXAMPLE 3 Finding the Greatest Common Factor by the Prime Factors Method

Find the greatest common factor of 80 and 63.

Solution

$$80 = 2^4 \cdot 5$$
$$63 = 3^2 \cdot 7$$

There are no primes in common here, so the GCF is 1. The number 1 is the largest number that will divide into both 80 and 63. ∎

Two numbers, such as 80 and 63, with a greatest common factor of 1 are called **relatively prime numbers**—that is, they are prime *relative* to one another. (They have no common factors other than 1.)

Another method of finding the greatest common factor involves dividing the numbers by common prime factors.

FINDING THE GREATEST COMMON FACTOR (DIVIDING BY PRIME FACTORS METHOD)

Step 1 Write the numbers in a row.

Step 2 Divide each of the numbers by a common prime factor. Try 2, then try 3, and so on.

Step 3 Divide the quotients by a common prime factor. Continue until no prime will divide into all the quotients.

Step 4 The product of the primes in Steps 2 and 3 is the greatest common factor.

EXAMPLE 4 **Finding the Greatest Common Factor by Dividing by Prime Factors**

Find the greatest common factor of 12, 18, and 60.

Solution

Write the numbers in a row and divide by 2.

$$\begin{array}{c|ccc} 2 & 12 & 18 & 60 \\ \hline & 6 & 9 & 30 \end{array}$$

The numbers 6, 9, and 30 are not all divisible by 2, but they are divisible by 3.

$$\begin{array}{c|ccc} 2 & 12 & 18 & 60 \\ \hline 3 & 6 & 9 & 30 \\ \hline & 2 & 3 & 10 \end{array}$$

No prime divides into 2, 3, and 10, so the greatest common factor of the numbers 12, 18, and 60 is given by the product of the primes on the left, 2 and 3.

$$\begin{array}{c|ccc} 2 & 12 & 18 & 60 \\ \hline 3 & 6 & 9 & 30 \\ \hline & 2 & 3 & 10 \end{array}$$

$$2 \cdot 3 = 6$$

The GCF of 12, 18, and 60 is 6. ∎

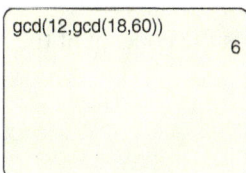

Another method of finding the greatest common factor of two numbers (but not more than two) is called the **Euclidean algorithm.** *

EXAMPLE 5 **Finding the Greatest Common Factor Using the Euclidean Algorithm**

Use the Euclidean algorithm to find the greatest common factor of 90 and 168.

Solution

Step 1 Begin by dividing the larger, 168, by the smaller, 90. Disregard the quotient, but note the remainder.

$$\begin{array}{r} 1 \\ 90\overline{)168} \\ 90 \\ \hline 78 \end{array}$$

Step 2 Divide the smaller of the two numbers by the remainder obtained in Step 1. Once again, note the remainder.

$$\begin{array}{r} 1 \\ 78\overline{)90} \\ 78 \\ \hline 12 \end{array}$$

Step 3 Continue dividing the successive remainders as many times as necessary to obtain a remainder of 0.

$$\begin{array}{r} 6 \\ 12\overline{)78} \\ 72 \\ \hline 6 \end{array}$$ Greatest common factor

Step 4 The *last positive remainder* in this process is the greatest common factor of 90 and 168. It can be seen that their GCF is 6.

$$\begin{array}{r} 2 \\ 6\overline{)12} \\ 12 \\ \hline 0 \end{array}$$

∎

*For a proof that this process does indeed give the greatest common factor, see *Elementary Introduction to Number Theory, Second Edition*, by Calvin T. Long, pp. 34–35.

The **greatest common divisor** is symbolized GCD or gcd. This screen uses the fact that

gcd(a, b, c) = gcd(a, gcd(b, c)).

Compare with **Example 4**.

gcd(12,gcd(18,60))
6

Pierre de Fermat (about 1601–1665), a government official who did not interest himself in mathematics until he was past 30, devoted leisure time to its study. He was a worthy scholar, best known for his work in number theory. His other major contributions involved certain applications in geometry and his original work in probability.

Much of Fermat's best work survived only on loose sheets or jotted, without proof, in the margins of works that he read.

The Euclidean algorithm is particularly useful if the two numbers are difficult to factor into primes. We summarize the algorithm here.

FINDING THE GREATEST COMMON FACTOR (EUCLIDEAN ALGORITHM)

To find the greatest common factor of two unequal numbers, divide the larger by the smaller. Note the remainder, and divide the previous divisor by this remainder. Continue the process until a remainder of 0 is obtained. The greatest common factor is the last positive remainder obtained.

Least Common Multiple

Closely related to the idea of the greatest common factor is the concept of the *least common multiple.*

LEAST COMMON MULTIPLE

The **least common multiple (LCM)** of a group of natural numbers is the smallest natural number that is a multiple of all the numbers in the group.

Example: 30 is the LCM of 15 and 10 because 30 is the smallest number that appears in both sets of multiples.

Multiples of 15: $\{15, 30, 45, 60, 75, 90, 105, \ldots\}$

Multiples of 10: $\{10, 20, 30, 40, 50, 60, 70, \ldots\}$

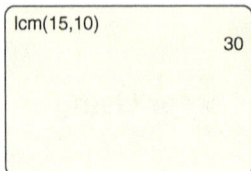

lcm(15,10)
                                30

The least common multiple of 15 and 10 is 30.

The set of natural numbers that are multiples of *both* 15 and 10 form the set of *common multiples:*

$$\{30, 60, 90, 120, \ldots\}.$$

While there are infinitely many common multiples, the *least* common multiple is observed to be 30.

A method similar to the first one given for the greatest common factor may be used to find the least common multiple of a group of numbers.

FINDING THE LEAST COMMON MULTIPLE (PRIME FACTORS METHOD)

Step 1 Write the prime factorization of each number.

Step 2 Choose all primes belonging to *any* factorization, with each prime raised to the power indicated by the *greatest* exponent that it has in any factorization.

Step 3 Form the product of all the numbers in Step 2. This product is the least common multiple.

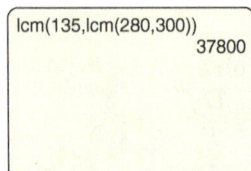

lcm(135,lcm(280,300))
                          37800

The least common multiple of 135, 280, and 300 is 37,800. Compare with **Example 6.**

EXAMPLE 6 **Finding the Least Common Multiple by the Prime Factors Method**

Find the least common multiple of 135, 280, and 300.

Solution

Write the prime factorizations:

$$135 = 3^3 \cdot 5, \quad 280 = 2^3 \cdot 5 \cdot 7, \quad \text{and} \quad 300 = 2^2 \cdot 3 \cdot 5^2.$$

Form the product of all the primes that appear in *any* of the factorizations. Use the *greatest* exponent from any factorization.

> Use the greatest exponents.

$$\text{LCM} = 2^3 \cdot 3^3 \cdot 5^2 \cdot 7 = 37{,}800$$

The least natural number divisible by 135, 280, and 300 is 37,800. ∎

The least common multiple of a group of numbers can also be found by dividing by prime factors. The process is slightly different from that for finding the GCF.

FINDING THE LEAST COMMON MULTIPLE (DIVIDING BY PRIME FACTORS METHOD)

Step 1 Write the numbers in a row.

Step 2 Divide each of the numbers by a common prime factor. Try 2, then try 3, and so on.

Step 3 Divide the quotients by a common prime factor. When no prime will divide all quotients, but a prime will divide some of them, divide where possible and bring any nondivisible quotients down. Continue until no prime will divide any two quotients.

Step 4 The product of all prime divisors from Steps 2 and 3 as well as all remaining quotients is the least common multiple.

EXAMPLE 7 Finding the Least Common Multiple by Dividing by Prime Factors

Find the least common multiple of 12, 18, and 60.

Solution

Proceed just as in **Example 4** to obtain the following.

$$
\begin{array}{r|rrr}
2 & 12 & 18 & 60 \\ \hline
3 & 6 & 9 & 30 \\ \hline
 & 2 & 3 & 10
\end{array}
$$

Now, even though no prime will divide 2, 3, and 10, the prime 2 will divide 2 and 10. Divide the 2 and the 10 and bring down the 3.

$$
\begin{array}{r|rrr}
2 & 12 & 18 & 60 \\ \hline
3 & 6 & 9 & 30 \\ \hline
2 & 2 & 3 & 10 \\ \hline
 & 1 & 3 & 5
\end{array}
$$

$2 \cdot 3 \cdot 2 \cdot 1 \cdot 3 \cdot 5 = 180$

The LCM of 12, 18, and 60 is 180. ∎

The least common multiple of two numbers *m* and *n* can be obtained by dividing their product by their greatest common factor.

FINDING THE LEAST COMMON MULTIPLE (FORMULA)

The least common multiple of *m* and *n* can be computed as follows.

$$\text{LCM} = \frac{m \cdot n}{\text{GCF of } m \text{ and } n}$$

(This method works only for two numbers, not for more than two.)

EXAMPLE 8 Finding the Least Common Multiple by Formula

Use the formula to find the least common multiple of 90 and 168.

Solution

In **Example 5** we used the Euclidean algorithm to find that the greatest common factor of 90 and 168 is 6. Therefore, the formula gives us

$$\text{Least common multiple of 90 and 168} = \frac{90 \cdot 168}{6} = 2520.$$ ∎

Problem-Solving Strategy

Problems that deal with questions such as "How many objects will there be in each group if each group contains the same number of objects?" and "When will two events occur at the same time?" can sometimes be solved using the ideas of greatest common factor and least common multiple.

EXAMPLE 9 Finding Common Starting Times of Movie Cycles

The King Theatre and the Star Theatre run movies continuously, and each starts its first feature at 1:00 P.M. If the movie shown at the King lasts 80 minutes and the movie shown at the Star lasts 2 hours, when will the two movies again start at the same time?

Solution

First, convert 2 hours to 120 minutes. The question can be restated as follows: "What is the smallest number of minutes it will take for the two movies to start at the same time again?" This is equivalent to asking, "What is the least common multiple of 80 and 120?"

Using any of the methods described in this section, we find that the least common multiple of 80 and 120 is 240. Therefore, it will take 240 minutes, or $\frac{240}{60} = 4$ hours, for the movies to start at the same time again. By adding 4 hours to 1:00 P.M., we find that they will start together again at 5:00 P.M. ∎

EXAMPLE 10 Finding the Greatest Common Size of Stacks of Cards

Joshua has 450 football cards and 840 baseball cards. He wants to place them in stacks on a table so that each stack has the same number of cards, and no stack has different types of cards within it. What is the greatest number of cards that he can have in each stack?

Solution

Here, we are looking for the greatest number that will divide evenly into 450 and 840. This is, of course, the greatest common factor of 450 and 840. Using any of the methods described in this section, we find that

greatest common factor of 450 and 840 = 30.

Therefore, the greatest number of cards he can have in each stack is 30. He will have 15 stacks of 30 football cards and 28 stacks of 30 baseball cards. ∎

WHEN | **Will I** | **Ever USE** | **This** | **?**

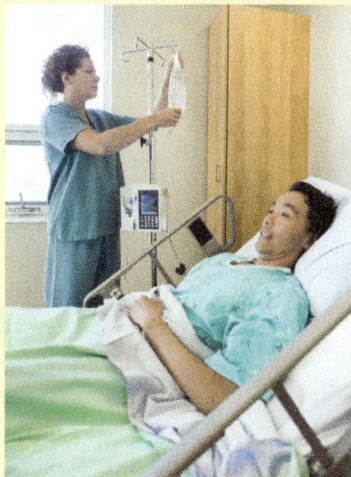

Addition and subtraction of fractions normally involve identifying the least common denominator, which is the LCM of all denominators. Suppose, for example, that a nurse needs the total displacement (volume) of the following quantities combined.

$$\frac{1}{5} \text{ teacup, } \frac{2}{3} \text{ cup, 5 teaspoons, and 3 tablespoons}$$

(In practice, the fractions would normally first be converted to decimal equivalents in metric units, but we illustrate the math here with the common fractions and non-metric units.)

The conversions necessary to obtain common units (in this case, ounces) are shown in the table.

1 teaspoon	=	$\frac{1}{6}$ ounce
1 tablespoon	=	$\frac{1}{2}$ ounce
1 teacup	=	6 ounces
1 cup	=	8 ounces

Thus
$$\frac{1}{5} \text{ teacup} = \frac{1}{5}(6 \text{ ounces}) = \frac{6}{5} \text{ ounces,}$$

$$\frac{2}{3} \text{ cup} = \frac{2}{3}(8 \text{ ounces}) = \frac{16}{3} \text{ ounces,}$$

$$5 \text{ teaspoons} = 5\left(\frac{1}{6} \text{ ounce}\right) = \frac{5}{6} \text{ ounces,}$$

$$3 \text{ tablespoons} = 3\left(\frac{1}{2} \text{ ounce}\right) = \frac{3}{2} \text{ ounces,}$$

and we need the sum $\frac{6}{5} + \frac{16}{3} + \frac{5}{6} + \frac{3}{2}$. One method of finding the least common denominator is dividing by prime factors.

$$
\begin{array}{c|cccc}
2 & 2 & 3 & 5 & 6 \\
3 & 1 & 3 & 5 & 3 \\
5 & 1 & 1 & 5 & 1 \\
\hline
 & 1 & 1 & 1 & 1
\end{array}
\quad \text{LCM} = 2 \cdot 3 \cdot 5 \cdot 1 \cdot 1 \cdot 1 \cdot 1 = 30
$$

Using 30 as the least common denominator, we obtain

$$
\frac{6}{5} + \frac{16}{3} + \frac{5}{6} + \frac{3}{2} = \frac{6}{5} \cdot \frac{6}{6} + \frac{16}{3} \cdot \frac{10}{10} + \frac{5}{6} \cdot \frac{5}{5} + \frac{3}{2} \cdot \frac{15}{15}
$$

$$
= \frac{36}{30} + \frac{160}{30} + \frac{25}{30} + \frac{45}{30}
$$

$$
= \frac{36 + 160 + 25 + 45}{30}
$$

$$
= \frac{266}{30} = 8\frac{26}{30}.
$$

The combined total is $8\frac{26}{30}$ ounces, or, in decimal form, about 8.867 ounces.

5.4 EXERCISES

Decide whether each statement is true *or* false.

1. No two even natural numbers can be relatively prime.

2. Two different prime numbers must be relatively prime.

3. If p is a prime number, then the greatest common factor of p and p^2 is p^2.

4. If p is a prime number, then the least common multiple of p and p^2 is p^3.

5. There is no prime number p such that the greatest common factor of p and 2 is 2.

6. The set of all common multiples of two given natural numbers is infinite.

7. Any two natural numbers have at least one common factor.

8. The least common multiple of two different primes is their product.

9. No two composite numbers can be relatively prime.

10. The product of any two natural numbers is equal to the product of their least common multiple and their greatest common factor.

Use the prime factors method to find the greatest common factor of each group of numbers.

11. 84 and 140   **12.** 315 and 90

13. 275 and 132   **14.** 264 and 504

15. 68, 102, and 425   **16.** 765, 780, and 990

Use the method of dividing by prime factors to find the greatest common factor of each group of numbers.

17. 150 and 260   **18.** 237 and 395

19. 600 and 90   **20.** 330 and 255

21. 84, 90, and 210   **22.** 585, 1680, and 990

Use the Euclidean algorithm to find the greatest common factor of each group of numbers.

23. 18 and 60   **24.** 77 and 84

25. 36 and 90   **26.** 72 and 90

27. 945 and 450   **28.** 200 and 350

Use the prime factors method to find the least common multiple of each group of numbers.

29. 48 and 60

30. 21 and 35

31. 81 and 45

32. 84 and 98

33. 20, 30, and 50

34. 15, 21, and 45

Use the method of dividing by prime factors to find the least common multiple of each group of numbers.

35. 27 and 36

36. 21 and 56

37. 63 and 99

38. 16, 120, and 216

39. 48, 54, and 60

40. 154, 165, and 2310

*Use the formula given in the text on **page 203** and the results of **Exercises 23–28** to find the least common multiple of each group of numbers.*

41. 18 and 60

42. 77 and 84

43. 36 and 90

44. 72 and 90

45. 945 and 450

46. 200 and 350

47. Explain in your own words how to find the greatest common factor of a group of numbers.

48. Explain in your own words how to find the least common multiple of a group of numbers.

49. If p, q, and r are different primes, and a, b, and c are natural numbers such that $a < b < c$,

(a) what is the greatest common factor of $p^a q^c r^b$ and $p^b q^a r^c$?

(b) what is the least common multiple of $p^b q^a$, $q^b r^c$, and $p^a r^b$?

50. Find (a) the greatest common factor and (b) the least common multiple of $2^{25} \cdot 5^{17} \cdot 7^{21}$ and $2^{28} \cdot 5^{22} \cdot 7^{13}$. Leave your answers in prime factored form.

It is possible to extend the Euclidean algorithm in order to find the greatest common factor of more than two numbers. For example, if we wish to find the greatest common factor of 150, 210, and 240, we can first use the algorithm to find the greatest common factor of two of these (say, for example, 150 and 210). Then we find the greatest common factor of that result and the third number, 240. The final result is the greatest common factor of the original group of numbers.

Use the Euclidean algorithm as just described to find the greatest common factor of each group of numbers.

51. 90, 105, and 315

52. 48, 315, and 450

53. 144, 180, and 192

54. 180, 210, and 630

55. Suppose that the least common multiple of p and q is pq. What can we say about p and q?

56. Suppose that the least common multiple of p and q is q. What can we say about p and q?

57. Recall some of your early experiences in mathematics (for example, in the elementary grade classroom). What topic involving fractions required the use of the least common multiple? Give an example.

58. Recall some of your experiences in elementary algebra. What topics required the use of the greatest common factor? Give an example.

*Refer to **Examples 9 and 10** to solve each problem.*

59. *Inspecting Calculators*   Jameel and Fahima work on an assembly line, inspecting electronic calculators. Jameel inspects the electronics of every sixteenth calculator, while Fahima inspects the workmanship of every thirty-sixth calculator. If they both start working at the same time, which calculator will be the first that they both inspect?

60. *Night Off for Security Guards*   Tomas and Jenny work as security guards at a factory. Tomas has every sixth night off, and Jenny has every tenth night off. If both are off on July 1, what is the next night that they will both be off together?

61. *Stacking Coins*   Suyín has 240 pennies and 288 nickels. She wants to place them all in stacks so that each stack has the same number of coins, and each stack contains only one denomination of coin. What is the greatest number of coins that she can place in each stack?

62. *Bicycle Racing*   Aki and Felipe are in a bicycle race, following a circular track. If they start at the same place and travel in the same direction, and Aki completes a revolution every 40 seconds, which Felipe takes 45 seconds to complete each revolution, how long will it take them before they reach the starting point again simultaneously?

63. *Selling Books*   Azad sold some books at $24 each and used the money to buy some concert tickets at $50 each. He had no money left over after buying the tickets. What is the least amount of money he could have earned from selling the books? What is the least number of books he could have sold?

64. *Sawing Lumber*   Terri has some pieces of two-by-four lumber. Some are 60 inches long, and some are 72 inches long. All of them must be sawn into shorter pieces. If all sawn pieces must be the same length, what is the longest such piece so that no lumber is left over?

OBJECTIVES

1 Work with the Fibonacci sequence.

2 Understand the golden ratio.

3 See relationships between the Fibonacci sequence and the golden ratio.

The solution of **Fibonacci's rabbit problem** is examined in **Chapter 1, pages 20–21.**

The **Fibonacci Association** is a research organization dedicated to investigation into the **Fibonacci sequence** and related topics. Check your library to see if it has the journal *Fibonacci Quarterly*. The first two journals of 1963 contain a basic introduction to the Fibonacci sequence.

The Fibonacci Sequence

One of the most famous problems in elementary mathematics comes from the book *Liber Abaci,* written in 1202 by Leonardo of Pisa, a.k.a. Fibonacci. The problem is as follows:

> A man put a pair of rabbits in a cage. During the first month the rabbits produced no offspring, but each month thereafter produced one new pair of rabbits. If each new pair thus produced reproduces in the same manner, how many pairs of rabbits will there be at the end of one year?

The solution of this problem leads to a sequence of numbers known as the **Fibonacci sequence.** Here are the first fifteen terms of the Fibonacci sequence:

$$1, 1, 2, 3, 5, 8, 13, 21, 34, 55, 89, 144, 233, 377, 610.$$

After the first two terms (both 1) in the sequence, each term is obtained by adding the two previous terms. For example, the third term is obtained by adding $1 + 1$ to get 2, the fourth term is obtained by adding $1 + 2$ to get 3, and so on. This can be described by a mathematical formula known as a **recursion formula.**

If F_n represents the Fibonacci number in the nth position in the sequence, then

$$F_1 = 1$$
$$F_2 = 1$$
$$F_n = F_{n-2} + F_{n-1}, \quad \text{for } n \geq 3.$$

Using the recursion formula $F_n = F_{n-2} + F_{n-1}$, we obtain

$$F_3 = F_1 + F_2 = 1 + 1 = 2, \quad F_4 = F_2 + F_3 = 1 + 2 = 3, \quad \text{and so on.}$$

The Fibonacci sequence exhibits many interesting patterns, and by inductive reasoning we can make many conjectures about these patterns. However, simply observing a finite number of examples does not provide a proof of a statement. Proofs of the properties of the Fibonacci sequence often involve mathematical induction (covered in college algebra texts). Here we simply observe some of the patterns and do not attempt to provide proofs.

EXAMPLE 1 Observing a Pattern of the Fibonacci Numbers

Find the sum of the squares of the first n Fibonacci numbers for $n = 1, 2, 3, 4, 5,$ and examine the pattern. Generalize this relationship.

Solution

$$1^2 = 1 = 1 \cdot 1 = F_1 \cdot F_2$$
$$1^2 + 1^2 = 2 = 1 \cdot 2 = F_2 \cdot F_3$$
$$1^2 + 1^2 + 2^2 = 6 = 2 \cdot 3 = F_3 \cdot F_4$$
$$1^2 + 1^2 + 2^2 + 3^2 = 15 = 3 \cdot 5 = F_4 \cdot F_5$$
$$1^2 + 1^2 + 2^2 + 3^2 + 5^2 = 40 = 5 \cdot 8 = F_5 \cdot F_6$$

The sum of the squares of the first n Fibonacci numbers seems to always be the product of F_n and F_{n+1}. This has been proved to be true, in general, using mathematical induction.

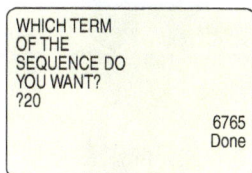

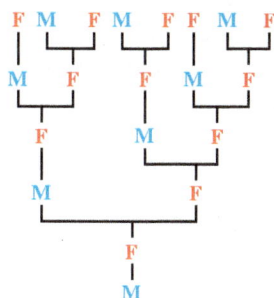

Figure 1

EXAMPLE 2   Observing the Fibonacci Sequence in a Long Division Problem

Observe the steps of the long-division algorithm used to find the first few decimal places of the reciprocal of 89, the eleventh Fibonacci number. Locate occurrences of the terms of the Fibonacci sequence in the algorithm.

Solution

```
        .011235 . . .
89)1.000000 . . .
   89
   110
    89
   210
   178
   320
   267
   530
   445
   850 . . .
```

Notice that after the 0 in the tenths place, the next five digits are the first five terms of the Fibonacci sequence. In addition, as indicated in color in the process, the digits 1, 1, 2, 3, 5, 8 appear in the division steps. Now, look at the digits next to the ones in color, beginning with the second "1"; they, too, are 1, 1, 2, 3, 5, ...

If the division process is continued past the final step shown above, the pattern seems to stop, because to ten decimal places, $\frac{1}{89} \approx 0.0112359551$. (The decimal representation actually begins to repeat later in the process, since $\frac{1}{89}$ is a rational number.) However, the sum below indicates how the Fibonacci numbers are actually "hidden" in this decimal.

$$0.01$$
$$0.001$$
$$0.0002$$
$$0.00003$$
$$0.000005$$
$$0.0000008$$
$$0.00000013$$
$$0.000000021$$
$$0.0000000034$$
$$0.00000000055$$
$$0.000000000089$$
$$\frac{1}{89} = 0.0112359550. \ldots$$

Fibonacci patterns have been found in numerous places in nature. For example, male honeybees (drones) hatch from eggs that have not been fertilized, so a male bee has only one parent, a female. On the other hand, female honeybees hatch from fertilized eggs, so a female has two parents, one male and one female. **Figure 1** shows several generations of ancestors for a male honeybee.

Notice that in the first generation, starting at the bottom, there is 1 bee, in the second there is 1 bee, in the third there are 2 bees, and so on. These are the terms of the Fibonacci sequence. Furthermore, beginning with the second generation, the numbers of female bees form the sequence, and beginning with the third generation, the numbers of male bees also form the sequence.

A fraction such as

$$1 + \cfrac{1}{1 + \cfrac{1}{1 + \cfrac{1}{1 + \cdots}}}$$

is called a **continued fraction**. This continued fraction can be evaluated as follows.

Let $\quad x = 1 + \cfrac{1}{1 + \cfrac{1}{1 + \cdots}}$

Then $\qquad x = 1 + \dfrac{1}{x}$

$$x^2 = x + 1$$

$$x^2 - x - 1 = 0.$$

By the quadratic formula from algebra,

$$x = \frac{1 \pm \sqrt{1 - 4(1)(-1)}}{2(1)}$$

$$x = \frac{1 \pm \sqrt{5}}{2}.$$

Notice that the positive solution is the **golden ratio.**

Successive terms in the Fibonacci sequence also appear in some plants. For example, the photo on the left below shows the double spiraling of a daisy head, with 21 clockwise spirals and 34 counterclockwise spirals. These numbers are successive terms in the sequence.

Most pineapples (see the photo on the right below) exhibit the Fibonacci sequence in the following way: Count the spirals formed by the "scales" of the cone, first counting from lower left to upper right. Then count the spirals from lower right to upper left. You should find that in one direction you get 8 spirals, and in the other you get 13 spirals, once again successive terms of the Fibonacci sequence. Many pinecones exhibit 5 and 8 spirals, and the cone of the giant sequoia has 3 and 5 spirals.

The Golden Ratio

If we consider the quotients of successive Fibonacci numbers, a pattern emerges.

$$\frac{1}{1} = 1 \qquad \frac{5}{3} = 1.666\ldots \qquad \frac{21}{13} \approx 1.615384615 \qquad \frac{89}{55} = 1.618181818\ldots$$

$$\frac{2}{1} = 2 \qquad \frac{8}{5} = 1.6 \qquad \frac{34}{21} \approx 1.619047619$$

$$\frac{3}{2} = 1.5 \qquad \frac{13}{8} = 1.625 \qquad \frac{55}{34} \approx 1.617647059$$

These quotients seem to be approaching some "limiting value" close to 1.618. In fact, as we go farther into the sequence, these quotients approach the number

$$\frac{1 + \sqrt{5}}{2}, \qquad \text{Golden ratio}$$

known as the **golden ratio** and often symbolized by ϕ, the Greek letter phi.

The golden ratio appears over and over in art, architecture, music, and nature. Its origins go back to the days of the ancient Greeks, who thought that a golden rectangle exhibited the most aesthetically pleasing proportions.

A **golden rectangle** is one that can be divided into a square and another (smaller) rectangle the same shape as the original rectangle. (See **Figure 2.**) If we let the smaller rectangle have length L and width W, as shown in the figure, then we see that the original rectangle has length $L + W$ and width L. Both rectangles (being "golden") have their lengths and widths in the golden ratio, ϕ, so we have

$$\frac{L}{W} = \frac{L + W}{L}$$

$$\frac{L}{W} = \frac{L}{L} + \frac{W}{L} \qquad \text{Write the right side as the sum of two fractions.}$$

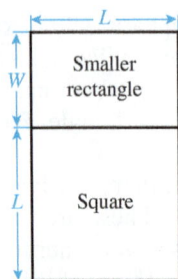

Figure 2

$$\phi = 1 + \frac{1}{\phi}$$ Substitute $\frac{L}{W} = \phi$, $\frac{L}{L} = 1$, and $\frac{W}{L} = \frac{1}{\phi}$.

$$\phi^2 = \phi + 1$$ Multiply both sides by ϕ.

$$\phi^2 - \phi - 1 = 0.$$ Write in standard quadratic form.

Using the quadratic formula from algebra, the positive solution of this equation is found to be $\frac{1 + \sqrt{5}}{2} \approx 1.618033989$, the golden ratio.

The Parthenon (see the photo), built on the Acropolis in ancient Athens during the fifth century B.C., is an example of architecture exhibiting many distinct golden rectangles.

To see an interesting connection between the terms of the Fibonacci sequence, the golden ratio, and a phenomenon of nature, we can start with a rectangle measuring 89 by 55 units. (See **Figure 3.**) This is a very close approximation to a golden rectangle. Within this rectangle a square is then constructed, 55 units on a side. The remaining rectangle is also approximately a golden rectangle, measuring 55 units by 34 units. Each time this process is repeated, a square and an approximate golden rectangle are formed.

As indicated in **Figure 3,** vertices of the squares may be joined by a smooth curve known as a *spiral*. This spiral resembles the outline of a cross section of the shell of the chambered nautilus, as shown in the photo.

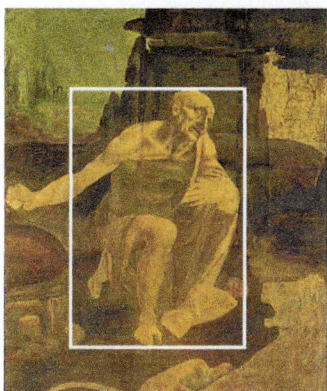

A Golden Rectangle in Art The rectangle outlining the figure in *St. Jerome* by Leonardo da Vinci is an example of a golden rectangle.

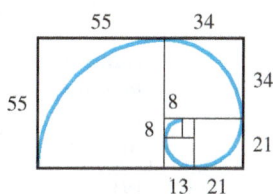

Figure 3

FOR FURTHER THOUGHT

Mathematical Animation

The 1959 animated film *Donald in Mathmagic Land* has endured for over 50 years as a classic. It provides a 25-minute trip with Donald Duck, led by the Spirit of Mathematics, through the world of mathematics. Several minutes of the film are devoted to the golden ratio (or, as it is termed there, the golden section).

Disney provides animation to explain the golden ratio in a way that the printed word simply cannot do. The golden ratio is seen in architecture, nature, and the human body.

For Group or Individual Investigation

1. Verify the following Fibonacci pattern in the conifer family. Obtain a pineapple, and count spirals formed by the "scales" of the cone, first counting from lower left to upper right. Then count the spirals from lower right to upper left. What do you find?

2. Two popular sizes of index cards are 3" by 5" and 5" by 8". Why do you think that these are industry-standard sizes?

3. Divide your height by the height to your navel. Find a class average. What value does this come close to?

Answer each question concerning the Fibonacci sequence or the golden ratio.

1. The fifteenth Fibonacci number is 610 and the seventeenth Fibonacci number is 1597. What is the value of the sixteenth Fibonacci number?

2. Recall that F_n represents the Fibonacci number in the nth position in the sequence. Find all values of n such that $F_n = n$.

3. Write the "parity" (o for odd, e for even) of each of the first twelve terms of the Fibonacci sequence.

4. From the pattern shown in the parity sequence, determine the parity of the one-hundredth term.

5. What is the exact value of the golden ratio?

6. What is the approximate value of the golden ratio to the nearest thousandth?

In each of Exercises 7–14, a pattern is established involving terms of the Fibonacci sequence. Use inductive reasoning to make a conjecture concerning the next equation in the pattern, and verify it. You may wish to refer to the first few terms of the sequence given in the text.

7. $1 = 2 - 1$
$1 + 1 = 3 - 1$
$1 + 1 + 2 = 5 - 1$
$1 + 1 + 2 + 3 = 8 - 1$

8. $1 = 2 - 1$
$1 + 3 = 5 - 1$
$1 + 3 + 8 = 13 - 1$
$1 + 3 + 8 + 21 = 34 - 1$

9. $1 = 1$
$1 + 2 = 3$
$1 + 2 + 5 = 8$
$1 + 2 + 5 + 13 = 21$

10. $1^2 + 1^2 = 2$
$1^2 + 2^2 = 5$
$2^2 + 3^2 = 13$
$3^2 + 5^2 = 34$

11. $2^2 - 1^2 = 3$
$3^2 - 1^2 = 8$
$5^2 - 2^2 = 21$
$8^2 - 3^2 = 55$

12. $2^3 + 1^3 - 1^3 = 8$
$3^3 + 2^3 - 1^3 = 34$
$5^3 + 3^3 - 2^3 = 144$
$8^3 + 5^3 - 3^3 = 610$

13. $1 = 1^2$
$1 - 2 = -1^2$
$1 - 2 + 5 = 2^2$
$1 - 2 + 5 - 13 = -3^2$

14. $1 - 1 = -1 + 1$
$1 - 1 + 2 = 1 + 1$
$1 - 1 + 2 - 3 = -2 + 1$
$1 - 1 + 2 - 3 + 5 = 3 + 1$

15. Every natural number can be expressed as a sum of Fibonacci numbers, where no number is used more than once. For example,

$$25 = 21 + 3 + 1.$$

Express each of the following in this way.
(a) 39   (b) 59   (c) 99

16. It has been shown that if m divides n, then F_m is a factor of F_n. Show that this is true for the following values of m and n.
(a) $m = 3, n = 6$ (b) $m = 4, n = 12$
(c) $m = 5, n = 15$

17. It has been shown that if the greatest common factor of m and n is r, then the greatest common factor of F_m and F_n is F_r. Show that this is true for the following values of m and n.
(a) $m = 10, n = 4$ (b) $m = 12, n = 6$
(c) $m = 14, n = 6$

18. For any prime number p except 2 or 5, either F_{p+1} or F_{p-1} is divisible by p. Show that this is true for the following values of p.
(a) $p = 3$ (b) $p = 7$ (c) $p = 11$

19. Earlier we saw that if a term of the Fibonacci sequence is squared and then the product of the terms on each side of the term is found, there will always be a difference of 1. Follow the steps below, choosing the sixth Fibonacci number, 8.
(a) Square 8. Multiply the terms of the sequence two positions away from 8 (i.e., 3 and 21). Subtract the smaller result from the larger, and record your answer.
(b) Square 8. Multiply the terms of the sequence three positions away from 8. Once again, subtract the smaller result from the larger, and record your answer.
(c) Repeat the process, moving four terms away from 8.
(d) Make a conjecture about what will happen when you repeat the process, moving five terms away. Verify your answer.

20. **A Number Trick** Here is a number trick that you can perform. Ask someone to pick any two numbers at random and to write them down. Ask the person to determine a third number by adding the first and second, a fourth number by adding the second and third, and so on, until ten numbers are determined. Then ask the person to add these ten numbers. You will be able to give the sum before the person even completes the list, because the sum will always be 11 times the seventh number in the list. Verify that this is true, by using x and y as the first two numbers arbitrarily chosen. (*Hint:* Remember the distributive property from algebra.)

Another Fibonacci-type sequence that has been studied by mathematicians is the **Lucas sequence,** *named after a French mathematician of the nineteenth century. The first nine terms of the Lucas sequence are*

$$1, 3, 4, 7, 11, 18, 29, 47, 76.$$

21. What is the tenth term of the Lucas sequence?

22. Choose any term of the Lucas sequence and square it. Then multiply the terms on either side of the one you chose. Subtract the smaller result from the larger. Repeat this for a different term of the sequence. Do you get the same result? Make a conjecture about this pattern.

23. The first term of the Lucas sequence is 1. Add the first and third terms. Record your answer. Now add the first, third, and fifth terms and record your answer. Continue this pattern, each time adding another term that is in an *odd* position in the sequence. What do you notice about all of your sums?

24. The second term of the Lucas sequence is 3. Add the second and fourth terms. Record your answer. Now add the second, fourth, and sixth terms and record your answer. Continue this pattern, each time adding another term that is in an *even* position of the sequence. What do you notice about all of your sums?

25. Many interesting patterns exist among the terms of the Fibonacci sequence and the Lucas sequence. Make a conjecture about the next equation that would appear in each of the lists, and then verify it.

(a) $1 \cdot 1 = 1$
$1 \cdot 3 = 3$
$2 \cdot 4 = 8$
$3 \cdot 7 = 21$
$5 \cdot 11 = 55$

(b) $1 + 2 = 3$
$1 + 3 = 4$
$2 + 5 = 7$
$3 + 8 = 11$
$5 + 13 = 18$

(c) $1 + 1 = 2 \cdot 1$
$1 + 3 = 2 \cdot 2$
$2 + 4 = 2 \cdot 3$
$3 + 7 = 2 \cdot 5$
$5 + 11 = 2 \cdot 8$

(d) $1 + 4 = 5 \cdot 1$
$3 + 7 = 5 \cdot 2$
$4 + 11 = 5 \cdot 3$
$7 + 18 = 5 \cdot 5$
$11 + 29 = 5 \cdot 8$

26. In the text we illustrate that the quotients of successive terms of the Fibonacci sequence approach the golden ratio. Make a similar observation for the terms of the Lucas sequence; that is, find the decimal approximations for the quotients

$$\frac{3}{1}, \frac{4}{3}, \frac{7}{4}, \frac{11}{7}, \frac{18}{11}, \frac{29}{18},$$

and so on, using a calculator. Then make a conjecture about what seems to be happening.

Recall the **Pythagorean theorem** *from geometry: If a right triangle has legs of lengths a and b and hypotenuse of length c, then*

$$a^2 + b^2 = c^2.$$

Suppose that we choose any four successive terms of the Fibonacci sequence. Multiply the first and fourth. Double the product of the second and third. Add the squares of the second and third. The three results obtained form a **Pythagorean triple** *(three numbers that satisfy the equation $a^2 + b^2 = c^2$). Find the Pythagorean triple obtained this way, using the four given successive terms of the Fibonacci sequence.*

27. 1, 1, 2, 3   **28.** 2, 3, 5, 8   **29.** 5, 8, 13, 21

30. Look at the values of the hypotenuse (c) in the answers to **Exercises 27–29.** What do you notice about each of them?

31. The following array of numbers is called **Pascal's triangle.**

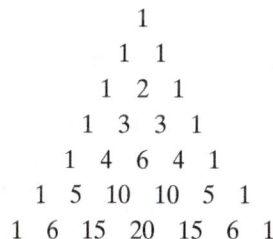

```
              1
            1   1
          1   2   1
        1   3   3   1
      1   4   6   4   1
    1   5   10  10   5   1
  1   6  15   20  15   6   1
```

This array is important in the study of counting techniques and probability (see later chapters) and appears in algebra in the binomial theorem. If the triangular array is written in a different form, as follows, and the sums along the diagonals as indicated by the dashed lines are found, there is an interesting occurrence. What do you find when the numbers are added?

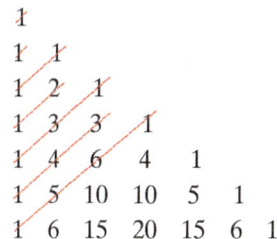

32. Write a paragraph explaining some of the occurrences of the Fibonacci sequence and the golden ratio in your everyday surroundings.

Exercises 33–36 require a scientific calculator.

33. The positive solution of $x^2 - x - 1 = 0$ is

$$\frac{1 + \sqrt{5}}{2},$$

as indicated in the text. The negative solution is

$$\frac{1 - \sqrt{5}}{2}.$$

Find the decimal approximations for both. What similarity do you notice between the two decimals?

34. In some cases, writers define the golden ratio to be the reciprocal of $\frac{1 + \sqrt{5}}{2}$. Find a decimal approximation for the reciprocal of $\frac{1 + \sqrt{5}}{2}$. What similarity do you notice between the decimals for $\frac{1 + \sqrt{5}}{2}$ and its reciprocal?

A remarkable relationship exists between the two solutions of the equation $x^2 - x - 1 = 0$, which are

$$\phi = \frac{1 + \sqrt{5}}{2} \quad and \quad \overline{\phi} = \frac{1 - \sqrt{5}}{2},$$

and the Fibonacci numbers. To find the nth Fibonacci number without using the recursion formula, use a calculator to evaluate

$$\frac{\phi^n - \overline{\phi}^n}{\sqrt{5}}.$$

Thus, to find the thirteenth Fibonacci number, evaluate

$$\frac{\left(\frac{1 + \sqrt{5}}{2}\right)^{13} - \left(\frac{1 - \sqrt{5}}{2}\right)^{13}}{\sqrt{5}}.$$

*This form is known as the **Binet form** of the nth Fibonacci number. Use the Binet form and a calculator to find the nth Fibonacci number for each of the following values of n.*

35. $n = 16$ **36.** $n = 24$

CHAPTER 5 SUMMARY

KEY TERMS

5.1
number theory
prime number
composite number
divisibility
divisibility tests
factor
divisor
multiple
factorization
prime factorization

Sieve of Eratosthenes
fundamental theorem
  of arithmetic
cryptography
modular systems

5.2
proof by contradiction
Mersenne number
Mersenne prime
partition

5.3
perfect number
deficient number
abundant number
amicable (friendly)
  numbers
Goldbach's conjecture
twin primes
primordial primes
Fermat's Last Theorem

5.4
greatest common factor (GCF)
relatively prime numbers
Euclidean algorithm
least common multiple (LCM)

5.5
Fibonacci sequence
Fibonacci numbers
golden ratio
golden rectangle

NEW SYMBOLS

$a \mid b$ *a divides b* ϕ the golden ratio

TEST YOUR WORD POWER

See how well you have learned the vocabulary in this chapter.

1. In mathematics, **number theory** in general deals with
  A. the properties of the rational numbers.
  B. operations on fractions.
  C. the properties of the natural numbers.
  D. only those numbers that are prime.

2. A **prime number** is
  A. divisible by 2, whereas a composite number is not.
  B. divisible by itself and 1 only.
  C. one of the finite set of numbers discovered by Marin Mersenne.
  D. any number expressible in the form $2^p - 1$, where p is a prime.

3. A **partition** of a set is
   A. a rule that separates half of its elements from the other half.
   B. a subset that contains some but not all of the elements of the set.
   C. a rule by which all elements of the set can be identified.
   D. a division of all elements of the set into disjoint subsets.

4. A natural number that is less than the sum of its **proper divisors** is
   A. an **abundant** number.
   B. a **perfect** number.
   C. a **deficient** number.
   D. a **Fermat number.**

5. An **amicable** pair consists of two numbers that are
   A. either both prime or both composite.
   B. either both odd or both even.
   C. each equal to the sum of the proper divisors of the other.
   D. one Fermat number and one Mersenne number.

6. **Goldbach's conjecture**
   A. was formulated over 300 years before it was proved.
   B. has to do with two nth powers summing to a third nth power.

C. has to do with expressing natural numbers as products of primes.
D. has to do with expressing natural numbers as sums of primes.

7. The so-called **twin primes**
   A. always occur in pairs that differ by 2.
   B. always occur in pairs that differ by 1.
   C. cannot exceed 10^{100}.
   D. have finally all been identified.

8. **Fermat's Last Theorem**
   A. was proved by Euler about 100 years after Fermat stated it.
   B. has never been proved, but most mathematicians believe it is true.
   C. was proved by Wiles in the 1990s.
   D. says that the equation $a^n + b^n = c^n$ cannot be true for $n = 2$.

ANSWERS
1. C 2. B 3. D 4. A
5. C 6. D 7. A 8. C

QUICK REVIEW

Concepts

Examples

5.1 Prime and Composite Numbers

The natural number a is divisible by the natural number b, or, equivalently, "b divides a" (denoted $b \mid a$), if there exists a natural number k such that $a = bk$.

$2 \mid 14$, since $14 = 2 \cdot 7$ ($k = 7$). 2 is a factor of 14, and 14 is a multiple of 2.

In the equation above, we say that b is a **factor** (or **divisor**) of a and that a is a multiple of b.

$2 \nmid 13$, since there is no suitable k. Here $13 = 2 \cdot 6.5$ (and $k = 6.5$ is *not* a natural number).

A **factorization** of a natural number expresses it as a product of some, or all, of its divisors.

$2 \cdot 12, \ 3 \cdot 8, \ 4 \cdot 6, \ 2 \cdot 2 \cdot 6, \ 2 \cdot 3 \cdot 4, \ 2 \cdot 2 \cdot 2 \cdot 3$
Factorizations of 24

A **prime number** is greater than 1 and has only itself and 1 as factors.

17 is prime. Its only factors are 1 and 17.

A **composite number** is greater than 1 and is *not* prime—that is, it has at least one factor other than itself and 1.

10 is composite. In addition to 1 and 10, it has the factors 2 and 5.

Divisibility tests can help determine the factors of a given natural number.

54,086,226 is divisible by 3 since the sum of its digits

$$5 + 4 + 0 + 8 + 6 + 2 + 2 + 6 = 33$$

is divisible by 3.

258,784 is *not* divisible by 3 since the sum of its digits

$$2 + 5 + 8 + 7 + 8 + 4 = 34$$

is *not* divisible by 3.

Concepts	*Examples*
A factorization with prime factors only is a number's **prime factorization.**	The prime factorization of 24 is $$2 \cdot 2 \cdot 2 \cdot 3.$$
The **fundamental theorem of arithmetic** states that the prime factorization of a number is unique (if the order of the factors is disregarded).	24 has no other prime factorization except rearrangements of the factors shown above, such as $2 \cdot 3 \cdot 2 \cdot 2$ and $3 \cdot 2 \cdot 2 \cdot 2$.
A natural number's prime factorization can be determined by using a **factor tree** or by using **repeated division by primes.**	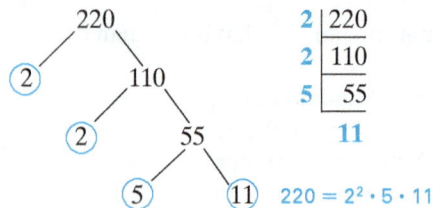 $220 = 2^2 \cdot 5 \cdot 11$

5.2 Large Prime Numbers

Modern **cryptography systems** are based on large prime numbers.	Cryptography systems, in practice, use primes so huge that the most powerful computers can multiply them but *cannot* factor their product.
Large primes are found today mainly among the **Mersenne numbers.**	The largest Mersenne prime discovered, as of summer 2014, is $2^{57,885,161} - 1$. It has 17,425,170 digits.

5.3 Selected Topics from Number Theory

The **proper divisors** of a natural number are all of its divisors except the number itself.	42 has proper divisors 1, 2, 3, 6, 7, 14, and 21.
A natural number is **perfect** if the sum of its proper divisors equals the number itself.	28 has proper divisors 1, 2, 4, 7, and 14. $$28 = 1 + 2 + 4 + 7 + 14$$
All even perfect numbers are of the form $2^{n-1}(2^n - 1)$, where $2^n - 1$ is prime.	The number $2^3 - 1 = 7$ is prime, so $$2^{3-1}(2^3 - 1) = 4(7) = 28 \quad \text{is perfect.}$$
A natural number is **abundant** if the sum of its proper divisors is greater than the number itself.	$1 + 2 + 3 + 6 + 7 + 14 + 21 = 54 > 42 \leftarrow$ Abundant
A natural number is **deficient** if the sum of its proper divisors is less than the number itself.	$1 + 2 + 4 + 8 = 15 < 16 \leftarrow$ Deficient
Two natural numbers are **amicable,** or **friendly,** if each equals the sum of the other's proper divisors.	The smallest amicable pair is 220 and 284. Proper divisors of 220: 1, 2, 4, 5, 10, 11, 20, 22, 44, 55, 110 Proper divisors of 284: 1, 2, 4, 71, 142 $$1 + 2 + 4 + 5 + 10 + 11 + 20 + 22 + 44 + 55 + 110 = 284$$ $$1 + 2 + 4 + 71 + 142 = 220$$
Goldbach's conjecture (which is not proved) states that every even number greater than 2 can be written as the sum of two prime numbers.	$12 = 5 + 7 \qquad 46 = 3 + 43$ Goldbach's conjecture
Twin primes are two primes that differ by 2.	17 and 19 29 and 31 Twin primes
Fermat's Last Theorem (proved in the 1990s) states that there are no triples of natural numbers (a, b, c) that satisfy the equation $a^n + b^n = c^n$ for any natural number $n \geq 3$.	$2^3 + 6^3 = 8 + 216 = 224$, but 224 is not a perfect cube.

Concepts

Examples

5.4 Greatest Common Factor and Least Common Multiple

The **greatest common factor (GCF)** of a group of natural numbers is the largest natural number that is a factor of all the numbers in the group.

Prime factors method:

$$450 = 2 \cdot 3^2 \cdot 5^2$$

$$660 = 2^2 \cdot 3 \cdot 5 \cdot 11$$

$$\text{GCF} = 2 \cdot 3 \cdot 5 = 30$$

Dividing by prime factors method:

$$\begin{array}{r|rr}
2 & 450 & 660 \\
3 & 225 & 330 \\
5 & 75 & 110 \\
\hline
 & 15 & 22
\end{array}$$

GCF = 2 · 3 · 5 = 30

Euclidean algorithm:

$$\begin{array}{r}
1 \\
450\overline{)660} \\
\underline{450} \\
210
\end{array}$$

$$\begin{array}{r}
2 \\
210\overline{)450} \\
\underline{420} \\
30 \quad \text{GCF}
\end{array}$$

$$\begin{array}{r}
7 \\
30\overline{)210} \\
\underline{210} \\
0
\end{array}$$

The **least common multiple (LCM)** of a group of natural numbers is the smallest natural number that is a multiple of all the numbers in the group.

Prime factors method:

$$450 = 2 \cdot 3^2 \cdot 5^2$$

$$660 = 2^2 \cdot 3 \cdot 5 \cdot 11$$

$$\text{LCM} = 2^2 \cdot 3^2 \cdot 5^2 \cdot 11 = 9900$$

Dividing by prime factors method:

$$\begin{array}{r|rr}
2 & 450 & 660 \\
2 & 225 & 330 \\
3 & 225 & 165 \\
3 & 75 & 55 \\
5 & 25 & 55 \\
 & 5 & 11
\end{array}$$

LCM = 2 · 2 · 3 · 3 · 5 · 5 · 11 = 9900

Formula:

$$\text{LCM} = \frac{450 \cdot 660}{\text{GCF of 450 and 660}} = \frac{297{,}000}{30} = 9900$$

5.5 The Fibonacci Sequence and the Golden Ratio

The **Fibonacci sequence** begins with two 1s, and each succeeding term is the sum of the preceding two terms.

1, 1, 2, 3, 5, 8, 13, 21, 34, 55, 89, 144, 233, 377, . . .

Concepts	Examples
The **golden ratio,** $$\frac{1+\sqrt{5}}{2},$$ arises in many ways. One is as the limit of the ratio of adjacent terms in the Fibonacci sequence.	The ratios (to three decimal places), $\frac{1}{1} = 1.000$, $\frac{2}{1} = 2.000$, $\frac{3}{2} = 1.500$, $\frac{5}{3} \approx 1.667$, $\frac{8}{5} = 1.600, \ldots$ are approaching $\frac{1+\sqrt{5}}{2} \approx 1.618$.

CHAPTER 5   TEST

In Exercises 1–6, decide whether each statement is true *or* false.

1. For all natural numbers n, 1 is a factor of n, and n is a multiple of n.

2. If a natural number is not perfect, then it must be abundant.

3. If a natural number is divisible by 9, then it must also be divisible by 3.

4. If p and q are different primes, 1 is their greatest common factor, and pq is their least common multiple.

5. No two prime numbers differ by 1.

6. There are infinitely many prime numbers.

7. Use divisibility tests to determine whether the number

$$540{,}487{,}608$$

is divisible by each of the following.
  (a) 2   **(b)** 3   **(c)** 4
  (d) 5   **(e)** 6   **(f)** 8
  (g) 9   **(h)** 10   **(i)** 12

8. Decide whether each number is prime, composite, or neither.
  (a) 91   **(b)** 101   **(c)** 1

9. Give the prime factorization of 2970.

10. In your own words state the fundamental theorem of arithmetic.

11. Decide whether each number is perfect, deficient, or abundant.
  (a) 18   **(b)** 32   **(c)** 28

12. Which of the following statements is false?
  A. There are no known odd perfect numbers.
  B. Every even perfect number must end in 8 or 26.
  C. Goldbach's conjecture for the number 12 is illustrated by the equation $12 = 7 + 5$.

13. Give a pair of twin primes between 60 and 80.

14. Find the greatest common factor of 99, 135, and 216.

15. Find the least common multiple of 91 and 154.

16. *Day Off for Fast-food Workers* Both Katherine and Josh work at a fast-food outlet. Katherine has every sixth day off, and Josh has every fourth day off. If they are both off on Wednesday of this week, what will be the day of the week that they are next off together?

17. The twenty-first Fibonacci number is 10,946 and the twenty-second Fibonacci number is 17,711. What is the twenty-third Fibonacci number?

18. Make a conjecture about the next equation in the following list, and verify it.

$$8 - (1 + 1 + 2 + 3) = 1$$
$$13 - (1 + 2 + 3 + 5) = 2$$
$$21 - (2 + 3 + 5 + 8) = 3$$
$$34 - (3 + 5 + 8 + 13) = 5$$

19. Choose the correct completion of this statement: If p is a prime number, then $2^p - 1$ is _____ prime.
  A. never   **B.** sometimes   **C.** always

20. Determine the eighth term of a Fibonacci-type sequence with first term 1 and second term 4.

21. Choose any term after the first in the sequence described in **Exercise 20.** Square it. Multiply the two terms on either side of it. Subtract the smaller result from the larger. Now repeat the process with a different term. Make a conjecture about what this process will yield for any term of the sequence.

22. Which one of the following is the *exact* value of the golden ratio?

  A. $\dfrac{1+\sqrt{5}}{2}$ **B.** $\dfrac{1-\sqrt{5}}{2}$ **C.** 1.6   **D.** 1.618

The Real Numbers and Their Representations

6

Unsettled Weather Ahead · weather.go
Friday, May 23, 2014 · 4:1

Western Cutoff Low Brings Unsettled Weather to the Area

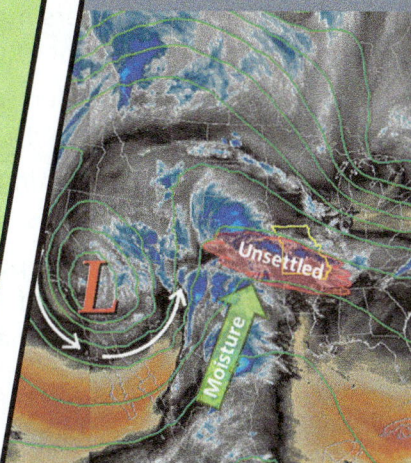

Today · Saturday · Sunc

Unsettled

L

Moisture

79° · 77° · 82

Rain Chance 40% Afternoon · Rain Chance 60% · Rain Cha 35%

Unsettled Weather
Expected into Next Wee

@NWSKansasCity · US National Weather
Service Pleasant Hill

In the March 12, 2014, issue of the *New Orleans Times Picayune*, columnist Jarvis DeBerry wrote the following in an article titled "Old Math or Common Core Math, Which Troubles You More?"

After a February board meeting for my alma mater's student newspaper, I told some students that I began Washington University in the engineering school. One asked if my previous focus on math and science ever proves useful in journalism. I joked: "Only when a colleague needs help figuring out the percent something has increased." Another board member who, like me, graduated from WashU and worked in journalism, responded with something close to awe: "You can do percentages? I mean, seriously?"

In the development of the real numbers (the topic of this chapter), the unit number 1 is the place at which everything starts. (0 showed up much later in our number system.) The concept of percent has become a standard by which portions of one whole thing are described.

Percent means "per one hundred," and the symbol % is used to represent percent. Thus,

1% means one one-hundredth, or $\frac{1}{100}$, or **0.01.**

One hundred percent, or 100%, means 1 whole thing.

As consumers, we are faced with the use of percent on a daily basis: advertisements for sale items, interest on loans and savings, poll numbers, chance of precipitation in daily weathercasts, and so on. Occupations that otherwise may not require formal mathematics often require the basic concept of percent.

- Retail managers marking up an item to sell might be required to find a given percent of the wholesale cost.

- The same retailer may need to use a percent to determine how much to subtract from the usual selling price to mark an item that is on sale.

- Teachers assigning grades may need to find a percent of a total number of points.

- A manager in a publishing company may need to determine the percent change of sales from one edition of a text to the next, or the percent of sales that a particular title captures in its market.

Some people, such as DeBerry, have a natural talent—or a learned ability—for determining percents, while others don't have a clue. Those who are proficient use many different methods to find simple percents. For example, if you are asked to find

20% of 50,

you might mentally calculate using one of the following methods.

1. You think, "Well, 20% means $\frac{1}{5}$, and to find $\frac{1}{5}$ of something I divide by 5, so 50 divided by 5 is 10. The answer is 10."

2. You think, "20% is twice 10%, and to find 10% of something I move the decimal point one place to the left. So 10% of 50 is 5.0, or just 5, and I simply double 5 to get 10. The answer is 10."

For those who don't think this way, there is the tried-and-true method of multiplying 50 by 0.20.

One whole dollar ($1.00) is the basic monetary unit in the United States. One one-hundredth of a dollar is a penny, or one cent. Notice the correlation between this concept and that of percent. One cent is represented $0.01, or just $.01. *Then how do we represent 99 cents?* The answer is **$.99.** But so many times we see signs such as the one here from Wendy's, indicating a price of **.99¢.** Because ¢ is a symbol for "cents," this posted price actually means that the spicy nuggets sell for less than one penny. Think about it. *Do you understand why?*

If you don't have the knack for computing percents using shortcuts, there are sound mathematical procedures for working with this concept. This chapter introduces the real number system, the operations of arithmetic with real numbers, and selected applications of fractions, decimals, and percents in occupations and trades.

OBJECTIVES

1 Represent a number on a number line.

2 Identify a number as positive, negative, or zero.

3 Identify a number as belonging to one or more sets of numbers.

4 Given two numbers *a* and *b,* determine whether $a = b$, $a < b$, or $a > b$.

5 Given a number *a,* determine its additive inverse and absolute value.

6 Interpret signed numbers in tables of economic and occupations data.

Sets of Real Numbers

The mathematician Leopold Kronecker (1823–1891) once made the statement, "God made the integers, all the rest is the work of man." The *natural numbers* are those numbers with which we count discrete objects. By including 0 in the set, we obtain the set of *whole numbers.*

NATURAL NUMBERS

$\{1, 2, 3, 4, \ldots\}$ is the set of **natural numbers.**

WHOLE NUMBERS

$\{0, 1, 2, 3, \ldots\}$ is the set of **whole numbers.**

These numbers, along with many others, can be represented on **number lines** like the one pictured in **Figure 1.** We draw a number line by locating any point on the line and labeling it 0. This is the **origin.** Choose any point to the right of 0 and label it 1. The distance between 0 and 1 gives a unit of measure used to locate other points, as shown in **Figure 1.** The numbers labeled and those continuing in the same way to the right correspond to the set of whole numbers.

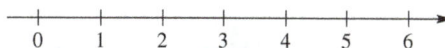

Figure 1

All the whole numbers starting with 1 are located to the right of 0 on the number line. But numbers may also be placed to the left of 0. These numbers, written −1, −2, −3, and so on, are shown in **Figure 2.** (The negative sign is used to show that the numbers are located to the *left* of 0.)

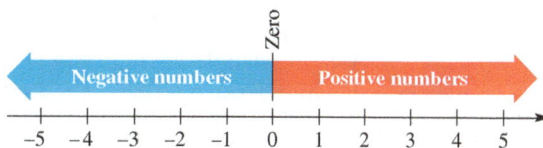

Figure 2

The numbers to the *left* of 0 are **negative numbers.** The numbers to the *right* of 0 are **positive numbers.** Positive numbers and negative numbers are called **signed numbers.** The number 0 itself is neither positive nor negative.

There are many practical applications of negative numbers.

- Temperatures may fall below zero. The lowest temperature ever recorded in meteorological records was −128.6°F at Vostok, Antarctica, on July 22, 1983.

- Altitudes below sea level can be represented by negative numbers. The shore surrounding the Dead Sea is 1312 feet below sea level, written −1312 feet.

The Origins of Zero The Mayan Indians of Mexico and Central America had one of the earliest numeration systems that included a symbol for zero. The very early Babylonians had a positional system, but they placed only a space between "digits" to indicate a missing power. When the Greeks absorbed Babylonian astronomy, they used the letter omicron, o, of their alphabet or ō to represent "no power," or "zero." The Greek numeration system was gradually replaced by the Roman numeration system.

The original Hindu word for zero was *sunya,* meaning "void." The Arabs adopted this word as *sifr,* or "vacant." The word *sifr* passed into Latin as *zephirum,* which over the years became *zevero, zepiro,* and finally, *zero.*

The natural numbers, their negatives, and zero make up the set of *integers*.

INTEGERS

$\{\ldots, -3, -2, -1, 0, 1, 2, 3, \ldots\}$ is the set of **integers.**

Not all numbers are integers. For example, $\frac{1}{2}$ is a number *halfway* between the integers 0 and 1. Several numbers that are not integers are *graphed* in **Figure 3**. The **graph** of a number is a point on the number line representing that number. Think of the graph of a set of numbers as a "picture" of the set. All the numbers indicated in **Figure 3** can be written as quotients of integers and are examples of *rational numbers*. An integer, such as 2, is also a rational number. This is true because $2 = \frac{2}{1}$.

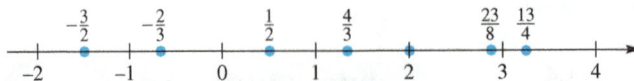

Figure 3

RATIONAL NUMBERS

$\{x \mid x$ is a quotient of two integers, with denominator not equal to $0\}$ is the set of **rational numbers.**

(Read the part in the braces as "the set of all numbers x such that x is a quotient of two integers, with denominator not equal to 0.")

The set symbolism used in the definition of rational numbers,

$$\{x \mid x \text{ has a certain property}\},$$

is called **set-builder notation.** This notation is convenient to use when it is not possible, or practical, to list all the elements of the set.

There are other numbers on the number line that are not rational. The first such number ever to be discovered was $\sqrt{2}$. The Pythagorean theorem, named for the Greek mathematician Pythagoras and covered in more detail in **Chapter 9,** says that in a right triangle, the length of the side opposite the right angle (the hypotenuse) is equal to the square root of the sum of the squares of the two perpendicular sides (the legs). In **Figure 4(a),** this is symbolized

$$c = \sqrt{a^2 + b^2}.$$

A right triangle having both legs a and b of length 1 is placed on a number line as shown in **Figure 4(b).** By the Pythagorean theorem, the length of the hypotenuse c is

$$c = \sqrt{1^2 + 1^2}$$

$$c = \sqrt{1 + 1}$$

$$c = \sqrt{2}.$$

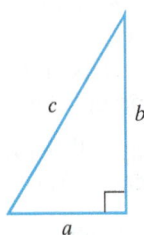

(a)

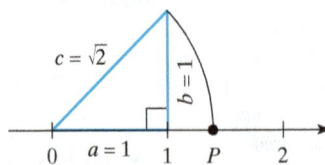

(b)

Figure 4

Figure 4(b) also shows an arc of a circle, centered at 0 with radius $\sqrt{2}$, intersecting the number line at point P. The coordinate of P is $\sqrt{2}$. The Greeks discovered that this number cannot be expressed as a quotient of integers—it is *irrational*.

IRRATIONAL NUMBERS

$\{x \mid x$ is a number on the number line that is not rational$\}$ is the set of **irrational numbers.**

Irrational numbers include $\sqrt{3}$, $\sqrt{7}$, $-\sqrt{10}$, and π, which is the ratio of the distance around a circle (its *circumference*) to the distance across it (its *diameter*). All numbers that can be represented by points on the number line (the union of the sets of rational and irrational numbers) are called *real numbers*.

REAL NUMBERS

$\{x \mid x$ is a number that can be represented by a point on the number line$\}$ is the set of **real numbers.**

Real numbers can be written as decimal numbers. Any rational number can be written as a decimal that will either come to an end (terminate) or repeat in a fixed "block" of digits. For example,

$$\frac{2}{5} = 0.4 \quad \text{and} \quad \frac{27}{100} = 0.27 \quad \text{are terminating decimals.}$$

$$\frac{1}{3} = 0.3333\ldots \quad \text{and} \quad \frac{3}{11} = 0.27272727\ldots \quad \text{are repeating decimals.}$$

The decimal representation of an irrational number neither terminates nor repeats. Decimal representations of rational and irrational numbers will be discussed further later in this chapter.

Figure 5 illustrates two ways to represent the relationships among the various sets of real numbers.

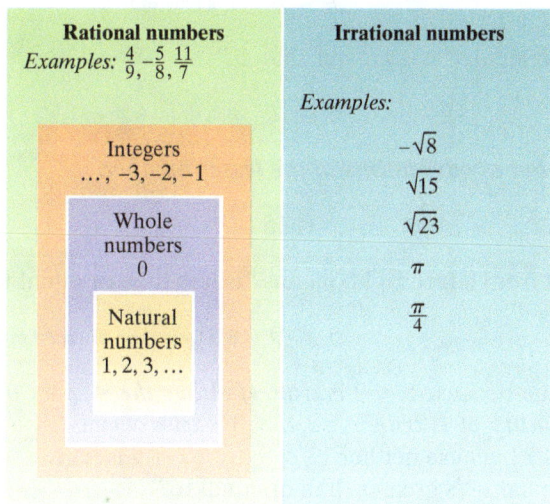

All numbers shown are real numbers.

(a)

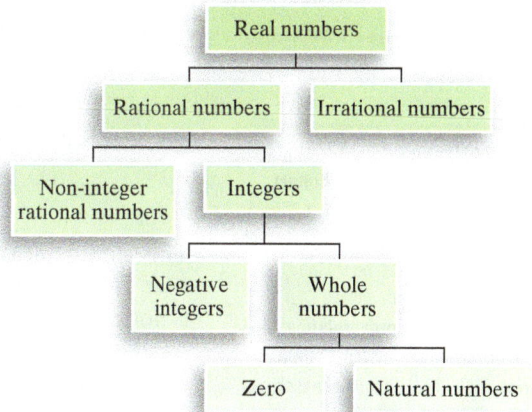

(b)

Figure 5

EXAMPLE 1 **Identifying Elements of a Set of Numbers**

List the numbers in the set that belong to each set of numbers.

$$\left\{-5, -\frac{2}{3}, 0, \sqrt{2}, \frac{13}{4}, 5, 5.8\right\}$$

(a) natural numbers   **(b)** whole numbers   **(c)** integers

(d) rational numbers   **(e)** irrational numbers   **(f)** real numbers

Solution

(a) The only natural number in the set is 5.

(b) The whole numbers consist of the natural numbers and 0, so the elements of the set that are whole numbers are 0 and 5.

(c) The integers in the set are -5, 0, and 5.

(d) The rational numbers are -5, $-\frac{2}{3}$, 0, $\frac{13}{4}$, 5, and 5.8, because each of these numbers *can* be written as the quotient of two integers. For example, $5.8 = \frac{58}{10} = \frac{29}{5}$.

(e) The only irrational number in the set is $\sqrt{2}$.

(f) All the numbers in the set are real numbers. ∎

Order in the Real Numbers

Suppose that a and b represent two real numbers. If their graphs on the number line are the same point, then a **is equal to** b. If the graph of a lies to the left of b, then a **is less than** b, and if the graph of a lies to the right of b, then a **is greater than** b. The **law of trichotomy** says that for two numbers a and b, one and only one of the following is true.

$$a = b, \quad a < b, \quad \text{or} \quad a > b$$

When read from left to right, the inequality $a < b$ is read "a is less than b."

$$7 < 8 \qquad \text{7 is less than 8.}$$

The inequality $a > b$ means "a is greater than b."

$$8 > 2 \qquad \text{8 is greater than 2.}$$

Notice that the symbol always points to the lesser number.

$$\text{Lesser number} \longrightarrow \quad 8 < 15$$

The symbol $\leq$, read from left to right, means "is less than or equal to."

$$5 \leq 9 \qquad \text{5 is less than or equal to 9.}$$

This statement is true because $5 < 9$ is true. *If either the < part or the = part is true, then the inequality $\leq$ is true.* Also, $8 \leq 8$ is true because $8 = 8$ is true. But it is not true that $13 \leq 9$ because neither $13 < 9$ nor $13 = 9$ is true.

The symbol $\geq$ means "is greater than or equal to."

$$9 \geq 5 \qquad \text{9 is greater than or equal to 5.}$$

This statement is true because $9 > 5$ is true.

EXAMPLE 2 Comparing Real Numbers

Determine whether each statement is *true* or *false*.

(a) $6 \neq 6$ **(b)** $5 < 19$ **(c)** $15 \leq 20$ **(d)** $25 \geq 30$ **(e)** $12 \geq 12$

Solution

(a) The statement $6 \neq 6$ is false because 6 *is equal to* 6.

(b) Because 5 is indeed less than 19, this statement is true.

(c) The statement $15 \leq 20$ is true because $15 < 20$.

(d) Both $25 > 30$ and $25 = 30$ are false, so $25 \geq 30$ is false.

(e) Because $12 = 12$, the statement $12 \geq 12$ is true.

Additive Inverses and Absolute Value

For any nonzero real number x, there is exactly one number on the number line the same distance from 0 as x but on the opposite side of 0. In **Figure 6,** the numbers 3 and -3 are the same distance from 0 but are on opposite sides of 0. Thus, 3 and -3 are called **additive inverses, negatives,** or **opposites** of each other.

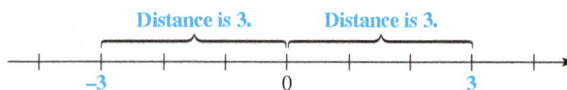

Figure 6

The additive inverse of the number 0 is 0 itself. This makes 0 the only real number that is its own additive inverse. Other additive inverses occur in distinct pairs. For example, 4 and -4 are additive inverses, as are 6 and -6. Several pairs of additive inverses are shown in **Figure 7.**

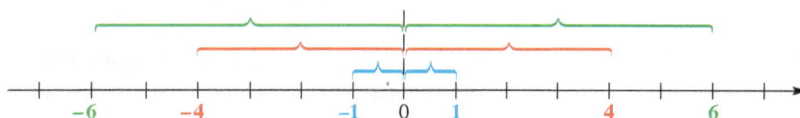

Figure 7

The additive inverse of a number can be indicated by writing the symbol $-$ in front of the number. With this symbol, the additive inverse of 7 is written -7. The additive inverse of -4 is written $-(-4)$ and can be read "the opposite of -4" or "the negative of -4." **Figure 7** suggests that 4 is an additive inverse of -4. A number can have only one additive inverse, so the symbols 4 and $-(-4)$ must represent the same number, which means that

$$-(-4) = 4.$$

DOUBLE NEGATIVE RULE

For any real number x, the following is true.

$$-(-x) = x$$

Table 1

Number	Additive Inverse
-4	$-(-4)$ or 4
0	0
19	-19
$-\dfrac{2}{3}$	$\dfrac{2}{3}$

Table 1 shows several numbers and their additive inverses. An important property of additive inverses will be studied later in this chapter:

$$a + (-a) = (-a) + a = 0, \quad \text{for all real numbers } a.$$

As mentioned above, additive inverses are numbers that are the same distance from 0 (but in opposite directions) on the number line. See **Figure 7** on the previous page. This idea can also be expressed by saying that a number and its additive inverse have the same *absolute value*. The **absolute value** of a real number can be defined as the undirected distance between 0 and the number on the number line.

The symbol for the absolute value of the number x is $|x|$, read **"the absolute value of x."** For example, the distance between 0 and 2 on the number line is 2 units, so

$$|2| = 2.$$

Because the distance between 0 and -2 on the number line is also 2 units,

$$|-2| = 2.$$

Absolute value is a measure of undirected distance, so *the absolute value of a number is never negative.* Because 0 is a distance of 0 units from 0, $|0| = 0.$

FORMAL DEFINITION OF ABSOLUTE VALUE

For any real number x, the absolute value of x is defined as follows.

$$|x| = \begin{cases} x & \text{if } x \geq 0 \\ -x & \text{if } x < 0 \end{cases}$$

If x is a positive number or 0, then its absolute value is x itself. For example, because 8 is a positive number, $|8| = 8$. By the second part of the definition, *if x is a negative number, then its absolute value is the additive inverse of x.* For example, if $x = -9$, then $|-9| = -(-9) = 9$, since the additive inverse of -9 is 9.

The formal definition of absolute value can be confusing if it is not read carefully. The "$-x$" in the second part of the definition *does not* represent a negative number. Since x is negative in the second part, $-x$ represents the opposite of a negative number, which is a positive number.

EXAMPLE 3 **Using Absolute Value**

Simplify by finding the absolute value.

(a) $|5|$ (b) $|-5|$ (c) $-|5|$

(d) $-|-14|$ (e) $|8-2|$ (f) $-|8-2|$

Solution

(a) $|5| = 5$ (b) $|-5| = -(-5) = 5$

(c) $-|5| = -(5) = -5$ (d) $-|-14| = -(14) = -14$

(e) $|8-2| = |6| = 6$ (f) $-|8-2| = -|6| = -6$ ∎

```
|(-5)|
              5
-|-14|
            -14
-|8-2|
             -6
```

This screen supports the results of
Example 3(b), (d), and (f).

Example 3(e) shows that absolute value bars also serve as grouping symbols. *Perform any operations that appear inside absolute value symbols before finding the absolute value.*

Applications of Real Numbers

EXAMPLE 4 Interpreting Signed Numbers in a Table

The consumer price index (CPI) measures the average change in prices of goods and services purchased by urban consumers in the United States. **Table 2** shows the percent change in the CPI for selected categories of goods and services from 2009 to 2010 and from 2010 to 2011. Use the table to answer each question.

Table 2

Category	Change from 2009 to 2010	Change from 2010 to 2011
Appliances	−4.5	−1.2
Education	4.4	4.2
Gasoline	18.4	26.4
Housing	−0.4	1.3
Medical care	3.4	3.0

Source: U.S. Bureau of Labor Statistics.

(a) Which category in which years represents the greatest percent decrease?

(b) Which category in which years represents the least change?

Solution

(a) We must find the negative number with the greatest absolute value. The number that satisfies this condition is −4.5, so the greatest percent decrease was shown by appliances from 2009 to 2010.

(b) We must find the number (either positive, negative, or zero) with the least absolute value. From 2009 to 2010, housing showed the least change, a decrease of 0.4 percent. ∎

In the next example, we see how the need for certain occupations during the upcoming years is changing, some for better and some for worse.

EXAMPLE 5 Comparing Occupational Rates of Change

Table 3 shows the total rates of change projected to occur in demand for some of the fastest-growing and some of the most rapidly-declining occupations from 2012 to 2022.

Table 3

Occupation (2012–2022)	Total Rate of Change (in percent)
Interpreters and translators	46.1
Personal care aides	48.8
Physicians' assistants	38.4
Word processors and typists	−25.1
Postal service clerks	−31.8
Locomotive firers	−42.0

Source: Bureau of Labor Statistics.

Which occupation in **Table 3** on the preceding page is expected to see

(a) the greatest change?   **(b)** the least change?

Solution

(a) We want the greatest change, without regard to whether the change is an increase or a decrease. Look for the number with the greatest absolute value. That number is for personal care aides:

$$|48.8| = 48.8.$$

(b) Using similar reasoning, the least change is for word processors and typists:

$$|-25.1| = 25.1.$$ ■

6.1 EXERCISES

Give a number that satisfies the given condition.

1. An integer between 4.5 and 5.5

2. A rational number between 2.8 and 2.9

3. A whole number that is not positive and is less than 1

4. A whole number greater than 4.5

5. An irrational number that is between $\sqrt{13}$ and $\sqrt{15}$

6. A real number that is neither negative nor positive

Decide whether each statement is true *or* false.

7. Every natural number is positive.

8. Every whole number is positive.

9. Every integer is a rational number.

10. Every rational number is a real number.

List all numbers from each set that are **(a)** *natural numbers;* **(b)** *whole numbers;* **(c)** *integers;* **(d)** *rational numbers;* **(e)** *irrational numbers;* **(f)** *real numbers.*

11. $\left\{ -9, -\sqrt{7}, -1\frac{1}{4}, -\frac{3}{5}, 0, \sqrt{5}, 3, 5.9, 7 \right\}$

12. $\left\{ -5.3, -5, -\sqrt{3}, -1, -\frac{1}{9}, 0, 1.2, 1.8, 3, \sqrt{11} \right\}$

Use an integer or decimal to express each number in bold-face print representing a change or measurement in each of the following.

13. *U.S. Population*   Between July 1, 2011, and July 1, 2012, the population of the United States increased by approximately **2,259,000**. (*Source:* U.S. Census Bureau.)

14. *Number of Movie Screens*   Between 2011 and 2012, the number of movie screens in the United States increased by **209**. (*Source:* Motion Picture Association of America.)

15. *World Series Attendance*   From 2011 to 2012, attendance at the first game of the World Series went from 46,406 to 42,982, a decrease of **3424**. (*Source:* Major League Baseball.)

16. *Number of U. S. Banks*   In 1935, there were 15,295 banks in the United States. By 2012, the number was 7083, representing a decrease of **8212**. (*Source:* Federal Deposit Insurance Corporation.)

17. *Dow Jones Average*   On Friday, August 23, 2013, the Dow Jones Industrial Average (DJIA) closed at 15,010.51. On the previous day, it had closed at 14,963.74. Thus, on Friday it closed up **46.77** points. (*Source: The Washington Post.*)

18. *NASDAQ*   On Tuesday, August 27, 2013, the NASDAQ closed at 3578.52. On the previous day, it had closed at 3657.57. Thus, on Tuesday it closed down **79.05** points. (*Source: The Washington Post.*)

19. *Height of Mt. Arenal in Costa Rica*   The height of Mt. Arenal, an active volcano in Costa Rica, is **5436** feet above sea level. (*Source: The New York Times Almanac.*)

20. *Elevation of New Orleans*   The city of New Orleans, devastated by Hurricane Katrina in 2005, lies **8** feet below sea level. (*Source:* U.S. Geological Survey, *Elevations and Distances in the United States.*)

21. *Melting Point of Fluorine*   The melting point of fluorine gas is **220°** below 0° Celsius.

22. *Boiling Point of Chlorine*   The boiling point of chlorine is approximately **30°** below 0° Fahrenheit.

Depths and Heights of Seas and Mountains   *The chart gives selected depths and heights of bodies of water and mountains.*

Bodies of Water	Average Depth in Feet (as a negative number)	Mountains	Altitude in Feet (as a positive number)
Pacific Ocean	−14,040	McKinley	20,237
South China Sea	−4802	Point Success	14,164
Gulf of California	−2375	Matlalcueyetl	14,636
Caribbean Sea	−8448	Ranier	14,416
Indian Ocean	−12,800	Steele	16,642

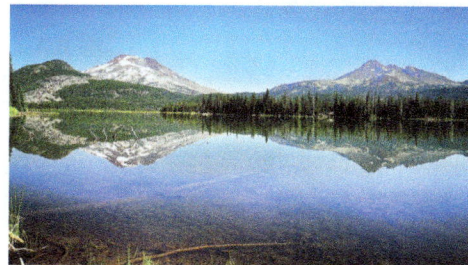

Source: The World Almanac and Book of Facts.

23. List the bodies of water in order, starting with the deepest and ending with the shallowest.

24. List the mountains in order, starting with the lowest and ending with the highest.

25. *True or false:* The absolute value of the depth of the Pacific Ocean is greater than the absolute value of the depth of the Indian Ocean.

26. *True or false:* The absolute value of the depth of the Gulf of California is greater than the absolute value of the depth of the Caribbean Sea.

Graph each group of numbers on a number line.

27. −2, −6, −4, 3, 4

28. −5, −3, −2, 0, 4

29. $\frac{1}{4}, 2\frac{1}{2}, -3\frac{4}{5}, -4, -1\frac{5}{8}$

30. $5\frac{1}{4}, 4\frac{5}{9}, -2\frac{1}{3}, 0, -3\frac{2}{5}$

31. Match each expression in Column I with its value in Column II. Some choices in Column II may not be used.

I	II
(a) $|-7|$	**A.** 7
(b) $-(-7)$	**B.** −7
(c) $-|-7|$	**C.** neither A nor B
(d) $-|-(-7)|$	**D.** both A and B

32. Fill in the blanks with the correct values: The opposite of −2 is ____, while the absolute value of −2 is ____. The additive inverse of −2 is ____, while the additive inverse of the absolute value of −2 is ____.

*Find **(a)** the additive inverse (or opposite) of each number and **(b)** the absolute value of each number.*

33. −2

34. −8

35. 6

36. 11

37. 7 − 4

38. 8 − 3

39. 7 − 7

40. 3 − 3

Select the lesser of the two given numbers.

41. −12, −4

42. −9, −14

43. −8, −1

44. −15, −16

45. 3, $|-4|$

46. 5, $|-2|$

47. $|-3|, |-4|$

48. $|-8|, |-9|$

49. $-|-6|, -|-4|$

50. $-|-2|, -|-3|$

51. $|5 - 3|, |6 - 2|$

52. $|7 - 2|, |8 - 1|$

Decide whether each statement is true *or* false.

53. $6 > -(-2)$

54. $-8 > -(-2)$

55. $-4 \le -(-5)$

56. $-6 \le -(-3)$

57. $|-6| < |-9|$

58. $|-12| < |-20|$

59. $-|8| > |-9|$

60. $-|12| > |-15|$

61. $-|-5| \ge -|-9|$

62. $-|-12| \le -|-15|$

63. $|6 - 5| \ge |6 - 2|$

64. $|13 - 8| \le |7 - 4|$

65. *Population Change* The table shows the percent change in population from 2000 to 2012 for selected metropolitan areas.

Metropolitan Area	Percent Change
Las Vegas	45.4
San Francisco	8.0
Chicago	4.7
New Orleans	−8.3
Phoenix	33.1
Detroit	−3.6

Source: U. S. Census Bureau.

(a) Which metropolitan area had the greatest percent change in population? What was this change? Was it an increase or a decrease?

(b) Which metropolitan area had the least change in population? What was this change? Was it an increase or a decrease?

66. *Trade Balance* The table gives the net trade balance, in millions of dollars, for selected U.S. trade partners for October 2013.

Country	Trade Balance (in millions of dollars)
India	−1967
China	−28,862
The Netherlands	2325
France	−1630
Turkey	396

Source: U.S. Census Bureau.

A negative balance means that imports to the United States exceeded exports from the United States. A positive balance means that exports exceeded imports.

(a) Which country had the greatest discrepancy between exports and imports? Explain.

(b) Which country had the least discrepancy between exports and imports? Explain.

(c) Which two countries were closest in their trade balance figures?

67. *CPI Data* Refer to **Table 2** in **Example 4.** Of the data for 2009 to 2010, which of the categories Appliances or Housing shows the greater change (without regard to sign)?

68. *Change in Occupations* Refer to **Table 3** in **Example 5.** Did the demand for physicians' assistants or for postal service clerks show the lesser change (without regard to sign)?

6.2	**OPERATIONS, PROPERTIES, AND APPLICATIONS OF REAL NUMBERS**

OBJECTIVES

1 Perform the operations of addition, subtraction, multiplication, and division of signed numbers.

2 Apply the rules for order of operations.

3 Identify and apply properties of addition and multiplication of real numbers.

4 Determine change in investment and meteorological data using subtraction and absolute value.

Operations on Real Numbers

The result of adding two numbers is called their **sum.** The numbers being added are called **addends** (or **terms**).

ADDING REAL NUMBERS

Like Signs Add two numbers with the *same* sign by adding their absolute values. The sign of the sum (either + or −) is the same as the sign of the two addends.

Unlike Signs Add two numbers with *different* signs by subtracting the lesser absolute value from the greater to find the absolute value of the sum. The sum is positive if the positive number has the greater absolute value. The sum is negative if the negative number has the greater absolute value.

For example, to add −12 and −8, first find their absolute values.

$$|-12| = 12 \quad \text{and} \quad |-8| = 8$$

−12 and −8 have the *same* sign, so add their absolute values: $12 + 8 = 20$. Give the sum the sign of the two numbers. Because both numbers are negative, the sum is negative and

$$-12 + (-8) = -20.$$

Find $-17 + 11$ by subtracting the absolute values, because these numbers have different signs.

$$|-17| = 17 \quad \text{and} \quad |11| = 11$$

$$17 - 11 = 6$$

Give the result the sign of the number with the larger absolute value.

$$-17 + 11 = -6$$

Negative because $|-17| > |11|$

EXAMPLE 1 Adding Signed Numbers

Find each sum.

(a) $-6 + (-3)$ **(b)** $-12 + (-4)$ **(c)** $4 + (-1)$

(d) $-9 + 16$ **(e)** $-16 + 12$ **(f)** $-4 + 4$

Solution

(a) $-6 + (-3) = -(6 + 3) = -9$

(b) $-12 + (-4) = -(12 + 4) = -16$

(c) $4 + (-1) = 3$

(d) $-9 + 16 = 7$

(e) $-16 + 12 = -4$

(f) $-4 + 4 = 0$ The sum of additive inverses is 0.   ∎

The result of subtracting two numbers is called their **difference.** In $a - b$, a is called the **minuend,** and b is called the **subtrahend.** Compare the two statements below.

$$7 - 5 = 2 \quad \text{and} \quad 7 + (-5) = 2$$

In a similar way,

$$9 - 3 = 9 + (-3).$$

That is, to subtract 3 from 9, add the additive inverse of 3 to 9. These examples suggest the following rule for subtraction.

DEFINITION OF SUBTRACTION

For all real numbers a and b,

$$a - b = a + (-b).$$

(Change the sign of the subtrahend and add.)

EXAMPLE 2 Subtracting Signed Numbers

Find each difference.

(a) $6 - 8$ **(b)** $-12 - 4$ **(c)** $-10 - (-7)$ **(d)** $15 - (-3)$

Solution

Change to addition.

Change sign of the subtrahend and add.

(a) $6 - 8 = 6 + (-8) = -2$

Change to addition.

Sign is changed.

(b) $-12 - 4 = -12 + (-4) = -16$

(c) $-10 - (-7) = -10 + [-(-7)]$ This step can be omitted.

$$= -10 + 7$$

$$= -3$$

(d) $15 - (-3) = 15 + 3 = 18$ ∎

The calculator supports the results of **Example 1(a), (c), and (e).**

```
-6+-3
            -9
4+-1
             3
-16+12
            -4
```

The calculator supports the results of **Example 2(a), (b), and (c).**

```
6-8
           -2
-12-4
          -16
-10-(-7)
           -3
```

Practical Arithmetic From the time of Egyptian and Babylonian merchants, practical aspects of arithmetic complemented mystical (or "Pythagorean") tendencies. This was certainly true in the time of **Adam Riese** (1489–1559), a "reckon master" influential when commerce was growing in Northern Europe. He championed new methods of reckoning using Hindu-Arabic numerals and quill pens. (The Roman methods then in common use moved counters on a ruled board.) Riese thus fulfilled Fibonacci's efforts 300 years earlier to supplant Roman numerals and methods.

The result of multiplying two numbers is called their **product.** The two numbers being multiplied are called **factors.**

Any rules for multiplication with negative real numbers should be consistent with the usual rules for multiplication of positive real numbers and zero. Observe the pattern of products below.

$$4 \cdot 5 = 20$$
$$4 \cdot 4 = 16$$
$$4 \cdot 3 = 12$$
$$4 \cdot 2 = 8$$ Products decrease by 4 in each line.
$$4 \cdot 1 = 4$$
$$4 \cdot 0 = 0$$
$$4 \cdot (-1) = ?$$

What number must be assigned as the product $4 \cdot (-1)$ so that the pattern is maintained? The numbers just to the left of the equality symbols decrease by 1 each time, and the products to the right decrease by 4 each time. To maintain the pattern, the number to the right in the bottom equation must be 4 less than 0, which is -4, so

$$4 \cdot (-1) = -4.$$

The pattern continues with

$$4 \cdot (-2) = -8$$
$$4 \cdot (-3) = -12$$
$$4 \cdot (-4) = -16,$$

and so on. In the same way,

$$-4 \cdot 2 = -8$$
$$-4 \cdot 3 = -12$$
$$-4 \cdot 4 = -16,$$

and so on. A similar observation can be made about the product of two negative real numbers. Look at the pattern that follows.

$$-5 \cdot 4 = -20$$
$$-5 \cdot 3 = -15$$
$$-5 \cdot 2 = -10$$ Products increase by 5 in each line.
$$-5 \cdot 1 = -5$$
$$-5 \cdot 0 = 0$$
$$-5 \cdot (-1) = ?$$

The numbers just to the left of the equality symbols decrease by 1 each time. The products on the right increase by 5 each time. To maintain the pattern, the product $-5 \cdot (-1)$ must be 5 more than 0, so it seems reasonable for the following to be true.

$$-5 \cdot (-1) = 5$$

Continuing this pattern gives the following.

$$-5 \cdot (-2) = 10$$
$$-5 \cdot (-3) = 15$$
$$-5 \cdot (-4) = 20$$
$$\vdots$$

$(+) \cdot (+) = +$

$(-) \cdot (-) = +$

$(+) \cdot (-) = -$

$(-) \cdot (+) = -$

MULTIPLYING REAL NUMBERS

Like Signs   Multiply two numbers with the *same* sign by multiplying their absolute values to find the absolute value of the product. The product is positive.

Unlike Signs   Multiply two numbers with *different* signs by multiplying their absolute values to find the absolute value of the product. The product is negative.

EXAMPLE 3 Multiplying Signed Numbers

Find each product.

(a) $-9 \cdot 7$ **(b)** $14 \cdot (-5)$ **(c)** $-8 \cdot (-4)$

Solution

(a) $-9 \cdot 7 = -63$ **(b)** $14 \cdot (-5) = -70$ **(c)** $-8 \cdot (-4) = 32$ ■

```
-9*7
              -63
14*-5
              -70
-8*-4
               32
```

The symbol (∗) represents multiplication on this screen. The display supports the results of **Example 3**.

The result of dividing two numbers is called their **quotient.** In the quotient $a \div b$ (or $\frac{a}{b}$), where $b \neq 0$, a is called the **dividend** (or **numerator**), and b is called the **divisor** (or **denominator**). For real numbers a, b, and c,

$$\text{if} \quad \frac{a}{b} = c, \quad \text{then} \quad a = b \cdot c.$$

To illustrate this, consider the quotient $\frac{10}{-2}$. The value of this quotient is obtained by asking, "What number multiplied by -2 gives 10?" From our discussion of multiplication, the answer to this question must be "-5." Therefore,

$$\frac{10}{-2} = -5, \quad \text{because} \quad 10 = -2 \cdot (-5).$$

Similarly,

$$\frac{-10}{2} = -5 \quad \text{and} \quad \frac{-10}{-2} = 5.$$

These facts, along with the fact that the quotient of two positive numbers is positive, lead to the following rules for division.

$(+)/(+) = +$

$(-)/(-) = +$

$(+)/(-) = -$

$(-)/(+) = -$

DIVIDING REAL NUMBERS

Like Signs   Divide two numbers with the *same* sign by dividing their absolute values to find the absolute value of the quotient. The quotient is positive.

Unlike Signs   Divide two numbers with *different* signs by dividing their absolute values to find the absolute value of the quotient. The quotient is negative.

EXAMPLE 4 Dividing Signed Numbers

Find each quotient.

(a) $\dfrac{15}{-5}$ **(b)** $\dfrac{-100}{-25}$ **(c)** $\dfrac{-60}{3}$

Solution

(a) $\dfrac{15}{-5} = -3$ **(b)** $\dfrac{-100}{-25} = 4$ **(c)** $\dfrac{-60}{3} = -20$

This is true because
$-5 \cdot (-3) = 15.$

```
15/-5
               -3
-100/-25
                4
-60/3
              -20
```

The division operation is represented by a slash (/). This screen supports the results of **Example 4**.

■

If 0 is divided by a nonzero number, the quotient is 0.

$$\frac{0}{a} = 0, \qquad \text{for } a \neq 0$$

This is true because $a \cdot 0 = 0$. However, we cannot divide by 0. There is a good reason for this. Whenever a division is performed, we want to obtain one and only one quotient. Now consider this division problem.

$$\frac{7}{0} = ?$$

We must ask ourselves,

"What number multiplied by 0 gives 7?"

There is no such number because the product of 0 and any number is zero. Now consider this quotient.

$$\frac{0}{0} = ?$$

There are infinitely many answers to the question, "What number multiplied by 0 gives 0?" Division by 0 does not yield a *unique* quotient, and thus it is not permitted.

```
ERR:DIVIDE BY 0
1: Quit
2: Goto
```

Dividing by zero leads to this message on the TI-83/84 Plus.

DIVISION INVOLVING ZERO

$\dfrac{a}{0}$ **is undefined for all** a. $\qquad$ $\dfrac{0}{a} = 0$ **for all nonzero** a.

Order of Operations

```
5+2*3
```

What result does the calculator give? The order of operations determines the answer. (See **Example 5(a)**.)

Given an expression such as

$$5 + 2 \cdot 3,$$

should 5 and 2 be added first, or should 2 and 3 be multiplied first? When an expression involves more than one operation, we use the following rules for **order of operations.**

The sentence **"Please excuse my dear Aunt Sally"** is often used to help us remember the rules for order of operations. The letters **P, E, M, D, A, S** are the first letters of the words of the sentence, and they stand for *parentheses, exponents, multiply, divide, add, subtract* (*Remember also that M and D have equal priority, as do A and S. Operations with equal priority are performed in order from left to right.*)

ORDER OF OPERATIONS

If parentheses or square brackets are present:

Step 1 Work separately above and below any **fraction bar.**

Step 2 Use the rules below within each set of **parentheses or square brackets.** Start with the innermost set and work outward.

If no parentheses or brackets are present:

Step 1 Apply any **exponents.**

Step 2 Do any **multiplications or divisions** in the order in which they occur, working from left to right.

Step 3 Do any **additions or subtractions** in the order in which they occur, working from left to right.

When evaluating an exponential expression that involves a negative sign, be aware that $(-a)^n$ and $-a^n$ do not necessarily represent the same quantity. For example, if $a = 2$ and $n = 6$, then

$$(-2)^6 = (-2)(-2)(-2)(-2)(-2)(-2) = 64 \qquad \text{The base is } -2.$$

while

$$-2^6 = -(2 \cdot 2 \cdot 2 \cdot 2 \cdot 2 \cdot 2) = -64. \qquad \text{The base is } 2.$$

EXAMPLE 5 Using the Order of Operations

Use the order of operations to simplify each expression.

(a) $5 + 2 \cdot 3$ 　　　　　**(b)** $4 \cdot 3^2 + 7 - (2 + 8)$

(c) $\dfrac{2(8 - 12) - 11(4)}{5(-2) - 3}$ 　　**(d)** -6^4

(e) $(-6)^4$ 　　　　　　**(f)** $(-8)(-3) - [4 - (3 - 6)]$

Solution 　　Be careful! Multiply first.

(a) $5 + 2 \cdot 3 = 5 + 6$ 　　Multiply.

$\qquad\qquad\quad = 11$ 　　Add.

(b) $4 \cdot 3^2 + 7 - (2 + 8) = 4 \cdot 3^2 + 7 - 10$ 　　Work within parentheses first.

3^2 means $3 \cdot 3$, not $3 \cdot 2$.

$\qquad\qquad\qquad\qquad = 4 \cdot 9 + 7 - 10$ 　　Apply the exponent.

$\qquad\qquad\qquad\qquad = 36 + 7 - 10$ 　　Multiply.

$\qquad\qquad\qquad\qquad = 43 - 10$ 　　Add.

$\qquad\qquad\qquad\qquad = 33$ 　　Subtract.

(c) $\dfrac{2(8 - 12) - 11(4)}{5(-2) - 3} = \dfrac{2(-4) - 11(4)}{5(-2) - 3}$ 　　Work separately above and below fraction bar.

$\qquad\qquad\qquad\qquad\quad = \dfrac{-8 - 44}{-10 - 3}$ 　　Multiply.

$\qquad\qquad\qquad\qquad\quad = \dfrac{-52}{-13}$ 　　Subtract.

$\qquad\qquad\qquad\qquad\quad = 4$ 　　Divide.

(d) $-6^4 = -(6 \cdot 6 \cdot 6 \cdot 6) = -1296$

The base is 6, not -6.

(e) $(-6)^4 = (-6)(-6)(-6)(-6) = 1296$

The base is -6 here.

(f) $-8(-3) - [4 - (3 - 6)] = -8(-3) - [4 - (-3)]$ 　　Work within parentheses.

Start here.

$\qquad\qquad\qquad\qquad\qquad = -8(-3) - [4 + 3]$ 　　Definition of subtraction

$\qquad\qquad\qquad\qquad\qquad = -8(-3) - 7$ 　　Work within brackets.

$\qquad\qquad\qquad\qquad\qquad = 24 - 7$ 　　Multiply.

$\qquad\qquad\qquad\qquad\qquad = 17$ 　　Subtract. ■

Properties of Addition and Multiplication of Real Numbers

PROPERTIES OF ADDITION AND MULTIPLICATION

For real numbers a, b, and c, the following properties hold.

Closure Properties $a + b$ **and** ab **are real numbers.**

Commutative Properties $a + b = b + a$ $ab = ba$

Associative Properties $(a + b) + c = a + (b + c)$
$$(ab)c = a(bc)$$

Identity Properties   There is a real number 0 such that
$$a + 0 = a \quad \text{and} \quad 0 + a = a.$$

There is a real number 1 such that
$$a \cdot 1 = a \quad \text{and} \quad 1 \cdot a = a.$$

Inverse Properties   For each real number a, there is a single real number $-a$ such that
$$a + (-a) = 0 \quad \text{and} \quad (-a) + a = 0.$$

For each nonzero real number a, there is a single real number $\frac{1}{a}$ such that
$$a \cdot \frac{1}{a} = 1 \quad \text{and} \quad \frac{1}{a} \cdot a = 1.$$

Distributive Property of Multiplication with Respect to Addition $a(b + c) = ab + ac$
$$(b + c)a = ba + ca$$

- The set of real numbers is said to be **closed** with respect to the operations of addition and multiplication. This means that the sum of two real numbers and the product of two real numbers are themselves real numbers.

- The **commutative** properties state that two real numbers may be added or multiplied in either order without affecting the result.

- The **associative** properties allow us to group terms or factors in any manner we wish without affecting the result.

- The number 0 is the **identity element for addition.** Adding 0 to a real number will always yield.

- The number 1 is the **identity element for multiplication.** Multiplying a real number by 1 will always yield that real number.

- Each real number a has an **additive inverse,** $-a$, such that the sum of a and its additive inverse is the additive identity element 0.

- Each nonzero real number a has a **multiplicative inverse,** or **reciprocal,** $\frac{1}{a}$, such that the product of a and its multiplicative inverse is the multiplicative identity element 1.

- The **distributive** property allows us to change certain products to sums and certain sums to products.

EXAMPLE 6   **Identifying Properties of Addition and Multiplication**

Identify the property illustrated in each statement.

(a) $5 + 7$ is a real number.   **(b)** $5 + (6 + 8) = (5 + 6) + 8$

(c) $8 + 0 = 8$ **(d)** $-4\left(-\dfrac{1}{4}\right) = 1$

(e) $4 + (3 + 9) = 4 + (9 + 3)$ **(f)** $5(x + y) = 5x + 5y$

Solution

(a) The statement that the sum of two real numbers is also a real number is an example of the *closure property of addition*.

(b) Because the grouping of the terms is different on the two sides of the equation, this illustrates the *associative property of addition*.

(c) Adding 0 to a number yields the number itself. This is an example of the *identity property of addition*.

(d) Multiplying a number by its reciprocal yields 1, and this illustrates the *inverse property of multiplication*.

(e) The order of the addends (terms) 3 and 9 is different, so this is justified by the *commutative property of addition*.

(f) The factor 5 is distributed to the terms x and y. This is an example of the *distributive property of multiplication with respect to addition*.   ■

Applications of Signed Numbers

Problem-Solving Strategy

When problems deal with gains and losses, the gains may be interpreted as positive numbers and the losses as negative numbers. The phrases "in the black" (for making money) and "in the red" (for losing money) go back to the days when bookkeepers used black ink to represent gains and red ink to represent losses.

Also, temperatures above 0° are positive, and those below 0° are negative. Altitudes above sea level are considered positive, and those below sea level are considered negative.

We must sometimes express a change from one number (quantity, measure, etc.) to another. For example, if the high temperature in New Orleans was 54°F yesterday and the high today is 68°F, we interpret this as an increase of 14°. The high temperature today is *greater than* it was yesterday, so there is an *increase*, represented by a *positive* number. That number can be found mathematically.

Subtract 54 from 68 to obtain 14.

If the opposite situation had occurred, similar reasoning would say that there was a *decrease*, which would be represented by a *negative* number.

Subtract 68 from 54 to obtain −14.

In this example, both measures are positive. However, if one or both of the measures are negative, determining the amount of change may not be as easy. We can mathematically define the magnitude (or amount) of change from one to the other by taking the absolute value of their difference. ***Because opposites have the same absolute value, the order in which we subtract does not matter.***

> **DETERMINING CHANGE FROM *a* TO *b***
>
> Suppose that *a* and *b* are real numbers. The amount (without regard to sign) of change *from a to b* is given by
>
> $$|a - b|, \quad \text{or, equivalently,} \quad |b - a|.$$
>
> The change is interpreted to be a **positive change** if $a < b$, while it is a **negative change** if $a > b$. If $a = b$, there is no change.

EXAMPLE 7 Interpreting Change from a Table

Suppose that you are an employee of a fund management company and you come across the results shown in **Figure 8** in a trade publication. Amounts are rounded to the nearest percent.

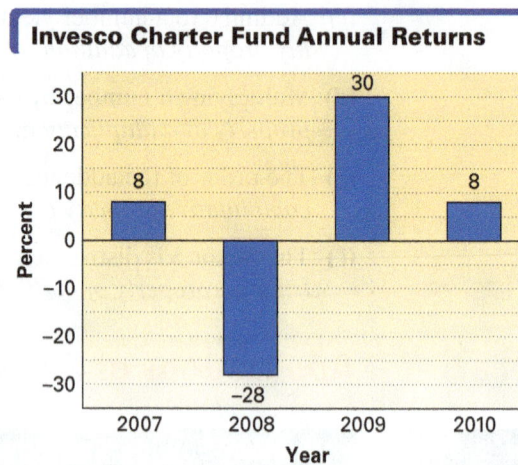

Invesco Charter Fund Annual Returns

Source: Morningstar, Invesco.

Figure 8

Illustrate by subtraction and absolute value how to determine the change in percent of annual returns from **(a)** 2007 to 2008 and **(b)** 2008 to 2009.

Solution

(a) From 2007 to 2008, the annual return went from 8 percent to −28 percent. This is a decrease, so our answer will be *negative*. The amount of the decrease is

$$|8 - (-28)| = |8 + 28| = |36| = 36.$$

(Note that we could have subtracted in the order $-28 - 8 = -36$, which has the same absolute value as 36.) The change from 2007 to 2008 is

$$- 36 \text{ percent.}$$

The negative sign indicates a decrease.   36 is the absolute value of the difference of the two amounts.

(b) To find the change from 2008 to 2009, which by inspection is an increase, find the absolute value of the difference of the two amounts in either order, and interpret it as a *positive* amount.

$$|-28 - 30| = |-58| = 58 \quad \text{or} \quad |30 - (-28)| = |58| = 58$$

The change from 2008 to 2009 is +58 percent.

The + sign is understood if not specifically indicated.

EXAMPLE 8 Interpreting Temperature Change from a Map

The daily highs for selected cities during a particularly cold two-day period in Minnesota are shown in **Figure 9.** Determine the change in temperature from Monday to Tuesday for each city listed.

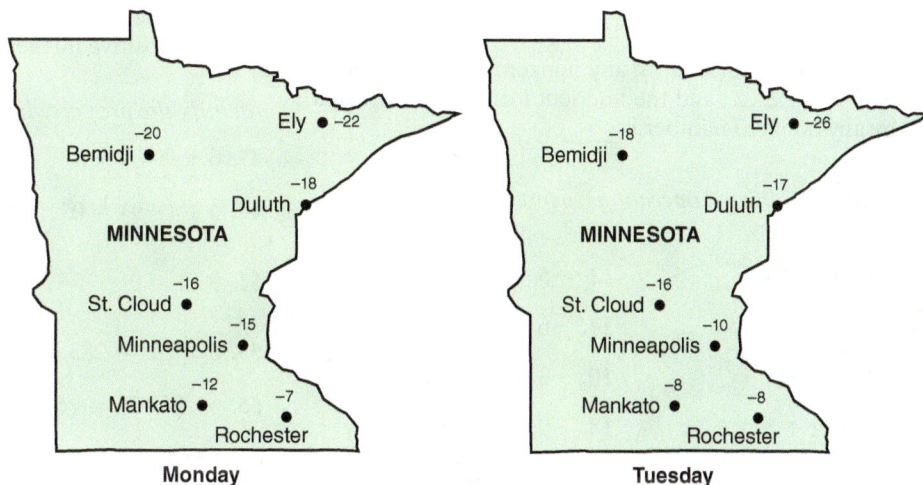

Figure 9

Solution

In each case we determine the sign of the change by observation. If the temperature on Tuesday *is greater than* that on Monday, the change is *positive*. If Tuesday's temperature *is less than* that on Monday, the change is *negative*. Equal temperatures indicate *no change*.

In Ely, the high temperature was less on Tuesday than on Monday because $-26 < -22$, so the change is negative. The amount is

$$|-26 - (-22)| = |-26 + 22| = |-4| = 4.$$

Thus the change is -4 degrees.

In Bemidji, the high temperature went from -20 to -18, which is an increase. Subtracting these temperatures in either order and taking the absolute value gives

$$|-20 - (-18)| = |-20 + 18| = |-2| = 2.$$

Thus, the change is $+2$ degrees. The other changes can be found similarly.

Duluth: +1 degree   St. Cloud: no change

Minneapolis: +5 degrees   Mankato: +4 degrees

Rochester: −1 degree

6.2 EXERCISES

Complete each statement and give an example.

1. The sum of two positive numbers is a _____ number.

2. The sum of two negative numbers is a _____ number.

3. The sum of a positive number and a negative number is negative if the negative number has the _____ absolute value.

4. The sum of a positive number and a negative number is positive if the positive number has the _____ absolute value.

5. The difference between two positive numbers is negative if _____.

6. The difference between two negative numbers is negative if _____.

7. The sum of a positive number and a negative number is 0 if the numbers are _____ .

8. The product of two numbers with the same sign is _____ .

9. The product of two numbers with different signs is _____ .

10. The quotient formed by any nonzero number divided by 0 is _____ , and the quotient formed by 0 divided by any nonzero number is _____ .

Perform the indicated operations, using the order of operations as necessary.

11. $-12 + (-8)$

12. $-5 + (-2)$

13. $12 + (-16)$

14. $-6 + 17$

15. $-12 - (-1)$

16. $-3 - (-8)$

17. $-5 + 11 + 3$

18. $-9 + 16 + 5$

19. $12 - (-3) - (-5)$

20. $15 - (-6) - (-8)$

21. $(-12)(-2)$

22. $(-3)(-5)$

23. $9(-12)(-4)(-1)(3)$

24. $-5(-17)(2)(-2)(4)$

25. $\dfrac{-18}{-3}$

26. $\dfrac{-100}{-50}$

27. $\dfrac{36}{-6}$

28. $\dfrac{52}{-13}$

29. $\dfrac{0}{12}$

30. $\dfrac{0}{-7}$

31. $-6 + [5 - (3 + 2)]$

32. $-8[4 + (7 - 8)]$

33. $-4 - 3(-2) + 5^2$

34. $-6 - 5(-8) + 3^2$

35. $\dfrac{2(-5) + (-3)(-2^2)}{-3^2 + 9}$

36. $\dfrac{3(-4) + (-5)(-2)}{2^3 - 2 + (-6)}$

37. $\dfrac{(-8 + 6) \cdot (-5)}{-5 - 5}$

38. $\dfrac{(-10 + 4) \cdot (-3)}{-7 - 2}$

39. $-8(-2) - [(4^2) + (7 - 3)]$

40. $-7(-3) - [2^3 - (3 - 4)]$

41. $\dfrac{(-6 + 3) \cdot (-4)}{-5 - 1} - \dfrac{(-9 + 6) \cdot (-3)}{-4 + 3}$

42. $\dfrac{2(-5 + 3)}{-2^2} - \dfrac{(-3^2 + 2)(3)}{3 - (-4)}$

43. $-\dfrac{1}{4}[3(-5) + 7(-5) + 1(-2)]$

44. $\dfrac{5 - 3\left(\dfrac{-5 - 9}{-7}\right) - 6}{-9 - 11 + 3 \cdot 7}$

45. Which of the following expressions are undefined?

A. $\dfrac{8}{0}$ **B.** $\dfrac{9}{6 - 6}$ **C.** $\dfrac{4 - 4}{5 - 5}$ **D.** $\dfrac{0}{-1}$

46. If you have no money in your pocket and you divide it equally among your three siblings, how much does each get? Use this situation to explain division of zero by a positive integer.

Identify the property illustrated by each statement.

47. $6 + 9 = 9 + 6$

48. $8 \cdot 4 = 4 \cdot 8$

49. $9 + (-9) = 0$

50. $12 + 0 = 12$

51. $9 \cdot 1 = 9$

52. $\left(\dfrac{1}{-3}\right) \cdot (-3) = 1$

53. $0 + 283 = 283$

54. $(3 \cdot 5) \cdot 4 = 4 \cdot (3 \cdot 5)$

55. $6 \cdot (4 \cdot 2) = (6 \cdot 4) \cdot 2$ **56.** $0 = -8 + 8$

57. $19 \cdot 12$ is a real number.

58. $19 + 12$ is a real number.

59. $7 + (2 + 5) = (7 + 2) + 5$

60. $2 \cdot (4 + 3) = 2 \cdot 4 + 2 \cdot 3$

61. $9 \cdot 6 + 9 \cdot 8 = 9 \cdot (6 + 8)$

Provide short answers in Exercises 62–66.

62. One of the authors received an email message from an old friend, Frank Capek. Frank said that his grandson had to evaluate

$$9 + 15 \div 3.$$

Frank and his wife, Barbara, said that the answer is 8, but the grandson said that the correct answer is 14. The grandson's reasoning is "There is a rule called the Order of Process so that you proceed from right to left rather than from left to right."

(a) What is the correct answer?

(b) Is his grandson's reasoning correct? Explain.

63. The following conversation actually took place between one of the authors of this text and his son, Jack, when Jack was four years old.

DADDY: "Jack, what is $3 + 0$?"
JACK: "3"
DADDY: "Jack, what is $4 + 0$?"
JACK: "4 ... and Daddy, *string* plus zero equals *string!*"

What property of addition of real numbers did Jack recognize?

64. Many everyday activities are commutative. The order in which they occur does not affect the outcome. For example, "putting on your shirt" and "putting on your pants" are commutative operations. Decide whether the given activities are commutative.

    (a) putting on your shoes; putting on your socks

    (b) getting dressed; taking a shower

    (c) combing your hair; brushing your teeth

65. Many everyday occurrences can be thought of as operations that have opposites or inverses. For example, the inverse operation for "going to sleep" is "waking up." For each of the given activities, specify its inverse activity.

    (a) cleaning up your room

    (b) earning money

    (c) increasing the volume on your MP3 player

66. The distributive property holds for multiplication with respect to addition. Does the distributive property hold for addition with respect to multiplication? That is, is

    $$a + (b \cdot c) = (a + b) \cdot (a + c)$$

    true for all values of a, b, and c? (*Hint:* Let $a = 2$, $b = 3$, and $c = 4$.)

Each expression in Exercises 67–74 is equal to either 81 *or* −81. *Decide which of these is the correct value in each case.*

67. -3^4 68. $-(3^4)$ 69. $(-3)^4$

70. $-(-3^4)$ 71. $-(-3)^4$ 72. $[-(-3)]^4$

73. $-[-(-3)]^4$ 74. $-[-(-3^4)]$

75. **S&P 500 Index Fund** The graph shows annual returns (to the nearest percent) for Class A shares of the Invesco S&P 500 Index Fund from 2009 to 2013. Use subtraction and absolute value to determine the change in returns from one year to the next. Give your answer as both a signed number (in percent) and a word description.

S&P 500 Index Fund

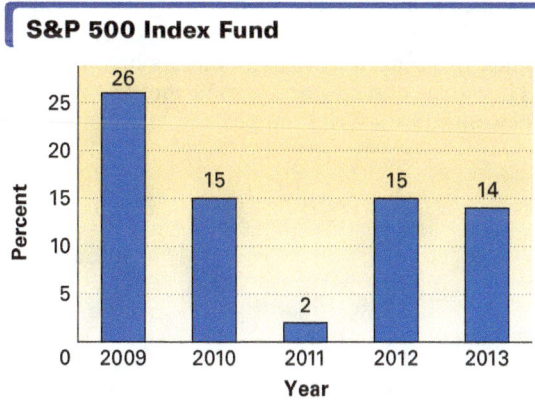

(a) 2009 to 2010   (b) 2010 to 2011

(c) 2011 to 2012   (d) 2012 to 2013

76. **Company Profits and Losses** The graph shows profits and losses for a private company for the years 2009 through 2013. Use subtraction and absolute value to determine the change in profit or loss from one year to the next. Give your answer as both a signed number (in thousands of dollars) and a word description.

Company Profits and Losses

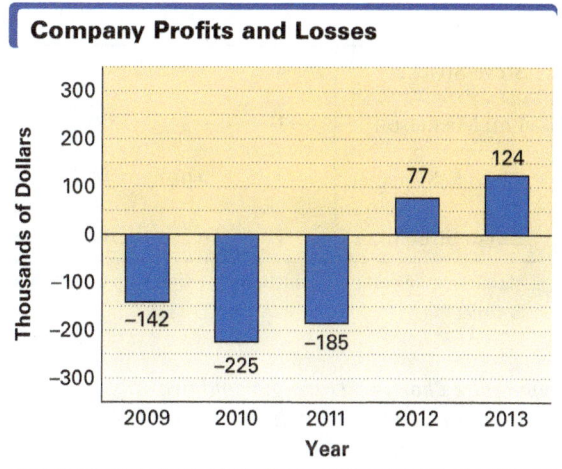

(a) 2009 to 2010   (b) 2010 to 2011

(c) 2011 to 2012   (d) 2012 to 2013

Heights of Mountains and Depths of Trenches *The table shows the heights of some selected mountains and the depths of some selected ocean trenches.*

Mountain	Height (in feet)	Trench	Depth (in feet, as a negative number)
Foraker	17,400	Philippine	−32,995
Wilson	14,246	Cayman	−24,721
Pikes Peak	14,110	Java	−23,376

Source: World Almanac and Book of Facts.

Use the information given to answer Exercises 77–82.

77. What is the difference between the height of Mt. Foraker and the depth of the Philippine Trench?

78. What is the difference between the height of Pikes Peak and the depth of the Java Trench?

79. How much deeper is the Cayman Trench than the Java Trench?

80. How much deeper is the Philippine Trench than the Cayman Trench?

81. How much higher is Mt. Wilson than Pikes Peak?

82. If Mt. Wilson and Pikes Peak were stacked one on top of the other, how much higher would they be than Mt. Foraker?

Golf Scores The table gives scores above or below par (that is, above or below the score "standard") for selected golfers during the 2013 PGA Tour Championship. Write a signed number that represents the total score for the four rounds (above or below par) for each golfer.

	Golfer	Round 1	Round 2	Round 3	Round 4
83.	Steve Stricker	−4	+1*	−2	−5
84.	Phil Mickelson	+1	−3	0	−2
85.	Charl Schwartzel	−2	+9	+7	−4
86.	Kevin Streelman	−1	+2	+4	−3

*Golf scoring commonly includes a + sign with a score over par.
Source: PGA.

Temperature Change During a cold two-day period in North Dakota, the daily highs for selected cities were as shown in the figure. Determine the change in temperature from Thursday to Friday for each city listed. **All temperatures in Exercises 87–96 are in degrees Fahrenheit.**

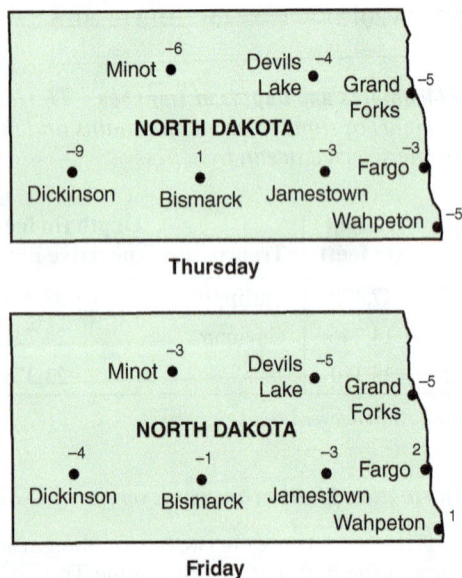

Thursday

Friday

87. Dickinson

88. Minot

89. Bismarck

90. Devils Lake

91. Jamestown

92. Grand Forks

93. Fargo

94. Wahpeton

95. *24-Hour Temperature Change* Television weather forecasters often show maps indicating how the current temperature compares to that of 24 hours earlier. Refer to the map for Thursday used in **Exercises 87–94,** and imagine that the temperatures listed indicated the change in temperature compared to 24 hours earlier. Use the table to determine the temperature in each city 24 hours earlier. (When no sign is shown on the map, + is understood.)

City	Current Temperature	Temperature 24 Hours Earlier
Dickinson	34	
Minot	28	
Bismarck	30	
Devils Lake	32	
Jamestown	35	
Grand Forks	31	
Fargo	34	
Wahpeton	36	

96. *24-Hour Temperature Change* Repeat **Exercise 95** for the Friday map used in **Exercises 87–94.**

97. *Breaching of Humpback Whales* Humpback whale researchers Mark and Debbie noticed that one of their favorite whales, "Pineapple," breached 15 feet above the surface of the ocean while her mate cruised 12 feet below the surface. What is the difference between these two levels?

98. *Altitude of Hikers* The surface, or rim, of a canyon is at altitude 0. On a hike down into the canyon, a party of hikers stop for a rest at 130 meters below the surface. They then descend another 54 meters. What is their new altitude? (Write the altitude as a signed number.)

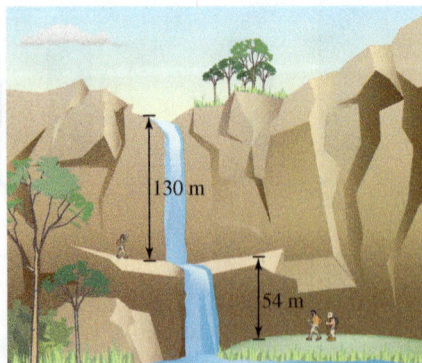

99. *Birth Date of a Greek Mathematician*   A certain Greek mathematician was born in 428 B.C. Her father was born 41 years earlier. In what year was her father born?

100. *Birth Date of a Roman Philosopher*   A certain Roman philosopher was born in 325 B.C. Her mother was born 35 years earlier. In what year was her mother born?

101. *Student Loans*   In 2005, undergraduate college students had an average of $4906 in student loans. This average increased by $788 by 2010 and then dropped $154 by 2012. What was the average amount of such loans in 2012? (*Source:* The College Board.)

102. *Entertainment Expenditures*   Among entertainment expenditures, the average annual spending per U.S. household on fees and admissions was $526 in 2001. This amount increased $80 by 2006 and then decreased $12 by 2011. What was the average household expenditure for fees and admissions in 2011? (*Source:* U.S. Bureau of Labor Statistics.)

103. *Record Low Temperatures*   The lowest temperature ever recorded in Illinois was −36°F. The lowest temperature ever recorded in Utah was 33°F lower than Illinois's record low. What is the record low temperature in Utah? (*Source:* National Climatic Data Center.)

104. *Record Low Temperatures*   The lowest temperature ever recorded in South Carolina was −19°F. The lowest temperature ever recorded in Wisconsin was 36°F lower than South Carolina's record low. What is the record low temperature in Wisconsin? (*Source:* National Climatic Data Center.)

105. *Temperature Extremes*   The lowest temperature ever recorded in Arkansas was −29°F. The highest temperature ever recorded there was 149°F more than the lowest. What is this highest temperature? (*Source:* National Climatic Data Center.)

106. *Drastic Temperature Change*   On January 23, 1943, the temperature rose 49°F in two minutes in Spearfish, South Dakota. If the starting temperature was −4°F, what was the temperature two minutes later? (*Source:* National Climatic Data Center.)

6.3   RATIONAL NUMBERS AND DECIMAL REPRESENTATION

OBJECTIVES

1 Define and identify rational numbers.

2 Write a rational number in lowest terms.

3 Add, subtract, multiply, and divide rational numbers in fraction form.

4 Solve a carpentry problem using operations with fractions.

5 Apply the density property and find arithmetic mean.

6 Convert a rational number in fraction form to a decimal number.

7 Convert a terminating or repeating decimal to a rational number in fraction form.

Definition and the Fundamental Property

Recall from **Section 6.1** that quotients of integers are called **rational numbers.** Think of the rational numbers as being made up of all the fractions (quotients of integers with denominator not equal to zero) and all the integers. Any integer can be written as a quotient of two integers.

For example, the integer 9 can be written as the quotient

$$\frac{9}{1}, \quad \text{or} \quad \frac{18}{2}, \quad \text{or} \quad \frac{27}{3}, \quad \text{and so on.}$$

Also, −5 can be expressed as a quotient of integers as

$$\frac{-5}{1}, \quad \text{or} \quad \frac{-10}{2}, \quad \text{and so on.}$$

(How can the integer 0 be written as a quotient of integers?)

RATIONAL NUMBERS

Rational numbers = $\{x \mid x \text{ is a quotient of two integers, with denominator not } 0\}$

A rational number is in **lowest terms** if the greatest common factor of the numerator (top number) and the denominator (bottom number) is 1. Rational numbers are written in lowest terms using the following property.

FUNDAMENTAL PROPERTY OF RATIONAL NUMBERS

If a, b, and k are integers with $b \neq 0$ and $k \neq 0$, then the following is true.

$$\frac{a \cdot k}{b \cdot k} = \frac{a}{b}$$

36/54▶Frac
$\frac{2}{3}$

The calculator gives 36/54 in lowest terms, as illustrated in **Example 1**.

EXAMPLE 1 **Writing a Fraction in Lowest Terms**

Write $\frac{36}{54}$ in lowest terms.

Solution

The greatest common factor of 36 and 54 is 18.

$$\frac{36}{54} = \frac{2 \cdot 18}{3 \cdot 18} = \frac{2}{3} \qquad \text{Use the fundamental property with } k = 18.$$

In **Example 1,** we see that $\frac{36}{54} = \frac{2}{3}$. If we multiply the numerator of the fraction on the left by the denominator of the fraction on the right, we obtain $36 \cdot 3 = 108$. If we multiply the denominator of the fraction on the left by the numerator of the fraction on the right, we obtain $54 \cdot 2 = 108$. The result is the same in both cases.

One way of determining whether two fractions are equal is to perform this test. If the product of the **"extremes"** (36 and 3 in this case) equals the product of the **"means"** (54 and 2), the fractions are equal. This is the **cross-product test.**

CROSS-PRODUCT TEST FOR EQUALITY OF RATIONAL NUMBERS

For rational numbers $\frac{a}{b}$ and $\frac{c}{d}$, where $b \neq 0$, $d \neq 0$, the following is true.

$$\frac{a}{b} = \frac{c}{d} \quad \text{if and only if} \quad a \cdot d = b \cdot c$$

Operations with Rational Numbers

The operation of addition of rational numbers can be illustrated by the sketches in **Figure 10.** The rectangle at the top left is divided into three equal portions, with one of the portions in color. The rectangle at the top right is divided into five equal parts, with two of them in color.

The total of the areas in color is represented by the sum

$$\frac{1}{3} + \frac{2}{5}.$$

To evaluate this sum, the areas in color must be redrawn in terms of a common unit. Since the least common multiple of 3 and 5 is 15, redraw both rectangles with 15 parts. See **Figure 11.** In the figure, 11 of the small rectangles are in color, so

$$\frac{1}{3} + \frac{2}{5} = \frac{5}{15} + \frac{6}{15} = \frac{11}{15}.$$

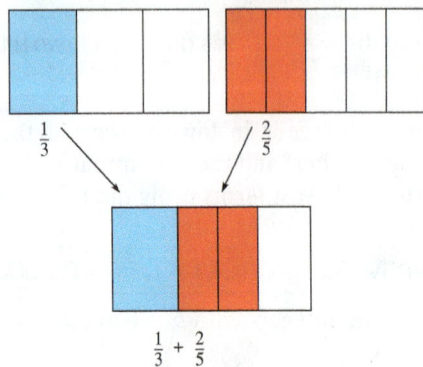

$\frac{1}{3}$ $\frac{2}{5}$

$\frac{1}{3} + \frac{2}{5}$

Figure 10

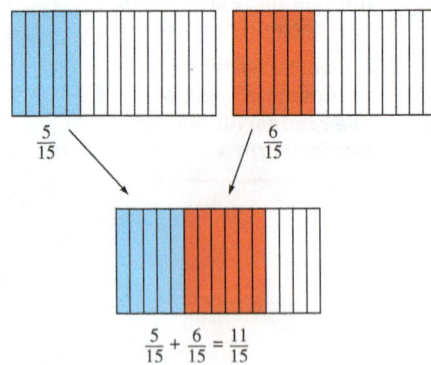

$\frac{5}{15}$ $\frac{6}{15}$

$\frac{5}{15} + \frac{6}{15} = \frac{11}{15}$

Figure 11

ADDING AND SUBTRACTING RATIONAL NUMBERS

If $\frac{a}{b}$ and $\frac{c}{d}$ are rational numbers, then the following are true.

$$\frac{a}{b} + \frac{c}{d} = \frac{ad + bc}{bd} \quad \text{and} \quad \frac{a}{b} - \frac{c}{d} = \frac{ad - bc}{bd}$$

This formal definition is seldom used in practice. We usually first rewrite the fractions with the least common multiple of their denominators, called the **least common denominator (LCD).**

EXAMPLE 2 Adding and Subtracting Rational Numbers

Perform each operation.

(a) $\dfrac{2}{15} + \dfrac{1}{10}$ (b) $\dfrac{173}{180} - \dfrac{69}{1200}$

Solution

(a) Since $30 \div 15 = 2$, $\quad \dfrac{2}{15} = \dfrac{2 \cdot 2}{15 \cdot 2} = \dfrac{4}{30}$

and since $30 \div 10 = 3$, $\quad \dfrac{1}{10} = \dfrac{1 \cdot 3}{10 \cdot 3} = \dfrac{3}{30}.$

The LCD is 30.

Thus, $\quad \dfrac{2}{15} + \dfrac{1}{10} = \dfrac{4}{30} + \dfrac{3}{30} = \dfrac{7}{30}.$

(b) The least common multiple of 180 and 1200 is 3600.

$$\frac{173}{180} - \frac{69}{1200} = \frac{3460}{3600} - \frac{207}{3600} = \frac{3460 - 207}{3600} = \frac{3253}{3600}$$

Fractions that are greater than 1 may be expressed in **mixed number** form.

The mixed number form of $\dfrac{3}{2}$ is $1\dfrac{1}{2}$.

It is understood that the whole number part and the fraction part of a mixed number are added (even though no addition symbol is shown).

EXAMPLE 3 Conversions Involving Mixed Numbers

Perform each conversion.

(a) $2\dfrac{5}{8}$ to a fraction (b) $\dfrac{86}{5}$ to a mixed number

Solution

(a) $2\dfrac{5}{8} = 2 + \dfrac{5}{8}$ The addition is understood.

$= \dfrac{16}{8} + \dfrac{5}{8}$ Write 2 as a fraction with denominator 8.

$= \dfrac{21}{8}$ Add the fractions.

2/15+1/10▶Frac
　　　　　7/30
173/180−69/1200▶Frac
　　　　　3253/3600

The results of **Example 2** are illustrated in this screen.

A shortcut for writing $2\frac{5}{8}$ as a fraction leads to the same result. We multiply the denominator (8) by the whole number part (2) and add the numerator (5) to find the numerator of the mixed number.

$$(8 \times 2) + 5 = 21$$

Then use the original denominator of the fraction part (8) as the denominator.

$$2\frac{5}{8} = \frac{21}{8}$$

(b) To convert $\frac{86}{5}$ to a mixed number, we use long division. The quotient is the whole number part of the mixed number. The remainder is the numerator of the fraction part, while the divisor is the denominator.

$$\begin{array}{r} 17 \leftarrow \text{Whole number part} \\ \text{Denominator of the fraction} \rightarrow 5\overline{)86} \\ 5 \\ \overline{36} \\ 35 \\ \overline{1} \leftarrow \text{Numerator of the fraction} \end{array}$$

Thus $\frac{86}{5} = 17\frac{1}{5}$. ∎

MULTIPLYING RATIONAL NUMBERS

If $\frac{a}{b}$ and $\frac{c}{d}$ are rational numbers, then the following is true.

$$\frac{a}{b} \cdot \frac{c}{d} = \frac{ac}{bd}$$

EXAMPLE 4 Multiplying Rational Numbers

Find each product. If it is greater than 1, write as a mixed number.

(a) $\frac{3}{4} \cdot \frac{7}{10}$ **(b)** $\frac{5}{18} \cdot \frac{3}{10}$ **(c)** $2\frac{1}{3} \cdot 1\frac{1}{2}$

Solution

(a) $\frac{3}{4} \cdot \frac{7}{10} = \frac{3 \cdot 7}{4 \cdot 10} = \frac{21}{40}$

(b) $\frac{5}{18} \cdot \frac{3}{10} = \frac{5 \cdot 3}{18 \cdot 10} = \frac{15}{180} = \frac{1 \cdot 15}{12 \cdot 15} = \frac{1}{12}$

In practice, a multiplication problem such as this is often solved by using slash marks to indicate that common factors have been divided out of the numerator and denominator.

$$\frac{\overset{1}{\cancel{5}}}{\underset{6}{\cancel{18}}} \cdot \frac{\overset{1}{\cancel{3}}}{\underset{2}{\cancel{10}}} = \frac{1}{6} \cdot \frac{1}{2}$$

3 is divided out of 3 and 18.
5 is divided out of 5 and 10.

$$= \frac{1}{12}$$

(3/4)*(7/10)▶Frac
                21/40
(5/18)*(3/10)▶Frac
                1/12

To illustrate the results of **Example 4(a) and (b),** we use parentheses around the fraction factors.

(c) $2\dfrac{1}{3}\cdot 1\dfrac{1}{2}=\dfrac{7}{3}\cdot\dfrac{3}{2}$ Convert mixed numbers to fractions.

$=\dfrac{7}{\underset{1}{3}}\cdot\dfrac{\overset{1}{3}}{2}$ Divide out the common factor 3.

$=\dfrac{7}{2}$ Multiply the fractions.

$=3\dfrac{1}{2}$ Write as a mixed number. ∎

A fraction bar indicates the operation of division. The multiplicative inverse of the nonzero number b is $\dfrac{1}{b}$. We define division using the multiplicative inverse.

DEFINITION OF DIVISION

If a and b are real numbers, where $b\neq 0$, then the following is true.

$$\frac{a}{b}=a\cdot\frac{1}{b}$$

You probably have heard the rule, "To divide fractions, invert the divisor and multiply." To illustrate this rule, suppose that you have $\dfrac{7}{8}$ of a gallon of milk and you wish to find how many quarts you have. A quart is $\dfrac{1}{4}$ of a gallon, so you must ask yourself, "*How many $\dfrac{1}{4}$s are there in $\dfrac{7}{8}$?*" This would be interpreted as

$$\frac{7}{8}\div\frac{1}{4},\quad\text{or}\quad\frac{\frac{7}{8}}{\frac{1}{4}}.$$

The fundamental property of rational numbers can be extended to rational number values of a, b, and k.

$$\frac{a}{b}=\frac{a\cdot k}{b\cdot k}=\frac{\frac{7}{8}\cdot 4}{\frac{1}{4}\cdot 4}=\frac{\frac{7}{8}\cdot 4}{1}=\frac{7}{8}\cdot\frac{4}{1}\quad\text{Let }a=\tfrac{7}{8},\,b=\tfrac{1}{4},\text{ and }k=4\left(\text{the reciprocal of }b=\tfrac{1}{4}\right).$$

To divide $\dfrac{7}{8}$ by $\dfrac{1}{4}$ is equivalent to multiplying $\dfrac{7}{8}$ by the reciprocal, $\dfrac{4}{1}$.

$$\frac{7}{8}\cdot\frac{4}{1}=\frac{28}{8}=\frac{7}{2}\quad\text{Multiply and reduce to lowest terms.}$$

Thus there are $\dfrac{7}{2}$, or $3\dfrac{1}{2}$, quarts in $\dfrac{7}{8}$ gallon.

DIVIDING RATIONAL NUMBERS

If $\dfrac{a}{b}$ and $\dfrac{c}{d}$ are rational numbers, where $\dfrac{c}{d}\neq 0$, then the following is true.

$$\frac{a}{b}\div\frac{c}{d}=\frac{a}{b}\cdot\frac{d}{c}=\frac{ad}{bc}$$

Early U.S. cents and **half cents** used fractions to denote their denominations. The half cent used $\tfrac{1}{200}$ and the cent used $\tfrac{1}{100}$. (See **Exercise 28** for a photo of an interesting error coin.)

The coins shown here were part of the collection of Louis E. Eliasberg, Sr. **Louis Eliasberg** was the only person ever to assemble a complete collection of United States coins. The Eliasberg gold coins were auctioned in 1982, while the copper, nickel, and silver coins were auctioned in two sales in 1996 and 1997. The half cent pictured sold for $506,000 and the cent sold for $27,500. The cent shown in **Exercise 28** went for a mere $2970.

EXAMPLE 5 Dividing Rational Numbers

Find each quotient.

(a) $\dfrac{3}{5} \div \dfrac{7}{15}$ **(b)** $\dfrac{-4}{7} \div \dfrac{3}{14}$ **(c)** $\dfrac{2}{9} \div 4$ **(d)** $-9 \div \dfrac{3}{5}$

Solution

(a) $\dfrac{3}{5} \div \dfrac{7}{15} = \dfrac{3}{5} \cdot \dfrac{15}{7} = \dfrac{45}{35} = \dfrac{9 \cdot 5}{7 \cdot 5} = \dfrac{9}{7}$

(b) $\dfrac{-4}{7} \div \dfrac{3}{14} = \dfrac{-4}{7} \cdot \dfrac{14}{3} = \dfrac{-56}{21} = \dfrac{-8 \cdot 7}{3 \cdot 7} = \dfrac{-8}{3} = -\dfrac{8}{3}$

> $\dfrac{-a}{b}, \dfrac{a}{-b},$ and $-\dfrac{a}{b}$ are all equal.

(c) $\dfrac{2}{9} \div 4 = \dfrac{2}{9} \div \dfrac{4}{1} = \dfrac{2}{9} \cdot \dfrac{1}{4} = \dfrac{\overset{1}{2}}{9} \cdot \dfrac{1}{\underset{2}{4}} = \dfrac{1}{18}$

(d) $-9 \div \dfrac{3}{5} = \dfrac{\overset{-3}{\cancel{-9}}}{1} \cdot \dfrac{5}{\underset{1}{\cancel{3}}} = -15$

(⁻4/7)/(3/14)▶Frac
$-\dfrac{8}{3}$

(2/9)/4▶Frac
$\dfrac{1}{18}$

This screen supports the results in **Example 4(b) and (c).**

EXAMPLE 6 Applying Operations with Fractions to Carpentry

A carpenter has a board 72 inches long that he must cut into 10 pieces of equal length. This will require 9 cuts. See **Figure 12.**

(a) If each cut causes a waste of $\dfrac{3}{16}$ inch, how many inches of actual board will remain after the cuts?

(b) What will be the length of each of the resulting pieces?

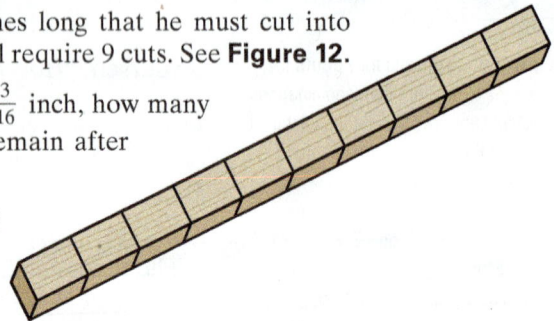

Figure 12

Solution

(a) We start with 72 inches and must subtract $\dfrac{3}{16}$ nine times.

Rather than writing out each of these, we represent the amount subtracted by $9\left(\dfrac{3}{16}\right)$.

$$72 - 9\left(\dfrac{3}{16}\right) = 72 - \left(\dfrac{9}{1} \cdot \dfrac{3}{16}\right) \qquad 9 = \dfrac{9}{1}$$

$$= 72 - \dfrac{27}{16} \qquad \text{Multiply the fractions.}$$

$$= \dfrac{1152}{16} - \dfrac{27}{16} \qquad \text{Use a common denominator.}$$

$$= \dfrac{1125}{16} \qquad \text{Subtract fractions.}$$

$$= 70\dfrac{5}{16} \qquad \text{Write as a mixed number.}$$

Thus $70\dfrac{5}{16}$ inches of actual board will remain.

(b) The $70\frac{5}{16}$, or $\frac{1125}{16}$, inches of board from part (a) will be divided into 10 pieces of equal length.

$$\frac{1125}{16} \div 10 = \frac{1125}{16} \cdot \frac{1}{10} \qquad \text{Definition of division}$$

$$= \frac{\overset{225}{\cancel{1125}}}{16} \cdot \frac{1}{\underset{2}{\cancel{10}}} \qquad \text{Divide out the common factor 5.}$$

$$= \frac{225}{32} \qquad \text{Multiply.}$$

$$= 7\frac{1}{32} \qquad \text{Convert to a mixed number.}$$

Each piece will measure $7\frac{1}{32}$ inches. ∎

Density and the Arithmetic Mean

There is no integer between two consecutive integers, such as 3 and 4. However, a rational number can always be found between any two distinct rational numbers. For this reason, the set of rational numbers is said to be *dense*.

> **DENSITY PROPERTY OF THE RATIONAL NUMBERS**
>
> If r and t are distinct rational numbers, with $r < t$, then there exists a rational number s such that
>
> $$r < s < t.$$

Repeated applications of the density property lead to the following conclusion.

> ***There are infinitely many rational numbers between two distinct rational numbers.***

One example of a rational number that is between two distinct rational numbers is the *arithmetic mean* of the two numbers. To find the **arithmetic mean**, or **average**, of n numbers, we add the numbers and then divide the sum by n. For two numbers, the number that lies halfway between them is their average.

Table 4

Academic Year	Amount of Tuition and Fees
2007–08	$21,427
2008–09	22,036
2009–10	21,908
2010–11	22,771
2011–12	23,479

Source: National Center for Education Statistics.

EXAMPLE 7 Finding an Arithmetic Mean (Average)

Table 4 shows the average amount of tuition and fees at private 4-year institutions for five academic years. What is the average cost for this period?

Solution

To find this average, divide the sum of the amounts by the number of amounts, **5**.

$$\frac{21,427 + 22,036 + 21,908 + 22,771 + 23,479}{5} = \frac{111,621}{5}$$

$$= 22,324\frac{1}{5}$$

The average amount for the period is $22,324 (rounded to the nearest dollar). ∎

EXAMPLE 8 Finding a Measurement in a Socket Wrench Set

A carpenter owns a socket wrench set that has measurements in the English system. As is the custom in carpentry, measurements are given in half-, quarter-, eighth-, and sixteenth-inches. He finds that his $\frac{3}{8}$-inch socket is just a bit too small for his job, while his $\frac{1}{2}$-inch socket is just a bit too large. He suspects that he will need to use the socket with the measure that is halfway between these (their arithmetic mean, or average). What measure socket must he use?

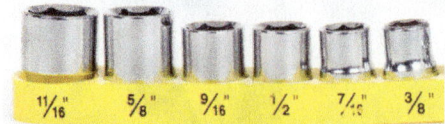

Solution

The carpenter needs to find the average of $\frac{3}{8}$ and $\frac{1}{2}$. To do this, find their sum and divide by 2.

$$\frac{3}{8} + \frac{1}{2} = \frac{3}{8} + \frac{4}{8} \qquad \text{Find a common denominator.}$$

$$= \frac{7}{8} \qquad \text{Add the numerators, and keep the same denominator.}$$

Now divide $\frac{7}{8}$ by 2.

$$\frac{7}{8} \div 2 = \frac{7}{8} \cdot \frac{1}{2} \qquad \text{Definition of division}$$

$$= \frac{7}{16} \qquad \text{Multiply the numerators and multiply the denominators.}$$

He must use the $\frac{7}{16}$-inch socket. ∎

Simon Stevin (1548–1620) worked as a bookkeeper in Belgium and became an engineer in the Netherlands army. He is usually given credit for the development of **decimals**.

Decimal Form of Rational Numbers

Rational numbers can be expressed as decimals. Decimal numerals have place values that are powers of 10. The place values are as shown here.

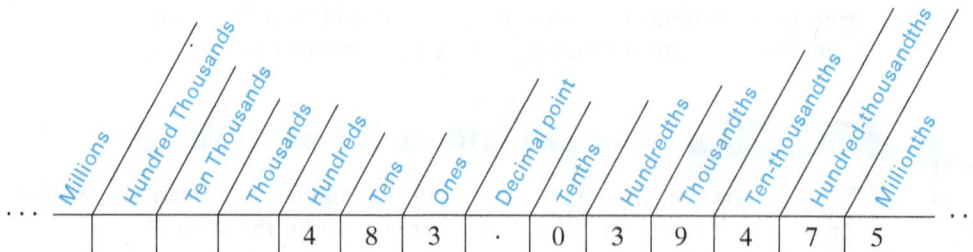

The decimal numeral 483.039475 is read "four hundred eighty-three and thirty-nine thousand, four hundred seventy-five millionths."

A rational number in the form $\frac{a}{b}$ can be expressed as a decimal most easily by entering it into a calculator. For example, to write $\frac{3}{8}$ as a decimal, enter 3, then enter the operation of division, then enter 8. Press the equals key to find the following equivalence.

$$\frac{3}{8} = 0.375$$

```
 0.375        0.3636...
8)3.000     11)4.00000...
 24            33
 ──            ──
 60            70
 56            66
 ──            ──
 40            40
 40            33
 ──            ──
  0            70
               66
               ──
               40
                ∴
```

This same result may be obtained by long division, as shown in the margin. By this result, the rational number $\frac{3}{8}$ is the same as the decimal 0.375. A decimal such as 0.375, which stops, is called a **terminating decimal.**

$$\frac{1}{4} = 0.25, \quad \frac{7}{10} = 0.7, \quad \text{and} \quad \frac{89}{1000} = 0.089 \quad \text{Examples of terminating decimals}$$

Not all rational numbers can be represented by terminating decimals. For example, convert $\frac{4}{11}$ into a decimal by dividing 11 into 4 using a calculator. The display shows

$$0.3636363636, \quad \text{or perhaps} \quad 0.363636364.$$

However, the long division process shown in the margin indicates that we will actually get 0.3636..., with the digits 36 repeating over and over indefinitely. To indicate this, we write a bar (called a *vinculum*) over the "block" of digits that repeats.

$$\frac{4}{11} = 0.\overline{36} \quad 0.\overline{36} \text{ means } 0.3636....$$

A decimal such as $0.\overline{36}$, which continues indefinitely, is called a **repeating decimal.**

$$\frac{5}{11} = 0.\overline{45}, \quad \frac{1}{3} = 0.\overline{3}, \quad \text{and} \quad \frac{5}{6} = 0.8\overline{3} \quad \text{Examples of repeating decimals}$$

```
2/3
         .6666666667
```

While 2/3 has a repeating decimal representation (2/3 = $0.\overline{6}$), the calculator rounds off in the final decimal place displayed.

While we distinguish between *terminating* and *repeating* decimals in this text, some mathematicians prefer to consider all rational numbers as repeating decimals. This can be justified by thinking this way: If the division process leads to a remainder of 0, then zeros repeat without end in the decimal form. For example, we can consider the decimal form of $\frac{3}{4}$ as follows.

$$\frac{3}{4} = 0.75\overline{0}$$

```
5/11
         .4545454545
1/3
         .3333333333
5/6
         .8333333333
```

Although only ten decimal digits are shown, all three fractions have decimals that repeat endlessly.

By considering the possible remainders that may be obtained when converting a quotient of integers to a decimal, we can draw an important conclusion about the decimal form of rational numbers. If the remainder is never zero, the division will produce a repeating decimal. This happens because each step of the division process must produce a remainder that is less than the divisor. Since the number of different possible remainders is less than the divisor, the remainders must eventually begin to repeat. This makes the digits of the quotient repeat, producing a repeating decimal.

DECIMAL REPRESENTATION OF RATIONAL NUMBERS

Any rational number can be expressed as either a terminating decimal or a repeating decimal.

To determine whether the decimal form of a quotient of integers will terminate or repeat, we use the following rule. Justification of this rule is based on the fact that the prime factors of 10 are 2 and 5, and the decimal system uses ten as its base.

CRITERIA FOR TERMINATING AND REPEATING DECIMALS

A rational number $\frac{a}{b}$ in *lowest terms* results in a **terminating decimal** if the only prime factor of the denominator is 2 or 5 (or both).

A rational number $\frac{a}{b}$ in *lowest terms* results in a **repeating decimal** if a prime other than 2 or 5 appears in the prime factorization of the denominator.

To find a baseball player's batting average, we divide the number of hits by the number of at-bats. A surprising paradox exists concerning averages. It is possible for Player A to have a higher batting average than Player B in each of two successive years, yet for the two-year period, Player B can have a higher total batting average. Look at the chart for Boston Red Sox players Mike Lowell and Jacoby Ellsbury.

Year	Mike Lowell	Jacoby Ellsbury
2007	$\frac{191}{589} = .324$	$\frac{41}{116} = .353$
2008	$\frac{115}{419} = .274$	$\frac{155}{554} = .280$
Two-year total	$\frac{306}{1008} = .304$	$\frac{196}{670} = .293$

In both individual years, Ellsbury had a higher average, but for the two-year period, Lowell had the higher average. This is an example of **Simpson's paradox** from statistics. (The authors thank Carol Merrigan for researching this information.)

EXAMPLE 9 Determining Whether a Decimal Terminates or Repeats

Determine whether the decimal form terminates or repeats.

(a) $\dfrac{7}{8}$ (b) $\dfrac{13}{150}$ (c) $\dfrac{6}{75}$

Solution

(a) The rational number $\frac{7}{8}$ is in lowest terms. Its denominator is 8, and since 8 factors as 2^3, the decimal form will terminate. No primes other than 2 or 5 divide the denominator.

(b) The rational number $\frac{13}{150}$ is in lowest terms with denominator

$$150 = 2 \cdot 3 \cdot 5^2.$$

Because **3** appears as a prime factor of the denominator, the decimal form will repeat.

(c) First write the rational number $\frac{6}{75}$ in lowest terms.

$$\frac{6}{75} = \frac{2}{25} \qquad \text{Denominator is 25.}$$

Because $25 = 5^2$, the decimal form will terminate. ■

We have seen that a rational number will be represented by either a terminating or a repeating decimal. Must a terminating decimal or a repeating decimal represent a rational number? The answer is *yes*. For example, the terminating decimal 0.6 represents a rational number.

$$0.6 = \frac{6}{10} = \frac{3}{5}$$

EXAMPLE 10 Writing Decimals as Quotients of Integers

Write each decimal as a quotient of integers.

(a) 0.437   (b) 8.2   (c) $0.\overline{85}$

Solution

```
.437▶Frac
               437
              ----
              1000
8.2▶Frac
               41
               --
               5
```

The results of **Example 10(a) and (b)** are supported in this screen.

(a) $0.437 = \dfrac{437}{1000}$ Read as "four hundred thirty-seven thousandths" and then write as a fraction.

(b) $\mathbf{8.2} = 8 + \dfrac{\mathbf{2}}{\mathbf{10}} = \dfrac{82}{10} = \dfrac{41}{5}$ Read as a decimal, write as a sum, and then add.

(c) To convert a repeating decimal to a quotient of integers, we use some simple algebra.

Step 1 Let $x = 0.\overline{85}$, so $x = 0.858585\ldots$.

Step 2 Multiply both sides of the equation $x = 0.858585\ldots$ by 100. (Use 100 since there are **two** digits in the part that repeats, and $100 = 10^2$.)

$$x = 0.858585\ldots$$
$$100x = 100(0.858585\ldots)$$
$$100x = 85.858585\ldots$$

Step 3   Subtract the expressions in Step 1 from the final expressions in Step 2.

$$100x = 85.858585\ldots$$

(Recall that $x = 1x$ and $100x - x = 99x$.)

$$\underline{\quad x = 0.858585\ldots}$$

$$99x = 85$$

Subtract.

Step 4   Solve the equation $99x = 85$ by dividing both sides by 99.

$$99x = 85$$

$$\frac{99x}{99} = \frac{85}{99}$$

Divide by 99.

$$x = \frac{85}{99}$$

$\frac{99x}{99} = x$

$$0.\overline{85} = \frac{85}{99}$$

$x = 0.\overline{85}$

When checking with a calculator, remember that the calculator will show only a finite number of decimal places and may round off in the final decimal place shown.   ■

1 = 0.99999⁹⁹⁹⁹⁹

Terminating or Repeating? One of the most baffling truths of elementary mathematics is the following:

$$1 = 0.9999\ldots.$$

Most people believe that $0.\overline{9}$ has to be less than 1, but this is not the case. The following argument shows why. Let $x = 0.9999\ldots.$ Then

$$10x = 9.9999\ldots$$

$$\underline{\quad x = 0.9999\ldots}$$

$$9x = 9$$ Subtract.

$$x = 1.$$ Divide.

Therefore, $1 = 0.9999\ldots.$ Similarly, it can be shown that any terminating decimal can be represented as a repeating decimal with an endless string of 9s. For example, $0.5 = 0.49999\ldots$ and $2.6 = 2.59999\ldots.$ This is a way of justifying that any rational number may be represented as a repeating decimal.

FOR FURTHER THOUGHT

The Influence of Spanish Coinage on Stock Prices

Until August 28, 2000, when decimalization of the U.S. stock market began, market prices were reported with fractions having denominators that were powers of 2, such as $17\frac{3}{4}$ and $112\frac{5}{8}$. Did you ever wonder why this was done?

    During the early years of the United States, prior to the minting of its own coinage, the Spanish eight-reales coin, also known as the Spanish milled dollar, circulated freely in the states. Its fractional parts, the four reales, two reales, and one real, were known as **pieces of eight** and were described as such in pirate and treasure lore. When the New York Stock Exchange was founded in 1792, it chose to use the Spanish milled dollar as its price basis, rather than the decimal base proposed by Thomas Jefferson that same year. All prices on the U.S. stock markets are now reported in decimals. (*Source:* "Stock price tables go to decimal listings," *The Times Picayune*, June 27, 2000.)

For Group or Individual Investigation

Consider this: Have you ever heard this old cheer? "Two bits, four bits, six bits, a dollar. All for the (home team), stand up and holler." The term **two bits** refers to 25 cents. Discuss how this cheer is based on the Spanish eight-reales coin.

6.3 EXERCISES

Basic Concepts of Fractions   *Complete each of the following, based on the discussion in this section.*

1. A number that can be represented as a quotient of integers with denominator not 0 is called a(n) _____.

2. In the fraction $\frac{7}{13}$, _____ is the numerator and _____ is the denominator.

3. A number of the type in **Exercise 1** is in lowest terms if its numerator and denominator have _____ as their greatest common factor.

4. The fundamental property of rational numbers allows us to write $\frac{18}{27}$ as the equivalent $\frac{2}{3}$ because there is a common factor of _____ in the numerator and the denominator.

5. The fractions $\frac{13}{27}$ and $\frac{221}{459}$ are equivalent because the product of the extremes, 13 and _____, is equal to the product of the means, 27 and _____. Both products are _____.

6. In the mixed number $3\frac{1}{7}$ there is a(n) _____ sign understood between the whole number part and the fraction part.

7. To divide $\frac{7}{12}$ by $\frac{3}{4}$, we multiply _____ by the reciprocal of the divisor. That reciprocal is _____.

8. The rational numbers exhibit the density property, which says that between any two distinct rational numbers there exists _____.

9. The integers do not exhibit the density property. For example, between 5 and _____ there is no other integer.

10. A rational number will have a decimal representation that will show one of two patterns: It will _____ or it will _____.

Choose the expression(s) that is (are) equivalent to the given rational number.

11. $\frac{4}{8}$

   A. $\frac{1}{2}$ **B.** $\frac{8}{4}$ **C.** 0.5   **D.** $0.5\overline{0}$ **E.** $0.\overline{55}$

12. $\frac{2}{3}$

   A. 0.67   **B.** $0.\overline{6}$ **C.** $\frac{20}{30}$ **D.** 0.666...   **E.** 0.6

13. $\frac{5}{9}$

   A. 0.56   **B.** 0.55   **C.** $0.\overline{5}$ **D.** $\frac{9}{5}$ **E.** $1\frac{4}{5}$

14. $\frac{1}{4}$

   A. 0.25   **B.** $0.24\overline{9}$ **C.** $\frac{25}{100}$ **D.** 4   **E.** $\frac{10}{400}$

Write each fraction in lowest terms.

15. $\frac{16}{48}$ 16. $\frac{21}{28}$ 17. $-\frac{15}{35}$ 18. $-\frac{8}{48}$

Write each fraction in three other ways.

19. $\frac{3}{8}$ 20. $\frac{9}{10}$

21. $-\frac{5}{7}$ 22. $-\frac{7}{12}$

23. **Fractional Parts**   Write a fraction in lowest terms that represents the portion of each figure that is in color.

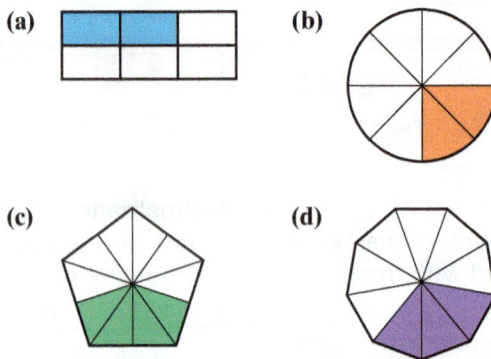

(a)   (b)

(c)   (d)

24. **Fractional Parts**   Write a fraction in lowest terms that represents the region described in parts (a)–(d).

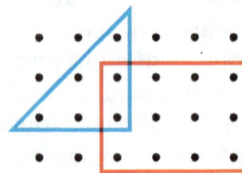

   (a) the dots in the rectangle as a part of the dots in the entire figure

   (b) the dots in the triangle as a part of the dots in the entire figure

   (c) the dots in the rectangle as a part of the dots in the union of the triangle and the rectangle

   (d) the dots in the intersection of the triangle and the rectangle as a part of the dots in the union of the triangle and the rectangle

25. *Fractional Parts* Refer to the figure for **Exercise 24**, and write a description of the region that is represented by the fraction $\frac{1}{12}$.

26. *Batting Averages* In a softball league, Paul got 8 hits in 20 at-bats, and Josh got 12 hits in 30 at-bats. Which player (if either) had the higher batting average?

27. *Batting Averages* After ten games, the following statistics were obtained.

Player	At-bats	Hits	Home Runs
Anne	40	9	2
Christine	36	12	3
Leah	11	5	1
Otis	16	8	0
Carol	20	10	2

Answer using estimation skills as necessary.

(a) Which player got a hit in exactly $\frac{1}{3}$ of his or her at-bats?

(b) Which player got a hit in just less than $\frac{1}{2}$ of his or her at-bats?

(c) Which player got a home run in just less than $\frac{1}{10}$ of his or her at-bats?

(d) Which player got a hit in just less than $\frac{1}{4}$ of his or her at-bats?

(e) Which two players got hits in exactly the same fractional parts of their at-bats? What was the fractional part, reduced to lowest terms?

28. *Error Coin* Refer to the margin note discussing the use of common fractions on early U.S. copper coinage. The photo here shows an error near the bottom that occurred on an 1802 large cent. Discuss the error and how it represents a mathematical impossibility.

Perform the indicated operations and express answers in lowest terms. Use the order of operations as necessary.

29. $\frac{3}{8} + \frac{1}{8}$ **30.** $\frac{7}{9} + \frac{1}{9}$ **31.** $\frac{5}{16} + \frac{7}{12}$

32. $\frac{1}{15} + \frac{7}{18}$ **33.** $\frac{2}{3} - \frac{7}{8}$ **34.** $\frac{13}{20} - \frac{5}{12}$

35. $\frac{5}{8} - \frac{3}{14}$ **36.** $\frac{19}{15} - \frac{7}{12}$ **37.** $\frac{3}{4} \cdot \frac{9}{5}$

38. $\frac{3}{8} \cdot \frac{2}{7}$ **39.** $-\frac{2}{3} \cdot \left(-\frac{5}{8}\right)$ **40.** $-\frac{2}{4} \cdot \frac{3}{9}$

41. $\frac{5}{12} \div \frac{15}{4}$ **42.** $\frac{15}{16} \div \frac{30}{8}$

43. $-\frac{9}{16} \div \left(-\frac{3}{8}\right)$ **44.** $-\frac{3}{8} \div \left(-\frac{5}{4}\right)$

45. $\left(\frac{1}{3} \div \frac{1}{2}\right) + \frac{5}{6}$ **46.** $\frac{2}{5} \div \left(-\frac{4}{5} \div \frac{3}{10}\right)$

Convert each mixed number to a fraction, and convert each fraction to a mixed number.

47. $4\frac{1}{3}$ **48.** $3\frac{7}{8}$ **49.** $2\frac{9}{10}$

50. $\frac{18}{5}$ **51.** $\frac{27}{4}$ **52.** $\frac{19}{3}$

Perform each operation and express your answer as a mixed number.

53. $3\frac{1}{4} + 2\frac{7}{8}$ **54.** $6\frac{1}{5} - 2\frac{7}{15}$

55. $-4\frac{7}{8} \cdot 3\frac{2}{3}$ **56.** $-4\frac{1}{6} \div 1\frac{2}{3}$

Carpenter Calculations *Carpentry applications often require calculations involving fractions and mixed numbers with denominators of 2, 4, 8, and 16. Perform each of the following, and express each answer as a mixed number.*

57. $\frac{9}{16} + 2\left(\frac{5}{8}\right) - \frac{1}{2}$ **58.** $\frac{5}{16} + 3\left(\frac{3}{8}\right) - \frac{1}{4}$

59. $9\left(2\frac{1}{2}\right) + 3\left(5\frac{1}{8}\right)$ **60.** $8\left(4\frac{1}{4}\right) - 3\left(7\frac{1}{2}\right)$

61. $5\left(\frac{1}{16}\right) + 3\left(2\frac{1}{2}\right) - \frac{7}{16}$ **62.** $8\left(\frac{3}{16}\right) + 2\left(6\frac{1}{4}\right) - \frac{13}{16}$

63. $4\left[3\left(\frac{1}{8}\right) + 2\left(\frac{7}{16}\right)\right] - 2\left(\frac{1}{2} - \frac{1}{16}\right)$

64. $6\left[2\left(\frac{5}{8}\right) + 3\left(\frac{3}{16}\right)\right] - 3\left(\frac{3}{8} - \frac{3}{16}\right)$

Solve each problem.

65. Socket Wrench Measurements A hardware store sells a 22-piece socket wrench set. The measure of the largest socket is $\frac{3}{4}$ in., while the measure of the smallest socket is $\frac{3}{16}$ in. What is the difference between these measures?

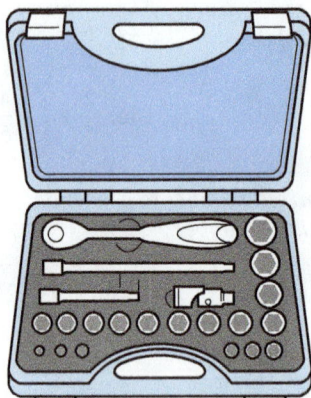

66. TV Guide First published in 1953, the digest-sized *TV Guide* has changed to a "full-sized" magazine, as shown in the figure. What is the difference in their heights? (*Source: TV Guide.*)

67. Desk Height The diagram appears in a woodworker's book. Find the height of the desk to the top of the writing surface.

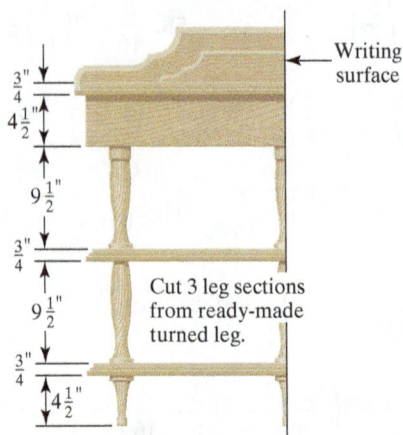

68. Legs of a Desk The desk in **Exercise 67** has four legs, each of which consists of three individual pieces. What is the total length of these twelve pieces?

69. Golf Tees The Pride Golf Tee Company, the only U.S. manufacturer of wooden golf tees, has created the Professional Tee system shown in the figure. (*Source: The Gazette.*)

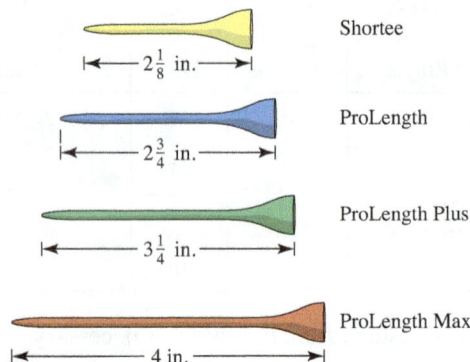

(a) Find the difference between the lengths of the ProLength Plus and the once-standard Shortee.

(b) The ProLength Max tee is the longest tee allowed by the U.S. Golf Association's Rules of Golf. How much longer is the ProLength Max than the Shortee?

70. Recipe for Grits The following chart appears on a package of Quaker® Quick Grits.

	Microwave		Stove Top		
Servings	**1**	**1**	**4**	**6**
Water	$\frac{3}{4}$ cup	1 cup	3 cups	4 cups
Grits	3 Tbsp	3 Tbsp	$\frac{3}{4}$ cup	1 cup
Salt (optional)	dash	dash	$\frac{1}{4}$ tsp	$\frac{1}{2}$ tsp

(a) How many cups of water would be needed for 6 microwave servings?

(b) How many cups of grits would be needed for 5 stove-top servings? (*Hint:* 5 is halfway between 4 and 6.)

71. Cuts on a Board A board is 48 in. long. It must be divided into four pieces of equal length, and each cut will cause a waste of $\frac{3}{16}$ in. How long will each piece be after the cuts are made?

72. Cuts on a Board A board is 72 in. long. It must be divided into six pieces of equal length, and each cut will cause a waste of $\frac{1}{8}$ in. How long will each piece be after the cuts are made?

73. Cake Recipe A cake recipe calls for $1\frac{3}{4}$ cups of sugar. A caterer has $15\frac{1}{2}$ cups of sugar on hand. How many cakes can he make?

74. Fabric It takes $2\frac{1}{4}$ yd of fabric to cover a chair of a particular design. How many chairs can be covered with $23\frac{2}{3}$ yd of fabric?

75. Fabric It takes $2\frac{3}{8}$ yd of fabric to make a costume for a school play. How much fabric would be needed for seven costumes?

76. Cookie Recipe A cookie recipe calls for $2\frac{2}{3}$ cups of sugar. How much sugar would be needed to make four batches of cookies?

77. Public 4-Year Institution Costs The table shows the average amount of tuition and fees at public 4-year institutions for in-state students for five academic years. What is the average cost for this period (rounded to the nearest dollar)? (*Source:* National Center for Education Statistics.)

Academic Year	Amount of Tuition and Fees
2007–08	$5943
2008–09	6312
2009–10	6695
2010–11	7136
2011–12	7701

78. Public 2-Year Institution Costs The table shows the average amount of tuition and fees at public 2-year institutions for in-state students for five academic years. What is the average cost for this period (rounded to the nearest dollar)? (*Source:* National Center for Education Statistics.)

Academic Year	Amount of Tuition and Fees
2007–08	$2061
2008–09	2136
2009–10	2285
2010–11	2439
2011–12	2647

Find the rational number halfway between the two given rational numbers.

79. $\dfrac{1}{2}, \dfrac{3}{4}$ **80.** $\dfrac{1}{3}, \dfrac{5}{12}$

81. $\dfrac{3}{5}, \dfrac{2}{3}$ **82.** $\dfrac{7}{12}, \dfrac{5}{8}$

83. $-\dfrac{2}{3}, -\dfrac{5}{6}$ **84.** $-\dfrac{5}{16}, -\dfrac{5}{2}$

*Use the method of **Example 9** to decide whether each rational number would yield a repeating or a terminating decimal. (Hint: Write in lowest terms before trying to decide.)*

85. $\dfrac{8}{15}$ **86.** $\dfrac{8}{35}$ **87.** $\dfrac{13}{125}$

88. $\dfrac{3}{24}$ **89.** $\dfrac{22}{55}$ **90.** $\dfrac{24}{75}$

Convert each rational number into either a repeating or a terminating decimal. Use a calculator if your instructor so allows.

91. $\dfrac{3}{4}$ **92.** $\dfrac{7}{8}$

93. $\dfrac{3}{16}$ **94.** $\dfrac{9}{32}$

95. $\dfrac{3}{11}$ **96.** $\dfrac{9}{11}$

97. $\dfrac{2}{7}$ **98.** $\dfrac{11}{15}$

Convert each decimal into a quotient of integers, written in lowest terms.

99. 0.4   **100.** 0.9

101. 0.85   **102.** 0.105

103. 0.934   **104.** 0.7984

105. $0.\overline{67}$ **106.** $0.\overline{53}$

107. $0.0\overline{42}$ **108.** $0.0\overline{86}$

109. $0.0\overline{1}$ **110.** $0.1\overline{2}$

111. Hard to Believe? Follow through on all parts of this exercise in order.

   (a) Find the decimal for $\frac{1}{3}$.

   (b) Find the decimal for $\frac{2}{3}$.

   (c) By adding the decimal expressions obtained in parts (a) and (b), obtain a decimal expression for $\frac{1}{3} + \frac{2}{3} = \frac{3}{3} = 1$.

   (d) State your result. Read the margin note on terminating and repeating decimals in this section, which refers to this idea.

112. Hard to Believe? It is a fact that $\frac{1}{3} = 0.333\ldots$. Multiply both sides of this equation by 3. Does your answer bother you? See the margin note on terminating and repeating decimals in this section.

6.4   IRRATIONAL NUMBERS AND DECIMAL REPRESENTATION

Definition and Basic Concepts

Every rational number has a decimal form that terminates or repeats, and every repeating or terminating decimal represents a rational number. However,

$$0.102001000200001000002\ldots$$

does not terminate and does not repeat. (It is true that there is a pattern in this decimal, but no single block of digits repeats indefinitely.)*

> **IRRATIONAL NUMBERS**
>
> **Irrational numbers** $= \{x \,|\, x$ is a number represented by a nonrepeating, nonterminating decimal$\}$
>   As the name implies, an irrational number cannot be represented as a quotient of integers.

The decimal number mentioned at the top of this page is an irrational number. Other irrational numbers include $\sqrt{2}$, $\frac{1+\sqrt{5}}{2}$ (ϕ, from **Section 5.5**), π (the ratio of the circumference of a circle to its diameter), and e (a constant *approximately equal to* 2.71828). There are infinitely many irrational numbers.

Irrationality of $\sqrt{2}$ and Proof by Contradiction

Figure 13 illustrates how a point with coordinate $\sqrt{2}$ can be located on a number line. (This was first mentioned in **Section 6.1**.)

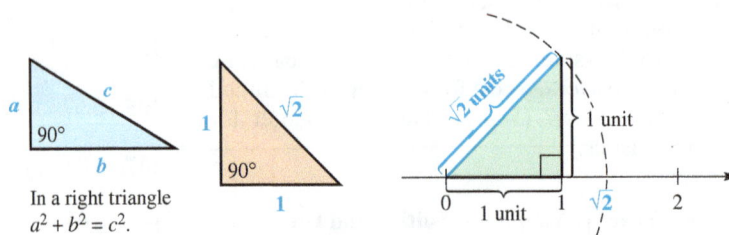

Figure 13

The irrational number $\sqrt{2}$ was discovered by the Pythagoreans in about 500 B.C. This discovery was a great setback to their philosophy that everything is based on the whole numbers. The Pythagoreans kept their findings secret, and legend has it that members of the group who divulged this discovery were sent out to sea, and, according to Proclus (410–485), "perished in a shipwreck, to a man."

The proof that $\sqrt{2}$ is irrational is a classic example of a **proof by contradiction.** We begin by assuming that $\sqrt{2}$ is rational, which leads to a contradiction, or absurdity. The method is also called **reductio ad absurdum** (Latin for "reduce to the absurd"). In order to understand the proof, we consider three preliminary facts:

1. When a rational number is written in lowest terms, the greatest common factor of the numerator and denominator is 1.

2. If an integer is even, then it has 2 as a factor and may be written in the form $2k$, where k is an integer.

3. If a perfect square is even, then its square root is even.

*In this section, we will assume that the digits of a number such as this continue indefinitely in the pattern established. The next few digits would be 000000100000002, and so on.

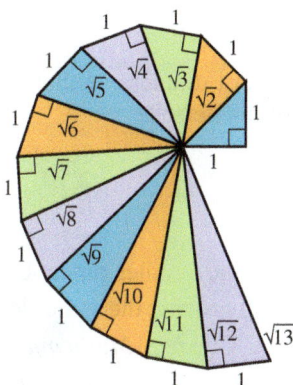

An interesting way to represent the lengths corresponding to $\sqrt{2}$, $\sqrt{3}$, $\sqrt{4}$, $\sqrt{5}$, and so on, is shown in the figure. Use the **Pythagorean theorem** to verify the lengths in the figure.

THEOREM

Statement: $\sqrt{2}$ is an irrational number.

Proof: Assume that $\sqrt{2}$ is a rational number. Then by definition,

$$\sqrt{2} = \frac{p}{q}, \quad \text{for some integers } p \text{ and } q.$$

Furthermore, assume that $\frac{p}{q}$ is the form of $\sqrt{2}$ that is written in lowest terms, so the greatest common factor of p and q is 1.

$$2 = \frac{p^2}{q^2} \quad \text{Square both sides of the equation.}$$

$$2q^2 = p^2 \quad \text{Multiply by } q^2.$$

The last equation, $2q^2 = p^2$, indicates that 2 is a factor of p^2. So p^2 is even, and thus p is even. Since p is even, it may be written in the form $2k$, where k is an integer. Now, substitute $2k$ for p in the last equation and simplify.

$$2q^2 = (2k)^2 \quad \text{Let } p = 2k.$$

$$2q^2 = 4k^2 \quad (2k)^2 = 2k \cdot 2k = 4k^2$$

$$q^2 = 2k^2 \quad \text{Divide by 2.}$$

Since q^2 has 2 as a factor, q^2 must be even, and thus q must be even. This leads to a contradiction: p and q cannot both be even because they would then have a common factor of 2. It was assumed that their greatest common factor is 1. The original assumption that $\sqrt{2}$ is rational has led to a contradiction, so $\sqrt{2}$ is irrational. ∎

Operations with Square Roots

In everyday mathematical work, nearly all of our calculations deal with rational numbers, usually in decimal form. However, we must sometimes perform operations with irrational numbers. ***Recall that $\sqrt{a}$, for $a \geq 0$, is the nonnegative number whose square is a. That is, $(\sqrt{a})^2 = a$.***

$$\sqrt{2}, \quad \sqrt{3}, \quad \text{and} \quad \sqrt{13} \quad \text{Examples of square roots that are irrational}$$

$$\sqrt{4} = 2, \quad \sqrt{36} = 6, \quad \text{and} \quad \sqrt{100} = 10 \quad \text{Examples of square roots that are rational}$$

If n is a positive integer that is not the square of an integer, then $\sqrt{n}$ is an irrational number.

A calculator with a square root key can give approximations of square roots of numbers that are not perfect squares. We use the $\approx$ symbol to indicate "is approximately equal to." Sometimes, for convenience, the $=$ symbol is used even if the statement is actually one of approximation, such as $\pi = 3.14$.

EXAMPLE 1 Using a Calculator to Approximate Square Roots

Use a calculator to verify the following approximations.

(a) $\sqrt{2} \approx 1.414213562$ **(b)** $\sqrt{6} \approx 2.449489743$ **(c)** $\sqrt{1949} \approx 44.14748011$

Solution

Use a calculator, such as the ones found on today's smartphones, to verify these approximations. Depending on the model, fewer or more digits may be displayed, and because of different rounding procedures, final digits may differ slightly. ∎

EXAMPLE 2 Applying a Formula for Pipe Flow

The following formula gives the number N of smaller pipes of diameter d that are necessary to supply the same total flow as one larger pipe of diameter D.

$$N = \sqrt{\left(\frac{D}{d}\right)^5}$$

This formula takes into account the extra friction caused by the smaller pipes. Use this formula to determine the number of $\frac{1}{2}$-inch pipes that will provide the same flow as one $1\frac{1}{2}$-inch pipe. (*Source:* Saunders, Hal M., and Robert A. Carman. *Mathematics for the Trades—A Guided Approach.* Pearson, 2015.)

Solution

Let $d = \frac{1}{2}$ and $D = 1\frac{1}{2} = \frac{3}{2}$ in the formula.

$$N = \sqrt{\left(\frac{\frac{3}{2}}{\frac{1}{2}}\right)^5} \qquad d = \frac{1}{2}, D = \frac{3}{2}$$

$$= \sqrt{3^5} \qquad \frac{3}{2} \div \frac{1}{2} = \frac{3}{2} \cdot \frac{2}{1} = 3$$

$$= \sqrt{243} \qquad 3^5 = 243$$

$$= 15.59 \qquad \text{Approximate with a calculator.}$$

In this case the answer must be a whole number, so we round 15.59 up to 16. A total of 16 pipes measuring $\frac{1}{2}$ inch will be needed. ∎

We will now look at some simple operations with square roots. Notice that

$$\sqrt{4} \cdot \sqrt{9} = 2 \cdot 3 = 6$$

and

$$\sqrt{4 \cdot 9} = \sqrt{36} = 6.$$

Thus, $\sqrt{4} \cdot \sqrt{9} = \sqrt{4 \cdot 9}$. This is a particular case of the following rule.

PRODUCT RULE FOR SQUARE ROOTS

For nonnegative real numbers a and b, the following is true.

$$\sqrt{a} \cdot \sqrt{b} = \sqrt{a \cdot b}$$

Just as every rational number $\frac{a}{b}$ can be written in *lowest terms* (by using the fundamental property of rational numbers), every square root radical has a *simplified form*.

CONDITIONS FOR A SIMPLIFIED SQUARE ROOT RADICAL

A square root radical is in **simplified form** if the following three conditions are met.

1. The number under the radical **(radicand)** has no factor (except 1) that is a perfect square.

2. The radicand has no fractions.

3. No denominator contains a radical.

Tsu Ch'ung-chih (about A.D. 500), the Chinese mathematician honored on the above stamp, investigated the digits of π. **Aryabhata,** his Indian contemporary, gave 3.1416 as the value.

EXAMPLE 3 Simplifying a Square Root Radical (Product Rule)

Simplify $\sqrt{27}$.

Solution

Since 9 is a factor of 27, and 9 is a perfect square, $\sqrt{27}$ is not in simplified form. The first condition for simplified form is not met. We simplify as follows.

$$\sqrt{27} = \sqrt{9 \cdot 3}$$

$$\boxed{\text{Simplified form}} \qquad = \sqrt{9} \cdot \sqrt{3} \qquad \text{Use the product rule.}$$

$$= 3\sqrt{3} \qquad \sqrt{9} = 3 \text{ because } 3^2 = 9.$$

Expressions such as $\sqrt{27}$ and $3\sqrt{3}$ represent the *exact value* of the square root of 27. If we use the square root key of a calculator, we find

$$\sqrt{27} \approx 5.196152423.$$

If we find $\sqrt{3}$ and then multiply the result by 3, we obtain

$$3\sqrt{3} \approx 3(1.732050808) \approx 5.196152423.$$

These approximations are the same, as we would expect. The work in **Example 3** provides the mathematical justification that they are indeed equal.

QUOTIENT RULE FOR SQUARE ROOTS

For nonnegative real numbers a and positive real numbers b, the following is true.

$$\frac{\sqrt{a}}{\sqrt{b}} = \sqrt{\frac{a}{b}}$$

EXAMPLE 4 Simplifying Square Root Radicals (Quotient Rule)

Simplify each radical.

(a) $\sqrt{\dfrac{25}{9}}$ **(b)** $\sqrt{\dfrac{3}{4}}$ **(c)** $\sqrt{\dfrac{1}{2}}$

Solution

(a) The radicand contains a fraction, so the radical expression is not simplified.

$$\sqrt{\frac{25}{9}} = \frac{\sqrt{25}}{\sqrt{9}} = \frac{5}{3} \qquad \text{Use the quotient rule.}$$

(b) $\sqrt{\dfrac{3}{4}} = \dfrac{\sqrt{3}}{\sqrt{4}} = \dfrac{\sqrt{3}}{2} \qquad \text{Use the quotient rule.}$

(c) $\sqrt{\dfrac{1}{2}} = \dfrac{\sqrt{1}}{\sqrt{2}} = \dfrac{1}{\sqrt{2}} \qquad \boxed{\text{This is not yet simplified.}}$

Condition 3 for simplified form is not met. To find an equivalent expression with no radical in the denominator, we **rationalize the denominator.**

Carl Gauss Square roots of negative numbers were called **imaginary numbers** by early mathematicians. Eventually the symbol *i* came to represent the **imaginary unit** $\sqrt{-1}$, and numbers of the form $a + bi$, where *a* and *b* are real numbers, were named **complex numbers**.

In about 1831, **Carl Gauss** was able to show that numbers of the form $a + bi$ can be represented as points on the plane just as real numbers are. He shared this contribution with **Robert Argand,** a bookkeeper in Paris, who wrote an essay on the geometry of the complex numbers in 1806. This went unnoticed at the time.

The complex number *i* has the property that its whole number powers repeat in a cycle, with four values.

$$i^0 = 1 \quad i^1 = i \quad i^2 = -1 \quad i^3 = -i$$

The pattern continues on and on this way.

To rationalize the denominator here, we multiply $\frac{1}{\sqrt{2}}$ by $\frac{\sqrt{2}}{\sqrt{2}}$, which is a form of 1, the identity element for multiplication.

$$\frac{1}{\sqrt{2}} = \frac{1}{\sqrt{2}} \cdot \frac{\sqrt{2}}{\sqrt{2}} = \frac{\sqrt{2}}{2} \qquad \sqrt{2} \cdot \sqrt{2} = 2$$

This is the simplified form of $\sqrt{\frac{1}{2}}$. ∎

Is $\sqrt{4} + \sqrt{9} = \sqrt{4+9}$ a true statement? The answer is *no*, because

$$\sqrt{4} + \sqrt{9} = 2 + 3 = 5, \quad \text{while} \quad \sqrt{4+9} = \sqrt{13}, \quad \text{and} \quad 5 \neq \sqrt{13}.$$

Square root radicals may be combined, however, if they have the same radicand. Such radicals are called **like radicals.** We add (and subtract) like radicals using the distributive property.

EXAMPLE 5   Adding and Subtracting Square Root Radicals

Add or subtract as indicated.

(a) $3\sqrt{6} + 4\sqrt{6}$ **(b)** $\sqrt{18} - \sqrt{32}$

Solution

(a) Since both terms contain $\sqrt{6}$, they are like radicals, and may be combined.

$$3\sqrt{6} + 4\sqrt{6} = (3+4)\sqrt{6} \qquad \text{Distributive property}$$

$$= 7\sqrt{6} \qquad \text{Add.}$$

(b) If we simplify $\sqrt{18}$ and $\sqrt{32}$, then this operation can be performed.

$$\sqrt{18} - \sqrt{32} = \sqrt{9 \cdot 2} - \sqrt{16 \cdot 2} \qquad \text{Factor so that perfect squares are in the radicands.}$$

$$= \sqrt{9} \cdot \sqrt{2} - \sqrt{16} \cdot \sqrt{2} \qquad \text{Product rule}$$

$$= 3\sqrt{2} - 4\sqrt{2} \qquad \text{Take square roots.}$$

$$= (3-4)\sqrt{2} \qquad \text{Distributive property}$$

$$= -1\sqrt{2} \qquad \text{Subtract.}$$

$$= -\sqrt{2} \qquad -1 \cdot a = -a \qquad ∎$$

Like radicals may be added or subtracted by adding or subtracting their coefficients (the numbers by which they are multiplied) and keeping the same radical.

Examples: $9\sqrt{7} + 8\sqrt{7} = 17\sqrt{7}$ (because $9 + 8 = 17$)

$4\sqrt{3} - 12\sqrt{3} = -8\sqrt{3}$ (because $4 - 12 = -8$)

The Irrational Numbers π, ϕ, and *e*

Figure 14 shows approximations for three important irrational numbers. The first, π, represents the ratio of the circumference of a circle to its diameter. The second, ϕ, is the Golden Ratio. Its exact value is $\frac{1 + \sqrt{5}}{2}$. The third is *e*, a fundamental number in our universe. It is the base of the *natural exponential* and *natural logarithmic* functions. The letter *e* was chosen to honor Leonhard Euler, who published extensive research on the number in 1746.

π	3.141592654
$(1+\sqrt{5})/2$	1.618033989
e	2.718281828

Figure 14

This poem, dedicated to Archimedes ("the immortal Syracusan"), allows us to learn the first 31 digits of the decimal representation of π. By replacing each word with the number of letters it contains, with a decimal point following the initial 3, the decimal is found. The poem was written by A. C. Orr, and it appeared in the *Literary Digest* in 1906.

Now I, even I, would celebrate
In rhymes unapt, the great
Immortal Syracusan, rivaled nevermore,
Who in his wondrous lore
Passed on before,
Left men his guidance
How to circles mensurate.

Pi (π)

$$\pi \approx 3.1415926535897932384626433832795$$

The computation of the digits of π has fascinated mathematicians since ancient times. Archimedes was the first to explore it extensively. As of August 5, 2014, its value had been computed to 12 trillion decimal places. Some of today's foremost researchers of the digits of π are Yasumasa Kanada, Gregory and David Chudnovsky, and the team of Alexander J. Yee and Shigeru Kondo.

A History of π, by Petr Beckmann, is a classic that is now in its third edition. Numerous Web sites, including the following, are devoted to the history and methods of computation of pi.

www.joyofpi.com/

www.math.utah.edu/~alfeld/Archimedes/Archimedes.html

www.super-computing.org/

www.pbs.org/wgbh/nova/sciencenow/3210/04.html

www.numberworld.org/misc_runs/pi-5t/details.html

One of the methods of computing pi involves the topic of *infinite series*.

EXAMPLE 6 Computing the Digits of Pi Using an Infinite Series

It is shown in higher mathematics that the *infinite series*

$$1 - \frac{1}{3} + \frac{1}{5} - \frac{1}{7} + \frac{1}{9} + \ldots \text{ "converges" to } \frac{\pi}{4}.$$

That is, as more and more terms are considered, its value becomes closer and closer to $\frac{\pi}{4}$. With a calculator, approximate the value of pi using twenty-one terms of this series.

Solution

Figure 15 shows the necessary calculation on the TI-83/84 Plus calculator. The sum of the first twenty-one terms is multiplied by 4, to obtain the approximation

$$3.189184782.$$

This series converges slowly. Although this is correct only to the first decimal place, better approximations are obtained using more terms of the series. ∎

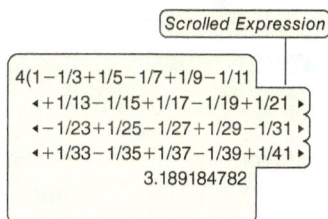

A rectangle that satisfies the condition that the ratio of its length to its width is equal to the ratio of the sum of its length and width to its length is called a **Golden Rectangle**. See **Figure 16**. This ratio is called the **Golden Ratio**. Its exact value is the irrational number $\frac{1+\sqrt{5}}{2}$. It is represented by the Greek letter ϕ (phi).

Scrolled Expression

4(1−1/3+1/5−1/7+1/9−1/11
◄+1/13−1/15+1/17−1/19+1/21 ►
◄−1/23+1/25−1/27+1/29−1/31 ►
◄+1/33−1/35+1/37−1/39+1/41 ►
3.189184782

Figure 15

$L + W$

L

W

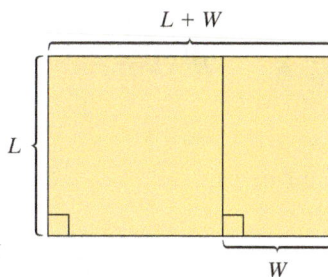

In a golden rectangle,

$$\frac{L+W}{L} = \frac{L}{W}.$$

Figure 16

Phi (ϕ)

$$\phi = \frac{1+\sqrt{5}}{2} \approx 1.6180339887498948482045868343656$$

In 1767 **J. H. Lambert** proved that π is irrational (and thus its decimal will never terminate and never repeat). Nevertheless, the 1897 Indiana state legislature considered a bill that would have *legislated* the value of π. In one part of the bill, the value was stated to be 4, and in another part, 3.2. Amazingly, the bill passed the House, but the Senate postponed action on the bill indefinitely.

Two books on phi are *The Divine Proportion, A Study in Mathematical Beauty,* by H. E. Huntley, and the more recent *The Golden Ratio,* by Mario Livio. Web sites devoted to this irrational number include the following.

www.mcs.surrey.ac.uk/Personal/R.Knott/Fibonacci/

www.goldennumber.net/

www.mathforum.org/dr.math/faq/faq.golden.ratio.html

www.geom.uiuc.edu/~demo5337/s97b/art.htm

EXAMPLE 7 Computing the Digits of Phi Using the Fibonacci Sequence

The first twelve terms of the Fibonacci sequence are

$$1, 1, 2, 3, 5, 8, 13, 21, 34, 55, 89, 144.$$

Each term after the first two terms is obtained by adding the two previous terms. Thus, the thirteenth term is $89 + 144 = 233$. As one goes farther out in the sequence, the ratio of a term to its predecessor gets closer to ϕ. How far out must one go in order to approximate ϕ so that the first five decimal places agree?

Solution

After 144, the next three Fibonacci numbers are 233, 377, and 610. **Figure 17** shows that $\frac{610}{377} \approx 1.618037135$, which agrees with ϕ to the fifth decimal place. ■

233/144	
	1.618055556
377/233	
	1.618025751
610/377	
	1.618037135

Figure 17

The irrational number e is a fundamental constant in mathematics.

e

$$e \approx 2.718281828459045235360287471353$$

Because of its nature, e is less well understood by the layperson than π (or even ϕ, for that matter). The 1994 book *e: The Story of a Number,* by Eli Maor, attempted to rectify this situation. The following Web sites also give information on e.

www.mathforum.org/dr.math/faq/faq.e.html

www-groups.dcs.st-and.ac.uk/~history/HistTopics/e.html

http://antwrp.gsfc.nasa.gov/htmltest/gifcity/e.1mil

www.math.toronto.edu/mathnet/answers/ereal.html

Example 8 illustrates another infinite series, this one converging to e.

EXAMPLE 8 Computing the Digits of e Using an Infinite Series

The infinite series

$$2 + \frac{1}{1 \cdot 2} + \frac{1}{1 \cdot 2 \cdot 3} + \frac{1}{1 \cdot 2 \cdot 3 \cdot 4} + \dots \text{ converges to } e.$$

Use a calculator to approximate e using the first seven terms of this series.

Solution

Figure 18 shows the sum of the first seven terms. (The denominators have all been multiplied out.) The sum is 2.718253968, which agrees with e to four decimal places. This series converges more rapidly than the one for π in **Example 6.** ■

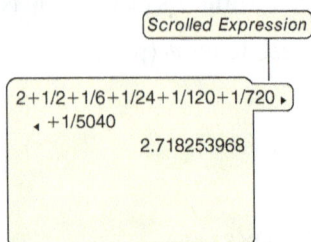

Scrolled Expression

2+1/2+1/6+1/24+1/120+1/720 ▸
 ◂ +1/5040
    2.718253968

Figure 18

Identify each number as rational *or* irrational.

1. $\dfrac{4}{9}$ 2. $\dfrac{7}{8}$ 3. $\sqrt{10}$ 4. $\sqrt{14}$

5. 1.618   6. 2.718   7. $0.\overline{41}$ 8. $0.\overline{32}$

9. π 10. e 11. 3.14159   12. $\dfrac{22}{7}$

13. 0.8787787777877778 . . .   14. $\dfrac{1 + \sqrt{5}}{2}$

In Exercises 15 and 16, work parts (a) and (b) in order.

15. (a) Find the sum.   0.272772777277772 . . .
           +0.616116111611116 . . .

   (b) Based on the result of part (a), we can conclude that the sum of two _____ numbers may be a(n) _____ number.

16. (a) Find the sum.   0.010110111011110 . . .
           +0.252552555255552 . . .

   (b) Based on the result of part (a), we can conclude that the sum of two _____ numbers may be a(n) _____ number.

Use a calculator to find a rational decimal approximation for each irrational number.

17. $\sqrt{39}$ 18. $\sqrt{44}$ 19. $\sqrt{15.1}$ 20. $\sqrt{33.6}$

21. $\sqrt{884}$ 22. $\sqrt{643}$ 23. $\sqrt{\dfrac{9}{8}}$ 24. $\sqrt{\dfrac{6}{5}}$

25. *Plumbing*   Use the formula from **Example 2,**

$$N = \sqrt{\left(\dfrac{D}{d}\right)^5},$$

to find the number of $\frac{3}{4}$-inch pipes that would be necessary to provide the same total flow as a single $1\frac{1}{2}$-inch pipe.

26. *Plumbing*   Use the formula from **Example 2** to find the number of $\frac{5}{8}$-inch pipes that would be necessary to provide the same total flow as a single 2-inch pipe.

Solve each problem. Use a calculator as necessary, and give approximations to the nearest tenth unless specified otherwise.

27. *Allied Health*   Body surface area (BSA) is used in the allied health field to calculate proper dosages of drugs based on a patient's height and weight. The formula

$$\text{BSA} = \sqrt{\dfrac{\text{weight in kg} \times \text{height in cm}}{3600}}$$

will give the patient's BSA in square meters. (*Source:* Saunders, Hal M., and Robert A. Carman. *Mathematics for the Trades—A Guided Approach, Tenth Edition.* Pearson, 2015.) What is the BSA for a person who weighs 100 kilograms and is 200 centimeters tall?

28. *Allied Health*   Use the formula from **Exercise 27** to find the BSA of a person who weighs 94 kilograms and is 190 centimeters tall.

29. *Diagonal of a Box*   The length of the diagonal of a box is given by

$$D = \sqrt{L^2 + W^2 + H^2},$$

where L, W, and H are the length, width, and height of the box. Find the length of the diagonal, D, of a box that is 4 feet long, 3 feet wide, and 2 feet high.

30. *Distance to the Horizon*   A meteorologist provided this answer to one of the readers of his newspaper column, a 6-foot man who lives 150 feet above the ground.

   To find the distance to the horizon in miles, take the square root of the height of your view and multiply that result by 1.224. That will give you the number of miles to the horizon.

   Assuming the viewer's eyes are 6 feet above his floor, the total height from the ground is $150 + 6 = 156$ feet. How far can he see to the horizon?

31. *Period of a Pendulum*   The period of a pendulum, in seconds, depends on its length, L, in feet, and is given by the formula

$$P = 2\pi\sqrt{\dfrac{L}{32}}.$$

   If a pendulum is 5.1 feet long, what is its period? Use 3.14 for π.

32. *Radius of an Aluminum Can*   The radius of the circular top or bottom of an aluminum can with surface area S in square cm and height h in cm is given by

$$r = \dfrac{-h + \sqrt{h^2 + 0.64S}}{2}.$$

What radius should be used to make a can with height 12 cm and surface area 400 square cm?

33. *Electronics Formula* The formula

$$I = \sqrt{\frac{2P}{L}}$$

relates the coefficient of self-induction L (in henrys), the energy P stored in an electronic circuit (in joules), and the current I (in amps). Find the value of I if $P = 120$ joules and $L = 80$ henrys.

34. *Law of Tensions* In the study of sound, one version of the law of tensions is

$$f_1 = f_2 \sqrt{\frac{F_1}{F_2}}.$$

Find f_1 to the nearest unit if $F_1 = 300$, $F_2 = 60$, and $f_2 = 260$.

Accident Reconstruction *Police sometimes use the following procedure to estimate the speed at which a car was traveling at the time of an accident. A police officer drives the car involved in the accident under conditions similar to those during which the accident took place and then skids to a stop. If the car is driven at 30 miles per hour, then the speed at the time of the accident is given by*

$$s = 30 \sqrt{\frac{a}{p}}.$$

Here, a is the length of the skid marks left at the time of the accident, and p is the length of the skid marks in the police test. Find s, to the nearest unit, for the following values of a and p.

35. $a = 862$ feet; $p = 156$ feet

36. $a = 382$ feet; $p = 96$ feet

37. $a = 84$ feet; $p = 26$ feet

38. $a = 90$ feet; $p = 35$ feet

39. *Area of the Bermuda Triangle* **Heron's formula** gives a method of finding the area of a triangle if the lengths of its sides are known. Suppose that a, b, and c are the lengths of the sides. Let s denote one-half of the perimeter of the triangle (called the **semiperimeter**); that is,

$$s = \frac{1}{2}(a + b + c).$$

Then the area $\mathcal{A}$ of the triangle is given by

$$\mathcal{A} = \sqrt{s(s - a)(s - b)(s - c)}.$$

Find the area of the Bermuda Triangle, if the "sides" of this triangle measure approximately 850 miles, 925 miles, and 1300 miles. Give your answer to the nearest thousand square miles.

40. *Area Enclosed by the Vietnam Veterans' Memorial* The Vietnam Veterans' Memorial in Washington, D.C., is in the shape of an unenclosed isosceles triangle with equal sides of length 246.75 feet. If the triangle were enclosed, the third side would have length 438.14 feet. Use Heron's formula from the previous exercise to find the area of this enclosure to the nearest hundred square feet. (*Source:* Information pamphlet obtained at the Vietnam Veterans' Memorial.)

41. *Perfect Triangles* A **perfect triangle** is a triangle whose sides have whole number lengths and whose area is numerically equal to its perimeter. Use Heron's formula to show that the triangle with sides of length 9, 10, and 17 is perfect.

42. *Heron Triangles* A **Heron triangle** is a triangle having integer sides and area. Use Heron's formula to show that each of the following is a Heron triangle.

(a) $a = 11, b = 13, c = 20$

(b) $a = 13, b = 14, c = 15$

(c) $a = 7, b = 15, c = 20$

*Use the methods of **Examples 3 and 4** to simplify each expression. Then, use a calculator to approximate both the given expression and the simplified expression. (Both should be the same.)*

43. $\sqrt{50}$ **44.** $\sqrt{32}$ **45.** $\sqrt{75}$

46. $\sqrt{150}$ **47.** $\sqrt{288}$ **48.** $\sqrt{200}$

49. $\dfrac{5}{\sqrt{6}}$ **50.** $\dfrac{3}{\sqrt{2}}$

51. $\sqrt{\dfrac{7}{4}}$ **52.** $\sqrt{\dfrac{8}{9}}$

53. $\sqrt{\dfrac{7}{3}}$ **54.** $\sqrt{\dfrac{14}{5}}$

*Use the method of **Example 5** to perform the indicated operations.*

55. $\sqrt{17} + 2\sqrt{17}$ **56.** $3\sqrt{19} + \sqrt{19}$

57. $5\sqrt{7} - \sqrt{7}$ **58.** $3\sqrt{27} - \sqrt{27}$

59. $3\sqrt{18} + \sqrt{2}$ **60.** $2\sqrt{48} - \sqrt{3}$

61. $-\sqrt{12} + \sqrt{75}$ **62.** $2\sqrt{27} - \sqrt{300}$

Irrational Investigations Exercises 63–78 deal with the irrational numbers π, ϕ, e, and $\sqrt{3}$. Use a calculator or computer as necessary.

63. Move one matchstick to make the equation approximately true. (*Source:* www.joyofpi.com)

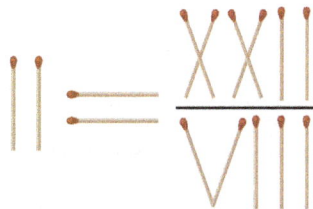

64. Find the square root of $\frac{2143}{22}$ using a calculator. Then find the square root of that result. Compare your result to the decimal given for π in the margin note. What do you notice?

65. Use a calculator to find the first eight digits in the decimal for $\frac{355}{113}$. Compare the result to the decimal for π given in the text. What do you notice?

66. You may have seen the statements

"Use $\frac{22}{7}$ for π"   and   "Use 3.14 for π."

Since $\frac{22}{7}$ is the quotient of two integers, and 3.14 is a terminating decimal, do these statements suggest that π is rational?

67. In the Bible (I Kings 7:23), a verse describes a circular pool at King Solomon's temple, about 1000 B.C. The pool is said to be ten cubits across, "and a line of 30 cubits did compass it round about." What value of π does this imply?

68. The ancient Egyptians used a method for finding the area of a circle that is equivalent to a value of 3.1605 for π. Write this decimal as a mixed number.

69. The computation of π has fascinated mathematicians and others for centuries. In the nineteenth century, the British mathematician William Shanks spent many years of his life calculating π to 707 decimal places. It turned out that only the first 527 were correct. Use an Internet search to find the 528th decimal digit of π (following the whole number part 3).

70. A **mnemonic device** is a scheme whereby one is able to recall facts by memorizing something completely unrelated to the facts. One way of learning the first few digits of the decimal for π is to memorize a sentence (or several sentences) and count the letters in each word of the sentence. For example,

"See, I know a digit,"

will give the first 5 digits of π: "See" has 3 letters, "I" has 1 letter, "know" has 4 letters, "a" has 1 letter, and "digit" has 5 letters. So the first five digits are 3.1415.

Verify that the following mnemonic devices work.

(a) "May I have a large container of coffee?"

(b) "See, I have a rhyme assisting my feeble brain, its tasks ofttimes resisting."

(c) "How I want a drink, alcoholic of course, after the heavy lectures involving quantum mechanics."

71. In the second season of the original *Star Trek* series, the episode "Wolf in the Fold" told the story of an alien entity that had taken over the computer of the starship *Enterprise*. To drive the entity out of the computer, Mr. Spock gave the alien the compulsory directive to compute π to the last digit. Explain why this strategy proved successful.

72. *Northern Exposure* ran between 1990 and 1995 on the CBS network. In the episode "Nothing's Perfect," the local disc jockey Chris Stevens meets and develops a relationship with a mathematician after accidentally running over her dog. Her area of research is computation of the decimal digits of pi. She mentions that a string of eight 8s appears in the decimal relatively early in the expansion. Use an Internet search to determine the position at which this string appears.

73. Use a calculator to find the decimal approximations for

$$\phi = \frac{1 + \sqrt{5}}{2} \quad \text{and its \textbf{conjugate,}} \quad \frac{1 - \sqrt{5}}{2}.$$

Comment on the similarities and differences in the two decimals.

74. In some literature, the Golden Ratio is defined to be the reciprocal of

$$\frac{1 + \sqrt{5}}{2}, \quad \text{which is} \quad \frac{2}{1 + \sqrt{5}}.$$

Use a calculator to find a decimal approximation for $\frac{2}{1 + \sqrt{5}}$, and compare it to ϕ as defined in this text. What do you observe?

75. Near the end of the 2008 movie *Harold & Kumar Escape from Guantanamo Bay,* Kumar (Kal Penn), recites a poem dealing with the square root of 3, written by the late David Feinberg. The text of the poem can be found on the Internet by searching

"David Feinberg Square Root of 3."

There are references to irrational numbers, integers, an approximation for the square root of 3, and the product rule for square root radicals. Why would Kumar prefer the 3 to be a 9?

76. See **Exercise 75.** What is the decimal approximation for $\sqrt{3}$ (to 4 decimal places) given in Feinberg's poem?

77. An approximation for e is

$$2.718281828.$$

A student noticed that there seems to be a repetition of four digits in this number (1, 8, 2, 8) and concluded that e is rational, because repeating decimals represent rational numbers. Was the student correct? Why or why not?

78. Use a calculator with an exponential key to find values for the following:

$$(1.1)^{10}, \quad (1.01)^{100}, \quad (1.001)^{1000},$$

$$(1.0001)^{10,000}, \quad \text{and} \quad (1.00001)^{100,000}.$$

Compare your results to the approximation given for e in this section. What do you find?

Roots Other Than Square Roots *The concept of square (second) root can be extended to* **cube (third) root, fourth root,** *and so on. For example,*

$$\sqrt[3]{8} = 2 \quad because \quad 2^3 = 8,$$

$$\sqrt[3]{1000} = 10 \quad because \quad 10^3 = 1000,$$

$$\sqrt[4]{81} = 3 \quad because \quad 3^4 = 81, \quad and \ so \ on.$$

If $n \geq 2$ and a is a nonnegative number, then $\sqrt[n]{a}$ represents the nonnegative number whose nth power is a.

Find each root.

79. $\sqrt[3]{64}$ **80.** $\sqrt[3]{125}$

81. $\sqrt[3]{343}$ **82.** $\sqrt[3]{729}$

83. $\sqrt[3]{216}$ **84.** $\sqrt[3]{512}$

85. $\sqrt[4]{1}$ **86.** $\sqrt[4]{16}$

87. $\sqrt[4]{256}$ **88.** $\sqrt[4]{625}$

89. $\sqrt[4]{4096}$ **90.** $\sqrt[4]{2401}$

Use a calculator to approximate each root. (Hint: To find the fourth root, find the square root of the square root.)

91. $\sqrt[3]{43}$ **92.** $\sqrt[3]{87}$

93. $\sqrt[3]{198}$ **94.** $\sqrt[4]{2107}$

95. $\sqrt[4]{10,265.2}$ **96.** $\sqrt[4]{863.5}$

97. $\sqrt[4]{968.1}$ **98.** $\sqrt[4]{12,966.4}$

6.5   APPLICATIONS OF DECIMALS AND PERCENTS

OBJECTIVES

1 Perform operations of arithmetic with decimal numbers.

2 Round whole numbers and decimals to a given place value.

3 Perform computations using percent.

4 Convert among forms of fractions, decimals, and percents.

5 Find percent increase and percent decrease.

6 Apply formulas involving fractions and decimals from the allied health industry.

Operations with Decimals

Because calculators have, for the most part, replaced paper-and-pencil methods for operations with decimals and percent, we will only briefly mention these latter methods. *We strongly suggest that the work in this section be done with a calculator at hand.*

ADDITION AND SUBTRACTION OF DECIMALS

To add or subtract decimal numbers, line up the decimal points in a column and perform the operation.

EXAMPLE 1   Adding and Subtracting Decimal Numbers

Find each of the following.

(a) $0.46 + 3.9 + 12.58$ **(b)** $12.1 - 8.723$

Solution

(a)
$$
\begin{array}{r}
0.46 \\
3.90 \\
+12.58 \\
\hline
16.94
\end{array}
$$
Line up decimal points.
Attach a zero as a placeholder.
← Sum

(b)
$$
\begin{array}{r}
12.100 \\
-8.723 \\
\hline
3.377
\end{array}
$$
Attach zeros.
← Difference

■

```
.46+3.9+12.58
                        16.94
12.1-8.723
                         3.377
```

This screen supports the results in
Example 1.

Recall that when two numbers are multiplied, the numbers are **factors** and the answer is the **product.** When two numbers are divided, the number being divided is the **dividend,** the number doing the dividing is the **divisor,** and the answer is the **quotient.**

MULTIPLICATION AND DIVISION OF DECIMALS

Multiplication To multiply decimals, multiply in the same manner as integers are multiplied. The number of decimal places to the right of the decimal point in the product is the *sum* of the numbers of places to the right of the decimal points in the factors.

Division To divide decimals, move the decimal point to the right the same number of places in the divisor and the dividend so as to obtain a whole number in the divisor. Divide in the same manner as integers are divided. The number of decimal places to the right of the decimal point in the quotient is the same as the number of places to the right of the decimal in the dividend.

EXAMPLE 2 **Multiplying and Dividing Decimal Numbers**

Find each of the following.

(a) 4.613×2.52 **(b)** $65.175 \div 8.25$

Solution

```
Normal Sci Eng
Float 0123456789
Radian Degree
Func Par Pol Seq
Connected Dot
Sequential Simul
Real a+bi re^θi
Full Horiz G-T
```

```
NORMAL  SCI  ENG
FLOAT 0123456789
RADIAN DEGREE
FUNC  PAR POL SEQ
CONNECTED DOT
SEQUENTIAL SIMUL
REAL a+bi re^θi
FULL  HORIZ  G-T
            ↓NEXT↓
```

Here the calculator is set to round the answer to two decimal places.

(a)
```
       4.613   ← 3 decimal places
   ×   2.52    ← 2 decimal places
     ─────────
       9226
      23065
       9226
   ─────────
    11.62476   ← 3 + 2 = 5 decimal places
```

(b)
$$8.25\overline{)65.175} \rightarrow$$

```
            7.9
    825)6517.5
        5775
        ────
        7425
        7425
        ────
           0
```

Bring the decimal point straight up in the answer.

Rounding Methods

To round, or approximate, a whole number or a decimal number to a given place value, use the following procedure. (You may wish to refer to the diagram in **Section 6.3** that describes place values determined by powers of 10.)

RULES FOR ROUNDING

Whole Numbers

Step 1 Locate the **place** to which the whole number is being rounded.

Step 2 Look at the next **digit to the right** of the place to which the number is being rounded.

Step 3A If the digit to the right is **less than 5,** replace it and all digits following it with zeros. Do not change the digit in the place to which the number is being rounded.

Step 3B If the digit to the right is **5 or greater,** replace it and all digits following it with zeros, but also add 1 to the digit in the place to which the number is being rounded. (If adding 1 to a 9, replace the 9 with 0 and add 1 to the next digit to the left. If that next digit is also 9, repeat this procedure; and so on, to the left.)

"**Technology** pervades the world outside school. There is no question that students will be expected to use calculators in other settings; this technology is now part of our culture . . . students no longer have the same need to perform these [paper-and-pencil] procedures with large numbers of lengthy expressions that they might have had in the past without ready access to technology."

From *Computation, Calculators, and Common Sense (A Position Paper of the National Council of Teachers of Mathematics).*

RULES FOR ROUNDING

Decimal Numbers

Steps 1 and 2 are the same as for whole numbers.

Step 3A   If the digit to the right is **less than 5,** drop all digits to the right of the place to which the number is being rounded. Do not change the digit in the place to which it is being rounded.

Step 3B   If the digit to the right is **5 or greater,** drop it and all digits following it, but also add 1 to the digit in the place to which the number is being rounded. (When adding 1 to a 9, use the same procedure as for whole numbers.)

Note: Some disciplines have guidelines specifying that in certain cases, a *downward* roundoff is made. We will not investigate these.

EXAMPLE 3 Rounding Numbers

Round to the place values indicated.

(a) 0.9175 to the nearest tenth, hundredth, and thousandth

(b) 1358 to the nearest thousand, hundred, and ten

Solution

(a) Round 0.9175 as follows: 0.9 to the nearest tenth,
0.92 to the nearest hundredth,
0.918 to the nearest thousandth.

(b) Round 1358 as follows: 1000 to the nearest thousand,
1400 to the nearest hundred,
1360 to the nearest ten.

> 0.9175
> 0.92
>
> The calculator rounds 0.9175 to the nearest hundredth.
>
> 0.9175
> 0.918
>
> The calculator rounds 0.9175 *up* to 0.918.

Percent

The word **percent** means **"per hundred."** The symbol **%** represents "percent."

PERCENT

$$1\% = \frac{1}{100} = 0.01$$

EXAMPLE 4 Converting Percents to Decimals

Convert each percent to a decimal.

(a) 98%   **(b)** 3.4%   **(c)** 0.2%   **(d)** 150%

Solution

(a) $98\% = 98(1\%) = 98(0.01) = 0.98$

(b) $3.4\% = 3.4(1\%) = 3.4(0.01) = 0.034$

(c) $0.2\% = 0.2(1\%) = 0.2(0.01) = 0.002$

(d) $150\% = 150(1\%) = 150(0.01) = 1.5$

EXAMPLE 5 Converting Decimals to Percents

Convert each decimal to a percent.

(a) 0.13   **(b)** 0.532   **(c)** 2.3   **(d)** 0.07

Solution

Reverse the procedure used in **Example 4.**

(a) $0.13 = 13(0.01) = 13(1\%) = 13\%$

(b) $0.532 = 53.2(0.01) = 53.2(1\%) = 53.2\%$

(c) $2.3 = 230(0.01) = 230(1\%) = 230\%$

(d) $0.07 = 7(0.01) = 7(1\%) = 7\%$ ∎

Examples 4 and 5 suggest the following shortcut methods for converting.

CONVERTING BETWEEN DECIMALS AND PERCENTS

To convert a percent to a decimal, drop the percent symbol (%) and move the decimal point two places to the left, inserting zeros as placeholders if necessary.

To convert a decimal to a percent, move the decimal point two places to the right, inserting zeros as placeholders if necessary, and attach the percent symbol (%).

EXAMPLE 6 Converting Fractions to Percents

Convert each fraction to a percent.

(a) $\dfrac{3}{5}$ **(b)** $\dfrac{14}{25}$ **(c)** $2\dfrac{7}{10}$

Solution

(a) First write $\dfrac{3}{5}$ as a decimal. Dividing 3 by 5 gives $\dfrac{3}{5} = 0.6 = 60\%$.

(b) $\dfrac{14}{25} = \dfrac{14 \cdot 4}{25 \cdot 4} = \dfrac{56}{100} = 0.56 = 56\%$

(c) $2\dfrac{7}{10} = 2.7 = 270\%$ ∎

CONVERTING A FRACTION TO A PERCENT

To convert a fraction to a percent, convert the fraction to a decimal, and then convert the decimal to a percent.

The percent symbol, %, probably evolved from a symbol introduced in an Italian manuscript of 1425. Instead of "per 100," "P 100," or "P cento," which were common at that time, the author used "P$\overset{R}{}$." By about 1650 the $\overset{R}{}$ had become $\frac{0}{0}$, so "per $\frac{0}{0}$" was often used. Finally the "per" was dropped, leaving $\frac{0}{0}$ or %.

(*Source: Historical Topics for the Mathematics Classroom,* the Thirty-first Yearbook of the National Council of Teachers of Mathematics, 1969.)

In the following examples involving percents, three methods are shown. They illustrate some basic ideas of solving equations. The second method in each case involves using cross-products. The third method involves the percent key of a basic calculator. (Keystrokes may vary among models.)

EXAMPLE 7 Finding a Percent of a Number

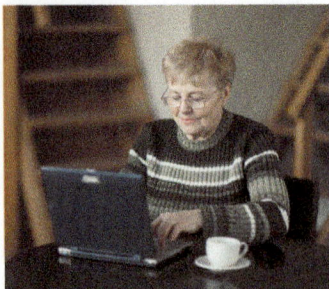

A publisher requires that an author reduce the page count in her next edition by 18%. The current edition has 250 pages. How many pages must the author cut from her book?

Solution

This problem can be reworded as follows: Find 18% of 250.

Method 1   The key word "of" translates as "times."

$$18\%(250) = 0.18(250) = 45$$

Method 2   Think "18 is to 100 as what (x) is to 250?" This translates as follows.

$$\frac{18}{100} = \frac{x}{250}$$

$$100x = 18 \cdot 250 \qquad \frac{a}{b} = \frac{c}{d} \text{ leads to } ad = bc.$$

$$x = \frac{18 \cdot 250}{100} \qquad \text{Divide by 100.}$$

$$x = 45 \qquad \text{Simplify.}$$

Method 3   Use the percent key on a calculator with the following keystrokes.

$$\boxed{2}\,\boxed{5}\,\boxed{0}\,\boxed{\times}\,\boxed{1}\,\boxed{8}\,\boxed{\%} \qquad \textbf{45} \quad \text{Final display}$$

Thus, 18% of 250 is 45. The author must reduce her page count by 45 pages. ∎

EXAMPLE 8 Finding What Percent One Number Is of Another

A kindergarten teacher has submitted 6 of the required 50 class preparations to the principal of her school. What percent of her requirement has she completed?

Solution

This problem can be reworded as follows: What percent of 50 is 6?

Method 1   Let the phrase "what percent" be represented by $x \cdot 1\%$ or $0.01x$. Again the word "of" translates as "times," while "is" translates as "equals."

$$0.01x \cdot 50 = 6$$

$$0.50x = 6 \qquad \text{Multiply on the left side.}$$

$$50x = 600 \qquad \text{Multiply by 100 to clear decimals.}$$

$$x = 12 \qquad \text{Divide by 50.}$$

Method 2   Think "What (x) is to 100 as 6 is to 50?"

$$\frac{x}{100} = \frac{6}{50}$$

$$50x = 600 \qquad \text{Cross-products}$$

$$x = 12 \qquad \text{Divide by 50.}$$

Method 3   Use the following keystrokes on a calculator.

$$\boxed{6}\,\boxed{\div}\,\boxed{5}\,\boxed{0}\,\boxed{\%} \qquad \textbf{12} \quad \text{Final display}$$

Thus, 6 is 12% of 50. She has completed 12% of her requirement. ∎

Cynthia (not her real name) had graduated in fashion merchandising and worked for eight years in an upscale women's clothing store in New Orleans. She decided to leave her position to go back to school with the goal of becoming a veterinarian.

Her curriculum required a course in college algebra, and Cynthia finished with the highest average in her class. Along the way she realized something that would have saved her a lot of time in her previous job. One day her instructor (one of the authors) made an offhand comment about percent increase. He said something like this:

If you want to mark up an item by a given percent, such as 15%, all you need to do is multiply the original price by 1.15. That's your final amount.

After class, she approached him and admitted that when she marked up items for retail sale, as her first step she would find the amount of markup on her calculator, and then in a separate step add it to the wholesale cost of the item.

For example, if the cost was $25.80, she would calculate **15% of $25.80** on her calculator in the first step, to obtain **$3.87**. Then in her second step, she would add

$$\textbf{\$3.87} + \$25.80 \quad \text{to obtain} \quad \$29.67. \quad \leftarrow \text{Selling price}$$

She only now realized that the selling price could be obtained in a single step by multiplying

$$\$25.80 \cdot 1.15 \quad \text{to obtain} \quad \$29.67. \quad \leftarrow \text{Selling price}$$

She realized that she could have saved time had she known this procedure.

The reason that this method works can be explained by the identity property of multiplication and the distributive property. The explanation below uses 15%, but any percent could be used with the same result.

Suppose that C represents the wholesale cost of an item. Then $0.15C$ represents 15% of C, the amount of the markup. The selling price S is found by adding wholesale cost to the amount of markup.

$$S = C + 0.15C \qquad \text{Selling price = wholesale cost + markup}$$
$$S = 1C + 0.15C \qquad \text{Identity property of multiplication}$$
$$S = (1 + 0.15)C \qquad \text{Distributive property}$$
$$S = 1.15C \qquad 1 + 0.15 = 1.15$$

A similar situation holds if a discount is to be applied. For example, to mark an item down 25%, just subtract from 100% to get 75%, and then multiply the original amount by 0.75. The result is the discounted price.

Cynthia went on to get her doctor's degree in veterinary medicine and open her own practice in the New Orleans area.

EXAMPLE 9 Finding a Number That Is a Given Percent of a Given Number

A government employee working for the county judicial system chooses 5% of a jury pool to question for possible service. This amounts to 38 people. What is the size of the entire jury pool?

Solution

This problem can be reworded as follows: 38 is 5% of what number?

Method 1

$$38 = 0.05x \qquad \text{Let } x \text{ represent the number.}$$

$$x = \frac{38}{0.05} \qquad \text{Divide by 0.05.}$$

$$x = 760 \qquad \text{Simplify.}$$

Method 2   Think "38 is to what number (x) as 5 is to 100?"

$$\frac{38}{x} = \frac{5}{100}$$

$$5x = 3800 \qquad \text{Cross-products}$$

$$x = 760 \qquad \text{Divide by 5.}$$

Method 3   Use the following keystrokes on a calculator.

| 3 | 8 | ÷ | 5 | % | **760**   ⟵ Final display

Each method shows us that 38 is 5% of 760. There are 760 in the jury pool. ∎

Consumers often encounter figures in the media that involve **percent change,** which includes **percent increase** and **percent decrease.** The following guidelines are helpful in understanding these concepts.

FINDING PERCENT INCREASE OR DECREASE

1. To find the **percent increase from *a* to *b*,** where $b > a$, subtract *a* from *b*, and divide this result by *a*. Convert to a percent.

   Example: The percent increase from 4 to 7 is $\frac{7-4}{4} = \frac{3}{4} = 75\%$.

2. To find the **percent decrease from *a* to *b*,** where $b < a$, subtract *b* from *a*, and divide this result by *a*. Convert to a percent.

   Example: The percent decrease from 8 to 6 is $\frac{8-6}{8} = \frac{2}{8} = \frac{1}{4} = 25\%$.

EXAMPLE 10 Solving Problems about Percent Change

Solve each problem involving percent change.

(a) An electronics store marked up a laptop computer from the store's cost of $1200 to a selling price of $1464. What was the percent markup?

(b) The enrollment in a community college declined from 12,750 during one school year to 11,350 the following year. Find the percent decrease to the nearest tenth of a percent.

Solution

(a) "Markup" is a name for an increase. Let x = the percent increase (as a decimal).

$$\text{percent increase} = \frac{\text{amount of increase}}{\text{original amount}}$$

Subtract to find the amount of increase.

$$x = \frac{1464 - 1200}{1200}$$ Substitute the given values. Use the original cost.

$$x = \frac{264}{1200}$$

$$x = 0.22$$ Use a calculator.

The computer was marked up 22%.

(b) Let x = the percent decrease (as a decimal).

$$\text{percent decrease} = \frac{\text{amount of decrease}}{\text{original amount}}$$

Subtract to find the amount of decrease.

$$x = \frac{12{,}750 - 11{,}350}{12{,}750}$$ Substitute the given values. Use the original enrollment.

$$x = \frac{1400}{12{,}750}$$

$$x \approx 0.110$$ Use a calculator.

The college enrollment decreased by about 11.0%.

*When calculating a percent increase or a percent decrease, use the original number (**before** the increase or decrease) as the base.* A common error is to use the final number (*after* the increase or decrease) in the denominator of the fraction. Applications of decimal numbers sometimes involve formulas.

EXAMPLE 11 Determining a Child's Body Surface Area

If a child weighs k kilograms, then the child's body surface area S in square meters (m²) is determined by the following formula.

$$S = \frac{4k + 7}{k + 90}$$

What is the body surface area of a child who weighs 40 pounds? (*Source:* Hegstad, Lorrie N., and Wilma Hayek. *Essential Drug Dosage Calculations*, 4th ed. Prentice Hall, 2001.)

Solution

Tables and Web sites for conversion between the metric and English systems are readily available. Using www.metric-conversions.org, we find that 40 pounds is approximately 18.144 kilograms. Use this value for k in the formula.

$$S = \frac{4(\mathbf{18.144}) + 7}{\mathbf{18.144} + 90}$$ Let k = 18.144 in the formula.

$$S = 0.74$$ Use a calculator. Round to the nearest hundredth.

The body surface area is approximately 0.74 m².

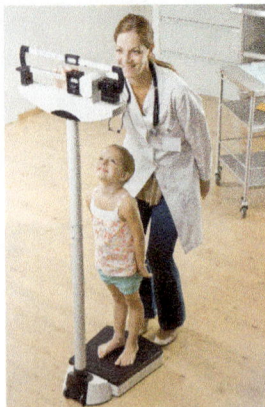

EXAMPLE 12 Determining a Child's Dose of a Drug

If D represents the usual adult dose of a drug, the corresponding child's dose C is calculated by the following formula.

$$C = \frac{\text{body surface area in square meters}}{1.7} \times D$$

Determine the appropriate dose for a child weighing 40 pounds if the usual adult dose is 50 milligrams. (*Source*: Hegstad, Lorrie N., and Wilma Hayek. *Essential Drug Dosage Calculations, 4th ed.* Prentice Hall, 2001.)

Solution

From **Example 11,** the body surface area of a child weighing 40 pounds is 0.74 m². Apply the formula for the child's dose.

$$C = \frac{0.74}{1.7} \times 50 \qquad \text{Body surface area} = 0.74, D = 50.$$

$$C = 22 \qquad \text{Use a calculator. Round to the nearest unit.}$$

The child's dose is 22 milligrams.

| 6.5 | **EXERCISES** |

Concepts of Percent   *Decide whether each statement is* true *or* false.

1. 300% of 12 is 36.

2. 25% of a quantity is the same as $\frac{1}{4}$ of that quantity.

3. To find 50% of a quantity, we may simply divide the quantity by 2.

4. A soccer team that has won 12 games and lost 8 games has a winning percentage of 60%.

5. If 70% is the lowest passing grade on a quiz that has 50 items of equal value, then answering at least 35 items correctly will assure you of a passing grade.

6. 30 is more than 40% of 120.

7. .99¢ = 99 cents

8. If an item usually costs $70.00 and it is discounted 10%, then the discount price is $7.00.

Calculate each of the following using either a calculator or paper-and-pencil methods, as directed by your instructor.

9. 8.53 + 2.785

10. 9.358 + 7.2137

11. 8.74 − 12.955

12. 2.41 − 3.997

13. 25.7 × 0.032

14. 45.1 × 8.344

15. 1019.825 ÷ 21.47

16. −262.563 ÷ 125.03

17. $\dfrac{118.5}{1.45 + 2.3}$

18. 2.45(1.2 + 3.4 − 5.6)

Personal Finance   *Solve each problem.*

19. Andrew has $48.35 in his checking account. He uses his debit card to make purchases of $35.99 and $20.00, which overdraws his account. His bank charges his account an overdraft fee of $28.50. He then deposits his paycheck for $66.27 from his part-time job at Arby's. What is the balance in his account?

20. Kayla has $37.60 in her checking account. She uses her debit card to make purchases of $25.99 and $19.34, which overdraws her account. Her bank charges her account an overdraft fee of $25.00. She then deposits her paycheck for $58.66 from her part-time job at Subway. What is the balance in her account?

21. Ahmad owes $382.45 on his Visa account. He returns two items costing $25.10 and $34.50 for credit. Then he makes purchases of $45.00 and $98.17.

(a) How much should his payment be if he wants to pay off the balance on the account?

(b) Instead of paying off the balance, he makes a payment of $300 and then incurs a finance charge of $24.66. What is the balance on his account?

22. Sabrina owes $237.59 on her MasterCard account. She returns one item costing $47.25 for credit and then makes two purchases of $12.39 and $20.00.

(a) How much should her payment be if she wants to pay off the balance on the account?

(b) Instead of paying off the balance, she makes a payment of $75.00 and incurs a finance charge of $32.06. What is the balance on her account?

23. Bank Account Balance In August, Kimberly began with a bank account balance of $904.89. Her withdrawals and deposits for August are given below:

Withdrawals	Deposits
$35.84	$85.00
$26.14	$120.76
$3.12	$205.00
$21.46	

Assuming no other transactions, what was her account balance at the end of August?

24. Bank Account Balance In September, David began with a bank account balance of $904.89. His withdrawals and deposits for September are given below:

Withdrawals	Deposits
$41.29	$80.59
$13.66	$276.13
$84.40	$550.00
$93.00	

Assuming no other transactions, what was his account balance at the end of September?

Rounding *Round each number to the place value indicated. For example, in part (a) of Exercise 25, round 54,793 to the nearest ten thousand. (Hint: Always round from the original number.)*

25. 54,793

(a) ten thousand   (b) thousand

(c) hundred   (d) ten

26. 453,258

(a) hundred thousand   (b) thousand

(c) hundred   (d) ten

27. 0.892451

(a) hundred-thousandth

(b) ten-thousandth

(c) thousandth

(d) hundredth

(e) tenth

(f) one or unit

28. 22.483956

(a) hundred-thousandth

(b) ten-thousandth

(c) thousandth

(d) hundredth

(e) tenth

(f) ten

Use the concept of percent in Exercises 29–32.

29. Match the fractions in Group II with their equivalent percents in Group I.

I
(a) 25%   (b) 10%
(c) 2%   (d) 20%
(e) 75%   (f) $33\frac{1}{3}\%$

II
A. $\frac{1}{3}$ B. $\frac{1}{50}$
C. $\frac{3}{4}$ D. $\frac{1}{10}$
E. $\frac{1}{4}$ F. $\frac{1}{5}$

30. Fill in each blank with the correct numerical response.

(a) 5% means _____ in every 100.

(b) 25% means 6 in every _____.

(c) 200% means _____ for every 4.

(d) 0.5% means _____ in every 100.

(e) _____ % means 12 for every 2.

31. The figures in **Exercise 23** of **Section 6.3** are reproduced here. Express the fractional parts represented by the shaded areas as percents.

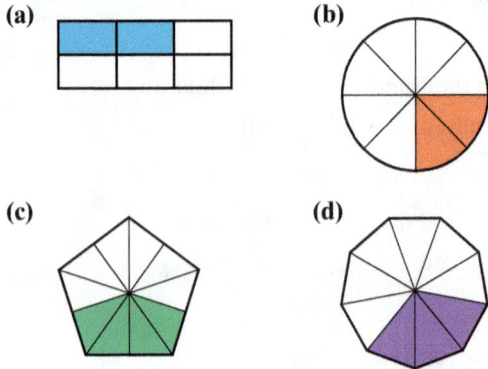

(a)

(b)

(c)

(d)

32. The Venn diagram shows the numbers of elements in the four regions formed.

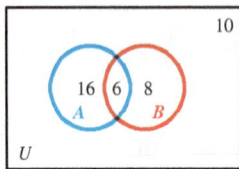

(a) What percent of the elements in the universe are in $A \cap B$?

(b) What percent of the elements in the universe are in A but not in B?

(c) What percent of the elements in $A \cup B$ are in $A \cap B$?

(d) What percent of the elements in the universe are in neither A nor B?

Convert each decimal to a percent.

33. 0.42   **34.** 0.87   **35.** 0.365

36. 0.792   **37.** 0.008   **38.** 0.0093

39. 2.1   **40.** 8.9

Convert each percent to a decimal.

41. 96%   **42.** 23%   **43.** 5.46%

44. 2.99%   **45.** 0.3%   **46.** 0.6%

47. 400%   **48.** 260%   **49.** $\frac{1}{2}\%$

50. $\frac{2}{5}\%$ **51.** $3\frac{1}{2}\%$ **52.** $8\frac{3}{4}\%$

Convert each fraction to a percent.

53. $\frac{1}{5}$ **54.** $\frac{2}{5}$ **55.** $\frac{1}{100}$ **56.** $\frac{1}{50}$

57. $\frac{3}{8}$ **58.** $\frac{5}{6}$ **59.** $\frac{3}{2}$ **60.** $\frac{7}{4}$

61. Explain the difference between $\frac{1}{2}$ of a quantity and $\frac{1}{2}\%$ of the quantity.

62. Explain the difference between 9% and .9%.

Work each problem involving percent.

63. What is 26% of 480?

64. What is 38% of 12?

65. What is 10.5% of 28?

66. What is 48.6% of 19?

67. What percent of 30 is 45?

68. What percent of 48 is 20?

69. 25% of what number is 150?

70. 12% of what number is 3600?

71. 0.392 is what percent of 28?

72. 78.84 is what percent of 292?

Solve each problem involving percent increase or decrease.

73. *Percent Increase*   After one year on the job, Grady got a raise from $10.50 per hour to $11.34 per hour. What was the percent increase in his hourly wage?

74. *Percent Discount*   Clayton bought a ticket to a rock concert at a discount. The regular price of the ticket was $70.00, but he paid only $59.50. What was the percent discount?

75. *Percent Decrease*   Between 2000 and 2012, the estimated population of Toledo, Ohio, declined from 313,619 to 284,012. What was the percent decrease to the nearest tenth? (*Source:* U.S. Census Bureau.)

76. *Percent Increase*   Between 2000 and 2012, the estimated population of Oklahoma City, Oklahoma, grew from 506,132 to 599,199. What was the percent increase to the nearest tenth? (*Source:* U.S. Census Bureau.)

77. *Percent Discount*   The DVD of the Original Broadway production of the musical *Memphis* was available at www.amazon.com for $16.22. The list price of this DVD was $17.98. To the nearest tenth, what was the percent discount? (*Source:* www.amazon.com)

78. *Percent Discount*   The Blu-ray of the movie *Django Unchained* had a list price of $39.99 and was for sale at www.amazon.com at $14.99. To the nearest tenth, what was the percent discount? (*Source:* www.amazon.com)

79. Value of 1916-D Mercury Dime The 1916 Mercury dime minted in Denver is quite rare. In 1979 its value in Extremely Fine condition was $625. The 2014 value had increased to $6200. What was the percent increase in the value of this coin from 1979 to 2014? (*Sources: A Guide Book of United States Coins; Coin World Coin Values.*)

80. Value of 1903-O Morgan Dollar In 1963, the value of a 1903 Morgan dollar minted in New Orleans in typical Uncirculated condition was $1500. Due to a discovery of a large hoard of these dollars late that year, the value plummeted. Its value in 2014 was $650. What was the percent decrease in its value from 1963 to 2014? (*Sources: A Guide Book of United States Coins; Coin World Coin Values.*)

Gasoline Prices *The line graph shows the average price, adjusted for inflation, that Americans paid for a gallon of gasoline for selected years between 1958 and 2008. Use this information in Exercises 81 and 82.*

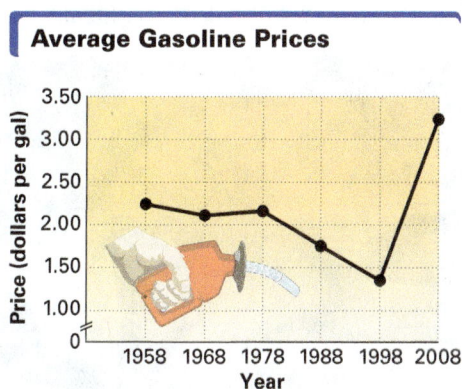

Average Gasoline Prices

Source: www.inflationdata.com

81. By what percent did prices increase from 1998 to 2008?

82. By what percent did prices decrease from 1978 to 1988?

Business Earnings Report *An article in the March 1, 2014, business section of* The Gazette *covered various facets of the earnings report of a major corporation. Answer each of the following questions about a particular phase of the report. Round to the nearest tenth if applicable.*

83. In 2013 profit fell from $1.56 billion to $1.23 billion. What was the percent decrease in profit?

84. Sales rose 2.3% to $8.7 billion. What were the sales prior to this rise?

85. Profit in one division dropped 24% to $680 million. What was the profit prior to this drop?

86. Shares of the company closed on one recent day at $17.98 per share, down $1.19. What percent decrease did this represent?

Use mental techniques to answer the questions in Exercises 87–92. Try to avoid using paper and pencil or a calculator.

87. Allowance Increase Carly's allowance was raised from $4.00 per week to $5.00 per week. What was the percent of the increase?

A. 25%   **B.** 20%   **C.** 50%   **D.** 30%

88. Boat Purchase and Sale Susan bought a boat five years ago for $5000 and sold it this year for $2000. What percent of her original purchase did she lose on the sale?

A. 40%   **B.** 50%   **C.** 20%   **D.** 60%

89. Population of Alabama A 2012 report indicated that the population of Alabama was 4,822,023, with 26.0% represented by African Americans. What is the best estimate of the African American population in Alabama? (*Source:* U.S. Census Bureau.)

A. 500,000   **B.** 1,500,000
C. 1,250,000   **D.** 750,000

90. Population of Hawaii A 2012 report indicated that the population of Hawaii was 1,392,313, with 19.4% of the population being of two or more races. What is the best estimate of this racial demographic population of Hawaii? (*Source:* U.S. Census Bureau.)

A. 280,000 **B.** 300,000 **C.** 21,400 **D.** 24,000

91. Discount and Markup Suppose that an item regularly costs $100.00 and is discounted 20%. If it is then marked up 20%, is the resulting price $100.00? If not, what is it?

92. Computing a Tip Suppose that you have decided that you will always tip 20% when dining in restaurants. By what whole number should you divide the bill to find the amount of the tip?

Body Surface Area *Use the formula for determining a child's body surface area*

$$S = \frac{4k + 7}{k + 90}$$

in **Example 11** *to work Exercises 93–96. Round to the nearest hundredth.*

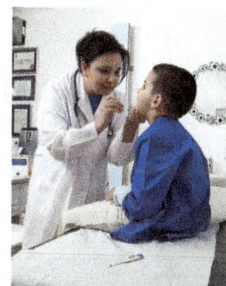

93. Find the body surface area S in square meters of a child who weighs 20 kilograms.

94. Find the body surface area S in square meters of a child who weighs 26 kilograms.

95. Find the body surface area S in square meters of a child who weighs 20 pounds. (Use $k = 9.07$.)

96. Find the body surface area S in square meters of a child who weighs 26 pounds. (Use $k = 11.79$.)

Child's Medication Dosage *Use the formula for determining a child's dose of a drug*

$$C = \frac{\text{body surface area in square meters}}{1.7} \times D$$

in *Example 12* to work *Exercises 97–100*. Round to the nearest unit.

97. If the usual adult dose D of a drug is 250 mg, what is the child's dose for a child weighing 20 kilograms? (*Hint:* Use the result of **Exercise 93.**)

98. If the usual adult dose D of a drug is 250 mg, what is the child's dose for a child weighing 26 kilograms? (*Hint:* Use the result of **Exercise 94.**)

99. If the usual adult dose D of a drug is 500 mg, what is the child's dose for a child weighing 20 pounds? (*Hint:* Use the result of **Exercise 95.**)

100. If the usual adult dose D of a drug is 500 mg, what is the child's dose for a child weighing 26 pounds? (*Hint:* Use the result of **Exercise 96.**)

Metabolic Units *One way to measure a person's cardio fitness is to calculate how many METs, or metabolic units, he or she can reach at peak exertion. One MET is the amount of energy used when sitting quietly. To calculate ideal METs, we can use one of the following formulas.*

$$\text{MET} = 14.7 - \text{age} \cdot 0.13$$

$$\text{MET} = 14.7 - \text{age} \cdot 0.11$$

(*Source: New England Journal of Medicine.*)

101. A 40-year-old woman wishes to find her ideal MET.

   (a) Write the expression using her age.

   (b) Calculate her ideal MET.

   (c) Researchers recommend that people reach about 85% of their MET when exercising. Calculate 85% of the ideal MET from part (b). Then refer to the following table. What activity can the woman do that is approximately this value?

Activity	METs	Activity	METs
Golf (with cart)	2.5	Skiing (water or downhill)	6.8
Walking (3 mph)	3.3	Swimming	7.0
Mowing lawn (power mower)	4.5	Walking (5 mph)	8.0
Ballroom or square dancing	5.5	Jogging	10.2

Source: Harvard School of Public Health.

102. Repeat **Exercise 101** for a 55-year-old man.

Exercises 103–106 are based on formulas found in Auto Math Handbook: Mathematical Calculations, Theory, and Formulas for Automotive Enthusiasts, *by John Lawlor (1991, HP Books).*

103. *Blood Alcohol Concentration* The blood alcohol concentration (BAC) of a person who has been drinking is given by the formula

$$\text{BAC} = \frac{(\text{ounces} \times \text{percent alcohol} \times 0.075)}{\text{body weight in lb}} - (\text{hours of drinking} \times 0.015).$$

Suppose a policeman stops a 190-pound man who, in two hours, has ingested four 12-ounce beers, each having a 3.2% alcohol content. The formula would then read

$$\text{BAC} = \frac{[(4 \times 12) \times 3.2 \times 0.075]}{190} - (2 \times 0.015).$$

   (a) Find this BAC.

   (b) Find the BAC for a 135-pound woman who, in three hours, has drunk three 12-ounce beers, each having a 4.0% alcohol content.

104. *Approximate Automobile Speed* The approximate speed of an automobile in miles per hour (mph) can be found in terms of the engine's revolutions per minute (rpm), the tire diameter in inches, and the overall gear ratio by the following formula.

$$\text{mph} = \frac{\text{rpm} \times \text{tire diameter}}{\text{gear ratio} \times 336}$$

If a certain automobile has an rpm of 5600, a tire diameter of 26 inches, and a gear ratio of 3.12, what is its approximate speed (mph)?

105. *Engine Horsepower* Horsepower can be found from mean effective pressure (mep) in pounds per square inch, engine displacement in cubic inches, and revolutions per minute (rpm) using the following formula.

$$\text{Horsepower} = \frac{\text{mep} \times \text{displacement} \times \text{rpm}}{792,000}$$

An engine has displacement of 302 cubic inches and indicated mep of 195 pounds per square inch at 4000 rpm. What is its approximate horsepower?

106. *Torque Approximation* To determine the torque at a given value of rpm, the following formula applies.

$$\text{Torque} = \frac{5252 \times \text{horsepower}}{\text{rpm}}$$

If the horsepower of a certain vehicle is 400 at 4500 rpm, what is the approximate torque?

Win-Loss Record *Exercises 107 and 108 deal with winning percentage in the standings of sports teams.*

107. At the end of the regular 2013 Major League Baseball season, the standings of the East Division of the American League were as shown. Winning percentage is commonly expressed as a decimal rounded to the nearest thousandth. To find the winning percentage of a team, divide the number of wins (W) by the total number of games played (W + L). Find the winning percentage of each team.

(a) Boston   (b) Tampa Bay

(c) New York Yankees   (d) Toronto

Team	W	L	Pct.
Boston	97	65	
Tampa Bay	92	71	
Baltimore	85	77	.525
New York Yankees	85	77	
Toronto	74	88	

Source: World Almanac and Book of Facts.

108. Repeat **Exercise 107** for the following standings for the East Division of the National League.

(a) Washington   (b) New York Mets

(c) Philadelphia   (d) Miami

Team	W	L	Pct.
Atlanta	96	66	.593
Washington	86	76	
New York Mets	74	88	
Philadelphia	73	89	
Miami	62	100	

Source: World Almanac and Book of Facts.

Tipping Procedure *It is customary in our society to "tip" the wait staff when dining in restaurants. One common rate for tipping is 15%. A quick way of figuring a tip that will give a close approximation of 15% follows.*

Step 1 *Round off the bill to the nearest dollar.*

Step 2 *Find 10% of this amount by moving the decimal point one place to the left.*

Step 3 *Take half of the amount obtained in Step 2 and add it to the result of Step 2.*

These steps will give approximately 15% of the bill. The amount obtained in Step 3 is 5%, and

$$10\% + 5\% = 15\%.$$

Use the method above to find an approximation of 15% for each restaurant bill.

109. $29.57   **110.** $38.32

111. $5.15   **112.** $7.89

For Excellent Service *Suppose that you get extremely good service and decide to tip 20%. You can use the first two steps listed earlier and then, in Step 3, double the amount you obtained in Step 2. Use this method to find an approximation of 20% for each restaurant bill.*

113. $59.96   **114.** $40.24

115. $180.43   **116.** $199.86

Postage Stamp Pricing *Refer to the margin note on decimal point abuse. At one time, the United States Postal Service sold rolls of 33-cent stamps that featured fruit berries. One such stamp is suggested on the left. On the right is a photo of the pricing information found on the cellophane wrapper of such a roll.*

100 STAMPS PSA
.33¢ ea. TOTAL $33.00
FRUIT BERRIES
ITEM 7757
BCA

117. Look at the second line of the pricing information. According to the price listed *per stamp,* how many stamps should you be able to purchase for one cent?

118. The total price listed is the amount the Postal Service actually charges. If you were to multiply the listed price *per stamp* by the number of stamps, what should the total price be?

Pricing of Pie and Coffee *The photos here were taken at a flea market near Natchez, Mississippi. The handwritten signs indicate that a piece of pie costs .10¢ and a cup of coffee ("ffee") costs .5¢. Assuming these are the actual prices, answer the questions in Exercises 119–122.*

119. How much will 10 pieces of pie and 10 cups of coffee cost?

120. How much will 20 pieces of pie and 10 cups of coffee cost?

121. How many pieces of pie can you get for $1.00?

122. How many cups of coffee can you get for $1.00?

123. *Producer Percent* In the 1967 movie *The Producers*, Leo Bloom (Gene Wilder) and Max Bialystock (Zero Mostel) scheme to make a fortune by overfinancing what they think will be a Broadway flop. After enumerating the percent of profits all of Max's little old ladies have been offered in the production, reality sets in. Watch the movie to see what percent of the profits was sold.

124. *Willy Wonka and Percent* There are several appearances of percent in the 1971 movie *Willy Wonka and the Chocolate Factory*. In one of them, upon preparing a mixture in his laboratory, Willy Wonka (Gene Wilder) states an impossible percent analysis as he drinks his latest concoction. Watch the movie to see what percent the ingredients total up to be.

CHAPTER 6   SUMMARY

KEY TERMS

6.1

natural numbers
whole numbers
number line
origin
negative numbers
positive numbers
signed numbers
integers
graph
rational numbers
irrational numbers
real numbers
additive inverses
    (negatives, opposites)
absolute value

6.2

sum
addends (terms)
difference
minuend
subtrahend
product
factors
quotient
dividend (numerator)
divisor (denominator)
positive change
negative change

6.3

extremes
means
cross-product test
    (for equality of fractions)
least common denominator
    (LCD)
mixed number
arithmetic mean (average)
terminating decimal
repeating decimal

6.4

irrational number
radicand
like radicals
Golden Rectangle
Golden Ratio

6.5

factors
product
dividend
divisor
quotient
percent

NEW SYMBOLS

$\{x \mid x \text{ has a certain property}\}$	set-builder notation	$a > b$	a is greater than b
%	percent	$a \geq b$	a is greater than or equal to b
¢	cents	$\lvert x \rvert$	absolute value of x
$a = b$	a is equal to b	π	irrational number $\pi \approx 3.14159$
$a < b$	a is less than b	ϕ	irrational number $\phi \approx 1.618$
$a \leq b$	a is less than or equal to b	e	irrational number $e \approx 2.71828$

TEST YOUR WORD POWER

See how well you have learned the vocabulary in this chapter.

1. The **absolute value** of a real number is
   A. the same as the opposite of the number in all cases.
   B. the same as the number itself in all cases.
   C. never zero.
   D. its undirected distance from zero on the number line.

2. The **sum** of two numbers is the result obtained by
   A. addition.   B. subtraction.
   C. multiplication.   D. division.

3. The **identity element for multiplication** is
   A. 0.   B. 1.   C. −1.   D. −a, for the real number a.

4. An example of an **irrational number** is
   A. ϕ.   B. e.
   C. π.   D. all of these.

5. An example of a rational number that has a **terminating decimal** representation is
   A. $\frac{1}{3}$.   B. $\frac{1}{7}$.   C. $\frac{1}{2}$.   D. $\frac{1}{9}$.

6. The irrational number π represents
   A. the circumference of a circle divided by its radius.
   B. the radius of a circle divided by its diameter.
   C. the diameter of a circle divided by its radius.
   D. the circumference of a circle divided by its diameter.

ANSWERS
1. D 2. A 3. B 4. D 5. C 6. D

QUICK REVIEW

Concepts	Examples

6.1 Real Numbers, Order, and Absolute Value

Sets of Numbers

Natural Numbers $\{1, 2, 3, 4, \ldots\}$ — 10, 25, 143 — Natural numbers

Whole Numbers $\{0, 1, 2, 3, 4, \ldots\}$ — 0, 8, 47 — Whole numbers

Integers $\{\ldots, -2, -1, 0, 1, 2, \ldots\}$ — −22, −7, 0, 4, 9 — Integers

Rational Numbers $\{x \mid x$ is a quotient of two integers, with denominator not equal to 0$\}$ — $-\frac{2}{3}, -0.14, 0, \frac{15}{8}, 6, 0.33333\ldots, \sqrt{4}$ — Rational numbers

Irrational Numbers $\{x \mid x$ is a number on the number line that is not rational$\}$ — $-\sqrt{22}, \sqrt{3}, \pi$ — Irrational numbers

Real Numbers $\{x \mid x$ is a number that can be represented by a point on the number line$\}$ — $-3, -\frac{2}{7}, 0.7, \pi, \sqrt{11}$ — Real numbers

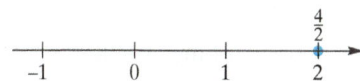

Order in the Real Numbers
Suppose that a and b are real numbers.

$a = b$ if they are represented by the same point on the number line.

$a < b$ if the graph of a lies to the left of the graph of b.

$a > b$ if the graph of a lies to the right of the graph of b.

$2 = \frac{4}{2}$ $0 < 1$ $2 > -1$

Additive Inverse
The additive inverse of a is −a. For all a,
$$a + (-a) = (-a) + a = 0.$$

$-(-5) = 5$ $-\left(-\frac{2}{3}\right) = \frac{2}{3}$
$5 + (-5) = 0$

Absolute Value
$$|x| = \begin{cases} x & \text{if } x \geq 0 \\ -x & \text{if } x < 0 \end{cases}$$

$|-1| = 1$ $|0| = 0$ $|2| = 2$

Concepts	Examples

6.2 Operations, Properties, and Applications of Real Numbers

Addition

Like Signs: Add the absolute values. The sum has the same sign as the given numbers.

Unlike Signs: Find the absolute values of the numbers, and subtract the lesser absolute value from the greater. The sum has the same sign as the number with the greater absolute value.

Add.
$$9 + 4 = 13$$
$$-8 + (-5) = -13$$
$$7 + (-12) = -5$$
$$-5 + 13 = 8$$

Subtraction

For all real numbers a and b,
$$a - b = a + (-b).$$

Subtract.
$$-3 - 4 = -3 + (-4) = -7$$
$$-2 - (-6) = -2 + 6 = 4$$
$$13 - (-8) = 13 + 8 = 21$$

Multiplication and Division

Like Signs: The product or quotient of two numbers with like signs is positive.

Unlike Signs: The product or quotient of two numbers with different signs is negative.

Multiply or divide.
$$6 \cdot 5 = 30 \quad -7(-8) = 56 \quad \frac{20}{4} = 5 \quad \frac{-24}{-6} = 4$$
$$-6(5) = -30 \quad 6(-5) = -30$$
$$\frac{-18}{9} = -2 \quad \frac{49}{-7} = -7$$

Division Involving 0

$\dfrac{a}{0}$ is undefined for all a. $\quad \dfrac{0}{a} = 0$ for all nonzero a.

$\dfrac{5}{0}$ is undefined. $\quad \dfrac{0}{5} = 0$

Order of Operations

If parentheses or square brackets are present:

Step 1 Work separately above and below any fraction bar.

Step 2 Use the rules below within each set of parentheses or square brackets. Start with the innermost set and work outward.

If no parentheses or brackets are present:

Step 1 Apply any exponents.

Step 2 Do any multiplications or divisions in the order in which they occur, working from left to right.

Step 3 Do any additions or subtractions in the order in which they occur, working from left to right.

Simplify.
$$(-6)[2^2 - (3 + 4)] + 3 \qquad \text{Work inside the innermost parentheses.}$$
$$= (-6)[2^2 - 7] + 3$$
$$= (-6)[4 - 7] + 3 \qquad \text{Work inside the brackets.}$$
$$= (-6)[-3] + 3$$
$$= 18 + 3 \qquad \text{Multiply.}$$
$$= 21 \qquad \text{Add.}$$

Properties of Real Numbers

If a and b are real numbers, then

Closure Properties

$a + b$ and ab are real numbers.

$3 + 4$ and $3 \cdot 4$ are real numbers.

Commutative Properties
$$a + b = b + a$$
$$ab = ba$$

$$7 + (-1) = -1 + 7$$
$$5(-3) = (-3)5$$

Associative Properties
$$(a + b) + c = a + (b + c)$$
$$(ab)c = a(bc)$$

$$(3 + 4) + 8 = 3 + (4 + 8)$$
$$[-2(6)]4 = -2[(6)4]$$

Concepts	Examples

Identity Properties
$$a + 0 = a \quad 0 + a = a$$
$$a \cdot 1 = a \quad 1 \cdot a = a$$

$$-7 + 0 = -7 \quad 0 + (-7) = -7$$
$$9 \cdot 1 = 9 \quad 1 \cdot 9 = 9$$

Inverse Properties
$$a + (-a) = 0 \quad -a + a = 0$$
$$a \cdot \frac{1}{a} = 1 \quad \frac{1}{a} \cdot a = 1 \quad (a \neq 0)$$

$$7 + (-7) = 0 \quad -7 + 7 = 0$$
$$-2\left(-\frac{1}{2}\right) = 1 \quad -\frac{1}{2}(-2) = 1$$

Distributive Property
$$a(b + c) = ab + ac$$
$$(b + c)a = ba + ca$$

$$5(4 + 2) = 5(4) + 5(2)$$
$$(4 + 2)5 = 4(5) + 2(5)$$

6.3 Rational Numbers and Decimal Representation

Fundamental Property of Rational Numbers
$$\frac{a \cdot k}{b \cdot k} = \frac{a}{b} \quad (b \neq 0, k \neq 0)$$

Write $\frac{8}{12}$ in lowest terms.
$$\frac{8}{12} = \frac{2 \cdot 4}{3 \cdot 4} = \frac{2}{3}$$

Cross-Product Test for Equality of Rational Numbers
$$\frac{a}{b} = \frac{c}{d} \text{ if and only if } a \cdot d = b \cdot c \quad (b \neq 0, d \neq 0)$$

Is $\frac{25}{36} = \frac{5}{6}$ a true statement?
$$25 \cdot 6 = 150 \quad \text{and} \quad 36 \cdot 5 = 180$$
Because $150 \neq 180$, the statement is false.

Adding and Subtracting Rational Numbers
$$\frac{a}{b} + \frac{c}{d} = \frac{ad + bc}{bd} \text{ and } \frac{a}{b} - \frac{c}{d} = \frac{ad - bc}{bd}$$

In practice, we usually find the least common denominator to add and subtract fractions.

Add or subtract.
$$\frac{2}{5} + \frac{7}{5} = \frac{2 + 7}{5} = \frac{9}{5}, \text{ or } 1\frac{4}{5}$$
$$\frac{2}{3} - \frac{1}{2} = \frac{4}{6} - \frac{3}{6} \quad \text{6 is the LCD.}$$
$$= \frac{1}{6}$$

Multiplying and Dividing Rational Numbers
$$\frac{a}{b} \cdot \frac{c}{d} = \frac{ac}{bd} \quad (b \neq 0, d \neq 0)$$
$$\frac{a}{b} \div \frac{c}{d} = \frac{a}{b} \cdot \frac{d}{c} = \frac{ad}{bc} \quad (b \neq 0, c \neq 0, d \neq 0)$$

Multiply or divide.
$$\frac{4}{3} \cdot \frac{5}{6} = \frac{20}{18} = \frac{10}{9}, \text{ or } 1\frac{1}{9}$$
$$\frac{6}{5} \div \frac{1}{4} = \frac{6}{5} \cdot \frac{4}{1} = \frac{24}{5}, \text{ or } 4\frac{4}{5}$$

Density Property of the Rational Numbers
If r and t are distinct rational numbers, with $r < t$, then there exists a rational number s such that
$$r < s < t.$$

Find the average, or mean, of $\frac{2}{3}$ and $\frac{3}{4}$.
$$\frac{2}{3} + \frac{3}{4} = \frac{8 + 9}{12} = \frac{17}{12} \quad \text{Add.}$$
$$\frac{1}{2} \cdot \frac{17}{12} = \frac{17}{24} \quad \text{Find half of the sum.} \leftarrow \text{Average of } \frac{2}{3} \text{ and } \frac{3}{4}$$

The decimal representation of a rational number will either terminate or will repeat indefinitely in a "block" of digits.

$$\frac{2}{3} = 0.\overline{6}, \text{ or } 0.666\ldots \quad \frac{3}{16} = 0.1875$$

Concepts	Examples

6.4 Irrational Numbers and Decimal Representation

Decimal Representation
The decimal for an irrational number neither terminates nor repeats.

$$\sqrt{2}, \quad \sqrt{10}, \quad 0.10110111011110\ldots$$

Approximations
Approximations for irrational numbers (for example, square roots of whole numbers that are not perfect squares) can be found with a calculator.

$$\sqrt{2} \approx 1.414213562 \qquad \sqrt{10} \approx 3.16227766$$

Product and Quotient Rules for Square Roots

$$\sqrt{a} \cdot \sqrt{b} = \sqrt{a \cdot b} \quad (a \geq 0, b \geq 0)$$

$$\frac{\sqrt{a}}{\sqrt{b}} = \sqrt{\frac{a}{b}} \quad (a \geq 0, b > 0)$$

Simplify.

$$\sqrt{54} = \sqrt{9 \cdot 6} = \sqrt{9} \cdot \sqrt{6} = 3\sqrt{6}$$

$$\frac{\sqrt{36}}{\sqrt{4}} = \sqrt{\frac{36}{4}} = \sqrt{9} = 3$$

Conditions Necessary for the Simplified Form of a Square Root Radical

1. The number under the radical (radicand) has no factor (except 1) that is a perfect square.

2. The radicand has no fractions.

3. No denominator contains a radical.

Simplify.

$$\sqrt{\frac{5}{6}} = \frac{\sqrt{5}}{\sqrt{6}} = \frac{\sqrt{5} \cdot \sqrt{6}}{\sqrt{6} \cdot \sqrt{6}} = \frac{\sqrt{30}}{6}$$

Three Important Irrational Numbers

$\pi \approx 3.141592653589793238462643383279$

$\phi = \dfrac{1 + \sqrt{5}}{2} \approx 1.618033988749894848204586834365$

$e \approx 2.718281828459045235360287471353$

The rational number $\frac{355}{113}$ gives an excellent approximation for the irrational number π.

The Golden Ratio ϕ appears throughout nature.

The irrational number e is important in higher mathematics.

6.5 Applications of Decimals and Percents

Addition and Subtraction of Decimals
To add or subtract decimal numbers, line up the decimal points in a column and perform the operation.

Add or subtract.

$1.2 + 36.158 + 9.26$

$$\begin{array}{r} 1.200 \\ 36.158 \\ 9.260 \\ \hline 46.618 \end{array}$$

$93.86 - 42.9142$

$$\begin{array}{r} 93.8600 \\ - 42.9142 \\ \hline 50.9458 \end{array}$$

Concepts	Examples

Multiplication and Division of Decimals

Multiplication   To multiply decimals, multiply in the same manner as integers are multiplied. The number of decimal places to the right of the decimal point in the product is the *sum* of the numbers of places to the right of the decimal points in the factors.

Division   To divide decimals, move the decimal point to the right the same number of places in the divisor and the dividend so as to obtain a whole number in the divisor. Divide in the same manner as integers are divided. The number of decimal places to the right of the decimal point in the quotient is the same as the number of places to the right of the decimal point in the dividend.

Multiply or divide.

51.6×2.3

$$\begin{array}{r} 51.6 \\ 2.3 \\ \hline 1548 \\ 1032 \\ \hline 118.68 \end{array}$$

$35.38 \div 6.1$

$$\begin{array}{r} 5.8 \\ 61\overline{)353.8} \\ 305 \\ \hline 488 \\ 488 \\ \hline 0 \end{array}$$

Rules for Rounding
See pages 269–270.

Round 745.2935 to the given place value.

hundreds: 700

ones or units: 745

tenths: 745.3

thousandths: 745.294

Percent
The word **percent** means "per hundred." The symbol % represents "percent."

$$1\% = \frac{1}{100} = 0.01$$

Converting between Decimals and Percents
To convert a percent to a decimal, drop the percent symbol and move the decimal point two places to the left, inserting zeros as placeholders if necessary.

To convert a decimal to a percent, move the decimal point two places to the right, inserting zeros as placeholders if necessary, and attach the percent symbol.

Convert 0.8% to a decimal.
$$0.8\% = 0.008$$

Convert 0.8 to a percent.
$$0.8 = 0.80 = 80\%$$

Converting a Fraction to a Percent
To convert a fraction to a percent, convert the fraction to a decimal, and then convert the decimal to a percent.

Convert $\frac{11}{4}$ to a percent.
$$\frac{11}{4} = 2\frac{3}{4} = 2.75 = 275\%$$

Finding Percent Increase or Decrease
To find the percent increase from a to b, where $b > a$, subtract a from b, and divide this result by a. Convert to a percent.

To find the percent decrease from a to b, where $b < a$, subtract b from a, and divide this result by a. Convert to a percent.

The sales of a textbook decreased from \$3,500,000 in its third edition to \$2,975,000 in its fourth edition. What was the percent decrease?

$$\text{Percent decrease} = \frac{\$3,500,000 - \$2,975,000}{\$3,500,000}$$

$$= \frac{\$525,000}{\$3,500,000}$$

$$= 0.15$$

The percent decrease was 15%.

CHAPTER 6	TEST

1. Consider $\{-4, -\sqrt{5}, -\frac{3}{2}, -0.5, 0, \sqrt{3}, 4.1, 12\}$. List the elements of the set that belong to each of the following.

   (a) natural numbers

   (b) whole numbers

   (c) integers

   (d) rational numbers

   (e) irrational numbers

   (f) real numbers

2. Match each set in (a)–(d) with the correct set-builder notation description in A–D.

   (a) $\{\ldots, -4, -3, -2, -1\}$

   (b) $\{3, 4, 5, 6, \ldots\}$

   (c) $\{1, 2, 3, 4, \ldots\}$

   (d) $\{-12, \ldots, -2, -1, 0, 1, 2, \ldots, 12\}$

   A. $\{x \mid x$ is an integer with absolute value less than or equal to 12$\}$

   B. $\{x \mid x$ is an integer greater than 2.5$\}$

   C. $\{x \mid x$ is a negative integer$\}$

   D. $\{x \mid x$ is a positive integer$\}$

3. Decide whether each statement is *true* or *false*.

   (a) The absolute value of a number must be positive.

   (b) $|-7| = -(-7)$

   (c) $\frac{2}{5}$ is an example of a real number that is not an integer.

   (d) Every real number is either positive or negative.

Perform the indicated operations. Use the order of operations as necessary.

4. $6^2 - 4(9-1)$

5. $\dfrac{(-8+3) - (5+10)}{7-9}$

6. $(-3)(-2) - [5 + (8-10)]$

7. $5(-6) - (3-8)^2 + 12$

8. **Temperature Extremes** The record high temperature in the United States was 134° Fahrenheit, recorded at Death Valley, California, in 1913. The record low was $-80°F$, at Prospect Creek, Alaska, in 1971. How much greater was the highest temperature than the lowest temperature? (*Source: The World Almanac and Book of Facts.*)

9. **Altitude of a Plane** The surface of the Dead Sea has altitude 1299 ft below sea level. A pilot is flying 80 ft above that surface. How much altitude must she gain to clear a 3852-ft pass by 225 ft? (*Source:. The World Almanac and Book of Facts.*)

10. Match each statement in (a)–(f) with the property that justifies it in A–F.

    (a) $7 \cdot (8 \cdot 5) = (7 \cdot 8) \cdot 5$

    (b) $3x + 3y = 3(x + y)$

    (c) $8 \cdot 1 = 1 \cdot 8 = 8$

    (d) $7 + (6 + 9) = (6 + 9) + 7$

    (e) $9 + (-9) = -9 + 9 = 0$

    (f) $5 \cdot 8$ is a real number.

    A. Distributive property

    B. Identity property

    C. Closure property

    D. Commutative property

    E. Associative property

    F. Inverse property

11. **Basketball Shot Statistics** Six players on the local high school basketball team had shooting statistics as shown in the table below. Answer each question, using estimation skills as necessary.

Player	Field Goal Attempts	Field Goals Made
Ed	40	13
Jack	10	4
Chuck	20	8
Ben	6	4
Charlie	7	2
Jason	7	6

   (a) Which players made more than half of their attempts?

   (b) Which players made just less than $\frac{1}{3}$ of their attempts?

   (c) Which player made exactly $\frac{2}{3}$ of his attempts?

   (d) Which two players made the same fractional parts of their attempts? What was the fractional part, reduced to lowest terms?

   (e) Which player made the greatest fractional part of his attempts?

Perform each operation. Write your answer in lowest terms.

12. $\dfrac{3}{16} + \dfrac{1}{2}$

13. $\dfrac{9}{20} - \dfrac{3}{32}$

14. $\dfrac{3}{8} \cdot \left(-\dfrac{16}{15}\right)$

15. $\dfrac{7}{9} \div 1\dfrac{3}{4}$

16. *Foreign-born Population* Approximately 40 million people living in the United States in 2010 were born in other countries. The graph gives the fractional number for each region of birth for these people.

U.S. Foreign-Born Population by Region of Birth

Other ?

Europe $\dfrac{3}{25}$

Latin America $\dfrac{27}{50}$

Asia $\dfrac{7}{25}$

Source: U.S. Census Bureau.

(a) What fractional part of the U.S. foreign-born population was from a region other than the ones specified?

(b) What fractional part of the foreign-born population was from Latin America or Asia?

(c) About how many people (in millions) were born in Europe?

17. Convert each rational number into a repeating or terminating decimal. Use a calculator if your instructor so allows.

(a) $\dfrac{9}{20}$

(b) $\dfrac{5}{12}$

18. Convert each decimal into a quotient of integers, reduced to lowest terms.

(a) 0.72

(b) $0.\overline{58}$

19. Identify each number as rational or irrational.

(a) $\sqrt{10}$

(b) $\sqrt{16}$

(c) 0.01

(d) $0.\overline{01}$

(e) $0.0101101110\ldots$

(f) π

*For each of the following, **(a)** use a calculator to find a decimal approximation and **(b)** simplify the radical according to the guidelines in this chapter.*

20. $\sqrt{150}$

21. $\dfrac{13}{\sqrt{7}}$

22. $2\sqrt{32} - 5\sqrt{128}$

23. *Rate of Return on an Investment* If an investment of P dollars grows to A dollars in two years, the annual rate of return on the investment is given by

$$r = \dfrac{\sqrt{A} - \sqrt{P}}{\sqrt{P}}.$$

What is the rate if an investment of $50,000 grows to $58,320?

24. Work each of the following using either a calculator or paper-and-pencil methods, as directed by your instructor.

(a) $4.6 + 9.21$

(b) $12 - 3.725 - 8.59$

(c) $86(0.45)$

(d) $236.439 \div (-9.73)$

25. Round 346.0449 to the given place values.

(a) tens

(b) hundredths

(c) thousandths

26. *Sale Price* A dress that originally sold for $100.00 is discounted 40% for a sale. Then the owner decides to offer an additional 10% off of the sale price. What is the final price of the dress?

27. Consider the figure.

(a) What percent of the total number of shapes are circles?

(b) What percent of the total number of shapes are not stars?

28. *Sales of Books* Use estimation techniques to answer the following: In 2013, Carol sold $300,000 worth of books. In 2014, she sold $900,000. Her 2014 sales were _____ of her 2013 sales.

A. 30%   **B.** $33\dfrac{1}{3}\%$ **C.** 3%   **D.** 300%

29. *Creature Comforts* From a list of "everyday items" often taken for granted, adults were asked to indicate those items they wouldn't want to live without. Complete the results shown in the table if 2400 adults were surveyed.

Item	Percent That Wouldn't Want to Live Without	Number That Wouldn't Want to Live Without
Toilet paper	69%	
Zipper	42%	
Frozen food		384
Self-stick note pads		144

(Other items included tape, hairspray, pantyhose, paper clips, and Velcro.)

Source: Market Facts for Kleenex Cottonelle.

30. *Child's Drug Dosage* If D represents the usual adult dose of a drug, the corresponding child's dose C is calculated by the following formula.

$$C = \frac{\text{body surface area in square meters}}{1.7} \times D$$

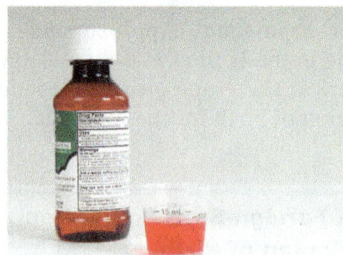

Determine the appropriate dose for a child with body surface area 0.85 m^2 if the usual adult dose is 150 mg.

Probability

11

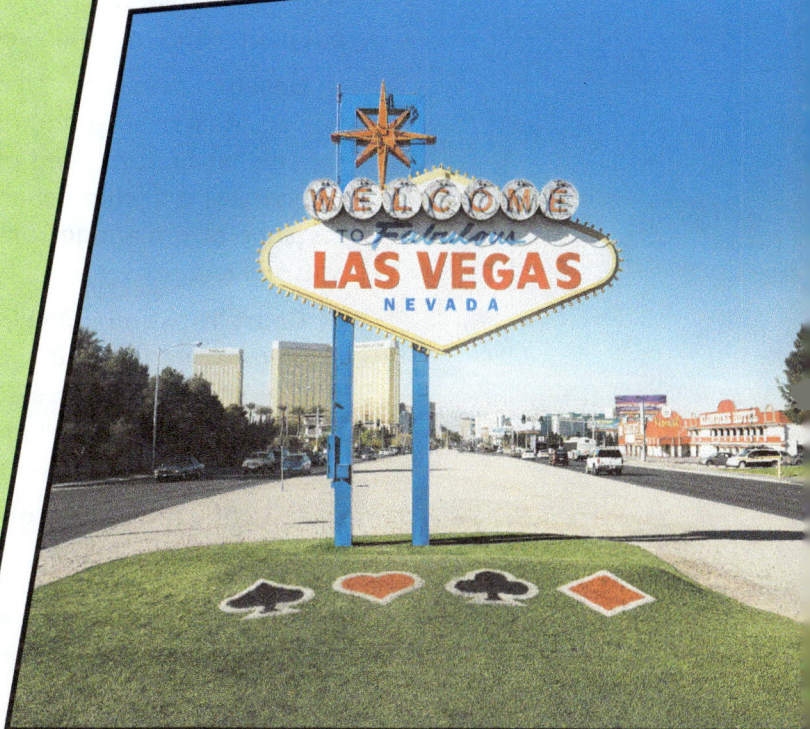

The mathematics of *probability* occurs in various occupations and experiences in daily life. An important feature distinguishes probability from other topics in this text. For example, an equation such as

$$4x + 6 = 14$$

has a specific solution (in this case, 2) that we can state without reservation. Results found in probability are often theoretical—that is, we cannot say for sure what will actually happen in a certain case. We can only say what trends will occur "in the long run."

Here are a few examples.

Weather forecasts Meteorologists state the chances of rain based on data collected on days where the atmospheric conditions have been the same in the past. If the chance of rain is 90%, it's a good bet that rain will occur, but there is still a 10% chance that no rain will fall.

Insurance rates   The rates that are charged on automobile or life insurance are based on data obtained from customers who fit the same basic profile. Actuarial science is devoted to analyzing data to provide information to insurance companies, which then make financial decisions based on those data.

Lotteries and casinos   The *odds* favor the lottery commission and the gaming industry—games of chance are designed so that in the long run, "the house wins." Once in a while, the player wins (see **page 626**), but over a lifetime, the average gambler will likely lose.

Baseball manager   History has shown that right-handed batters fare better against left-handed pitchers, and left-handed batters fare better against right-handed pitchers. Late in a close game, a manager will often make a pitching change based on this probability. This is only one instance of a manager or head coach making a decision based on the *law of averages*.

Court litigation   One interesting application of probability, called the *calculus of negligence*, was mentioned in the cult movie *Fight Club.* A typical situation involves determination by a manufacturer whether to recall a product that has proved defective. The manufacturer may decide that it is financially feasible not to recall the product if

(the probability of loss in court) × (the amount of that loss)

is less than what it would cost to conduct the recall.

11.1   BASIC CONCEPTS

OBJECTIVES

1 Understand the basic terms in the language of probability.

2 Work simple problems involving theoretical and empirical probability.

3 Understand the law of large numbers (law of averages).

4 Find probabilities related to flower colors as described by Mendel in his genetics research.

5 Determine the odds in favor of an event and the odds against an event.

The Language of Probability

If you go to a supermarket and select five pounds of peaches at $2.49 per pound, you can easily predict the amount you will be charged at the checkout counter.

$$5 \cdot \$2.49 = \$12.45$$

This is an example of a **deterministic phenomenon.** It can be predicted exactly on the basis of obtainable information, namely, in this case, number of pounds and cost per pound.

On the other hand, consider the problem faced by the produce manager of the market, who must order peaches to have on hand each day without knowing exactly how many pounds customers will buy during the day. Customer demand is an example of a **random phenomenon.** It fluctuates in such a way that its value on a given day cannot be predicted exactly with obtainable information.

The study of probability is concerned with such random phenomena. Even though we cannot be certain whether a given result will occur, we often can obtain a good measure of its *likelihood*, or **probability.** This chapter discusses various ways of finding and using probabilities.

Any observation, or measurement, of a random phenomenon is an **experiment.** The possible results of the experiment are **outcomes,** and the set of all possible outcomes is the **sample space.**

Usually we are interested in some particular collection of the possible outcomes. Any such subset of the sample space is an **event.** See the Venn Diagram in **Figure 1.** Outcomes that belong to the event are "favorable outcomes," or "successes." Any time a success is observed, we say that the event has "occurred." The probability of an event, being a numerical measure of the event's likelihood, is determined in one of two ways: either *theoretically* (mathematically) or *empirically* (experimentally). We use the notation

$$P(E) \text{ to represent the probability of event } E.$$

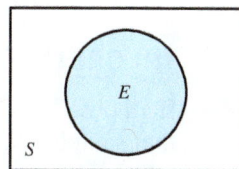

Every event *E* is a subset of the sample space *S*.

Figure 1

Examples in Probability

EXAMPLE 1 Finding Probability When Tossing a Coin

If a single coin is tossed, find the probability that it will land heads up.

Solution

There is no apparent reason for one side of a coin to land up any more often than the other (in the long run), so we assume that heads and tails are equally likely.

The experiment here is the tossing of a single fair coin, the sample space is $S = \{h, t\}$, and the event whose probability we seek is $E = \{h\}$. Since one of the two equally likely outcomes is a head, the probability of heads is the quotient of 1 and 2.

$$\text{Probability (heads)} = \frac{1}{2}, \quad \text{written} \quad P(h) = \frac{1}{2} \quad \text{or} \quad P(E) = \frac{1}{2}. \quad \blacksquare$$

EXAMPLE 2 Finding Probability When Tossing a Cup

If a Styrofoam cup is tossed, find the probability that it will land on its top.

Solution

Intuitively, it seems that such a cup will land on its side much more often than on its top or its bottom. But just how much more often is not clear. To get an idea, we performed the experiment of tossing such a cup 50 times. It landed on its side 44 times, on its top 5 times, and on its bottom just 1 time. By the frequency of "success" in this experiment, we concluded that for the cup we used,

$$P(\text{top}) \approx \frac{5}{50} = \frac{1}{10}. \quad \boxed{\text{Write in lowest terms.}} \quad \blacksquare$$

The Birth of Probability The basic ideas of probability arose largely in the context of games and gambling. In 1654 two French mathematicians, **Pierre de Fermat** (about 1601–1665) and **Blaise Pascal** (1623–1662) corresponded with each other regarding a problem posed by the Chevalier de Méré, a gambler and member of the aristocracy.

If the two players of a game are forced to quit before the game is finished, how should the pot be divided?

Pascal and Fermat solved the problem by developing basic methods of determining each player's chance, or probability, of winning.

In his 2010 book *The Unfinished Game: Pascal, Fermat, and the Seventeenth-Century Letter That Made the World Modern*, Keith Devlin describes how the two "struggled for several weeks" to solve the unfinished-game problem. In fact, this is no exception, but rather the rule, even for the greatest mathematicians. The reams of scratch work behind the elegant results are seldom seen and rarely published.

In **Example 1** involving the tossing of a fair coin, the number of possible outcomes was obviously two, both were equally likely, and one of the outcomes was a head. No actual experiment was required. The desired probability was obtained *theoretically.* Theoretical probabilities apply to dice rolling, card games, roulette, lotteries, and so on, and apparently to many phenomena in nature.

Laplace, in his famous *Analytic Theory of Probability,* published in 1812, gave a formula that applies to any such theoretical probability, as long as the sample space S is finite and all outcomes are equally likely. It is often referred to as the *classical definition of probability.*

THEORETICAL PROBABILITY FORMULA

If all outcomes in a sample space S are equally likely, and E is an event within that sample space, then the **theoretical probability** of event E is given by the following formula.

$$P(E) = \frac{\text{number of favorable outcomes}}{\text{total number of outcomes}} = \frac{n(E)}{n(S)}$$

WHEN Will I Ever USE This ?

The traditional coin flip is a tried-and-true method of making a decision when only two people (players) are involved. It is easy to see that each outcome of heads and tails has a probability of $\frac{1}{2}$. We sometimes say that "the chances are 50–50," meaning that each player has a 50% chance of winning. But suppose that we are faced with determining a winner among *three* players? *Can we use coin flipping in such a way that each player has an equal probability of winning?* The answer is yes.

The method is sometimes called "odd man wins." Each player flips a coin, and then the results are examined. If all three players show heads or all show tails, the experiment is repeated. (We are eliminating the outcomes hhh and ttt in our sample space this way.) There are 6 other possible outcomes, and each one shows 1 tail or 1 head. If a player has this "odd" result, that player is the winner. The tree diagram here shows that each player has a

$$\frac{2}{6}, \text{ or } \frac{1}{3}, \text{ probability of winning.}$$

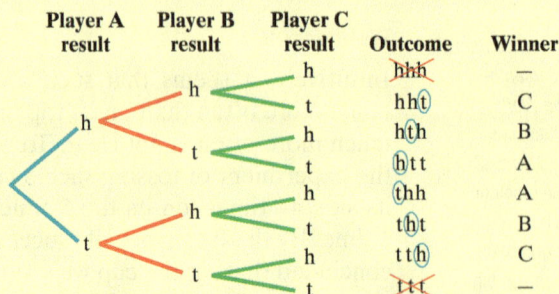

Of the 6 equally likely outcomes considered, each player wins 2 times.

On the other hand, **Example 2** involved the tossing of a cup, where the likelihoods of the various outcomes were not intuitively clear. It took an actual experiment to arrive at a probability value of $\frac{1}{10}$, and that value, based on a portion of all possible tosses of the cup, should be regarded as an approximation of the true theoretical probability. The value was found according to the *experimental*, or *empirical*, probability formula.

EMPIRICAL PROBABILITY FORMULA

If E is an event that may happen when an experiment is performed, then an **empirical probability** of event E is given by the following formula.

$$P(E) = \frac{\text{number of times event } E \text{ occurred}}{\text{number of times the experiment was performed}}$$

Usually it is clear in applications which probability formula should be used.

EXAMPLE 3 Finding the Probability of Having Daughters

Kathy wants to have exactly two daughters. Assuming that boy and girl babies are equally likely, find her probability of success for the following cases.

(a) She has a total of two children.

(b) She has a total of three children.

Solution

(a) The equal-likelihood assumption allows the use of theoretical probability. We can determine the number of favorable outcomes and the total number of possible outcomes by using a tree diagram (see **Section 10.1**) to enumerate the possibilities, as shown in **Figure 2**. From the outcome column we obtain the sample space $S = \{gg, gb, bg, bb\}$. Only one outcome, marked with an arrow, is favorable to the event of exactly two daughters: $E = \{gg\}$.

$$P(E) = \frac{n(E)}{n(S)} = \frac{1}{4} \qquad \text{Theoretical probability formula}$$

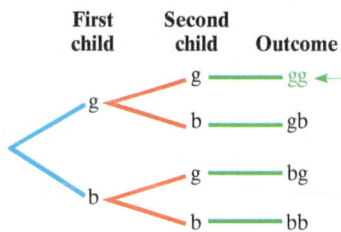

Exactly two girls among two children
Figure 2

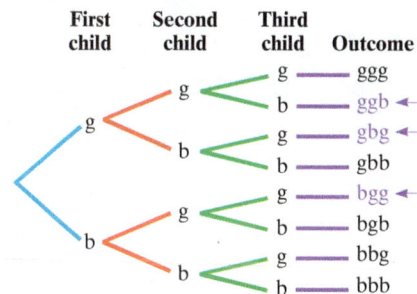

Exactly two girls among three children
Figure 3

(b) For three children altogether, we construct another tree diagram, as shown in **Figure 3**. In this case, we see that

$$S = \{ggg, ggb, gbg, gbb, bgg, bgb, bbg, bbb\} \quad \text{and} \quad E = \{ggb, gbg, bgg\},$$

so

$$P(E) = \frac{3}{8}.$$

When dealing or drawing cards, as in the next example, the dealing is generally done "without replacement." Once dealt, a card is *not* replaced in the deck. So all cards in a hand are distinct. Repetitions are *not* allowed. In many cases, such as building three-digit numbers, repetition of digits *is* allowed. For example, 255 is a legitimate three-digit number. So digit selection is done "with replacement."

EXAMPLE 4 Finding Probability When Dealing Cards

Find the probability of being dealt each of the following hands in five-card poker. Use a calculator to obtain answers to eight decimal places.

(a) a full house (three of one denomination and two of another)

(b) a royal flush (the five highest cards—ace, king, queen, jack, ten—of a single suit)

Solution

(a) **Table 1** summarizes the various possible kinds of five-card hands as first seen in **Section 10.3.** Because the **2,598,960** possible individual hands all are equally likely, we can enter the appropriate numbers from the table into the theoretical probability formula.

$$P(\text{full house}) = \frac{3744}{2{,}598{,}960} = \frac{6}{4165} \approx 0.00144058$$

(b) The table shows that there are four royal flush hands, one for each suit.

$$P(\text{royal flush}) = \frac{4}{2{,}598{,}960} = \frac{1}{649{,}740} \approx 0.00000154$$

Examples 3 and 4 both utilized the theoretical probability formula because we were able to enumerate all possible outcomes and all were equally likely. In **Example 3,** however, the equal likelihood of girl and boy babies was *assumed.* While male births typically occur a little more frequently, there usually are more females living at any given time, due to higher infant mortality rates among males and longer female life expectancy in general. **Example 5** shows a way of incorporating such empirical information.

Table 1

Number of Poker Hands in 5-Card Poker; Nothing Wild

Event E	Number of Outcomes Favorable to E
Royal flush	4
Straight flush	36
Four of a kind	624
Full house	3744
Flush	5108
Straight	10,200
Three of a kind	54,912
Two pairs	123,552
One pair	1,098,240
No pair	1,302,540
Total	2,598,960

EXAMPLE 5 Finding the Probability of the Gender of a Resident

According to the United States Census figures, on July 1, 2012, there were an estimated 154.5 million males and 159.4 million females. If a person were selected randomly from the population in that year, what is the probability that the person would be a male? (*Source:* www.census.gov)

Solution

In this case, we calculate the empirical probability from the given experimental data.

$$P(\text{male}) = \frac{\text{number of males}}{\text{total number of persons}}$$
$$= \frac{154.5 \text{ million}}{154.5 \text{ million} + 159.4 \text{ million}}$$
$$\approx 0.492$$

The **law of large numbers** (or **law of averages**) also can be stated as follows.

A theoretical probability really says nothing about one, or even a few, repetitions of an experiment, but only about the proportion of successes we would expect over the long run.

The Law of Large Numbers (or Law of Averages)

Recall the cup of **Example 2.** If we tossed it 50 more times, we would have 100 total tosses on which to base an empirical probability of the cup landing on its top. The new value would likely be slightly different from what we obtained before. It would still be an empirical probability, but it would be "better" in the sense that it was based on a larger set of outcomes.

If, as we increase the number of tosses, the resulting empirical probability values approach some particular number, that number can be defined as the theoretical probability of that particular cup landing on its top. We could determine this "limiting" value only as the actual number of observed tosses approached the total number of possible tosses of the cup. There are potentially an infinite number of possible tosses, so we could never actually find the theoretical probability. But we can still assume such a number exists. And as the number of actual observed tosses increases, the resulting empirical probabilities should tend ever closer to the theoretical value.

This principle is known as the **law of large numbers** or the **law of averages.**

LAW OF LARGE NUMBERS (LAW OF AVERAGES)

As an experiment is repeated more and more times, the proportion of outcomes favorable to any particular event will tend to come closer and closer to the theoretical probability of that event.

EXAMPLE 6   Graphing a Sequence of Proportions

A fair coin was tossed 35 times, producing the following sequence of outcomes.

tthhh,   ttthh,   hthtt,   hhthh,   ttthh,   thttt,   hhthh

Calculate the ratio of heads to total tosses after the first toss, the second toss, and so on through all 35 tosses, and plot these ratios on a graph.

Solution

After the first toss, we have 0 heads out of 1 toss, for a ratio of $\frac{0}{1} = 0.00$. After two tosses, we have $\frac{0}{2} = 0.00$. After three tosses, we have $\frac{1}{3} \approx 0.33$. Verify that the first six ratios are

0.00,   0.00,   0.33,   0.50,   0.60,   0.50.

The 35 ratios are plotted in **Figure 4.** The fluctuations away from 0.50 become smaller as the number of tosses increases, and the ratios appear to approach 0.50 toward the right side of the graph, in keeping with the law of large numbers.

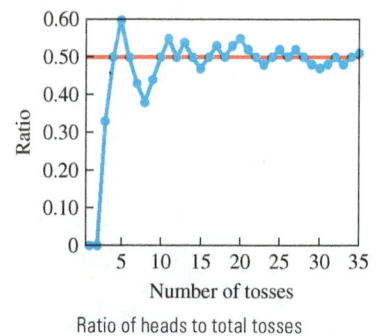

Ratio of heads to total tosses
Figure 4

COMPARING EMPIRICAL AND THEORETICAL PROBABILITIES

A series of repeated experiments provides an **empirical probability** for an event, which, by *inductive reasoning*, is an *estimate* of the event's *theoretical probability*. Increasing the number of repetitions increases the reliability of the estimate.

Likewise, an established **theoretical probability** for an event enables us, by *deductive reasoning*, to *predict* the proportion of times the event will occur in a series of repeated experiments. The prediction should be more accurate for larger numbers of repetitions.

Taking Your Chances

Smoking 1.4 cigarettes
Spending 1 hour in a coal mine
Living 2 days in New York or Boston
Eating 40 teaspoons of peanut butter
Living 2 months with a cigarette smoker
Flying 1000 miles in a jet
Traveling 300 miles in a car
Riding 10 miles on a bicycle

Risk is the probability that a harmful event will occur. Almost every action or substance exposes a person to some risk, and the assessment and reduction of risk account for a great deal of study and effort in our world. The list above, from *Calculated Risk*, by J. Rodricks, contains activities that increase a person's annual risk of death by one chance in a million.

Gregor Johann Mendel (1822–1884) came from a peasant family who managed to send him to school. By 1847 he had been ordained and was teaching at the Abbey of St. Thomas. He finished his education at the University of Vienna and returned to the abbey to teach mathematics and natural science.

Mendel began to carry out experiments on plants in the abbey garden, notably pea plants, whose distinct traits (unit characters) he had puzzled over. In 1865 he published his results. His work was not appreciated at the time, even though he had laid the foundation of **classical genetics.**

Probability in Genetics

Probabilities, both empirical and theoretical, have been valuable tools in many areas of science. An important early example was the work of the Austrian monk Gregor Mendel, who used the idea of randomness to help establish the study of genetics.

In an effort to understand the mechanism of character transmittal from one generation to the next in plants, Mendel counted the number of occurrences of various characteristics. He found that the flower color in certain pea plants obeyed this scheme:

Pure red crossed with pure white produces red.

Mendel theorized that red is "dominant," symbolized in this explanation with the capital letter R, while white is "recessive," symbolized with the lowercase letter r. The pure red parent carried only genes for red (R), and the pure white parent carried only genes for white (r). The offspring would receive one gene from each parent, hence one of the four combinations shown in the body of **Table 2.** Because every offspring receives one gene for red, that characteristic dominates, and each offspring exhibits the color red.

Table 2 First to Second Generation

		Second Parent	
		r	**r**
First Parent	**R**	Rr	Rr
	R	Rr	Rr

Table 3 Second to Third Generation

		Second Parent	
		R	**r**
First Parent	**R**	RR	Rr
	r	rR	rr

Now each of these second-generation offspring, though exhibiting the color red, still carries one of each gene. So when two of them are crossed, each third-generation offspring will receive one of the gene combinations shown in **Table 3.** Mendel theorized that each of the four possibilities would be equally likely, and he produced experimental counts that were close enough to support this hypothesis.

EXAMPLE 7 Finding Probabilities of Flower Colors

Referring to **Table 3,** determine the probability that a third-generation offspring will exhibit each flower color. Base the probabilities on the sample space of equally likely outcomes.

$$S = \{RR, Rr, rR, rr\}$$

(a) red   (b) white

Solution

(a) Since red dominates white, any combination with at least one gene for red (R) will result in red flowers. Since three of the four possibilities meet this criterion,

$$P(\text{red}) = \frac{3}{4}.$$

(b) Only the combination rr has no gene for red, so

$$P(\text{white}) = \frac{1}{4}.$$

Super Bowl Coin Tosses During the pregame coin flip of Super Bowl XLIV in 2010, the New Orleans Saints of the National Football Conference (NFC) called "heads" and won the toss. The probability of this, of course, was $\frac{1}{2}$. At the time, however, an interesting situation existed. For the twelve previous Super Bowls, the NFC representative in the Super Bowl had won the toss over the American Football Conference (AFC). The Saints continued this streak, and thus the NFC had won the toss for thirteen straight years.

Immediately following the toss, the television announcer indicated that the odds against such a run of thirteen straight wins were

"about 8100 to 1."

Where did this information come from? Someone had done the mathematics ahead of time and evaluated

$$\left(\frac{1}{2}\right)^{13} \quad \text{to obtain} \quad \frac{1}{8192}.$$

Using the concepts described in this section, the actual theoretical odds are 8191 to 1 against a run of thirteen straight wins by one conference.

Odds

Whereas probability compares the number of favorable outcomes to the total number of outcomes, **odds** compare the number of favorable outcomes to the number of unfavorable outcomes. Odds are commonly quoted, rather than probabilities, in horse racing, lotteries, and most other gambling situations. And the odds quoted normally are odds "against" rather than odds "in favor."

ODDS

If all outcomes in a sample space are equally likely, a of them are favorable to the event E, and the remaining b outcomes are unfavorable to E, then the

odds in favor of E are a to b, and the **odds against E** are b to a.

EXAMPLE 8 Finding the Odds of Getting an Intern Position

Theresa has been promised one of six jobs, three of which would be intern positions at the state capitol. If she has equal chances for all six jobs, find the odds *in favor* of her getting one of the intern positions.

Solution

Since three possibilities are favorable and three are not, the odds of *becoming an intern* at the capitol are 3 to 3 (or 1 to 1 in reduced terms). Odds of 1 to 1 are often termed

"even odds," or a "50–50 chance." ∎

EXAMPLE 9 Finding the Odds of Winning a Raffle

Bob has purchased 12 tickets for an office raffle in which the winner will receive an iPad. If 104 tickets were sold altogether and each has an equal chance of winning, what are the odds *against* Bob's winning the iPad?

Solution

Bob has 12 chances to win and $104 - 12 = 92$ chances to lose, so the odds *against* winning are

92 to 12.

In practice, we often use the fact that each number can be divided by their greatest common factor, to express the odds in lower terms. Dividing each of 92 and 12 by 4 gives the equivalent odds of

23 to 3. Lowest terms ∎

CONVERTING BETWEEN PROBABILITY AND ODDS

Let E be an event.

- If $P(E) = \frac{a}{b}$, then the odds in favor of E are a to $(b-a)$.
- If the odds in favor of E are a to b, then $P(E) = \frac{a}{a+b}$.

EXAMPLE 10   Converting from Probability to Odds

There is a 30% chance of rain tomorrow. Give this information in terms of odds.

Solution

$$P(\text{rain}) = 0.30 = \frac{30}{100} = \frac{3}{10}$$

> Convert the decimal fraction to a quotient of integers and reduce.

By the first conversion formula, if $P(E) = \frac{a}{b}$, then the odds *in favor* of rain tomorrow are a to $(b - a)$—that is,

$$3 \text{ to } (10 - 3)$$

or

$$3 \text{ to } 7. \quad \text{Odds in favor of rain}$$

We can also say that the odds *against* rain tomorrow are

$$7 \text{ to } (10 - 7)$$

or

$$7 \text{ to } 3. \quad \text{Odds against rain} \qquad \blacksquare$$

EXAMPLE 11   Converting from Odds to Probability

In a certain sweepstakes, your odds of winning are 1 to 99,999. What is the probability that you will win?

Solution

Use the second conversion formula $P(E) = \frac{a}{a + b}$.

$$P(\text{win}) = \frac{1}{1 + 99,999} = \frac{1}{100,000} = 0.00001 \qquad \blacksquare$$

11.1   EXERCISES

In Exercises 1–4, give the probability that the spinner shown would land on **(a)** *red,* **(b)** *yellow, and* **(c)** *blue.*

1.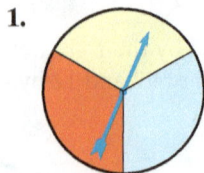

2.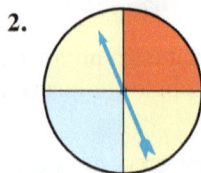

3.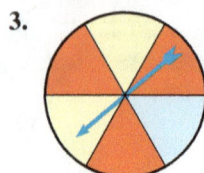

4.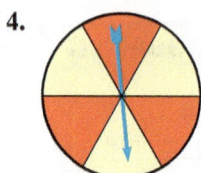

Solve each probability problem.

5. *Using Spinners to Generate Numbers*   Suppose the spinner shown here is spun once, to determine a single-digit number, and we are interested in the event E that the resulting number is odd. Give each of the following.

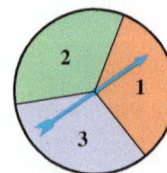

(a) the sample space

(b) the number of favorable outcomes

(c) the number of unfavorable outcomes

(d) the total number of possible outcomes

(e) the probability of an odd number

(f) the odds in favor of an odd number

6. *Lining Up Preschool Children* Kim's group of pre-school children includes nine girls and seven boys. If she randomly selects one child to be first in line, with *E* being the event that the one selected is a girl, give each of the following.

   (a) the total number of possible outcomes

   (b) the number of favorable outcomes

   (c) the number of unfavorable outcomes

   (d) the probability of event *E*

   (e) the odds in favor of event *E*

7. *Using Spinners to Generate Numbers* The spinner of **Exercise 5** is spun twice in succession to determine a two-digit number. Give each of the following.

   (a) the sample space

   (b) the probability of an odd number

   (c) the probability of a number with repeated digits

   (d) the probability of a number greater than 30

   (e) the probability of a prime number

8. *Probabilities in Coin Tossing* Two fair coins are tossed (say a dime and a quarter). Give each of the following.

   (a) the sample space

   (b) the probability of heads on the dime

   (c) the probability of heads on the quarter

   (d) the probability of getting both heads

   (e) the probability of getting the same outcome on both coins

9. *Random Selection of Fifties Music* Butch has fifty vinyl records from the fifties, including exactly one by Smiley Lewis, two by The Drifters, three by Bobby Darin, four by The Coasters, and five by Fats Domino. If he randomly selects one from his collection of fifty, find the probability it will be by each of the following.

   (a) Smiley Lewis

   (b) The Drifters

   (c) Bobby Darin

   (d) The Coasters

   (e) Fats Domino

10. *Probabilities in Coin Tossing* Three fair coins are tossed.

    (a) Write out the sample space.

    Determine the probability of each event.

    (b) no heads   (c) exactly one head

    (d) exactly two heads   (e) three heads

11. *Number Sums for Rolling Two Dice* The sample space for the rolling of two fair dice appeared in **Table 2** of **Section 10.1.** Reproduce that table, but replace each of the 36 equally likely ordered pairs with its corresponding sum (for the two dice). Then find the probability of rolling each sum.

    (a) 2   (b) 3   (c) 4

    (d) 5   (e) 6   (f) 7

    (g) 8   (h) 9   (i) 10

    (j) 11   (k) 12

12. *Probabilities of Two Daughters among Four Children* In **Example 3,** what would be Kathy's probability of having exactly two daughters if she were to have four children altogether? (You may want to use a tree diagram to construct the sample space.)

*Probabilities of Poker Hands In 5-card poker, find the probability of being dealt each of the following. Give each answer to eight decimal places. (Refer to **Table 1**.)*

13. a straight flush

14. two pairs

15. four of a kind

16. four queens

17. a hearts flush (*not* a royal flush or a straight flush)

18. no pair

In Exercises 19 and 20, give answers to three decimal places.

19. *Probability of Seed Germination* In a hybrid corn research project, 200 seeds were planted, and 175 of them germinated. Find the empirical probability that any particular seed of this type will germinate.

20. *Probability of Forest Land in California* According to *The World Almanac and Book of Facts,* California has 155,959 square miles of land area, 51,250 square miles of which are forested. Find the probability that a randomly selected location in California will be forested.

Selecting Class Reports Assuming that Ben, Jill, and Pam are three of the 26 members of the class, and that three of the class members will be chosen randomly to deliver their reports during the next class meeting, find the probability (to six decimal places) of each event.

21. Ben, Jill, and Pam are selected, in that order.

22. Ben, Jill, and Pam are selected, in any order.

23. *Probabilities in Olympic Curling*
In Olympic curling, the scoring area (shown here) consists of four concentric circles on the ice with radii of 6 inches, 2 feet, 4 feet, and 6 feet. If a team member lands a (43-pound) stone *randomly* within the scoring area, find the probability that it ends up centered on the given color.

(a) red   (b) white   (c) blue

24. *Probabilities in Dart Throwing*
If a dart hits the square target shown here at random, what is the probability that it will hit in a colored region? (*Hint:* Compare the area of the colored regions to the total area of the target.)

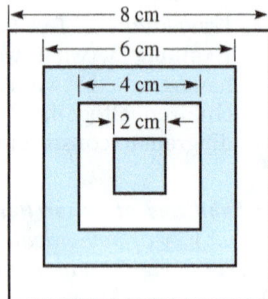

Genetics in Snapdragons *Mendel found no dominance in snapdragons (in contrast to peas) with respect to red and white flower color. When pure red and pure white parents are crossed (see* **Table 2***), the resulting* Rr *combination (one of each gene) produces second-generation offspring with pink flowers. These second-generation pinks, however, still carry one red and one white gene, so when they are crossed, the third generation is still governed by* **Table 3.***

Find each probability for third-generation snapdragons.

25. *P*(red)   **26.** *P*(pink)   **27.** *P*(white)

Genetics in Pea Plants *Mendel also investigated various characteristics besides flower color. For example, round peas are dominant over recessive wrinkled peas. First, second, and third generations can again be analyzed using* **Tables 2 and 3,** *where* R *represents round and* r *represents wrinkled.*

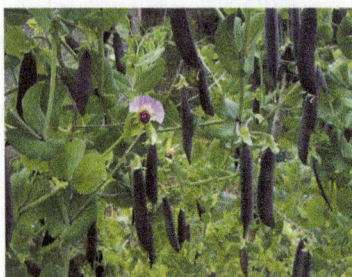

28. Explain why crossing pure round and pure wrinkled first-generation parents will always produce round peas in the second-generation offspring.

29. When second-generation round pea plants (each of which carries both R and r genes) are crossed, find the probability that a third-generation offspring will have the following.

(a) round peas   (b) wrinkled peas

Genetics of Cystic Fibrosis *Cystic fibrosis is one of the most common inherited diseases in North America (including the United States), occurring in about 1 of every 2000 Caucasian births and about 1 of every 250,000 non-Caucasian births. Even with modern treatment, victims usually die from lung damage by their early twenties.*

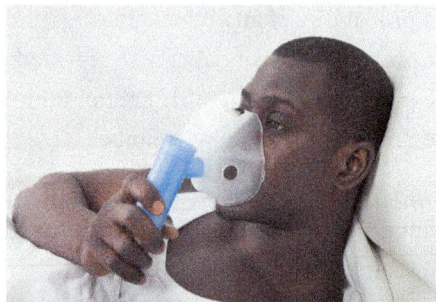

If we denote a cystic fibrosis gene with a c *and a disease-free gene with a* C *(since the disease is recessive), then only a* cc *person will actually have the disease. Such persons would ordinarily die before parenting children, but a child can also inherit the disease from two* Cc *parents (who themselves are healthy—that is, have no symptoms but are "carriers" of the disease). This is like a pea plant inheriting white flowers from two red-flowered parents that both carry genes for white.*

30. Find the empirical probability (to four decimal places) that cystic fibrosis will occur in a randomly selected infant birth among U.S. Caucasians.

31. Find the empirical probability (to six decimal places) that cystic fibrosis will occur in a randomly selected infant birth among U.S. non-Caucasians.

32. Among 150,000 North American Caucasian births, about how many occurrences of cystic fibrosis would you expect?

Suppose that both partners in a marriage are cystic fibrosis carriers (a rare occurrence). Construct a chart similar to **Table 3** *and determine the probability of each of the following events.*

33. Their first child will have the disease.

34. Their first child will be a carrier.

35. Their first child will neither have nor carry the disease.

Suppose a child is born to one cystic fibrosis carrier parent and one non-carrier parent. Find the probability of each of the following events.

36. The child will have cystic fibrosis.

37. The child will be a healthy cystic fibrosis carrier.

38. The child will neither have nor carry the disease.

Genetics of Sickle-Cell Anemia *Sickle-cell anemia occurs in about 1 of every 500 black baby births and about 1 of every 160,000 non-black baby births. It is ordinarily fatal in early childhood. There is a test to identify carriers. Unlike cystic fibrosis, which is recessive, sickle-cell anemia is* **codominant.** *This means that inheriting two sickle-cell genes causes the disease, while inheriting just one sickle-cell gene causes a mild (non-fatal) version (which is called* **sickle-cell trait***). This is similar to a snapdragon plant manifesting pink flowers by inheriting one red gene and one white gene.*

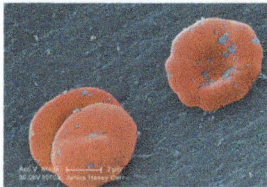

In Exercises 39 and 40, find the empirical probabilities of the given events.

39. A randomly selected black baby will have sickle-cell anemia. (Give your answer to three decimal places.)

40. A randomly selected non-black baby will have sickle-cell anemia. (Give your answer to six decimal places.)

41. Among 80,000 births of black babies, about how many occurrences of sickle-cell anemia would you expect?

In Exercises 42–44, find the theoretical probability of each condition in a child both of whose parents have sickle-cell trait.

42. The child will have sickle-cell anemia.

43. The child will have sickle-cell trait.

44. The child will be healthy.

Drawing Balls from an Urn Anne Kelly randomly chooses a single ball from the urn shown here. Find the odds against each event.

45. red

46. yellow

47. blue

48. red or yellow

49. yellow or blue

50. red or blue

Random Selection of Club Officers *Five people (Alan, Bill, Cathy, David, and Evelyn) form a club:* $N = \{A, B, C, D, E\}$. *Cathy and Evelyn are women, and the others are men. If they choose a president randomly, find the odds against each of the following becoming president.*

51. Cathy

52. a woman

53. a person whose name begins with a consonant

54. a man

In Exercises 55 and 56, assume that the probability of an event E is

$$P(E) = 0.37.$$

Find each of the following.

55. the odds in favor of E

56. the odds against E

Make the requested conversions in Exercises 57 and 58.

57. If the odds in favor of event E are 12 to 19, find $P(E)$.

58. If the odds against event E are 10 to 3, find $P(E)$.

59. Women's 100-Meter Run In the history of track and field, no woman has broken the 10-second barrier in the 100-meter run.

(a) From the statement above, find the empirical probability that a woman runner will break the 10-second barrier next year.

(b) Can you find the theoretical probability for the event of part (a)?

(c) Is it possible that the event of part (a) will occur?

60. On page 27 of their book *Descartes' Dream*, Philip Davis and Reuben Hersh ask the question, "Is probability real or is it just a cover-up for ignorance?" What do you think?

The remaining exercises require careful thought to determine n(E) and n(S). (In some cases, you may want to employ counting methods from Chapter 10, such as the fundamental counting principle, permutations, or combinations.)

Probabilities of Seating Arrangements *Six people (three married couples) arrange themselves randomly in six consecutive seats in a row. In Exercises 61–64,*

(a) *determine the number of ways in which the described event can occur, and*

(b) *determine the probability of the event.*

(Hint: In each case the denominator of the probability fraction will be 6! = 720, the total number of ways to arrange six items.)

61. Each man will sit immediately to the left of his wife.

62. Each man will sit immediately to the left of a woman.

63. The women will be in three adjacent seats.

64. The women will be in three adjacent seats, as will the men.

Solve each problem.

65. *Altered Dice*   A six-sided die has been altered so that the side that had been a single dot is now a blank face. Another die has a blank face instead of the face with four dots. What is the probability that a sum of 7 is rolled when the two dice are thrown? (*Mathematics Teacher* calendar problem)

66. *Location in a Tunnel*   Mr. Davis is driving through a tunnel that is eight miles long. At this instant, what is the probability that he is at least six miles from one end of the tunnel? (*Mathematics Teacher* calendar problem)

67. *Slopes*   Two lines, neither of which is vertical, are perpendicular if and only if their slopes are negative reciprocals (as are $\frac{2}{3}$ and $-\frac{3}{2}$). If two distinct numbers are chosen randomly from the set

$$\left\{-2, -\tfrac{4}{3}, -\tfrac{1}{2}, 0, \tfrac{1}{2}, \tfrac{3}{4}, 3\right\},$$

find the probability that they will be the slopes of two perpendicular lines.

68. *Drawing Cards*   When drawing cards without replacement from a standard 52-card deck, find the maximum number of cards you could possibly draw and still get the following.

(a) fewer than three black cards

(b) fewer than six spades

(c) fewer than four face cards

(d) fewer than two kings

69. *Student Course Schedules*   A student plans to take three courses next term. If he selects them randomly from a list of twelve courses, five of which are science courses, what is the probability that all three courses selected will be science courses?

70. *Symphony Performances*   Rhonda randomly selects three symphony performances to attend this season from a schedule of ten performances, three of which feature works by Beethoven. Find the probability that she will select all of the Beethoven programs.

71. *Racing Bets*   A "trifecta" is a particular horse race in which you win by picking the "win," "place," and "show" horses (the first-, second-, and third-place winners), in their proper order. If five horses of equal ability are entered in a trifecta race, and Tracy selects an entry, what is the probability that she will be a winner?

72. *Random Selection of Prime Numbers*   If two distinct prime numbers are randomly selected from among the first eight prime numbers, what is the probability that their sum will be 24?

73. *Random Sums*   Two integers are randomly selected from the set $\{1, 2, 3, 4, 5, 6, 7, 8, 9\}$ and are added together. Find the probability that their sum is 11 if they are selected as described.

(a) with replacement

(b) without replacement

74. *Numbers from Sets of Digits*   The digits 1, 2, 3, 4, and 5 are randomly arranged to form a five-digit number. Find the probability of each event.

(a) The number is even.

(b) The first and last digits of the number both are even.

75. *Divisibility of Random Products*   When a fair six-sided die is tossed on a tabletop, the bottom face cannot be seen. What is the probability that the product of the numbers on the five faces that can be seen is divisible by 6? (*Mathematics Teacher* calendar problem)

76. *Random Sums and Products*   Tamika selects two different numbers at random from the set $\{8, 9, 10\}$ and adds them. Carlos takes two different numbers at random from the set $\{3, 5, 6\}$ and multiplies them. What is the probability that Tamika's result is greater than Carlos's result? (*Mathematics Teacher* calendar problem)

77. *Classroom Demographics*   A high school class consists of 6 seniors, 10 juniors, 12 sophomores, and 4 freshmen. Exactly 2 of the juniors are female, and exactly 2 of the sophomores are males. If two students are randomly selected from this class, what is the probability that the pair consists of a male junior and a female sophomore? (*Mathematics Teacher* calendar problem)

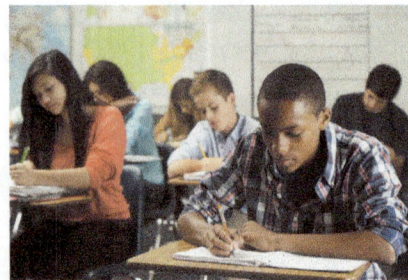

78. *Fractions from Dice Rolls*   Lisa has one red die and one green die, which she rolls to make up fractions. The green die is the numerator, and the red die is the denominator. Some of the fractions have terminating decimal representations. How many different terminating decimal results can these two dice represent? What is the probability of rolling a fraction with a terminating decimal representation? (*Mathematics Teacher* calendar problem)

Palindromic Numbers   *Numbers that are **palindromes** read the same forward and backward.*

> *30203 is a five-digit palindrome.*

If a single number is chosen randomly from each set, find the probability that it will be palindromic.

79. the set of all two-digit numbers

80. the set of all three-digit numbers

Six people, call them

> *A, B, C, D, E, and F,*

are randomly divided into three groups of two. Find the probability of each event. (Do not impose unwanted ordering among groups.)

81. *A* and *B* are in the same group, as are *C* and *D*.

82. *E* and *F* are in the same group.

11.2   EVENTS INVOLVING "NOT" AND "OR"

OBJECTIVES

1 Know that the probability of an event is a real number between 0 and 1, inclusive of both, and know the meanings of the terms *impossible event* and *certain event*.

2 Understand the correspondences among set theory, logic, and arithmetic.

3 Determine the probability of "not *E*" given the probability of *E*.

4 Determine the probability of "*A* or *B*" given the probabilities of *A*, *B*, and "*A* and *B*."

Properties of Probability

Recall that an empirical probability, based upon experimental observation, may be the best value available but still is only an approximation to the ("true") theoretical probability. For example, no human has ever been known to jump higher than 8.5 feet vertically, so the empirical probability of such an event is zero. Observing the rate at which high-jump records have been broken, we suspect that the event is, in fact, possible and may one day occur. Hence it must have some nonzero theoretical probability, even though we have no way of assessing its exact value.

Recall also that the theoretical probability formula,

$$P(E) = \frac{n(E)}{n(S)},$$

is valid only when all outcomes in the sample space *S* are equally likely. For the experiment of tossing two fair coins, we can write $S = \{hh, ht, th, tt\}$ and compute

$$P(\text{both heads}) = \frac{1}{4}. \quad \text{Correct}$$

However, if we define the sample space with non-equally likely outcomes as $S = \{\text{both heads}, \text{both tails}, \text{one of each}\}$, we are led to

$$P(\text{both heads}) = \frac{1}{3}. \quad \text{Incorrect}$$

To convince yourself that $\frac{1}{4}$ is a better value than $\frac{1}{3}$, toss two fair coins 100 times or so, to see what the empirical fraction seems to approach.

For any event *E* within a sample space *S*, we know that $0 \le n(E) \le n(S)$. Dividing all members of this inequality by $n(S)$ gives

$$\frac{0}{n(S)} \le \frac{n(E)}{n(S)} \le \frac{n(S)}{n(S)}, \quad \text{or} \quad 0 \le P(E) \le 1.$$

In words, the probability of any event is a number from 0 through 1, inclusive.

If event *E* is *impossible* (cannot happen), then $n(E)$ must be 0 (*E* is the empty set), so

$$P(E) = 0. \quad E \text{ is impossible.}$$

If event *E* is *certain* (cannot help but happen), then $n(E) = n(S)$, so

$$P(E) = \frac{n(E)}{n(S)} = \frac{n(S)}{n(S)} = 1. \quad E \text{ is certain.}$$

More on the Birth of Probability
The modern mathematical theory of probability came mainly from the Russian scholars **P. L. Chebyshev** (1821–1922), **A. A. Markov** (1856–1922), and **Andrei Nikolaevich Kolmogorov** (1903–1987). The Dutch mathematician and scientist **Christiaan Huygens** (1629–1695) wrote a formal treatise on probability. It appeared in 1657 and was based on the Pascal–Fermat correspondence. One of the first to apply probability to matters other than gambling was the French mathematician **Pierre Simon de Laplace** (1749–1827), who is usually credited with being the "father" of probability theory.

Pierre Simon de Laplace (1749–1827) began in 1773 to solve the problem of why Jupiter's orbit seems to shrink and Saturn's orbit seems to expand. Eventually Laplace worked out a complete theory of the solar system. *Celestial Mechanics* resulted from almost a lifetime of work. In five volumes, it was published between 1799 and 1825 and gained for Laplace the reputation "Newton of France."

Laplace's work on probability was actually an adjunct to his celestial mechanics. He needed to demonstrate that probability is useful in interpreting scientific data.

PROPERTIES OF PROBABILITY

Let E be an event within the sample space S. That is, E is a subset of S. Then the following properties hold.

1. $0 \leq P(E) \leq 1$ The probability of an event is a number from 0 through 1, inclusive.

2. $P(\emptyset) = 0$ The probability of an impossible event is 0.

3. $P(S) = 1$ The probability of a certain event is 1.

Probabilities are often expressed in terms of percents in the media. We will, however, give them in terms of fractions or decimals.

EXAMPLE 1 Finding Probability When Rolling a Die

When a single fair die is rolled, find the probability of each event.

(a) The number 2 is rolled. **(b)** A number other than 2 is rolled.

(c) The number 7 is rolled. **(d)** A number less than 7 is rolled.

Solution

(a) Since one of the six possibilities is a 2, $P(2) = \frac{1}{6}$.

(b) There are five such numbers, 1, 3, 4, 5, and 6, so $P(\text{a number other than 2}) = \frac{5}{6}$.

(c) None of the possible outcomes is 7. Thus, $P(7) = \frac{0}{6} = 0$.

(d) Since all six of the possible outcomes are less than 7,

$$P(\text{a number less than 7}) = \frac{6}{6} = 1.$$

Refer to the box preceding **Example 1.** No probability in that example was less than 0 or greater than 1, which illustrates Property 1. The "impossible" event of part (c) had probability 0, illustrating Property 2. The "certain" event of part (d) had probability 1, illustrating Property 3.

Events Involving "Not"

Table 4 repeats the information of **Table 8** of **Section 10.5,** with a third correspondence added in row 3. The correspondences shown in **Table 4** are the basis for the probability rules. For example, the probability of an event *not* happening involves the *complement* and *subtraction*, according to row 1 of the table.

Table 4 Set Theory/Logic/Arithmetic Correspondences

	Set Theory	Logic	Arithmetic
1. Operation or Connective (Symbol)	Complement ($'$)	Not ($\sim$)	Subtraction ($-$)
2. Operation or Connective (Symbol)	Union ($\cup$)	Or ($\vee$)	Addition ($+$)
3. Operation or Connective (Symbol)	Intersection ($\cap$)	And ($\wedge$)	Multiplication ($\cdot$)

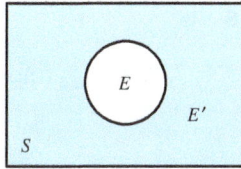

The logical connective "not" corresponds to "complement" in set theory.

$$P(not\,E) = P(S) - P(E)$$
$$= 1 - P(E)$$

Figure 5

The rule for the probability of a complement follows and is illustrated in **Figure 5.**

PROBABILITY OF A COMPLEMENT (FOR THE EVENT "NOT E")

The probability that an event E will *not* occur is equal to 1 minus the probability that it *will* occur.

$$P(not\,E) = 1 - P(E)$$

Notice that the events of **Examples 1(a) and (b)**, namely "2" and "not 2," are complements of one another, and that their probabilities add up to 1. This illustrates the above probability rule. The equation

$$P(E) + P(E') = 1$$

is a rearrangement of the formula for the probability of a complement. Another form of the equation that is also useful at times follows.

$$P(E) = 1 - P(E')$$

EXAMPLE 2 Finding the Probability of a Complement

When a single card is drawn from a standard 52-card deck, what is the probability that it will not be a king?

Solution

$$P(\text{not a king}) = 1 - P(\text{king}) = 1 - \frac{4}{52} = \frac{48}{52} = \frac{12}{13}$$

Remember to write in lowest terms.

EXAMPLE 3 Finding the Probability of a Complement

If five fair coins are tossed, find the probability of obtaining at least two heads.

Solution

A tree diagram will show that there are $2^5 = 32$ possible outcomes for the experiment of tossing five fair coins. Most include at least two heads. In fact, only the outcomes

ttttt,   htttt,   thttt,   tthtt,   tttht,   and   tttth   6 of these

do *not* include at least two heads. If E denotes the event "at least two heads," then E' is the event "not at least two heads,"

$$P(E) = 1 - P(E') = 1 - \frac{6}{32} = \frac{26}{32} = \frac{13}{16}$$

Events Involving "Or"

Examples 2 and 3 showed how the probability of an event can be approached *indirectly*, by first considering the complement of the event. Another indirect approach is to break the event into simpler component events. Row 2 of **Table 4** indicates that the probability of one event *or* another should involve the *union* and *addition*.

Mary Somerville (1780–1872) is associated with Laplace because of her brilliant exposition of his *Celestial Mechanics.*

Somerville studied Euclid thoroughly and perfected her Latin so she could read Newton's *Principia.* In about 1816 she went to London and soon became part of its literary and scientific circles.

Somerville's book on Laplace's theories came out in 1831 to great acclaim. Then followed a panoramic book, *Connection of the Physical Sciences* (1834). A statement in one of its editions suggested that irregularities in the orbit of Uranus might indicate that a more remote planet, not yet seen, existed. This caught the eye of the scientists who worked out the calculations for Neptune's orbit.

EXAMPLE 4 Selecting from a Set of Numbers

If one number is selected randomly from the set $\{1, 2, 3, 4, 5, 6, 7, 8, 9, 10\}$, find the probability of each of the following events.

(a) The number is odd or a multiple of 4.

(b) The number is odd or a multiple of 3.

Solution

Define the following events.

$$S = \{1, 2, 3, 4, 5, 6, 7, 8, 9, 10\} \quad \text{Sample space}$$

$$A = \{1, 3, 5, 7, 9\} \quad \text{Odd outcomes; } P(A) = \tfrac{5}{10}$$

$$B = \{4, 8\} \quad \text{Multiples of 4; } P(B) = \tfrac{2}{10}$$

$$C = \{3, 6, 9\} \quad \text{Multiples of 3; } P(C) = \tfrac{3}{10}$$

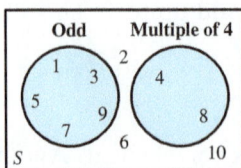

Figure 6

(a) **Figure 6** shows the positioning of the 10 integers within the sample space and within the pertinent sets A and B. The composite event "A or B" corresponds to the set $A \cup B = \{1,3,4,5,7,8,9\}$. $A \cup B$ has **seven** elements. By the theoretical probability formula,

$$P(A \text{ or } B) = \frac{7}{10}. \qquad \text{Of 10 total outcomes, seven are favorable.}$$

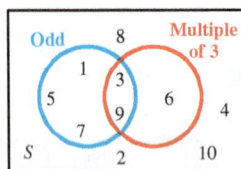

Figure 7

(b) **Figure 7** shows the situation.   Of 10 total outcomes, six are favorable.

$$P(A \text{ or } C) = \frac{6}{10} = \frac{3}{5}$$

Would an addition formula have worked in **Example 4?** Let's check.

Part (a): $P(A \text{ or } B) = P(A) + P(B) = \dfrac{5}{10} + \dfrac{2}{10} = \dfrac{7}{10}$ Correct

Part (b): $P(A \text{ or } C) = P(A) + P(C) = \dfrac{5}{10} + \dfrac{3}{10} = \dfrac{8}{10} = \dfrac{4}{5}$ Incorrect

The trouble in part (b) is that A and C are not disjoint sets. They have outcomes in common. Just as with the additive counting principle in **Chapter 10,** an adjustment must be made here to compensate for counting the common outcomes twice.

$$P(A \text{ or } C) = P(A) + P(C) - P(A \text{ and } C)$$

$$= \frac{5}{10} + \frac{3}{10} - \frac{2}{10} = \frac{6}{10} = \frac{3}{5} \quad \text{Correct}$$

In probability theory, events that are disjoint sets are called *mutually exclusive events.*

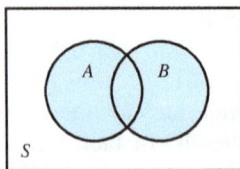

The logical connective "or" corresponds to "union" in set theory.

$P(A \text{ or } B)$
$= P(A) + P(B) - P(A \text{ and } B)$

Figure 8

MUTUALLY EXCLUSIVE EVENTS

Two events A and B are **mutually exclusive events** if they have no outcomes in common. Mutually exclusive events cannot occur simultaneously.

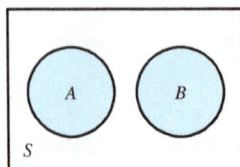

When A and B are mutually exclusive,

$P(A \text{ or } B) = P(A) + P(B).$

Figure 9

The results observed in **Example 4** are generalized as follows. The two possibilities are illustrated in **Figures 8 and 9.**

ADDITION RULE OF PROBABILITY (FOR THE EVENT "A OR B")

If A and B are any two events, then the following holds.

$$P(A \text{ or } B) = P(A) + P(B) - P(A \text{ and } B)$$

If A and B are mutually exclusive, then the following holds.

$$P(A \text{ or } B) = P(A) + P(B)$$

Actually, the first formula in the addition rule applies in all cases. The third term on the right drops out when A and B are mutually exclusive, because $P(A \text{ and } B) = 0$. Still, it is good to remember the second formula in the preceding box for the many cases where the component events are mutually exclusive. In this section, we consider only cases where the event "A and B" is simple. We deal with more involved composites involving "and" in the next section.

EXAMPLE 5 Finding the Probability of an Event Involving "Or"

If a single card is drawn from a standard 52-card deck, what is the probability that it will be a spade or a red card?

Solution

First note that "spade" and "red" cannot both occur, because there are no red spades. All spades are black. Therefore, we can use the formula for mutually exclusive events. There are 13 spades and 26 red cards in the deck.

$$P(\text{spade or red}) = P(\text{spade}) + P(\text{red}) = \frac{13}{52} + \frac{26}{52} = \frac{39}{52} = \frac{3}{4} \quad \blacksquare$$

Two examples of "success" in Example 5

We often need to consider composites of more than two events. When each event involved is mutually exclusive of all the others, we extend the addition rule to the appropriate number of components.

EXAMPLE 6 Treating Unions of Several Components

Amy plans to spend from 1 to 6 hours on her homework. If x represents the number of hours to be spent on a given night, then the probabilities of the various values of x, rounded to the nearest hour, are as shown in **Table 5**. Find the probabilities that Amy will spend each of the following amounts of time.

(a) fewer than 3 hours

(b) more than 2 hours

(c) more than 1 but no more than 5 hours

(d) fewer than 5 hours

Table 5

x	$P(x)$
1	0.05
2	0.10
3	0.20
4	0.40
5	0.10
6	0.15

Solution

Because the time periods in **Table 5** are mutually exclusive of one another, we can simply add the appropriate component probabilities.

(a) $P(\text{fewer than 3}) = P(1 \textbf{ or } 2)$ Fewer than 3 means 1 or 2.

$= P(1) + P(2)$ Addition rule

$= 0.05 + 0.10$ Substitute values from **Table 5.**

$= 0.15$ Add.

(b) $P(\text{more than }2) = P(3 \textbf{ or } 4 \textbf{ or } 5 \textbf{ or } 6)$ More than 2 means 3, 4, 5, or 6.

$\qquad\qquad\qquad = P(3) + P(4) + P(5) + P(6)$ Addition rule

$\qquad\qquad\qquad = 0.20 + 0.40 + 0.10 + 0.15$ Substitute values from **Table 5**.

$\qquad\qquad\qquad = 0.85$ Add.

(c) $P(\text{more than 1 but no more than 5})$

$\qquad = P(2 \textbf{ or } 3 \textbf{ or } 4 \textbf{ or } 5)$ 2, 3, 4, and 5 are more than 1 and no more than 5.

$\qquad = P(2) + P(3) + P(4) + P(5)$ Addition rule

$\qquad = 0.10 + 0.20 + 0.40 + 0.10$ Substitute values from **Table 5**.

$\qquad = 0.80$ Add.

(d) Although we could take a direct approach here, as in parts (a), (b), and (c), we will combine the complement rule with the addition rule.

$\qquad P(\text{fewer than 5}) = 1 - P(\text{not fewer than 5})$ Complement rule

$\qquad\qquad\qquad = 1 - P(5 \text{ or more})$ 5 or more is equivalent to not fewer than 5.

$\qquad\qquad\qquad = 1 - P(5 \textbf{ or } 6)$ 5 or more means 5 or 6.

$\qquad\qquad\qquad = 1 - [P(5) + P(6)]$ Addition rule

$\qquad\qquad\qquad = 1 - (0.10 + 0.15)$ Substitute values from **Table 5**.

$\qquad\qquad\qquad = 1 - 0.25$ Add inside the parentheses first.

$\qquad\qquad\qquad = 0.75$

Table 5 in **Example 6** lists all possible time intervals so the corresponding probabilities add up to 1, a necessary condition for the way part (d) was done. The time spent on homework here is an example of a **random variable.** It is "random" because we cannot predict which of its possible values will occur.

A listing like **Table 5,** which shows all possible values of a random variable, along with the probabilities that those values will occur, is a **probability distribution** for that random variable. *All* possible values are listed, so they make up the entire sample space, and thus the listed probabilities must add up to 1 (by probability Property 3). Probability distributions will be discussed in later sections.

EXAMPLE 7 Finding the Probability of an Event Involving "Or"

Find the probability that a single card drawn from a standard 52-card deck will be a diamond or a face card.

Solution

The component events "diamond" and "face card" can occur simultaneously. (The jack, queen, and king of diamonds belong to both events.) So, we must use the first formula of the addition rule. We let D denote "diamond" and F denote "face card."

$\qquad P(D \text{ or } F) = P(D) + P(F) - P(D \text{ and } F)$ Addition rule

$\qquad\qquad = \dfrac{13}{52} + \dfrac{12}{52} - \dfrac{3}{52}$ There are 13 diamonds, 12 face cards, and 3 that are both.

$\qquad\qquad = \dfrac{22}{52}$ Add and subtract.

$\qquad\qquad = \dfrac{11}{26}$ Write in lowest terms.

One example of a card that is both a diamond and a face card

EXAMPLE 8 Finding the Probability of an Event Involving "Or"

Of 20 elective courses, Emily plans to enroll in one, which she will choose by throwing a dart at the schedule of courses. If 8 of the courses are recreational, 9 are interesting, and 3 are both recreational and interesting, find the probability that the course she chooses will have at least one of these two attributes.

Solution

If R denotes "recreational" and I denotes "interesting," then

$$P(R) = \frac{8}{20}, \quad P(I) = \frac{9}{20}, \quad \text{and} \quad P(R \text{ and } I) = \frac{3}{20}.$$

R and I are not mutually exclusive.

$$P(R \text{ or } I) = \frac{8}{20} + \frac{9}{20} - \frac{3}{20}$$

$$= \frac{14}{20} \qquad \text{Addition rule}$$

$$= \frac{7}{10} \qquad \text{Write in lowest terms.} \qquad \blacksquare$$

11.2   EXERCISES

Probabilities for Rolling a Die   For the experiment of rolling a single fair die, find the probability of each event. (Hint: Recall that 1 is neither prime nor composite.)

1. not prime
2. not less than 2
3. even or prime
4. odd or less than 5
5. less than 3 or greater than 4
6. odd or even

Probability and Odds for Drawing a Card   For the experiment of drawing a single card from a standard 52-card deck, find (a) the probability of each event, and (b) the odds in favor of each event.

7. king or queen
8. not an ace
9. spade or face card
10. club or heart
11. neither a heart nor a 7
12. not a heart, or a 7

Number Sums for Rolling a Pair of Dice   For the experiment of rolling an ordinary pair of dice, find the probability that the sum will be each of the following. (You may want to use a table showing the sum for each of the 36 equally likely outcomes.)

13. even or a multiple of 3
14. 11 or 12
15. less than 3 or greater than 9
16. odd or greater than 9

Prime Results   Find the probability of getting a prime number in each case.

17. A number is chosen randomly from the set $\{1, 2, 3, 4, \ldots, 12\}$.
18. Two dice are rolled and the sum is observed.
19. A single digit is randomly chosen from the number 578.
20. A single digit is randomly chosen from the number 1234.

21. **Determining Whether Events Are Mutually Exclusive** Amanda has three office assistants. If A is the event that at least two of them are men and B is the event that at least two of them are women, are A and B mutually exclusive?

22. **Determining Whether Events Are Mutually Exclusive** Jeanne earned her college degree several years ago. Consider the following four events.

    Her alma mater is in the East.
    Her alma mater is a private college.
    Her alma mater is in the Northwest.
    Her alma mater is in the South.

    Are these events all mutually exclusive of one another?

Probabilities of Poker Hands *If you are dealt a 5-card hand (this implies without replacement) from a standard 52-card deck, find the probability of getting each of the following. Refer to* **Table 1** *of* **Section 11.1,** *and give answers to six decimal places.*

23. a full house or a straight

24. a black flush or two pairs

25. nothing any better than two pairs

26. a flush or three of a kind

Probabilities in Golf Scoring *The table gives Josh's probabilities of scoring in various ranges on a par-70 course. In a given round, find the probability of each event in Exercises 27–31.*

x	P(x)
Below 60	0.04
60–64	0.06
65–69	0.14
70–74	0.30
75–79	0.23
80–84	0.09
85–89	0.06
90–94	0.04
95–99	0.03
100 or above	0.01

27. par or above   **28.** in the 80s

29. less than 90   **30.** not in the 70s, 80s, or 90s

31. 95 or higher

32. What are the odds of Josh's scoring below par?

33. **Probability Distribution** Let x denote the sum of two distinct numbers selected randomly from the set of numbers $\{1, 2, 3, 4, 5\}$. Construct the probability distribution for the random variable x.

34. **Probability Distribution** Anne Kelly randomly chooses a single ball from the urn shown here, and x represents the color of the ball chosen. Construct a complete probability distribution for the random variable x.

Comparing Empirical and Theoretical Probabilities for Rolling Dice *Roll a pair of dice 50 times, keeping track of the number of times the sum is "less than 3 or greater than 9" (that is 2, 10, 11, or 12).*

35. From your results, calculate an empirical probability for the event "less than 3 or greater than 9."

36. By how much does your answer differ from the *theoretical* probability of **Exercise 15?**

37. Explain the difference between the two formulas in the addition rule of probability, illustrating each one with an appropriate example.

38. Suppose, for a given experiment, A, B, C, and D are events, all mutually exclusive of one another, such that

$$A \cup B \cup C \cup D = S \text{ (the sample space)}.$$

By extending the addition rule of probability to this case and utilizing probability Property 3, what statement can you make?

Complemental Probability *For Exercises 39–41, let A be an event within the sample space S, and let*

$$n(A) = a \quad and \quad n(S) = s.$$

39. Use the complements principle of counting to find an expression for $n(A')$.

40. Use the theoretical probability formula to express $P(A)$ and $P(A')$.

41. Evaluate and simplify $P(A) + P(A')$.

42. What rule have you proved?

The remaining exercises require careful thought for the determination of n(E) and n(S). (In some cases, you may want to employ counting methods from Chapter 10, such as the fundamental counting principle, permutations, or combinations.)

Building Numbers from Sets of Digits *Suppose we want to form three-digit numbers using the set of digits*

$$\{0, 1, 2, 3, 4, 5\}.$$

For example, 501 and 224 are such numbers, but 035 is not.

43. How many such numbers are possible?

44. How many of these numbers are multiples of 10?

45. How many of these numbers are multiples of 5?

46. If one three-digit number is chosen at random from all those that can be made from the above set of digits, find the probability that the one chosen is not a multiple of 5.

47. **Drawing Colored Marbles from Boxes** A bag contains fifty blue and fifty green marbles. Two marbles at a time are randomly selected. If both are green, they are placed in box A; if both are blue, in box B; if one is green and the other is blue, in box C. After all marbles are drawn, what is the probability that the numbers of marbles in box A and box B are the same? (*Mathematics Teacher* calendar problem)

48. *Multiplying Numbers Generated by Spinners* An experiment consists of spinning both spinners shown here and multiplying the resulting numbers together. Find the probability that the resulting product will be even.

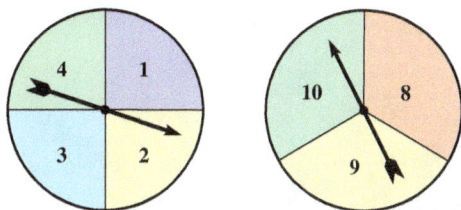

49. Luka and Janie are playing a coin toss game. If the coin lands heads up, Luka earns a point; otherwise, Janie earns a point. The first player to reach 20 points wins the game. If 19 of the first 37 tosses have been heads, what is the probability that Janie wins the game? (*Mathematics Teacher* calendar problem)

50. *Random Births on the Same Day of the Week* What is the probability that, of three people selected at random, at least two were born on the same day of the week? (*Mathematics Teacher* calendar problem)

11.3 CONDITIONAL PROBABILITY AND EVENTS INVOLVING "AND"

OBJECTIVES

1 Apply the conditional probability formula.

2 Determine whether two events are independent.

3 Apply the multiplication rule for the event "*A* and *B*."

Conditional Probability

Sometimes the probability of an event must be computed using the knowledge that some other event has happened (or is happening, or will happen—the timing is not important). This type of probability is called *conditional probability*.

CONDITIONAL PROBABILITY

The probability of event *B*, computed on the assumption that event *A* has happened, is called the **conditional probability of *B* given *A*** and is denoted

$$P(B|A).$$

EXAMPLE 1 **Selecting from a Set of Numbers**

From the sample space

$$S = \{1, 2, 3, 4, 5, 6, 7, 8, 9, 10\},$$

a single number is to be selected randomly. Find each probability given the events

   A: The selected number is odd,   and   *B*: The selected number is a multiple of 3.

(a) $P(B)$ **(b)** $P(A \text{ and } B)$ **(c)** $P(B|A)$

Solution

(a) $B = \{3, 6, 9\}$, so

$$P(B) = \frac{n(B)}{n(S)} = \frac{3}{10}.$$

(b) *A* and *B* is the set $A \cap B = \{1, 3, 5, 7, 9\} \cap \{3, 6, 9\} = \{3, 9\}$.

$$P(A \text{ and } B) = \frac{n(A \cap B)}{n(S)} = \frac{2}{10} = \frac{1}{5}$$

> Two elements

(c) The given condition, that *A* occurs, effectively reduces the sample space from *S* to *A*, and the elements of the new sample space *A* that are also in *B* are the elements of $A \cap B$.

$$P(B|A) = \frac{n(A \cap B)}{n(A)} = \frac{2}{5}$$

WHEN Will I Ever USE This ?

Monty Hall

The probability is low that anyone reading this will ever be on a televised game show. But who knows? A former popular game show host, Monty Hall, has a rather famous problem named after him, and knowing the solution to the problem might help a contestant win a car instead of a goat.

Suppose that there are three doors, and you are asked to choose one of them. One of the doors hides a new car, and behind the two other doors are goats. You pick a door, say Door A. Then the host, who knows what's behind all three doors, opens another door, say Door C, and a goat appears. The host asks you whether you want to change your choice to Door B. *Is it to your advantage to do so?*

One way to look at the problem, given that the car is *not* behind Door C, is that Doors A and B are now equally likely to contain the car. Thus, switching doors will neither help nor hurt your chances of winning the car.

However, there is another way to look at the problem. When you picked Door A, the probability was $\frac{1}{3}$ that it contained the car. Being shown the goat behind Door C doesn't really give you any new information; after all, you knew that there was a goat behind at least one of the other doors. So seeing the goat behind Door C does nothing to change your assessment of the probability that Door A has the car. It remains $\frac{1}{3}$. But because Door C has been ruled out, the probability that Door B has the car is now $\frac{2}{3}$. Thus, you should switch.

Analysis of this problem depends on the psychology of the host. If we suppose that the host must *always* show you a losing door and then give you an option to switch, then you should switch. This was not specifically stated in the problem but has been pointed out by many mathematicians.

There are several Web sites that simulate the Monty Hall problem. One of them is http://stayorswitch.com/

(The authors wish to thank David Berman of the University of New Orleans for his assistance with this explanation.)

A **cosmic impact,** the collision of a meteor, comet, or asteroid with Earth could be as catastrophic as full-scale nuclear war, killing a billion or more people.

The Spaceguard Survey has discovered more than half of the estimated number of near-Earth asteroids (NEAs) 1 kilometer or more in diameter and hopes to locate 90% of them in the next decade. Although the risk of finding one on a collision course with the Earth is slight, it is anticipated that, if we did, we would be able to deflect it before impact.

The photo above shows a crater in Arizona, 4000 feet in diameter and 570 feet deep. It is thought to have been formed 20,000 to 50,000 years ago by a meteorite about 50 meters across, hitting the ground at several kilometers per second. (See http://en.wikipedia.org/wiki/Meteor_Crater.)

Example 1 illustrates some important points. First, because

$$\frac{n(A \cap B)}{n(A)} = \frac{\frac{n(A \cap B)}{n(S)}}{\frac{n(A)}{n(S)}} \qquad \text{Multiply numerator and denominator by } \tfrac{1}{n(S)}.$$

$$= \frac{P(A \cap B)}{P(A)}, \qquad \text{Theoretical probability formula}$$

the final line of the example gives the following convenient formula.

CONDITIONAL PROBABILITY FORMULA

The **conditional probability of B given A** is calculated as follows.

$$P(B|A) = \frac{P(A \cap B)}{P(A)} = \frac{P(A \text{ and } B)}{P(A)}$$

A second observation from **Example 1** is that the conditional probability of B given A was $\frac{2}{5}$, whereas the "unconditional" probability of B (with no condition given) was $\frac{3}{10}$, so the condition *did* make a difference.

EXAMPLE 2 **Finding Probabilities of Boys and Girls in a Family**

Given a family with two children, find the probability of each event.

(a) Both are girls, given that at least one is a girl.

(b) Both are girls, given that the older child is a girl.

(Assume boys and girls are equally likely.)

Solution

We define the following events. (The older child's gender appears first.)

$$S = \{gg, gb, bg, bb\} \qquad \text{Sample space } S; \ n(S) = 4$$
$$A = \{gg\} \qquad \text{Both are girls. } P(A) = \tfrac{1}{4}$$
$$B = \{gg, gb, bg\} \qquad \text{At least one is a girl. } P(B) = \tfrac{3}{4}$$
$$C = \{gg, gb\} \qquad \text{The older one is a girl. } P(C) = \tfrac{2}{4}$$

Note that $A \cap B = \{gg\}$, and $A \cap C = \{gg\}$ as well. Thus

$$P(A \cap B) = P(A \cap C) = \frac{1}{4}.$$

(a) $P(A|B) = \dfrac{P(A \text{ and } B)}{P(B)} = \dfrac{\frac{1}{4}}{\frac{3}{4}} = \dfrac{1}{4} \div \dfrac{3}{4} = \dfrac{1}{4} \cdot \dfrac{4}{3} = \dfrac{1}{3}$

(b) $P(A|C) = \dfrac{P(A \text{ and } C)}{P(C)} = \dfrac{\frac{1}{4}}{\frac{2}{4}} = \dfrac{1}{4} \div \dfrac{2}{4} = \dfrac{1}{4} \cdot \dfrac{4}{2} = \dfrac{1}{2}$ ∎

Independent Events

Sometimes a conditional probability is no different from the corresponding unconditional probability, in which case we call the two events *independent*.

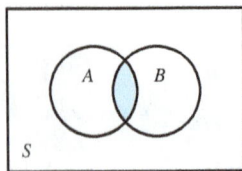

One example of a card that is both a face card and black

Two events A and B are **independent events** if knowledge about the occurrence of one of them has no effect on the probability of the other one occurring. A and B are independent if

$$P(B \mid A) = P(B), \quad \text{or, equivalently,} \quad P(A \mid B) = P(A).$$

EXAMPLE 3 **Checking Events for Independence**

A single card is to be drawn from a standard 52-card deck. (The sample space S has 52 elements.) Find each of the following, given the events

    A: The selected card is a face card,   and B: The selected card is black.

(a) $P(B)$ **(b)** $P(B \mid A)$

(c) Use the results of parts (a) and (b) to determine whether events A and B are independent.

Solution

(a) There are 26 black cards in the 52-card deck.

$$P(B) = \frac{26}{52} = \frac{1}{2} \qquad \text{Theoretical probability formula}$$

(b) $P(B \mid A) = \dfrac{P(B \text{ and } A)}{P(A)}$ Conditional probability formula

$$= \frac{\dfrac{6}{52}}{\dfrac{12}{52}} \qquad \text{Of 52 cards, 12 are face cards and 6 are black face cards.}$$

$$= \frac{6}{52} \cdot \frac{52}{12} \qquad \text{To divide, multiply by the reciprocal of the divisor.}$$

$$= \frac{1}{2} \qquad \text{Calculate and write in lowest terms.}$$

(c) Because $P(B \mid A) = P(B)$, events A and B are independent. ■

Events Involving "And"

If we multiply both sides of the conditional probability formula by $P(A)$, we obtain an expression for $P(A \cap B)$, which applies to events of the form "A and B." The resulting formula is related to the fundamental counting principle of **Chapter 10.** It is illustrated in **Figure 10.**

     Just as the calculation of $P(A \text{ or } B)$ is simpler when A and B are mutually exclusive, the calculation of $P(A \text{ and } B)$ is simpler when A and B are independent.

The logical connective "and" corresponds to "intersection" in set theory.
$P(A \text{ and } B) = P(A) \cdot P(B \mid A)$

Figure 10

MULTIPLICATION RULE OF PROBABILITY (FOR THE EVENT "A AND B")

If A and B are *any two events*, then

$$P(A \text{ and } B) = P(A) \cdot P(B \mid A).$$

If A and B are *independent*, then

$$P(A \text{ and } B) = P(A) \cdot P(B).$$

The first formula in the multiplication rule actually applies in all cases, because $P(B|A) = P(B)$ when A and B are independent. The independence of the component events is clear in many cases, so it is good to remember the second formula.

EXAMPLE 4 Selecting from a Set of Books

Each year, Jacqui adds to her book collection a number of new publications that she believes will be of lasting value and interest. She has categorized each of her twenty acquisitions for 2015 as hardcover or paperback and as fiction or nonfiction. The numbers of books in the various categories are shown in **Table 6.**

Table 6 Year 2015 Books

	Fiction (F)	Nonfiction (N)	Totals
Hardcover (H)	3	5	8
Paperback (P)	8	4	12
Totals	11	9	20

Reduced sample space in part (b)

If Jacqui randomly chooses one of these 20 books, find the probability that it will be each of the following.

(a) hardcover   (b) fiction, given it is hardcover   (c) hardcover and fiction

Solution

(a) **Eight** of the **20** books are hardcover, so $P(H) = \frac{8}{20} = \frac{2}{5}$.

(b) The given condition that the book is hardcover reduces the sample space to **eight** books. Of those eight, just three are fiction, so $P(F|H) = \frac{3}{8}$.

(c) $P(H \text{ and } F) = P(H) \cdot P(F|H) = \frac{2}{5} \cdot \frac{3}{8} = \frac{3}{20}$ Multiplication rule

It is easier here if we simply notice, directly from **Table 6,** that 3 of the 20 books are "hardcover and fiction." This verifies that the general multiplication rule of probability did give us the correct answer. ∎

EXAMPLE 5 Selecting from a Set of Planets

Table 7 lists the eight planets of our solar system, together with their mean distances from the sun, in millions of kilometers. (Data is from *The World Almanac and Book of Facts.*) Carrie must choose two distinct planets to cover in her astronomy report. If she selects randomly, find the probability that the first one selected is closer to the sun than Mars and the second is closer to the sun than Saturn.

Solution

We define the following events.

   A: The first is closer than Mars. B: The second is closer than Saturn.

Then $P(A) = \frac{3}{8}$ because three of the original eight choices are favorable to event A. If the planet selected first is closer than Mars, it is also closer than Saturn, and since that planet is no longer available, four of the remaining seven are favorable to event B. Thus $P(B|A) = \frac{4}{7}$. Now apply the multiplication rule.

$$P(A \text{ and } B) = P(A) \cdot P(B|A) = \frac{3}{8} \cdot \frac{4}{7} = \frac{3}{14} \approx 0.214$$ Multiplication rule ∎

Table 7 Mean Distance of Planets from the Sun

Mercury	58
Venus	108
Earth	150
Mars	228
Jupiter	778
Saturn	1430
Uranus	2870
Neptune	4500

Even **a rare occurrence** can sometimes cause widespread controversy. When Mattel Toys marketed a new talking Barbie doll a few years ago, some of the Barbies were programmed to say "Math class is tough." The National Council of Teachers of Mathematics (NCTM), the American Association of University Women (AAUW), and numerous consumers voiced complaints about the damage such a message could do to the self-confidence of children and to their attitudes toward school and mathematics. Mattel subsequently agreed to erase the phrase from the microchip to be used in future doll production.

   Originally, each Barbie was programmed to say four different statements, randomly selected from a pool of 270 prerecorded statements. Therefore, the probability of getting a Barbie that said "Math class is tough" was only

$$\frac{1 \cdot {}_{269}C_3}{{}_{270}C_4} \approx 0.015 = 1.5\%.$$

Other messages included in the pool were "I love school, don't you?," "I'm studying to be a doctor," and "Let's study for the quiz."

In **Example 5,** the condition that A had occurred changed the probability of B, since the selection was done, in effect, without replacement. (Repetitions were not allowed.) Events A and B were not independent. On the other hand, in the next example the same events, A and B, are independent.

EXAMPLE 6 Selecting from a Set of Planets

Carrie must again select two planets, but this time one is for an oral report, the other is for a written report, and they need not be distinct. Here, the same planet may be selected for both reports. Again, find the probability that if she selects randomly, the first is closer than Mars and the second is closer than Saturn.

Solution

Defining events A and B as in **Example 5,** we have $P(A) = \frac{3}{8}$, just as before. But the selection is now done *with* replacement because repetitions *are* allowed. Event B is independent of event A, so we can use the second form of the multiplication rule.

$$P(A \text{ and } B) = P(A) \cdot P(B) = \frac{3}{8} \cdot \frac{5}{8} = \frac{15}{64} \approx 0.234 \quad \text{Answer is different than in Example 5.} \quad \blacksquare$$

EXAMPLE 7 Selecting from a Deck of Cards

A single card is drawn from a standard 52-card deck. Let B denote the event that the card is black, and let D denote the event that it is a diamond. Are events B and D

(a) independent?   (b) mutually exclusive?

One example of a card that is black and one example of a card that is a diamond

(No single card is both black and a diamond.)

Solution

(a) For the unconditional probability of D, we get $P(D) = \frac{13}{52} = \frac{1}{4}$ because 13 of the 52 cards are diamonds. But for the conditional probability of D given B, we have $P(D \mid B) = \frac{0}{26} = 0$. None of the 26 black cards are diamonds. Since the conditional probability $P(D \mid B)$ is different from the unconditional probability $P(D)$, **B and D are not independent.**

(b) Mutually exclusive events are events that cannot both occur for a given performance of an experiment. Since no card in the deck is both black and a diamond, **B and D are mutually exclusive.** $\blacksquare$

EXAMPLE 8 Selecting from an Urn of Balls

Anne is still drawing balls from the same urn, as shown at the side. This time she draws three balls, without replacement. Find the probability that she gets red, yellow, and blue balls, in that order.

Solution

Using appropriate letters to denote the colors, and subscripts to indicate first, second, and third draws, the event can be symbolized "R_1 and Y_2 and B_3."

$$P(R_1 \text{ and } Y_2 \text{ and } B_3) = P(R_1) \cdot P(Y_2 \mid R_1) \cdot P(B_3 \mid R_1 \text{ and } Y_2)$$

$$= \frac{4}{11} \cdot \frac{5}{10} \cdot \frac{2}{9} = \frac{4}{99} \approx 0.0404 \qquad \blacksquare$$

EXAMPLE 9 Selecting from a Deck of Cards

If five cards are drawn without replacement from a standard 52-card deck, find the probability that they all are hearts.

Solution

Each time a heart is drawn, the number of available cards decreases by one and the number of hearts decreases by one.

$$P(\text{all hearts}) = \frac{13}{52} \cdot \frac{12}{51} \cdot \frac{11}{50} \cdot \frac{10}{49} \cdot \frac{9}{48} = \frac{33}{66,640} \approx 0.000495$$

One example of five cards that are all hearts

We saw in **Chapter 10** that the problem of **Example 9** can also be solved by using the theoretical probability formula and combinations. The total possible number of 5-card hands, drawn without replacement, is $_{52}C_5$, and the number of those containing only hearts is $_{13}C_5$.

$$P(\text{all hearts}) = \frac{_{13}C_5}{_{52}C_5} = \frac{\dfrac{13!}{5! \cdot 8!}}{\dfrac{52!}{5! \cdot 47!}} \approx 0.000495 \quad \text{Use a calculator.}$$

EXAMPLE 10 Using Both Addition and Multiplication Rules

The local garage employs two mechanics, Ray and Tom. Your consumer club has found that Ray does twice as many jobs as Tom, Ray does a good job three out of four times, and Tom does a good job only two out of five times. If you plan to take your car in for repairs, find the probability that a good job will be done.

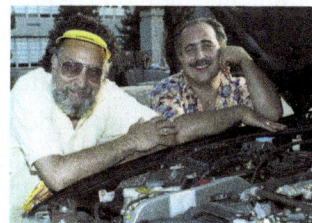

Solution

We define the following events.

R: work done by Ray T: work done by Tom G: good job done

Since Ray does twice as many jobs as Tom, the unconditional probabilities of events R and T are, respectively, $\frac{2}{3}$ and $\frac{1}{3}$. Because Ray does a good job three out of four times, the probability of a good job, given that Ray did the work, is $\frac{3}{4}$. And because Tom does well two out of five times, the probability of a good job, given that Tom did the work, is $\frac{2}{5}$. (These last two probabilities are conditional.) These four values can be summarized.

$$P(R) = \frac{2}{3}, \quad P(T) = \frac{1}{3} \quad P(G \mid R) = \frac{3}{4}, \quad \text{and} \quad P(G \mid T) = \frac{2}{5}$$

Event G can occur in two mutually exclusive ways: Ray could do the work and do a good job ($R \cap G$), or Tom could do the work and do a good job ($T \cap G$).

$$P(G) = P(R \cap G) + P(T \cap G) \qquad \text{Addition rule}$$
$$= P(R) \cdot P(G \mid R) + P(T) \cdot P(G \mid T) \qquad \text{Multiplication rule}$$

Multiply first, then add.

$$= \frac{2}{3} \cdot \frac{3}{4} + \frac{1}{3} \cdot \frac{2}{5} \qquad \text{Substitute the values.}$$

$$= \frac{1}{2} + \frac{2}{15} = \frac{19}{30} \approx 0.633$$

The **search for extraterrestrial intelligence (SETI)** may have begun in earnest as early as 1961, when Dr. Frank Drake presented an equation for estimating the number of possible civilizations in the Milky Way galaxy whose communications we might detect. Over the years, the effort has been advanced by many scientists, including the late astronomer and exobiologist Carl Sagan, who popularized the issue in TV appearances and in his book *The Cosmic Connection: An Extraterrestrial Perspective* (Dell Paperback). "There must be other starfolk," said Sagan. In fact, some astronomers have estimated the odds against life on Earth being the only life in the universe at one hundred billion billion to one.

Other experts disagree. Freeman Dyson, a noted mathematical physicist and astronomer, says (in his book *Disturbing the Universe*) that after considering the same evidence and arguments, he believes it is just as likely as not (even odds) that there never was any other intelligent life out there.

The tree diagram in **Figure 11** shows a graphical way to organize the work of **Example 10.** Use the given information to draw the tree diagram; then find the probability of a good job by adding the probabilities from the two indicated branches of the tree.

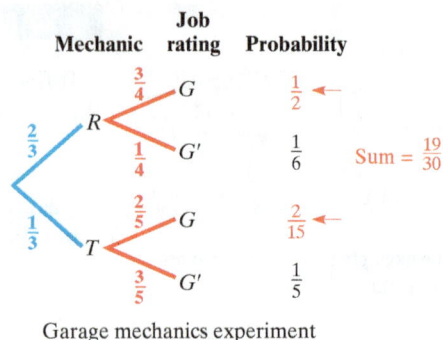

Garage mechanics experiment

Figure 11

The Birthday Problem

A classic problem (with a surprising result) involves the probability that a given group of people will include at least one pair of people with the same birthday (the same day of the year, not necessarily the same year). This problem can be analyzed using the probability of a complement formula (**Section 11.2**) and the multiplication rule of probability from this section. Suppose there are three people in the group.

P (at least one duplication of birthdays)

$= 1 - P$(no duplications)   Complement formula

$= 1 - P$(2nd is different than 1st, and 3rd is different than 1st and 2nd)

$= 1 - \dfrac{364}{365} \cdot \dfrac{363}{365}$ Multiplication rule

$\approx 1 - 0.992$

$= 0.008$

(To simplify the calculations, we have assumed 365 possible birth dates, ignoring February 29.)

By doing more calculations like the one above, we find that the smaller the group, the smaller the probability of a duplication. The larger the group, the larger the probability of a duplication. The table below shows the probability of at least one duplication for numbers of people from 2 through 52.

For Group or Individual Investigation

1. Based on the data shown in the table, what are the odds in favor of a duplication in a group of 30 people?

2. Estimate from the table the least number of people for which the probability of duplication is at least $\frac{1}{2}$.

3. How small a group is required for the probability of a duplication to be *exactly* 0?

4. How large a group is required for the probability of a duplication to be *exactly* 1?

The answers are on **page 613.**

Number of People	Probability of at Least One Duplication	Number of People	Probability of at Least One Duplication	Number of People	Probability of at Least One Duplication
2	0.003	19	0.379	36	0.832
3	0.008	20	0.411	37	0.849
4	0.016	21	0.444	38	0.864
5	0.027	22	0.476	39	0.878
6	0.040	23	0.507	40	0.891
7	0.056	24	0.538	41	0.903
8	0.074	25	0.569	42	0.914
9	0.095	26	0.598	43	0.924
10	0.117	27	0.627	44	0.933
11	0.141	28	0.654	45	0.941
12	0.167	29	0.681	46	0.948
13	0.194	30	0.706	47	0.955
14	0.223	31	0.730	48	0.961
15	0.253	32	0.753	49	0.966
16	0.284	33	0.775	50	0.970
17	0.315	34	0.795	51	0.974
18	0.347	35	0.814	52	0.978

The **search for extraterrestrial
intelligence (SETI)** has been mainly
accomplished in recent years through
SETI@HOME, a very large distributed
computing program. Most of the data
are collected by the world's largest radio
telescope, built into a 20-acre natural
bowl in Aricebo, Puerto Rico (pictured
above), and processed by millions of
personal computers around the world.

To learn more, or for a chance to be
the first to "contact" an extraterrestrial
civilization, check out
www.setiathome.ssl.berkeley.edu.

EXAMPLE 11 Selecting Door Prizes

Rob is among five door prize winners at a Christmas party. The
five winners are asked to choose, without looking, from a bag
that, they are told, contains five tokens, four of them redeemable for candy canes and one specific token redeemable for a
$100 gift certificate. Can Rob improve his chance of getting the
gift certificate by drawing first among the five people?

Solution

We denote candy cane by C, gift certificate by G, and first draw, second draw, and
so on by subscripts 1, 2,.... If Rob draws first, his probability of getting the gift
certificate is

$$P(G_1) = \frac{1}{5}.$$

If he draws second, his probability of getting the gift certificate is

$$P(G_2) = P(C_1 \text{ and } G_2)$$
$$= P(C_1) \cdot P(G_2 | C_1)$$
$$= \frac{4}{5} \cdot \frac{1}{4} = \frac{1}{5}. \qquad \text{Same result as above}$$

For the third draw,

$$P(G_3) = P(C_1 \text{ and } C_2 \text{ and } G_3)$$
$$= P(C_1) \cdot P(C_2 | C_1) \cdot P(G_3 | C_1 \text{ and } C_2)$$
$$= \frac{4}{5} \cdot \frac{3}{4} \cdot \frac{1}{3} = \frac{1}{5}. \qquad \text{Same result as above}$$

The probability of getting the gift certificate is also $\frac{1}{5}$ when drawing fourth or fifth.
The order in which the five winners draw does not affect Rob's chances. ∎

11.3 EXERCISES

From the sample space

$$S = \{1, 2, 3, 4, \ldots, 15\},$$

*a single number is to be selected at random. Given the
events*

    A: *The selected number is even,*

    B: *The selected number is a multiple of 4,*

    C: *The selected number is a prime number,*

find each probability in Exercises 1–10.

1. $P(A)$ **2.** $P(B)$

3. $P(C)$ **4.** $P(A \text{ and } B)$

5. $P(A \text{ and } C)$ **6.** $P(B \text{ and } C)$

7. $P(A | B)$ **8.** $P(B | A)$

9. $P(C | A)$ **10.** $P(A | C)$

*Given a family with three children, find the probability of
each event.*

11. All are girls.   **12.** All are boys.

13. The oldest two are boys, given that there are at least
two boys.

14. The youngest is a girl, given that there are at least two
girls.

15. The oldest is a boy and the youngest is a girl, given that there are at least one boy and at least one girl.

16. The youngest two are girls, given that there are at least two girls.

17. The oldest two are boys, given that the oldest is a boy.

18. The oldest is a boy, given that there is at least one boy.

19. The youngest is a boy, given that the birth order alternates between girls and boys.

20. The oldest is a girl, given that the middle child is a girl.

For each experiment, determine whether the two given events are independent.

21. **Coin Tosses**   A fair coin is tossed twice. The events are "head on the first" and "head on the second."

22. **Dice Rolls**   A pair of dice are rolled. The events are "even on the first" and "odd on the second."

23. **Planets' Mean Distances from the Sun**   Two planets are selected, without replacement, from the list in **Table 7**. The events are "the first selected planet is closer than Jupiter" and "the second selected planet is farther than Mars."

24. **Mean Distances from the Sun**   Two planets are selected, with replacement, from the list in **Table 7**. The events are "the first selected planet is closer than Earth" and "the second selected planet is farther than Uranus."

25. **Answers on a Multiple-choice Test**   The answers are all guessed on a twenty-question multiple-choice test. The events are "the first answer is correct" and "the last answer is correct."

26. **Committees of U.S. Senators**   A committee of five is randomly selected from the 100 U.S. senators. The events are "the first member selected is a Republican" and "the second member selected is a Republican." (Assume that there are both Republicans and non-Republicans in the Senate.)

Gender and Career Motivation of College Students   *One hundred college seniors attending a career fair at a university were categorized according to gender and according to primary career motivation. See the table for the results.*

	Primary Career Motivation			
	Money	**Allowed to be Creative**	**Sense of Giving to Society**	**Total**
Male	19	15	14	48
Female	12	23	17	52
Total	31	38	31	100

If one of these students is to be selected at random, find the probability that the student selected will satisfy each condition in Exercises 27–32.

27. female

28. motivated primarily by creativity

29. not motivated primarily by money

30. male and motivated primarily by money

31. male, given that primary motivation is a sense of giving to society

32. motivated primarily by money or creativity, given that the student is female

Pet Selection   *A pet store has seven puppies, including four huskies, two terriers, and one retriever. If Rebecka and Aaron, in that order, each select one puppy at random, with replacement (they may both select the same one), find the probability of each event in Exercises 33–36.*

33. Both select a husky.

34. Rebecka selects a retriever; Aaron selects a terrier.

35. Rebecka selects a terrier; Aaron selects a retriever.

36. Both select a retriever.

Pet Selection   *Suppose two puppies are selected as earlier, but this time without replacement (Rebecka and Aaron cannot both select the same puppy). Find the probability of each event in Exercises 37–42.*

37. Both select a husky.

38. Aaron selects a terrier, given that Rebecka selects a husky.

39. Aaron selects a retriever, given that Rebecka selects a husky.

40. Rebecka selects a retriever.

41. Aaron selects a retriever, given that Rebecka selects a retriever.

42. Both select a retriever.

Card Dealing *Let two cards be dealt successively, without replacement, from a standard 52-card deck. Find the probability of each event in Exercises 43–49.*

43. spade dealt second, given a spade dealt first

44. club dealt second, given a diamond dealt first

45. two face cards   **46.** no face cards

47. The first card is a jack and the second is a face card.

48. The first card is the ace of hearts and the second is black.

49. The first card is black and the second is red.

50. Proof of Formulas Given events A and B within the sample space S, the following sequence of steps establishes formulas that can be used to compute conditional probabilities. Justify each statement.

(a) $P(A \text{ and } B) = P(A) \cdot P(B \mid A)$

(b) Therefore, $P(B \mid A) = \dfrac{P(A \text{ and } B)}{P(A)}$.

(c) Therefore, $P(B \mid A) = \dfrac{n(A \text{ and } B)/n(S)}{n(A)/n(S)}$.

(d) Therefore, $P(B \mid A) = \dfrac{n(A \text{ and } B)}{n(A)}$.

Conditions in Card Drawing *Use the results of **Exercise 50** to find each probability when a single card is drawn from a standard 52-card deck.*

51. $P(\text{queen} \mid \text{face card})$ **52.** $P(\text{face card} \mid \text{queen})$

53. $P(\text{red} \mid \text{diamond})$ **54.** $P(\text{diamond} \mid \text{red})$

Property of P(A and B) *Complete Exercises 55 and 56 to discover a general property of the probability of an event of the form "A and B."*

55. If one number is chosen randomly from the integers 1 through 10, the probability of getting a number that is *odd and prime,* by the multiplication rule, is

$$P(\text{odd}) \cdot P(\text{prime} \mid \text{odd}) = \frac{5}{10} \cdot \frac{3}{5} = \frac{3}{10}.$$

Compute the product $P(\text{prime}) \cdot P(\text{odd} \mid \text{prime})$, and compare to the product above.

56. What does **Exercise 55** imply, in general, about the probability of an event of the form "*A and B*"?

57. Gender in Sequences of Babies Two authors of this book each have three sons and no daughters. Assuming boy and girl babies are equally likely, what is the probability of this event?

58. Dice Rolls Three dice are rolled. What is the probability that the numbers shown will all be different? (*Mathematics Teacher* calendar problem)

The remaining exercises, and groups of exercises, may require concepts from earlier sections, such as the complements principle of counting and addition rules, as well as the multiplication rule of this section.

Warehouse Grocery Shopping Therese manages a grocery warehouse that encourages volume shopping on the part of its customers. Therese has discovered that on any given weekday, 70% of the customer sales amount to more than $100. That is, any given sale on such a day has a probability of 0.70 of being for more than $100. (Actually, the conditional probabilities throughout the day would change slightly, depending on earlier sales, but this effect would be negligible for the first several sales of the day, so we can treat them as independent.)

Find the probability of each event in Exercises 59–62. (Give answers to three decimal places.)

59. The first two sales on Wednesday are both for more than $100.

60. The first three sales on Wednesday are all for more than $100.

61. None of the first three sales on Wednesday is for more than $100.

62. Exactly one of the first three sales on Wednesday is for more than $100.

Pollution from the Space Shuttle Launch Site *One problem encountered by developers of the space shuttle program was air pollution in the area surrounding the launch site. A certain direction from the launch site was considered critical in terms of hydrogen chloride pollution from the exhaust cloud. It has been determined that weather conditions would cause emission cloud movement in the critical direction only 5% of the time.*

In Exercises 63–66, find the probability for each event. Assume that probabilities for a particular launch in no way depend on the probabilities for other launches. (Give answers to two decimal places.)

63. A given launch will not result in cloud movement in the critical direction.

64. No cloud movement in the critical direction will occur during any of 5 launches.

65. Any 5 launches will result in at least one cloud movement in the critical direction.

66. Any 10 launches will result in at least one cloud movement in the critical direction.

Job Interviews *Three men and three women are waiting to be interviewed for jobs. If they are all selected in random order, find the probability of each event in Exercises 67–70.*

67. All the women will be interviewed first.

68. The first interviewee will be a man, and the remaining ones will alternate gender.

69. The first three chosen will all be the same gender.

70. No man will be interviewed until at least two women have been interviewed.

Garage Door Opener *Kevin installed a certain brand of automatic garage door opener that utilizes a transmitter control with six independent switches, each one set on or off. The receiver (wired to the door) must be set with the same pattern as the transmitter. (Exercises 71–74 are based on ideas similar to those of the "birthday problem" in the* **For Further Thought** *feature in this section.)*

71. How many different ways can Kevin set the switches?

72. If one of Kevin's neighbors also has this same brand of opener, and both of them set the switches randomly, what is the probability, to four decimal places, that they will use the same settings?

73. If five neighbors with the same type of opener set their switches independently, what is the probability of at least one pair of neighbors using the same settings? (Give your answer to four decimal places.)

74. What is the minimum number of neighbors who must use this brand of opener before the probability of at least one duplication of settings is greater than $\frac{1}{2}$?

Weather Conditions on Successive Days *In November, the rain in a certain valley tends to fall in storms of several days' duration. The unconditional probability of rain on any given day of the month is 0.500. But the probability of rain on a day that follows a rainy day is 0.800, and the probability of rain on a day following a nonrainy day is 0.300. Find the probability of each event in Exercises 75–78. Give answers to three decimal places.*

75. rain on two randomly selected consecutive days in November

76. rain on three randomly selected consecutive days in November

77. rain on November 1 and 2, but not on November 3 (The October 31 weather is unknown.)

78. rain on the first four days of November, given that October 31 was clear all day

Engine Failures in a Vintage Aircraft *In a certain four-engine vintage aircraft, now quite unreliable, each engine has a 10% chance of failure on any flight, as long as it is carrying its one-fourth share of the load. But if one engine fails, then the chance of failure increases to 20% for each of the other three engines. And if a second engine fails, each of the remaining two has a 30% chance of failure.*

Assuming that no two engines ever fail simultaneously, and that the aircraft can continue flying with as few as two operating engines, find each probability for a given flight of this aircraft. (Give answers to four decimal places.)

79. no engine failures

80. exactly one engine failure (any one of four engines)

81. exactly two engine failures (any two of four engines)

82. a failed flight

Fair Decisions from Biased Coins *Many everyday decisions, like who will drive to lunch or who will pay for the coffee, are made by the toss of a (presumably fair) coin and using the criterion "heads, you will; tails, I will." This criterion is not quite fair, however, if the coin is biased (perhaps due to slightly irregular construction or wear). John von Neumann suggested a way to make perfectly fair decisions, even with a possibly biased coin. If a coin, biased so that*

$$P(\text{h}) = 0.5200 \quad \text{and} \quad P(\text{t}) = 0.4800,$$

is tossed twice, find each probability. Give answers to four decimal places.

83. $P(\text{hh})$

84. $P(\text{ht})$

85. $P(\text{th})$

86. $P(\text{tt})$

87. Having completed **Exercises 83–86,** can you suggest what von Neumann's scheme may have been?

One-and-one Free Throw Shooting in Basketball *In basketball, "one-and-one" free throw shooting (commonly called foul shooting) is done as follows: If the player makes the first shot (1 point), she is given a second shot. If she misses the first shot, she is not given a second shot (see the tree diagram).*

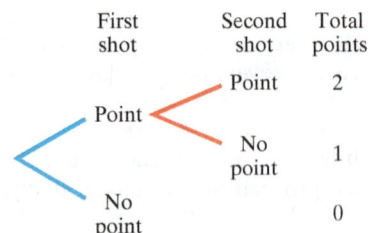

First shot	Second shot	Total points
	Point	2
Point		
	No point	1
No point		0

Christine, a basketball player, has a 70% free throw record. (She makes 70% of her free throws.) Find the probability that, on a given one-and-one free throw shooting opportunity, Christine will score each number of points.

88. no points   **89.** one point   **90.** two points

91. *Gender in Sequences of Babies*   Assuming boy and girl babies are equally likely, find the probability that it would take

   (a) at least three births to obtain two girls

   (b) at least four births to obtain two girls

   (c) at least five births to obtain two girls.

92. *Card Choices*   Cards are drawn, without replacement, from an ordinary 52-card deck.

   (a) How many must be drawn before the probability of obtaining at least one face card is greater than $\frac{1}{2}$?

   (b) How many must be drawn before the probability of obtaining at least one king is greater than $\frac{1}{2}$?

93. *A Two-Headed Coin?*   A gambler has two coins in his pocket—one fair coin and one two-headed coin. He selects a coin at random and flips it twice. If he gets two heads, what is the probability that he selected the fair coin? (*Mathematics Teacher* calendar problem)

94. *Cube Cuts*   A $4'' \times 4'' \times 4''$ cube is painted and then cut into sixty-four $1'' \times 1'' \times 1''$ cubes. A unit cube is then randomly selected and rolled. What is the probability that the top face of the rolled cube is painted? Express your answer as a common fraction. (*Mathematics Teacher* calendar problem)

95. *Anne Continues*   In **Example 8,** where Anne draws three balls without replacement, what would be her probability of getting one of each color, where the order does not matter?

96. *Card Choices*   There are three cards, one that is green on both sides, one that is red on both sides, and one that is green on one side and red on the other. One of the three cards is selected randomly and laid on the table. If it happens that the card on the table has a red side up, what is the probability that it is also red on the other side?

97. *Empirical and Theoretical Probabilities in Dice Rolling*   Roll a pair of dice until a sum of seven appears, keeping track of how many rolls it took. Repeat the process a total of 50 times, each time recording the number of rolls it took to get a sum of seven.

   (a) Use your experimental data to compute an empirical probability to two decimal places that it would take at least three rolls to get a sum of seven.

   (b) Find the theoretical probability to two decimal places that it would take at least three rolls to obtain a sum of seven.

98. Go to the Web site http://www.planetary.org/explore/ space-topics/ and select the topic *Asteroids and Comets*. Write a report on the threat to humanity of cosmic impacts. Include an explanation of the acronym *NEO*.

11.4   BINOMIAL PROBABILITY

OBJECTIVES

1 Construct a simple binomial probability distribution.

2 Apply the binomial probability formula for an experiment involving Bernoulli trials.

Binomial Probability Distribution

Suppose the spinner in **Figure 12** on the next page is spun twice. We are interested in the number of times a 2 is obtained. (Assume that 1, 2, and 3 all are equally likely on a given spin.) We can think of the outcome 2 as a "success," while a 1 or a 3 would be a "failure."

When the outcomes of an experiment are divided into just two categories, success and failure, the associated probabilities are called "binomial." (The prefix *bi* means *two*.) Repeated performances of such an experiment, where the probability of success remains constant throughout all repetitions, are also known as repeated **Bernoulli trials** (after James Bernoulli). If we use an ordered pair to represent the result of each pair of spins, then the sample space for this experiment is

$$S = \{(1,1), (1,2), (1,3), (2,1), (2,2), (2,3), (3,1), (3,2), (3,3)\}.$$

The nine outcomes in *S* are all equally likely. This follows from the numbers 1, 2, and 3 being equally likely on a particular spin.

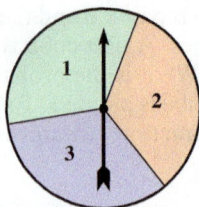

Figure 12

Table 8 Probability Distribution for the Number of 2s in Two Spins

x	$P(x)$
0	$\frac{4}{9}$
1	$\frac{4}{9}$
2	$\frac{1}{9}$
Sum: $\frac{9}{9} = 1$	

James Bernoulli (1654–1705) is also known as Jacob or Jacques. He was charmed away from theology by the writings of Leibniz, became his pupil, and later headed the mathematics faculty at the University of Basel. His results in probability are contained in the *Art of Conjecture*, which was published in 1713, after his death, and which also included a reprint of the earlier Huygens paper. Bernoulli also made many contributions to calculus and analytic geometry.

If x denotes the number of 2s occurring on each pair of spins, then x is an example of a *random variable*. Although we cannot predict the result of any particular pair of spins, we can find the probabilities of various events from the sample space listing. In S, the number of 2s is 0 in four cases, 1 in four cases, and 2 in one case, as reflected in **Table 8**. Because the table includes all possible values of x, together with their probabilities, it is an example of a *probability distribution*. In this case, we have a **binomial probability distribution**. The probability column in **Table 8** has a sum of 1, in agreement with probability Property 3 in **Section 11.2**.

In order to develop a general formula for binomial probabilities, we can consider another way to obtain the probability values in **Table 8**. The various spins of the spinner are independent of one another, and on each spin the probability of success (S) is $\frac{1}{3}$ and the probability of failure (F) is $\frac{2}{3}$. We will denote success on the first spin by S_1, failure on the second by F_2, and so on.

$$P(x = 0) = P(F_1 \text{ and } F_2)$$

$$= P(F_1) \cdot P(F_2) \qquad \text{Multiplication rule}$$

$$= \frac{2}{3} \cdot \frac{2}{3} \qquad \text{Substitute values.}$$

$$= \frac{4}{9} \qquad \text{Multiply.}$$

$$P(x = 1) = P[(S_1 \text{ and } F_2) \text{ or } (F_1 \text{ and } S_2)] \qquad \text{2 ways to get } x = 1$$

$$= P(S_1 \text{ and } F_2) + P(F_1 \text{ and } S_2) \qquad \text{Addition rule}$$

$$= P(S_1) \cdot P(F_2) + P(F_1) \cdot P(S_2) \qquad \text{Multiplication rule}$$

$$= \frac{1}{3} \cdot \frac{2}{3} + \frac{2}{3} \cdot \frac{1}{3} \qquad \text{Substitute values.}$$

$$= \frac{2}{9} + \frac{2}{9} \qquad \text{Multiply.}$$

$$= \frac{4}{9} \qquad \text{Add.}$$

$$P(x = 2) = P(S_1 \text{ and } S_2)$$

$$= P(S_1) \cdot P(S_2) \qquad \text{Multiplication rule}$$

$$= \frac{1}{3} \cdot \frac{1}{3} \qquad \text{Substitute values.}$$

$$= \frac{1}{9} \qquad \text{Multiply.}$$

Notice the following pattern in the above calculations. There is only one way to get $x = 0$ (namely, F_1 and F_2). And there is only one way to get $x = 2$ (namely, S_1 and S_2). But there are two ways to get $x = 1$. One way is S_1 and F_2; the other is F_1 and S_2. There are two ways because the one success required can occur on the first spin or on the second spin. How many ways can exactly one success occur in two repeated trials? This question is equivalent to

How many size-one subsets are there of the set of two trials?

The answer is $_2C_1 = 2$. The expression $_2C_1$ denotes "combinations of 2 things taken 1 at a time." Each of the two ways to get exactly one success has a probability equal to $\frac{1}{3} \cdot \frac{2}{3}$, the probability of success times the probability of failure.

EXAMPLE 1 **Constructing a Probability Distribution**

Use a tree diagram to construct a probability distribution for the number of 2s in three spins of the spinner in **Figure 12**.

Solution

If the spinner is spun three times rather than two, then x, the number of successes (2s) could have values of 0, 1, 2, or 3. Then the number of ways to get exactly 1 success is $_3C_1 = 3$. They are

$$S_1 \text{ and } F_2 \text{ and } F_3, \quad F_1 \text{ and } S_2 \text{ and } F_3, \quad F_1 \text{ and } F_2 \text{ and } S_3.$$

The probability of each of these three ways is $\frac{1}{3} \cdot \frac{2}{3} \cdot \frac{2}{3} = \frac{4}{27}$.

$$P(x = 1) = 3 \cdot \frac{4}{27} = \frac{12}{27} = \frac{4}{9}$$

Figure 13 shows all possibilities for three spins, and **Table 9** gives the associated probability distribution. In the tree diagram, the number of ways of getting two successes in three trials is 3, in agreement with the fact that $_3C_2 = 3$. Also the sum of the $P(x)$ column in **Table 9** is again 1.

Table 9 Probability Distribution for the Number of 2s in Three Spins

x	$P(x)$
0	$\frac{8}{27}$
1	$\frac{12}{27}$
2	$\frac{6}{27}$
3	$\frac{1}{27}$
Sum:	$\frac{27}{27} = 1$

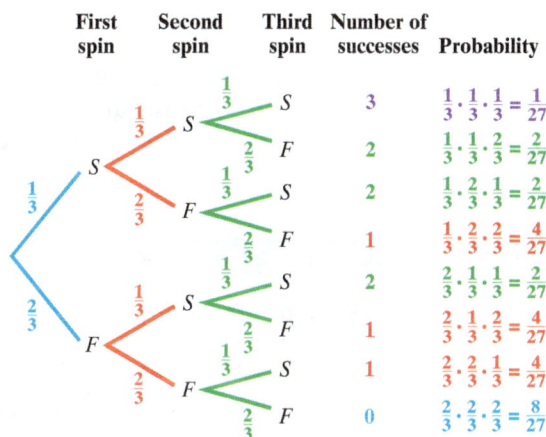

Tree diagram for three spins

Figure 13

Problem-Solving Strategy

One of the various problem-solving strategies from **Chapter 1** was "Look for a pattern." Having constructed complete probability distributions for binomial experiments with 2 and 3 repeated trials (and probability of success $\frac{1}{3}$), we can now generalize the observed pattern to any binomial experiment, as shown next.

Binomial Probability Formula

Define the following quantities.

n = the number of repeated trials
p = the probability of success on any given trial
$q = 1 - p$ = the probability of failure on any given trial
x = the number of successes that occur

Note that p remains fixed throughout all n trials. This means that all trials are independent of one another. The random variable x representing the number of successes can have any integer value from 0 through n. In general, x successes can be assigned among n repeated trials in $_nC_x$ different ways, because this is the number of different subsets of x positions among a set of n positions. Also, regardless of which x of the trials result in successes, there will always be x successes and $n - x$ failures, so we multiply x factors of p and $n - x$ factors of q together.

BINOMIAL PROBABILITY FORMULA

When n independent repeated trials occur, where

$$p = \text{probability of success} \quad \text{and} \quad q = \text{probability of failure}$$

with p and q (where $q = 1 - p$) remaining constant throughout all n trials, the probability of exactly x successes is calculated as follows.

$$P(x) = {_nC_x}\, p^x q^{n-x} = \frac{n!}{x!(n-x)!} p^x q^{n-x}$$

Binomial probabilities for particular values of n, p, and x can be found directly using tables, statistical software, and some handheld calculators. In the following examples, we use the formula derived above.

From the DISTR menu

binompdf(5,.5,3)
.3125
Ans▸Frac
$\frac{5}{16}$

The TI-83/84 Plus calculator will find the probability discussed in **Example 2.**

EXAMPLE 2 Finding Probability in Coin Tossing

Find the probability of obtaining exactly three heads in five tosses of a fair coin.

Solution

Let heads be "success." Then this is a binomial experiment with

$$n = 5, \quad p = \frac{1}{2}, \quad q = \frac{1}{2}, \quad \text{and} \quad x = 3.$$

$$P(x = 3) = {_5C_3}\left(\frac{1}{2}\right)^3\left(\frac{1}{2}\right)^{5-3} = 10 \cdot \frac{1}{8} \cdot \frac{1}{4} = \frac{5}{16} \qquad \text{Binomial probability formula} \quad \blacksquare$$

binompdf(6,1/6,2)
.200938786

This screen supports the answer in **Example 3.**

EXAMPLE 3 Finding Probability in Dice Rolling

Find the probability of obtaining exactly two 5s in six rolls of a fair die.

Solution

Let 5 be "success." Then $n = 6$, $p = \frac{1}{6}$, $q = \frac{5}{6}$, and $x = 2$.

$$P(x = 2) = {_6C_2}\left(\frac{1}{6}\right)^2\left(\frac{5}{6}\right)^4 = 15 \cdot \frac{1}{36} \cdot \frac{625}{1296} = \frac{3125}{15{,}552} \approx 0.201 \qquad \blacksquare$$

In the case of repeated independent trials, when an event involves more than one specific number of successes, we can employ the binomial probability formula along with the complement and/or addition rules.

EXAMPLE 4 **Finding the Probability of Female Children**

A couple plans to have 5 children. Find the probability that they will have more than 3 girls. (Assume girl and boy babies are equally likely.)

Solution

Let a girl be "success." Then $n = 5$, $p = q = \frac{1}{2}$, and $x > 3$.

$$P(x > 3) = P(x = 4 \text{ or } 5) \qquad \text{More than 3 means 4 or 5.}$$
$$= P(4) + P(5) \qquad \text{Addition rule}$$
$$= {}_5C_4\left(\frac{1}{2}\right)^4\left(\frac{1}{2}\right)^1 + {}_5C_5\left(\frac{1}{2}\right)^5\left(\frac{1}{2}\right)^0 \qquad \text{Binomial probability formula}$$
$$= 5 \cdot \frac{1}{16} \cdot \frac{1}{2} + 1 \cdot \frac{1}{32} \cdot 1 \qquad \text{Simplify.}$$
$$= \frac{5}{32} + \frac{1}{32} = \frac{6}{32} = \frac{3}{16} = 0.1875$$

binompdf(5,.5,4)+binompdf▸
(5,.5,5)
                      .1875
Ans▸Frac
                       $\frac{3}{16}$

This screen supports the answer in
Example 4.

EXAMPLE 5 **Finding the Probability of Hits in Baseball**

Gary Bell, a pitcher who is also an excellent hitter, has a well-established career batting average of .300. Suppose that he bats ten times. Find the probability that he will get more than two hits in the ten at-bats.

Solution

This "experiment" involves $n = 10$ repeated Bernoulli trials, with probability of success (a hit) given by $p = 0.3$ (which implies $q = 1 - 0.3 = 0.7$). Since, in this case, "more than 2" means

"3 or 4 or 5 or 6 or 7 or 8 or 9 or 10"   (eight different possibilities),

it will be less work to apply the complement rule.

$$P(x > 2) = 1 - P(x \le 2) \qquad \text{Complement rule}$$
$$= 1 - P(x = 0 \text{ or } 1 \text{ or } 2) \qquad \text{Only three different possibilities}$$
$$= 1 - [P(0) + P(1) + P(2)] \qquad \text{Addition rule}$$
$$= 1 - [{}_{10}C_0(0.3)^0(0.7)^{10} \qquad \text{Binomial probability formula}$$
$$+ {}_{10}C_1(0.3)^1(0.7)^9 + {}_{10}C_2(0.3)^2(0.7)^8]$$
$$\approx 1 - [0.0282 + 0.1211 + 0.2335] \qquad \text{Simplify.}$$
$$= 1 - 0.3828 = 0.6172$$

1−(binompdf(10,.3,0)+bin▸
ompdf(10,.3,1)+binompdf▸
(10,.3,2))
                  .6172172136

This screen supports the answer in
Example 5.

11.4 **EXERCISES**

For Exercises 1–22, give all numerical answers as common fractions reduced to lowest terms. For Exercises 23–56, give all numerical answers to three decimal places.

Coin Tossing *If three fair coins are tossed, find the probability of each number of heads.*

1. 0

2. 1

3. 2

4. 3

5. 1 or 2

6. at least 1

7. no more than 1

8. fewer than 3

9. *Relating Pascal's Triangle to Coin Tossing* Pascal's triangle was shown in **Table 5** of **Section 10.4.** Explain how the probabilities in **Exercises 1–4** here relate to row 3 of the "triangle." (We will refer to the topmost row of the triangle as "row number 0" and to the leftmost entry of each row as "entry number 0.")

10. Generalize the pattern in **Exercise 9** to complete the following statement. If *n* fair coins are tossed, the probability of exactly *x* heads is the fraction whose numerator is entry number _____ of row number _____ in Pascal's triangle, and whose denominator is the sum of the entries in row number _____.

Binomial Probability Applied to Tossing Coins *Use the pattern noted in* **Exercises 9 and 10** *to find the probabilities of each number of heads when seven fair coins are tossed.*

11. 0	**12.** 1	**13.** 2	**14.** 3
15. 4	**16.** 5	**17.** 6	**18.** 7

Binomial Probability Applied to Rolling Dice *A fair die is rolled three times. A 4 is considered "success," while all other outcomes are "failures." Find the probability of each number of successes.*

19. 0	**20.** 1	**21.** 2	**22.** 3

For n repeated independent trials, with constant probability of success p for all trials, find the probability of exactly x successes in each of Exercises 23–26.

23. $n = 5$, $p = \frac{1}{3}$, $x = 4$ **24.** $n = 10$, $p = \frac{7}{10}$, $x = 5$

25. $n = 20$, $p = 0.125$, $x = 2$

26. $n = 30$, $p = 0.6$, $x = 22$

For Exercises 27–30, refer to **Example 5.**

27. ***Batting Averages in Baseball*** If Gary's batting average is exactly .300 going into the series based on exactly 1200 career hits out of 4000 previous times at bat, what is the greatest his average could possibly be when he goes up to bat the tenth time of the series?

28. ***Batting Averages in Baseball*** Refer to **Exercise 27.** What is the least his average could possibly be when he goes up to bat the tenth time of the series?

29. ***Batting Averages in Baseball*** Does Gary's probability of a hit really remain constant at exactly 0.300 through all ten times at bat? Explain your reasoning.

30. Do you think the use of the binomial probability formula was justified in **Example 5,** even though *p* is not strictly constant? Explain your reasoning.

Random Selection of Answers on a Multiple-choice Test *Beth is taking a ten-question multiple-choice test for which each question has three answer choices, only one of which is correct. Beth decides on answers by rolling a fair die and marking the first answer choice if the die shows 1 or 2, the second if it shows 3 or 4, and the third if it shows 5 or 6. Find the probability of each event in Exercises 31–34.*

31. exactly four correct answers

32. exactly seven correct answers

33. fewer than three correct answers

34. at least seven correct answers

Side Effects of Prescription Drugs *It is known that a certain prescription drug produces undesirable side effects in 35% of all patients who use it. Among a random sample of eight patients using the drug, find the probability of each event in Exercises 35–38.*

35. None have undesirable side effects.

36. Exactly one has undesirable side effects.

37. Exactly two have undesirable side effects.

38. More than two have undesirable side effects.

Likelihood of Capable Students Attending College *It has been shown that 60% of the high school graduates who are capable of college work actually enroll in colleges. Find the probability that, among nine capable high school graduates in a state, each number will enroll in college.*

39. exactly 4	**40.** from 4 through 6
41. all 9	**42.** at least 3

43. ***Ball Choices*** A bag contains only white balls and black balls. Let *p* be the probability that a ball selected at random is black. Each time a ball is selected, it is placed back in the bag before the next ball is selected. Four balls are selected at random. What is the probability that two of the four balls are black and two are white? (*Mathematics Teacher* calendar problem) (*Hint*: Use the binomial probability formula to express the probability in terms of *p*.)

44. ***Ball Choices*** Evaluate the probability of **Exercise 43** in the case where the bag actually contains 15 black balls and 25 white balls.

45. ***Frost Survival among Orange Trees*** If it is known that 65% of all orange trees will survive a hard frost, then what is the probability that at least half of a group of six trees will survive such a frost?

46. ***Rate of Favorable Media Coverage of an Incumbent President*** During a presidential campaign, 64% of the political columns in a certain group of major newspapers were favorable to the incumbent president. If a sample of fifteen of these columns is selected at random, what is the probability that exactly ten of them will be favorable?

47. *Student Ownership of Personal Computers* At a large midwestern university, 90% of all students have their own personal computers. If five students at that university are selected at random, find the probability that exactly three of them have their own computers.

Taking a Random Walk *Abby is parked at a mile marker on an east-west country road. She decides to toss a fair coin 10 times, each time driving 1 mile east if it lands heads up and 1 mile west if it lands tails up. The term "random walk" applies to this process, even though she drives rather than walks. It is a simplified model of Brownian motion, which is mentioned in a margin note in **Section 11.1**.*

In each of Exercises 48–55, find the probability that Abby's "walk" will end as described.

48. 6 miles east of the start

49. 10 miles east of the start

50. 5 miles west of the start

51. 6 miles west of the start

52. at least 2 miles east of the start

53. 2 miles east of the start

54. exactly at the start

55. at least 2 miles from the start

11.5 EXPECTED VALUE AND SIMULATION

OBJECTIVES

1 Determine expected value of a random variable and expected net winnings in a game of chance.

2 Determine if a game of chance is a fair game.

3 Use expected value to make business and insurance decisions.

4 Use simulation in genetic processes such as flower color and birth gender.

Table 10

x	$P(x)$
1	0.05
2	0.10
3	0.20
4	0.40
5	0.10
6	0.15

Expected Value

We repeat the beginning of **Example 6** in **Section 11.2.**

Amy plans to spend from 1 to 6 hours on her homework. If x represents the number of hours to be spent on a given night, then the probabilities of the various values of x, rounded to the nearest hour, are as shown in **Table 10**.

If Amy's friend Tara asks her how many hours her studies will take, what would be her best guess? Six different time values are possible, with some more likely than others. One thing Amy could do is calculate a "weighted average" by multiplying each possible time value by its probability and then adding the six products.

$$1(0.05) + 2(0.10) + 3(0.20) + 4(0.40) + 5(0.10) + 6(0.15)$$
$$= 0.05 + 0.20 + 0.60 + 1.60 + 0.50 + 0.90$$
$$= 3.85$$

Thus 3.85 hours is the **expected value** (or the **mathematical expectation**) of the quantity of time to be spent.

> **EXPECTED VALUE**
>
> If a random variable x can have any of the values $x_1, x_2, x_3, \ldots, x_n$, and the corresponding probabilities of these values occurring are $P(x_1)$, $P(x_2)$, $P(x_3), \ldots, P(x_n)$, then $E(x)$, the **expected value of x,** is calculated as follows.
>
> $$E(x) = x_1 \cdot P(x_1) + x_2 \cdot P(x_2) + x_3 \cdot P(x_3) + \cdots + x_n \cdot P(x_n)$$

EXAMPLE 1 Finding the Expected Number of Boys

Find the expected number of boys for a three-child family—that is, the expected value of the number of boys. Assume girls and boys are equally likely.

Solution

The sample space for this experiment is

$$S = \{ggg, ggb, gbg, bgg, gbb, bgb, bbg, bbb\}.$$

The probability distribution is shown in **Table 11,** along with the products and their sum, which gives the expected value.

Table 11

Number of Boys x	Probability $P(x)$	Product $x \cdot P(x)$
0	$\frac{1}{8}$	0
1	$\frac{3}{8}$	$\frac{3}{8}$
2	$\frac{3}{8}$	$\frac{6}{8}$
3	$\frac{1}{8}$	$\frac{3}{8}$

Expected value: $E(x) = \frac{12}{8} = \frac{3}{2}$

The expected number of boys is $\frac{3}{2}$, or 1.5. This result seems reasonable. Boys and girls are equally likely, so "half" the children are expected to be boys. ■

The expected value for the number of boys in the family could never actually occur. It is only a kind of long-run average of the various values that *could* occur. If we record the number of boys in many different three-child families, then by the law of large numbers, or law of averages, as the number of observed families increases, the observed average number of boys should approach the expected value.

Games and Gambling

EXAMPLE 2 **Finding Expected Winnings**

A player pays $3 to play the following game: He tosses three fair coins and receives back "payoffs" of $1 if he tosses no heads, $2 for one head, $3 for two heads, and $4 for three heads. Find the player's expected net winnings for this game.

Solution

See **Table 12.** For each possible event, "net winnings" are "gross winnings" (payoff) minus cost to play. Probabilities are derived from the sample space.

$$S = \{ttt, htt, tht, tth, hht, hth, thh, hhh\}$$

The expected net loss of 50 cents is a long-run average only. On any particular play of this game, the player would lose $2 or lose $1 or break even or win $1. Over a long series of plays, say 100, there would be some wins and some losses, but the total net result would likely be around a $100 \cdot (\$0.50) = \$50$ *loss*.

Table 12

Number of Heads	Payoff	Net Winnings x	Probability $P(x)$	Product $x \cdot P(x)$
0	$1	−$2	$\frac{1}{8}$	$-\$\frac{2}{8}$
1	2	−1	$\frac{3}{8}$	$-\frac{3}{8}$
2	3	0	$\frac{3}{8}$	0
3	4	1	$\frac{1}{8}$	$\frac{1}{8}$

Expected value: $E(x) = -\$\frac{1}{2} = -\$0.50$

■

Eisenhower Dollar Coins

A game in which the expected net winnings are zero is called a **fair game.** The game in **Example 2** has negative expected net winnings, so it is unfair against the player. A game with positive expected net winnings is unfair in favor of the player.

EXAMPLE 3 Finding the Fair Cost to Play a Game

The $3 cost to play the game of **Example 2** makes the game unfair against the player (since the player's expected net winnings are negative). What cost would make this a fair game?

Solution

We already computed, in **Example 2,** that the $3 cost to play resulted in an expected net loss of $0.50. Therefore, we can conclude that the $3 cost was 50 cents too high. A fair cost to play the game would then be $3 - $0.50 = $2.50. ∎

The result in **Example 3** can be verified. Disregard the cost to play, and find the expected *gross* winnings by summing the products of payoff times probability.

$$E(\text{gross winnings}) = \$1 \cdot \frac{1}{8} + \$2 \cdot \frac{3}{8} + \$3 \cdot \frac{3}{8} + \$4 \cdot \frac{1}{8} = \frac{\$20}{8} = \$2.50$$

Expected gross winnings (payoff) are $2.50, so this amount is a fair cost to play.

EXAMPLE 4 Finding the Fair Cost to Play a Game

In a certain state lottery, a player chooses three digits, in a specific order. Leading digits may be 0, so numbers such as 028 and 003 are legitimate entries. The lottery operators randomly select a three-digit sequence, and any player matching that selection receives a payoff of $600. What is a fair cost to play this game?

Solution

In this case, no cost has been proposed, so we have no choice but to compute expected *gross* winnings. The probability of selecting all three digits correctly is $\frac{1}{10} \cdot \frac{1}{10} \cdot \frac{1}{10} = \frac{1}{1000}$, and the probability of not selecting all three correctly is $1 - \frac{1}{1000} = \frac{999}{1000}$. The expected gross winnings are

$$E(\text{gross winnings}) = \$600 \cdot \frac{1}{1000} + \$0 \cdot \frac{999}{1000} = \$0.60.$$

Thus the fair cost to play this game is 60 cents. (In fact, the lottery charges $1 to play, so players should expect to lose 40 cents per play *on the average.*) ∎

State lotteries must be unfair against players because they are designed to help fund benefits (such as the state's school system), as well as to cover administrative costs and certain other expenses. Among people's reasons for playing may be a willingness to support such causes, but most people undoubtedly play for the chance to "beat the odds" and be one of the few net winners.

Gaming casinos are major business enterprises, by no means designed to break even. The games they offer are always unfair in favor of the house. The bias does not need to be great, however, because even relatively small average losses per player, multiplied by large numbers of players, can result in huge profits for the house.

Roulette ("little wheel") was invented in France in the seventeenth or early eighteenth century. It has been a featured game of chance in the gambling casino of Monte Carlo.

The disk is divided into red and black alternating compartments numbered 1 to 36 (but not in that order). There is a compartment also for 0 (and for 00 in the United States). In roulette, the wheel is set in motion, and an ivory ball is thrown into the bowl opposite to the direction of the wheel. When the wheel stops, the ball comes to rest in one of the compartments—the number and color determine who wins.

The players bet against the banker (person in charge of the pool of money) by placing money or equivalent chips in spaces on the roulette table corresponding to the wheel's colors or numbers. Bets can be made on one number or several, on odd or even, on red or black, or on combinations. The banker pays off according to the odds against the particular bet(s). For example, the classic payoff for a winning single number is $36 for each $1 bet.

EXAMPLE 5 Finding Expected Winnings in Roulette

One simple type of *roulette* is played with an ivory ball and a wheel set in motion. The wheel contains thirty-eight compartments. Eighteen of the compartments are black, eighteen are red, one is labeled "zero," and one is labeled "double zero." (These last two are neither black nor red.) In this case, assume the player places $1 on either red or black. If the player picks the correct color of the compartment in which the ball finally lands, the payoff is $2. Otherwise, the payoff is zero. Find the expected net winnings.

Solution

By the expected value formula, expected net winnings are

$$E(\text{net winnings}) = (\$1)\frac{18}{38} + (-\$1)\frac{20}{38} = -\$\frac{1}{19}.$$

The expected net *loss* here is $\$\frac{1}{19}$, or about 5.3¢, per play.

FOR FURTHER THOUGHT

Expected Value of Games of Chance

Slot machines are a popular game for those who want to lose their money with very little mental effort. We cannot calculate an expected value applicable to all slot machines since payoffs vary from machine to machine. But we can calculate the "typical expected value."

A player operates a classic slot machine by pulling a handle after inserting a coin or coins. (We assume one coin in this discussion.) Reels inside the machine then rotate, and come to rest in some random order. (Most *modern* machines employ an electronic equivalent rather than actual rotating reels.) Assume that three reels show the pictures listed in **Table 13**. For example, of the 20 pictures on the first reel, 2 are cherries, 5 are oranges, 5 are plums, 2 are bells, 2 are melons, 3 are bars, and 1 is the number 7.

This Cleveland Indians fan hit four 7s in a row on a progressive nickel slot machine at the Sands Casino in Las Vegas in 1988.

A picture of cherries on the first reel, but not on the second, leads to a payoff of 3 coins (*net winnings*: 2 coins); a picture of cherries on the first two reels, but not the third, leads to a payoff of 5 coins (*net* winnings: 4 coins). These and all other winning combinations are listed in **Table 14**.

Since, according to **Table 13**, there are 2 ways of getting cherries on the first reel, 15 ways of *not* getting cherries on the second reel, and 20 ways of getting anything on the third reel, we have a total of $2 \cdot 15 \cdot 20 = 600$ ways of getting a net payoff of 2. Since there are 20 pictures per reel, there are a total of $20 \cdot 20 \cdot 20 = 8000$ possible outcomes. Hence, the probability of receiving a net payoff of 2 coins is $\frac{600}{8000}$.

Table 14 takes into account all *winning* outcomes, with the necessary products for finding expectation added in the last column. However, since a *nonwinning* outcome can occur in

$8000 - 988 = 7012$ ways (with winnings of -1 coin),

the product $(-1) \cdot 7012/8000$ must also be included. Hence, the expected value of this particular slot machine is

$$\frac{6318}{8000} + (-1) \cdot \frac{7012}{8000} \approx -0.087 \text{ coin.}$$

Table 13 Pictures on Reels

Pictures	Reels		
	1	**2**	**3**
Cherries	2	5	4
Oranges	5	4	5
Plums	5	3	3
Bells	2	4	4
Melons	2	1	2
Bars	3	2	1
7s	1	1	1
Totals	20	20	20

Table 14 Calculating Expected Loss on a Three-Reel Slot Machine

Winning Combinations	Number of Ways	Probability	Number of Coins Received	Net Winnings (in coins)	Probability Times Net Winnings
1 cherry (on first reel)	$2 \cdot 15 \cdot 20 = 600$	600/8000	3	2	1200/8000
2 cherries (on first two reels)	$2 \cdot 5 \cdot 16 = 160$	160/8000	5	4	640/8000
3 cherries	$2 \cdot 5 \cdot 4 = 40$	40/8000	10	9	360/8000
3 oranges	$5 \cdot 4 \cdot 5 = 100$	100/8000	10	9	900/8000
3 plums	$5 \cdot 3 \cdot 3 = 45$	45/8000	14	13	585/8000
3 bells	$\_\cdot\_\cdot\_=\_$	___/8000	18	___	___/8000
3 melons (jackpot)	$\_\cdot\_\cdot\_=\_$	___/8000	100	___	___/8000
3 bars (jackpot)	$\_\cdot\_\cdot\_=\_$	___/8000	200	___	___/8000
3 7s (jackpot)	$\_\cdot\_\cdot\_=\_$	___/8000	500	___	___/8000
Totals	___				6318/8000

On a machine costing one dollar per play, the expected *loss* (per play) is about

$$(0.087)(1 \text{ dollar}) = 8.7 \text{ cents.}$$

Actual slot machines vary in expected loss per dollar of play. But author Hornsby was able to beat a Las Vegas slot machine in 1988.

Table 15 comes from an article by Andrew Sterrett in *The Mathematics Teacher* (March 1967), in which he discusses rules for various games of chance and calculates their expected values. He uses expected values to find expected times it would take to lose $1000 if you played continually at the rate of $1 per play and one play per minute.

For Group or Individual Investigation

1. Explain why the entries of the "Net Winnings" column of **Table 14** are all one fewer than the corresponding entries of the "Number of Coins Received" column.

2. Find the 29 missing values in **Table 14** and verify that the final result, 6318/8000, is correct. (Refer to **Table 13** for the values in the "Number of Ways" column.)

3. In order to make your money last as long as possible in a casino, which game should you play? (Refer to **Table 15**.)

Table 15 Expected Time to Lose $1000

Game	Expected Value	Days	Hours	Minutes
Roulette (with one 0)	−$0.027	25	16	40
Roulette (with 0 and 00)	−$0.053	13	4	40
Chuck-a-luck	−$0.079	8	19	46
Keno (one number)	−$0.200	3	11	20
Numbers	−$0.300	2	7	33
Football pool (4 winners)	−$0.375	1	20	27
Football pool (10 winners)	−$0.658	1	1	19

Investments

EXAMPLE 6 **Finding Expected Investment Profits**

Nick has $5000 to invest and will commit the whole amount, for six months, to one of three technology stocks. A number of uncertainties could affect the prices of these stocks, but Nick is confident, based on his research, that one of only several possible profit scenarios will prove true of each one at the end of the six-month period. His complete analysis is shown in **Table 16**. For example, stock ABC could lose $400, gain $800, or gain $1500. Find the expected profit (or loss) for each of the three stocks, and select Nick's optimum choice based on these calculations.

Table 16

Company ABC		Company RST		Company XYZ	
Profit or Loss x	Probability $P(x)$	Profit or Loss x	Probability $P(x)$	Profit or Loss x	Probability $P(x)$
−$400	0.2	$500	0.8	$0	0.4
800	0.5	1000	0.2	700	0.3
1500	0.3			1200	0.1
				2000	0.2

Solution

Apply the expected value formula.

ABC: $-\$400 \cdot (0.2) + \$800 \cdot (0.5) + \$1500 \cdot (0.3) = \$770$ ← Largest

RST: $\$500 \cdot (0.8) + \$1000 \cdot (0.2) = \$600$

XYZ: $\$0 \cdot (0.4) + \$700 \cdot (0.3) + \$1200 \cdot (0.1) + \$2000 \cdot (0.2) = \$730$

The largest expected profit is $770. Nick should invest in stock ABC. ■

Of course, by investing in stock ABC, Nick may in fact *lose* $400 over the six months. The "expected" return of $770 is only a long-run average over many identical situations. Because this particular investment situation may never occur again, you may argue that using expected values is not the best approach for Nick to use.

An optimist would ignore most possibilities and focus on the *best* that each investment could do, while a pessimist would focus on the *worst* possibility.

EXAMPLE 7 **Choosing Stock Investments**

Decide which stock of **Example 6** Nick would pick in each case.

(a) He is an optimist.   **(b)** He is a pessimist.

Solution

(a) Disregarding the probabilities, he would focus on the best case for each stock. Since ABC could return as much as $1500, RST as much as $1000, and XYZ as much as $2000, the optimum is $2000. He would buy stock XYZ (the best of the three *best* cases).

(b) In this situation, he would focus on the worst possible cases. Since ABC might return as little as −$400 (a $400 loss), RST as little as $500, and XYZ as little as $0, he would buy stock RST (the best of the three *worst* cases). ■

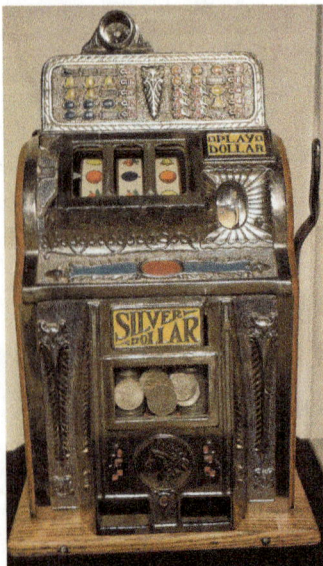

The first **Silver Dollar Slot Machine** was fashioned in 1929 by the Fey Manufacturing Company, San Francisco, inventors of the 3-reel, automatic payout machine (1895).

Business and Insurance

EXAMPLE 8   Finding Expected Lumber Revenue

Mike, a lumber wholesaler, is considering the purchase of a (railroad) carload of varied dimensional lumber. He calculates that the probabilities of reselling the load for $10,000, for $9000, and for $8000 are 0.22, 0.33, and 0.45, respectively. In order to ensure an *expected* profit of at least $3000, how much can Mike pay for the load?

Solution

The expected revenue (or income) from resales can be found as in **Table 17.**

Table 17 Expected Lumber Revenue

Income x	Probability $P(x)$	Product $x \cdot P(x)$
$10,000	0.22	$2200
9000	0.33	2970
8000	0.45	3600
	Expected revenue:	**$8770**

In general, we have the relationship

$$\text{profit} = \text{revenue} - \text{cost}.$$

Therefore, in terms of expectations,

$$\text{expected profit} = \text{expected revenue} - \text{cost}.$$

So

$$\$3000 = \$8770 - \text{cost}$$

$$\text{cost} = \$8770 - \$3000 \quad \text{Add cost and subtract \$3000.}$$

$$\text{cost} = \$5770. \quad \text{Subtract.}$$

Mike can pay up to $5770 and still maintain an expected profit of at least $3000. ■

EXAMPLE 9   Analyzing an Insurance Decision

Jeff, a farmer, will realize a profit of $150,000 on his wheat crop, unless there is rain before harvest, in which case he will realize only $40,000. The long-term weather forecast assigns rain a probability of 0.16. (The probability of no rain is $1 - 0.16 = 0.84$.) An insurance company offers crop insurance of $150,000 against rain for a premium of $20,000. Should he buy the insurance?

Solution

In order to make a wise decision, Jeff computes his expected profit under both options: to insure and not to insure. The complete calculations are summarized in the two "expectation" **Tables 18 and 19** on the next page.

For example, if insurance is purchased and it rains, Jeff's net profit is

$$\begin{bmatrix} \text{Insurance} \\ \text{proceeds} \end{bmatrix} + \begin{bmatrix} \text{Reduced} \\ \text{crop profit} \end{bmatrix} - \begin{bmatrix} \text{Insurance} \\ \text{premium} \end{bmatrix} \qquad \text{Net profit}$$

$$\$150,000 + \$40,000 - \$20,000 = \$170,000.$$

Table 18 Expectation When Insuring

	Net Profit x	Probability $P(x)$	Product $x \cdot P(x)$
Rain	$170,000	0.16	$27,200
No rain	130,000	0.84	109,200
		Expected profit:	**$136,400**

Table 19 Expectation When Not Insuring

	Net Profit x	Probability $P(x)$	Product $x \cdot P(x)$
Rain	$40,000	0.16	$6400
No rain	150,000	0.84	126,000
		Expected profit:	**$132,400**

By comparing expected profits (**136,400** > **132,400**), we conclude that Jeff is better off buying the insurance. ∎

Pilots, astronauts, race car drivers, and others train in **simulators.** Some of these devices, which may be viewed as very technical, high-cost versions of video games, imitate conditions to be encountered later in the "real world." A simulator session allows estimation of the likelihood, or probability, of different responses that the learner would display under actual conditions. Repeated sessions help the learner to develop more successful responses before actual lives and equipment are put at risk.

Simulation

An important area within probability theory is the process called **simulation.** It is possible to study a complicated, or unclear, phenomenon by *simulating*, or imitating, it with a simpler phenomenon involving the same basic probabilities. **Simulation methods** (also called **Monte Carlo methods**) require huge numbers of random digits, so computers are used to produce them. A computer, however, cannot toss coins. It must use an algorithmic process, programmed into the computer, that is called a *random number generator.* It is very difficult to avoid all nonrandom patterns in the results, so the digits produced are called "pseudorandom" numbers. They must pass a battery of tests of randomness before being "approved for use."

Computer scientists and physicists have been encountering unexpected difficulties with even the most sophisticated random number generators. Therefore, these must be carefully checked along with each new simulation application proposed.

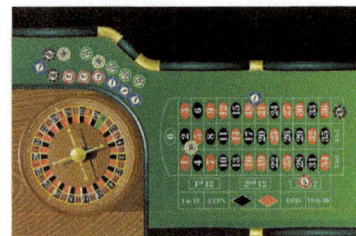

Simulating Genetic Traits

	Second Parent	
	R	r
First R	RR	Rr
Parent r	rR	rr

Recall from **Section 11.1** Mendel's discovery that when two Rr pea plants (red-flowered but carrying genes for both red and white flowers) are crossed, the offspring will have red flowers if an R gene is received from either parent or from both. This is because red is dominant and white is recessive. **Table 3,** reproduced here in the margin, shows that three of the four equally likely possibilities result in red-flowered offspring.

Now suppose we want to estimate the probability that three offspring in a row will have red flowers. It is much easier (and quicker) to toss coins than to cross pea plants. And the equally likely outcomes, heads and tails, can be used to simulate the transfer of the equally likely genes, R and r. If we toss two coins (say a nickel and a penny), then we can interpret the results as follows.

A:
B:

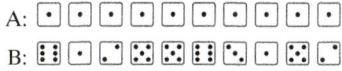

Two **possible sequences** of outcomes, when a fair die is rolled ten times in succession, are

A: 1–1–1–1–1–1–1–1–1–1 and
B: 6–1–2–5–5–6–3–1–5–2.

Which of the following options would you choose?

(a) Result A is more likely.
(b) Result B is more likely.
(c) They are equally likely.

Even though the specific sequences A and B are, in fact, equally likely (choice (c)), 67% (rounded) of the students in a class of college freshman business majors, not previously trained in probability theory, initially selected choice (b), as shown here.

Answer Choice	Pre-activity Survey	Post-activity Survey
(a)	4%	0%
(b)	67%	42%
(c)	29%	58%

Note that, even after a classroom session of hands-on simulation activities, 42% of the students persisted in their faulty reasoning.

(*Source: Mathematics Teacher*, September 2014, page 126.)

hh $\Rightarrow$ RR $\Rightarrow$ red gene from first parent and red gene from second parent

$\Rightarrow$ red flowers

ht $\Rightarrow$ Rr $\Rightarrow$ red gene from first parent and white gene from second parent

$\Rightarrow$ red flowers

th $\Rightarrow$ rR $\Rightarrow$ white gene from first parent and red gene from second parent

$\Rightarrow$ red flowers

tt $\Rightarrow$ rr $\Rightarrow$ white gene from first parent and white gene from second parent

$\Rightarrow$ white flowers

Although nothing is certain for a few tosses, the law of large numbers indicates that larger and larger numbers of tosses should become better and better indicators of general trends in the genetic process.

EXAMPLE 10 Simulating Genetic Processes

Toss two coins 50 times and use the results to approximate the probability that the crossing of Rr pea plants will produce three successive red-flowered offspring.

Solution

We actually tossed two coins 50 times and got the following sequence.

th, hh, th, tt, th, hh, ht, th, ht, th, hh, hh, tt, th, hh, ht, ht, ht, ht, th, hh, hh, hh, tt, ht, tt, hh, ht, ht, hh, tt, tt, tt, th, tt, tt, hh, ht, ht, ht, hh, tt, th, hh, tt, hh, ht, tt, tt, tt

By the color interpretation described above, this gives the following sequence of flower colors in the offspring. (Only "both tails" gives white.)

red–red–red–white–red–red–red–red–red–red–red–red–white–

red–red–red–red–red–red–red–red–red–red–white–red–white–

red–red–red–red–white–white–white–red–white–white–red–red–

red–red–red–white–red–red–white–red–red–white–white–white

We now have an experimental list of 48 sets of three successive plants, the 1st, 2nd, and 3rd entries, then the 2nd, 3rd, and 4th entries, and so on. Do you see why there are 48 in all?

Now we just count up the number of these sets of three that are "red-red-red." Since there are 20 of those, our empirical probability of three successive red offspring, obtained through simulation, is $\frac{20}{48} = \frac{5}{12}$, or about 0.417. By applying the multiplication rule of probability (with all outcomes independent of one another), we find that the theoretical value is $\left(\frac{3}{4}\right)^3 = \frac{27}{64}$, or about 0.422, so our approximation obtained by simulation is very close. ∎

Simulating Other Phenomena

In human births, boys and girls are (essentially) equally likely. Therefore, an individual birth can be simulated by tossing a fair coin, letting a head correspond to a girl and a tail to a boy.

Table 20

→ 51592
77876
36500
40571
04822
→ 53033
92080
01587
36006
63698
→ 17297
22841
→ 91979
96480
74949
76896
47588
45521
02472
55184
40177
84861
86937
20931
22454
→ 73219
→ 55707
48007
→ 65191
06772
94928
→ 15709
39922
96365
14655
65587
76905
12369
54219
89329
90060
06975
05050
69774
→ 78351
11464
84086
→ 51497
12307
68009

EXAMPLE 11 Simulating Births with Coin Tossing

A sequence of 40 actual coin tosses produced the results below.

bbggb, gbbbg, gbgbb, bggbg, bbbbg, gbbgg, gbbgg, bgbbg

(For every head we have written g, for girl. For every tail, b, for boy.)

(a) How many pairs of two successive births are represented by the sequence?

(b) How many of those pairs consist of both boys?

(c) Find the empirical probability, based on this simulation, that two successive births both will be boys. Give your answer to three decimal places.

Solution

(a) Beginning with the 1st–2nd pair and ending with the 39th–40th pair, there are 39 pairs.

(b) Observing the above sequence, we count 11 pairs of two consecutive boys.

(c) Utilizing parts (a) and (b), we have $\frac{11}{39} \approx 0.282$. ∎

Another way to simulate births, and other phenomena, is with random numbers. The spinner in **Figure 14** can be used to obtain a table of random digits, as in **Table 20**. The 250 random digits generated have been grouped conveniently so that we can easily follow down a column or across a row to carry out a simulation.

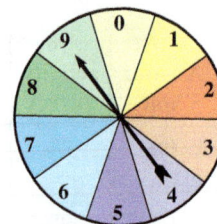

Figure 14

EXAMPLE 12 Simulating Births with Random Numbers

A couple plans to have five children. Use random number simulation to estimate the probability that they will have more than three boys.

Solution

Let each sequence of five digits, as they appear in **Table 20**, represent a family with five children, and (arbitrarily) associate odd digits with boys, even digits with girls. (Recall that 0 is even.) Verify that, of the fifty families simulated, only the ten marked with arrows have more than 3 boys (4 boys or 5 boys).

$$P(\text{more than 3 boys}) = \frac{10}{50} = 0.20 \quad \text{Estimated (empirical) probability} \quad ∎$$

The theoretical value for the probability in **Example 12** above would be the same as that obtained in **Example 4** of **Section 11.4**. It was 0.1875. Our estimate above was fairly close. In light of the law of large numbers, a larger sampling of random digits (more than 50 simulated families) would likely yield a closer approximation.

EXAMPLE 13 Simulating Card Drawing with Random Numbers

Use random number simulation to estimate the probability that two cards drawn from a standard deck, with replacement, both will be of the same suit.

Solution

Use this correspondence: 0 and 1 mean clubs, 2 and 3 mean diamonds, 4 and 5 mean hearts, 6 and 7 mean spades, 8 and 9 are disregarded. Now refer to **Table 20**.

If we (arbitrarily) use the first digit of each five-digit group, omitting 8s and 9s, we obtain the sequence

$$5\text{-}7\text{-}3\text{-}4\text{-}0\text{-}5\text{-}0\text{-}3\text{-}6\text{-}1\text{-}2\text{-}7\text{-}7\text{-}4\text{-}4\text{-}0\text{-}5\text{-}4\text{-}2\text{-}2\text{-}$$
$$7\text{-}5\text{-}4\text{-}6\text{-}0\text{-}1\text{-}3\text{-}1\text{-}6\text{-}7\text{-}1\text{-}5\text{-}0\text{-}0\text{-}6\text{-}7\text{-}1\text{-}5\text{-}1\text{-}6.$$

First digits of all groups

This 40-digit sequence of digits yields the sequence of suits shown next.

5 gives hearts, 7 gives spades, 3 gives diamonds, and so on.

hearts–spades–diamonds–hearts–clubs–hearts–clubs–diamonds–spades–
clubs–diamonds–spades–spades–hearts–hearts–clubs–hearts–hearts–
diamonds–diamonds–spades–hearts–hearts–spades–clubs–clubs–
diamonds–clubs–spades–spades–clubs–hearts–clubs–clubs–spades–
spades–clubs–hearts–clubs–spades

Verify that of the 39 successive pairs of suits (hearts–spades, spades–diamonds, diamonds–hearts, etc.), 9 of them are pairs of the same suit. This makes the estimated probability $\frac{9}{39} \approx 0.23$. (For comparison, the theoretical value is 0.25.) ■

FOR FURTHER THOUGHT

Assessing Randomness

Observe the two listings that follow, each of which simulates 200 tosses of a coin. Read from left to right across the top row, then from left to right across the second row, and so on, to the bottom row. One of the listings was obtained by actual tossing, and the other was made up by a student.

Make an educated guess about which is which, before reading the paragraph that follows the listings.

B

t h h t h	h t h h h	t h t t t	t t t t t	t h h t h
h h h t t	t h t t h	t t h t h	t t t t t	h h t t h
h t h h h	t t h t h	t h h t t	t t t t h	t h h t h
t h h h t	h t h h h	t t t h t	t h h t t	h t h h t
h t h h t	h h h t h	t h t h h	h t h t t	t t h h t
h t t t t	t h h h h	t h h h h	h h t h h	t h t h h
h t t h h	t h h t t	t t h t h	t t t t t	h t h t t
t t h h t	t h t h h	h t h h h	h t h t h	h t h t h

Consider the following fact of probability. When a fair coin is tossed 200 times, there is a 97% chance that there will be a run of six consecutive heads or six consecutive tails. Now look at the listings again, and based on this *almost* certain outcome, decide whether your earlier choice is correct. The answer is in the margin above. (The authors thank Marty Triola for his input into this For Further Thought feature.)

A

t h h t t	t h h t t	h t h h h	t t h t h	h t t t h
h t h h t	t h t t t	t h h t h	t t h h t	h t h h h
t h h h t	t h t t t	t h h h t	h t h t h	h t t h h
h t t h h	h t t h h	t h t t t	h h t h t	t h h h t
t h t t h	h t h h h	h t t h t	h h t t h	h t h t h
h t t t h	h t h h h	h t t h t	h h t t t	t h h t h
h t t h h	h t h t t	t t t h t	h h t t h	h h t t h
t h h t t	t h h t h	h t h t h	t h h t t	h t t h h

---

1. ***Coin Tosses*** Five fair coins are tossed. Find the expected number of heads.

2. ***Card Choices*** Two cards are drawn, with replacement, from a standard 52-card deck. Find the expected number of diamonds.

Expected Winnings in a Die-rolling Game *A game consists of rolling a single fair die and pays off as follows: $3 for a 6, $2 for a 5, $1 for a 4, and no payoff otherwise.*

3. Find the expected winnings for this game.

4. What is a fair price to pay to play this game?

Expected Winnings in a Die-rolling Game For Exercises 5 and 6, consider a game consisting of rolling a single fair die, with payoffs as follows. If an even number of spots turns up, you receive as many dollars as there are spots up. But if an odd number of spots turns up, you must pay as many dollars as there are spots up.

5. Find the expected net winnings of this game.

6. Is this game fair, or unfair against the player, or unfair in favor of the player?

7. *Expected Winnings in a Coin-tossing Game* A certain game involves tossing 3 fair coins, and it pays 10¢ for 3 heads, 5¢ for 2 heads, and 3¢ for 1 head. Is 5¢ a fair price to pay to play this game? That is, does the 5¢ cost to play make the game fair?

8. *Expected Winnings in Roulette* In a form of roulette slightly different from that in **Example 5,** a more generous management supplies a wheel having only thirty-seven compartments, with eighteen red, eighteen black, and one zero. Find the expected net winnings if you bet $1 on red in this game.

9. *Expected Number of Absences in a Math Class* In a certain mathematics class, the probabilities have been empirically determined for various numbers of absentees on any given day. These values are shown in the table below. Find the expected number of absentees on a given day. Give the answer to two decimal places.

Number absent	0	1	2	3	4
Probability	0.18	0.26	0.29	0.23	0.04

10. *Expected Profit of an Insurance Company* An insurance company will insure a $200,000 home for its total value for an annual premium of $650. If the company spends $25 per year to service such a policy, the probability of total loss for such a home in a given year is 0.002, and you assume that either total loss or no loss will occur, what is the company's expected annual gain (or profit) on each such policy?

Profits from a College Foundation Raffle A college foundation raises funds by selling raffle tickets for a new car worth $36,000.

11. If 600 tickets are sold for $120 each, determine each of the following.

(a) The expected *net* winnings of a person buying one of the tickets

(b) The total profit for the foundation, assuming that it had to purchase the car

(c) The total profit for the foundation, assuming that the car was donated

12. For the raffle described in **Exercise 11,** if 720 tickets are sold for $120 each, determine each of the following.

(a) The expected *net* winnings of a person buying one of the tickets

(b) The total profit for the foundation, assuming that it had to purchase the car

(c) The total profit for the foundation, assuming that the car was donated

Winnings and Profits of a Raffle Five thousand raffle tickets are sold. One first prize of $1000, two second prizes of $500 each, and three third prizes of $100 each will be awarded, with all winners selected randomly.

13. If you purchased one ticket, what are your expected gross winnings?

14. If you purchased ten tickets, what are your expected gross winnings?

15. If the tickets were sold for $1 each, how much profit goes to the raffle sponsor?

16. *Expected Sales at a Theater Snack Bar* A children's theater found in a random survey that 58 customers bought one snack bar item, 49 bought two items, 31 bought three items, 4 bought four items, and 8 avoided the snack bar altogether. Use this information to find the expected number of snack bar items per customer. Round your answer to the nearest tenth.

17. *Expected Number of Children to Attend an Amusement Park* An amusement park, considering adding some new attractions, conducted a study over several typical days and found that, of 10,000 families entering the park, 1020 brought just one child (defined as younger than age twelve), 3370 brought two children, 3510 brought three children, 1340 brought four children, 510 brought five children, 80 brought six children, and 170 brought no children at all. Find the expected number of children per family attending this park. Round your answer to the nearest tenth.

18. *Expected Sums of Randomly Selected Numbers* Four cards are numbered 1 through 4. Two of these cards are chosen randomly without replacement, and the numbers on them are added. Find the expected value of this sum.

19. *Prospects for Electronics Jobs in a City* In a certain California city, projections for the next year are that there is a 20% chance that electronics jobs will increase by 200, a 50% chance that they will increase by 300, and a 30% chance that they will decrease by 800. What is the expected change in the number of electronics jobs in that city in the next year?

20. *Expected Winnings in Keno* In one version of the game *keno*, the house has a pot containing 80 balls, numbered 1 through 80. A player buys a ticket for $1 and marks on it one number from 1 to 80. The house then selects 20 of the 80 numbers at random. If the number selected by the player is among the 20 selected by the management, the player is paid $3.20. Find the expected net winnings for this game.

Contractor Decisions Based on Expected Profits *Lori, a commercial building contractor, will commit her company to one of three projects, depending on her analysis of potential profits or losses as shown here.*

Project A		Project B		Project C	
Profit or Loss x	Probability $P(x)$	Profit or Loss x	Probability $P(x)$	Profit or Loss x	Probability $P(x)$
$60,000	0.10	$0	0.20	$40,000	0.65
180,000	0.60	210,000	0.35	340,000	0.35
250,000	0.30	290,000	0.45		

In Exercises 21–23, determine which project Lori should choose according to each approach.

21. expected values

22. the optimist viewpoint

23. the pessimist viewpoint

24. Refer to **Examples 6 and 7.** Considering the three different approaches (expected values, optimist, and pessimist), which one seems most reasonable to you, and why?

Expected Winnings in a Game Show *A game show contestant is offered the option of receiving a computer system worth $2300 or accepting a chance to win either a luxury vacation worth $5000 or a boat worth $8000. If the second option is chosen the contestant's probabilities of winning the vacation and of winning the boat are 0.20 and 0.15, respectively.*

25. If the contestant were to turn down the computer system and go for one of the other prizes, what would be the expected winnings?

26. Purely in terms of monetary value, what is the contestant's wiser choice?

Insurance Purchase *David, the promoter of an outdoor concert, expects a gate profit of $100,000 unless it rains, which would reduce the gate profit to $30,000. The probability of rain is 0.20. For a premium of $25,000 David can purchase insurance coverage that would pay him $100,000 in case of rain.*

Use this information for Exercises 27–30.

27. Find the expected net profit when the insurance is purchased.

28. Find the expected net profit when the insurance is not purchased.

29. Based on expected values, which is David's wiser choice in this situation?

30. If you were the promoter, would you base your decision on expected values? Explain your reasoning.

Expected Values in Book Sales *Jessica, an educational publisher representative, presently has five accounts, and her manager is considering assigning her three additional accounts. The new accounts would bring potential volume to her business, and some of her present accounts have potential for growth as well. See the accompanying table.*

1	2	3	4	5	6
Account Number	Existing Volume	Potential Additional Volume	Probability of Getting Additional Volume	Expected Value of Additional Volume	Existing Volume plus Expected Value of Additional Volume
1	$10,000	$10,000	0.40	$4000	$14,000
2	30,000	0	—	—	30,000
3	25,000	15,000	0.20	3000	
4	35,000	0	—	—	
5	15,000	5,000	0.30		
6	0	30,000	0.10		
7	0	25,000	0.70		
8	0	45,000	0.60		

Use the preceding table to work Exercises 31–35.

31. Compute the four missing expected values in column 5.

32. Compute the six missing amounts in column 6.

33. What is Jessica's total "expected" additional volume?

34. If Jessica achieved her expected additional volume in all accounts, what would be the total volume of all her accounts?

35. If Jessica achieved her expected additional volume in all accounts, by what percentage (to the nearest tenth of a percent) would she increase her total volume?

36. *Expected Winnings in Keno* Recall that in the game keno of **Exercise 20,** the house randomly selects 20 numbers from the counting numbers 1–80. In the variation called 6-spot keno, the player pays 60¢ for his ticket and marks 6 numbers of his choice. If the 20 numbers selected by the house contain at least 3 of those chosen by the player, he gets a payoff according to this scheme.

3 of the player's numbers among the 20	$0.35
4 of the player's numbers among the 20	2.00
5 of the player's numbers among the 20	60.00
6 of the player's numbers among the 20	1250.00

Find the player's expected net winnings in this game.

[*Hint:* The four probabilities required here can be found using combinations **(Section 10.3),** the fundamental counting principle **(Section 10.2),** and the theoretical probability formula **(Section 11.1).**]

37. *Pea Plant Reproduction Simulation with Coin Tossing* Use the sequence of flower colors of **Example 10** to approximate the probability that *four* successive offspring all will have red flowers.

38. *Pea Plant Reproduction Simulation with Coin Tossing* Explain why, in **Example 10,** fifty tosses of the coins produced only 48 sets of three successive offspring.

39. *Boy and Girl Simulation with Random Numbers* Use **Table 20** to simulate fifty families with three children. Let 0–4 correspond to boys and 5–9 to girls, and use the middle three digits of the 5-digit groupings (159, 787, 650, and so on). Estimate the probability of exactly two boys in a family of three children. Compare with the theoretical probability, which is $\frac{3}{8} = 0.375$.

40. *Empirical Probability of Successive Girls* Simulate 40 births by tossing coins yourself, and obtain an empirical probability for two successive girls.

41. *Likelihoods of Girl and Boy Births* Should the probability of two successive girl births be any different from that of two successive boy births?

One-and-One Free Throw Shooting *In Exercises 88–90 of Section 11.3, Christine, who had a 70% foul-shooting record, had probabilities of scoring 0, 1, or 2 points of 0.30, 0.21, and 0.49, respectively.*

Use **Table 20** *(with digits 0–6 representing hit and 7–9 representing miss) to simulate 50 one-and-one shooting opportunities for Christine. Begin at the top left (5, 7, 3, etc., to the bottom), then move to the second column (1, 7, 6, etc.), going until 50 one-and-one opportunities are obtained. (Some "opportunities" involve one shot and one random digit, while others involve two shots and two random digits.) Keep a tally of the numbers of times 0, 1, and 2 points are scored.*

Number of Points	Tally
0	
1	
2	

From the tally, find the empirical probability (to two decimal places) of each event.

42. no points

43. 1 point

44. 2 points

Simulation with Coin Tossing *A coin was actually tossed 200 times, producing the following sequence of outcomes. Read from left to right across the top row, then from left to right across the second row, and so on, to the bottom row.*

```
h h t h t    t h t h h    t t t t t    t t h t t    h t h h h
t h t t h    h h h h h    h t t h h    h t h h t    h h t t h
t h t h t    h h t t t    h h h h h    t t h t h    t h t h h
t h h h h    h h h h t    t h t t h    t h h t h    t h h t t
t h h t t    t h t t t    h t h h t    t h h h t    h t t t t
h t t h h    h t t t t    t t t h t    t t t t h    t h t h h
h h h h h    t t h h t    t t h h t    h t h t t    h h h h t
h t h t t    h t t t t    h h t t h    t t t h h    t h t t h
```

Find the empirical probability of each of the following. Round to three decimal places.

45. two consecutive heads

46. two consecutive tails

47. three consecutive tosses of the same outcome

48. six consecutive tosses of the same outcome

Path of a Random Walk Using a Die and a Coin *Exercises 49–56 of Section 11.4 illustrated a simple version of the idea of a "random walk." Atomic particles released in nuclear fission also move in a random fashion. During World War II, John von Neumann and Stanislaw Ulam used simulation with random numbers to study particle motion in nuclear reactions. Von Neumann coined the name "Monte Carlo" for the methods used.*

The figure suggests a model for random motion in two dimensions. Assume that a particle moves in a series of 1-unit "jumps," each one in a random direction, any one of 12 equally likely possibilities. One way to choose directions is to roll a fair die and toss a fair coin. The die determines one of the directions 1–6, coupled with heads on the coin. Tails on the coin reverses the direction of the die, so that the die coupled with tails gives directions 7–12. So 3h (meaning 3 with the die and heads with the coin) gives direction 3; 3t gives direction 9 (opposite to 3); and so on.

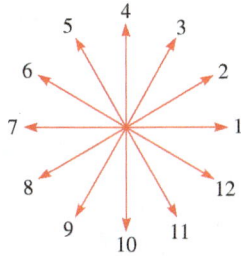

49. Simulate the motion described above with 10 rolls of a die and tosses of a coin. Draw the 10-jump path you get. Make your drawing accurate enough so you can estimate (by measuring) how far from its starting point the particle ends up.

50. Repeat the experiment of **Exercise 49** four more times. Measure distance from start to finish for each of the 5 "random trips." Add these 5 distances and divide the sum by 5, to arrive at an "expected net distance" for such a trip.

In Exercises 51 and 52, consider another two-dimensional random walk governed by the following conditions.

- *Start out from a given street corner, and travel one block north. At each intersection:*
- *Turn left with probability $\frac{1}{6}$.*
- *Go straight with probability $\frac{2}{6}\left(=\frac{1}{3}\right)$.*
- *Turn right with probability $\frac{3}{6}\left(=\frac{1}{2}\right)$.*
  (Never turn around.)

51. *A Random Walk Using a Random Number Table* Use **Table 20** to simulate this random walk. For every 1 encountered in the table, turn left and proceed for another block. For every 2 or 3, go straight and proceed for another block. For every 4, 5, or 6, turn right and proceed for another block. Disregard all other digits—that is, 0s, 7s, 8s, and 9s. (Do you see how this scheme satisfies the probabilities given above?) This time, begin at the upper right corner of the table, running down the column 2, 6, 0, and so on, to the bottom. When this column of digits is used up, stop the "walk." Describe, in terms of distance and direction, where you have ended up relative to your starting point.

52. *A Random Walk Using a Fair Die* Explain how a fair die could be used to simulate this random walk.

Buffon's Needle Problem *The following problem was posed by Georges Louis Leclerc, Comte de Buffon (1707–1788) in his* Histoire Naturelle *in 1777. A large plane area is ruled with equidistant parallel lines, the distance between two consecutive lines of the series being "a". A thin needle of length*

$$\ell < a$$

is tossed randomly onto the plane. What is the probability that the needle will intersect one of these lines?

The answer to this problem is found using integral calculus, and the probability p is shown to be $p=\frac{2\ell}{\pi a}$. Solving for π gives the formula

$$\pi = \frac{2\ell}{pa},$$

which can be used to approximate the value of π experimentally. This was first observed by Pierre Simon de Laplace, and such an experiment was carried out by Johann Wolf, a professor of astronomy at Bern, in about 1850. (Source: Burton, David M. History of Mathematics: An Introduction. *Wm. C. Brown, 1995.)*

53. On a sheet of paper, draw a series of parallel lines evenly spaced across it. Then find a thin needle, or needlelike object, with a length less than the distance between adjacent parallel lines on the paper.

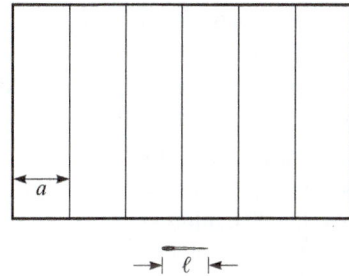

Carry out the following steps in order.

(a) Measure and record the distance between lines (a) and the length of the needle (ℓ), using the same units for both.

(b) Drop the needle onto the paper 100 times. Determine whether the needle "hits" a line or not, and keep a tally of hits and misses.

(c) Now calculate $p =$ (number of hits)/100. Is this probability value theoretical or empirical?

(d) Enter the calculated value of p and the measured values of a and ℓ into the formula to obtain an approximation for π. Round this value to four decimal places.

(e) The correct value of π, to four decimal places, is 3.1416. What value did your simulation give? How far off is it?

54. *Dynamic Simulation* See www.metablake.com/pi.swf for a dynamic illustration of the Buffon Needle Problem. (One of the authors of this text used 21,734 iterations, to get 13,805 intersections, for an empirical value of π to be 3.148569358927.)

CHAPTER 11   SUMMARY

KEY TERMS

11.1

deterministic phenomenon
random phenomenon
probability
experiment
outcome
sample space
event
theoretical probability

empirical probability
law of large numbers
   (law of averages)
odds in favor
odds against

11.2

mutually exclusive events
random variable
probability distribution

11.3

conditional probability of
   B given A
independent events

11.4

Bernoulli trials
binomial probability
   distribution

11.5

expected value
   (mathematical expectation)
simulation
simulation methods
   (Monte Carlo methods)
random number generator

NEW SYMBOLS

$P(E)$ probability of event E

$P(B|A)$ probability of event B, given that event A has occurred

$E(x)$ expected value of random variable x

TEST YOUR WORD POWER

See how well you have learned the vocabulary in this chapter.

1. The **sample space** of an experiment is
   A. the set of all possible events of the experiment.
   B. the set of all possible outcomes of the experiment.
   C. the number of possible events of the experiment.
   D. the number of possible outcomes of the experiment.

2. The **probability of an event E**
   A. must be a number greater than 1.
   B. may be a negative number.
   C. must be a number between 0 and 1, inclusive of both.
   D. cannot be 0.

3. If E is an event, the **probability of "not E"** must be
   A. 0.
   B. equal to 1 plus the probability of E.
   C. equal to 1 minus the probability of E.
   D. less than the probability of E.

4. A **probability distribution**
   A. shows all possible values of a random variable, along with the probabilities that those values will occur.
   B. is a listing of empirical values of an experiment.
   C. is a number between 0 and 1, inclusive of both.
   D. is an application of the distributive property to probability.

ANSWERS

1. B   2. C   3. C   4. A

QUICK REVIEW

Concepts

Examples

11.1 Basic Concepts

Theoretical Probability Formula

If all outcomes in a sample space S are equally likely, and E is an event within that sample space, then the **theoretical probability** of event E is given by the following formula.

$$P(E) = \frac{\text{number of favorable outcomes}}{\text{total number of outcomes}} = \frac{n(E)}{n(S)}$$

The spinner shown here is spun twice. Find the probability that both spins yield the same number.

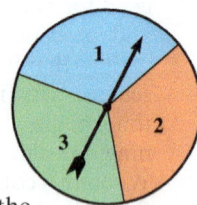

The sample space S is $\{(1,1), (1,2), (1,3), (2,1), (2,2), (2,3), (3,1), (3,2)(3,3)\}$. Of these, three indicate the same number: $E = \{(1,1), (2,2), (3,3)\}$. Therefore,

$$P(E) = \frac{n(E)}{n(S)} = \frac{3}{9} = \frac{1}{3}.$$

Concepts	Examples

Empirical Probability Formula

If E is an event that may happen when an experiment is performed, then an **empirical probability** of event E is given by the following formula.

$$P(E) = \frac{\text{number of times event } E \text{ occurred}}{\text{number of times the experiment was performed}}$$

A coin was actually tossed 100 times, producing the following sequence of outcomes. If E represents two consecutive tosses of heads (h), what is the empirical probability of E?

thhht   thhht   hthth   thhhh   thttt
ttttt   tthht   htthh   tttth   tttth
hhthh   thttt   thttt   thtth   thhht
hthhh   ttthh   hhhht   ththt   hhhht

Law of Large Numbers (Law of Averages)

As an experiment is repeated more and more times, the proportion of outcomes favorable to any particular event will tend to come closer and closer to the theoretical probability of that event.

There are 99 cases of two consecutive tosses, and 24 of them show both heads, so

$$P(E) = \frac{24}{99} \approx 0.242. \quad \text{Empirical probability}$$

In the experiment above, what is the theoretical probability that two consecutive heads will occur? How does the empirical result compare?

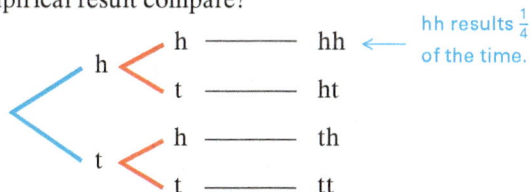

The theoretical probability is $\frac{1}{4} = 0.25$. This is very close to 0.242, the empirical value obtained above.

Odds

If all outcomes in a sample space are equally likely, a of them are favorable to the event E, and the remaining b outcomes are unfavorable to E, then

the **odds in favor** of E are a to b,

and   the **odds against** E are b to a.

A single card is drawn from a standard deck of 52 cards.

(a) What is the probability that the card is a queen?
There are 4 queens (and thus 48 non-queens).

$$P(\text{queen}) = \frac{4}{52} = \frac{1}{13} \quad \text{Probability of a queen}$$

Converting between Probability and Odds

Let E be an event.

- If $P(E) = \frac{a}{b}$, then the odds in favor of E are a to $(b-a)$.
- If the odds in favor of E are a to b, then $P(E) = \frac{a}{a+b}$.

(b) What are the odds in favor of drawing a queen?
4 to 48,   or   1 to 12   Odds in favor of a queen

(c) What are the odds against drawing a queen?
48 to 4,   or   12 to 1   Odds against a queen

11.2 Events Involving "Not" and "Or"

Properties of Probability

Let E be an event within the sample space S. That is, E is a subset of S. Then the following properties hold.

1. $0 \leq P(E) \leq 1$ (The probability of an event is a number from 0 through 1, inclusive.)
2. $P(\varnothing) = 0$ (The probability of an impossible event is 0.)
3. $P(S) = 1$ (The probability of a certain event is 1.)

If 6 fair coins are tossed, what is the probability that at most 5 will show tails?

There are $2^6 = 64$ possible outcomes. Only one of them,

tttttt,

does not show at most 5 tails, so there are $64 - 1 = 63$ favorable outcomes.

$$P(\text{at most 5 tails}) = \frac{64-1}{64} = \frac{63}{64}$$

Concepts	Examples

Probability of a Complement (for the Event "Not E")

The probability that an event E will *not* occur is equal to 1 minus the probability that it will occur.

$$P(\text{not } E) = 1 - P(E)$$

If a single card is drawn from a standard 52-card deck, find the probability that it is the following.

(a) a heart or a black card   **(b)** a heart or a 6

$P(\text{heart or black}) = P(\text{heart}) + P(\text{black})$ These are mutually exclusive.

Mutually Exclusive Events

Two events A and B are **mutually exclusive events** if they have no outcomes in common. Mutually exclusive events cannot occur simultaneously.

$$= \frac{13}{52} + \frac{26}{52}$$
$$= \frac{39}{52}$$
$$= \frac{3}{4}$$

Addition Rule of Probability (for the Event "A or B")

If A and B are any two events, then the following holds.

$$P(A \text{ or } B) = P(A) + P(B) - P(A \text{ and } B)$$

If A and B are mutually exclusive, then the following holds.

$$P(A \text{ or } B) = P(A) + P(B)$$

$P(\text{heart or a 6}) = P(\text{heart}) + P(6) - P(6 \text{ of hearts})$ These are not mutually exclusive.

$$= \frac{13}{52} + \frac{4}{52} - \frac{1}{52}$$
$$= \frac{16}{52}$$
$$= \frac{4}{13}$$

11.3 Conditional Probability and Events Involving "And"

From the sample space

$$S = \{1, 2, 3, 4, 5, 6, 7, 8, 9, 10, 11, 12\},$$

a single number is to be selected randomly. Find each probability given the events

Conditional Probability

The probability of event B, computed on the assumption that event A has happened, is the **conditional probability of B given A** and is denoted $P(B|A)$.

A: The selected number is even.

B: The selected number is a multiple of 3.

(a) $P(B)$

$B = \{3, 6, 9, 12\}$, so

Conditional Probability Formula

The **conditional probability of B given A** is calculated as

$$P(B|A) = \frac{P(A \cap B)}{P(A)} = \frac{P(A \text{ and } B)}{P(A)}.$$

$$P(B) = \frac{n(B)}{n(S)} = \frac{4}{12} = \frac{1}{3}.$$

(b) $P(A \text{ and } B)$

$A = \{2, 4, 6, 8, 10, 12\}$, so $A \cap B = \{6, 12\}$.

$$P(A \text{ and } B) = P(A \cap B) = \frac{2}{12} = \frac{1}{6}$$

Independent Events

Two events A and B are **independent events** if knowledge about the occurrence of one of them has no effect on the probability of the other one occurring. A and B are independent if

$$P(B|A) = P(B), \quad \text{or, equivalently,} \quad P(A|B) = P(A).$$

(c) $P(B|A)$ Are A and B independent?

In the reduced sample space of A, 2 of the 6 outcomes belong to B.

$$P(B|A) = \frac{2}{6} = \frac{1}{3}$$

Multiplication Rule of Probability (for the Event "A and B")

If A and B are *any two events*, then

$$P(A \text{ and } B) = P(A) \cdot P(B|A).$$

If A and B are *independent*, then

$$P(A \text{ and } B) = P(A) \cdot P(B).$$

Equivalently, $P(B|A) = \dfrac{P(A \cap B)}{P(A)} = \dfrac{\frac{1}{6}}{\frac{1}{2}} = \dfrac{1}{6} \cdot \dfrac{2}{1} = \dfrac{1}{3}.$

Because $P(B) = P(B|A)$, A and B are independent.

Concepts	Examples

11.4 Binomial Probability

Binomial Probability Formula

When n independent repeated trials occur, where

p = probability of success and q = probability of failure

with p and q (where $q = 1 - p$) remaining constant throughout all n trials, the probability of exactly x successes is calculated as follows.

$$P(x) = {_nC_x}p^x q^{n-x} = \frac{n!}{x!(n-x)!}p^x q^{n-x}$$

A die is rolled ten times. Find the probability that exactly four of the tosses result in a 3.

The probability of rolling a 3 on one roll is $p = \frac{1}{6}$, so the probability of not rolling a 3 is

$$q = 1 - \frac{1}{6} = \frac{5}{6}.$$

Here $n = 10$ and $x = 4$.

$$P(\text{exactly four 3s}) = {_{10}C_4}\left(\frac{1}{6}\right)^4\left(\frac{5}{6}\right)^{10-4}$$

$$= 210\left(\frac{1}{1296}\right)\left(\frac{15,625}{46,656}\right)$$

$$\approx 0.054$$

11.5 Expected Value and Simulation

Expected Value

If a random variable x can have any of the values $x_1, x_2, x_3, \ldots, x_n$, and the corresponding probabilities of these values occurring are $P(x_1), P(x_2), P(x_3), \ldots, P(x_n)$, then $E(x)$, the **expected value of x,** is calculated as follows.

$$E(x) = x_1 \cdot P(x_1) + x_2 \cdot P(x_2) + x_3 \cdot P(x_3)$$

$$+ \cdots + x_n \cdot P(x_n)$$

In a certain lottery, a player chooses four digits in a specific order. The leading digit may be 0, so a number such as 0460 is legitimate. A four-digit sequence is randomly chosen, and any player matching that selection receives a payoff of $10,000. Find the expected gross winnings.

The probability of selecting the correct four digits is $\left(\frac{1}{10}\right)^4 = \frac{1}{10,000}$, and the probability of not selecting them is $1 - \left(\frac{1}{10}\right)^4 = \frac{9999}{10,000}$. The expected gross winnings are

$$E(\text{gross winnings}) = \$10,000 \cdot \frac{1}{10,000} + \$0 \cdot \frac{9999}{10,000}$$

$$= \$1.00. \quad \text{A fair cost to play this game}$$

CHAPTER 11   TEST

Numbers from Sets of Digits *Two numbers are randomly selected without replacement from the set*

$$\{1, 2, 3, 4, 5\}.$$

Find the probability of each event.

1. Both numbers are even.

2. Both numbers are prime.

3. The sum of the two numbers is odd.

4. The product of the two numbers is odd.

Days Off for Pizza Parlor Workers *The manager of a pizza parlor (which operates seven days a week) allows each of three employees to select one day off next week. Assuming the selection is done randomly and independently, find the probability of each event.*

5. All three select different days.

6. All three select the same day, given that all three select a day beginning with the same letter.

7. Exactly two of them select the same day.

Genetics of Cystic Fibrosis *The chart represents genetic transmission of cystic fibrosis. C denotes a normal gene, and c denotes a cystic fibrosis gene. (Normal is dominant.) Both parents in this case are Cc, which means that they inherited one of each gene. Therefore, they are both carriers but do not have the disease.*

		Second Parent	
		C	c
First Parent	C		Cc
	c		

8. Complete the chart, showing all four equally likely gene arrangements.

9. Find the probability that a child of these parents will also be a carrier without the disease.

10. What are the odds that a child of these parents actually will have cystic fibrosis?

Drawing Cards *A single card is chosen at random from a standard 52-card deck. Find the odds against its being each of the following.*

11. a heart

12. a red queen

13. a king or a black face card

Selecting Committees *A three-member committee is selected randomly from a group consisting of three men and two women.*

14. Let x denote the number of men on the committee, and complete the probability distribution table.

x	$P(x)$
0	0
1	
2	
3	

15. Find the probability that the committee members are not all men.

16. Find the expected number of men on the committee.

Rolling Dice *A pair of dice are rolled. Find the following.*

17. the probability of "doubles" (the same number on both dice)

18. the odds in favor of a sum greater than 2

19. the odds against a sum of "7 or 11"

20. the probability of a sum that is even and less than 5

Making Par in Golf *Ted has a 0.78 chance of making par on each hole of golf that he plays. Today he plans to play just three holes. Find the probability of each event. Round answers to three decimal places.*

21. He makes par on all three holes.

22. He makes par on exactly two of the three holes.

23. He makes par on at least one of the three holes.

24. He makes par on the first and third holes but not on the second.

Card Choices *Two cards are drawn, without replacement, from a standard 52-card deck. Find the probability of each event.*

25. Both cards are red.

26. Both cards are the same color.

27. The second card is a queen, given that the first card is an ace.

28. The first card is a face card and the second is black.

29. **Gender in Sequences of Babies** Assuming boy and girl babies are equally likely, find the probability that a family with three children will have exactly two boys.

30. **Simulation of Pea Plant Reproduction with Coin Tossing** Use the sequence of flower colors in **Example 10** of **Section 11.5** to approximate, to three decimal places, the probability that three successive offspring will all have white flowers.

Simulation with Coin Tossing *A coin was actually tossed 100 times, producing the following sequence of outcomes. Read from left to right across the top row, then from left to right across the second row, and so on, to the bottom row.*

tttth	ththh	htttt	hhhth	ththh
hhhth	httht	hhhht	thttt	thtth
httth	hhhhh	hhhhh	ttht	ttttt
htthh	tttth	httth	tthhh	tthhh

Find, to three decimal places, the empirical probability of each event.

31. Two consecutive tails (tt) occur.

32. Three consecutive heads (hhh) occur.

Statistics

12

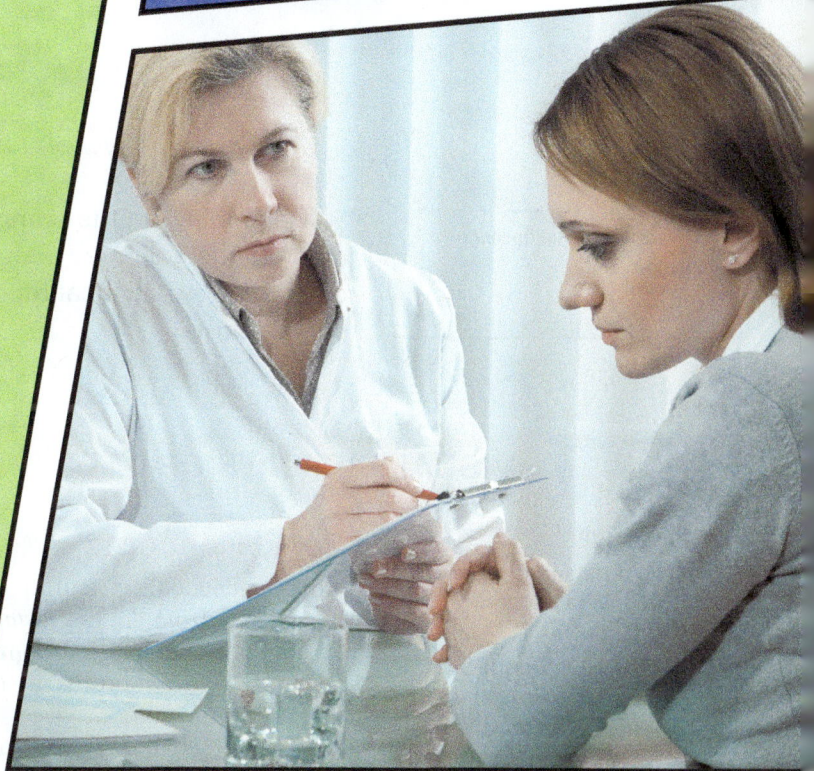

Suppose you're a psychological therapist who specializes in helping patients with various early-stage addictions. You have noted that under your treatment plan, 50% of your clients have been able to quit drinking (their cumulative probability of remission is 0.50) by 10 years after their dependency began. ("Remission" is defined as quitting for 1 year.)

Compare these results with the general situation, as reported in a long-term study and shown in the graph on the next page. Complete the following statements.

1. According to the long-term study, the more general cumulative probability of remission is 0.50 after about _____ years.
2. Your own results for 0.50 probability of remission are better than the average by about _____ years.

(Check your answers with those given on page 646.)

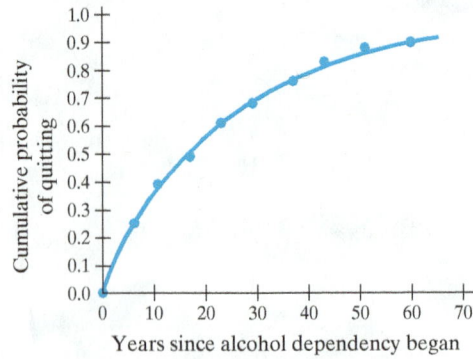

Data Source: G. Heyman, *Annual Review of Clinical Psychology,* March 28, 2013.

Depending on the comparison between the general results and your own results, you may want to either reconsider your treatment protocol or prepare a statistical report on your results for publication in a professional journal.

12.1   VISUAL DISPLAYS OF DATA

OBJECTIVES

1 Understand the basic concepts of data displays.
2 Work with frequency distributions.
3 Work with grouped frequency distributions.
4 Construct and interpret stem-and-leaf displays.
5 Work with bar graphs, circle graphs, and line graphs.

Basic Concepts

Governments collect and analyze an amazing quantity of "statistics." The word itself comes from the Latin *statisticus,* meaning "of the state."

In statistical work, a **population** includes *all* items of interest, and a **sample** includes *some,* but ordinarily not all, of the items in the population. See the Venn diagram in the margin.

To predict the outcome of an approaching presidential election, we may be interested in a population of many millions of voter preferences (those of all potential voters in the country). As a practical matter, however, even national polling organizations with considerable resources obtain only a relatively small sample, say 2000, of those preferences.

The study of statistics is divided into two main areas. **Descriptive statistics** has to do with collecting, organizing, summarizing, and presenting data (information). **Inferential statistics** has to do with drawing inferences or conclusions (making conjectures) about populations on the basis of information from samples.

The photos below show two random samples drawn from a large bowl of 10,000 colored beads. The 25-bead sample contains 9 green beads, from which we infer, by inductive reasoning, that the bowl (the population) must contain about

$$\frac{9}{25}, \text{ or } 36\%, \text{ that is, about 3600 green beads.}$$

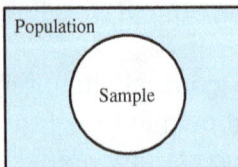

Population

Sample

A population of 10,000

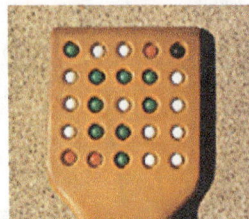

A random sample of 25

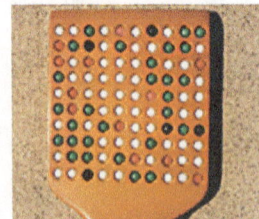

A random sample of 100

The 100-bead sample contains 28 green beads, leading to the inference that the population must contain about

$$\frac{28}{100}, \text{ or } 28\%, \text{ that is, about 2800 green beads.}$$

This estimate, based on a larger sample, should be more accurate. In fact it is, since the bowl actually contains

$$30\%, \text{ or } 3000 \text{ green beads,}$$

and 2800 is closer to 3000 than 3600 is. (The "error" is one-third as much.)

Summarizing, if we know what a population is like, then probability theory enables us to predict what is likely to happen in a sample (deductive reasoning). If we know what a sample is like, then inferential statistics enables us to infer estimates about the population (inductive reasoning).

Information that has been collected but not yet organized or processed is called **raw data.** It is often **quantitative** (or **numerical**), but it can also be **qualitative** (or **non-numerical**), as illustrated in **Table 1.**

Table 1 Examples of Raw Data

Quantitative data: The number of siblings in ten different families: 3, 1, 2, 1, 5, 4, 3, 3, 8, 2

Qualitative data: The makes of six different automobiles: Toyota, Ford, Kia, Toyota, Chevrolet, Honda

Quantitative data are generally more useful when they are **ranked,** or arranged in numerical order. In ranked form, the first list in **Table 1** appears as follows.

$$1, 1, 2, 2, 3, 3, 3, 4, 5, 8$$

Frequency Distributions

When a data set includes many repeated items, it can be organized into a **frequency distribution,** which lists the distinct data values (x) along with their frequencies (f). The frequency designates the number of times the corresponding item occurred in the data set.

It is also helpful to show the **relative frequency** of each distinct item. This is the fraction, or percentage, of the data set represented by the item. If n denotes the total number of items, and a given item, x, occurred f times, then the relative frequency of x is

$$\frac{f}{n}.$$

Example 1 illustrates these ideas.

EXAMPLE 1 **Constructing Frequency and Relative Frequency Distributions**

The 25 members of a psychology class were polled about the number of siblings in their individual families. Construct a frequency distribution and a relative frequency distribution for their responses, which are shown here.

$$2, 3, 1, 3, 3, 5, 2, 3, 3, 1, 1, 4, 2, 4, 2, 5, 4, 3, 6, 5, 1, 6, 2, 2, 2$$

Solution

The data range from a low of 1 to a high of 6. The frequencies (obtained by inspection) and relative frequencies are shown in **Table 2** on the next page.

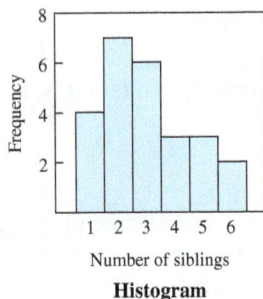

Number of siblings
Histogram

Figure 1

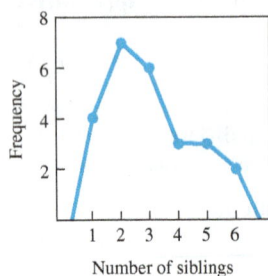

Number of siblings
Frequency polygon

Figure 2

Table 2 Frequency and Relative Frequency Distributions for Numbers of Siblings

Number x	Frequency f	Relative Frequency $\frac{f}{n}$
1	4	$\frac{4}{25} = 16\%$
2	7	$\frac{7}{25} = 28\%$
3	6	$\frac{6}{25} = 24\%$
4	3	$\frac{3}{25} = 12\%$
5	3	$\frac{3}{25} = 12\%$
6	2	$\frac{2}{25} = 8\%$
Total: $n = 25$		

Problem-Solving Strategy

Three of the strategies from **Chapter 1** are "Make a table or chart," "Look for a pattern," and "Draw a sketch." The techniques of this section apply all three of these to present raw data in more meaningful visual forms.

The numerical data of **Table 2** can be interpreted more easily with the aid of a **histogram.** A series of rectangles, whose lengths represent the frequencies, are placed next to one another as shown in **Figure 1.** On each axis, horizontal and vertical, a label and the numerical scale should be shown.

The information shown in the histogram in **Figure 1** can also be conveyed by a **frequency polygon,** as in **Figure 2.** Simply plot a single point at the appropriate height for each frequency, connect the points with a series of connected line segments, and complete the polygon with segments that trail down to the axis beyond 1 and 6.

The frequency polygon is an instance of the more general *line graph*, which was first introduced in **Chapter 1.**

Grouped Frequency Distributions

Data sets containing large numbers of items are often arranged into groups, or **classes.** All data items are assigned to their appropriate classes, and then a **grouped frequency distribution** can be set up and a graph displayed. Although there are no fixed rules for establishing the classes, most statisticians agree on a few general guidelines.

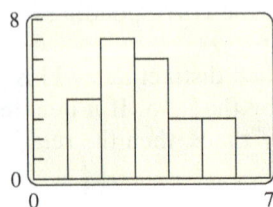

This histogram was generated with a graphing calculator using the data in **Table 2.** Compare with **Figure 1** above.

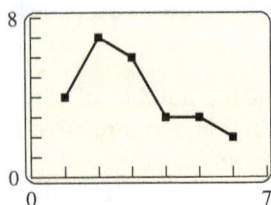

This line graph resembles the frequency polygon in **Figure 2** above. It was generated with a graphing calculator using the data in **Table 2.**

GUIDELINES FOR THE CLASSES OF A GROUPED FREQUENCY DISTRIBUTION

1. Make sure each data item will fit into one, and only one, class.
2. Try to make all classes the same width.
3. Make sure the classes do not overlap.
4. Use from 5 to 12 classes. Too few or too many classes can obscure the tendencies in the data.

EXAMPLE 2 **Constructing a Histogram and a Frequency Polygon**

Forty students, selected randomly in the school cafeteria one morning, were asked to estimate the number of hours they had spent studying in the past week (including both in-class and out-of-class time). Their responses are recorded here.

$$18 \quad 60 \quad 72 \quad 58 \quad 33 \quad 15 \quad 12 \quad 36 \quad 16 \quad 29$$

$$26 \quad 41 \quad 45 \quad 25 \quad 32 \quad 24 \quad 22 \quad 55 \quad 30 \quad 31$$

$$55 \quad 39 \quad 29 \quad 44 \quad 29 \quad 14 \quad 36 \quad 31 \quad 45 \quad 62$$

$$36 \quad 52 \quad 47 \quad 38 \quad 36 \quad 23 \quad 33 \quad 44 \quad 17 \quad 24$$

Tabulate a grouped frequency distribution and a grouped relative frequency distribution, and construct a histogram and a frequency polygon for the given data.

Solution

The data range from a low of 12 to a high of 72—that is, over a range of

$$72 - 12 = 60 \text{ units.}$$

The widths of the classes should be uniform (by Guideline 2), and there should be from 5 to 12 classes (by Guideline 4). Five classes would imply a class width of about $\frac{60}{5} = 12$, while twelve classes would imply a class width of about $\frac{60}{12} = 5$. A class width of 10 will be convenient. We let our classes run from 10 through 19, from 20 through 29, and so on up to 70 through 79, for a total of seven classes. All four guidelines are met.

Next go through the data set, tallying each item into the appropriate class. The tally totals produce class frequencies, which in turn produce relative frequencies, as shown in **Table 3** below. The histogram is displayed in **Figure 3**.

In **Table 3** (and **Figure 3**) the numbers 10, 20, 30, and so on are called the **lower class limits.** They are the smallest possible data values within the respective classes. The numbers 19, 29, 39, and so on are called the **upper class limits.** The common **class width** for the distribution is the difference of any two successive lower class limits (such as $30 - 20$) or of any two successive upper class limits (such as $59 - 49$). The class width for this distribution is 10, as noted earlier.

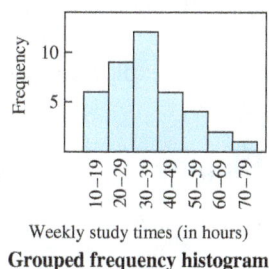

Grouped frequency histogram

Figure 3

Table 3 Grouped Frequency and Relative Frequency Distributions for Weekly Study Times

Class Limits	Tally	Frequency f	Relative Frequency $\frac{f}{n}$
10–19	卌 l	6	$\frac{6}{40} = 15.0\%$
20–29	卌 llll	9	$\frac{9}{40} = 22.5\%$
30–39	卌 卌 ll	12	$\frac{12}{40} = 30.0\%$
40–49	卌 l	6	$\frac{6}{40} = 15.0\%$
50–59	llll	4	$\frac{4}{40} = 10.0\%$
60–69	ll	2	$\frac{2}{40} = 5.0\%$
70–79	l	1	$\frac{1}{40} = 2.5\%$
	Total:	$n = 40$	

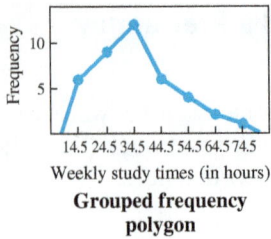

Weekly study times (in hours)

Grouped frequency polygon

Figure 4

Table 4

Grouped Frequency Distribution for Weekly Study Times

Class Limits	Frequency
10–19	6
20–29	9
30–39	12
40–49	6
50–59	4
60–69	2
70–79	1

To construct a frequency polygon, notice that in a *grouped* frequency distribution, the data items in a given class are generally not all the same. We can obtain the "middle" value, or **class mark,** by adding the lower and upper class limits and dividing this sum by 2. We locate all the class marks along the horizontal axis and plot points above the class marks. The heights of the plotted points represent the class frequencies. The resulting points are connected just as for a nongrouped frequency distribution. The result is shown in **Figure 4.**

Stem-and-Leaf Displays

In **Table 3,** the tally marks give a good visual impression of how the data are distributed. In fact, the tally marks are almost like a histogram turned on its side. Nevertheless, once the tallying is done, the tally marks are usually dropped, and the grouped frequency distribution is presented as in **Table 4.**

The pictorial advantage of the tally marks is now lost. Furthermore, we cannot tell, from the grouped frequency distribution itself (or from the tally marks either, for that matter), what any of the original items were. We only know, for example, that there were six items in the class 40–49. We do not know specifically what any of them were.

One way to avoid these shortcomings is to employ a tool of exploratory data analysis, the **stem-and-leaf display,** as shown in **Example 3.**

EXAMPLE 3 **Constructing a Stem-and-Leaf Display**

Present the study times data of **Example 2** in a stem-and-leaf display.

Solution

See **Example 2** for the original raw data. We arrange the numbers in **Table 5.** The tens digits, to the left of the vertical line, are the "stems," while the corresponding ones digits are the "leaves." We have entered all items from the first row of the original data, from left to right, and then the items from the second row through the fourth row.

Table 5 Stem-and-Leaf Display for Weekly Study Times

1	8	5	2	6	4	7					
2	9	6	5	4	2	9	9	3	4		
3	3	6	2	0	1	9	6	1	6	8	6	3
4	1	5	4	5	7	4					
5	8	5	5	2							
6	0	2									
7	2										

Notice that the stem-and-leaf display of **Example 3** conveys at a glance the same pictorial impressions that a histogram would convey, without the need for constructing the drawing. It also preserves the exact data values.

Bar Graphs, Circle Graphs, and Line Graphs

A frequency distribution of non-numerical observations can be presented in the form of a **bar graph,** which is similar to a histogram except that the rectangles (bars) usually are not touching one another and sometimes are arranged horizontally rather than vertically. The bar graph of **Figure 5** shows the frequencies of occurrence of the vowels A, E, I, O, and U in this paragraph.

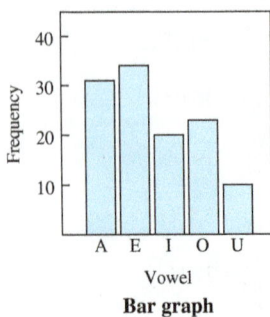

Vowel

Bar graph

Figure 5

EXAMPLE 4 Interpreting a Bar Graph

Refer to the bar graph in **Figure 6** to answer these questions.

(a) What was the total annual antibiotic usage?

(b) About what percentage was used in livestock production?

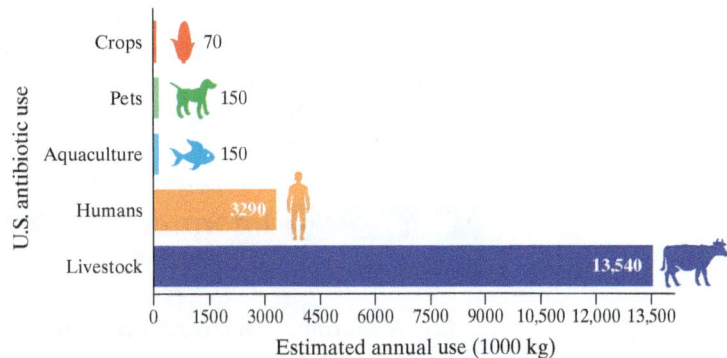

Source: A. Hollis and Z. Ahmed, *New England Journal of Medicine,* December 26, 2013.

U.S. antibiotic use (2013)

Figure 6

Solution

(a) $70{,}000 + 150{,}000 + 150{,}000 + 3{,}290{,}000 + 13{,}540{,}000 = 17{,}200{,}000$

The total antibiotic usage was 17,200,000 kilograms.

(b) $\dfrac{13{,}540{,}000}{17{,}200{,}000} \approx 0.787$ Livestock production was about 79% of total usage. ■

Table 6

Portfolio Allocations

Investment Category	Percent of Total
Stocks	55%
Bonds	20
Cash	15
Metals	10

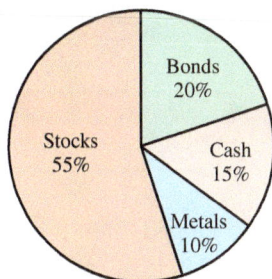

Investment allocations

Figure 7

A graphical alternative to the bar graph is the **circle graph,** or **pie chart,** which uses a circle to represent the total of all the categories and divides the circle into sectors, or wedges (like pieces of pie), whose sizes show the relative magnitudes of the categories. The angle around the entire circle measures 360°. For example, a category representing 20% of the whole should correspond to a sector whose central angle is 20% of 360°, that is,

$$0.20(360°) = 72°.$$

A circle graph shows, at a glance, the relative magnitudes of various categories.

EXAMPLE 5 Constructing a Circle Graph

Filipe's investment advisor recommends the portfolio allocations shown in **Table 6.** Present this information in a circle graph.

Solution

The central angles for the sectors are as follows.

Stocks: $0.55(360°) = 198°$

Bonds: $0.20(360°) = 72°$

Cash: $0.15(360°) = 54°$

Metals: $0.10(360°) = 36°$

Draw a circle and mark off the angles with a protractor. (See **Figure 7.**) ■

To demonstrate how a quantity *changes*, say with respect to time, use a **line graph.** Connect a series of line segments that rise and fall with time, according to the magnitude of the quantity being illustrated. To compare the patterns of change for two or more quantities, we can even plot multiple line graphs together in a "comparison line graph." (A line graph looks somewhat like a frequency polygon, but the quantities graphed are not necessarily frequencies.)

In December 2013, the U.S. Food and Drug Administration initiated a voluntary program phasing out antibiotics used to make livestock grow bigger (see **Figure 6** in **Example 4**) because it was thought that such usage was a cause of bacterial resistance to drugs in humans.

EXAMPLE 6 Interpreting a Comparison Line Graph

Answer the following questions by referring to **Figure 8.**

(a) What time frame is reflected in the figure?

(b) In which years was resistance to ciprofloxacin greater than resistance to ceftriaxone?

(c) Over this time span, what was the highest resistance to ceftriaxone achieved, and when did it occur?

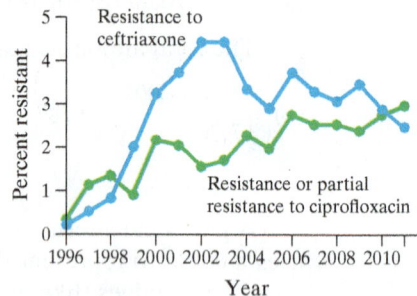

Source: U.S. CDC.

Antibiotic resistance of Salmonella infections in humans

Figure 8

Solution

(a) 1996 through 2011

(b) 1997, 1998, and 2011

(c) About 4.5%, in 2002 and 2003

FOR FURTHER THOUGHT

Expected and Observed Frequencies

When fair coins are tossed, the results on particular tosses cannot be reliably predicted. As more and more coins are tossed, however, the proportions of heads and tails become more predictable. This is a consequence of the "law of large numbers."

For example, if five coins are tossed, then the resulting number of heads, denoted x, is a "random variable," whose possible values are

0, 1, 2, 3, 4, and 5.

If the five coins are tossed repeatedly, say 64 separate times, then the binomial probability formula can be used to get **expected frequencies** (or **theoretical frequencies**), as shown in the table below. The first two columns of the table comprise the **expected frequency distribution** for 64 tosses of five fair coins.

Number of Heads x	Expected Frequency e	Observed Frequency o
0	2	
1	10	
2	20	
3	20	
4	10	
5	2	

In an actual experiment, we could obtain **observed frequencies** (or **empirical frequencies**), which would most likely differ somewhat from the expected frequencies. But 64 repetitions of the experiment should be enough to provide fair consistency between expected and observed values.

For Group or Individual Investigation

Toss five coins a total of 64 times, keeping a record of the results.

1. Enter your experimental results in the third column of the accompanying table, producing an **observed frequency distribution**.
2. Compare the second and third column entries.
3. Construct two histograms, one from the expected frequency distribution and one from your observed frequency distribution.
4. Compare the two histograms.

12.1   EXERCISES

In Exercises 1 and 2, use the given data to do the following:

(a) *Construct frequency and relative frequency distributions, in a table similar to* **Table 2.**

(b) *Construct a histogram.*

(c) *Construct a frequency polygon.*

1. *Preparation for Summer*   According to *Newsmax* (May 2010, page 76), the following are five popular "maintenance" activities performed as summer approaches.

   1. Prep the car for road trips.
   2. Clean up the house or apartment.
   3. Groom the garden.
   4. Exercise the body.
   5. Organize the wardrobe.

   The following data are the responses of 30 people who were asked, on June 1, how many of the five activities they had accomplished.

   ```
   1 1 3 1 0 3 0 0 2 1
   2 2 0 0 5 3 4 0 1 0
   4 2 0 2 0 1 0 1 2 3
   ```

2. *Favorite Numbers*   The following data are the "favorite numbers" (single digits, 1 through 9) of the 28 pupils of a second grade class.

   ```
   4 7 2 7 6 3 1
   7 2 9 8 5 6 1
   4 3 8 9 5 4 5
   9 2 6 5 2 5 7
   ```

In Exercises 3–6, use the given data to do the following:

(a) *Construct grouped frequency and relative frequency distributions, in a table similar to* **Table 3.** *(Follow the suggested guidelines for class limits and class width.)*

(b) *Construct a histogram.*

(c) *Construct a frequency polygon.*

3. *Exam Scores*   The scores of the 48 members of a sociology lecture class on a 50-point exam were as follows.

   ```
   40 43 44 33 41 40 33 39
   42 36 43 42 45 42 23 41
   43 40 37 47 44 48 31 46
   38 31 46 38 39 42 47 40
   47 41 48 42 45 29 32 38
   45 28 49 46 47 36 48 46
   ```

   Use six classes with a uniform class width of 5 points, and use a lower limit of 21 points for the first class.

4. *Charge Card Account Balances*   The following raw data represent the monthly account balances (to the nearest dollar) for a sample of 50 brand-new charge card users.

   ```
    78 175   46 138   79 118 90 163   88 107
   126 154   85  60   42  54 62 128  114  73
    67 119  116 145  129 130 81 105   96  71
   100 145  117  60  125 130 94  88  136 112
    85 165  118  84   74  62 81 110  108  71
   ```

   Use seven classes with a uniform width of 20 dollars, where the lower limit of the first class is 40 dollars.

5. **Daily High Temperatures** The following data represent the daily high temperatures (in degrees Fahrenheit) for the month of June in a town in the Sacramento Valley of California.

79	84	88	96	102	104	99	97	92	94
85	92	100	99	101	104	97	108	106	106
90	82	74	72	83	107	98	102	97	94

Use eight classes with a uniform width of 5 degrees, where the lower limit of the first class is 70 degrees.

6. **IQ Scores of College Freshmen** The following data represent IQ scores of a group of 50 college freshmen.

113	109	118	92	130	112	114	117	122	115
127	107	108	113	124	112	111	106	116	118
121	107	118	118	110	124	115	103	100	114
104	124	116	123	104	135	121	126	116	111
96	134	98	129	102	103	107	113	117	112

Use nine classes with a uniform width of 5, where the lower limit of the first class is 91.

In each of Exercises 7–10, construct a stem-and-leaf display for the given data. In each case, treat the ones digits as the leaves. For any single-digit data, use a stem of 0.

7. **Games Won in the National Basketball Association** Approaching midseason, the teams in the National Basketball Association had won the following numbers of games.

27	20	29	11	26	11	12	7	26	18
22	19	14	13	22	9	25	11	10	15
38	10	22	23	31	8	24	15	24	15

8. **Accumulated College Units** The students in a biology class were asked how many college units they had accumulated to date. Their responses are shown below.

12	4	13	12	21	22	15	17	33	24
32	42	26	11	53	62	42	25	13	8
54	18	21	14	19	17	38	17	20	10

9. **Distances to School** The following data are the daily round-trip distances to school (in miles) for 30 randomly chosen students attending a community college in California.

16	30	10	11	18	26	34	18	8	12
21	14	5	22	4	25	9	10	6	21
12	18	9	16	44	23	4	13	36	8

10. **Yards Gained in the National Football League** The following data represent net yards gained per game by National Football League running backs who played during a given week of the season.

25	19	36	73	37	88	67	33	54	123	79
19	39	45	22	58	7	73	30	43	24	36
65	43	33	55	40	29	112	60	94	86	62
52	29	18	25	41	3	49	102	16	32	46

Federal Revenue and Spending *The graph shows U.S. government receipts and outlays (both on-budget and off-budget) for 2001–2012. Refer to the graph for Exercises 11–15.*

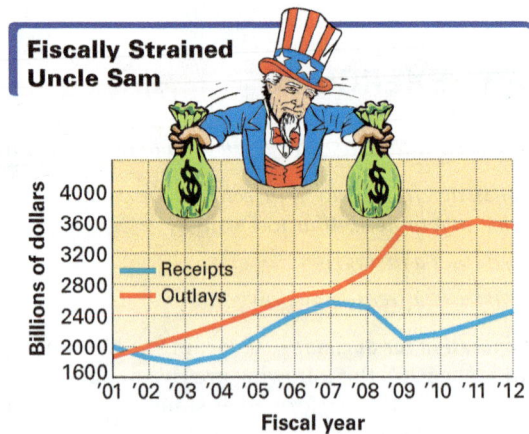

Source: Department of the Treasury, Office of Management and Budget.

11. For the period 2001–2012, list all years when receipts exceeded outlays.

12. Identify each of the following amounts, and indicate when it occurred.

    (a) the greatest one-year drop in receipts

    (b) the greatest one-year rise in outlays

13. In what years did receipts appear to climb faster than outlays?

14. About what was the greatest deficit (outlays minus receipts), and in what year did it occur?

15. Plot a point for each year and draw a line graph showing the federal surplus (+) or deficit (−) over the years 2001–2012.

Reading Bar Graphs of Economic Indicators *The bar graphs here show trends in several economic indicators over the period 2007–2012.*

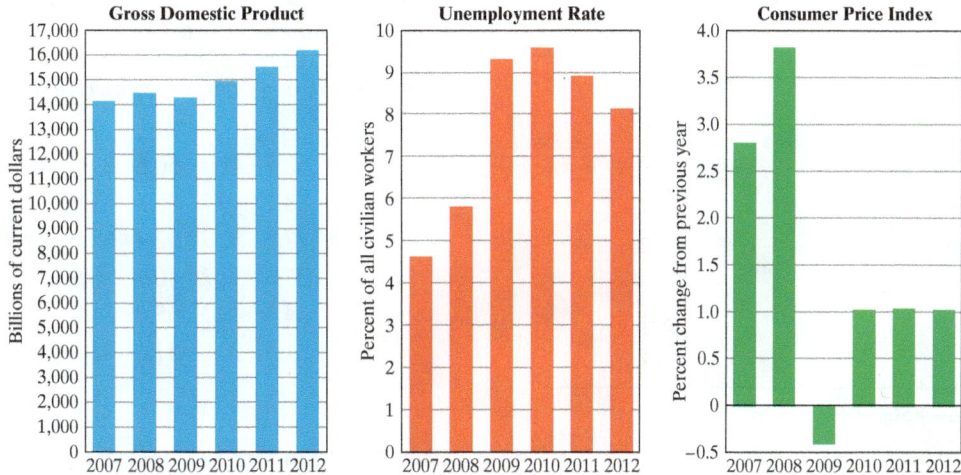

Selected Key U.S. Economic Indicators

Sources: U.S. Department of Commerce, Bureau of Economic Analysis.
U.S. Department of Labor, Bureau of Labor Statistics.

Refer to the above graphs for Exercises 16–20.

16. About what was the gross domestic product in 2012?

17. Over the six-year period, about what was the greatest change in consumer price index, and when did it occur?

18. What was the greatest year-to-year change in the unemployment rate, and when did it occur?

19. Observing these graphs, what would you say was the most unusual occurrence during the six years represented?

20. Explain why the gross domestic product would generally increase when the unemployment rate decreases.

Reading a Circle Graph of Job-Training Sources *The circle graph in the next column shows how a surveyed group of American workers were trained for their jobs. Refer to it for Exercises 21 and 22.*

21. What is the greatest single training source? To the nearest degree, what is the central angle of that category's sector?

22. What percentage of job training was provided directly from employers, either formally or informally?

23. Sources of U.S. Budget Receipts The table at the right gives the major sources of government income for 2012. Use the information to draw a circle graph.

Graph for Exercises 21 and 22

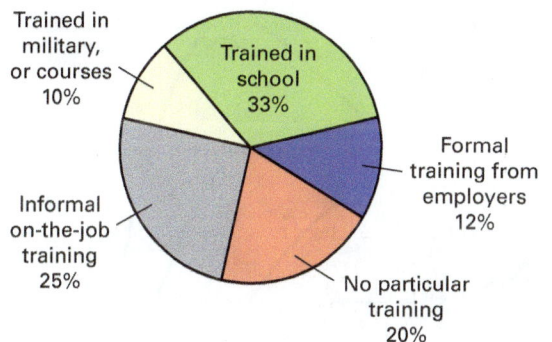

Job Training Sources

Trained in military, or courses 10%
Trained in school 33%
Formal training from employers 12%
No particular training 20%
Informal on-the-job training 25%

Table for Exercise 23

Source of Receipts	Approximate Percentage of Total
Individual income taxes	46%
Corporate income taxes	10
Social insurance and retirement receipts	35
Excise taxes	3
Other receipts	6

Source: Congressional Budget Office; *Budget of the U.S. Government,* Office of Management and Budget, Executive Office of the President.

24. *Correspondence between Education and Earnings* Data for a recent year showed that the average annual earnings of American workers corresponded to educational level as shown in the table below. Draw a bar graph that shows this information.

Educational Level	Median Weekly Earnings
Less than a high school diploma	$453
High school graduate	618
Some college, no degree	699
Associate degree	757
Bachelor's degree	1012
Master's degree	1233
Professional degree	1531
Doctoral degree	1561

Source: Bureau of Labor Statistics.

Net Worth of Retirement Savings Saura, wishing to retire at age 60, is studying the comparison line graph here, which shows (under certain assumptions) how the net worth of her retirement savings (initially $500,000 at age 60) will change as she gets older and as she withdraws living expenses from savings. Refer to the graph for Exercises 25–28.

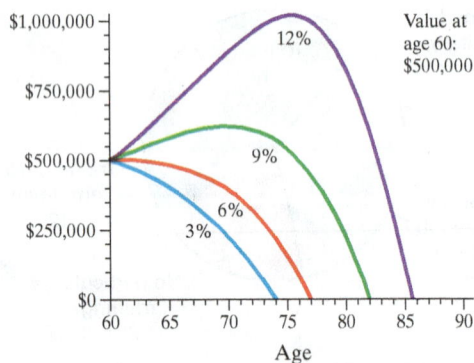

25. Assuming Saura can maintain an average annual return of 9%, how old will she be when her money runs out?

26. If she could earn an average of 12% annually, what maximum net worth would Saura achieve? At about what age would the maximum occur?

27. Suppose Saura reaches age 70 in good health, and the average annual return has proved to be 6%.

    (a) About how much longer can she expect her money to last?

    (b) What options might she consider in order to extend that time?

28. At age 71, about how many times more will Saura's net worth be if she averages a 12% return than if she averages a 3% return?

Sample Masses in a Geology Laboratory Stem-and-leaf displays can be modified in various ways in order to obtain a reasonable number of stems. The following data, representing the measured masses (in grams) of thirty mineral samples in a geology lab, are shown in a *double-stem* display in **Table 7.**

60.7	41.4	50.6	39.5	46.4
58.1	49.7	38.8	61.6	55.2
47.3	52.7	62.4	59.0	44.9
35.6	36.2	40.6	56.9	42.6
34.7	48.3	55.8	54.2	33.8
51.3	50.1	57.0	42.8	43.7

Table 7 Stem-and-Leaf Display for Mineral Sample Masses

(30–34)	3	4.7	3.8			
(35–39)	3	9.5	8.8	5.6	6.2	
(40–44)	4	1.4	4.9	0.6	2.6	2.8	3.7
(45–49)	4	6.4	9.7	7.3	8.3	
(50–54)	5	0.6	2.7	4.2	1.3	0.1
(55–59)	5	8.1	5.2	9.0	6.9	5.8	7.0
(60–64)	6	0.7	1.6	2.4		

29. Describe how the stem-and-leaf display of **Table 7** was constructed.

30. Explain why **Table 7** is called a "double-stem" display.

31. In general, how many stems (total) are appropriate for a stem-and-leaf display? Explain your reasoning.

32. *Record Temperatures* According to the National Climatic Data Center, U.S. Department of Commerce, the highest temperatures (in degrees Fahrenheit) ever recorded in the 50 states (as of May 22, 2013) were as follows.

112	100	128	120	134	114	106	110	109	112
100	118	117	116	118	121	114	114	105	109
107	112	115	115	118	117	118	125	106	110
122	108	110	121	113	120	119	111	104	113
120	113	120	117	107	110	118	112	114	115

Present these data in a double-stem display.

33. *Letter Occurrence Frequencies in the English Language* The table below shows commonly accepted percentages of occurrence for the various letters in English language usage. (Code breakers have carefully analyzed these percentages as an aid in deciphering secret codes.)

For example, notice that E is the most commonly occurring letter, followed by T, A, O, N, and so on. The letters Q and Z occur least often. Referring to **Figure 5** in the text, would you say that the relative frequencies of occurrence of the vowels in the associated paragraph were typical or unusual? Explain your reasoning.

Letter	Percent	Letter	Percent
E	13	L	$3\frac{1}{2}$
T	9	C, M, U	3
A, O	8	F, P, Y	2
N	7	W, G, B	$1\frac{1}{2}$
I, R	$6\frac{1}{2}$	V	1
S, H	6	K, X, J	$\frac{1}{2}$
D	4	Q, Z	$\frac{1}{5}$

Frequencies and Probabilities of Letter Occurrence The percentages shown in **Exercise 33** are based on a very large sampling of English language text. Since they are based on experiment, they are "empirical" rather than "theoretical." By converting each percent in that table to a decimal fraction, you can produce an **empirical probability distribution**.

For example, if a single letter is randomly selected from a randomly selected passage of text, the probability that it will be an E is 0.13. The probability that a randomly selected letter will be a vowel (A, E, I, O, or U) is

$$(0.08 + 0.13 + 0.065 + 0.08 + 0.03) = 0.385.$$

34. Rewrite the distribution shown in **Exercise 33** as an empirical probability distribution. Give values to three decimal places. Note that the 26 probabilities in this distribution—one for each letter of the alphabet—should add up to 1 (except for, perhaps, a slight round-off error).

35. **(a)** From your distribution in **Exercise 34,** construct an empirical probability distribution just for the vowels A, E, I, O, and U. (*Hint:* Divide each vowel's probability, from **Exercise 34,** by 0.385 to obtain a distribution whose five values add up to 1.) Give values to three decimal places.

(b) Construct an appropriately labeled bar graph from your distribution in part (a).

36. Based on the occurrences of vowels in the paragraph represented by **Figure 5,** construct a probability distribution for the vowels. Give probabilities to three decimal places. The frequencies are

A–31,   E–34,   I–20,   O–23,   U–10.

37. Is the probability distribution in **Exercise 36** theoretical or empirical? Is it different from the distribution in **Exercise 35**? Which one is more accurate? Explain your reasoning.

38. *Frequencies and Probabilities of Study Times* Convert the grouped frequency distribution in **Table 3** to an empirical probability distribution, using the same classes and giving probability values to three decimal places.

39. *Probabilities of Study Times* Recall that the distribution in **Exercise 38** was based on weekly study times for a sample of 40 students. Suppose one of those students was selected randomly. Using your distribution, find the probability that the study time in the past week for the student selected would have been in each of the following ranges.

(a) 20–29 hours   **(b)** 50–69 hours

(c) fewer than 40 hours   **(d)** at least 60 hours

Favorite Sports among Recreation Students The 40 members of a recreation class were asked to name their favorite sports. The table shows the numbers who responded in various ways.

Sport	Number of Class Members
Sailing	7
Archery	4
Snowboarding	10
Bicycling	8
Rock climbing	8
Rafting	3

Use this information in Exercises 40–42.

40. If a member of this class is selected at random, what is the probability that the favorite sport of the person selected is sailing?

41. **(a)** Based on the data in the table, construct a probability distribution, giving probabilities to three decimal places.

(b) Do you think the distribution in part (a) is theoretical, or empirical?

(c) Explain your answer to part (b).

42. Explain the distinction between a relative frequency distribution and a probability distribution.

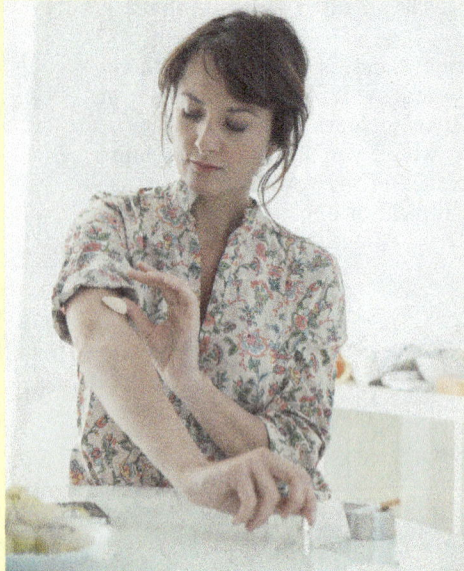

Imagine that you are the therapist mentioned in the chapter opener and that, again, you want to compare the results you have achieved in your own practice with the broader results. In this case, you are interested in tobacco dependency and remission. The study referred to earlier produced the following data for smokers, where

x = years since dependency began, and

y = cumulative probability of quitting (for a year).

x	y	x	y	x	y
3.3	0.08	22.9	0.39	45.8	0.63
7.9	0.17	28.0	0.45	50.9	0.69
14.0	0.26	33.2	0.52	56.1	0.78
18.2	0.32	37.9	0.57	62.6	0.84

1. To clarify the meaning of the data, plot the twelve points and sketch a smooth curve that approximately fits the points.

2. Use the graph you have prepared to complete the following statement:

   It takes about _____ years for 50% of smokers to quit the habit.

3. Compare with the graph on **page 646,** and complete the same statement for drinkers:

   It takes about _____ years before 50% of drinkers quit the habit.

   (Check your answers with those given below.)

4. Give some possible reasons for the discrepancy between the results for smokers and those for drinkers, as reflected in your answers to Questions 2 and 3 above.

Answers: 2. 30 3. 17 4. Answers will vary.

OBJECTIVES

1 Find and interpret means.
2 Find and interpret medians.
3 Find and interpret modes.
4 Infer central tendency measures from stem-and-leaf displays.
5 Discern symmetry in data sets.
6 Compare measures of central tendency.

A city issued the following numbers of building permits over a six-month period.

$$305, \quad 285, \quad 240, \quad 376, \quad 198, \quad 264$$

A single number that is in some sense representative of this whole set of numbers, a kind of "middle" value, would be a **measure of central tendency.**

Mean

The most common measure of central tendency is the **mean** (or **arithmetic mean**). The mean of a sample is denoted $\bar{x}$ (read "x bar"), while the mean of a complete population is denoted μ (the lowercase Greek letter *mu*). For our purposes here, data sets are considered to be samples, so we use $\bar{x}$.

The mean of a set of data items is found by adding up all the items and then dividing the sum by the number of items. (The mean is what most people associate with the word "average.") Since adding up, or summing, a list of items is a common procedure in statistics, we use the symbol for "summation," Σ (the capital Greek letter *sigma*). Therefore, the sum of n items—say $x_1, x_2, \ldots, x_n$—can be denoted

$$\Sigma x = x_1 + x_2 + \cdots + x_n.$$

Many calculators find the **mean** (as well as other statistical measures) automatically when a set of data items are entered. To recognize these calculators, look for a key marked $\boxed{\bar{x}}$, or perhaps $\boxed{\mu}$, or look in a menu such as "LIST" for a listing of mathematical measures.

MEAN

The **mean** of n data items $x_1, x_2, \ldots, x_n$, is calculated as follows.

$$\bar{x} = \frac{\Sigma x}{n}$$

We find the central tendency of the monthly building permit figures as follows.

mean({305,285,240,376,198, ▸
  ◂264})
                     278

A calculator can find the mean of items in a list. This screen supports the text discussion of monthly building permits.

$$\text{Mean} = \bar{x} = \frac{\Sigma x}{n}$$

$$= \frac{305 + 285 + 240 + 376 + 198 + 264}{6} \quad \text{Add the monthly numbers.}$$
Divide by the number of months.

$$= \frac{1668}{6}, \quad \text{or} \quad 278$$

The mean value (the "average monthly permits issued") is 278.

EXAMPLE 1 Finding the Mean of a List of Sales Figures

Last year's annual sales for eight different online businesses were as follows.

$374,910 $321,872 $242,943 $351,147
$382,740 $412,111 $334,089 $262,900

Find the mean annual sales for the eight businesses.

Solution

Scrolled Expression

mean({374910,321872,242 ▸
  ◂943,351147,382740,412111, ▸
  ◂334089,262900})
                     335339

This screen supports the result in Example 1.

$$\bar{x} = \frac{\Sigma x}{n} = \frac{2{,}682{,}712}{8} = 335{,}339 \quad \text{Add the sales.}$$
Divide by the number of businesses.

The mean annual sales amount is $335,339.

The following table shows the units and grades earned by one student last term.

Course	Grade	Units
Mathematics	A	3
History	C	3
Chemistry	B	5
Art	B	2
PE	A	1

In one common method of defining **grade-point average,** an A grade is assigned 4 points, with 3 points for B, 2 points for C, and 1 for D. Compute grade-point average as follows.

Step 1 Multiply the number of units for a course and the number assigned to each grade.

Step 2 Add these products.

Step 3 Divide by the total number of units.

Course	Grade	Grade Points	Units	(Grade Points) · (Units)
Mathematics	A	4	3	12
History	C	2	3	6
Chemistry	B	3	5	15
Art	B	3	2	6
PE	A	4	1	4
			Totals: 14	43

$$\text{Grade-point average} = \frac{43}{14} = 3.07 \text{ (rounded)}$$

The calculation of a grade-point average is an example of a **weighted mean,** because the grade points for each course grade must be weighted according to the number of units for the course. (For example, five units of A is better than two units of A.) The number of units is called the **weighting factor.**

WEIGHTED MEAN

The **weighted mean** of n numbers, $x_1, x_2, \ldots, x_n$ that are weighted by the respective factors $f_1, f_2, \ldots, f_n$ is calculated as follows.

$$\overline{w} = \frac{\Sigma(x \cdot f)}{\Sigma f}$$

In words, the weighted mean of a group of (weighted) items is the sum of all products of items times weighting factors, divided by the sum of all weighting factors.

The weighted mean formula is commonly used to find the mean for a frequency distribution. In this case, the weighting factors are the frequencies.

Salary x	Number of Employees f
$12,000	8
$16,000	11
$18,500	14
$21,000	9
$34,000	2
$50,000	1

Scrolled Expression

mean({12000,16000,18500, ▸
◂21000,34000,50000},{8,11, ▸
◂14,9,2,1})
                18622.22222

In this screen supporting **Example 2**, the first list contains the salaries, and the second list contains their frequencies.

EXAMPLE 2 **Finding the Mean of a Frequency Distribution of Salaries**

Find the mean salary for a small company that pays annual salaries to its employees as shown in the frequency distribution in the margin.

Solution

According to the weighted mean formula, we can set up the work as follows.

Salary x	Number of Employees f	Salary · Number $x \cdot f$
$12,000	8	$96,000
$16,000	11	$176,000
$18,500	14	$259,000
$21,000	9	$189,000
$34,000	2	$68,000
$50,000	1	$50,000
Totals:	45	$838,000

$$\text{Mean salary} = \frac{\Sigma(x \cdot f)}{\Sigma f} = \frac{\$838,000}{45} = \$18,622 \quad (\text{rounded})$$ ∎

For some data sets the mean can be a misleading indicator of average. Consider a small business that employs five workers at the following annual salaries.

$$\$16,500, \quad \$16,950, \quad \$17,800, \quad \$19,750, \quad \$20,000$$

The employees, struggling to survive on these salaries, approach the business owner, Aiguo, with the following computation.

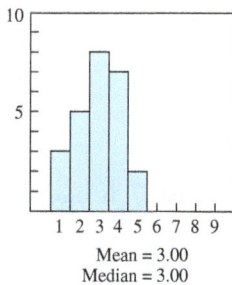
Mean = 3.00
Median = 3.00

$$\bar{x} = \frac{\$16,500 + \$16,950 + \$17,800 + \$19,750 + \$20,000}{5}$$

$$= \frac{\$91,000}{5}, \quad \text{or} \quad \$18,200 \quad \text{Mean salary (employees)}$$

Aiguo responds with his own calculations, for *all* workers (including his own salary of $188,000).

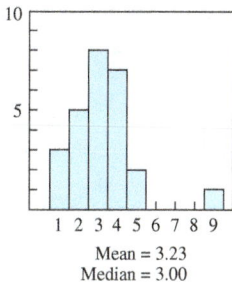
Mean = 3.23
Median = 3.00

The introduction of a single "outlier" above increased the mean by 8 percent but left the median unaffected.

Outliers should usually be considered as *possible* errors in the data.

$$\bar{x} = \frac{\$16,500 + \$16,950 + \$17,800 + \$19,750 + \$20,000 + \$188,000}{6}$$

$$= \frac{\$279,000}{6}, \quad \text{or} \quad \$46,500 \quad \text{Mean salary (including Aiguo's)}$$

This mean is quite different, and it looks much more respectable.

The employees, of course, would argue that when Aiguo included his own salary in the calculation, it caused the mean to be a misleading indicator of average. This was so because Aiguo's salary is not typical. It lies a good distance away from the general grouping of the items (salaries). An extreme value like this is referred to as an **outlier**. Since a single outlier can have a significant effect on the value of the mean, we say that the mean is "highly sensitive to extreme values."

Median

Another measure of central tendency, which is not so sensitive to extreme values, is the **median.** This measure divides a group of numbers into two parts, with half the numbers below the median and half above it.

MEDIAN

Find the **median** of a group of items as follows.

Step 1 Rank the items (that is, arrange them in numerical order from least to greatest).

Step 2 If the number of items is *odd*, the median is the middle item in the list.

Step 3 If the number of items is *even*, the median is the mean of the two middle items.

For Aiguo's business, all salaries (including his own), arranged in numerical order, are shown here.

$$\$16{,}500, \quad \$16{,}950, \quad \$17{,}800, \quad \$19{,}750, \quad \$20{,}000, \quad \$188{,}000$$

Thus, $\quad\text{median} = \dfrac{\$17{,}800 + \$19{,}750}{2} = \dfrac{\$37{,}550}{2} = \$18{,}775.$

This figure is a representative average, based on all six salaries, that the employees would probably agree is reasonable.

EXAMPLE 3 **Finding Medians of Lists of Numbers**

Find the median of each list of numbers.

(a) 6, 7, 12, 13, 18, 23, 24   **(b)** 17, 15, 9, 13, 21, 32, 41, 7, 12

(c) 147, 159, 132, 181, 174, 253

Solution

(a) This list is already ranked. The number of values in the list, 7, is odd, so the median is the middle value, or 13.

(b) First rank the items.

$$7, 9, 12, 13, \textbf{15}, 17, 21, 32, 41$$
$$\uparrow$$
$$\text{Median}$$

The middle number can now be picked out. The median is 15.

(c) First rank the items.

$$132, 147, \textbf{159}, \textbf{174}, 181, 253$$

Since the list contains an even number of items, namely 6, there is no single middle item. Find the median by taking the mean of the two middle items, 159 and 174.

$$\frac{159 + 174}{2} = \frac{333}{2} = 166.5 \;\leftarrow\; \text{Median}$$

Scrolled Expression

median({6,7,12,13,18,23,24})
                13
median({147,159,132,181,174,253})
              166.5

The calculator can find the median of the entries in a list. This screen supports the results in **Examples 3(a) and (c).**

Locating the middle item (the median) of a frequency distribution is a bit more involved. First find the total number of items in the set by adding the frequencies ($n = \Sigma f$). Then the median is the item whose *position* is given by the following formula.

POSITION OF THE MEDIAN IN A FREQUENCY DISTRIBUTION

$$\text{Position of median} = \frac{n+1}{2} = \frac{\Sigma f + 1}{2}$$

This formula gives only the position, and not the actual value, of the median.

WHEN Will I Ever USE This ?

There is considerable controversy over how important professional intervention is, in general, for addiction remission. But it seems clear that for those who *do* need help with their follow-up recovery, a complete quarterly checkup is much more effective than just a brief interview in getting them back into treatment.

The graph below reflects the results of an extensive study, *Drug and Alcohol Dependence 2012*, reported by M. L. Dennis and C. K. Scott. Conveying information graphically, and interpreting information from graphs, are important in the work of psychologists and other social scientists.

Effects of follow-up on readmission to treatment

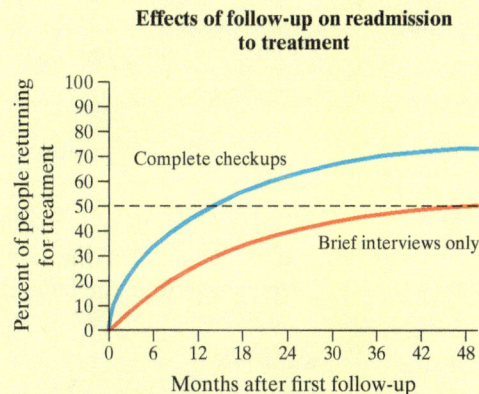

Data Source: M. L. Dennis and C. K. Scott, *Drug and Alcohol Dependence 2012.*

The median time to readmission is the number of months by which half the people who needed help with their follow-up recovery had returned for treatment. Complete the following statements.

1. The median time for those having checkups was about _____ months.

2. The median time for those not having checkups was about _____ months.

3. For those who had complete quarterly checkups, the median time to readmission was reduced by about _____ months.

EXAMPLE 4　**Finding Medians for Frequency Distributions**

Find the medians for the following distributions.

(a)

Value	1	2	3	4	5	6
Frequency	1	3	2	4	8	2

(b)

Value	2	4	6	8	10
Frequency	5	8	10	6	6

Solution

(a) Arrange the work as follows. Tabulate the values and frequencies, and the **cumulative frequencies,** which tell, for each different value, how many items have that value or a lesser value.

Value	Frequency	Cumulative Frequency	
1	1	1	1 item 1 or less
2	3	4	$1 + 3 = 4$ items 2 or less
3	2	6	$4 + 2 = 6$ items 3 or less
4	4	10	$6 + 4 = 10$ items 4 or less
5	8	18	$10 + 8 = 18$ items 5 or less
6	2	20	$18 + 2 = 20$ items 6 or less

　　　　Total: 20

Adding the frequencies shows that there are

20 items total.

(The final entry in the cumulative frequency column also shows the total number of items.)

$$\text{position of median} = \frac{20 + 1}{2} = \frac{21}{2} = 10.5$$

The median, then, is the average of the tenth and eleventh items. To find these items, make use of the cumulative frequencies. Since the value 4 has a cumulative frequency of 10, the tenth item is 4 and the eleventh item is 5, making the median

$$\frac{4 + 5}{2} = \frac{9}{2} = 4.5.$$

(b)

Value	Frequency	Cumulative Frequency
2	5	5
4	8	13
6	10	23
8	6	29
10	6	35

　　　　Total: 35

There are 35 items total.

$$\text{position of median} = \frac{35 + 1}{2} = \frac{36}{2} = 18$$

From the cumulative frequency column, the fourteenth through the twenty-third items are all 6s. This means the eighteenth item is a 6, so the median is 6. ■

median({1,2,3,4,5,6},{1,3,2, ▸
    ◂4,8,2})
　　　　　　　　　　4.5

median({2,4,6,8,10},{5,8,10, ▸
    ◂6,6})
　　　　　　　　　　6

These two screens support the results in
Example 4.

Mode

The third important measure of central tendency is the **mode.** Suppose ten students earned the following scores on a business law examination.

74, 81, 39, 74, 82, 80, 100, 92, 74, 85

Notice that more students earned the score 74 than any other score.

MODE

The **mode** of a data set is the value that occurs most often.

EXAMPLE 5 Finding Modes for Sets of Data

Find the mode for each set of data.

(a) 51, 32, 49, 49, 74, 81, 92   **(b)** 482, 485, 483, 485, 487, 487, 489

(c) 10,708, 11,519, 10,972, 17,546, 13,905, 12,182

(d)

Value	19	20	22	25	26	28
Frequency	1	3	8	7	4	2

Solution

(a) 51, 32, 49, 49, 74, 81, 92

The number 49 occurs more times than any other. Therefore, 49 is the mode. *The numbers do not need to be in numerical order when you look for the mode.*

(b) 482, 485, 483, 485, 487, 487, 489

Both 485 and 487 occur twice. This list is said to have *two* modes, or to be **bimodal.**

(c) No number here occurs more than once. This list has no mode.

(d)

Value	Frequency	
19	1	
20	3	
22	8	← Greatest frequency
25	7	
26	4	
28	2	

The frequency distribution shows that the most frequently occurring value (and, thus, the mode) is 22.

It is traditional to include the mode as a measure of *central tendency*, because many important kinds of data sets do have their most frequently occurring values "centrally" located. However, there is no reason why the mode cannot be one of the least values in the set or one of the greatest. In such a case, the mode really is not a good measure of "central tendency."

When the data items being studied are non-numerical, the mode may be the only usable measure of central tendency. For example, the bar graph of **Figure 5** in **Section 12.1** showed frequencies of occurrence of vowels in a sample paragraph. Since A, E, I, O, and U are not numbers, they cannot be added, nor can they be numerically ordered. Thus, neither their mean nor their median exists. The mode, however, does exist. As the bar graph shows, the mode is the letter E.

Sometimes a distribution is bimodal, as in **Example 5(b).** In a large distribution, this term is commonly applied even when the two modes do not have exactly the same frequency. Three or more different items sharing the highest frequency of occurrence rarely offer useful information. We say that such a distribution has *no* mode.

Central Tendency from Stem-and-Leaf Displays

As shown in **Section 12.1,** data are sometimes presented in a stem-and-leaf display in order to give a visual impression of their distribution. We can also calculate measures of central tendency from a stem-and-leaf display. The median and mode are more easily identified when the "leaves" are ranked on their "stems."

In **Table 8,** we have ranked the leaves of **Table 5** in **Section 12.1** (which showed the weekly study times from **Example 2** of that section).

Table 8   Stem-and-Leaf Display for Weekly Study Times, with Leaves Ranked

1	2	4	5	6	7	8					
2	2	3	4	4	5	6	9	9	9		
3	0	1	1	2	3	3	6	6	6	6	8	9
4	1	4	4	5	5	7					
5	2	5	5	8							
6	0	2									
7	2										

EXAMPLE 6   **Finding the Mean, Median, and Mode from a Stem-and-Leaf Display**

For the data in **Table 8,** find the following.

(a) the mean

(b) the median

(c) the mode

Solution

(a) A calculator with statistical capabilities will automatically compute the mean. Otherwise, add all items (reading from the stem-and-leaf display) and divide by $n = 40$.

$$\text{mean} = \frac{12 + 14 + 15 + \cdots + 60 + 62 + 72}{40}$$

$$= \frac{1414}{40}$$

$$= 35.35$$

(b) In this case, $n = 40$ (an even number), so the median is the average of the twentieth and twenty-first items, in order. Counting leaves, we see that these will be the fifth and sixth items on the stem 3.

$$\text{median} = \frac{33 + 33}{2} = 33$$

(c) By inspection, we see that 36 occurred four times and no other value occurred that often.

$$\text{mode} = 36$$

■

Symmetry in Data Sets

The most useful way to analyze a data set often depends on whether the distribution is **symmetric** or **nonsymmetric.** In a "symmetric" distribution, as we move out from the central point, the pattern of frequencies is the same (or nearly so) to the left and to the right. In a "nonsymmetric" distribution, the patterns to the left and right are significantly different.

Figure 9 shows several types of symmetric distributions, while **Figure 10** shows some nonsymmetric distributions. A nonsymmetric distribution with a tail extending out to the left, shaped like a J, is **skewed to the left.** If the tail extends out to the right, the distribution is **skewed to the right.** Notice that a bimodal distribution may be either symmetric or nonsymmetric.

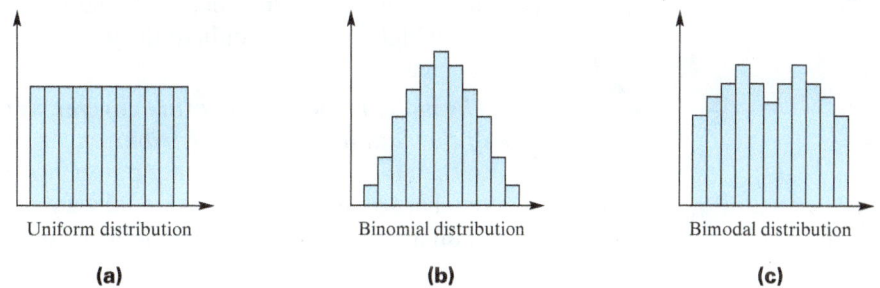

Uniform distribution	Binomial distribution	Bimodal distribution
(a)	(b)	(c)

Some symmetric distributions

Figure 9

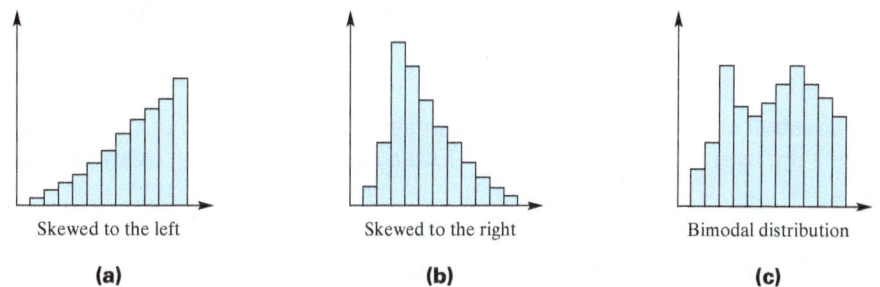

Skewed to the left	Skewed to the right	Bimodal distribution
(a)	(b)	(c)

Some nonsymmetric distributions

Figure 10

Summary

We conclude this section with a summary of the measures presented and a brief discussion of their relative advantages and disadvantages.

SUMMARY OF THE COMMON MEASURES OF CENTRAL TENDENCY

The **mean** of a set of numbers is found by adding all the values in the set and dividing by the number of values.

The **median** is a kind of "middle" number. To find the median, first rank the values. For an *odd* number of values, the median is the middle value in the list. For an *even* number of values, the median is the mean of the two middle values.

The **mode** is the value that occurs with the greatest frequency. Some sets of numbers have two most frequently occurring values and are **bimodal.** Other sets have no mode at all (if no value occurs more often than the others or if more than two values occur most often).

Some helpful points of comparison follow.

1. For distributions of numerical data, the mean and median will always exist, while the mode may not exist. On the other hand, for non-numerical data, it may be that none of the three measures exists, or that only the mode exists.

2. Because even a single change in the data may cause the mean to change significantly, while the median and mode may not be affected at all, *the mean is the most "sensitive" measure.*

3. In a symmetric distribution, the mean, median, and mode (if a single mode exists) will all be equal. In a nonsymmetric distribution, the mean is often unduly affected by relatively few extreme values and, therefore, may not be a good representative measure of central tendency. For example, distributions of salaries, family incomes, or home prices often include a few values that are much higher than the bulk of the items. In such cases, the median is a more useful measure.

4. *The mode is the only measure covered here that must always be equal to one of the data items of the distribution.* In fact, more of the data items are equal to the mode than to any other number. A fashion shop planning to stock only one hat size for next season would want to know the mode (the most common) of all hat sizes among its potential customers.

FOR FURTHER THOUGHT

Simpson's Paradox

In baseball statistics, a player's "batting average" gives the average number of hits per time at bat. For example, a player who has gotten 84 hits in 250 times at bat has a batting average of $\frac{84}{250} = .336$.

This "average" can be interpreted as the empirical probability of that player's getting a hit the next time at bat.

The following are actual comparisons of hits and at-bats for two major league players in the 2007, 2008, and 2009 seasons. The numbers illustrate a puzzling statistical occurrence known as **Simpson's paradox.** The example below, involving Mike Lowell and Jacoby Ellsbury, was provided by the alert Boston Red Sox fan Carol Merrigan.

For Group or Individual Investigation

1. Fill in the twelve blanks in the table, giving batting averages to three decimal places.

2. Which player had a better average in 2007?

3. Which player had a better average in 2008?

4. Which player had a better average in 2009?

5. Which player had a better average in 2007, 2008, and 2009 combined?

6. Did the results above surprise you? How can it be that one player's batting average leads another's for each of three years, and yet trails the other's for the combined years?

	Mike Lowell			Jacoby Ellsbury		
	Hits	**At-bats**	**Batting Average**	**Hits**	**At-bats**	**Batting Average**
2007	191	589	_____	41	116	_____
2008	115	419	_____	155	554	_____
2009	129	445	_____	188	624	_____
Combined (2007–2009)	_____	_____	_____	_____	_____	_____

12.2   EXERCISES

For each list of data, calculate **(a)** *the mean (to the nearest tenth),* **(b)** *the median, and* **(c)** *the mode or modes (if any).*

1. 6, 9, 12, 14, 21

2. 20, 27, 42, 45, 53, 62, 62, 64

3. 218, 230, 196, 224, 196, 233

4. 26, 31, 46, 31, 26, 29, 31

5. 3.1, 4.5, 6.2, 7.1, 4.5, 3.8, 6.2, 6.3

6. 14,322, 16,959, 17,337, 15,474

7. 128, 131, 136, 125, 139, 128, 125, 127

8. 8.97, 5.64, 2.31, 1.02, 4.35, 7.68

Answer each question.

9. *Gymnasts' Scores* An Olympic gymnast can earn an average score of 16.5 points for each of two vaults. Another can earn an average score of 15.5 points. The first gymnast falters, losing 2.375 points on her first vault and 1.575 points on her second vault. What average score must the second gymnast earn to win the gold? (*Mathematics Teacher* calendar problem, August 2, 2013)

10. *Quiz Grades* The average of 5 quiz grades is 10. When the lowest grade is dropped and the new average is calculated, it turns out to be 11. What was the score of the dropped grade? (*Mathematics Teacher* calendar problem, February 7, 2014)

Airline Fatalities in the United States *The table pertains to scheduled commercial carriers. Fatalities data include those on the ground except for the September 11, 2001 terrorist attacks. Use this information for Exercises 11–16.*

U.S. Airline Safety, 1999–2008

Year	Departures (in millions)	Fatal Accidents	Fatalities
1999	10.9	2	12
2000	11.1	2	89
2001	10.6	6	531
2002	10.3	0	0
2003	10.2	2	22
2004	10.8	1	13
2005	10.9	3	22
2006	10.6	2	50
2007	10.7	0	0
2008	10.6	0	0

Source: The World Almanac and Book of Facts 2010.

For each category in Exercises 11–16, find **(a)** *the mean (to the nearest tenth),* **(b)** *the median, and* **(c)** *the mode or modes (if any).*

11. departures

12. fatal accidents

13. fatalities

The year 2001 was clearly an anomaly. If the data for that year are reduced by 4 fatal accidents and 265 fatalities, which of the three measures change and what are their new values for each of the following?

14. **Exercise 12**   15. **Exercise 13**

16. Following 2001, in what year did airline departures start to increase again?

Leading U.S. Businesses *The table shows the top five U.S. businesses, ranked by 2012 revenue.*

Business	Revenue (in billions of dollars)	Profit (in billions of dollars)
Wal-Mart Stores	$469.2	$17.0
ExxonMobil	449.9	44.9
Chevron	233.9	26.2
Phillips 66	169.6	4.1
Berkshire Hathaway	162.5	14.8

Sources: The World Almanac and Book of Facts 2014; Fortune magazine.

Find each of the following quantities for these five businesses.

17. the mean revenue   18. the median profit

Measuring Elapsed Times *While doing an experiment, a physics student recorded the following sequence of elapsed times (in seconds) in a lab notebook.*

2.16, 22.2, 2.96, 2.20, 2.73, 2.28, 2.39

19. Find the mean.   20. Find the median.

The student from **Exercises 19 and 20,** *when reviewing the calculations later, decided that the entry 22.2 should have been recorded as 2.22 and made that change in the listing.*

21. Find the mean for the new list.

22. Find the median for the new list.

23. Which measure, the mean or the median, was affected more by correcting the error?

24. In general, which measure, mean or median, is affected less by the presence of an extreme value in the data?

Scores on Management Examinations *Thao earned the following scores on her six management exams last semester.*

$$79, \ 81, \ 44, \ 89, \ 79, \ 90$$

25. Find the mean, the median, and the mode for Thao's scores.

26. Which of the three averages probably is the best indicator of Thao's ability?

27. If Thao's instructor gives her a chance to replace her score of 44 by taking a "make-up" exam, what must she score on the make-up exam to get an overall average (mean) of 85?

For each of the following frequency distributions, find **(a)** *the mean (to the nearest tenth),* **(b)** *the median, and* **(c)** *the mode or modes (if any).*

28.

Value	Frequency
6	3
7	1
8	8
9	4

29.

Value	Frequency
615	13
540	7
605	9
579	14
586	7
600	5

30. *Average Employee Salary* A company has

15 employees with a salary of $21,500,

11 employees with a salary of $23,000,

17 employees with a salary of $25,800,

2 employees with a salary of $31,500,

4 employees with a salary of $38,900,

1 employee with a salary of $147,500.

Find the mean salary for the employees (to the nearest hundred dollars).

Grade-Point Averages *Find the grade-point average for each of the following students. Assume* $A = 4$, $B = 3$, $C = 2$, $D = 1$, *and* $F = 0$. *Round to the nearest hundredth.*

31.

Units	Grade
8	A
3	B
5	C

32.

Units	Grade
4	C
3	B
7	A
3	F

Federal Budget Totals *The table gives federal receipts and outlays for the years 2010–2014. (The 2013 and 2014 figures are estimates.) Use this information for Exercises 33 and 34.*

Fiscal Year	Receipts (in billions of dollars)	Outlays (in billions of dollars)
2010	$2162.7	$3457.1
2011	2303.5	3603.1
2012	2450.2	3537.1
2013	2712.0	3684.9
2014	3033.6	3777.8

Sources: The World Almanac and Book of Facts 2014; Budget of the U.S. Government, Fiscal Year 2014, Office of Management and Budget, Executive Office of the President.

Over the five-year period, find **(a)** *the mean and* **(b)** *the median for each of the following.*

33. receipts

34. outlays

World Cell Phone Use *In 2012, just the top six countries accounted for about 48% of cell phone subscriptions worldwide. Use the data in the table for Exercises 35 and 36.*

Country	Cell Phone Subscriptions (in thousands)
China	1100.0
India	864.7
United States	310.0
Indonesia	282.0
Russia	261.9
Brazil	248.3

Sources: The World Almanac and Book of Facts 2014; International Telecommunications Union.

35. Find the approximate mean number of cell phone subscriptions for these six countries in 2012.

36. The United States accounted for about 4.835% of worldwide subscriptions. About how many subscriptions were active in the world that year?

Crew, Passengers, and Hijackers on 9/11 Airliners The table shows, for each hijacked flight on September 11, 2001, the numbers of crew members, passengers, and hijackers (not included as passengers). For each quantity in Exercises 37–39, find **(a)** the mean, and **(b)** the median.

Flight	Crew	Passengers	Hijackers
American #11	11	76	5
United #175	9	51	5
American #77	6	53	5
United #93	7	33	4

Source: www.911research.wtc7.net

37. number of crew members per plane

38. number of passengers per plane

39. total number of persons per plane

Olympic Medal Standings The top ten medal-winning nations in the 2014 Winter Olympics at Sochi, Russia, are shown in the table. Use the given information for Exercises 40–43.

Medal Standings for the 2014 Winter Olympics

Place	Nation	Gold	Silver	Bronze	Total
1	Russian Federation	13	11	9	33
2	United States	9	7	12	28
3	Norway	11	5	10	26
4	Canada	10	10	5	25
5	Netherlands	8	7	9	24
6	Germany	8	6	5	19
7	Austria	4	8	5	17
8	France	4	4	7	15
9	Sweden	2	7	6	15
10	Switzerland	6	3	2	11

Source: www.nbcolympics.com

Calculate the following for all nations shown.

40. the mean number of gold medals

41. the median number of bronze medals

42. the mode, or modes, for the number of silver medals

43. each of the following for the total number of medals
  (a) mean
  (b) median
  (c) mode or modes

In Exercises 44 and 45, use the given stem-and-leaf display to identify **(a)** *the mean,* **(b)** *the median, and* **(c)** *the mode or modes (if any) for the data represented.*

44. *Online Sales* The display here represents prices (to the nearest dollar) charged by 23 different online sellers for a new car alternator. Give answers to the nearest dollar.

9	9
10	2 3
10	5 8 9
11	1 2 3 3
11	5 6 7 8 8
12	0 2 4
12	5 6 7 9
13	4

45. *Scores on a Biology Exam* The display here represents scores achieved on a 100-point biology exam by the 34 members of a class.

4	7
5	1 3 6
6	2 5 5 6 7 8 8
7	0 4 5 6 7 7 8 8 8 8 9
8	0 1 1 3 4 5 5
9	0 0 0 1 6

46. *Calculating a Missing Test Score* Katie's Business professor lost his computer memory, which contained her five test scores for the course. A summary of the scores (each of which was an integer from 0 to 100) indicates the following:

The mean was 88.

The median was 87.

The mode was 92.

(The data set was not bimodal.) What is the least possible number among the missing scores?

47. Explain what an "outlier" is and how it affects measures of central tendency.

48. *Scores on a Math Quiz* The following are scores earned by 15 college students on a 20-point math quiz.

0, 1, 3, 14, 14, 15, 16, 16, 17, 17, 18, 18, 18, 19, 20

(a) Calculate the mean, median, and mode values.

(b) Which measure in part (a) is most representative of the data?

49. *Consumer Preferences in Food Packaging* A food processing company that packages individual cups of instant soup wishes to find out the best number of cups to include in a package. A survey of 22 consumers revealed that five prefer a package of 1, five prefer a package of 2, three prefer a package of 3, six prefer a package of 4, and three prefer a package of 6.

(a) Calculate the mean, median, and mode values for preferred package size.

(b) Which measure in part (a) should the food processing company use?

(c) Explain your answer to part (b).

In Exercises 50–53, begin a list of the given numbers, in order, starting with the least one. Continue the list only until the median of the listed numbers is a multiple of 4. Stop at that point and find (a) *the number of numbers listed, and* (b) *the mean of the listed numbers (to two decimal places).*

50. counting numbers

51. prime numbers

52. Fibonacci numbers (see **page 208**)

53. triangular numbers (see **pages 13–14**)

Answer each question in Exercises 54–56.

54. Seven consecutive even whole numbers add up to 168. What is the result when their mean is subtracted from their median?

55. Bajir wants to include a fifth counting number, n, along with the numbers 2, 5, 8, and 9 so that the mean and median of the five numbers will be equal. How many choices does he have for the number n, and what are those choices?

56. If the mean, median, and mode are all equal for the set

$$\{70, 110, 80, 60, x\},$$

find the value of x.

For Exercises 57–59, refer to the grouped frequency distribution shown here.

Class Limits	Frequency f
21–25	5
26–30	3
31–35	8
36–40	12
41–45	21
46–50	38
51–55	35
56–60	20

57. Is it possible to identify, on the basis of the data shown in the table, any specific data items that occurred in this sample?

58. Is it possible to compute the actual mean for this sample?

59. Describe how you might approximate the mean for this sample. Justify your procedure.

60. *Average Employee Salaries* Refer to the salary data of **Example 2,** specifically the dollar amounts given in the salary column of the table. Explain what is wrong with simply calculating the mean salary by adding those six numbers and dividing the result by 6.

12.3 MEASURES OF DISPERSION

OBJECTIVES
1 Find the range of a data set.
2 Calculate the standard deviation of a data set.
3 Interpret measures of dispersion.
4 Calculate the coefficient of variation.

The mean is a good indicator of the central tendency of a set of data values, but it does not completely describe the data. Compare distribution A with distribution B in **Table 9** on the next page.

Both distributions have the same mean and the same median, but they are quite different. In the first, 7 is a fairly typical value, but in the second, most of the values differ considerably from 7. What is needed here is some measure of the **dispersion,** or *spread,* of the data.

Range

The **range** of a data set is a straightforward measure of dispersion.

RANGE

For any set of data, the **range** of the set is defined as follows.

Range = (greatest value in the set) − (least value in the set)

Table 9

	A	B
	5	1
	6	2
	7	7
	8	12
	9	13
Mean	7	7
Median	7	7

For a short list of data, calculation of the range is simple. For a more extensive list, it is more difficult to be sure you have accurately identified the greatest and least values.

EXAMPLE 1 **Finding and Comparing Range Values**

Find the ranges for distributions A and B in **Table 9,** and describe what they imply.

Solution

In distribution A, the greatest value is 9 and the least is 5.

$$\text{Range} = \text{greatest} - \text{least} = 9 - 5 = 4$$

Distribution B is handled similarly.

$$\text{Range} = 13 - 1 = 12$$

We can say that even though the two distributions have identical averages, distribution B exhibits three times more dispersion, or *spread,* than distribution A. ∎

Table 10

Quiz	Max	Molly
1	28	27
2	22	27
3	21	28
4	26	6
5	18	27
Mean	23	23
Median	22	27
Range	10	22

The range can be misleading if it is interpreted unwisely. For example, look at the points scored by Max and Molly on five different quizzes, as shown in **Table 10.** The ranges for the two students make it tempting to conclude that Max is more consistent than Molly. However, Molly is actually more consistent, with the exception of one very poor score. That score, 6, is an outlier which, if not actually recorded in error, must surely be due to some special circumstance. (Notice that the outlier does not seriously affect Molly's median score, which is more typical of her overall performance than is her mean score.)

Standard Deviation

One of the most useful measures of dispersion, the *standard deviation,* is based on *deviations from the mean* of the data values.

Once the data are entered, a calculator with statistical functions may actually show the range (among other things), or at least sort the data and identify the minimum and maximum items. (The associated symbols may be something like $\boxed{\text{MIN } \Sigma}$ and $\boxed{\text{MAX } \Sigma}$, or $\min X$ and $\max X$.) Given these two values, a simple subtraction produces the range.

EXAMPLE 2 **Finding Deviations from the Mean**

Find the deviations from the mean for all data values in the following sample.

$$32, 41, 47, 53, 57$$

Solution

Add these values and divide by the total number of values, 5. The mean is 46. To find the deviations from the mean, subtract 46 from each data value.

Data value	32	41	47	53	57
Deviation	−14	−5	1	7	11

$32 - 46 = -14$ $57 - 46 = 11$

To check your work, add the deviations. ***The sum of the deviations for a set of data is always 0.*** ∎

We cannot obtain a measure of dispersion by finding the mean of the deviations, because this number is always 0, since the positive deviations just cancel out the negative ones. To avoid this problem of positive and negative numbers canceling each other, we *square* each deviation.

The following chart shows the squares of the deviations for the data in **Example 2.**

Data value	32	41	47	53	57
Deviation	−14	−5	1	7	11
Square of deviation	196	25	1	49	121

$$(-14) \cdot (-14) = 196 \qquad\qquad 11 \cdot 11 = 121$$

An average of the squared deviations could now be found by dividing their sum by the number of data values n (5 in this case), which we would do if our data values composed a population. However, since we are considering the data to be a sample, we divide by $n - 1$ instead.*

The average that results is itself a measure of dispersion, called the **variance,** but a more common measure is obtained by taking the square root of the variance. This compensates, in a way, for squaring the deviations earlier and gives a kind of average of the deviations from the mean, which is called the sample **standard deviation.** It is denoted by the letter s. (The standard deviation of a population is denoted σ, the lowercase Greek letter *sigma*.)

Continuing our calculations from the chart above, we obtain

Most calculators find square roots, such as $\sqrt{98}$, to as many digits as you need using a key like $\boxed{\sqrt{x}}$. In this text, we normally give from two to four significant figures for such calculations.

$$s = \sqrt{\frac{196 + 25 + 1 + 49 + 121}{4}} = \sqrt{\frac{392}{4}} = \sqrt{98} \approx 9.90.$$

$$\underset{n-1}{\uparrow}$$

The algorithm (process) described above for finding the sample standard deviation can be summarized as follows.

stdDev({32,41,47,53,57})
            9.899494937
√98
            9.899494937

This screen supports the text discussion. Note that the standard deviation reported agrees with the approximation for $\sqrt{98}$.

CALCULATION OF STANDARD DEVIATION

Let a sample of n numbers $x_1, x_2, \ldots, x_n$ have mean $\bar{x}$. Then the **sample standard deviation, s,** of the numbers is calculated as follows.

$$s = \sqrt{\frac{\Sigma(x - \bar{x})^2}{n - 1}}$$

The individual steps involved in this calculation follow.

Step 1 Calculate $\bar{x}$, the mean of the numbers.

Step 2 Find the deviations from the mean.

Step 3 Square each deviation.

Step 4 Sum the squared deviations.

Step 5 Divide the sum in Step 4 by $n - 1$.

Step 6 Take the square root of the quotient in Step 5.

---

*Although the reasons cannot be explained at this level, dividing by $n - 1$ rather than n produces a sample measure that is more accurate for purposes of inference. In most cases, the results using the two divisors are only slightly different.

The preceding description helps show why standard deviation measures the amount of spread in a data set. For actual calculation purposes, we recommend the use of a scientific calculator, or a statistical calculator, that does all the detailed steps automatically. We illustrate both methods in **Example 3.**

EXAMPLE 3 Finding a Sample Standard Deviation

Find the standard deviation of the following sample by using **(a)** the step-by-step process, and **(b)** the statistical functions of a calculator.

$$7, 9, 18, 22, 27, 29, 32, 40$$

Solution

(a) Carry out the six steps summarized above.

Step 1 Find the mean of the values.

$$\frac{7 + 9 + 18 + 22 + 27 + 29 + 32 + 40}{8} = 23$$

Step 2 Find the deviations from the mean.

Data value	7	9	18	22	27	29	32	40
Deviation	−16	−14	−5	−1	4	6	9	17

Step 3 Square each deviation.

Squares of deviations: 256 196 25 1 16 36 81 289

Step 4 Sum the squared deviations.

$$256 + 196 + 25 + 1 + 16 + 36 + 81 + 289 = 900$$

Step 5 Divide by $n - 1 = 8 - 1 = 7$: $\quad \frac{900}{7} \approx 128.57$

Step 6 Take the square root: $\quad \sqrt{128.57} \approx 11.3$

(b) Enter the eight data values. (The key for entering data may look something like $\boxed{\Sigma +}$. Find out which key it is on your calculator.) Then press the key for standard deviation. It may look like one of these.

$$\boxed{\text{STDEV}} \quad \text{or} \quad \boxed{\text{SD}} \quad \text{or} \quad \boxed{S_{n-1}} \quad \text{or} \quad \boxed{\sigma_{n-1}}$$

If your calculator also has a key that looks like σ_n, it is probably for *population* standard deviation, which involves dividing by n rather than by $n - 1$, as mentioned earlier.

The result should again be 11.3. ← *If you mistakenly used the population standard deviation key, the result would be 10.6.*

For data given in the form of a frequency distribution, some calculators allow entry of both values and frequencies, or each value can be entered separately the number of times indicated by its frequency. Then press the standard deviation key.

The following example is included only to strengthen your understanding of frequency distributions and standard deviation, not as a practical algorithm for calculating.

L1   L2   L3   1
7   ------ ------
9
18
22
27
29
32
L1(1)=7

The sample in **Example 3** is stored in a list. (The last entry, 40, is not shown here.)

1–Var Stats
→ x̄=23
  Σx=184
  Σx²=5132
→ Sx=11.33893419
→ σx=10.60660172
  ↓n=8

The arrows point to the mean and the sample and population standard deviations. See **Example 3.**

Table 11

Value	Frequency
2	5
3	8
4	10
5	2

stdDev({2,3,4,5},{5,8,10,2})
　　　　　　　　　　.9073771726
√19.76/24
　　　　　　　　　　.9073771726

The screen supports the result in
Example 4.

Higher
average

Lower
dispersion

The more desirable basket depends on
your objective.

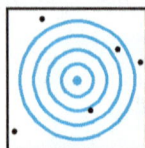

Good average,
poor
consistency

Good
consistency,
poor average

In this case, good consistency (lesser
dispersion) is more desirable than a good
average (central tendency).

EXAMPLE 4 Finding the Standard Deviation of a Frequency Distribution

Find the sample standard deviation for the frequency distribution shown in **Table 11**.

Solution

Complete the calculations as shown in **Table 12** below. To find the numbers in the "Deviation" column, first find the mean, and then subtract the mean from the numbers in the "Value" column.

Table 12

Value	Frequency	Value Times Frequency	Deviation	Squared Deviation	Squared Deviation Times Frequency
2	5	10	−1.36	1.8496	9.2480
3	8	24	−0.36	0.1296	1.0368
4	10	40	0.64	0.4096	4.0960
5	2	10	1.64	2.6896	5.3792
Sums	25	84			19.76

$$\bar{x} = \frac{84}{25} = 3.36 \qquad s = \sqrt{\frac{19.76}{24}} \approx \sqrt{0.8233} \approx 0.91$$

Central tendency and dispersion (or "spread tendency") are different and independent aspects of a set of data. Which one is more critical can depend on the specific situation.

For example, suppose tomatoes sell by the basket. Each basket costs the same, and each contains one dozen tomatoes. If you want the most fruit possible per dollar spent, you would look for the basket with the highest average weight per tomato (regardless of the dispersion of the weights). On the other hand, if the tomatoes are to be served on an hors d'oeuvre tray where "presentation" is important, you would look for a basket with uniform-sized tomatoes, that is a basket with the lowest weight dispersion (regardless of the average of the weights). See the illustration at the side.

Another situation involves target shooting (also illustrated at the side). The five hits on the top target are, *on average*, very close to the bull's-eye. If the square grid is 10 by 10 units and the concentric circles have radii 2, 4, 6, and 8, then the hits shown are at

$$(-9, -7), \ (-5, 8), \ (2, -4), \ (6, 4), \ \text{and} \ (9, 2),$$

so the average position is the point

$$\left(\frac{-9 - 5 + 2 + 6 + 9}{5}, \frac{-7 + 8 - 4 + 4 + 2}{5} \right) = \left(\frac{3}{5}, \frac{3}{5} \right),$$

which is about 0.85 unit from the bull's-eye. However, the large dispersion (spread) of the hits—the lack of consistency—implies that improvement will require much effort on the part of the shooter. On the other hand, the bottom target exhibits a poorer *average*. (The hits in this case have an approximate average position at $(-3, 6)$, which is about 6.71 units from the bull's-eye.) But in this case the poor average is balanced by a small dispersion—that is, good consistency. Improvement will require only a minor adjustment of the gun sights. In general, consistent errors can be corrected more easily than more dispersed errors.

Interpreting Measures of Dispersion

A main use of dispersion measures is to compare the amounts of spread in two (or more) data sets as we did with distributions A and B at the beginning of this section. A common technique in inferential statistics is to draw comparisons between populations by analyzing samples that come from those populations.

EXAMPLE 5 Comparing Populations Based on Samples

Two companies, *A* and *B*, sell 12-ounce jars of instant coffee. Five jars of each were randomly selected from markets, and the contents were carefully weighed, with the following results.

$$A:\quad 12.02,\quad 12.08,\quad 11.99,\quad 11.96,\quad 11.99$$

$$B:\quad 12.40,\quad 12.21,\quad 12.36,\quad 12.22,\quad 12.27$$

Find **(a)** which company provides more coffee in its jars, and **(b)** which company fills its jars more consistently.

Solution

The mean and standard deviation values for both samples are shown in **Table 13**.

(a) Since $\bar{x}_B$ is greater than $\bar{x}_A$, we *infer* that Company B most likely provides more coffee (greater mean) per jar.

(b) Since s_A is less than s_B, we *infer* that Company A seems more consistent (smaller standard deviation).

Table 13

Sample *A*	Sample *B*
$\bar{x}_A = 12.008$	$\bar{x}_B = 12.292$
$s_A = 0.0455$	$s_B = 0.0847$

The conclusions drawn in **Example 5** are tentative, because the samples were small. We could place more confidence in our inferences if we used larger samples, for then it would be more likely that the samples were accurate representations of their respective populations.

It is clear that a larger dispersion value means more "spread" than a smaller one. But it is difficult to say exactly what a single dispersion value says about a data set. *It is impossible* (though it would be nice) to make a general statement like: "Exactly half of the items of any distribution lie within one standard deviation of the mean of the distribution." Such a statement can be made only of specialized kinds of distributions. (See, for example, **Section 12.5** on the normal distribution.) There is, however, one useful result that does apply to all data sets, no matter what their distributions are like. This result is named for the Russian mathematician Pafnuty Lvovich Chebyshev.

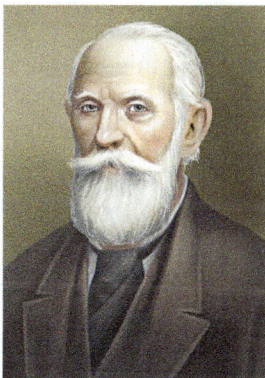

Pafnuty Lvovich Chebyshev (1821–1894) was a Russian mathematician known mainly for his work on the theory of prime numbers. Chebyshev and French mathematician and statistician **Jules Bienaymé** (1796–1878) independently developed an important inequality of probability now known as the Bienaymé–Chebyshev inequality.

CHEBYSHEV'S THEOREM

For any set of numbers, regardless of how they are distributed, the fraction of them that lie within k standard deviations of their mean (where $k > 1$) is *at least*

$$1 - \frac{1}{k^2}.$$

Be sure to notice the words *at least* in the theorem statement. In certain distributions the fraction of items within k standard deviations of the mean may be more than $1 - \frac{1}{k^2}$, but in no case will it ever be less. The theorem is meaningful for any value of k greater than 1 (integer or noninteger).

EXAMPLE 6 Applying Chebyshev's Theorem

What is the minimum percentage of the items in a data set that lie within 3 standard deviations of the mean?

Solution

With $k = 3$, we calculate as follows.

$$1 - \frac{1}{3^2} = 1 - \frac{1}{9} = \frac{8}{9} \approx 0.889 = 88.9\% \quad \leftarrow \text{Minimum percentage}$$

$3^2 = 3 \cdot 3, \text{not } 3 \cdot 2$

Coefficient of Variation

Look again at the top target pictured on **page 676.** The dispersion, or spread, among the five bullet holes may not be especially impressive if the shots were fired from 100 yards, but it would be much more so at, say, 300 yards. There is another measure, the *coefficient of variation*, that takes this distinction into account. It is not strictly a measure of dispersion, as it combines central tendency and dispersion. It expresses the standard deviation as a percentage of the mean. ***Often this is a more meaningful measure than a straight measure of dispersion, especially when we are comparing distributions whose means are appreciably different.***

COEFFICIENT OF VARIATION

For any set of data, the **coefficient of variation** is calculated as follows.

$$V = \frac{s}{\bar{x}} \cdot 100\% \quad \text{for a sample} \qquad \text{or} \qquad V = \frac{\sigma}{\mu} \cdot 100\% \quad \text{for a population}$$

EXAMPLE 7 Comparing Samples

Compare the dispersions in the two samples A and B.

$$A: 12, 13, 16, 18, 18, 20 \qquad B: 125, 131, 144, 158, 168, 193$$

Table 14

Sample A	Sample B
$\bar{x}_A = 16.167$	$\bar{x}_B = 153.167$
$s_A = 3.125$	$s_B = 25.294$
$V_A = 19.3$	$V_B = 16.5$

Solution

Using a calculator, we apply the first formula of the previous definition to obtain the values shown in **Table 14.** From the calculated values, we see that sample B has a much larger dispersion (standard deviation) than sample A. But sample A actually has the larger *relative* dispersion (coefficient of variation). The dispersion within sample A is larger as a percentage of that sample's mean.

FOR FURTHER THOUGHT

Measuring Skewness in a Distribution

Section 12.2 included a discussion of "symmetry in data sets." Here we present a common method of measuring the amount of "skewness," or nonsymmetry, inherent in a distribution.

In a skewed distribution, the mean will be farther out toward the tail than the median, as shown in the sketch.

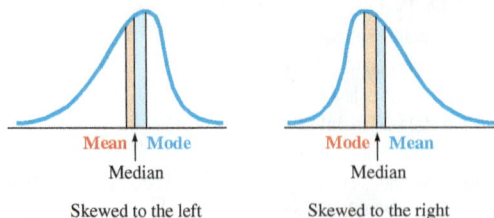

Mean ↑ Mode
Median
Skewed to the left

Mode ↑ Mean
Median
Skewed to the right

The degree of skewness can be measured by the **skewness coefficient**, which involves both central tendency and dispersion and is calculated as follows.

$$SK = \frac{3 \cdot (\text{mean} - \text{median})}{\text{standard deviation}}$$

For Group or Individual Investigation

1. Under what conditions would the skewness coefficient be each of the following?

  (a) positive **(b)** negative

2. Explain why the mean of a skewed distribution is always farther out toward the tail than the median.

3. In a skewed distribution, how many standard deviations apart are the mean and median in each case?

  (a) $SK = \frac{1}{2}$ **(b)** $SK = 1$ **(c)** $SK = 3$

4. Suppose $SK = -6$. Complete the following statement. The mean is _____ standard
                       (how many?)
deviations _____ the median.
         (above/below)

12.3 EXERCISES

1. If your calculator finds both the sample standard deviation and the population standard deviation, which of the two will be a larger number for a given set of data? (*Hint:* Recall the difference in the ways the two standard deviations are calculated.)

2. If your calculator finds only one kind of standard deviation, explain how you could determine, without the calculator instructions, whether it is sample or population standard deviation.

Find **(a)** *the range, and* **(b)** *the standard deviation for each sample in Exercises 3–10. If necessary, round answers to the nearest hundredth.*

3. 2, 4, 5, 8, 9, 11, 16

4. 6, 12, 10, 8, 9, 20, 22, 16, 5

5. 34, 27, 22, 41, 30, 15, 31

6. 62, 81, 55, 63, 75, 61, 88, 72, 65

7. 74.96, 74.60, 74.58, 74.48, 74.72, 75.62, 75.03, 75.10, 74.53

8. 314.3, 310.4, 309.3, 312.1, 310.8, 313.5, 310.6, 310.5, 311.0, 314.2

9.

Value	Frequency
9	2
12	5
8	6
3	4
1	2

10.

Value	Frequency
25	6
16	8
23	5
20	6
18	12
22	2
26	5

Use Chebyshev's theorem for Exercises 11–22. Give answers as common fractions reduced to lowest terms.

Find the least possible fraction of the numbers in a data set lying within the given number of standard deviations of the mean.

11. 2

12. 4

13. $\frac{5}{2}$

14. $\frac{7}{4}$

In a certain distribution of numbers, the mean is 50 and the standard deviation is 5. At least what fraction of the numbers are between the following pairs of numbers?

15. 40 and 60

16. 35 and 65

17. 30 and 70

18. 25 and 75

In a distribution with mean 80 and standard deviation 8, find the largest fraction of the numbers that could meet the following requirements.

19. less than 64 or more than 96

20. less than 60 or more than 100

21. less than 52 or more than 108

22. less than 62 or more than 98

Travel Accommodation Costs *Gabriel and Lucia took a road trip across the country. The room costs, in dollars, for their overnight stays are listed here.*

| 99 | 105 | 120 | 165 | 185 | 178 |
| 110 | 245 | 134 | 134 | 120 | 260 |

Use this distribution of costs for Exercises 23–28.

23. Find the mean of the distribution.

24. Find the standard deviation of the distribution.

25. How many of the cost amounts are within one standard deviation of the mean?

26. How many of the cost amounts are within two standard deviations of the mean?

27. What does Chebyshev's theorem say about the number of the amounts that are within two standard deviations of the mean?

28. Explain any discrepancy between your answers for **Exercises 26 and 27.**

In Exercises 29 and 30, two samples are given. In each case, (a) find both sample standard deviations, (b) find both sample coefficients of variation, (c) decide which sample has the higher dispersion, and (d) decide which sample has the higher relative dispersion.

29. *A:* 3, 7, 4, 3, 8   *B:* 10, 8, 10, 6, 7, 3, 5

30. *A:* 68, 72, 69, 65, 71, 72, 68, 71, 67, 67

   B: 26, 35, 30, 28, 31, 36, 38, 29, 34, 33

Utilize the following sample for Exercises 31–36.

13, 14, 17, 19, 21, 22, 25

31. Compute the mean and standard deviation for the sample (each to the nearest hundredth).

32. Now add 5 to each item of the given sample, and compute the mean and standard deviation for the new sample.

33. Go back to the original sample. This time subtract 10 from each item, and compute the mean and standard deviation of the new sample.

34. Based on your answers for **Exercises 31–33,** make conjectures about what happens to the mean and standard deviation when all items of the sample have the same constant k added or subtracted.

35. Go back to the original sample again. This time multiply each item by 3, and compute the mean and standard deviation of the new sample.

36. Based on your answers for **Exercises 31 and 35,** make conjectures about what happens to the mean and standard deviation when all items of the sample are multiplied by the same constant k.

37. **Comparing Water Heater Lifetimes** Two brands of electric water heaters, both carrying 6-year warranties, were sampled and tested under controlled conditions. Five of each brand failed after the numbers of months shown here.

Brand A: 74, 65, 70, 64, 71

Brand B: 69, 70, 62, 72, 60

(a) Calculate both sample means.

(b) Calculate both sample standard deviations.

(c) Which brand apparently lasts longer?

(d) Which brand has the more consistent lifetime?

Lifetimes of Engine Control Modules *Chin manages the service department of a trucking company. Each truck in the fleet utilizes an electronic engine control module. Long-lasting modules are desirable. A preventive replacement program also avoids costly breakdowns. For this purpose it is desirable that the modules be fairly consistent in their lifetimes, so that preventive replacements can be timed efficiently.*

Chin tested a sample of 20 Brand A modules, and they lasted 48,560 highway miles on the average (mean), with a standard deviation of 2116 miles. The listing below shows how long each of another sample of 20 Brand B modules lasted. Use these data for Exercises 38–40.

44,660	51,300	45,680	48,840	47,510
61,220	49,100	48,660	47,790	47,210
48,050	49,920	47,420	45,880	50,110
52,910	47,930	45,800	46,690	49,240

38. According to the sampling, which brand of module has the longer average life (in highway miles)?

39. Which brand of module apparently has a more consistent (or uniform) length of life (in highway miles)?

40. If Brands A and B are the only modules available, which one should Chin purchase for the maintenance program? Explain your reasoning.

41. In **Section 12.2** we showed that the mean, as a measure of central tendency, is highly sensitive to extreme values. Which measure of dispersion, covered in this section, would be more sensitive to extreme values? Illustrate your answer with one or more examples.

A Cereal-Marketing Survey   A food distribution company conducted a survey to determine whether a proposed premium to be included in boxes of its cereal was appealing enough to generate new sales. Four cities were used as test markets, where the cereal was distributed with the premium, and four cities as control markets, where the cereal was distributed without the premium. The eight cities were chosen on the basis of their similarity in terms of population, per capita income, and total cereal purchase volume. The results are shown in the following table.

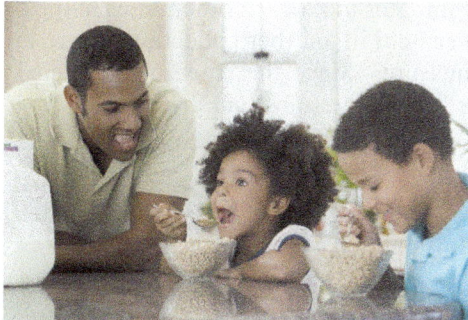

Percent Change in Average Market Share per Month

Test cities	1	+18
	2	+15
	3	+7
	4	+10
Control cities	1	+1
	2	−8
	3	−5
	4	0

42. Find the mean of the percent change in market share for the four test cities.

43. Find the mean of the percent change in market share for the four control cities.

44. Find the standard deviation of the percent change in market share for the test cities.

45. Find the standard deviation of the percent change in market share for the control cities.

46. Find the difference between the means of the test cities and the control cities. This difference represents the estimate of the percent change in sales due to the premium.

47. The two standard deviations from the test cities and the control cities were used to calculate an "error" of ± 7.95 for the estimate in **Exercise 46.** With this amount of error, what are the least and greatest estimates of the increase in sales?

(On the basis of the interval estimate of **Exercise 47,** the company decided to mass produce the premium and distribute it nationally.)

*For Exercises 48–50, refer to the grouped frequency distribution shown below. (Also refer to **Exercises 57–59** in Section 12.2.)*

Class Limits	Frequency f
21–25	5
26–30	3
31–35	8
36–40	12
41–45	21
46–50	38
51–55	35
56–60	20

48. Is it possible to identify any specific data items that occurred in this sample?

49. Is it possible to compute the actual standard deviation for this sample?

50. Describe how you might approximate the standard deviation for this sample. Justify your procedure.

51. Suppose the frequency distribution of **Example 4** involved 50 or 100 (or even more) distinct data values, rather than just four. Explain why the procedure of that example would then be very inefficient.

52. A "J-shaped" distribution can be skewed either to the right or to the left. (When skewed right, it is sometimes called a "reverse J" distribution.)

   (a) In a J-shaped distribution skewed to the right, which data item would be the mode, the greatest or the least item?

   (b) In a J-shaped distribution skewed to the left, which data item would be the mode, the greatest or the least item?

   (c) Explain why the mode is a weak measure of central tendency for a J-shaped distribution.

12.4 MEASURES OF POSITION

OBJECTIVES

1 Understand the *z*-score.
2 Compute and interpret percentiles.
3 Compute and interpret deciles and quartiles.
4 Work with box plots.

Lists of **the top ten jobs of 2014** may vary somewhat, depending on which organization's ratings are consulted. The list below is based generally on (a) job prospects, (b) salaries, (c) stress levels, and (d) "work-life balance." The number in parentheses is the rating (out of a possible 10 points), and the dollar amount is the median salary for 2012. See the Web site below for detailed information.

1. Software developer (8.4) $90,060
2. Computer systems
   analyst (8.2) $79,680
3. Dentist (8.2) $145,240
4. Nurse practitioner (8.1) $89,960
5. Pharmacist (8.1) $116,670
6. Registered nurse (8.0) $65,470
7. Physical therapist (7.9) $79,860
8. Physician (7.8) $123,300
9. Web developer (7.8) $62,500
10. Dental hygienist (7.7) $70,210

(*Source:* www.money.usnews.com)

Measures of central tendency and measures of dispersion give us an effective way of characterizing an overall set of data. Central tendency indicates where, along a number scale, the overall data set is centered. Dispersion indicates how much the data set is spread out from the center point. And Chebyshev's theorem, stated in the previous section, tells us in a general sense what portions of the data set may be dispersed different amounts from the center point.

In some cases, we are interested in certain individual items within a data set, rather than in that set as a whole. So we would like to measure how an item fits into the collection, how its placement compares to those of other items in the collection, or even how it compares to another item in another collection. There are several common ways of creating such measures. Since they measure an item's position within the data set, they usually are called **measures of position.**

The *z*-Score

Each individual item in a sample can be assigned a **z-score,** which is the first measure of position that we consider. It is defined as follows.

THE *z*-SCORE

If x is a data item in a sample with mean $\bar{x}$ and standard deviation s, then the **z-score** of x is calculated as follows.

$$z = \frac{x - \bar{x}}{s}$$

Because $x - \bar{x}$ gives the amount by which x differs (or deviates) from the mean $\bar{x}$, the z-score simply reflects the number of standard deviations by which x differs from $\bar{x}$. Notice that z will be positive if x is greater than $\bar{x}$ but negative if x is less than $\bar{x}$. Chebyshev's theorem assures us that in any distribution whatsoever, at least 89% (roughly) of the items will lie within three standard deviations of the mean. That is, at least 89% of the items will have z-scores between -3 and 3. In fact, many common distributions, especially symmetric ones, have considerably more than 89% of their items within three standard deviations of the mean (as we will see in the next section). Hence, a z-score greater than 3 or less than -3 is rare.

EXAMPLE 1 Comparing Positions Using *z*-Scores

Two friends, Cyrus and Aisha, who take different history classes, had midterm exams on the same day. Cyrus's score was 86, and Aisha's was only 78. Based on z-scores, which student did relatively better, given the class data shown here?

	Cyrus	Aisha
Class mean	73	69
Class standard deviation	8	5

Solution

Calculate as follows.

$$\text{Cyrus: } z = \frac{86 - 73}{8} = 1.625 \qquad \text{Aisha: } z = \frac{78 - 69}{5} = 1.8$$

Because Aisha's z-score is higher, she was positioned relatively higher within her class than Cyrus was within his class.

Percentiles

When you take the Scholastic Aptitude Test (SAT), or any other standardized test taken by large numbers of students, your raw score usually is converted to a **percentile** score, which is defined as follows.

PERCENTILE

If approximately n percent of the items in a distribution are less than the number x, then x is the **nth percentile** of the distribution, denoted P_n.

For example, if you scored at the eighty-third percentile on the SAT, it means that you outscored approximately 83% of all those who took the test. (It does *not* mean that you got 83% of the answers correct.) Since the percentile score gives the position of an item within the data set, it is another "measure of position." The following example approximates percentiles for a fairly small collection of data.

EXAMPLE 2 Finding Percentiles

The following are the numbers of dinner customers served by a restaurant on 40 consecutive days. The numbers have been ranked least to greatest.

46	51	52	55	56	56	58	59	59	59
60	61	62	62	63	63	64	64	64	65
66	66	66	67	67	67	68	68	69	69
70	70	71	71	72	75	79	79	83	88

For this data set, find **(a)** the sixty-fifth percentile, and **(b)** the eighty-eighth percentile.

Solution

(a) The sixty-fifth percentile can be taken as the item below which 65% of the items are ranked. Since 65% of 40 is $0.65(40) = 26$, we take the twenty-seventh item, or 68, as the sixty-fifth percentile.

(b) Since 88% of 40 is $0.88(40) = 35.2$, we round *up* and take the eighty-eighth percentile to be the thirty-sixth item, or 75. ■

Technically, percentiles originally were conceived as a set of 99 values P_1, P_2, P_3, ..., P_{99} (not necessarily data items) along the scale that would divide the data set into 100 equal-sized parts. They were computed only for very large data sets. With smaller data sets, as in **Example 2,** dividing the data into 100 parts would necessarily leave many of those parts empty. However, the modern techniques of exploratory data analysis seek to apply the percentile concept to even small data sets. Thus, we use approximation techniques as in **Example 2.** Another option is to divide the data into a lesser number of equal-sized (or nearly equal-sized) parts, say ten parts, or just four parts, as we discuss next.

Deciles and Quartiles

Deciles are the nine values (denoted $D_1, D_2, \ldots, D_9$) along the scale that divide a data set into ten (approximately) equal-sized parts, and **quartiles** are the three values ($Q_1, Q_2,$ and Q_3) that divide a data set into four (approximately) equal-sized parts. Since deciles and quartiles serve to position particular items within portions of a distribution, they also are "measures of position." We can evaluate deciles by finding their equivalent percentiles.

$$D_1 = P_{10}, \quad D_2 = P_{20}, \quad D_3 = P_{30}, \quad \ldots, \quad D_9 = P_{90}$$

EXAMPLE 3 Finding Deciles

Find the fourth decile for the dinner customer data of **Example 2.**

Solution

Refer to the ranked data table. The fourth decile is the fortieth percentile, and 40% of 40 is $0.40(40) = 16$. We take the fourth decile to be the seventeenth item, or 64. ■

Although the three quartiles also can be related to corresponding percentiles, notice that the second quartile, Q_2, also is equivalent to the median, a measure of central tendency introduced in **Section 12.2.** A common convention for computing quartiles goes back to the way we computed the median.

FINDING QUARTILES

For any set of data (ranked in order from least to greatest):

The **second quartile, Q_2,** is just the median, the middle item when the number of items is odd, or the mean of the two middle items when the number of items is even.

The **first quartile, Q_1,** is the median of all items below Q_2.

The **third quartile, Q_3,** is the median of all items above Q_2.

EXAMPLE 4 Finding Quartiles

Find the three quartiles for the data of **Example 2.**

Solution

Refer to the ranked data. The two middle data items are 65 and 66.

$$Q_2 = \frac{65 + 66}{2} = 65.5$$

The least 20 items (an even number) are all below Q_2, and the two middle items in that set are 59 and 60.

$$Q_1 = \frac{59 + 60}{2} = 59.5$$

The greatest 20 items are above Q_2.

$$Q_3 = \frac{69 + 70}{2} = 69.5$$ ■

The Box Plot

A **box plot,** or **box-and-whisker plot,** involves the median (a measure of central tendency), the range (a measure of dispersion), and the first and third quartiles (measures of position), all incorporated into a simple visual display.

BOX PLOT

For a given set of data, a **box plot** (or **box-and-whisker plot**) consists of a rectangular box positioned above a numerical scale, extending from Q_1 to Q_3, with the value of Q_2 (the median) indicated within the box, and with "whiskers" (line segments) extending to the left and right from the box out to the minimum and maximum data items.

EXAMPLE 5   **Constructing a Box Plot**

Construct a box plot for the forty weekly study times of **Example 2** in **Section 12.1.**

Solution

To determine the quartiles and the minimum and maximum values more easily, we use the stem-and-leaf display (with leaves ranked), given in **Table 8** of **Section 12.2.**

1	2 4 5 6 7 8
2	2 3 4 4 5 6 9 9 9
3	0 1 1 2 3 3 6 6 6 6 8 9
4	1 4 4 5 5 7
5	2 5 5 8
6	0 2
7	2

The median (determined earlier in **Example 6** of **Section 12.2**) is

$$\frac{33 + 33}{2} = 33.$$

From the stem-and-leaf display,

$$Q_1 = \frac{24 + 25}{2} = 24.5 \quad \text{and} \quad Q_3 = \frac{44 + 45}{2} = 44.5.$$

The minimum and maximum items are evident from the stem-and-leaf display. They are 12 and 72. The box plot is shown in **Figure 11.**

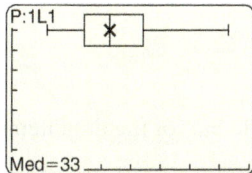

Weekly study times (in hours)

Box plot

Figure 11

The box plot in **Figure 11** conveys the following important information:

1. central tendency (the location of the median);
2. the location of the middle half of the data (the extent of the box);
3. dispersion (the range is the extent of the whiskers); and
4. skewness (the nonsymmetry of both the box and the whiskers).

1-Var Stats
↑n=40
  minX=12
  Q₁=24.5
  Med=33
  Q₃=44.5
  maxX=72

This screen supports the results of **Example 5.**

P:1L1

Med=33

This box plot corresponds to the results of **Example 5.** It indicates the median in the display at the bottom. The TRACE function of the TI-83/84 Plus will locate the minimum, maximum, and quartile values as well.

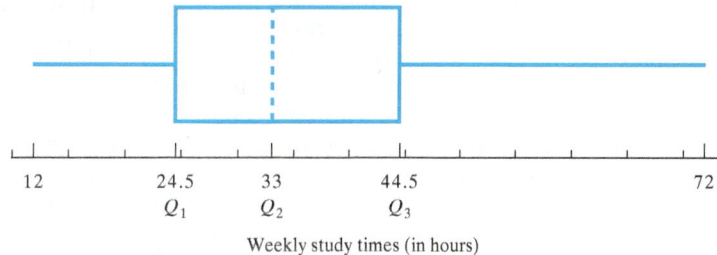

12.4   **EXERCISES**

Numbers of Restaurant Customers   *Refer to the dinner customers data of* ***Example 2.*** *Approximate each of the following. Use the methods illustrated in this section.*

1. the fifteenth percentile

2. the eighty-fifth percentile

3. the second decile   4. the seventh decile

In Exercises 5–8, make use of z-scores.

5. ***Relative Positions on Sociology Quizzes***   In a sociology class, Neil scored 5 on a quiz for which the class mean and standard deviation were 4.6 and 2.1, respectively. Janet scored 6 on another quiz for which the class mean and standard deviation were 4.9 and 2.3, respectively. Relatively speaking, which student did better?

6. **Relative Performances in Track Events** In Saturday's track meet, Edgar, a high jumper, jumped 6 feet 3 inches. Conference high jump marks for the past season had a mean of 6 feet even and a standard deviation of 3.5 inches. Kurt, Edgar's teammate, achieved 18 feet 4 inches in the long jump. In that event the conference season average (mean) and standard deviation were 16 feet 6 inches and 1 foot 10 inches, respectively. Relative to this past season in this conference, which athlete had a better performance on Saturday?

7. **Relative Lifetimes of Tires** The lifetimes of Brand A tires are distributed with mean 45,000 miles and standard deviation 4500 miles, while Brand B tires last for only 38,000 miles on the average (mean) with standard deviation 2080 miles. Nicole's Brand A tires lasted 37,000 miles, and Yvette's Brand B tires lasted 35,000 miles. Relatively speaking, within their own brands, which driver got the better wear?

8. **Relative Ratings of Fish Caught** In a certain lake, the trout average 12 inches in length with a standard deviation of 2.75 inches. The bass average 4 pounds in weight with a standard deviation of 0.9 pound. If Tobi caught an 18-inch trout and Katrina caught a 6-pound bass, then relatively speaking, which catch was the better trophy?

World's Largest Energy Producers and Consumers *The table includes only countries in the top ten in 2011 for both production and consumption of energy. (Energy units are quadrillion Btu.) Population is for midyear 2013, in millions. Use this information for Exercises 9–20.*

Country	Population	Production	Consumption
China	1350	97.8	109.6
United States	317	78.0	97.5
Russia	143	54.9	32.8
Canada	35	18.8	13.5
India	1221	16.0	23.6
Brazil	201	9.9	11.7

Source: The World Almanac and Book of Facts 2014.

Compute z-scores (accurate to one decimal place) for Exercises 9–12.

9. Russia's population

10. U.S. production

11. China's consumption

12. Brazil's consumption

In each of Exercises 13–16, determine which country occupied the given position.

13. the twenty-fifth percentile in population

14. the fourth decile in production

15. the third quartile in consumption

16. the ninth decile in production

17. Determine which was relatively higher: Canada in production or India in consumption.

18. Construct box plots for both production and consumption, one above the other in the same drawing.

19. What does your box plot of **Exercise 18** *for consumption* indicate about the following characteristics of the consumption data?
    (a) the central tendency
    (b) the dispersion
    (c) the location of the middle half of the data items

20. Comparing your two box plots of **Exercise 18,** what can you say about energy among the world's top producers and consumers of 2011?

21. The text stated that for *any* distribution of data, at least 89% of the items will be within three standard deviations of the mean. Why couldn't we just move some items farther out from the mean to obtain a new distribution that would violate this condition?

22. Describe the basic difference between a measure of central tendency and a measure of position.

This chapter has introduced three major characteristics, central tendency, dispersion, and position, and has developed various ways of measuring them in numerical data. In each of Exercises 23–26, a new measure is described. In each case indicate which of the three characteristics you think it would measure, and explain why.

23. **Midrange** $= \dfrac{\text{minimum item} + \text{maximum item}}{2}$

24. **Midquartile** $= \dfrac{Q_1 + Q_3}{2}$

25. **Interquartile range** $= Q_3 - Q_1$

26. **Semi-interquartile range** $= \dfrac{Q_3 - Q_1}{2}$

27. The "skewness coefficient" was defined in **For Further Thought** in the previous section, and it is calculated as follows.

$$SK = \frac{3 \cdot (\bar{x} - Q_2)}{s}$$

Is this a measure of individual data items or of the overall distribution?

28. For the energy data preceding **Exercise 9,** calculate the skewness coefficient for the following.

    (a) production   (b) consumption

29. From **Exercise 28,** how would you compare the skewness of production versus consumption?

30. In a national standardized test, Kimberly scored at the ninety-second percentile. If 67,500 individuals took the test, about how many scored higher than Kimberly?

31. Let the three quartiles (from least to greatest) for a large population of scores be denoted Q_1, Q_2, and Q_3.

    (a) Is it necessarily true that
    $$Q_2 - Q_1 = Q_3 - Q_2?$$

    (b) Explain your answer to part (a).

In Exercises 32–35, answer yes *or* no *and explain your answer. (Consult* **Exercises 23–26** *for definitions.)*

32. Is the midquartile necessarily the same as the median?

33. Is the midquartile necessarily the same as the midrange?

34. Is the interquartile range necessarily half the range?

35. Is the semi-interquartile range necessarily half the interquartile range?

36. *Relative Positions on a Standardized Chemistry Test*   Omer and Alessandro participated in the standardization process for a new statewide chemistry test. Within the large group participating, their raw scores and corresponding z-scores were as shown here.

	Raw Score	**z-Score**
Omer	60	0.69
Alessandro	72	1.67

Find the overall mean and standard deviation of the distribution of scores (to two decimal places).

Rating Passers in the National Football League   *Since the National Football League began keeping official statistics in 1932, the passing effectiveness of quarterbacks has been rated by several different methods. The current system, adopted in 1973, is based on four performance components: completions, touchdowns, yards gained, and interceptions, as percentages of the number of passes attempted. The computation can be accomplished using the formula in the next column.*

$$\text{Rating} = \frac{\left(250 \cdot \frac{C}{A}\right) + \left(1000 \cdot \frac{T}{A}\right) + \left(12.5 \cdot \frac{Y}{A}\right) + 6.25 - \left(1250 \cdot \frac{I}{A}\right)}{3},$$

where A = attempted passes,
C = completed passes,
T = touchdown passes,
Y = yards gained passing,

and I = interceptions.

In addition to the weighting factors (coefficients) appearing in the formula, the four category ratios are limited to non-negative values with the following maximums.

$$0.775 \text{ for } \frac{C}{A}, \quad 0.11875 \text{ for } \frac{T}{A}, \quad 12.5 \text{ for } \frac{Y}{A}, \quad 0.095 \text{ for } \frac{I}{A}$$

These limitations are intended to prevent any one component of performance from having an undue effect on the overall rating. They are not often invoked, but in special cases they can have a significant effect.

   The formula rates all passers against the same performance standard and is applied, for example, after a single game, an entire season, or a career. The ratings for the ten leading passers in the league for 2013 regular season play are ranked in the following table.

Rank	NFL Passer	Rating Points
1	Nick Foles, Philadelphia	119.2
2	Peyton Manning, Denver	115.1
3	Josh McCown, Chicago	109.0
4	Philip Rivers, San Diego	105.5
5	Aaron Rodgers, Green Bay	104.9
6	Drew Brees, New Orleans	104.7
7	Russell Wilson, Seattle	101.2
8	Tony Romo, Dallas	96.7
9	Ben Roethlisberger, Pittsburgh	92.0
10	Colin Kaepernick, San Francisco	91.6

Source: www.espn.go.com

Find the measures (to one decimal place) in Exercises 37–42.

37. the three quartiles

38. the sixth decile

39. the ninety-fifth percentile

40. the midrange (See **Exercise 23.**)

41. the midquartile (see **Exercise 24.**)

42. the interquartile range (See **Exercise 25.**)

43. Construct a box plot for the rating points data.

44. Peyton Manning's brother, Eli, quarterbacking for the New York Giants, completed the 2013 season at rank 35, with a rating of 69.4. His component statistics were as follows.

$$A = 551, \ C = 317, \ T = 18, \ Y = 3818, \ I = 27$$

If Eli had managed to avoid all 27 interceptions (a tall order), without changing any of his other numbers, by what amount would his rating points have *increased?*

45. With no interceptions, would Eli have risen into the top ten rankings?

46. Given Eli Manning's 551 attempts and 27 interceptions, suppose he had completed 20 more passes, for 420 additional yards. In order to make it into the top ten, how many more of his passes would have needed to result in touchdowns?

Archie Manning, father of NFL quarterbacks Peyton and Eli, signed this photo for author Hornsby's son, Jack.

47. Steve Young set a full-season rating record of 112.8 in 1994 and held that record until Peyton Manning achieved a rating of 121.1 in 2004. *His* record was broken in 2011 by Aaron Rodgers, with a rating of 122.5. If, in 2011, Rodgers had 343 completions, 45 touchdowns, and 4643 yards, in 502 attempts, how many times was he intercepted that year?

48. Use the passer rating formula to find the highest rating possible (considered a "perfect" passer rating).

*Given our method of finding quartiles, **Example 4** shows that quartiles are not necessarily data items. For each data set in Exercises 49–52, **(a)** find all three quartiles, and **(b)** state how many of them are data items.*

49. 1 2 3 4 5 6 7 8   **50.** 1 2 3 4 5 6 7 8 9

51. 1 2 3 4 5 6 7 8 9 10

52. 1 2 3 4 5 6 7 8 9 10 11

53. The pattern of **Exercises 49–52** holds in general. Complete the following statements for any data set with n distinct items (no item occurs more than once).

   (a) If $n = 4k$ for some counting number k, then _____ of the quartiles are data items.

   (b) If $n = 4k + 1$ for some counting number k, then _____ of the quartiles are data items.

   (c) If $n = 4k + 2$ for some counting number k, then _____ of the quartiles are data items.

   (d) If $n = 4k + 3$ for some counting number k, then _____ of the quartiles are data items.

54. Suppose a data set contains n distinct items (no items occur more than once). Determine how many of the three quartiles are data items in each case.

   (a) $n = 47$ **(b)** $n = 128$

   (c) $n = 101$ **(d)** $n = 166$

12.5   THE NORMAL DISTRIBUTION

OBJECTIVES

1 Understand the distinction between discrete and continuous random variables.

2 Understand the properties of a normal curve.

3 Use a table of standard normal curve areas.

4 Interpret normal curve areas in several ways.

Discrete and Continuous Random Variables

A random variable that can take on only certain fixed values is called a **discrete random variable.** For example, the number of heads in 5 tosses of a coin is discrete because its only possible values are

$$0, 1, 2, 3, 4, \text{ and } 5.$$

A variable whose values are not restricted in this way is a **continuous random variable.** For example, the diameter of camellia blossoms would be a continuous variable, spread over a scale perhaps from 5 to 25 centimeters. The values would not be restricted to whole numbers, or even to tenths, or hundredths, etc. A discrete random variable takes on only a countable number of values. A continuous random variable takes on an uncountable number of values.

The normal curve was first developed by **Abraham De Moivre** (1667–1754), but his work went unnoticed for many years. It was independently redeveloped by Pierre de Laplace (1749–1827) and Carl Friedrich Gauss (1777–1855). Gauss found so many uses for this curve that it is sometimes called the *Gaussian curve.*

Most distributions discussed in this chapter were *empirical* (based on observation). The distributions covered in this section are *theoretical* (based on theoretical probabilities). A knowledge of theoretical distributions enables us to identify when actual observations are inconsistent with stated assumptions.

The theoretical probability distribution for the discrete random variable "number of heads" when 5 fair coins are tossed is shown in **Table 15. Figure 12** shows the corresponding histogram. The probability values can be found using the binomial probability formula (**Section 11.4**) or using Pascal's triangle (**Section 10.4**).

Table 15 Probability Distribution

x	P(x)
0	0.03125
1	0.15625
2	0.31250
3	0.31250
4	0.15625
5	0.03125
Sum:	1.00000

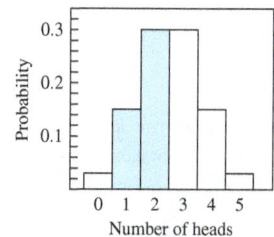

Figure 12

Since each rectangle in **Figure 12** is 1 unit wide, the *area* of the rectangle is also equal to the probability of the corresponding number of heads. The area, and thus the probability, for the event "1 head or 2 heads" is shaded in the figure. The graph consists of 6 distinct rectangles since "number of heads" is a *discrete* variable with 6 possible values. The sum of the 6 rectangular areas is exactly 1 square unit.

A probability distribution for camellia blossom diameters cannot be tabulated or graphed in quite the same way, since this variable is *continuous*. The graph would be smeared out into a "continuous" bell-shaped curve (rather than a set of rectangles) as shown in **Figure 13**. The vertical scale on the graph in this case shows what we call "probability density," the probability per unit along the horizontal axis.

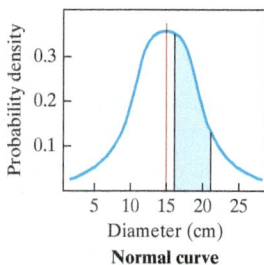

Figure 13

Definition and Properties of a Normal Curve

The camellia blossom curve is highest at a diameter value of 15 cm, its center point, and drops off rapidly and equally toward a zero level in both directions. Such a symmetric, bell-shaped curve is called a **normal curve.** Any random variable whose graph has this characteristic shape is said to have a **normal distribution.**

The area under the curve along a certain interval is numerically equal to the probability that the random variable will have a value in the corresponding interval. The area of the shaded region in **Figure 13** is equal to the probability of a randomly chosen blossom having a diameter in the interval from the left extreme, say 16.4, to the right extreme, say 21.2. Normal curves are very important in the study of statistics because *a great many continuous random variables have normal distributions, and many discrete variables are distributed approximately normally.*

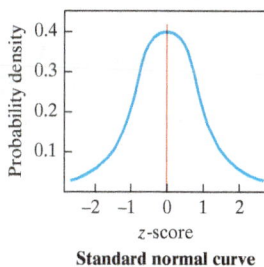

Figure 14

Each point on the horizontal scale of a normal curve lies some number of standard deviations from the mean (positive to the right, negative to the left). This number is the "standard score" for that point. It is the same as the z-score defined in **Section 12.4.** By relabeling the horizontal axis, as in **Figure 14,** we obtain the **standard normal curve,** which we can use to analyze *any* normal (or approximately normal) distribution. We relate the random variable value, x, to its z-score by

$$z = \frac{x - \bar{x}}{s}.$$

Close but Never Touching When a curve approaches closer and closer to a line, without ever actually meeting it (as a normal curve approaches the horizontal axis), the line is called an **asymptote,** and the curve is said to approach the line **asymptotically.**

Figure 15 shows several of infinitely many possible normal curves. Each is completely characterized by its mean and standard deviation. Only one of these, the one marked S, is the *standard* normal curve. That one has mean 0 and standard deviation 1.

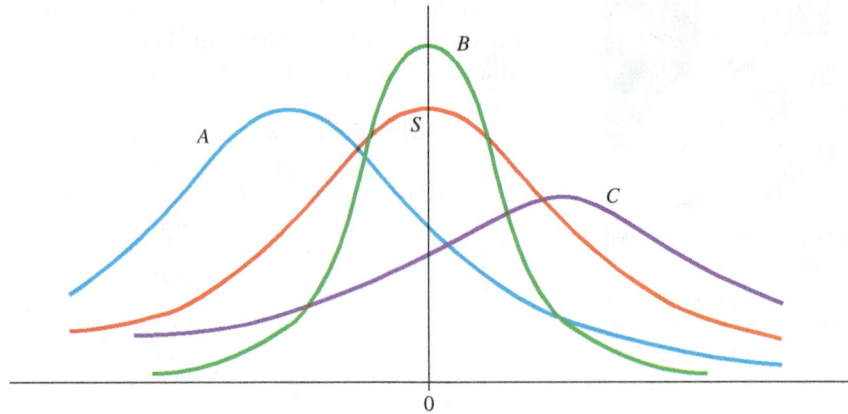

Normal curve S is standard, with mean = 0 and standard deviation = 1.
Normal curve A has mean < 0 and standard deviation = 1.
Normal curve B has mean = 0 and standard deviation < 1.
Normal curve C has mean > 0 and standard deviation > 1.

Figure 15

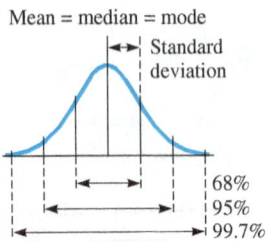

Figure 16

Several properties of normal curves are summarized below and are illustrated in **Figure 16.**

PROPERTIES OF NORMAL CURVES

The graph of a normal curve is bell-shaped and symmetric about a vertical line through its center.

The mean, median, and mode of a normal curve are all equal and occur at the center of the distribution.

Empirical Rule About 68% of all data values of a normal curve lie within 1 standard deviation of the mean (in both directions), about 95% within 2 standard deviations, and about 99.7% within 3 standard deviations.

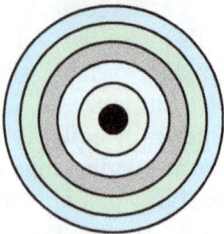

The empirical rule indicates that a very small percentage of the items in a normal distribution will lie more than 3 standard deviations from the mean (approximately 0.3%, divided equally between the upper and lower tails of the distribution). As we move away from the center, the curve *never* actually touches the horizontal axis. No matter how far out we go, there is always a chance of an item occurring even farther out. Theoretically then, the range of a true normal distribution is infinite.

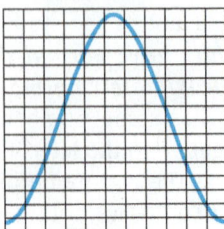

A normal distribution occurs in darts if the player, always aiming at the bull's-eye, tosses a fairly large number of times, and the aim on each toss is affected only by independent random errors.

EXAMPLE 1 Applying the Empirical Rule

Suppose that 300 chemistry students take a midterm exam and the distribution of their scores can be treated as normal. Find the number of scores falling into each of the following intervals.

(a) Within 1 standard deviation of the mean

(b) Within 2 standard deviations of the mean

Solution

(a) By the empirical rule, 68% of all scores lie within one standard deviation of the mean. There is a total of 300 scores.

$$0.68(300) = 204 \quad \text{68\%} = 0.68$$

(b) A total of 95% of all scores lie within 2 standard deviations of the mean.

$$0.95(300) = 285 \quad \text{95\%} = 0.95 \qquad ■$$

A Table of Standard Normal Curve Areas

Most questions we need to answer about normal distributions involve regions other than those within 1, 2, or 3 standard deviations of the mean. We might need the percentage of items within $1\frac{1}{2}$ or $2\frac{1}{5}$ standard deviations of the mean, or perhaps the area under the curve from 0.8 to 1.3 standard deviations above the mean.

In such cases, we need more than the empirical rule. The traditional approach is to refer to a table of area values, such as **Table 16** on the next page. Computer software designed for statistical uses generally will produce the required values on command, and some advanced calculators also have this capability. Those tools are recommended. As an option, we illustrate the use of **Table 16** here.

The table gives the fraction of all scores in a normal distribution that lie between the mean and z standard deviations from the mean. ***Because of the symmetry of the normal curve, the table can be used for values above the mean or below the mean.*** All of the items in the table can be thought of as corresponding to the area under the curve. The total area is arranged to be 1.000 square unit, with 0.500 square unit on each side of the mean. The table shows that at 3.30 standard deviations from the mean, essentially all of the area is accounted for. Whatever remains beyond is so small that it does not appear in the first three decimal places.

Carl Friedrich Gauss (1777–1855) was one of the greatest mathematical thinkers of history. In his *Disquisitiones Arithmeticae,* published in 1798, he pulled together work by predecessors and enriched and blended it with his own into a unified whole. The book is regarded by many as the beginning of the modern theory of numbers (**Chapter 5**).

Of his many contributions to science, the statistical method of least squares is the most widely used today in astronomy, biology, geodesy, physics, and the social sciences. Gauss took special pride in his contributions to developing the method. Despite an aversion to teaching, he taught an annual course in the method for the last twenty years of his life.

It has been said that Gauss was the last person to have mastered all of the mathematics known in his day.

EXAMPLE 2 **Applying the Normal Curve Table**

Use **Table 16** to find the percent of all scores that lie between the mean and the following values.

(a) 1 standard deviation above the mean

(b) 2.45 standard deviations below the mean

Solution

(a) Here $z = 1.00$ (the number of standard deviations, written as a decimal to the nearest hundredth). Refer to **Table 16.** Find 1.00 in the z column. The table entry is 0.341, so 34.1% of all values lie between the mean and 1 standard deviation above the mean.

Another way of looking at this is to say that the area in color in **Figure 17** represents 34.1% of the total area under the normal curve.

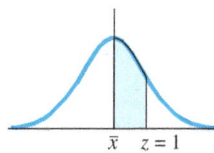

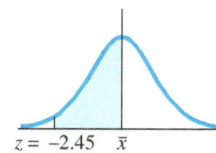

| **Figure 17** | **Figure 18** |

(b) Even though we go *below* the mean here (to the left), **Table 16** still works since the normal curve is symmetric about its mean. Find 2.45 in the z column. A total of 0.493, or 49.3%, of all values lie between the mean and 2.45 standard deviations below the mean. This region is colored in **Figure 18.** ■

The column under *A* gives the **fraction of the area under the entire curve** that is between *z* = 0 and a positive value of *z*.

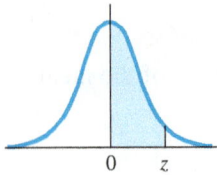

Because the curve is symmetric about the 0-value, the area between *z* = 0 and a *negative* value of *z* can be found by using the corresponding positive value of *z*.

Table 16   Areas under the Standard Normal Curve

z	A	z	A	z	A	z	A	z	A	z	A
.00	.000	.56	.212	1.12	.369	1.68	.454	2.24	.487	2.80	.497
.01	.004	.57	.216	1.13	.371	1.69	.454	2.25	.488	2.81	.498
.02	.008	.58	.219	1.14	.373	1.70	.455	2.26	.488	2.82	.498
.03	.012	.59	.222	1.15	.375	1.71	.456	2.27	.488	2.83	.498
.04	.016	.60	.226	1.16	.377	1.72	.457	2.28	.489	2.84	.498
.05	.020	.61	.229	1.17	.379	1.73	.458	2.29	.489	2.85	.498
.06	.024	.62	.232	1.18	.381	1.74	.459	2.30	.489	2.86	.498
.07	.028	.63	.236	1.19	.383	1.75	.460	2.31	.490	2.87	.498
.08	.032	.64	.239	1.20	.385	1.76	.461	2.32	.490	2.88	.498
.09	.036	.65	.242	1.21	.387	1.77	.462	2.33	.490	2.89	.498
.10	.040	.66	.245	1.22	.389	1.78	.462	2.34	.490	2.90	.498
.11	.044	.67	.249	1.23	.391	1.79	.463	2.35	.491	2.91	.498
.12	.048	.68	.252	1.24	.393	1.80	.464	2.36	.491	2.92	.498
.13	.052	.69	.255	1.25	.394	1.81	.465	2.37	.491	2.93	.498
.14	.056	.70	.258	1.26	.396	1.82	.466	2.38	.491	2.94	.498
.15	.060	.71	.261	1.27	.398	1.83	.466	2.39	.492	2.95	.498
.16	.064	.72	.264	1.28	.400	1.84	.467	2.40	.492	2.96	.498
.17	.067	.73	.267	1.29	.401	1.85	.468	2.41	.492	2.97	.499
.18	.071	.74	.270	1.30	.403	1.86	.469	2.42	.492	2.98	.499
.19	.075	.75	.273	1.31	.405	1.87	.469	2.43	.492	2.99	.499
.20	.079	.76	.276	1.32	.407	1.88	.470	2.44	.493	3.00	.499
.21	.083	.77	.279	1.33	.408	1.89	.471	2.45	.493	3.01	.499
.22	.087	.78	.282	1.34	.410	1.90	.471	2.46	.493	3.02	.499
.23	.091	.79	.285	1.35	.411	1.91	.472	2.47	.493	3.03	.499
.24	.095	.80	.288	1.36	.413	1.92	.473	2.48	.493	3.04	.499
.25	.099	.81	.291	1.37	.415	1.93	.473	2.49	.494	3.05	.499
.26	.103	.82	.294	1.38	.416	1.94	.474	2.50	.494	3.06	.499
.27	.106	.83	.297	1.39	.418	1.95	.474	2.51	.494	3.07	.499
.28	.110	.84	.300	1.40	.419	1.96	.475	2.52	.494	3.08	.499
.29	.114	.85	.302	1.41	.421	1.97	.476	2.53	.494	3.09	.499
.30	.118	.86	.305	1.42	.422	1.98	.476	2.54	.494	3.10	.499
.31	.122	.87	.308	1.43	.424	1.99	.477	2.55	.495	3.11	.499
.32	.126	.88	.311	1.44	.425	2.00	.477	2.56	.495	3.12	.499
.33	.129	.89	.313	1.45	.426	2.01	.478	2.57	.495	3.13	.499
.34	.133	.90	.316	1.46	.428	2.02	.478	2.58	.495	3.14	.499
.35	.137	.91	.319	1.47	.429	2.03	.479	2.59	.495	3.15	.499
.36	.141	.92	.321	1.48	.431	2.04	.479	2.60	.495	3.16	.499
.37	.144	.93	.324	1.49	.432	2.05	.480	2.61	.495	3.17	.499
.38	.148	.94	.326	1.50	.433	2.06	.480	2.62	.496	3.18	.499
.39	.152	.95	.329	1.51	.434	2.07	.481	2.63	.496	3.19	.499
.40	.155	.96	.331	1.52	.436	2.08	.481	2.64	.496	3.20	.499
.41	.159	.97	.334	1.53	.437	2.09	.482	2.65	.496	3.21	.499
.42	.163	.98	.336	1.54	.438	2.10	.482	2.66	.496	3.22	.499
.43	.166	.99	.339	1.55	.439	2.11	.483	2.67	.496	3.23	.499
.44	.170	1.00	.341	1.56	.441	2.12	.483	2.68	.496	3.24	.499
.45	.174	1.01	.344	1.57	.442	2.13	.483	2.69	.496	3.25	.499
.46	.177	1.02	.346	1.58	.443	2.14	.484	2.70	.497	3.26	.499
.47	.181	1.03	.348	1.59	.444	2.15	.484	2.71	.497	3.27	.499
.48	.184	1.04	.351	1.60	.445	2.16	.485	2.72	.497	3.28	.499
.49	.188	1.05	.353	1.61	.446	2.17	.485	2.73	.497	3.29	.499
.50	.191	1.06	.355	1.62	.447	2.18	.485	2.74	.497	3.30	.500
.51	.195	1.07	.358	1.63	.448	2.19	.486	2.75	.497	3.31	.500
.52	.198	1.08	.360	1.64	.449	2.20	.486	2.76	.497	3.32	.500
.53	.202	1.09	.362	1.65	.451	2.21	.486	2.77	.497	3.33	.500
.54	.205	1.10	.364	1.66	.452	2.22	.487	2.78	.497	3.34	.500
.55	.209	1.11	.367	1.67	.453	2.23	.487	2.79	.497	3.35	.500

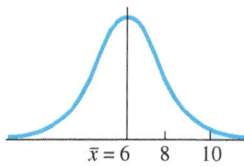

Figure 19

A Basic Consumer (T)issue It all
started when a reporter on consumer
issues for a Midwest TV station received
a complaint that rolls of Brand X toilet
paper manufactured by Company Y did not
have the number of sheets claimed on the
wrapper. Brand X is supposed to have 375
sheets, but three rolls of it were found by
reporters to have 360, 361, and 363.

Shocked Company Y executives said
that the **odds against** six rolls having
fewer than 375 sheets each are 1 billion
to 1. They counted sheets and found
that several rolls of Brand X actually had
380 sheets each (machines count the
sheets only in 10s). TV reporters made
an independent count, and their results
agreed with Company Y.

What happened the first time?
Well, the reporters hadn't actually
counted sheets but, rather, had measured
rolls and divided the length of a roll by the
length of one sheet. Small variations in
length can be expected, which can add up
over a roll, giving false results.

This true story perhaps points
out the distinction in probability and
statistics between **discrete values** and
continuous values.

EXAMPLE 3 Finding Probabilities of Phone Call Durations

The time lengths of phone calls placed through a certain company are distributed
normally with mean 6 minutes and standard deviation 2 minutes. If one call is randomly selected from phone company records, what is the probability that it will
have lasted more than 10 minutes?

Solution

Here 10 minutes is 2 standard deviations above the mean. The probability of such a
call is equal to the area of the colored region in **Figure 19.**

From **Table 16,** the area between the mean and 2 standard deviations above
is 0.477 ($z = 2.00$). The total area to the right of the mean is 0.500. Find the area
from $z = 2.00$ to the right by subtracting.

$$0.500 - 0.477 = 0.023$$

The probability of a call exceeding 10 minutes is 0.023, or 2.3%. ∎

EXAMPLE 4 Finding Areas under the Normal Curve

Find the total area that is shaded in each of **Figures 20** and **21.**

Solution

For **Figure 20,** find the area from 1.45 standard deviations below the mean to
2.71 standard deviations above the mean. From **Table 16,** $z = 1.45$ leads to an
area of 0.426, while $z = 2.71$ leads to 0.497. The total area is the sum of these, or
$0.426 + 0.497 = 0.923$.

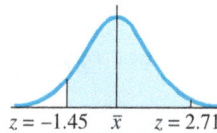

Figure 20

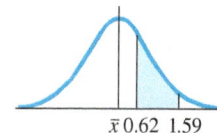

Figure 21

To find the area indicated in **Figure 21,** refer to **Table 16.** From the table,
$z = 0.62$ leads to an area of 0.232, and $z = 1.59$ gives 0.444. To get the area between
these two values of z, subtract the areas.

$$0.444 - 0.232 = 0.212$$ ∎

Interpreting Normal Curve Areas

Examples 2–4 emphasize the *equivalence* of three quantities, as follows.

MEANING OF NORMAL CURVE AREAS

In the standard normal curve, the following three quantities are equivalent.

1. **Percentage** (of total items that lie in an interval)
2. **Probability** (of a randomly chosen item lying in an interval)
3. **Area** (under the normal curve along an interval)

Which quantity we think of depends on how a particular question is formulated.
They all can be evaluated by using A-values from **Table 16.**

In general, when we use **Table 16**, z is the z-score of a particular data item x. When one of these values is known and the other is required, we use the formula

$$z = \frac{x - \bar{x}}{s}.$$

EXAMPLE 5 Applying the Normal Curve to Driving Distances

In one area, the distribution of monthly miles driven by motorists has mean 1200 miles and standard deviation 150 miles. Assume that the number of miles is closely approximated by a normal curve, and find the percent of all motorists driving the following distances.

(a) Between 1200 and 1600 miles per month

(b) Between 1000 and 1500 miles per month

Solution

(a) Start by finding how many standard deviations 1600 miles is above the mean. Use the formula for z.

$$z = \frac{1600 - 1200}{150} = \frac{400}{150} \approx 2.67$$

From **Table 16**, 0.496, or 49.6%, of all motorists drive between 1200 and 1600 miles per month.

(b) As shown in **Figure 22**, values of z must be found for both 1000 and 1500.

$$\text{For 1000:} \quad z = \frac{1000 - 1200}{150} = \frac{-200}{150} \approx -1.33$$

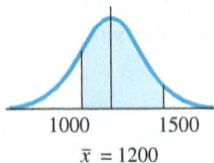

1000   1500
$\bar{x} = 1200$

Figure 22

$$\text{For 1500:} \quad z = \frac{1500 - 1200}{150} = \frac{300}{150} = 2.00$$

From **Table 16**, $z = -1.33$ leads to an area of 0.408, while $z = 2.00$ gives 0.477. This means a total of

$$0.408 + 0.477 = 0.885, \quad \text{or} \quad 88.5\%,$$

of all motorists drive between 1000 and 1500 miles per month. ■

EXAMPLE 6 Identifying a Data Value within a Normal Distribution

A particular normal distribution has mean $\bar{x} = 81.7$ and standard deviation $s = 5.21$. What data value from the distribution would correspond to $z = -1.35$?

Solution

Solve for x.
$$z = \frac{x - \bar{x}}{s} \qquad \text{\textit{z}-score formula}$$

$$-1.35 = \frac{x - 81.7}{5.21} \qquad \text{Substitute the given values for } z, \bar{x}, \text{ and } s.$$

$$-1.35(5.21) = \frac{x - 81.7}{5.21}(5.21) \qquad \text{Multiply each side by 5.21 to clear the fraction.}$$

$$-7.0335 = x - 81.7 \qquad \text{Simplify.}$$

$$74.6665 = x \qquad \text{Add 81.7.}$$

Rounding to the nearest tenth, the required data value is 74.7. ■

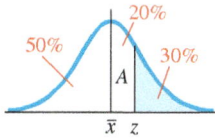

20%
50%
A
30%
$\bar{x}$ z

Figure 23

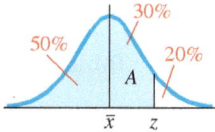

30%
50%
A
20%
$\bar{x}$ z

Figure 24

| EXAMPLE 7 | **Finding z-Values for Given Areas under the Normal Curve** |

Assuming a normal distribution, find the z-value meeting each condition.

(a) 30% of the total area is to the right of z.

(b) 80% of the total area is to the left of z.

Solution

(a) Because 50% of the area lies to the right of the mean, there must be 20% between the mean and z. (See **Figure 23**.) In **Table 16**, $A = 0.200$ corresponds to $z = 0.52$ or 0.53, or we could average the two: $z = 0.525$.

(b) This situation is shown in **Figure 24**. The 50% to the left of the mean plus 30% additional makes up the 80%. From **Table 16**, $A = 0.300$ implies $z = 0.84$. ■

12.5 EXERCISES

Note: For problems requiring the calculation of z-scores or A-values, our answers are based on **Table 16,** *or the empirical rule when specified. By using a calculator or computer package, you will sometimes obtain a slightly more precise answer.*

Identify each variable quantity as discrete *or* continuous.

1. the number of heads in 30 rolled dice

2. the number of babies born in one day at a certain hospital

3. the average weight of babies born in a week

4. the heights of seedling fir trees at two years of age

5. the time as shown on a digital watch

6. the time as shown on a watch with a sweep hand

Measuring the Mass of Ore Samples *Suppose 100 geology students measure the mass of an ore sample. Due to human error and limitations in the reliability of the balance, not all the readings are equal. The results are found to closely approximate a normal curve, with mean 75 g and standard deviation 2 g.*

Use the symmetry of the normal curve and the empirical rule to estimate the number of students reporting readings in the following ranges.

7. more than 75 g

8. more than 73 g

9. between 71 and 79 g

10. between 69 and 77 g

Distribution of IQ Scores *On standard IQ tests, the mean is 100, with a standard deviation of 15. The results come very close to fitting a normal curve. Suppose an IQ test is given to a very large group of people. Use the empirical rule to find the percentage of people whose IQ scores fall into each category.*

11. more than 100

12. less than 85

13. between 70 and 130

14. less than 115

Find the percentage of area under a normal curve between the mean and the given number of standard deviations from the mean. (Note that positive indicates above the mean, while negative indicates below the mean.)

15. 1.60

16. 0.76

17. -1.68

18. -2.45

Find the percentage of the total area under a normal curve between the given values of z.

19. $z = 1.31$ and $z = 1.73$

20. $z = -1.56$ and $z = -1.02$

21. $z = -3.02$ and $z = 2.03$

22. $z = -1.42$ and $z = 0.98$

Find a value of z such that each condition is met.

23. 20% of the total area is to the right of z.

24. 23% of the total area is to the left of z.

25. 78% of the total area is to the left of z.

26. 63% of the total area is to the right of z.

Weights of Peaches *A fruit-packing company produced peaches last summer whose weights were normally distributed with mean 14 ounces and standard deviation 0.6 ounce. Among a sample of 1000 of those peaches, about how many could be expected to have weights as follows?*

27. at least 14 ounces

28. between 14 and 15.2 ounces

29. between 14.9 and 15.5 ounces

30. less than 13.4 ounces

31. more than 12 ounces

32. between 13.5 and 15.5 ounces

IQs of Employees *A large company employs workers whose IQs are distributed normally with mean 105 and standard deviation 12.5. Management uses this information to assign employees to projects that will be challenging, but not too challenging. What percentage of the employees would have IQs that satisfy the following criteria?*

33. less than 90   **34.** more than 125

35. between 100 and 120   **36.** between 80 and 100

Net Weight of Cereal Boxes *A certain dry cereal is packaged in 12-oz boxes. The machine that fills the boxes is set so that, on the average a box contains 12.4 oz. The machine-filled boxes have content weight that can be closely approximated by a normal curve. What is the probability that a randomly selected box will be underweight (net weight less than 12 oz) if the standard deviation is as follows?*

37. 0.5 oz   **38.** 0.4 oz   **39.** 0.3 oz   **40.** 0.2 oz

41. Recommended Daily Vitamin Allowances In nutrition, the recommended daily allowance of vitamins is a number set by the government to guide an individual's daily vitamin intake. Actually, vitamin needs vary drastically from person to person, but the needs are closely approximated by a normal curve. To calculate the recommended daily allowance, the government first finds the average (mean) need for vitamins among people in the population and the standard deviation. The **recommended daily allowance** is then defined as the mean plus 2.5 times the standard deviation. What fraction of the population would receive adequate amounts of vitamins under this plan?

Recommended Daily Vitamin Allowances *Find the recommended daily allowance for each vitamin if the mean need and standard deviation are as follows. (See **Exercise 41**.)*

42. mean need $= 1600$ units;
standard deviation $= 140$ units

43. mean need $= 146$ units;
standard deviation $= 8$ units

Assume the following distributions are all normal, and use the areas under the normal curve given in **Table 16** *to find the appropriate areas.*

44. Assembling Cell Phones The times taken by workers to assemble a certain kind of cell phone are normally distributed with mean 18.5 minutes and standard deviation 3.1 minutes. Find the probability that one such phone will require less than 12.8 minutes in assembly.

45. Finding Blood Clotting Times The mean clotting time of blood is 7.47 sec, with a standard deviation of 3.6 sec. What is the probability that an individual's blood-clotting time will be less than 7 sec or greater than 8 sec?

46. Sizes of Fish The average length of the fish caught in Vernal Lake is 10.6 in., with a standard deviation of 3.2 in. Find the probability that a fish caught there will be longer than 16 in.

47. Size Grading of Eggs To be graded extra large, an egg must weigh at least 2.2 oz. If the average weight for an egg is 1.5 oz, with a standard deviation of 0.4 oz, how many of five dozen randomly chosen eggs would you expect to be extra large?

Distribution of Student Grades *Professor Cheng teaches a marketing course. He uses the following grading system.*

Grade	Score in Class
A	Greater than $\bar{x} + 1.5s$
B	$\bar{x} + 0.5s$ to $\bar{x} + 1.5s$
C	$\bar{x} - 0.5s$ to $\bar{x} + 0.5s$
D	$\bar{x} - 1.5s$ to $\bar{x} - 0.5s$
F	Below $\bar{x} - 1.5s$

What percent of the students receive the following grades?

48. A   **49.** B   **50.** C

51. Do you think that Professor Cheng's system would be more likely to be fair in a large first-year lecture class or in a graduate seminar of five students? Why?

Normal Distribution of Student Grades *A teacher gives a test to a large group of students. The results are closely approximated by a normal curve. The mean is 72 with a standard deviation of 5. The teacher wishes to give As to the top 8% of the students and Fs to the bottom 8%. A grade of B is given to the next 15%, with Ds given similarly. All other students get Cs. Find the bottom point cutoffs (rounded to the nearest whole number) for the following grades. (Hint: Use* **Table 16** *to find z-scores from known A-values.)*

52. A   **53.** B   **54.** C   **55.** D

56. ***Selecting Industrial Rollers*** A manufacturer makes carts with industrial swivel rollers that must support a load of at least 125 pounds. Supplier A can provide rollers supporting 130 pounds on the average, with standard deviation 2.2 pounds, and Supplier B's rollers will support 135 pounds on the average, with standard deviation 4.2 pounds.

(a) If both distributions can be assumed normal, which supplier offers the smaller percentage of unsatisfactory rollers?

(b) Give your reasoning.

A normal distribution has mean 76.8 and standard deviation 9.42. Follow the method of ***Example 6*** *and find data values corresponding to the following values of z. Round to the nearest tenth.*

57. $z = 1.44$ **58.** $z = 0.72$

59. $z = -3.87$ **60.** $z = -2.39$

61. What percentage of the items lie within 1.25 standard deviations of the mean

(a) in any distribution (using the results of Chebyshev's theorem)?

(b) in a normal distribution (by **Table 16**)?

62. Explain the difference between the answers to **Exercise 61,** parts (a) and (b).

CHAPTER 12   SUMMARY

KEY TERMS

12.1

population
sample
descriptive statistics
inferential statistics
raw data
quantitative (numerical)
    data
qualitative (non-numerical)
    data
ranked data
frequency distribution
relative frequency
    distribution
histogram
frequency polygon
classes
grouped frequency
    distribution

lower class limits
upper class limits
class width
stem-and-leaf display
bar graph
circle graph (pie chart)
line graph
expected (theoretical)
    frequencies
observed (empirical)
    frequencies

12.2

measure of central
    tendency
mean (arithmetic mean)
weighted mean
weighting factor
outlier

median
cumulative frequency
mode
bimodal
symmetry in
    data sets
skewed to the left
skewed to the right
Simpson's paradox

12.3

dispersion (spread) of
    data
range
variance
standard deviation
Chebyshev's theorem
coefficient of variation
skewness coefficient

12.4

measures of position
z-score
percentile
decile
quartile
box plot (box-and-whisker
    plot)

12.5

discrete random variable
continuous random
    variable
normal curve
normal distribution
standard normal curve
asymptote
empirical rule

TEST YOUR WORD POWER

See how well you have learned the vocabulary in this chapter.

1. Using a **sample** to gain knowledge about a **population** is part of
   A. descriptive statistics.
   B. inferential statistics.
   C. qualitative data analysis.
   D. calculus.

2. A **class mark** is the
   A. average grade of all the class members.
   B. least item in a grouped frequency distribution.
   C. middle of a class in a grouped frequency distribution.
   D. number of items in the class.

3. A common property of **bar graphs, circle graphs, line graphs,** and **stem-and-leaf displays** is that they provide
   A. a visual representation of a data set.
   B. the range value of a data set.
   C. a way of comparing two data sets.
   D. clear divisions among the various classes.

4. The measure that, when it exists, is always equal to a data item is the
   A. mean.   B. standard deviation.
   C. median.   D. mode.

5. The **median** is an example of a
   A. weighted mean.
   B. measure of central tendency.
   C. measure of dispersion.
   D. measure of position.

6. For a frequency distribution, the formula $\frac{\Sigma f + 1}{2}$ gives
   A. the total number of data items.
   B. one of the measures of position.
   C. the position of the mean.
   D. the position of the median.

7. **Measures of dispersion** include
   A. mean, median, and mode.
   B. range, variance, and percentile.
   C. range, variance, and standard deviation.
   D. percentiles, quartiles, and deciles.

8. The **z-score** determines
   A. how many items must lie within 1 standard deviation of the mean.
   B. how many standard deviations from the mean a data item is.
   C. the ratio of the standard deviation to the mean.
   D. the percentage of items that lie below the given item.

9. A **bimodal** distribution will never be
   A. normal.
   B. skewed.
   C. symmetric.
   D. spread out more than 3 standard deviations from its mean.

10. In any approximately normal distribution, an item with a negative *z*-score will be
    A. negative.
    B. less than the distribution range.
    C. between the first and third quartiles.
    D. less than the distribution mean.

ANSWERS

1. B   **2.** C   **3.** A   **4.** D   **5.** B
6. D   **7.** C   **8.** B   **9.** A   **10.** D

QUICK REVIEW

Concepts	Examples

12.1 Visual Displays of Data

Qualitative raw data are non-numerical.

Category	Number of Positions Available	
Surgeons	3	
GP physicians	5	Qualitative data
Nurse practitioners	7	
Registered nurses	12	
Licensed vocational nurses	8	

Quantitative raw data are numerical.

32, 41, 18, 22, 32, 41, 32, 38, 18, 28, 41, 28, 38, 28, 41, 18, 22, 32, 28, 28   Quantitative data

Concepts	Examples

Quantitative data can be **ranked** (arranged in numerical order).

18, 18, 18, 22, 22, 28, 28, 28, 28, 28, 32, 32, 32, 32, 38, 38, 41, 41, 41, 41   *Quantitative data (ranked)*

A **frequency distribution** makes a data presentation more concise.

A **relative frequency distribution** shows the fraction of the total number of items represented by each value.

x	f	$\dfrac{f}{n}$
18	3	0.15
22	2	0.10
28	5	0.25
32	4	0.20
38	2	0.10
41	4	0.20

Frequency distribution and relative frequency distribution

Total: $n = 20$

A **grouped frequency distribution** collects multiple items into separate classes.

Class limits are the least (or greatest) items that can occur in each class.

Class marks are the center points of the classes.

A **histogram** is a graphical presentation of a grouped frequency distribution.

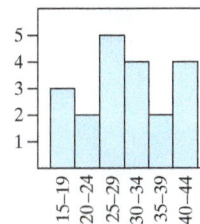

A **frequency polygon** is a sequence of line segments representing a frequency distribution.

Histogram

Frequency Polygon

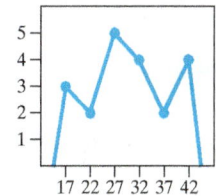

Lower class limits	15 20 25 30 35 40	
Upper class limits	19 24 29 34 39 44	
Class marks	17 22 27 32 37 42	

A **stem-and-leaf display** shows all data items in a visual arrangement.

(15–19) 1	8 8 8
(20–24) 2	2 2
(25–29) 2	8 8 8 8 8
(30–34) 3	2 2 2 2
(35–39) 3	8 8
(40–44) 4	1 1 1 1

A (double stem) stem-and-leaf display for the above data

Qualitative data can be effectively presented in a number of ways, including a bar graph, a circle graph, and a line graph.

Number of flu cases treated at a clinic during five days:

Mon	Tue	Wed	Thu	Fri
5	7	12	20	28

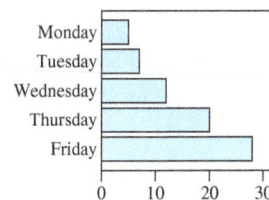

A **bar graph** is a visual depiction of a distribution, appropriate for either qualitative or quantitative data.

Concepts	*Examples*

A **circle graph** represents classes of data as sectors of a circle, so it conveys the relative sizes of the classes.

A **line graph** is useful especially for showing how a quantity changes over time, and it can also show comparisons of two or more quantities.

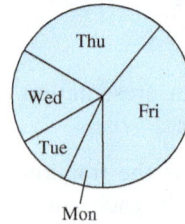

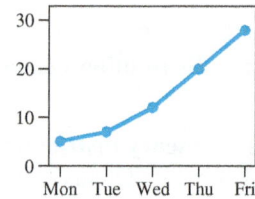

12.2 Measures of Central Tendency

The **mean,** the most common measure of central tendency, is computed using the formula

$$\bar{x} = \frac{\Sigma x}{n},$$

or, for frequency distributions,

$$\bar{x} = \frac{\Sigma (x \cdot f)}{\Sigma f}.$$

Find the mean for the distribution 2, 6, 3, 7, 4.

$$\bar{x} = \frac{\Sigma x}{n} = \frac{2 + 6 + 3 + 7 + 4}{5} = \frac{22}{5} = 4.4$$

Find the mean for the frequency distribution.

x	f	$x \cdot f$
3	2	6
4	5	20
5	3	15
Totals:	10	41

$$\bar{x} = \frac{\Sigma (x \cdot f)}{\Sigma f} = \frac{41}{10} = 4.1$$

A **weighted mean** is computed as

$$\bar{w} = \frac{\Sigma (x \cdot f)}{\Sigma f},$$

where f represents the "weighting factors." In the case of the mean of a frequency distribution, the mean is actually a weighted mean where the weighting factors are the frequencies.

Find the grade-point average. (Grade-point average is a weighted mean, where the weighting factors are the numbers of units for which each grade is assigned.)

Grade Earned	Grade Points Assigned, x	Number of Units, f	$x \cdot f$
A	4	3	12
B	3	6	18
C	2	4	8
		Totals: 13	38

$$\text{Grade-point average} = \bar{w} = \frac{\Sigma (x \cdot f)}{\Sigma f} = \frac{38}{13} \approx 2.92$$

The **median** is the "center" item of a distribution, or the mean of the two center items if there are an even number of items.

Find the median of the data set 2, 6, 3, 7, 4.

First rank the data: 2, 3, 4, 6, 7.

The median is the middle item, 4.

The **mode,** when it exists, is the item that occurs in the distribution more times than any other. A distribution having two distinct items with higher frequencies is **bimodal.**

Find the mode for the frequency distribution for which the mean (4.1) was computed above.

The mode is 4, since there were more 4s than there were of any other item.

Concepts	**Examples**

Concepts

It is possible to find measures of central tendency for a distribution presented in a stem-and-leaf display.

Symmetry and **skewness** are important properties of a distribution.

- If the graph of a distribution drops off about equally in both directions from the center, it is symmetric.

- If it trails off farther to the left (or right), it is "skewed" left (or right).

Examples

Consider this stem-and-leaf display.

```
5 | 2  4  7  8  8
6 | 1  2  5  5  6
7 | 1  1  2  3  6  6  9    Counting leaves gives n = 28.
8 | 2  2  4  5  8  8
9 | 0  0  0  2  4
```

$$\Sigma x = 52 + 54 + 57 + 58 + 58 + 61 + \cdots + 94 = 2081$$

$$\bar{x} = \frac{\Sigma x}{n} = \frac{2081}{28} \approx 74.3 \quad \text{Mean}$$

$$\frac{73 + 76}{2} = 74.5 \quad \text{The median is the mean of the 14th and 15th items.}$$

Examining the leaves reveals that the mode is 90.

Symmetry and skewness are illustrated by these distributions.

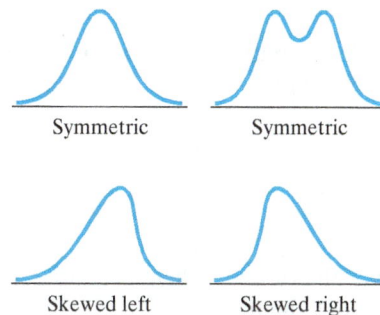

Symmetric   Symmetric

Skewed left   Skewed right

12.3 Measures of Dispersion

Range is the distance from the least item to the greatest item of the data set.

Range = (greatest value in the set) − (least value in the set)

Standard deviation s is a sort of "average" amount by which all items in a distribution differ from their mean.

$$s = \sqrt{\frac{\Sigma(x - \bar{x})^2}{n - 1}}$$

Chebyshev's Theorem
In any distribution, the fraction of all items that lie within k standard deviations of the mean (where $k > 1$) is *at least*

$$1 - \frac{1}{k^2}.$$

The **coefficient of variation** for a distribution expresses standard deviation as a percentage of the mean.

$$V = \frac{s}{\bar{x}} \cdot 100\%$$

Consider again the distribution

$$2, 6, 3, 7, 4; \quad \text{or, when ranked,} \quad 2, 3, 4, 6, 7.$$

$$\text{Range} = 7 - 2 = 5$$

Since the mean is 4.4, the standard deviation is as follows.

$$s = \sqrt{\frac{(2 - 4.4)^2 + (3 - 4.4)^2 + (4 - 4.4)^2 + (6 - 4.4)^2 + (7 - 4.4)^2}{5 - 1}}$$

$$\approx 2.07$$

For $k = 2$, at least

$$1 - \frac{1}{2^2}, \text{ or } 1 - \frac{1}{4} = \frac{3}{4},$$

of the items will lie within 2 standard deviations of the mean; that is, at least 3.75 items will lie between 0.26 and 8.54. In fact, in this case all 5 items lie within that range.

Find the coefficient of variation for this distribution.

$$V = \frac{2.07}{4.4} \cdot 100\% \approx 47\%$$

Concepts	**Examples**

12.4 Measures of Position

The **z-score** of an item gives the number of standard deviations that item is from the mean.

$$z = \frac{x - \bar{x}}{s}$$

Earlier, we saw a stem-and-leaf display for the data set

52 54 57 58 58 61 62 65 65 66 71 71 72 73
76 76 79 82 82 84 85 88 88 90 90 90 92 94.

The mean was 74.3. The standard deviation was 12.9.

Find the z-score for the item 84.

$$z = \frac{84 - 74.3}{12.9} \approx 0.75$$

84 is three-fourths of a standard deviation *above* the mean.

Find the z-score for the item 62.

$$z = \frac{62 - 74.3}{12.9} \approx -0.95$$

62 is $\frac{95}{100}$ of a standard deviation *below* the mean.

The **percentile** corresponding to an item tells roughly the percentage of all items that are below that item. There are ninety-nine percentiles, the first through the ninety-ninth.

For the above data, find the ninety-fifth percentile.

Since 95% of 28 (the number of items) is 26.6, we round up. The twenty-seventh item is 92.

For any numerical distribution, there are nine **deciles,** the first through the ninth.

Find the fourth decile (the same as the fortieth percentile).

40% of 28 is 11.2.

Therefore, take the twelfth item, which is 71.

For any numerical distribution, there are three **quartiles,** the first through the third.

Find the three quartiles for the data.

The median (Q_2) was found to be

74.5 (the mean of 73 and 76).

Q_1 is the median of the lower half of the items.

$$\frac{62 + 65}{2} = 63.5$$

Q_3 is the median of the upper half of the items.

$$\frac{85 + 88}{2} = 86.5$$

A **box plot** (or **box-and-whisker plot**) for a distribution is a visual representation that shows the least value, the greatest value, and all three quartiles.

Construct a box plot for this distribution of twenty-eight items.

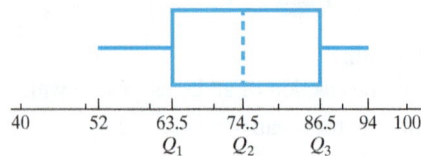

Concepts   *Examples*

12.5 **The Normal Distribution**

Many quantities are **distributed approximately "normally"**—that is, according to a characteristic bell-shaped curve. Their distributions may have any mean and any (positive) standard deviation, but we can calculate their properties by relating them to the so-called standard normal curve, which has mean 0 and standard deviation 1. The total area under the standard normal curve, from $-\infty$ to $+\infty$, is 1 square unit.

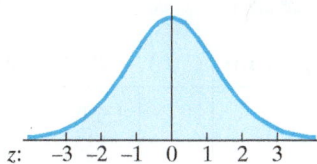

For any two items in a normal distribution, say x_1 and x_2, if we calculate their corresponding z-scores,

$$z_1 = \frac{x_1 - \bar{x}}{s} \quad \text{and} \quad z_2 = \frac{x_2 - \bar{x}}{s},$$

then **the following three quantities are equivalent.**

1. **The percentage of items in the original distribution that are between x_1 and x_2**

2. **The probability that a randomly chosen item from the original distribution will lie between x_1 and x_2**

3. **The area under the standard normal curve between z_1 and z_2**

Suppose the heights of the trainees on a large military base are normally distributed with mean 70 inches and standard deviation 5 inches.

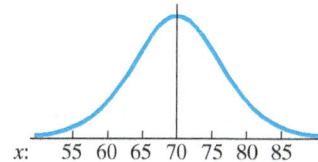

1. Find the *percentage* of trainees with heights between $x_1 = 70$ inches and $x_2 = 74$ inches.

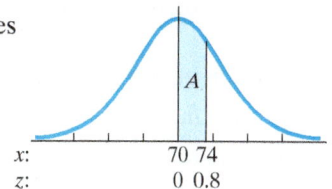

Since x_1 is at the mean, we need only calculate

$$z_2 = \frac{x_2 - \bar{x}}{s} = \frac{74 - 70}{5} = 0.8.$$

From the standard normal curve table,

$$A = 0.288, \quad \text{and the required percentage is} \quad 28.8\%.$$

2. Find the *probability* that a randomly selected trainee will have a height between $x_1 = 67$ inches and $x_2 = 77$ inches.

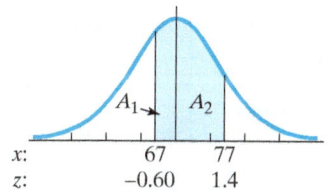

$$z_1 = \frac{67 - 70}{5} = -0.60, \quad \text{so} \quad A_1 = 0.226.$$

$$z_2 = \frac{77 - 70}{5} = 1.4, \quad \text{so} \quad A_2 = 0.419.$$

The required probability is

$$A_1 + A_2 = 0.226 + 0.419 = 0.645.$$

3. Find the *area* under the standard normal curve between the z-scores corresponding to trainee heights between 60 inches and 68 inches.

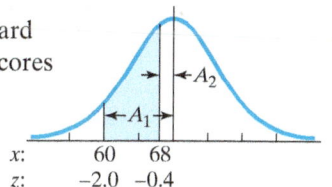

$$z_1 = \frac{60 - 70}{5} = -2.0 \quad \text{and} \quad z_2 = \frac{68 - 70}{5} = -0.4,$$

so $\qquad A_1 = 0.477 \quad \text{and} \quad A_2 = 0.155.$

The area under the curve between these two points is

$$A_1 - A_2 = 0.477 - 0.155 = 0.322.$$

CHAPTER 12 | TEST

1. **Visits to Social Networking Sites** The table shows the number of persons age 2 and older in any U.S. location, in millions, who visited the listed Web sites at least once in June 2013.

Rank	Web Site	Visitors
1	Facebook	144.7
2	LinkedIn	50.6
3	Twitter	39.1
4	Tumblr	35.9
5	Pinterest	27.9

Source: comScore Media Metrix; comScore qSearch

Use this information to determine each of the following.

(a) the mean number of visitors per site

(b) the range

(c) the standard deviation

(d) the coefficient of variation

2. **Cheaters Never Learn** The table here shows the results of an educational study of university physics students, comparing exam scores with students' rates of copying others' homework. The numbers in the table approximate letter grades on a 4-point scale (4.0 is an A, 3.0 is a B, and so on). Answer the questions in terms of copy rate.

Copy Rate	Pretest	Exam 1	Exam 2	Exam 3	Final Exam
<10%	2.70	2.75	2.90	2.80	2.95
10% to 30%	2.50	2.45	2.35	2.40	2.30
30% to 50%	2.45	2.43	2.30	2.10	2.00
>50%	2.40	2.05	1.70	1.80	1.60

Source: Table created using data from research reported in Physics Review–Special Topics–Physics Education *by David J. Palazzo, Young-Jin Lee, Rasil Warnakulasooriya, and David E. Pritchard of the Massachusetts Institute of Technology (MIT) physics faculty.*

(a) Which students generally improved their exam performance over the course of the semester?

(b) Which students did better on exam 3 than on exam 2?

(c) Which students had lower scores consistently from one exam to the next throughout the semester?

(d) Do you think that copying homework is generally a *cause* of lower exam scores? Explain.

Champion Trees *The table below lists the 9 largest national champion trees, based on the formula*

$$T = G + H + 0.25C,$$

where T = total points,

G = girth (*circumference of trunk 4.5 feet above the ground, measured in inches*)

H = height, measured in feet

and C = average crown spread, measured in feet

Tree Type	G (in.)	H (ft)	C (ft)	T	Location
Giant sequoia	1020	274	107	1321	Sequoia National Park, CA
Coast redwood	950	321	75	1290	Jedediah Smith Redwoods State Park, CA
Coast redwood	895	307	83	1223	Jedediah Smith Redwoods State Park, CA
Coast redwood	844	349	89	1216	Redwoods National Park, CA
Coast redwood	867	299	101	1191	Prairie Creek Redwoods State Park, CA
Western red cedar	761	159	45	931	Olympic National Park, WA
Sitka spruce	668	191	96	883	Olympic National Park, WA
Coast Douglas fir	505	281	71	804	Olympic National Forest, WA
Port Orford cedar	522	242	35	773	Coos, OR

Source: The World Almanac and Book of Facts 2014.

Use this information for Exercises 3 and 4.

3. For the nine trees listed, find the following.

(a) the median height

(b) the first quartile in girth

(c) the seventh decile in total points

4. The tenth-ranking tree in the country is a common bald cypress on Cat Island, Louisiana, with $G = 647$ inches, $H = 96$ feet, and $C = 74$ feet. For this tree, answer the following questions.

   (a) Find its total points.

   (b) Where would it have ranked based on girth alone?

   (c) How much greater would its girth need to be to exceed the seventh-ranked girth figure?

   (d) Assuming a roughly circular cross section at 4.5 feet above the ground, approximate the diameter of the trunk (to the nearest foot) at that height.

College Endowment Assets   *The table shows the top six U.S. colleges for 2012 endowment assets, in billions of dollars.*

Rank	College/University	Endowment Assets
1	Harvard University	30.4
2	Yale University	19.3
3	University of Texas System	18.3
4	Stanford University	17.0
5	Princeton University	17.0
6	Massachusetts Institute of Technology	10.1

Source: World Almanac and Book of Facts 2014.

5. Construct a bar graph for these data.

6. What percentage (to one decimal place) of the total was held by Yale?

7. What was the median value?

Pediatrics Patients   *Khalida worked the 22 weekdays of last month in the pediatrics clinic and checked in the following numbers of patients.*

14	24	26	11	16	25	10	12	8	18	13
15	20	18	22	24	17	19	19	7	15	30

8. Construct grouped frequency and relative frequency distributions. Use five uniform classes of width 5 where the first class has a lower limit of 6. (Round relative frequencies to two decimal places.)

9. From your frequency distribution of **Exercise 8,** construct **(a)** a histogram and **(b)** a frequency polygon. Use appropriate scales and labels.

10. For the data above, how many uniform classes would be required if the first class had limits 7–9?

In Exercises 11–14, find the indicated measures for the following frequency distribution.

Value	8	10	12	14	16	18
Frequency	3	8	10	8	5	1

11. the mean

12. the median

13. the mode

14. the range

15. *Exam Scores in a Criminal Justice Class*   The following data are scores achieved by 30 students on a criminal justice exam. Arrange the data into a stem-and-leaf display with leaves ranked.

52	43	84	65	77	70	79	61	66	80
55	48	68	78	71	38	45	64	67	73
33	49	61	91	84	77	50	67	79	72

Use the stem-and-leaf display below for Exercises 16–21.

2	3 3 4
2	6 7 8 9 9
3	0 1 1 2 3 3 3 4
3	5 6 7 8 8 9
4	1 2 2 4
4	5 7 9
5	2 4
5	8
6	0

Compute the measures required in Exercises 16–20.

16. the median

17. the mode(s), if any

18. the range

19. the third decile

20. the eighty-fifth percentile

21. Construct a box plot for the given data, showing values for the five important quantities on the numerical scale.

22. *Test Scores in a Training Institute*   A certain training institute gives a standardized test to large numbers of applicants nationwide. The resulting scores form a normal distribution with mean 80 and standard deviation 5. Find the percent of all applicants with scores as follows. (Use the empirical rule.)

   (a) between 70 and 90

   (b) greater than 95 or less than 65

   (c) less than 75

   (d) between 85 and 90

Triple Jump Champions　*All Summer Olympic champion-ship distances in the Triple Jump from 1896 through 2012 averaged 16.38 meters, with standard deviation 1.34 meters. (As of 2015, the all-time greatest distance was 18.17 meters, achieved by Mike Conley of the United States in 1992.) Assuming the distribution of distances was approximately normal, find (to the nearest thousandth) the probability that a randomly selected distance from the distribution would fall into each of the following intervals.*

23. between 16.8 meters and 17.9 meters

24. between 14.6 meters and 16.8 meters

Season Statistics in Major League Baseball　*The tables below show the 2013 statistics on games won for all three divisions of both major baseball leagues. In each case,*

$n =$ *number of teams in the division,*

$\bar{x} =$ *average (mean) number of games won,*

and　$s =$ *standard deviation of number of games won.*

American League

East Division	Central Division	West Division
$n = 5$	$n = 5$	$n = 5$
$\bar{x} = 86.6$	$\bar{x} = 80.0$	$\bar{x} = 77.4$
$s = 8.7$	$s = 14.4$	$s = 17.8$

National League

East Division	Central Division	West Division
$n = 5$	$n = 5$	$n = 5$
$\bar{x} = 78.2$	$\bar{x} = 84.2$	$\bar{x} = 79.8$
$s = 13.1$	$s = 13.5$	$s = 7.3$

Refer to the preceding tables for Exercises 25–28.

25. Overall, who had the greatest winning average, the East teams, the Central teams, or the West teams?

26. Overall, where were the teams the least "consistent" in number of games won: East, Central, or West?

27. Find (to the nearest tenth) the average number of games won for all East Division teams.

28. The Boston Red Sox, in the American League East, and the St. Louis Cardinals, in the National League Central, each won 97 games. Use *z*-scores to determine which of these two teams did relatively better within its own division of 5 teams.

Remission Data for Cocaine and Marijuana Addiction　*The chapter opener and the **When Will I Ever Use This?** feature in **Section 12.1** showed research results for alcohol and tobacco addiction and recovery. (A 0.5 cumulative probability of remission required about 17 years for alcoholics, and about 30 years for smokers.) This table gives data inferred from the same study for cocaine and marijuana addiction.*

Cocaine		Marijuana	
x	y	x	y
1.9	0.27	2.1	0.22
6.1	0.64	6.1	0.55
9.8	0.79	10.3	0.69
14.0	0.89	14.0	0.79
21.0	0.94	19.2	0.84

Data source: G. Heyman, Annual Review of Clinical Psychology 2013.

29. Plot the data points from the table, draw two smooth curves, and then estimate the 0.5 cumulative probability of remission for:

(a) cocaine.　　　　**(b)** marijuana.

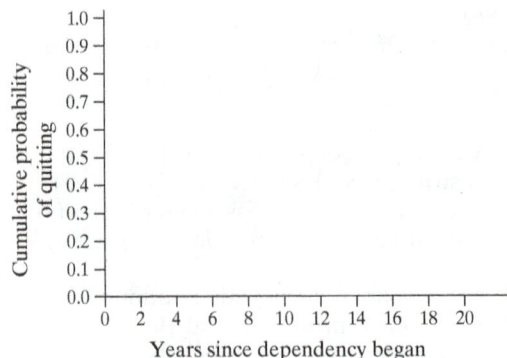

30. Make a conjecture about why remission periods are this much shorter for cocaine and marijuana than for alcohol and tobacco.

Personal Financial Management

13

Financial planners advise their clients in making decisions that will help them reach their financial goals. Many people find it beneficial to consult a professional who can offer a knowledgeable and objective perspective in this area. Financial planners can help people establish financial goals, increase their financial security, and even ask the right questions.

1. *Am I spending my income wisely?*
2. *Do I need a budget?*
3. *What are my financial priorities?*
4. *Should I buy a home, or is renting a better idea?*
5. *Am I saving enough for retirement or for my children's college education?*
6. *Am I taking advantage of available tax savings?*
7. *What are the best investment vehicles for me?*

Questions like these are within the purview of good financial planners, and many of the topics discussed in this chapter are related to them.

13.1   THE TIME VALUE OF MONEY

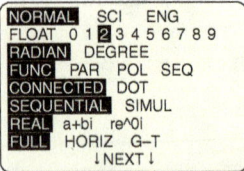

Interest

To determine the value of money, we consider not only the amount (number of dollars) but also the particular point in time that the value is to be determined. If we borrow an amount of money today, we will repay a larger amount later. This increase in value is known as **interest.** The money *gains value over time.*

The amount of a loan or a deposit is called the **principal.** The interest is usually computed as a percent of the principal. This percent is called the **rate of interest** (or the **interest rate,** or simply the **rate**). The rate of interest is always assumed to be an annual rate unless otherwise stated.

Interest calculated only on principal is called **simple interest.** Interest calculated on principal plus any previously earned interest is called **compound interest.**

Simple Interest

Simple interest is calculated according to the following formula.

```
NORMAL   SCI   ENG
FLOAT  0 1 2 3 4 5 6 7 8 9
RADIAN  DEGREE
FUNC   PAR  POL  SEQ
CONNECTED  DOT
SEQUENTIAL  SIMUL
REAL  a+bi  re^θi
FULL  HORIZ  G-T
       ↓NEXT↓
```

Because the calculations in this chapter involve money, answers often will be rounded to the nearest $0.01. The TI-83/84 will display answers to the nearest 0.01 with the setting shown above.

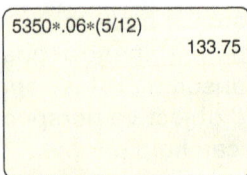

SIMPLE INTEREST

If P = principal, r = annual interest rate, and t = time (in years), then the **simple interest** I is calculated as follows.

$$I = Prt$$

EXAMPLE 1 Finding Simple Interest

Find the simple interest paid to borrow $5350 for 5 months at 6%.

Solution

```
5350*.06*(5/12)
              133.75
```

This is the computation required to solve **Example 1.**

$$I = Prt \qquad \text{Simple interest formula}$$

$5 \text{ months is } \frac{5}{12} \text{ of a year.}$

$$= \$5350(0.06)\left(\frac{5}{12}\right) \qquad P = \$5350, r = 6\% = 0.06$$

$$= \$133.75 \qquad \text{Calculate.}$$ ∎

Future Value and Present Value

In **Example 1,** at the end of 5 months the borrower would have to repay

Principal   Interest

$\$5350 \quad + \quad \$133.75 = \$5483.75.$

The total amount repaid is sometimes called the **maturity value** (or simply the **value**) of the loan. We will generally refer to it as the **future value,** or **future amount,** because when a loan is being set up, repayment will be occurring in the future. We use A to denote future amount (or value). The original principal, denoted P, can also be thought of as **present value.**

Future value depends on principal (present value) and interest as follows.

$$A = P + I = P + Prt = P(1 + rt)$$

> ### FUTURE VALUE FOR SIMPLE INTEREST
>
> If a principal P is borrowed at simple interest for t years at an annual interest rate of r, then the **future value** of the loan, denoted A, is calculated as follows.
>
> $$A = P(1 + rt)$$

EXAMPLE 2 Finding Future Value for Simple Interest

James took out a simple interest loan for $210 to purchase textbooks and school supplies. If the annual interest rate is 6% and he must repay the loan after 6 months, find the future value (the maturity value) of the loan.

Solution

$$A = P(1 + rt) \qquad \text{Future value formula}$$

$$= \$210\left[1 + 0.06\left(\frac{6}{12}\right)\right] \qquad P = \$210,\ r = 6\% = 0.06,\ t = \frac{6}{12}$$

$$= \$216.30 \qquad \text{Calculate.}$$

At the end of 6 months, James will need to repay $216.30. ∎

```
210(1+.06(6/12))
                  216.3
```

This is the computation required to solve **Example 2.**

Sometimes the future value is known, and we need to compute the present value. For this purpose, we solve the future value formula for P.

$$A = P(1 + rt) \qquad \text{Future value formula for simple interest}$$

$$P = \frac{A}{1 + rt} \qquad \text{Solve the formula for } P.$$

EXAMPLE 3 Finding Present Value for Simple Interest

Suppose that James (**Example 2**) is granted a 6-month *deferral* of the $210 payment rather than a loan. That is, instead of incurring interest for 6 months, he will have to pay just the $210 at the end of that period. If he has extra money right now, and if he can earn 2% simple interest on savings, what lump sum must he deposit now so that its value will be $210 after 6 months?

Solution

$$P = \frac{A}{1 + rt} \qquad \text{Future value formula solved for } P$$

$$P = \frac{\$210}{1 + (0.02)\left(\frac{6}{12}\right)} = \$207.92 \qquad \text{Substitute known values and calculate.}$$

```
210/(1+.02(6/12))
                  207.92
```

This is the computation required to solve **Example 3.**

A deposit of $207.92 now, growing at 2% simple interest, will grow to $210 over a 6-month period. ∎

Compound Interest

Interest paid on principal plus interest is called *compound interest*. After a certain period, the interest earned so far is *credited* (added) to the account, and the sum (principal plus interest) then earns interest during the next period.

Compare simple and compound interest for a $1000 deposit at 2% interest for 5 years.

Solution

$$A = P(1 + rt)$$ Future value formula

$$= \$1000(1 + 0.02 \cdot 5)$$ $P = \$1000$, $r = 2\%$, $t = 5$

$$= \mathbf{\$1100}$$ Calculate.

After 5 years, $1000 grows to $1100 subject to 2% simple interest.
The result of compounding (annually) for 5 years is shown in **Table 1.**

```
1000 → X
              1000.00
X+.02X → X
              1020.00
              1040.40
              1061.21
              1082.43
```

This screen supports the Ending Balance figures in **Table 1.**

Table 1 A $1000 Deposit at 2% Interest Compounded Annually

Year	Beginning Balance	Interest Earned $I = Prt$	Ending Balance
1	$1000.00	$1000.00(0.02)(1) = $20.00	$1020.00
2	$1020.00	$1020.00(0.02)(1) = $20.40	$1040.40
3	$1040.40	$1040.40(0.02)(1) = $20.81	$1061.21
4	$1061.21	$1061.21(0.02)(1) = $21.22	$1082.43
5	$1082.43	$1082.43(0.02)(1) = $21.65	**$1104.08**

Under annual compounding for 5 years, $1000 grows to $1104.08, which is $4.08 more than under simple interest. ∎

Based on the compounding pattern of **Example 4,** we now develop a future value formula for compound interest. In practice, earned interest can be credited to an account at time intervals other than 1 year (usually more often). For example, it can be done semiannually, quarterly, monthly, or daily. This time interval is called the **compounding period** (or simply the **period**). Start with the following definitions:

P = original principal deposited, $\qquad$ r = annual interest rate,

n = number of periods per year, $\qquad$ m = total number of periods.

During each individual compounding period, interest is earned according to the simple interest formula, and as interest is added, the beginning principal increases from one period to the next. During the first period, the interest earned is given by

$$\text{Interest} = P(r)\left(\frac{1}{n}\right) \qquad \begin{array}{l}\text{Interest} = (\text{Principal})(\text{rate})(\text{time}),\\ \text{one period} = \frac{1}{n}\,\text{year}\end{array}$$

$$= P\left(\frac{r}{n}\right). \qquad \text{Rewrite } (r)\left(\frac{1}{n}\right) \text{ as } \left(\frac{r}{n}\right).$$

At the end of the first period, the account then contains

$$\text{Ending amount} = \underset{\substack{\uparrow\\ \text{Beginning amount}}}{P} + \underset{\substack{\uparrow\\ \text{Interest}}}{P\left(\frac{r}{n}\right)}$$

$$= P\left(1 + \frac{r}{n}\right). \qquad \text{Factor } P \text{ from both terms.}$$

Now during the second period, the interest earned is given by

$$\text{Interest} = \left[P\left(1 + \frac{r}{n}\right)\right](r)\left(\frac{1}{n}\right) \qquad \textcolor{teal}{\text{Interest} = [\,\text{Principal}\,](\text{rate})(\text{time})}$$

$$= P\left(1 + \frac{r}{n}\right)\left(\frac{r}{n}\right),$$

so the account ends the second period containing

$$\text{Ending amount} = \overset{\textcolor{teal}{\text{Beginning amount}}}{P\left(1 + \frac{r}{n}\right)} + \overset{\textcolor{teal}{\text{Interest}}}{P\left(1 + \frac{r}{n}\right)\left(\frac{r}{n}\right)}$$

$$= P\left(1 + \frac{r}{n}\right)\left[1 + \frac{r}{n}\right] \qquad \textcolor{teal}{\text{Factor } P\left(1 + \frac{r}{n}\right) \text{ from both terms.}}$$

$$= P\left(1 + \frac{r}{n}\right)^2. \qquad \textcolor{teal}{a \cdot a = a^2}$$

Consider one more period, namely, the third. The interest earned is

$$\text{Interest} = \left[P\left(1 + \frac{r}{n}\right)^2\right](r)\left(\frac{1}{n}\right) \qquad \textcolor{teal}{\text{Interest} = [\,\text{Principal}\,](\text{rate})(\text{time})}$$

$$= P\left(1 + \frac{r}{n}\right)^2\left(\frac{r}{n}\right), \qquad \textcolor{teal}{\text{Multiply.}}$$

so the account ends the third period containing

$$\text{Ending amount} = \overset{\textcolor{teal}{\text{Beginning amount}}}{P\left(1 + \frac{r}{n}\right)^2} + \overset{\textcolor{teal}{\text{Interest}}}{P\left(1 + \frac{r}{n}\right)^2\left(\frac{r}{n}\right)}$$

$$= P\left(1 + \frac{r}{n}\right)^2\left[1 + \frac{r}{n}\right] \qquad \textcolor{teal}{\text{Factor } P\left(1 + \frac{r}{n}\right)^2 \text{ from both terms.}}$$

$$= P\left(1 + \frac{r}{n}\right)^3. \qquad \textcolor{teal}{a^2 \cdot a = a^3}$$

Table 2 summarizes the preceding results.

Table 2 Compound Amount

Period Number	Beginning Amount	Interest Earned During Period	Ending Amount
1	P	$P\left(\frac{r}{n}\right)$	$P\left(1 + \frac{r}{n}\right)$
2	$P\left(1 + \frac{r}{n}\right)$	$P\left(1 + \frac{r}{n}\right)\left(\frac{r}{n}\right)$	$P\left(1 + \frac{r}{n}\right)^2$
3	$P\left(1 + \frac{r}{n}\right)^2$	$P\left(1 + \frac{r}{n}\right)^2\left(\frac{r}{n}\right)$	$P\left(1 + \frac{r}{n}\right)^3$
$\vdots$	$\vdots$	$\vdots$	$\vdots$
m	$P\left(1 + \frac{r}{n}\right)^{m-1}$	$P\left(1 + \frac{r}{n}\right)^{m-1}\left(\frac{r}{n}\right)$	$P\left(1 + \frac{r}{n}\right)^{m}$

King Hammurabi tried to hold interest rates at 20% for both silver and gold, but moneylenders ignored his decrees.

The quantity

$$\left(1 + \frac{r}{n}\right)^m$$

can be evaluated using a key such as y^x on a scientific calculator or $\wedge$ on a graphing calculator.

The lower right entry of **Table 2** provides the following formulas.

FUTURE VALUE FOR COMPOUND INTEREST

If P dollars are deposited at an annual interest rate of r, compounded n times per year, and the money is left on deposit for a total of m periods (or $t = \frac{m}{n}$ years), then the **future value, A** (the final amount on deposit), is calculated as follows.

$$A = P\left(1 + \frac{r}{n}\right)^m \quad \text{or} \quad A = P\left(1 + \frac{r}{n}\right)^{nt}$$

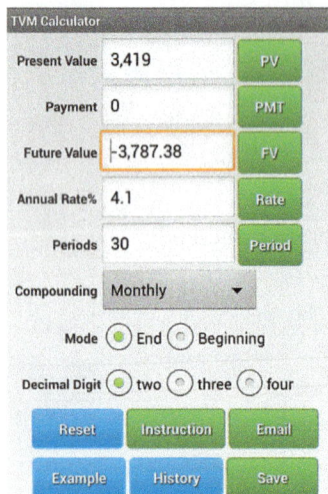

TVM Calculator

Present Value	3,419	PV
Payment	0	PMT
Future Value	-3,787.38	FV
Annual Rate%	4.1	Rate
Periods	30	Period
Compounding	Monthly ▼	

Mode ⦿ End ○ Beginning

Decimal Digit ⦿ two ○ three ○ four

Reset | Instruction | Email
Example | History | Save

The highlighted field in the screenshot above shows the future value calculated in **Example 5(a).** The negative value in the Present Value field indicates a negative cash flow for the investor at the time of deposit.

Several **financial calculator apps** are available for smartphones. This one is the (free) TVM Calculator from www.fncalculator.com.

```
N=20
I%=3
PV=-12450
PMT=0
→  FV=14456.74257
P/Y=4
C/Y=4
PMT: END  BEGIN
```

The financial functions of the TI-83/84 Plus calculator allow the user to solve for a missing quantity related to the time value of money. The arrow indicates the display that supports the future value answer in **Example 5(b).**

EXAMPLE 5 Finding Future Value for Compound Interest

Find the future value (final amount on deposit) and the amount of interest earned for the following deposits.

(a) $3419 at 4.1% compounded monthly for 30 months

(b) $12,450 at 3% compounded quarterly for 5 years

Solution

(a) Here $P = \$3419$, $r = 4.1\% = 0.041$, $n = 12$, and $m = 30$.

$$A = P\left(1 + \frac{r}{n}\right)^m$$

$$A = \$3419\left(1 + \frac{0.041}{12}\right)^{30} = \$3787.38$$

Future value   Present value

Interest earned $= \$3787.38 \;-\; \$3419$

$= \$368.38$

(b) Here $P = \$12,450$, $r = 3\% = 0.03$, $n = 4$, and $t = 5$.

$$A = P\left(1 + \frac{r}{n}\right)^{nt}$$

$$A = \$12,450\left(1 + \frac{0.03}{4}\right)^{(4)(5)} = \$14,456.74$$

Future value   Present value

Interest earned $= \$14,456.74 \;-\; \$12,450$

$= \$2006.74$ ∎

Compound interest problems sometimes require that we solve the formula for P to compute the present value when the future value is known.

$$A = P\left(1 + \frac{r}{n}\right)^m \qquad \text{Future value for compound interest}$$

$$P = \frac{A}{\left(1 + \frac{r}{n}\right)^m} \qquad \text{Solve the formula for } P.$$

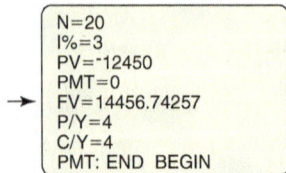

EXAMPLE 6   Finding Present Value for Compound Interest

Ashley will need $40,000 in 5 years to help pay for her college education. What lump sum, deposited today at 4% compounded quarterly, will produce the necessary amount?

Solution

This question requires that we find the present value P based on the following.

Future value:	$A = \$40{,}000$
Annual rate:	$r = 4\% = 0.04$
Periods per year:	$n = 4$
Total number of periods:	$m = (5)(4) = 20$

$$P = \frac{A}{\left(1 + \frac{r}{n}\right)^m} \qquad \text{Future value formula solved for } P$$

$$P = \frac{\$40{,}000}{\left(1 + \frac{0.04}{4}\right)^{20}} \qquad \text{Substitute known values.}$$

$$P = \$32{,}781.78 \qquad \text{Calculate.}$$

Assuming interest of 4% compounded quarterly can be maintained, $40,000 can be attained 5 years in the future by depositing $32,781.78 today. ∎

The next example uses logarithms. See **Section 8.6** or your calculator manual.

EXAMPLE 7   Finding the Time Required to Double a Principal Deposit

In a savings account paying 3% interest, compounded daily, when will the amount in the account be twice the original principal?

Solution

The future value must equal two times the present value.

$$2P = P\left(1 + \frac{r}{n}\right)^{nt} \qquad \text{Substitute } 2P \text{ for } A \text{ in the future value formula.}$$

$$2 = \left(1 + \frac{r}{n}\right)^{nt} \qquad \text{Divide both sides by } P.$$

$$2 = \left(1 + \frac{0.03}{365}\right)^{365t} \qquad \text{Substitute values of } r \text{ and } n.$$

$$\log 2 = \log\left(1 + \frac{0.03}{365}\right)^{365t} \qquad \text{Take the logarithm of both sides.}$$

$$\log 2 = 365t \log\left(1 + \frac{0.03}{365}\right) \qquad \text{Use the power property of logarithms.}$$

$$t = \frac{\log 2}{365 \log\left(1 + \frac{0.03}{365}\right)} \qquad \text{Solve for } t.$$

$$t = 23.106 \qquad \text{Round to the nearest thousandth of a year.}$$

This equates to 23 years, 39 days (ignoring leap years). ∎

The 1954 musical *The Pajama Game* was revived on Broadway twice, most recently in 2006 starring Harry Connick, Jr. The musical tells the story of the Sleeptite Pajama Factory workers attempting to secure a raise of $7\frac{1}{2}$ cents per hour.

The musical features a number called **"Seven and a Half Cents"** (after the novel that inspired the musical), with lyrics claiming that a small raise of $7\frac{1}{2}$ cents per hour will amount to $852.74 in 5 years, $1705.48 in 10 years, and $3411.96 in 20 years. Assuming that the figure for 5 years is correct, is the one for 10 years also correct? How about the one for 20 years? Oops! Judging by the first two figures, were the workers planning to invest their "extra" wages?

The power of compound interest is being put to use in the town of Union City, Michigan. Eli Hooker, chairman of the local Bicentennial Committee in 1976, saw that there was not enough money to put on a proper celebration that year. So, in order to help his town prepare for the tricentennial, in 2076, he collected twenty-five dollars apiece from 42 patriotic residents and deposited the money in a local bank. Compounded at 7%, that money would grow to a million dollars by 2076.

Unfortunately, the million dollars won't be worth as much then as we might think. If the community decides to hire people to parade around in historical costumes, it might have paid them $3 per hour in 1976. The going wage in 2076, assuming 5% annual inflation for a hundred years, would be nearly $400 per hour.

Example 8 shows the importance of starting early to maximize the long-term advantage of compounding.

EXAMPLE 8   Comparing Retirement Plans

Compare the results at age 65 for the following two retirement plans. Both plans earn 8% annual interest throughout the account-building period.

> *Plan A:*   Angie begins saving at age 20, deposits $2000 on every birthday from age 21 to age 30 (10 deposits, or $20,000 total), and thereafter makes no additional contributions.

> *Plan B:*   Barry waits until age 30 to start saving and then makes deposits of $2000 on every birthday from age 31 to age 65 (35 deposits, or $70,000 total).

Solution

Table 3 shows how both accounts build over the years. $20,000, deposited earlier, produces $83,744 more than $70,000, deposited later.

Table 3 Retirement Plans Compared

Age	Plan A	Plan B
20	0	0
25	11,733	0
30	28,973	0
35	42,571	11,733
40	62,551	28,973
45	91,908	54,304
50	135,042	91,524
55	198,422	146,212
60	291,547	226,566
65	428,378	344,634

In **Example 8,** the *early* deposits of Plan A outperformed the *greater number* of deposits of Plan B because the interest rate was high enough to result in a Plan A balance at age 30 that Plan B could never overtake, despite additional Plan B deposits from then on. However, if the interest rate had been 6% rather than 8%, Plan B would have overtaken Plan A (at age 57) and at age 65 would have come out ahead by $20,253 (to the nearest dollar).

Effective Annual Yield

Banks, credit unions, and other financial institutions often advertise two rates: first, the actual annualized interest rate, or **nominal rate** (the "named" or "stated" rate), and second, the rate that would produce the same final amount, or future value, at the end of 1 year if the interest being paid were simple rather than compound. This is called the "effective rate," or, more commonly the **effective annual yield.** (It may be denoted **APY** for **"annual percentage yield."**) Because the interest is normally compounded multiple times per year, the effective annual yield will usually be somewhat higher than the nominal rate.

EXAMPLE 9 Finding Effective Annual Yield

What is the effective annual yield of an account paying a nominal rate of 2.50%, compounded quarterly?

Solution

From the given data, $r = 0.025$ and $n = 4$. Suppose we deposited $P = \$1$ and left it for 1 year ($m = 4$). Then the compound future value formula gives

$$A = 1 \cdot \left(1 + \frac{0.025}{4}\right)^4 \approx 1.0252.$$

The initial deposit of $1, after 1 year, has grown to $1.0252.

$$\text{Interest earned} = 1.0252 - 1 \qquad \text{Interest} = \text{future value} - \text{present value}$$

$$= 0.0252, \quad \text{or} \quad 2.52\%$$

A nominal rate of 2.50% results in an effective annual yield of 2.52%. ∎

Generalizing the procedure of **Example 9** gives the following formula.

> **EFFECTIVE ANNUAL YIELD**
>
> A nominal interest rate of r, compounded n times per year, is equivalent to the following **effective annual yield.**
>
> $$Y = \left(1 + \frac{r}{n}\right)^n - 1$$

When shopping for loans or savings opportunities, a borrower should seek the least yield available, while a depositor should look for the greatest.

EXAMPLE 10 Comparing Savings Rates

Yolanda wants to deposit $2800 into a savings account and has narrowed her choices to the three institutions represented here. Which is the best choice?

Institution	Rate on Deposits of $1000 to $5000
Friendly Credit Union	2.08% annual rate, compounded monthly
Premier Savings	2.09% annual yield
Neighborhood Bank	2.05% compounded daily

Solution

Compare the effective annual yields for the three institutions.

Friendly: $Y = \left(1 + \dfrac{0.0208}{12}\right)^{12} - 1 = 0.0210 = 2.10\%$

Premier: $Y = 2.09\%$

Neighborhood: $Y = \left(1 + \dfrac{0.0205}{365}\right)^{365} - 1 = 0.0207 = 2.07\%$

The best of the three yields, 2.10%, is offered by Friendly Credit Union. ∎

▶Eff(2.50,4)

         2.523535309

Financial calculators are programmed to compute effective interest rate. Compare to the result in **Example 9.**

Table 4

Consumer Price Index
(CPI-U)*

Year	Average CPI-U	Percent Change in CPI-U
1980	82.4	13.5
1981	90.9	10.3
1982	96.5	6.2
1983	99.6	3.2
1984	103.9	4.3
1985	107.6	3.6
1986	109.6	1.9
1987	113.6	3.6
1988	118.3	4.1
1989	124.0	4.8
1990	130.7	5.4
1991	136.2	4.2
1992	140.3	3.0
1993	144.5	3.0
1994	148.2	2.6
1995	152.4	2.8
1996	156.9	3.0
1997	160.5	2.3
1998	163.0	1.6
1999	166.6	2.2
2000	172.2	3.4
2001	177.1	2.8
2002	179.9	1.6
2003	184.0	2.3
2004	188.9	2.7
2005	195.3	3.4
2006	201.6	3.2
2007	207.3	2.8
2008	215.3	3.8
2009	214.5	−0.4
2010	218.1	1.6
2011	224.9	3.2
2012	229.6	2.1
2013	233.0	1.5

Source: Bureau of Labor Statistics.
*The period 1982 to 1984: 100

Inflation

Interest reflects how money *gains value over time* when it is borrowed or lent. On the other hand, in terms of the equivalent number of goods or services that a given amount of money will buy, money normally *loses value over time*. This results in a periodic increase in the cost of living, which is called **price inflation.**

In the United States, the Bureau of Labor Statistics publishes **consumer price index (CPI)** figures, which reflect the prices of certain items purchased by large numbers of people. The items include such things as food, housing, automobiles, fuel, and clothing. Items such as yachts and expensive jewelry are not included.

Current data are published regularly at www.bls.gov/cpi. **Table 4** gives values of the primary index representing "all urban consumers" (the CPI-U). The value shown for a given year is actually the average (arithmetic mean) of the twelve monthly figures for that year. The table shows, for example, that the average CPI-U for 2007 (207.3) was 2.8% greater than the average value for 2006 (201.6).

Deflation is a *decrease* in price levels from one year to the next. A brief period of minor deflation, along with a general economic slowdown, usually is called a **recession.** (For example, in **Table 4,** note the 2009 rate of −0.4.)

Unlike account values under interest compounding, which make sudden jumps at just certain points in time (such as quarterly, monthly, or daily), price levels tend to fluctuate gradually over time. Thus, it is appropriate, for inflationary estimates, to use the formula for continuous compounding (introduced in **Section 8.6**).

FUTURE VALUE FOR CONTINUOUS COMPOUNDING

If an initial deposit of P dollars earns continuously compounded interest at an annual rate r for a period of t years, then the **future value, A,** is calculated as follows.

$$A = Pe^{rt}$$

EXAMPLE 11 Predicting Inflated Salary Levels

In 2012, the median salary for athletic trainers and exercise physiologists was $42,690 (*Source:* Bureau of Labor Statistics, U.S. Dept. of Labor). Approximately what salary would a person in this field need in 2032 to maintain purchasing power, assuming the inflation rate were to persist at each of the following levels?

(a) 3% (approximately the 2011 level)

(b) 13% (approximately the 1980 level)

Solution

(a) In this case we can use the continuous compounding future value formula with $P = \$42,690$, $r = 0.03$, and $t = 20$ (since 20 years pass between 2012 and 2032), along with the e^x key on a calculator.

$$A = Pe^{rt} = \$42,690\, e^{(0.03)(20)} = \$77,786.25$$

The required salary in 2032 would be about $78,000.

(b) For this level of inflation, we have the following.

$$A = Pe^{rt} = \$42,690\, e^{(0.13)(20)} = \$574,766.98$$

The required salary in 2032 would be about $575,000. ∎

$42690e^{0.03*20}$

$42690e^{0.13*20}$

77786.25

574766.98

These are the computations required to solve the two parts of **Example 11** on the previous page.

To compare equivalent general price levels in any 2 years, we can use the proportion below. (A proportion is a statement that says that two ratios are equal. See **Section 7.3.**)

> ### INFLATION PROPORTION
>
> For a given consumer product or service subject to average inflation, prices in two different years are related as follows.
>
> $$\frac{\text{Price in year A}}{\text{Price in year B}} = \frac{\text{CPI in year A}}{\text{CPI in year B}}$$

EXAMPLE 12   Comparing a Tuition Increase to Average Inflation

The average cost of tuition and fees at U.S. institutions of higher learning increased from $1626 for the 1982–83 academic year to $10,683 for the 2012–13 academic year (*Source:* National Center for Education Statistics, U.S. Dept. of Education). Compare these increases to average inflation over the same period.

Solution

Since an academic year spans the last half of one calendar year and the first half of the next, use the average of the CPI-U figures from **Table 4** over each two-year period.

1982–83 academic year: $\dfrac{96.5 + 99.6}{2} = 98.05$ Average of CPI-U figures for 1982 and 1983

2012–13 academic year: $\dfrac{229.6 + 233.0}{2} = 231.3$ Average of CPI-U figures for 2012 and 2013

Let x represent what we would expect tuition and fees to cost in 2012–13 if their cost had increased at the average rate since the 1982–83 academic year.

$$\frac{\text{Cost in } 2012-13}{\text{Cost in } 1982-83} = \frac{\text{CPI for } 2012-13}{\text{CPI for } 1982-83} \qquad \text{Inflation proportion}$$

$$\frac{x}{\$1626} = \frac{231.3}{98.05} \qquad \text{Substitute CPI values.}$$

$$x = \frac{231.3}{98.05} \cdot \$1626 \qquad \text{Solve for } x.$$

$$x \approx \$3836$$

Now compare the actual 2012–13 cost of $10,683 with the expected figure, $3836.

$$\frac{\$10,683}{\$3836} = 2.78 \qquad 2.78 = 1 + 1.78 = 100\% + 178\%$$

Over the period of interest, the cost of tuition and fees at institutions of higher learning in the United States increased at a rate approximately 178% greater than the average CPI-U rate.   ■

Monetary inflation devalues the currency just as **price inflation** does, but it does so through a direct increase in the money supply within the economy. Examples in the United States include the issuance in 2009 and 2010 of massive government debt and massive government spending in "stimulus" programs, as well as the Federal Reserve's purchasing of government and mortgage bonds through "quantitative easing," which continues into 2014.

Debate has persisted among economists about whether actions of the government and central bank would accomplish their goals.

Would economic recovery continue, or would major deflation occur?

Most everyone agreed that the long-term result of monetary inflation would be more and more price inflation. Some deny that any real distinction exists; they maintain that because of complex interrelated economic forces, monetary inflation and price inflation are essentially the same thing.

When working with quantities, such as inflation, where continual fluctuations and inexactness prevail, we often develop rough "rules of thumb" for obtaining quick estimates. One example is the estimation of the **years to double,** which is the number of years it takes for the general level of prices to double for a given annual rate of inflation.

We can derive an estimation rule as follows.

$$A = Pe^{rt} \qquad \text{Future value formula}$$

$$2P = Pe^{rt} \qquad \text{Prices are to double.}$$

$$2 = e^{rt} \qquad \text{Divide both sides by } P.$$

$$\ln 2 = rt \qquad \text{Take the natural logarithm of both sides.}$$

$$t = \frac{\ln 2}{r} \qquad \text{Solve for } t.$$

$$t = \frac{100 \ln 2}{100r} \qquad \text{Multiply numerator and denominator by 100.}$$

$$\text{years to double} \approx \frac{70}{\text{annual inflation rate}} \qquad 100 \ln 2 \approx 70$$

(Because r is the inflation rate as a *decimal*, $100r$ is the inflation rate as a *percent*.)

The result above usually is called the **rule of 70.** The value it produces, if not a whole number, should be rounded *up* to the next whole number of years.

```
(100*ln(2))/2.3
                30.13683394
70/2.3
                30.43478261
```

Because 100 ln 2 ≈ 70, the two results shown here are approximately equal. See **Example 13** and the preceding discussion.

EXAMPLE 13 **Estimating Years to Double by the Rule of 70**

Estimate the years to double for an annual inflation rate of 2.3%.

Solution

$$\text{Years to double} \approx \frac{70}{2.3} \approx 30.43 \qquad \text{Rule of 70}$$

With a sustained inflation rate of 2.3%, prices would double in about 31 years. ■

13.1 EXERCISES

In the following exercises, assume whenever appropriate that, unless otherwise known, there are 12 months per year, 30 days per month, and 365 days per year.

Find the simple interest owed for each loan.

1. $600 at 8% for 1 year

2. $5000 at 3% for 1 year

3. $920 at 7% for 9 months

4. $5400 at 5% for 4 months

5. $2675 at 5.4% for $2\frac{1}{2}$ years

6. $2620 at 4.82% for 32 months

Find the future value of each deposit if the account pays **(a)** *simple interest, and* **(b)** *interest compounded annually.*

7. $700 at 2% for 6 years

8. $4000 at 4% for 5 years

9. $2500 at 3% for 3 years

10. $3000 at 5% for 4 years

Solve each interest-related problem.

11. **Simple Interest on a Late Property Tax Payment** Andrew was late on his property tax payment to the county. He owed $7500 and paid the tax 4 months late. The county charges a penalty of 5% simple interest. Find the amount of the penalty.

12. **Simple Interest on a Loan for Equipment** Jonathan purchased new equipment for his yard maintenance business. He borrowed $2325 to pay for the equipment, and he agreed to pay off the loan in 9 months at 6% simple interest. Find the amount of interest he will owe.

13. **Simple Interest on a Small Business Loan** Tony opened a hot dog stand last April. He borrowed $6500 to pay for the stand and startup inventory, and he agreed to pay off the loan in 10 months at 7% simple interest. Find the *total amount* required to repay the loan.

14. **Simple Interest on a Tax Overpayment** Paul is owed $530 by the Internal Revenue Service for overpayment of last year's taxes. The IRS will repay the amount at 4% simple interest. Find the *total amount* Paul will receive if the interest is paid for 8 months.

Find the missing final amount (future value) and/or interest earned.

	Principal	Rate	Compounded	Time	Final Amount	Compound Interest
15.	$975	4%	quarterly	4 years	$1143.26	_____
16.	$1150	7%	semiannually	6 years	$1737.73	_____
17.	$480	6%	semiannually	9 years	_____	$337.17
18.	$2370	5%	quarterly	5 years	_____	_____
19.	$7500	$3\frac{1}{2}$%	annually	25 years	_____	_____
20.	$3450	2.4%	semiannually	10 years	_____	_____

For each deposit, find the future value (that is, the final amount on deposit) when compounding occurs (a) annually, (b) semiannually, and (c) quarterly.

	Principal	Rate	Time
21.	$2000	4%	4 years
22.	$5000	2%	6 years
23.	$18,000	1%	3 years
24.	$10,000	3%	12 years

Occasionally a savings account may actually pay interest compounded continuously. For each deposit, find the interest earned if interest is compounded (a) semiannually, (b) quarterly, (c) monthly, (d) daily, and (e) continuously.

	Principal	Rate	Time
25.	$950	1.6%	4 years
26.	$2650	2.8%	33 months (In parts (d) and (e), assume 1003 days.)

27. Describe the effect of interest being compounded more and more often. In particular, how good is continuous compounding?

Solve each interest-related problem.

28. Finding the Amount Borrowed in a Simple Interest Loan Brandon takes out a 7% simple interest loan today that will be repaid 15 months from now with a payoff amount of $815.63. What amount is he borrowing?

29. Finding the Amount Borrowed in a Simple Interest Loan What is the maximum amount Carole can borrow today if it must be repaid in 4 months with simple interest at 8% and she knows that at that time she will be able to repay no more than $1500?

30. In the development of the future value formula for compound interest in the text, at least four specific problem-solving strategies were employed. Identify (name) as many of them as you can, and describe their use in this case.

Find the present value for each future amount.

31. $1000 (6% compounded annually for 5 years)

32. $14,000 (4% compounded quarterly for 2.5 years)

33. $9860 (8% compounded semiannually for 12 years)

34. $15,080 (5% compounded monthly for 6 years)

Finding the Present Value of a Compound Interest Retirement Account *Tanja wants to establish an account that will supplement her retirement income beginning 30 years from now. For each interest rate, find the lump sum she must deposit today so that $750,000 will be available at time of retirement.*

35. 5% compounded quarterly

36. 6% compounded quarterly

37. 5% compounded daily

38. 6% compounded daily

Finding the Effective Annual Yield in a Savings Account *Suppose a savings and loan pays a nominal rate of 2% on savings deposits. Find the effective annual yield if interest is compounded as stated in Exercises 39–45. (Give answers to the nearest thousandth of a percent.)*

39. annually   **40.** semiannually

41. quarterly   **42.** monthly

43. daily   **44.** 1000 times per year

45. 10,000 times per year

46. Judging from **Exercises 39–45,** what do you suppose is the effective annual yield if a nominal rate of 2% is compounded continuously? Explain your reasoning.

Comparing CD Rates and Yields *A certificate of deposit (CD) is similar to a savings account, but deposits are generally held to maturity. They usually offer higher rates than savings accounts because they "tie up" money for a period of time and often require a minimum deposit. The table shows the two best 5-year CD rates available on a certain online listing on April 18, 2014. Use this information in Exercises 47 and 48.*

5-Year CD	Rate	Yield
CiT Bank	2.274%	2.300%
GE Capital Retail Bank	2.225%	2.250%

47. If you deposit $100,000 with CiT Bank, how much will you have at the end of the 5 years?

48. How often does compounding occur in the GE Capital CD: daily, monthly, quarterly, or semiannually?

Solve each problem.

49. **Finding Years to Double** How long would it take to double your money in an account paying 4% compounded quarterly? (Answer in years plus days, ignoring leap years.)

50. **Comparing Principal and Interest Amounts** After what time period would the interest earned equal the original principal in an account paying 2% compounded daily? (Answer in years plus days, ignoring leap years.)

51. Solve the effective annual yield formula for r to obtain a general formula for nominal rate in terms of yield and the number of compounding periods per year.

52. **Finding the Nominal Rate of a Savings Account** Ridgeway Savings compounds interest monthly. The effective annual yield is 1.95%. What is the nominal rate?

53. **Comparing Bank Savings Rates** Bank A pays a nominal rate of 1.800% compounded daily on deposits. Bank B produces the same annual yield as Bank A but compounds interest only quarterly and pays no interest on funds deposited for less than an entire quarter.

   (a) What nominal rate does Bank B pay (to the nearest thousandth of a percent)?

   (b) Which bank should Nancy choose if she has $2000 to deposit for 10 months? How much more interest will she earn than in the other bank?

   (c) Which bank should Dara choose for a deposit of $5000 for one year? How much interest will be earned?

Estimating the Years to Double by the Rule of 70 *Use the rule of 70 to estimate the years to double for each annual inflation rate.*

54. 1%   **55.** 2%

56. 8%   **57.** 9%

Estimating the Inflation Rate by the Rule of 70 *Use the rule of 70 to estimate the annual inflation rate (to the nearest tenth of a percent) that would cause the general level of prices to double in each time period.*

58. 5 years   **59.** 7 years

60. 16 years   **61.** 22 years

62. Derive a rule for estimating the "years to triple"—that is, the number of years it would take for the general levels of prices to triple for a given annual inflation rate.

Estimating Future Prices for Constant Annual Inflation *The year 2014 prices of several items are given below. Find the estimated future prices required to fill the blanks in the chart. (Give a number of significant figures consistent with the 2014 price figures provided.)*

	Item	2014 Price	2020 Price 2% Inflation	2030 Price 2% Inflation	2020 Price 10% Inflation	2030 Price 10% Inflation
63.	Fast-food meal	$6.39	_____	_____	_____	_____
64.	House	$262,000	_____	_____	_____	_____
65.	New car	$30,500	_____	_____	_____	_____
66.	Gallon of gasoline	$3.53	_____	_____	_____	_____

Estimating Future Prices for Variable Annual Inflation   *Assume that prices for the items below increased at the average annual inflation rates shown in* **Table 4.** *Use the inflation proportion to find the missing prices in the last column of the chart. Round to the nearest dollar.*

	Item	Price	Year Purchased	Price in 2013
67.	Evening dress	$175	2000	_____
68.	Desk	$450	2002	_____
69.	Lawn tractor	$1099	1996	_____
70.	Designer puppy	$250	1998	_____

Solve each interest-related problem.

71. *Finding the Present Value of a Future Equipment Purchase* Human gene sequencing is a major research area of biotechnology. PE Biosystems leased 300 of its sequencing machines to a sibling company, Celera. If Celera was able to earn 7% compounded quarterly on invested money, what lump sum did it need to invest in order to purchase those machines 18 months later at a price of $300,000 each? (*Source: Forbes,* February 21, 2000, p. 102.)

72. The CPI-U for medical care expenses was 260.8 in 2000 and 433.4 in March 2014. The overall CPI-U was 172.2 in 2000 and 236.3 in March 2014.

(a) What would we expect medical care expenses to be in March 2014 if they had increased at the average rate of inflation since 2000?

(b) Compare actual increases in medical care expenses to average inflation over this time period.

13.2   CONSUMER CREDIT

OBJECTIVES

1 Identify the different types of consumer credit.

2 Determine the interest, payments, and cost for installment loans.

3 Determine account balances and finance charges for revolving loans.

Types of Consumer Credit

Consumer credit is the loan extended to people who borrow money to finance the purchase of cars, furniture, appliances, jewelry, electronics, and other things. Technically, **real estate mortgages,** loans to finance home purchases, are consumer credit, but they usually involve much larger amounts and longer repayment periods.

In this section we discuss two common types of consumer credit. The first type, **installment loans,** or **closed-end** credit, involves borrowing a set amount up front and paying a series of equal installments (payments) until the loan is paid off. Furniture, appliances, and cars commonly are financed through closed-end credit.

With the second type of consumer credit, **revolving loans,** or **open-end** credit, there is no fixed number of installments—the consumer continues paying until no balance is owed. With revolving loans, additional credit often is extended before the initial amount is paid off. Examples of open-end credit include most department store charge accounts and bank charge cards such as MasterCard and VISA.

Installment Loans

Installment loans, set up under closed-end credit, often are based on **add-on interest.** This means that interest is calculated by the simple interest formula $I = Prt$, and we simply "add on" this amount of interest to the principal borrowed to arrive at the total debt (or amount to be repaid).

$$\text{Amount to be repaid} = P + Prt$$
$$= P(1 + rt)$$

The total debt is then paid in equal periodic installments (usually monthly) to be made over the t years.

EXAMPLE 1 **Repaying an Installment Loan**

Laura buys $5400 worth of furniture and appliances for her first apartment. She pays $1100 down and agrees to pay the balance at a 7% add-on rate for 2 years. Find

(a) the total amount to be repaid,   **(b)** the monthly payment, and

(c) the total cost of the purchases, including finance charges.

Solution

(a) Amount to be financed = $5400 − $1100

Purchase amount → $5400   Down payment → $1100

$$= \$4300$$

Amount to be repaid $= P(1 + rt)$ Future value formula

$$= \$4300(1 + 0.07 \cdot 2)$$ Substitute known values.

$$= \$4902$$ Calculate.

(b) Monthly payment $= \dfrac{\$4902}{24}$ ← Amount to be repaid
← Number of payments

$$= \$204.25$$

(c) Total cost of purchases = $1100 + $4902

Down payment ↓   Loan repayment ↓

$$= \$6002$$

Laura will end up paying $6002, which is 11.1% more than the price tag total. ■

Notice that the repayment amount in **Example 1,** $P(1 + rt)$, was the same as the final amount A in a savings account paying a simple interest rate r for t years. (See **Section 13.1.**) But in the case of savings, the bank keeps all of your money for the entire time period. In **Example 1,** Laura did not keep the full principal amount for the full time period. She repaid it in 24 monthly installments. The 7% add-on rate turns out to be equivalent to a much higher "true annual interest" rate.

Revolving Loans

With a typical department store account, or bank card, a **credit limit** is established initially and the consumer can make many purchases during a month (up to the credit limit). The required monthly payment can vary from a set minimum (which may depend on the account balance) up to the full balance.

For purchases made in a store, the customer may authorize the charge by signing on a paper "charge slip," or on an electronic signature pad. Authorization also can be given by telephone or Internet for mail orders and other expenditures.

At the end of each billing period (normally once a month), the customer receives an **itemized billing,** a statement listing purchases and cash advances, the total balance owed, the minimum payment required, and perhaps other account information. Any charges beyond cash advanced and cash prices of items purchased are called **finance charges.** Finance charges may include interest, an annual fee, credit insurance coverage, a time payment differential, or carrying charges.

Most revolving credit plans calculate finance charges by the **average daily balance method.** The average daily balance is the *weighted mean* (see **Section 12.2**) of the various amounts of credit utilized for different parts of the billing period, with the weighting factors being the number of days that each credit amount applied.

Outstanding debt among students has become a growing concern in recent years. Though credit card balances among students have decreased since new policies were enacted in 2010, debt from student loans has continued to rise, roughly doubling from about $550 billion in 2007 to over $1 trillion in 2014.

By the end of 2012, the average balance of the nearly 40 million student loans in the United States was just under $25,000. And of the federally guaranteed loans in their payoff period, about 21% were delinquent.

A survey of graduating college seniors in 2013 showed an average of $35,200 in college-related debt. (*Sources:* www.federalreserve.gov *and* www.money.cnn.com)

SEPTEMBER
30

Learning the following rhyme will help you in problems like the one found in **Example 2.**

Thirty days hath September,

April, June, and November.

All the rest have thirty-one,

Save February which has twenty-eight days clear,

And twenty-nine in each leap year.

What's *not* in your wallet? Some consumers carry high balances on their credit card(s) and fall into the habit of paying only the **minimum payment** each month. But they should read their statement carefully. The **Credit Card Act of 2009** requires that credit card issuers disclose "minimum payment" information.

For example, a VISA card charging 12.9% APR with a balance of $2994.82 has an initial minimum payment of just $60. This may sound good, but making only the minimum payment each month would result in total charges of $5689.00 over a 16-year period. Ouch!

1% Back!

Other features of revolving credit accounts can be at least as important as the stated interest rate. For example:

1. Is an annual fee charged? If so, how much is it?

2. Is a special "introductory" rate offered? If so, how long will it last?

3. Are there other incentives, such as rebates, credits toward certain purchases, "free" airline miles, or return of interest charges for long-time use?

EXAMPLE 2 Using the Average Daily Balance Method

The activity in Carmen's MasterCard account for one billing period is shown below. If the previous balance (on March 3) was $209.46, and the bank charges 1.3% per month on the average daily balance, find

(a) the average daily balance for the next billing (April 3),

(b) the finance charge to appear on the April 3 billing, and

(c) the account balance on April 3.

March 3	Billing date	
March 12	Payment	$50.00
March 17	Clothes	$28.46
March 20	Mail order	$31.22
April 1	Auto parts	$59.10

Solution

(a) First make a table that shows the beginning date of the billing period and the dates of all transactions in the billing period. Along with each of these, compute the running balance on that date.

Date	Running Balance
March 3	$209.46
March 12	$209.46 − $50 = $159.46
March 17	$159.46 + $28.46 = $187.92
March 20	$187.92 + $31.22 = $219.14
April 1	$219.14 + $59.10 = $278.24

Next, tabulate the running balance figures, along with the number of days until the balance changed. Multiply each balance amount by the number of days. The sum of these products gives the "sum of the daily balances."

Date	Running Balance	Number of Days Until Balance Changed	$\left(\begin{array}{c}\text{Running}\\\text{Balance}\end{array}\right) \cdot \left(\begin{array}{c}\text{Number}\\\text{of Days}\end{array}\right)$
March 3	$209.46	9	$1885.14
March 12	$159.46	5	$797.30
March 17	$187.92	3	$563.76
March 20	$219.14	12	$2629.68
April 1	$278.24	2	$556.48
		Totals: 31	$6432.36

$$\text{Average daily balance} = \frac{\text{Sum of daily balances}}{\text{Days in billing period}} = \frac{\$6432.36}{31} = \$207.50$$

Carmen will pay a finance charge based on the average daily balance of $207.50.

(b) The finance charge for the April 3 billing will be

$$1.3\% \text{ of } \$207.50 = 0.013 \cdot \$207.50 \qquad \text{1.3\% = 1.3 × 0.01 = 0.013}$$

$$= \$2.70.$$

(c) The account balance on the April 3 billing will be the latest running balance plus the finance charge.

$$\$278.24 + \$2.70 = \$280.94$$

Credit cards and other revolving loans are relatively expensive credit. The monthly rate of 1.3% in **Example 2** is typical and is equivalent to an annual rate of 15.6%. A single card (like VISA or MasterCard), however, can be a great convenience.

The best practice is *not* to carry a balance from month to month, but to pay the entire new balance by the due date each month. Because purchases made during the month are not billed until the next billing date, and the payment due date may be 20 days or more after the billing date, items can often be charged without actually paying for them for nearly two months. To obtain this "free credit," resist the temptation to buy more than you can pay for by the next payment date.

13.2 EXERCISES

Round all monetary answers to the nearest cent unless directed otherwise. Assume that, unless otherwise known, there are 12 months per year, 30 days per month, and 365 days per year.

Financing an Appliance Purchase *Neema bought appliances costing $3795 at a store charging 6% add-on interest. She made a $1000 down payment and agreed to monthly payments over two years.*

1. Find the total amount to be financed.

2. Find the total interest to be paid.

3. Find the total amount to be repaid.

4. Find the monthly payment.

5. Find the total cost, for appliances plus interest.

Financing a New Car Purchase *Suppose you want to buy a new car that costs $18,500. You have no cash—only your old car, which is worth $2000 as a trade-in.*

6. How much do you need to finance to buy the new car?

7. The dealer says the interest rate is 4% add-on for 4 years. Find the total interest.

8. Find the total amount to be repaid.

9. Find the monthly payment.

10. Find your total cost, for the new car plus interest.

In Exercises 11–16, use the add-on method of calculating interest to find the total interest and the monthly payment.

	Amount of Loan	Length of Loan	Interest Rate
11.	$4500	3 years	4.5%
12.	$2700	2 years	4%
13.	$750	18 months	3.7%
14.	$2450	30 months	4.6%
15.	$535	16 months	5.1%
16.	$798	29 months	2.9%

Work each problem.

17. **Finding the Monthly Payment for an Add-On Interest Furniture Loan** The Giordanos buy $8500 worth of furniture for their new home. They pay $3000 down. The store charges 10% add-on interest. The Giordanos will pay off the furniture in 30 monthly payments ($2\frac{1}{2}$ years). Find the monthly payment.

18. **Finding the Monthly Payment for an Add-On Interest Auto Loan** Find the monthly payment required to pay off an auto loan of $15,780 over 3 years if the add-on interest rate is 3.9%.

19. **Finding the Monthly Payment for an Add-On Interest Home Electronics Loan** The total purchase price of a new home entertainment system is $14,240. If the down payment is $2900 and the balance is to be financed over 48 months at 6% add-on interest, what is the monthly payment?

20. **Finding the Monthly Payment for an Add-On Interest Loan** What are the monthly payments that Holly pays on a loan of $1680 for a period of 10 months if 9% add-on interest is charged?

21. **Finding the Amount Borrowed for an Add-On Interest Car Loan** Kyle has misplaced the sales contract for his car and cannot remember the amount he originally financed. He does know that the add-on interest rate was 4.9% and the loan required a total of 48 monthly payments of $314.65 each. How much did Kyle borrow (to the nearest dollar)?

22. **Finding an Add-On Interest Rate** Ella is making monthly payments of $257.49 to pay off a $3\frac{1}{2}$-year loan of $9400. What is her add-on interest rate (to the nearest tenth of a percent)?

23. **Finding the Term of an Add-On Interest Loan** How long (in years) will it take Hiroshi to pay off an $11,000 loan with monthly payments of $240.17 if the add-on interest rate is 6.2%?

24. **Finding the Number of Payments for an Add-On Interest Loan** How many monthly payments must Jawann make on a $10,000 loan if he pays $417.92 per month and the add-on interest rate is 10.15%?

Finding Finance Charges *Find the finance charge for each charge account. Assume interest is calculated on the average daily balance of the account.*

	Average Daily Balance	Monthly Interest Rate
25.	$249.94	1.4%
26.	$350.75	1.5%
27.	$419.95	1.38%
28.	$450.21	1.26%
29.	$1073.40	1.425%
30.	$1320.42	1.375%

Finding Finance Charges and Account Balances Using the Average Daily Balance Method *For each credit card account, assume one month between billing dates (with the appropriate number of days) and interest of 1.3% per month on the average daily balance. Find* **(a)** *the average daily balance,* **(b)** *the monthly finance charge, and* **(c)** *the account balance for the next billing.*

31. Previous balance: $728.36

May 9	Billing date	
May 17	Payment	$200.00
May 30	Dinner	$46.11
June 3	Theater tickets	$64.50

32. Previous balance: $514.79

January 27	Billing date	
February 9	Candy	$11.08
February 13	Returns	$26.54
February 20	Payment	$59.00
February 25	Repairs	$71.19

33. Previous balance: $462.42

June 11	Billing date	
June 15	Returns	$106.45
June 20	Jewelry	$115.73
June 24	Car rental	$74.19
July 3	Payment	$115.00
July 6	Flowers	$68.49

34. Previous balance: $983.25

August 17	Billing date	
August 21	Mail order	$14.92
August 23	Returns	$25.41
August 27	Beverages	$31.82
August 31	Payment	$108.00
September 9	Returns	$71.14
September 11	Concert tickets	$110.00
September 14	Cash advance	$100.00

Finding Finance Charges *In Exercises 35 and 36, assume no purchases or returns are made.*

35. At the beginning of a 31-day billing period, Alicia has an unpaid balance of $720 on her credit card. Three days before the end of the billing period, she pays $600. Find her finance charge at 1.4% per month using the average daily balance method.

36. Vincent's VISA bill dated April 14 shows an unpaid balance of $1070. Five days before May 14, the end of the billing period, Vincent makes a payment of $900. Find his finance charge at 1.32% per month using the average daily balance method.

Analyzing a "90 Days Same as Cash" Offer *One version of the "90 Days Same as Cash" promotion was offered by a "major purchase card," which established an account charging 1.3167% interest per month on the account balance. Interest charges are added to the balance each month, becoming part of the balance on which interest is computed the next month. If you pay off the original purchase charge within 3 months, all interest charges are cancelled. Otherwise, you are liable for all the interest. Suppose you purchase* $2900 *worth of carpeting under this plan.*

37. Find the interest charge added to the account balance at the end of

  (a) the first month,

  (b) the second month,

  (c) the third month.

38. Suppose you pay off the account 1 day late (3 months plus 1 day). What total interest amount must you pay? (Do not include interest for the one extra day.)

39. Treating the 3 months as $\frac{1}{4}$ year, find the equivalent simple interest rate for this purchase (to the nearest tenth of a percent).

Various Charges of a Bank Card Account *Aimee's bank card account charges 1.1% per month on the average daily balance, as well as the following special fees:*

Cash advance fee:	2% *(neither less than $2 nor more than $10)*
Late payment fee:	$15
Over-the-credit-limit fee:	$5

In the month of June, Aimee's average daily balance was $1846. *She was on vacation during the month and did not get her account payment in on time, which resulted in a late payment and, as a consequence, charges accumulated to a sum above her credit limit. She also used her card for six* $100 *cash advances while on vacation. Find the following based on account transactions in that month.*

40. interest charges to the account

41. special fees charged to the account

Write out your response to each of the following.

42. Is it possible to use a bank credit card for your purchases without paying anything for credit? If so, explain how.

43. Obtain applications or descriptive brochures for several different bank card programs, compare their features (including those in fine print), and explain which deal would be best for you, and why.

44. Research and explain the difference, if any, between a "credit" card and a "debit" card.

45. Many charge card offers include the option of purchasing credit insurance coverage, whereby the insurer would make your monthly payments if you became disabled and could not work and/or would pay off the account balance if you died. Find out the details on at least one such offer, and discuss why you would or would not accept it.

46. Make a list of "special incentives" offered by bank cards you are familiar with, and briefly describe the pros and cons of each one.

47. One bank offered a card with a "low introductory rate" of 5.9%, good through the end of the year. And furthermore, you could receive back a percentage (up to 100%!) of all interest you pay, as shown in the table at the top of the next column.

Use your card for:	2 years	5 years	10 years	15 years	20 years
Get back:	10%	25%	50%	75%	100%

(As soon as you take a refund, the time clock starts over.) Because you can eventually claim all your interest payments back, is this card a good deal? Explain.

48. Recall a car-buying experience you have had, or visit a new-car dealer and interview a salesperson. Describe the procedure involved in purchasing a car on credit.

Comparing Bank Card Accounts *Ruby is considering two bank card offers that are the same in all respects except for the following:*

Bank A charges no annual fee and charges monthly interest of 1.18% on the unpaid balance.

Bank B charges a $30 annual fee and monthly interest of 1.01% on the unpaid balance.

From her records, Ruby has found that the unpaid balance she tends to carry from month to month is quite consistent and averages $900.

49. Estimate her total yearly cost to use the card if she chooses the card from

   (a) Bank A.   **(b)** Bank B.

50. Which card is her better choice?

13.3   TRUTH IN LENDING

OBJECTIVES

1 Determine the annual percentage rate for different types of loans.

2 Calculate unearned interest.

Free consumer information is available from the Federal Trade Commission at www.consumer.ftc.gov. For example, you can find out about "Money & Credit," "Homes & Mortgages," "Jobs & Making Money," "Privacy & Identity," and many other consumer topics. You can also file consumer complaints and report identity theft using online forms.

Annual Percentage Rate (APR)

The Consumer Credit Protection Act, passed in 1968, has commonly been known as the **Truth in Lending Act.** The law addressed two major issues:

1. How can we tell the true annual interest rate a lender is charging?

2. How much of the finance charge are we entitled to save if we decide to pay off a loan sooner than originally scheduled?

Question 1 above arose because lenders were computing and describing the interest they charged in several different ways. For example, how does 1.9% per month at Sears compare to 9% per year add-on interest at a furniture store? Truth in Lending standardized the so-called true annual interest rate, or **annual percentage rate,** commonly denoted **APR.** All sellers (car dealers, stores, banks, insurance agents, credit card companies, and the like) must disclose the APR when asked, and the written contract must state the APR in all cases. This enables a borrower to more easily compare the true costs of different loans.

    Theoretically, a borrower should not need to calculate APR, but it is possible to verify the value stated by the lender. Since the formulas for finding APR are quite involved, it is easiest to use a table provided by the Federal Reserve Bank. We show an abbreviated version in **Table 5** on the next page. It identifies APR values to the nearest half percent from 6.0% to 14.0%, for loans requiring *monthly* payments and extending over the most common lengths for consumer loans from 6 to 60 months.

Table 5 relates the following three quantities.

APR = true annual interest rate (shown across the top)

n = total number of scheduled monthly payments (shown down the left side)

h = finance charge per $100 of amount financed (shown in the body of the table)

Table 5 Annual Percentage Rate (APR) for Monthly Payment Loans

Number of Monthly Payments (n)	Annual Percentage Rate (APR)																
	6.0%	6.5%	7.0%	7.5%	8.0%	8.5%	9.0%	9.5%	10.0%	10.5%	11.0%	11.5%	12.0%	12.5%	13.0%	13.5%	14.0%
	Finance Charge per $100 of Amount Financed ($h$)																
6	$1.76	$1.90	$2.05	$2.20	$2.35	$2.49	$2.64	$2.79	$2.94	$3.08	$3.23	$3.38	$3.53	$3.68	$3.83	$3.97	$4.12
12	3.28	3.56	3.83	4.11	4.39	4.66	4.94	5.22	5.50	5.78	6.06	6.34	6.62	6.90	7.18	7.46	7.74
18	4.82	5.22	5.63	6.04	6.45	6.86	7.28	7.69	8.10	8.52	8.93	9.35	9.77	10.19	10.61	11.03	11.45
24	6.37	6.91	7.45	8.00	8.55	9.09	9.64	10.19	10.75	11.30	11.86	12.42	12.98	13.54	14.10	14.66	15.23
30	7.94	8.61	9.30	9.98	10.66	11.35	12.04	12.74	13.43	14.13	14.83	15.54	16.24	16.95	17.66	18.38	19.10
36	9.52	10.34	11.16	11.98	12.81	13.64	14.48	15.32	16.16	17.01	17.86	18.71	19.57	20.43	21.30	22.17	23.04
48	12.73	13.83	14.94	16.06	17.18	18.31	19.45	20.59	21.74	22.90	24.06	25.23	26.40	27.58	28.77	29.97	31.17
60	16.00	17.40	18.81	20.23	21.66	23.10	24.55	26.01	27.48	28.96	30.45	31.96	33.47	34.99	36.52	38.06	39.61

EXAMPLE 1 Finding the APR for an Add-On Loan

Recall that Laura (in **Example 1** of **Section 13.2**) paid $1100 down on a $5400 purchase and agreed to pay the balance at a 7% add-on rate for 2 years. Find the APR for her loan.

Solution

As shown previously, the total amount financed was

$$\underset{\substack{\uparrow \\ \text{Purchase} \\ \text{price}}}{\$5400} - \underset{\substack{\uparrow \\ \text{Down} \\ \text{payment}}}{\$1100} = \$4300.$$

The finance charge (interest) was

$$I = Prt = \$4300 \cdot 0.07 \cdot 2 = \$602.$$

Next find the finance charge per $100 of the amount financed. To do this, divide the finance charge by the amount financed, and then multiply by $100.

$$\left(\begin{array}{c}\text{Finance charge per} \\ \text{\$100 financed}\end{array}\right) = \frac{\text{Finance charge}}{\text{Amount financed}} \cdot \$100$$

$$= \frac{\$602}{\$4300} \cdot \$100, \quad \text{or} \quad \$14$$

This amount, $14, represents h, the finance charge per $100 of the amount financed. Because the loan was to be paid over 24 months, look down to the "24 monthly payments" row of **Table 5** ($n = 24$). Then look across the table for the h-value closest to $14.00, which is $14.10. From that point, read up the column to find the APR, 13.0% (to the nearest half percent). In this case, a 7% add-on rate is equivalent to an APR of 13.0%. ■

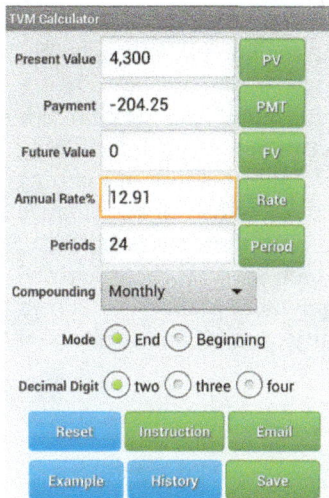

TVM Calculator

Present Value	4,300	PV
Payment	-204.25	PMT
Future Value	0	FV
Annual Rate%	12.91	Rate
Periods	24	Period
Compounding	Monthly ▾	
Mode	● End ○ Beginning	
Decimal Digit	● two ○ three ○ four	

Reset Instruction Email

Example History Save

Financial calculator apps give another option (besides tables) for calculating APR. For **Example 1,** dividing the total amount to be repaid by the number of payments gives the monthly payment.

$$\frac{\$4300 + \$602}{24} = \$204.25$$

This amount can then be entered (as negative cash flow) into the financial calculator, and the rate calculated. We see here that the APR for Laura's loan is 12.91%.(App used: TVM Calculator)

EXAMPLE 2 Finding the APR for a Car Loan

After a down payment on her new car, Vanessa still owed $12,378. She agreed to repay the balance in 48 monthly payments of $294 each. What is the APR on her loan?

Solution

First find the finance charge.

Total payments   Amount financed

$$\text{Finance charge} = 48 \cdot \$294 - \$12{,}378$$

$$= \$1734 \quad \boxed{\text{Multiply first. Then subtract.}}$$

Now find the finance charge per $100 financed as in **Example 1.**

$$\left(\begin{array}{c}\text{Finance charge per}\\ \$100 \text{ financed}\end{array}\right) = \frac{\text{Finance charge}}{\text{Amount financed}} \cdot \$100$$

$$= \frac{\$1734}{\$12{,}378} \cdot \$100, \quad \text{or} \quad \$14.01$$

Find the "48 payments" row of **Table 5,** read across to find the number closest to 14.01, which is 13.83. From there read up to find the APR, which is 6.5%. ∎

```
N=48
▪I%=6.579891685
 PV=12378
 PMT=‑294
 FV=0
 P/Y=12
 C/Y=12
 PMT:  ███  BEGIN
```

Unknown APR may be found using the TI-84's TVM Solver. Enter the number of payments (N), the present value of the loan (PV), the payment amount (PMT), the number of payments per year, and the number of compoundings per year. Then locate the cursor in the I% field and press [ALPHA] [SOLVE]. The screen above confirms the approximate value found in **Example 2** using **Table 5.**

Unearned Interest

Question 2 at the beginning of this section arises when a borrower decides to pay off an installment loan earlier than originally scheduled. In such a case, it turns out that the lender has not loaned as much money for as long as planned, and so he has not really "earned" the full finance charge originally disclosed.

If a loan is paid off early, the amount by which the original finance charge is reduced is called the **unearned interest.** We will discuss two methods of calculating unearned interest, the **actuarial method** and the **rule of 78.** The Truth in Lending Act requires that the method for calculating this refund (or reduction) of finance charge (in case of an early payoff) be disclosed at the time the loan is initiated. Whichever method is used, the borrower may not, in fact, save all the unearned interest, since the lender is entitled to impose an **early payment penalty** to recover certain costs. A lender's intention to impose such a penalty in case of early payment also must be disclosed at initiation of the loan.

Rights and responsibilities apply to all credit accounts, and the consumer should read all disclosures provided by the lender. *The Fair Credit Billing Act* and *The Fair Credit Reporting Act* regulate, among other things, procedures for billing and for disputing bills, and for providing and disputing personal information on consumers.

If you have ever applied for a charge account, a personal loan, insurance, or a job, then information about where you work and live, how you pay your bills, and whether you've been sued or arrested or have filed for bankruptcy, appears in the files of Consumer Reporting Agencies (CRAs), which sell that information to creditors, employers, insurers, and other businesses.

For more detailed information on these and many other consumer issues, you may want to consult the Web site www.usa.gov/consumer.

UNEARNED INTEREST—ACTUARIAL METHOD

For an installment loan requiring *monthly* payments, which is paid off earlier than originally scheduled, let

R = regular monthly payment,

k = remaining number of scheduled payments (*after* current payment), and

h = finance charge per $100, corresponding to a loan with the same APR and k monthly payments.

Then the **unearned interest, u,** is calculated as follows.

$$u = kR\left(\frac{h}{\$100 + h}\right)$$

Once the unearned interest u is calculated (by any method), the amount required to pay off the loan early is easily found. It consists of the present regular payment due, plus k additional future payments, minus the unearned interest.

PAYOFF AMOUNT

An installment loan requiring regular monthly payments R can be paid off early, along with the current payment. If the original loan had k additional payments scheduled (after the current payment), and the unearned interest is u, then, disregarding any possible prepayment penalty, the **payoff amount** is calculated as follows.

$$\text{Payoff amount} = (k + 1)R - u$$

EXAMPLE 3   Finding Early Payoff Amount (Actuarial Method)

Vanessa got an unexpected pay raise and wanted to pay off her car loan of **Example 2** at the end of 3 years rather than paying for 4 years as originally agreed.

(a) Find the unearned interest (the amount she will save by retiring the loan early).

(b) Find the "payoff amount" (the amount required to pay off the loan at the end of 3 years).

Solution

(a) From **Example 2,** recall that $R = \$294$ and APR $= 6.5\%$. The current payment is payment number 36, so $k = 48 - 36 = 12$. Use **Table 5,** with 12 payments and APR 6.5%, to obtain $h = \$3.56$. Then use the actuarial method formula.

$$u = 12 \cdot \$294\left(\frac{\$3.56}{\$100 + \$3.56}\right) = \$121.28$$

By this method, Vanessa will save $121.28 in interest by retiring the loan early.

(b) The payoff amount is found by using the appropriate formula.

$$\text{Payoff amount} = (12 + 1)\$294 - \$121.28 = \$3700.72$$

The required payoff amount at the end of 3 years is $3700.72.   ∎

The actuarial method, being based on the APR value, probably is the better method. However, some lenders historically used a second method, the **rule of 78.**

UNEARNED INTEREST—RULE OF 78

For a closed-end loan requiring *monthly* payments, which is paid off earlier than originally scheduled, let

F = original finance charge,

n = number of payments originally scheduled, and

k = remaining number of scheduled payments (*after* current payment).

Then the **unearned interest, u,** is calculated as follows.

$$u = \frac{k(k + 1)}{n(n + 1)} \cdot F$$

```
12*294(3.56/(100+3.56))
                  121.28
(12+1)*294-121.28
                 3700.72
```

These are the computations required in the solution of **Example 3.**

EXAMPLE 4 Finding Early Payoff Amount (Rule of 78)

Again assume that the loan in **Example 2** is paid off at the time of the thirty-sixth monthly payment. This time, however, instead of the actuarial method, use the rule of 78 to find

(a) the unearned interest, and

(b) the payoff amount.

Solution

(a) From **Example 2,** the original finance charge is $F = \$1734$. Also, $n = 48$ and $k = 12$.

$$u = \frac{12(12+1)}{48(48+1)} \cdot \$1734 = \$115.01 \quad \text{Rule of 78}$$

By the rule of 78, Vanessa will save $115.01 in interest, which is $6.27 *less* than her savings by the actuarial method.

(b) $$\text{Payoff amount} = (12+1)\$294 - \$115.01 = \$3706.99$$

The payoff amount is $3706.99, which is $6.27 *more* than the payoff amount calculated by the actuarial method in **Example 3.**

(12*(12+1))/(48(48+1))*1734
115.01
(12+1)*294−115.01
3706.99

These are the computations required in the solution of **Example 4.**

When the rule of 78 was first introduced into financial law (by the Indiana legislature in 1935), loans were ordinarily written for 1 year or less, interest rates were relatively low, and loan amounts were less than they tend to be today. For these reasons the rule of 78 was acceptably accurate then. Today, however, with very accurate tables and/or calculators readily available, the rule of 78 is used much less often than previously.

Suppose we want to compute unearned interest accurately (so we don't trust the rule of 78), but the APR value, or the number of scheduled payments, or the number of remaining payments (or at least one of the three) is not included in **Table 5.** Then what? Actually, in the actuarial method, we can evaluate h (the finance charge per $100 financed) using the same formula that was used to generate **Table 5.**

FINANCE CHARGE PER $100 FINANCED

If an installment loan requires n equal monthly payments and APR denotes the true annual interest rate for the loan (as a decimal), then h, the **finance charge per $100 financed,** is calculated as follows.

$$h = \frac{n \cdot \frac{\text{APR}}{12} \cdot \$100}{1 - \left(1 + \frac{\text{APR}}{12}\right)^{-n}} - \$100$$

EXAMPLE 5 Finding Unearned Interest and Early Payoff Amount

Mark borrowed $4000 to pay for music equipment for his band. His loan contract states an APR of 9.8% and stipulates 28 monthly payments of $160.39 each. Mark decides to pay the loan in full at the time of his nineteenth scheduled payment. Find

(a) the unearned interest, and   **(b)** the payoff amount.

(9*(.098/12)*100)/(1−(1+
.098/12)⁻⁹)−100
　　　　　　4.13
9*160.39*4.13/(100+4.13)
　　　　　　57.25

These are the computations required in the solution of **Example 5(a).**

Solution

(a) First find h from the finance charge formula just given.

> Remember to use the remaining number of payments, $28 - 19 = 9$, as the value of n.

$$h = \frac{9\left(\frac{0.098}{12}\right)(\$100)}{1 - \left(1 + \frac{0.098}{12}\right)^{-9}} - \$100 = \$4.13$$

Next use the actuarial formula for unearned interest u.

Regular monthly payment: $\qquad R = \$160.39$

Remaining number of payments: $\quad k = 28 - 19 = 9$

Finance charge per \$100: $\qquad h = \$4.13$

$$u = 9 \cdot \$160.39 \cdot \frac{\$4.13}{\$100 + \$4.13} = \$57.25$$

The amount of interest Mark will save is \$57.25.

(b) $\qquad$ Payoff amount $= (9 + 1)(\$160.39) - \$57.25 = \$1546.65.$

To pay off the loan at the time of his nineteenth scheduled payment, Mark must pay \$1546.65. ■

13.3 EXERCISES

Round all monetary answers to the nearest cent unless otherwise directed.

Finding True Annual Interest Rate *Find the APR (true annual interest rate), to the nearest half percent, for each loan.*

	Amount Financed	Finance Charge	Number of Monthly Payments
1.	$1000	$38	12
2.	$1700	$202	24
3.	$6600	$750	30
4.	$5900	$1150	48

Finding the Monthly Payment *Find the monthly payment for each loan.*

	Purchase Price	Down Payment	Finance Charge	Number of Monthly Payments
5.	$3500	$500	$275	24
6.	$4750	$450	$750	36
7.	$3850	$300	$800	48
8.	$9500	$1500	$1400	60

Finding True Annual Interest Rate *Find the APR (true annual interest rate), to the nearest half percent, for each loan.*

	Purchase Price	Down Payment	Add-On Interest Rate	Number of Payments
9.	$4190	$390	3.5%	12
10.	$3250	$750	7%	36
11.	$7480	$2200	5%	18
12.	$12,800	$4500	3.5%	48

Unearned Interest by the Actuarial Method *Each loan was paid in full before its due date.* **(a)** *Obtain the value of h from* **Table 5.** *Then* **(b)** *use the actuarial method to find the amount of unearned interest, and* **(c)** *find the payoff amount.*

	Regular Monthly Payment	APR	Remaining Number of Scheduled Payments after Payoff
13.	$346.70	11.0%	18
14.	$783.50	8.5%	12
15.	$595.80	6.5%	6
16.	$314.50	10.0%	24

Finding Finance Charge and True Annual Interest Rate *For each loan, find* **(a)** *the finance charge, and* **(b)** *the APR.*

17. Jesse financed a $1990 computer with 24 monthly payments of $91.50 each.

18. Margey bought a horse trailer for $6090. She paid $1240 down and paid the remainder at $175.50 per month for $2\frac{1}{2}$ years.

19. Walter still owed $2000 on his new garden tractor after the down payment. He agreed to pay monthly payments for 18 months at 4.6% add-on interest.

20. Daniel paid off a $15,000 car loan over 3 years with monthly payments of $466.59 each.

Comparing the Actuarial Method and the Rule of 78 for Unearned Interest *Each loan was paid off early. Find the unearned interest by* **(a)** *the actuarial method, and* **(b)** *the rule of 78.*

	Amount Financed	Regular Monthly Payments	Total Number of Payments Scheduled	Remaining Number of Scheduled Payments after Payoff
21.	$3310	$201.85	18	6
22.	$10,230	$277.00	48	12
23.	$29,850	$641.58	60	12
24.	$16,730	$539.82	36	18

Unearned Interest by the Actuarial Method *Each loan was paid in full before its due date.* **(a)** *Obtain the value of h from the appropriate formula. Then* **(b)** *use the actuarial method to find the amount of unearned interest, and* **(c)** *find the payoff amount.*

	Regular Monthly Payment	APR	Remaining Number of Scheduled Payments after Payoff
25.	$212	8.6%	4
26.	$575	5.9%	8

Comparing Loan Choices *Phan needs to borrow $5000 to pay for NBA season tickets for her family. She can borrow the amount from a finance company (at 6.5% add-on interest for 3 years) or from the credit union (36 monthly payments of $164.50 each). Use this information for Exercises 27–31.*

27. Find the APR (to the nearest half percent) for each loan, and decide which one is Phan's better choice.

28. Phan takes the credit union loan. At the time of her thirtieth payment she pays it off. If the credit union uses the rule of 78 for computing unearned interest, how much will she save by paying in full now?

29. What would Phan save in interest if she paid in full at the time of the thirtieth payment and the credit union used the actuarial method for computing unearned interest?

30. Under the conditions of **Exercise 29,** what amount must Phan come up with to pay off her loan?

31. Suppose the credit union uses the rule of 78 and imposes an early payoff penalty of 1% of the original finance charge per month of early payment. For example, if Phan pays the loan off 9 months early, then the penalty will be 9% of the original finance charge. Find the least value of k that would result in any savings for Phan.

32. Describe why, in **Example 1,** the APR and the add-on rate differ. Which one is more legitimate? Why?

Approximating the APR of an Add-On Rate *To convert an add-on interest rate to its corresponding APR, some people recommend using the formula*

$$APR = \frac{2n}{n+1} \cdot r,$$

where r is add-on rate and n is total number of payments.

33. Apply the given formula to calculate the APR (to the nearest half percent) for the loan of **Example 1** ($r = 0.07, n = 24$).

34. Compare your APR value in **Exercise 33** to the value in **Example 1.** What do you conclude?

The Rule of 78 with Prepayment Penalty *A certain retailer's credit contract designates the rule of 78 for computing unearned interest and imposes a "prepayment penalty." In case of any payoff earlier than the due date, the lender will charge an additional 10% of the original finance charge. Find the least value of k (remaining payments after payoff) that would result in any net savings in each case.*

35. 24 payments originally scheduled

36. 36 payments originally scheduled

37. 48 payments originally scheduled

38. 60 payments originally scheduled

Write out your response to each exercise.

39. Why might a lender be justified in imposing a prepayment penalty?

40. Discuss reasons why a borrower might want to pay off a loan early.

41. Find out what federal agency you can contact if you have questions about compliance with the Truth in Lending Act. (Any bank, or retailer's credit department, should be able to help you with this, or you could try a Web search.)

42. Study the table below, which pertains to a 12-month loan. The column-3 entries are designed so that they are in the same ratios as the column-2 entries but will add up to 1 because their denominators are all equal to

$$1 + 2 + 3 + 4 + 5 + \ldots + 12 = \frac{12 \cdot 13}{2} = 78.$$

(This is the origin of the term "rule of 78.")

Month	Fraction of Loan Principal Used by Borrower	Fraction of Finance Charge Owed
1	12/12	12/78
2	11/12	11/78
3	10/12	10/78
4	9/12	9/78
5	8/12	8/78
6	7/12	7/78
7	6/12	6/78
8	5/12	5/78
9	4/12	4/78
10	3/12	3/78
11	2/12	2/78
12	1/12	1/78
		78/78 = 1

Suppose the loan is paid in full after eight months. Use the table to determine the unearned fraction of the total finance charge.

43. Find the fraction of unearned interest of **Exercise 42** by using the rule of 78 formula.

Understanding the Actuarial Method   *The actuarial method of computing unearned interest assumes that, throughout the life of the loan, the borrower is paying interest at the rate given by APR for money actually being used by the borrower. When contemplating complete payoff along with the current payment, think of k future payments as applying to a separate loan with the same APR and of h as being the finance charge per $100 of that loan. Refer to the following formula.*

$$u = kR \left(\frac{h}{\$100 + h} \right)$$

44. Describe in words the quantity represented by

$$\frac{h}{\$100 + h}.$$

45. Describe in words the quantity represented by kR.

46. Explain why the product of the two quantities above represents unearned interest.

13.4 **THE COSTS AND ADVANTAGES OF HOME OWNERSHIP**

OBJECTIVES

1 Understand the characteristics of fixed-rate mortgages.

2 Understand the characteristics of adjustable-rate mortgages.

3 Be familiar with the closing costs associated with mortgages.

4 Be familiar with the recurring costs of home ownership.

Fixed-Rate Mortgages

For many decades, home ownership has been considered a centerpiece of the "American dream." For most people, a home represents the largest purchase of their lifetime, and it is certainly worth careful consideration.

A loan for a substantial amount, extending over a lengthy time interval (typically up to 30 years), for the purpose of buying a home or other property or real estate, and for which the property is pledged as security for the loan, is called a **mortgage.** (In some areas, a mortgage may also be called a **deed of trust** or a **security deed.**)

- The time until final payoff is called the **term** of the mortgage.
- The portion of the purchase price of the home that the buyer pays initially is called the **down payment.**
- The **principal amount of the mortgage** (the amount borrowed) is found by subtracting the down payment from the purchase price.

With a **fixed-rate mortgage,** the interest rate will remain constant throughout the term, and the initial principal balance, together with interest due on the loan, is repaid to the lender through regular (constant) periodic (we assume monthly) payments. This is called **amortizing** the loan. The regular monthly payment needed to amortize a loan depends on the amount financed, the term of the loan, and the interest rate, according to the formula on the next page.

Heating a house is another cost that may get you involved with banks and interest rates after you finally get a roof over your head. The roof you see above does more than keep off the rain. It holds solar panels, part of the solar heating system in the building.

REGULAR MONTHLY PAYMENT

The **regular monthly payment** required to repay a loan of P dollars, together with interest at an annual rate r, over a term of t years, is calculated as follows.

$$R = \frac{P\left(\frac{r}{12}\right)}{1 - \left(\frac{12}{12 + r}\right)^{12t}}$$

EXAMPLE 1 Using a Formula to Find a Monthly Mortgage Payment

Find the monthly payment necessary to amortize a $21,000 mortgage at 3.0% annual interest for 15 years.

Solution

$$R = \frac{\$21{,}000\left(\frac{0.030}{12}\right)}{1 - \left(\frac{12}{12 + 0.030}\right)^{(12)(15)}} = \$145.02$$

With a programmable or financial calculator, you can store the monthly payment formula and minimize the work.

Another option is shown in **Table 7** on **page 734,** which gives payment values (per $1000 principal) for typical ranges of mortgage terms and interest rates. Table entries are given to five decimal places for accuracy to the nearest cent. ■

```
N=180
I%=3
PV=21000
▪PMT=-145.02
FV=0
P/Y=12
C/Y=12
PMT: END  BEGIN
```

The arrow indicates a payment of $145.02, supporting the result of **Example 1.**

WHEN Will I Ever USE This ?

Suppose you are a salesperson for a solar energy company. A couple who are potential customers contact you. To make your presentation more compelling, you provide cost comparisons of

"going solar"   and   "staying in the dark."

The customers give you permission to contact the power company and request a record of their power consumption over the last year, and they agree to meet the next day to go over the numbers.

At this meeting, you explain that the government is providing a tax incentive in the form of a tax credit equal to 30% of the cost of the system. Your company adds to the incentive by offering a 12-month "same-as-cash" loan for that 30%, and a 15-year fixed-rate mortgage at 3.0% for the other 70% of the cost. This allows the customers to have zero "out-of-pocket" expense for the purchase and installation of the system. They can use their tax refund to pay off the 30% "same-as-cash" loan before any interest charges apply.

Analysis of this couple's power consumption for the last year shows that their average monthly bill was $200 (totaling $2400 for the year), and a $30,000 solar system would initially produce enough energy to match their consumption. Energy prices are projected to increase by 5% per year. A conservative estimate of solar panel degradation is 0.5% per year, and all equipment will be under warranty for 20 years. A cost comparison can now be provided for a 20-year period using the following calculations.

Initial principal amount of 3% loan = 0.7 · $30,000 = $21,000

Monthly payment = $145.02   **Example 1** calculation

First-year energy costs with solar ≈ 12 · $145 = $1740

First-year energy costs without solar = $2400 · 1.05 = $2520   5% price increase

Second-year energy costs without solar = $2400 · 1.05^2 = $2646   5% price increase

Second-year costs with solar = 0.005($2646) + $1740 ≈ $1753   0.5% degradation

Continuing these calculations results in the analysis of **Table 6.**

Cost of system:	$30,000	Interest rate:	3.0%
Same-as-cash loan:	$9,000	Term of loan (years):	15
12-yr fixed-rate mortgage:	$21,000	Monthly payment:	$145.02

Table 6 Analysis of Cost/Savings of Solar Energy

Year	Annual Cost without Solar	Energy Bills		Monthly Payments on Loan		Annual Cost with Solar
1	$2520	$0	+	$1740	=	$1740
2	2646	13	+	1740	=	1753
3	2778	28	+	1740	=	1768
4	2917	44	+	1740	=	1784
5	3063	61	+	1740	=	1801
6	3216	80	+	1740	=	1820
7	3377	101	+	1740	=	1841
8	3546	124	+	1740	=	1864
9	3723	149	+	1740	=	1889
10	3909	176	+	1740	=	1916
11	4105	205	+	1740	=	1945
12	4310	237	+	1740	=	1977
13	4526	272	+	1740	=	2012
14	4752	309	+	1740	=	2049
15	4989	349	+	1740	=	2089
16	5239	393	+	0	=	393
17	5501	440	+	0	=	440
18	5776	491	+	0	=	491
19	6065	546	+	0	=	546
20	6368	605	+	0	=	605
Totals:	**$83,326**					**$30,723**

20-Year Savings: $83,326 − $30,723 = $52,603

As you (and your customers) can see, the projected savings over the next 20 years are substantial.

Table 7 Monthly Payments to Repay Principal and Interest on a $1000 Mortgage

Annual rate (r)	5	10	15	20	25	30	40
2.0%	$17.52776	$9.20135	$6.43509	$5.05883	$4.23854	$3.69619	$3.02826
2.5%	17.74736	9.42699	6.66789	5.29903	4.48617	3.95121	3.29778
3.0%	17.96869	9.65607	6.90582	5.54598	4.74211	4.21604	3.57984
3.5%	18.19174	9.88859	7.14883	5.79960	5.00624	4.49045	3.87391
4.0%	18.41652	10.12451	7.39688	6.05980	5.27837	4.77415	4.17938
4.5%	18.64302	10.36384	7.64993	6.32649	5.55832	5.06685	4.49563
5.0%	18.87123	10.60655	7.90794	6.59956	5.84590	5.36822	4.82197
5.5%	19.10116	10.85263	8.17083	6.87887	6.14087	5.67789	5.15770
6.0%	19.33280	11.10205	8.43857	7.16431	6.44301	5.99551	5.50214
6.5%	19.56615	11.35480	8.71107	7.45573	6.75207	6.32068	5.85457
7.0%	19.80120	11.61085	8.98828	7.75299	7.06779	6.65302	6.21431
7.5%	20.03795	11.87018	9.27012	8.05593	7.38991	6.99215	6.58071
8.0%	20.27639	12.13276	9.55652	8.36440	7.71816	7.33765	6.95312
8.5%	20.51653	12.39857	9.84740	8.67823	8.05227	7.68913	7.33094
9.0%	20.75836	12.66758	10.14267	8.99726	8.39196	8.04623	7.71361
9.5%	21.00186	12.93976	10.44225	9.32131	8.73697	8.40854	8.10062
10.0%	21.24704	13.21507	10.74605	9.65022	9.08701	8.77572	8.49146
10.5%	21.49390	13.49350	11.05399	9.98380	9.44182	9.14739	8.88570
11.0%	21.74242	13.77500	11.36597	10.32188	9.80113	9.52323	9.28294

N=300.00
I%=4.50
PV=160000.00
→ ▪PMT=-889.33
FV=0.00
P/Y=12.00
C/Y=12.00
PMT: END BEGIN

Under the conditions of **Example 2,** the monthly payment is $889.33. Compare with the table method.

EXAMPLE 2 **Using a Table to Find a Monthly Mortgage Payment**

Find the monthly payment necessary to amortize a $160,000 loan at 4.5% for 25 years.

Solution

In **Table 7,** read down to the 4.5% row and across to the column for 25 years, to find the entry 5.55832. As this is the monthly payment amount needed to amortize a loan of $1000, and our loan is for $160,000, our required monthly payment is

$$160 \cdot \$5.55832 = \$889.33.$$

So that the borrower pays interest only on the money actually owed in a month, interest on real-estate loans is computed on the decreasing balance of the loan. Each equal monthly payment is first applied toward interest for the previous month. The remainder of the payment is then applied toward reduction of the principal amount owed.

Payments in the early years of a real-estate loan are mostly interest (typically 80% or more); only a small amount goes toward reducing the principal. The amount of interest decreases each month, so that larger and larger amounts of the payment will apply to the principal. During the last years of the loan, most of the monthly payment is applied toward the principal. (See **Table 9** on **page 736.**)

Longer-term mortgages have become more popular in recent years. In fact, 40-year mortgages are not uncommon today. Extending the term may help a buyer "afford" a more expensive home. For example, the monthly payments for principal and interest on a $200,000 mortgage at 4% APR would be $954.83 for a 30-yr term, but only $835.88 for a 40-yr term. In some cases, this may be advantageous, but the long-term effects of a longer term loan should be considered. Over the lives of the two loans mentioned above, the total payments would be as follows.

    30-yr term: $343,739.01
    40-yr term: $401,220.93

Which loan would you prefer?

Once the regular monthly payment has been determined, as in **Examples 1 and 2,** an **amortization schedule** (or **repayment schedule**) can be generated. It will show the allotment of payments for interest and principal, and the principal balance, for one or more months during the life of the loan. Tables showing these breakdowns are available from lenders or can be produced on a computer spreadsheet. The following steps demonstrate how the computations work.

Step 1 Interest for the month $= \left(\dfrac{\text{Old balance}}{\text{of principal}}\right)\left(\dfrac{\text{Annual}}{\text{interest rate}}\right)\left(\dfrac{1}{12}\,\text{year}\right)$

Step 2 Payment on principal $= \left(\dfrac{\text{Monthly}}{\text{payment}}\right) - \left(\dfrac{\text{Interest for}}{\text{the month}}\right)$

Step 3 New balance of principal $= \left(\dfrac{\text{Old balance}}{\text{of principal}}\right) - \left(\dfrac{\text{Payment on}}{\text{principal}}\right)$

This sequence of steps is done for the end of each month. The new balance obtained in Step 3 becomes the Step 1 old balance for the next month.

EXAMPLE 3   Preparing an Amortization Schedule

The Petersons have a $120,000 mortgage with a term of 30 years and a 4.0% interest rate. Prepare an amortization schedule for the first 2 months of their mortgage.

Solution

First get the monthly payment. We use **Table 7.** (You could also use the formula.)

$$R = 120 \quad \cdot \quad \$4.77415 = \$572.90$$

Mortgage amount in $1000s   Intersection of 4.0% row with 30-year column in **Table 7**

Now apply Steps 1–3.

Step 1 Interest for the month $= \$120{,}000(0.04)\left(\frac{1}{12}\right) = \$400$

Step 2 Payment on principal $= \$572.90 - \$400 = \$172.90$

Step 3 New balance of principal $= \$120{,}000 - \$172.90 =$ **$119,827.10**

Starting with an old balance of $119,827.10, repeat the steps for the second month.

Step 1 Interest for the month $=$ **$119,827.10**$(0.04)\left(\frac{1}{12}\right) = \$399.42$

Step 2 Payment on principal $= \$572.90 - \$399.42 = \$173.48$

Step 3 New balance of principal $= \$119{,}827.10 - \$173.48 =$ **$119,653.62**

These calculations are summarized in **Table 8.**

The book *You Can Do the Math*, by Ron Lipsman of the University of Maryland, is a practical resource for most aspects of **personal financial management.** The associated Web site, www.math.umd .edu/~rll/cgi-bin/finance.html, provides "calculators" with which you can easily input your own values to get the results of many different financial computations.

Table 8 Amortization Schedule

Payment Number	Interest Payment	Principal Payment	Balance of Principal
			$120,000.00
1	$400.00	$172.90	$119,827.10
2	$399.42	$173.48	$119,653.62

Table 9 Amortization Schedules for a $120,000, 30-Year Mortgage

4.0% Interest Monthly Payment: $572.90				12.5% Interest Monthly Payment: $1280.71			
Payment Number	Interest Payment	Principal Payment	Balance of Principal	Payment Number	Interest Payment	Principal Payment	Balance of Principal
Initially →			120,000.00	Initially →			120,000.00
1	400.00	172.90	119,827.10	1	1,250.00	30.71	119,969.29
2	399.42	173.48	119,653.62	2	1,249.68	31.03	119,938.26
3	398.85	174.05	119,479.57	3	1,249.36	31.35	119,906.91
12	393.55	179.35	117,886.76	12	1,246.29	34.42	119,609.63
60	362.49	210.41	108,537.02	60	1,224.11	56.60	117,458.11
152	287.12	285.78	85,851.63	152	1,133.87	146.84	108,704.57
153	286.17	286.73	85,564.90	153	1,132.34	148.37	108,556.20
230	202.43	370.47	60,358.56	230	951.19	329.52	90,984.42
231	201.20	371.70	59,986.86	231	947.75	332.96	90,651.46
294	114.50	458.40	33,890.53	294	641.10	639.61	60,905.68
295	112.97	459.93	33,430.60	295	634.43	646.28	60,259.40
296	111.44	461.46	32,969.13	296	627.70	653.01	59,606.39
359	3.80	569.10	571.00	359	26.27	1254.44	1267.51
360	1.90	571.00	0.00	360	13.20	1267.51	0.00
Totals:	86,243.41	120,000.00		**Totals:**	341,055.35	120,000.00	

Prevailing mortgage interest rates have varied considerably over the years. **Table 9** shows portions of the amortization schedule for the Petersons' loan of **Example 3** and shows what the corresponding values would have been had their interest rate been 12.5%. (Rates that high have not been seen for many years.) Notice how much interest is involved in this home mortgage. At the (low) 4.0% rate, $86,243.41 in interest was paid, along with the $120,000 principal. At a rate of 12.5%, the interest alone would total the huge sum of $341,055.35, which is about 2.8 times the mortgage principal.

Adjustable-Rate Mortgages

The lending industry uses many variations on the basic fixed-rate mortgage. An **adjustable-rate mortgage (ARM),** also known as a **variable-rate mortgage (VRM),** generally starts out with a lower rate than similar fixed-rate loans, but the rate changes periodically, reflecting changes in prevailing rates.

Amortization Schedule				
No.	Amount	Interest	Principal	Balance
1	1,280.71	1,250.00	30.71	119,969.29
2	1,280.71	1,249.68	31.03	119,938.26
3	1,280.71	1,249.36	31.35	119,906.91

Amortizations to Go There are many free (and premium) apps for smartphones that will generate amortization tables. This screenshot is from the Loan Calculator feature of the fncalculator.com app. It agrees with the amortization of the 12.5% loan in **Table 9.**

QUANTITIES GOVERNING ADJUSTABLE-RATE MORTGAGES (ARMS)

- **Adjustment period**—Time interval between rate adjustments (typically 1, 3, or 5 years)

- **Index**—Standard fluctuating average that is the basis for the new, adjusted rate (typically the 1-, 3-, or 5-year U.S. Treasury security rate or a national or regional "cost of funds" index)

A Note on Margin When determining the margin on a loan, the lender considers the creditworthiness of the borrower. Those with lower credit scores are a higher risk to the lender, so the margin is increased to cover that risk.

Credit scores are determined in part by comparing the amount of credit a borrower has to the amount they use, as well as their record of making payments on time. The Fair Credit Reporting Act permits consumers to request a free credit report once per year from each of the three major credit-reporting agencies (Equifax, TransUnion, and Experian).

Potential homebuyers should establish good credit to qualify for lower interest rates. For sound advice on building good credit, check out http://www.fdic.gov/consumers/consumer/ccc/savvy.html

QUANTITIES GOVERNING ARMS, CONTINUED

- **Margin**—Additional amount added to the index by the lender (typically a few percentage points)

- **Discount**—Amount by which the *initial* rate may be less than the sum of the index and the margin (typically arranged between seller and lender)

- **Interest rate cap**—Limits on (interest) rate increases

- **Periodic cap**—Limit on rate increase per adjustment period (typically about 1% per 6 months or 2% per year)

- **Overall cap**—Limit on rate increases over the life of the loan (typically about 5% total)

- **Payment cap**—Limit on how much the payment can increase at each adjustment

- **Negative amortization**—An increasing loan principal (perhaps caused by a payment cap preventing the payment from covering a higher interest rate)

- **Convertibility feature**—A contractual ability to convert to a fixed-rate mortgage (usually at certain designated points in time)

- **Prepayment penalty**—Charges imposed by the lender if payments are made early

EXAMPLE 4   Comparing ARM Payments Before and After a Rate Adjustment

We pay $40,000 down on a $200,000 house and take out a 3/1 ARM (meaning that the initial rate stays fixed for the first three years and then adjusts each year after that to reflect the current index) for a 30-year term. The lender uses the 1-year Treasury index (at 0.2%) and a 2.8% margin.

(a) Find the monthly payment for the first three years.

(b) Suppose that after three years, the 1-year Treasury index has increased to 1.4%. Find the monthly payment for the fourth year.

Solution

(a)   Cost of house ↓   Down payment ↓

Mortgage amount = $200,000 − $40,000 = $160,000

The interest rate for the first three years will be

ARM interest rate = Index rate + Margin = 0.2% + 2.8% = 3%.

Now from **Table 7** (using 3% over 30 years) we obtain 4.21604.

Initial monthly payment = 160 · $4.21604 = $674.57

(b) During the first three years, some of the mortgage principal has been paid, so in effect we will now have a new "mortgage amount." (Also, the term will now be three years less than the original term.) The amortization schedule for the first three years (not shown here) yields a loan balance, after the thirty-sixth monthly payment, of $149,670.66.

For the fourth year,

ARM interest rate = Index rate + Margin = 1.4% + 2.8% = 4.2%.

Because 4.2% is not included in **Table 7,** we use the regular monthly payment formula with the new mortgage balance and 27 years for the remaining term.

$$\text{Fourth-year monthly payment} = \frac{P\left(\frac{r}{12}\right)}{1 - \left(\frac{12}{12 + r}\right)^{12t}} \qquad \text{Regular payment formula}$$

$$= \frac{\$149{,}670.66\left(\frac{0.042}{12}\right)}{1 - \left(\frac{12}{12 + 0.042}\right)^{(12)(27)}} \qquad \text{Substitute known values.}$$

$$= \$773.07$$

The first ARM interest rate adjustment has caused the fourth-year monthly payment to rise to $773.07, which is an increase of $98.50 over the initial monthly payment. ∎

A "seller buydown" occurs when the seller (a new-home builder, for example) pays the lender an amount in order to discount the buyer's loan. This reduces the initial rate and monthly payments, thereby giving the buyer a better chance to qualify for the loan. But it may be combined with higher initial fees or home price.

EXAMPLE 5 Discounting a Mortgage Rate

In **Example 4,** suppose that a seller buydown discounts our initial rate by 0.5%. Find the initial and fourth-year monthly payments.

Solution

The initial interest rate is discounted to 2.5% (rather than the 3% of **Example 4**), so the **Table 7** entry is 3.95121.

$$\text{Initial monthly payment} = 160(\$3.95121) = \$632.19$$

The amortization schedule shows a balance at the end of three years of $148,839.36. The discount now expires, and the index has increased to 1.4%, so for the fourth year,

$$\text{ARM interest rate} = \text{Index rate} + \text{Margin} = 1.4\% + 2.8\% = 4.2\%.$$

(This is just as in **Example 4.**) Using the monthly payment formula, with $r = 4.2\%$ and $t = 27$,

$$\text{Fourth-year monthly payment} = \frac{\$148{,}839.36\left(\frac{0.042}{12}\right)}{1 - \left(\frac{12}{12 + 0.042}\right)^{(12)(27)}} \qquad \begin{array}{l}\text{Substitute values in the}\\\text{monthly payment formula.}\end{array}$$

$$= \$768.78.$$

The initial monthly payment of $632.19 looks better than the $674.57 of **Example 4,** but at the start of year four, monthly payments jump by $136.59. ∎

Closing Costs

Apart from principal and interest payments, buying a home involves a variety of one-time expenses called **closing costs,** or **settlement charges,** which are imposed when the loan is finalized (at "closing"). Often a buyer agrees to establish an **escrow account,** or **reserve account,** maintained by the lender for the purpose of paying property taxes and homeowner's insurance premiums. In such cases, closing costs will include an "escrow deposit" to fund the escrow account.

Because escrow deposits increase closing costs, some buyers decide to pay property taxes and insurance premiums directly. However, some lenders lower the interest rate slightly for borrowers who use an escrow account, and many home-owners prefer the simplicity of letting the lender process these payments. A buyer is entitled to a "good faith estimate" (GFE) of closing costs from the lender and, if desired, may shop for alternative providers of settlement services.

EXAMPLE 6   Computing Total Closing Costs

A buyer borrowed $130,000 at 4% interest to purchase a home. At the "closing" of the loan on June 19, the following closing costs were charged.

Loan origination fee (1% of mortgage amount)	$ ___?___
Broker loan fee	1625
Lender document and underwriting fees	534
Transfer taxes (for state and local fees on mortgages)	180
Title services and lender's title insurance	401
Title insurance	457
Government recording charges	55
Escrow deposit	1032
Daily interest charges ($14.25 per day for 12 days)	___?___

Compute the total closing costs for this mortgage.

Solution

"Loan origination fees" are sometimes referred to as **points.** Each "point" amounts to 1% of the mortgage amount. By imposing points, the lender can effectively raise the interest rate without raising monthly payments (because points are normally paid at closing rather than over the life of the loan).

In this case, "one point" amounts to **$1300**. Because mortgage payments are typically made on the first of each month, and this loan closed on June 19, twelve days will pass before July 1. Interest on the $130,000 balance for these twelve days is

$$12 \cdot \$130,000 \cdot \frac{0.04}{365} \approx \$171.$$

Adding these amounts to those listed gives total closing costs of $5755.   ■

Recurring Costs of Home Ownership

The primary considerations for most homebuyers are the following.

1. Accumulating the down payment
2. Having sufficient cash and income to qualify for the loan
3. Making the monthly mortgage payments

Three additional expenses of homeownership, listed below, should be anticipated realistically. These are **recurring costs** (as opposed to the one-time closing costs), and they can be significant.

1. **Property taxes** are collected by a county or other local government. Depending on the location and value of the home, taxes can range up to several thousand dollars annually. Property taxes and mortgage interest are income tax deductible. Therefore, money expended for those items will decrease income taxes. This is one way in which the government has historically, through the tax code, encouraged home ownership.

Government-backed mortgages, including FHA (Federal Housing Administration) and VA (Veterans Administration) loans, carry a government guarantee to protect the lender in case the borrower fails to repay the loan. Those who obtain a conventional loan, instead, are usually required to buy private mortgage insurance (PMI), unless they are able to make a down payment of at least 20% of the purchase price of the property.

This feature can be a surprise recurring expense to first-time home purchasers. It was introduced to protect lenders but indirectly also protects the buyers, who may lose a great deal if some catastrophe makes it impossible for them to make their mortgage payments.

2. **Homeowner's insurance** usually covers losses due to fire, storm damage, and other casualties. Some types, such as earthquake, flood, or hurricane coverage, may be unavailable or very expensive, depending on location.

3. All homes require **maintenance,** but these costs can vary greatly, depending mainly on the size, construction type, age, and condition of the home.

Insurance and maintenance expenses, though necessary to protect the home investment, are not generally tax deductible.

EXAMPLE 7 **Determining the Affordability of Home Ownership**

Greg and Tamara Jamieson have household income that places them in a 30% combined state and local tax bracket. They can afford a net average monthly expenditure of $1700 for a home. The home of their dreams is priced at $300,000. They have saved a 20% down payment and are prequalified for a 20-year, $240,000 fixed-rate mortgage at 4.5%. Property taxes are estimated to be $3900 annually, homeowner's insurance premiums would be $600 per year, and annual maintenance costs are estimated to be $1200. Can the Jamiesons afford their dream home?

Solution

Let's "do the math."

$$\text{Regular mortgage payment} = \underset{\substack{\text{Mortgage amount} \\ \text{in \$1000s}}}{240} \cdot \underset{\substack{\text{Value from} \\ \textbf{Table 7}}}{\$6.32649} \approx \$1518$$

$$\text{Monthly property taxes} = \frac{\$3900 \quad \leftarrow \text{Annual taxes}}{12 \quad \leftarrow \text{Months per year}}$$

$$= \$325$$

$$\text{Monthly insurance and maintenance cost} = \frac{\$600 + \$1200}{12} = \$150$$

$$\text{Total monthly expense} = \underset{\text{Mortgage payment}}{\$1518} + \underset{\text{Property taxes}}{\$325} + \underset{\substack{\text{Insurance and} \\ \text{maintenance}}}{\$150}$$

$$= \mathbf{\$1993}$$

Because $1993 > $1700, it seems that the Jamiesons cannot afford this home. But wait—remember that mortgage interest and property taxes are both income tax deductible. Let's consider further.

$$\text{Monthly interest} = \$240{,}000(0.045)\left(\frac{1}{12}\right) \qquad \textit{Interest} = Prt$$

$$= \$900$$

Of the $1518 mortgage payment, $900 would be interest (initially).

$$\text{Monthly deductible expenses} = \underset{\text{Mortgage interest}}{\$900} + \underset{\text{Property tax}}{\$325} = \$1225$$

$$\text{Monthly income tax savings} = \underset{\text{Deductible expense}}{\$1225} \cdot \underset{\text{Tax bracket}}{30\%} \approx \$368$$

$$\text{Net monthly cost of home} = \underset{\text{Gross cost}}{\$1993} - \underset{\text{Tax savings}}{\$368} = \mathbf{\$1625}$$

By considering the effect of taxes, we see that the net monthly cost of $1625 is indeed within the Jamiesons' $1700 affordability limit. ■

Example 7 raises some reasonable questions.

1. Is it wise to buy a house close to your spending limit, in light of future uncertainties?

2. What about the fact that the interest portion of mortgage payments, and therefore the tax savings, will decrease over time?

3. Won't taxes, insurance, and maintenance costs probably increase over time, making it more difficult to keep up the payments?

Some possible responses follow.

1. The Jamiesons may have built in sufficient leeway when they decided on their limit of $1700 per month.

2. (a) Look again at **Table 9** to see how slowly the interest portion drops.

   (b) Most people find that their income over time increases faster than expenses for a home that carries a fixed-rate mortgage. (A variable-rate mortgage is more risky and should have its initial rate locked in for as long as possible.)

   (c) Prevailing interest rates rise and fall over time. While rising rates will not affect a fixed-rate mortgage, falling rates may offer the opportunity to reduce mortgage expenses by refinancing. (For guidelines, see the margin note on refinancing.)

3. Here again, increases in income will probably keep pace. Although most things, including personal income, tend to follow inflation, the fixed-rate mortgage ensures that mortgage amortization, a major item, will stay constant.

WHEN Will I Ever USE This ?

Suppose you are a certified financial planner (CFP) with clients who are relocating and need help in deciding whether to purchase a home or rent. Their household income puts them in a 25% combined state and federal income tax bracket. If they purchase a home, they would plan to stay in it for at least 10 years. They are interested in a $280,000 home and are prequalified for a 30-year fixed mortgage at 4.5%. There are comparable homes in the same area that rent for $1,000 per month. Home values are projected to increase by 3% per year, and rents are also projected to increase by 3% per year.

Would it be better for them to buy or rent?

To answer this question, you lay out a side-by-side comparison of the housing costs over a ten-year period. See **Table 10** on the next page.

Explanation of the Costs of Buying

Because the purchase price is $280,000 and your clients want to avoid purchasing private mortgage insurance, they will be making a down payment of $56,000 (20% of the purchase price). Closing costs—including loan origination fees, discount points, documentation fees, and prorated interest and escrow deposits—typically come to about 3% of the purchase price, or $8,400. The down payment and closing costs make up the one-time "up-front" costs.

Table 10 Comparing Buying and Renting

Buying		Renting	
Purchase Price: $280,000			
Up-front Costs of Purchasing		**Up-front Costs of Renting**	
Down payment	$56,000	Rental deposit (one month's rent)	$1,000
Closing costs (about 3% of purchase price)	$8,400		
Recurring Costs (30-yr fixed, 4.5%)		**Recurring Costs**	
Mortgage principal	$44,600	Rent (increasing 3% per year)	$137,567
Mortgage interest (incl. 25% tax deduction)	$68,698	Renter's insurance (1.32% of annual rent)	$1,816
Property Taxes (incl. 25% tax deduction)	$30,093		
Maintenance (0.5% of purchase price, rising with inflation)	$15,330		
Insurance (0.2% of home value)	$6,420		
Opportunity Costs		**Opportunity Costs**	
Down payment/closing costs	$33,244	Initial rent deposit	$516
Yearly costs	$40,455	Yearly costs	$32,083
Selling Costs		**Terminating Rental Agreement**	
Selling costs (6% of selling price)	$22,578	Return of initial deposit	−$1,000
Remaining principal	$179,400		
Proceeds from home sale	−$376,297		
Total Housing Expenditures for 10 Years	**$128,921**	**Total Housing Expenditures for 10 Years**	**$171,982**

The recurring costs of home ownership include monthly principal and interest payments, property taxes (paid from an escrow account), maintenance costs, and homeowner's insurance. The amount of $44,600 for principal paid during the first ten years on the mortgage was calculated by amortizing the mortgage and adding the principal payments for 120 months. Your clients are in a 25% income tax bracket, so only 75% of the interest payments are considered "costs." Thus amortizing interest for the first 120 months, adding up these interest payments, and multiplying by 0.75 will yield the $68,698 figure for interest.

Property taxes in the county where the house is located are 1.25% of the value of the home. Thus taxes for the first year are

$$0.0125 \cdot \$280,000 = \$3,500.$$

Because property values are expected to increase by 3% per year, the property taxes in the second year are projected to be

$$\$3,500 \cdot 1.03 = \$3,605.$$

Continuing this pattern yields the property tax amounts shown in **Table 11**. But property taxes are income tax deductible, so the actual expense for property taxes for the first ten years (considering the 25% tax bracket) is

$$(1 - 0.25) \cdot \$40,124 = 0.75 \cdot \$40,124 = \$30,093.$$

Maintenance costs are assumed to be 0.5% of the purchase price and grow at the rate of inflation, which is projected to be 2%.

$$\text{First-year maintenance costs} = 0.005 \cdot \$280,000 = \$1,400$$

$$\text{Second-year maintenance costs} = \$1,400(1.02) = \$1,428$$

Table 11

Year	Prop Tax
1	$3,500
2	3,605
3	3,713
4	3,825
5	3,939
6	4,057
7	4,179
8	4,305
9	4,434
10	4,567
Total	**$40,124**

Table 12

Year	Maintenance
1	$1,400
2	1,428
3	1,457
4	1,486
5	1,515
6	1,546
7	1,577
8	1,608
9	1,640
10	1,673
Total	**$15,330**

Table 13

Year	Insurance
1	$560
2	577
3	594
4	612
5	630
6	649
7	669
8	689
9	709
10	731
Total	**$6,420**

The cost increase due to inflation continues throughout the ten years, resulting in the amounts shown in **Table 12**.

Homeowner's insurance costs are assumed to be about 0.2% of the value of the home (which is projected to increase by 3% per year).

$$\text{First-year insurance costs} = 0.002 \cdot \$280,000 = \$560$$

$$\text{Second-year insurance costs} = \$560(1.03) \approx \$577$$

Table 13 shows the insurance costs projected for the ten-year period.

Opportunity cost is incurred when a choice is made to forgo an investment opportunity. For example, if your clients decide to purchase the home, they will be spending $56,000 on a down payment and $8,400 for closing costs, rather than investing it in some other investment vehicle. For the purposes of your calculations, you have assumed that monies they have that are not spent on housing will instead be invested in the stock market at a 5% annual return. It is also assumed that the capital gains from these investments will be taxed at 15% (rather than the 25% income tax rate). Thus the "after-tax" effective growth rate will be

$$(1 - 0.15)(0.05) = 0.85(0.05) = 0.0425 = 4.25\%.$$

Therefore, the net gain (opportunity cost) over ten years on the up-front costs of purchasing the home would be

$$\$64,400(1 + 0.0425)^{10} - \$64,400 \approx \$33,244.$$

To calculate the opportunity cost of the yearly expenses of home ownership, you begin with the monthly mortgage payments. The amortization of the 30-year fixed mortgage at 4.5% and beginning principal of $224,000 shows that the first month's principal payment will be $295 and the interest will be $840.

Because the interest is tax deductible, the actual "cost" of the first month's mortgage payment would be

$$\$295 + 0.75(\$840) = \$925.$$

Investing this amount over ten years would yield an "after-tax" gain of

$$\$925(1 + 0.0425)^{10} - \$925 \approx \$477.$$

Performing similar calculations for the remaining 119 months of payments, as well as for the yearly costs of property taxes, maintenance, and insurance, yields the total opportunity cost of the yearly expenses of $40,455.

If your clients were to sell the home in ten years, the value (having increased by 3% per year) would be

$$\$280,000(1.03)^{10} \approx \$376,297,$$

which amounts to the proceeds from the sale of the home. Closing costs (for realtor services, home inspection, and other necessities in selling a home) can be expected to be about 6% of the selling price.

$$\text{Closing costs} = 0.06(\$376,297) = \$22,578$$

The remaining principal on the loan after ten years can be calculated as follows.

$$\text{Original loan amount} - \text{sum of principal payments} = \text{remaining principal}$$

$$\$224,000 - \$44,600 = \$179,400$$

Explanation of the Costs of Renting

Landlords typically charge a deposit equal to one month's rent. In your clients' case, this would be $1,000. The rent is expected to increase 3% per year over the next ten years. Rental agreements are often one-year leases, so the rent payment will be fixed for 12 months before it increases. All 120 projected rent payments are added to give the $137,567 figure for the rent payments over the ten-year period.

Renter's insurance in the area where your clients would like to live typically amount to 1.32% of the annual rent, which increases by 3% per year. Projected annual premiums for the ten years total $1,816.

Opportunity costs for rental expenses are calculated in a fashion similar to those for home ownership, taking into account the 15% tax rate on capital gains.

When all the calculations are made, you are able to show your clients that the difference between rental costs and ownership costs would be approximately

Rental Ownership
costs costs
↓ ↓

$$\$171{,}982 - \$128{,}921 = \mathbf{\$43{,}061.}$$

Therefore, they can save, on average, about $4,306 per year by buying rather than renting.

13.4   EXERCISES

Round all monetary answers to the nearest cent unless directed otherwise.

Monthly Payment on a Fixed-Rate Mortgage *Find the monthly payment needed to amortize principal and interest for each fixed-rate mortgage. You can use either the regular monthly payment formula or* **Table 7,** *as appropriate.*

	Loan Amount	Interest Rate	Term			Loan Amount	Interest Rate	Term
1.	$100,000	8.0%	20 years		**5.**	$227,750	10.5%	25 years
2.	$23,000	6.0%	15 years		**6.**	$195,450	5.5%	10 years
3.	$135,000	5.7%	25 years		**7.**	$42,500	4.6%	5 years
4.	$95,000	6.9%	30 years		**8.**	$200,000	3.5%	10 years

Amortization of a Fixed-Rate Mortgage *Complete the first one or two months (as required) of each amortization schedule for a fixed-rate mortgage.*

9. Mortgage: $58,500
Interest rate: 10.0%
Term of loan: 30 years

Amortization Schedule

Payment Number	Total Payment	Interest Payment	Principal Payment	Balance of Principal
1	(a) _____	(b) _____	(c) _____	(d) _____

10. Mortgage: $95,000
Interest rate: 3.5%
Term of loan: 20 years

Amortization Schedule

Payment Number	Total Payment	Interest Payment	Principal Payment	Balance of Principal
1	(a) _____	(b) _____	(c) _____	(d) _____

11. Mortgage: $143,200

     Interest rate: 6.5%

     Term of loan: 15 years

Amortization Schedule

Payment Number	Total Payment	Interest Payment	Principal Payment	Balance of Principal
1	(a) _____	(b) _____	(c) _____	(d) _____
2	(e) _____	(f) _____	(g) _____	(h) _____

12. Mortgage: $124,750

     Interest rate: 3.0%

     Term of loan: 25 years

Amortization Schedule

Payment Number	Total Payment	Interest Payment	Principal Payment	Balance of Principal
1	(a) _____	(b) _____	(c) _____	(d) _____
2	(e) _____	(f) _____	(g) _____	(h) _____

13. Mortgage: $113,650

     Interest rate: 8.2%

     Term of loan: 10 years

Amortization Schedule

Payment Number	Total Payment	Interest Payment	Principal Payment	Balance of Principal
1	(a) _____	(b) _____	(c) _____	(d) _____
2	(e) _____	(f) _____	(g) _____	(h) _____

14. Mortgage: $150,000

     Interest rate: 4.25%

     Term of loan: 16 years

Amortization Schedule

Payment Number	Total Payment	Interest Payment	Principal Payment	Balance of Principal
1	(a) _____	(b) _____	(c) _____	(d) _____
2	(e) _____	(f) _____	(g) _____	(h) _____

Finding Monthly Mortgage Payments   *Find the total monthly payment, including taxes and insurance.*

	Mortgage	Interest Rate	Term of Loan	Annual Taxes	Annual Insurance
15.	$62,300	5%	20 years	$610	$220
16.	$51,800	6%	25 years	$570	$145
17.	$89,560	3.5%	10 years	$915	$409
18.	$72,890	5.5%	15 years	$1850	$545

Comparing Total Principal and Interest on a Mortgage   *Suppose* $140,000 *is owed on a house. The monthly payment for principal and interest at 8.5% for 30 years is*

$$140 \cdot \$7.68913 = \$1076.48.$$

19. How many monthly payments will be made over the 30-year period?

20. What is the total amount that will be paid for principal and interest?

21. The total interest charged is the total amount paid minus the amount financed. What is the total interest?

22. Which is more—the amount financed or the total interest paid? By how much?

Comparing Mortgage Options   *Suppose you want to purchase a $300,000 home, and you have the required $60,000 down payment in savings. Complete the table below to compare different mortgage options presented by a mortgage broker. (Source: Bankrate.com, accessed May 20, 2014)*

	Principal	Interest Rate	Term (years)	Payment	Total of Payments over Life of Loan	Total Interest Paid over Life of Loan
23.	$240,000	4.250%	30	_____	_____	_____
24.	$240,000	3.875%	20	_____	_____	_____
25.	$240,000	3.250%	15	_____	_____	_____
26.	$240,000	3.000%	10	_____	_____	_____

27. Compare the monthly payments, total of payments, and total interest paid for the loans in **Exercises 23 and 25.** Which of these two loan options would you choose? Why?

28. Compare the monthly payments, total of payments, and total interest paid for the loans in **Exercises 24 and 26.** Which of these two loan options would you choose? Why?

Long-Term Effect of Interest Rates   *You may remember seeing home mortgage interest rates fluctuate widely in a period of not too many years. The following exercises show the effect of changing rates. Refer to **Table 9,** which compared the amortization of a $120,000, 30-year mortgage for rates of 4.0% and 12.5%. Give values of each of the following for (a) a 4.0% rate, and (b) a 12.5% rate.*

29. monthly payments

30. percentage of first monthly payment that is principal

31. balance of principal after 1 year

32. the number of payments required to pay down half of the original principal

33. the first monthly payment that includes more toward principal than toward interest

34. amount of interest included in final monthly payment of mortgage

The Effect of the Term on Total Amount Paid   *Suppose a $60,000 mortgage is to be amortized at 7.5% interest. Find the total amount of interest that would be paid for each term.*

35. 10 years

36. 20 years

37. 30 years

38. 40 years

The Effect of Adjustable Rates on the Monthly Payment   *For each adjustable-rate mortgage, find (a) the initial monthly payment, (b) the monthly payment for the second adjustment period, and (c) the change in monthly payment at the first adjustment.*

	Beginning Balance	Term	Initial Index Rate	Margin	Adjustment Period	Adjusted Index Rate	Adjusted Balance
39.	$75,000	20 years	6.5%	2.5%	1 year	8.0%	$73,595.52
40.	$44,500	30 years	7.2%	2.75%	3 years	6.6%	$43,669.14

(The "adjusted balance" is the principal balance at the time of the first rate adjustment. Assume no caps apply.)

The Effect of Rate Caps on Adjustable-Rate Mortgages   *Millie has a 1-year ARM for $110,000 over a 20-year term. The margin is 2%, and the index rate starts out at 0.5% and increases to 3.0% at the first adjustment. The balance of principal at the end of the first year is $105,706.34. The ARM includes a periodic rate cap of 2% per adjustment period. Use this information for Exercises 41–44.*

41. Find **(a)** the interest owed and **(b)** the monthly payment due for the first month of the first year.

42. Find **(a)** the interest owed and **(b)** the monthly payment due for the first month of the second year. (Remember the periodic rate cap!)

43. What is the monthly payment adjustment at the end of the first year?

44. If the index rate has dropped slightly at the end of the second year, will the third-year monthly payments necessarily drop? Why or why not?

Closing Costs of a Mortgage *For Exercises 45–48, refer to the following list of closing costs for the purchase of a $175,000 house requiring a 20% down payment, and find each requested amount.*

Title insurance premium	$400
Document recording fee	60
Loan fee (two points)	_____
Appraisal fee	400
Prorated property taxes	685
Prorated fire insurance premium	295

45. mortgage amount   **46.** loan fee

47. total closing costs

48. total amount of cash required of the buyer at closing (including down payment)

*Consider the scenario of **Example 7**. Recalling that mortgage interest is income tax deductible, find (to the nearest dollar) the additional initial net monthly savings resulting from each strategy. In each case, only the designated item changes. All other features remain the same.*

49. Change the mortgage term from 20 years to 30 years.

50. Change the mortgage from fixed at 4.5% to an ARM with an initial rate of 3.5%.

For each of Exercises 51–54, find all of the following quantities for a $200,000 fixed-rate mortgage. (Give answers to the nearest dollar.)

(a) *Monthly mortgage payment (principal and interest)*

(b) *Monthly house payment (including property taxes and insurance)*

(c) *Initial monthly interest*

(d) *Income tax deductible portion of initial house payment*

(e) *Net initial monthly cost for the home (considering tax savings)*

	Term of Mortgage	Interest Rate	Annual Property Tax	Annual Insurance	Owner's Income Tax Bracket
51.	15 years	5.5%	$960	$480	20%
52.	20 years	6.0%	$840	$420	25%
53.	10 years	6.5%	$1092	$540	30%
54.	30 years	7.5%	$1260	$600	40%

*Exercises 55 and 56 refer to the **When Will I Ever Use This?** feature at the end of this section. Suppose your clients can afford to buy a $200,000 home and have the required 20% down payment in savings. Comparable rentals are available for $900 per month. Assume a combined income tax bracket of 25%, and a return on investments of 4%, with a 15% tax rate on capital gains. Compare the costs of buying and renting under the given circumstances.*

55. Your clients plan to move in five years, so they would take out a 5/1 ARM at 3.5%, and then plan to sell the home and pay off the loan before the interest rate adjusts. The table reflects an increase in home values of 3% per year, and a 2% annual increase in rents. Determine

(a) the opportunity costs that would result from the up-front costs of purchasing the home,

(b) the total costs of five years of ownership, and

(c) the difference between the costs of renting and the costs of owning over five years.

Buying		Renting	
Up-front Costs of Purchasing		**Up-front Costs of Renting**	
Down payment	$40,000	Rental deposit	$900
Closing costs	$6,000		
Recurring Costs		**Recurring Costs**	
Mortgage payments	$36,452	Rent	$56,204
Other	$17,685	Renter's insurance	$1,124
Opportunity Costs		**Opportunity Costs**	
Up-front costs	_____	Initial rent deposit	$164
Yearly costs	$5,121	Yearly costs	$5,094
Selling Costs		**Terminating Agreement**	
Selling costs	$13,911	Return of deposit	−$900
Remaining principal	$143,515		
Proceeds from sale	−$231,855		
Total Costs for 5 Years _____		**Total Costs for 5 Years $62,586**	

56. Seeing the savings associated with home ownership, your clients now are considering staying in the area for 20 years. If they decided to purchase the home, they would take out a 30-year fixed mortgage at 4.5%. If home values increase by 3% per year, and rents also increase 3% per year, determine

(a) the proceeds from the sale of the same home in 20 years,

(b) the total costs of 20 years of ownership, and

(c) the difference between the costs of renting and the costs of owning over 20 years.

Buying		Renting	
Up-front Costs of Purchasing		**Up-front Costs of Renting**	
Down Payment	$40,000	Rental deposit	$900
Closing Costs	$6,000		
Recurring Costs		**Recurring Costs**	
Mortgage payments	$166,369	Rent	$290,200
Other	$87,425	Renter's insurance	$5,804
Opportunity Costs		**Opportunity Costs**	
Up-front costs	$43,778	Initial rent deposit	$857
Yearly costs	$104,715	Yearly costs	$112,388
Selling Costs		**Terminating Agreement**	
Selling costs	$21,673	Return of deposit	−$900
Remaining principal	$78,224		
Proceeds from sale	_____		
Total Costs for 20 Years _____		**Total Costs for 20 Years $409,249**	

On the basis of material in this section, or your own research, give brief written responses to each problem.

57. Suppose your ARM allows conversion to a fixed-rate loan at each of the first five adjustment dates. Describe circumstances under which you would want to convert.

58. Should a home buyer always pay the smallest down payment that will be accepted? Explain.

59. Should a borrower always choose the shortest term available in order to minimize the total interest expense? Explain.

60. Under what conditions would an ARM probably be a better choice than a fixed-rate mortgage?

61. Why are fourth-year monthly payments (slightly) less in **Example 5** than in **Example 4** even though the term, 27 years, and the interest rate, 4.2%, are the same in both cases?

62. What tends to make ownership less expensive than renting, especially over many years?

63. Discuss the term "payment shock" mentioned at the end of **Example 5.**

64. Find out what is meant by each term and describe some of the features of each.

FHA-backed mortgage
VA-backed mortgage
Conventional mortgage

13.5   FINANCIAL INVESTMENTS

OBJECTIVES

1 Become familiar with stocks.
2 Become familiar with bonds.
3 Become familiar with mutual funds.
4 Evaluate investment returns.
5 Explore strategies for building a nest egg.

Stocks

In a general sense, an *investment* is a way of putting resources to work so that (one hopes) they will grow. Our main emphasis here will be on the basic mathematical aspects of a restricted class of financial investments—namely, *stocks, bonds,* and *mutual funds.*

Buying stock in a corporation makes you a part owner of the corporation. You then share in any profits the company makes, and your share of profits is called a **dividend.** If the company prospers (or if increasing numbers of investors believe it will prosper in the future), your stock will be attractive to others, so that you may, if you wish, sell your shares at a profit. The profit you make by selling for more than you paid is called a **capital gain.** A negative gain, or **capital loss,** results if you sell for less than you paid. By **return on investment,** we mean the net difference between what you receive (including your sale price and any dividends received) and what you paid (your purchase price plus any other expenses of buying and selling the stock).

EXAMPLE 1 Finding the Return on Ownership of Stock

Sarah Patterson bought 100 shares of stock in Company A on January 15, 2015, paying $30 per share. On January 15, 2016, she received a dividend of $0.50 per share, and the stock price had risen to $30.85 per share. (Ignore any costs other than the purchase price.) Find the following.

(a) Sarah's total cost for the stock

(b) The total dividend amount

(c) Sarah's capital gain if she sold the stock on January 15, 2016

(d) Sarah's total return on her one year of ownership of this stock

(e) The percentage return

Solution

(a) Cost = ($30 per share) · (100 shares) = $3000

She paid $3000 total.

(b) Dividend = ($0.50 per share) · (100 shares) = $50

The dividend amount was $50.

(c)

Change in price per share   Number of shares

Capital gain = ($30.85 − $30.00) · 100 = $0.85 · 100 = $85

The capital gain was $85.

(d)

Dividend   Capital gain

Total return = $50 + $85 = $135

Total return on the investment was $135.

(e) Percentage return = $\dfrac{\text{Total return}}{\text{Total cost}} \cdot 100\% = \dfrac{\$135}{\$3000} \cdot 100\% = 4.5\%$

The percentage return on the investment was 4.5%. ■

The **NASDAQ (National Association of Securities Dealers Advanced Quotations)**, unlike the NYSE and other exchanges that actually carry out trading at specific locations, is an electronic network of brokerages established in 1971 to facilitate the trading of over-the-counter (unlisted) stocks. Most of the high-tech company stocks, which grew very rapidly as a group in the 1980s and especially the 1990s (and many of which declined as rapidly in 2000–2002), are (or were) listed on the Nasdaq market.

The price of a share of stock is determined by the law of supply and demand at institutions called **stock exchanges.** In the United States, the oldest and largest exchange is the New York Stock Exchange (NYSE), established in 1792 and located on Wall Street in New York City. Members of the public buy their stock through stockbrokers, people who have access to the exchange. Stockbrokers do business in offices throughout the country. They charge a fee for buying or selling stock.

Current prices and other information about particular stocks are published daily in many newspapers and on various Web sites, where information may be updated every few minutes. **Table 14** below shows examples of trading on the NYSE, on May 22, 2014. The exchange opens at 9:30 A.M. and closes at 4:00 P.M., Eastern time.

Observe the column headings and the numbers for the first company listed, Allstate, with market symbol ALL. The first four numbers show that the first sale of Allstate shares after 9:30 A.M. was for $58.40 per share, the highest sale during the trading day was for $58.52, the lowest was for $58.26, and the last sale before 4:00 P.M. was for $58.35 per share.

Moving across to the right, the next two numbers show that the closing price for the day was $0.07 lower than that of the previous day—that is, 0.12% lower. The volume number indicates that 1,797,900 shares of Allstate stock were sold that day.

Table 14 Selected Stock Quotes (May 22, 2014)

Company	Symbol	Open	High	Low	Close	Net Change	% Change	Volume	52-Wk High	52-Wk Low	Annual Div	Dividend Yield	PE	YTD % Change
Allstate Corp	ALL	58.40	58.52	58.26	58.35	−0.07	−0.12%	1,797,900	58.69	45.60	$1.12	1.92%	10.34	6.99%
Bank of America	BAC	14.57	14.72	14.55	14.71	0.10	0.68%	51,929,398	18.03	12.13	$0.04	0.27%	17.39	−5.52%
Carmax Inc	KMX	44.26	44.92	44.22	44.53	0.19	0.43%	943,500	53.08	42.21	$0.00	0.00%	19.71	−5.30%
Caterpillar Inc	CAT	102.90	103.88	102.51	103.08	0.20	0.19%	3,112,300	108.21	80.86	$2.40	2.33%	17.00	13.51%
Chevron Corp	CVX	124.10	124.41	123.56	123.63	−0.53	−0.43%	3,262,700	127.83	109.27	$4.28	3.46%	12.09	−1.02%
Deere & Co	DE	90.00	90.50	89.60	89.99	0.01	0.01%	1,657,700	94.89	79.50	$2.04	2.27%	9.86	−1.47%
Duke Energy	DUK	70.01	70.64	70.01	70.55	0.62	0.89%	1,834,700	75.13	64.16	$3.12	4.42%	15.54	2.23%
FedEx Corp	FDX	138.84	139.69	138.34	139.33	0.28	0.20%	1,550,700	144.39	94.60	$0.60	0.43%	21.52	−3.09%
General Mills	GIS	53.50	53.80	53.31	53.74	0.26	0.49%	1,693,900	54.78	46.18	$1.64	3.05%	19.96	7.67%
Harley-Davidson	HOG	71.65	72.25	71.35	71.64	−0.45	−0.62%	1,037,600	74.13	49.15	$1.10	1.54%	20.66	3.47%
Hewlett-Packard	HPQ	32.45	33.36	31.21	31.78	−0.74	−2.28%	32,458,801	33.90	20.25	$0.58	1.83%	8.93	13.58%
Kimberly-Clark	KMB	109.91	110.37	109.71	110.13	0.11	0.10%	889,000	113.09	91.44	$3.36	3.05%	19.07	5.43%
Pearson Plc	PSO	19.21	19.37	19.09	19.28	0.07	0.36%	252,300	22.40	16.58	$0.79	4.10%	0.00	−13.93%
Walt Disney	DIS	82.19	82.79	82.04	82.35	0.16	0.19%	4,651,900	83.65	60.41	$0.86	1.04%	20.81	7.79%
Yum! Brands	YUM	74.63	75.14	74.35	74.87	0.20	0.27%	1,761,800	79.70	64.08	$1.48	1.98%	23.78	−0.98%

The next two numbers show that, over the last year, the greatest and least prices paid for ALL were $58.69 and $45.60, respectively. The following two numbers show that Allstate's current dividend is $1.12 per share and that this represents a current yield of 1.92% (of the current stock closing price for the day).

PE is the **price-to-earnings ratio,** the current price per share divided by the earnings per share over the past 12 months. For ALL, this ratio is 10.34. Finally, the year-to-date percent change means that Allstate's price per share is now 6.99% higher than it was at the start of this calendar year.

EXAMPLE 2 Reading a Stock Table

Use the stock table (**Table 14**) to find the required amounts below.

(a) The highest price for the last 52 weeks for FedEx (FDX)

(b) The dividend for Hewlett-Packard (HPQ)

Solution

(a) Find the correct line from the stock table for FedEx.

The highest price for FedEx for the last 52 weeks was $144.39.

(b) The annual dividend value of $0.58 means that HPQ paid a dividend of $0.58 per share. ∎

EXAMPLE 3 Finding the Cost of a Stock Purchase

Find the cost for 100 shares of Deere (DE) stock, purchased at the low price for the day.

Solution

From **Table 14,** the low price for the day for Deere was $89.60 per share.

$$\underset{\text{share}}{\underset{\text{Price per}}{}} \qquad \underset{\text{shares}}{\underset{\text{Number of}}{}}$$

$$\text{Total cost} = \$89.60 \ \cdot \ 100 = \$8960.00$$

One hundred shares of this stock would cost $8960.00, plus any broker's fees. ∎

Most stock purchases and sales are done through a broker, who has representatives at the exchange to execute a buy or sell order. The broker will charge a **commission** (the broker's fee) for executing an order. Before 1974, commission rates were set by stock exchange rules and were uniform from broker to broker. Since then, however, rates are competitive and vary considerably among brokers.

Full-service brokers, who offer research, professional opinions on buying and selling individual issues, and various other services, tend to charge the highest commissions on transactions they execute. Many **discount brokers,** on the other hand, merely buy and sell stock for their clients, offering little in the way of additional services.

During the 1990s, when it often seemed that almost any reasonable choice of stock would pay off for almost any investor, a number of online companies appeared, offering much cheaper transactions than were available through conventional brokers. Subsequently, most conventional brokerage firms, as well as the major discount brokerages, introduced their own automated (online) services.

Commissions normally are a percentage of the value of a purchase or sale (called the **principal**). Some firms charge additional amounts in certain cases. For example, rates may depend on whether the order is for a **round lot** of shares (a multiple of 100) or an **odd lot** (any portions of an order for fewer than 100 shares). On the odd-lot portion, you may be charged an **odd-lot differential** (say, 10¢ per share).

Stock Quote
11/03/2014 01:02:03

Enter a symbol   + ▶

S&P 500   **2018.05**

Table 14 Portfolio - US ($) Li

Name	Symbol	Pr
X Allstate Corporat	ALL	6
X Bank of America C	BAC	1
X CarMax Inc	KMX	5
X Caterpillar Inc.	CAT	10
X Chevron Corporati	CVX	11
X Deere & Company C	DE	8
X Duke Energy Corpo	DUK	8
X FedEx Corporation	FDX	16
X General Mills In	GIS	5
X Harley-Davidson	HOG	6

Pocket Portfolios Today, investors can use smartphones to check their portfolios, issue buy and sell orders, and view the performance history of stocks on multiple exchanges. This screenshot (using the Stock Quote app from BISHINEWS) shows part of the portfolio from **Table 14.**

Also, it is possible to place a **limit order,** where the broker is instructed to execute a buy or sell if and when a stock reaches a predesignated price. An extra fee may be added to the commission on a limit order. Many discount brokerage firms do not charge these special fees; instead, they charge different commissions depending on whether the transactions are **automated trades** or **broker-assisted trades.**

Table 15 Typical Discount Brokerage Commissions

	Automated (Online)	Automated (Phone)	Broker-Assisted
Brokerage Firm A	$8.95	$13.95	$33.95
Brokerage Firm B	$9.99	$34.99	$44.99
Brokerage Firm C	$7.95	$12.95	$32.95

Also, the Securities and Exchange Commission (SEC), a federal agency that regulates stock markets, supports its own activities by charging the exchanges, based on volume of transactions. Typically this charge is passed on to investors, through brokers, in the form of an **SEC fee** assessed on stock sales only (not on purchases). As of February 2014, this fee was $22.10 per million dollars of principal (rounded *up* to the next cent). For example, to find the fee for a sale of $1600, first divide $1600 by $1,000,000 and then multiply by $22.10.

$$\text{SEC fee} = \frac{\$1600}{\$1,000,000} \cdot \$22.10 = \$0.03536, \quad \text{which is rounded up to } \$0.04.$$

EXAMPLE 4 Finding Total Cost of Stock Purchases

Carina has a brokerage account with Brokerage Firm A, and on May 22, 2014, she executed a broker-assisted purchase of 250 shares of Harley-Davidson (HOG) stock at the average price of the day. She purchased 100 more shares online on September 12 for $73.50 per share. Find Carina's total cost for purchasing the 350 shares of HOG.

Solution

Daily high   Daily low   (**Table 14** values)

$$\text{Average price for first purchase} = \frac{\$72.25 + \$71.35}{2} \quad \text{Arithmetic mean}$$

$$= \mathbf{\$71.80}$$

Price per share   Number of shares

$$\text{Basic cost of first purchase} = \$71.80 \cdot 250$$

$$= \mathbf{\$17,950.00}$$

Because the purchase was a broker-assisted transaction, Firm A charged a commission of $33.95 (as shown in **Table 15**).

Basic cost   Commission

$$\text{Total cost of first purchase} = \$17,950.00 + \$33.95$$

$$= \mathbf{\$17,983.95}$$

Price per share   Number of shares

$$\text{Basic cost of second purchase} = \$73.50 \cdot 100$$

$$= \mathbf{\$7350.00}$$

The second purchase was automated (online), so Firm A charged a commission of $8.95.

Basic cost   Commission

Total cost of second purchase = $7350.00 + $8.95

= **$7358.95**

Total cost for both transactions = $17,983.95 + $7358.95

= **$25,342.90** ← Total cost   ■

EXAMPLE 5 Finding the Return on a Stock Investment

Suppose Carina (from **Example 4**) sold all 350 shares of HOG stock using an online transaction on December 19, 2014, at a price of $74.28 per share. Find the return on the investment. (Assume that dividends were paid on September 5.)

Solution

The basic price of the stock sold is

$$(\$74.28 \text{ per share}) \cdot (350 \text{ shares}) = \$25,998.00.$$

Because Carina owned only 250 shares when dividends were paid, her dividend receipt was

$$(\$1.10 \text{ per share}) \cdot (250 \text{ shares}) = \$275.00.$$

The sale was executed online, so Firm A charged a commission of $8.95 (see **Table 15**). Find the SEC fee as described before **Example 4.**

$$\frac{\$25,998.00}{\$1,000,000} \cdot \$22.10 = \$0.574556$$

Round up to obtain an SEC fee of $0.58. Then

Basic price   Commission   SEC fee

$$\text{Net proceeds from sale} = \$25,998.00 - \$8.95 - \$0.58$$

$$= \$25,988.47.$$

Finally, the return on investment will be the sum of dividends and proceeds from the sale, less the total cost of purchasing the stock (from **Example 4**).

$$\text{Return on investment} = \$275.00 + \$25,988.47 - \$25,342.90$$

$$= \$920.57 \quad \leftarrow \text{Final return}$$

Carina's return on the investment in HOG shares was $920.57.   ■

Bonds

Rather than using your capital (money) to buy shares of stock in a company, you may prefer merely to *lend* money to the company, receiving an agreed-upon rate of interest for the use of your money. In this case you would buy a bond from the company rather than purchase stock. The bond (loan) is issued with a stated term (life span), after which the bond "matures" and the principal (or **face value**) is paid back to you. Over the term, the company pays you a fixed rate of interest, which depends on prevailing interest rates at the time of issue (and to some extent on the company's credit rating), rather than on the underlying value of the company.

A corporate bondholder has no stake in company profits but is (quite) certain to receive timely interest payments. If a company is unable to pay both bond interest and stock dividends, the bondholders must be paid first.

EXAMPLE 6 **Finding the Return on a Bond Investment**

Roy invests $10,000 in a 5-year corporate bond paying 3% annual interest, paid semiannually. Find the total return on this investment, assuming Roy holds the bond to maturity (for the entire 5 years). Ignore any broker's fees.

Solution

By the simple interest formula (see **Section 13.1**),

$$I = Prt = \$10{,}000(0.03)(5) = \$1500.$$

Notice that this amount, $1500, is the *total* return, over the 5-year period, not the annual return. Also, the fact that interest is paid twice per year (1.5% of the face value each time) has no bearing here since compounding does not occur. ■

Mutual Funds

A **mutual fund** is a pool of money collected by an investment company from many individuals and/or institutions and invested in many stocks, bonds, or money market instruments. As of 2013, in the United States alone, $15 trillion was invested in more than 7000 mutual funds. Nearly 57 million U.S. households held mutual funds, with the median mutual fund assets of these households approximately $100,000. Over a third ($6.5 trillion) of mutual fund assets were in retirement (IRA and deferred compensation plan) accounts.

The **Wall Street reform bill of 2010** put forth "the most sweeping set of changes to the financial regulatory system since the 1930s." Prompted by the failure of a number of large financial and other companies and a steep international recession beginning in 2007, the bill created a new Consumer Financial Protection Bureau within the Federal Reserve, headed by a presidentially appointed "independent" regulator, and with authority to write new rules to (theoretically) protect consumers from unfair or abusive practices in mortgages and credit cards. In fact, the bill creates vast bureaucratic control over virtually all financial areas discussed in this chapter.

A decade earlier, in 1999, congressional repeal of many earlier legal and regulatory barriers had allowed large financial conglomerates to offer one-stop shopping for services in the areas of insurance, banking, and investments.

It took over a year for the Democrat-controlled Congress to bring the 2010 bill to a vote. According to the watchdog group Public Citizen, over that period the financial industry spent nearly $600 million to hire some 1000 lobbyists to promote their interests. Debate raged about whether the bill was really to help consumers or to control business (or perhaps a combination of the two).

(*Source:* www.CNNMoney.com)

MUTUAL FUNDS VERSUS INDIVIDUAL STOCKS AND BONDS

Advantages of Mutual Funds

1. *Simplicity*
   Let someone else do the work.

2. *Diversification*
   Being part of a large pool makes it easier to own interests in many different stocks, which decreases vulnerability to large losses in one particular stock or one particular sector of the economy.

3. *Access to New Issues*
   Initial public offerings (IPOs) can sometimes be very profitable. Whereas individual investors find it difficult to get in on these, the large overall assets of a mutual fund give its managers much more access.

4. *Economies of Scale*
   Large stock purchases usually incur smaller expenses per dollar invested.

5. *Professional Management*
   Professional managers may have the expertise to pick stocks and time purchases to better achieve the stated objectives of the fund.

6. *Indexing*
   With minimal management, an index fund can maintain a portfolio that mimics a popular index (such as the S&P 500). This makes it easy for individual fund investors to achieve returns at least close to those of the index.

MUTUAL FUNDS VERSUS INDIVIDUAL STOCKS AND BONDS, CONTINUED

Disadvantages of Mutual Funds

1. *Impact of One-Time Charges and Recurring Fees*
   Sales charges, management fees, "12b-1" fees (used to pay sales representatives), and fund expenses can mount and can be difficult to identify. And paying management fees does not guarantee getting *quality* management.

2. *Hidden Cost of Brokerage*
   Recurring fees and expenses are added to give the *expense ratio* of a fund. But commissions paid by the fund for stock purchases and sales are in addition to the expense ratio. (Since 1995, funds have been required to disclose their average commission costs in their annual reports.)

3. *Some Hidden Risks of Fund Ownership*

   (a) In the event of a market crash, getting out of the fund may mean accepting securities rather than cash, and these may be difficult to redeem for a fair price.

   (b) Managers may stray from the stated strategies of the fund.

   (c) Since tax liability is passed on to fund investors and depends on purchases and sales made by fund managers, investors may be unable to avoid inheriting unwanted tax basis.

Source for advantages and disadvantages of mutual funds:
Forbes Guide to the Markets, John Wiley & Sons, Inc.

The risks and rewards associated with stocks and bonds are reflected in the mutual funds that own shares of those stocks and bonds. Historically, the rate of return over most time periods has been greater for stocks than for bonds. Accordingly, stock mutual funds have more potential for higher returns (on average) than bond mutual funds. On the other hand, there are fewer uncertainties associated with bonds, and so investors holding shares of bond mutual funds are exposed to less risk than those invested in stock funds. "Mixed asset" (or blended) mutual funds are designed to help investors balance their portfolios by dealing in both stocks and bonds, thus increasing potential earnings with stock holdings, but reducing volatility by virtue of bond holdings.

Because a mutual fund owns many different stocks, each share of the fund represents a fractional interest in each of those companies. In an "open-end" fund (by far the most common type), new shares are created and issued to buyers while the fund company absorbs the shares of sellers. Similarly, a bond mutual fund can invest in many different bonds of varying maturities, buying and selling according to market conditions. Every stock or bond held by a fund at any given time contributes to the total value of the fund. Each day, the value of a share in the fund is called the *net asset value, or* **NAV.**

An abundance of information about **mutual funds** is available on the Internet. For example, the Investment Company Institute (ICI), the national association of the American investment company industry, provides material at the site www.ici.org/.

NET ASSET VALUE OF A MUTUAL FUND

If A = Total fund assets, L = Total fund liabilities, and N = Number of shares outstanding, then the **net asset value** is calculated as follows.

$$\text{NAV} = \frac{A - L}{N}$$

EXAMPLE 7 Finding the Number of Shares in a Mutual Fund Purchase

Suppose, on a given day, a mutual fund has $500 million worth of stock, $500,000 in cash (not invested), and $300,000 in other assets. Total liabilities amount to $4 million, and there are 25 million shares outstanding. If Joanne invests $50,000 in this fund, how many shares will she obtain?

Solution

$$A = \$500,000,000 + \$500,000 + \$300,000$$

$$= \$500,800,000$$

$$L = \$4,000,000$$

$$N = 25,000,000$$

$$\text{NAV} = \frac{A - L}{N} \qquad \text{Formula}$$

$$= \frac{\$500,800,000 - \$4,000,000}{25,000,000} \qquad \text{Substitute values of } A, L, \text{ and } N.$$

$$= \$19.872 \qquad \text{Calculate.}$$

Because $\frac{\$50,000}{\$19.872} \approx 2516$, Joanne's $50,000 investment will purchase (to the nearest whole number) 2516 shares. ∎

Evaluating Investment Returns

Regardless of the type of investment you have or are considering (stocks, bonds, mutual funds, or others), it is important to be able to accurately evaluate and compare returns (profits or losses). Even though past performance is never a guarantee of future returns, it is crucial information. A wealth of information on stock and bond markets and mutual funds is available in many publications (for example, the *Wall Street Journal, Investor's Business Daily,* and *Barron's Weekly*) and at numerous Internet sources (for example, www.nasdaq.com).

The most important measure of performance of an investment is the **annual rate of return.** This is not *necessarily* the same as percentage return, as calculated in **Example 1,** because annual rate of return depends on the time period involved. Some commonly reported periods are daily, seven-day, monthly, month-to-date, quarterly, quarter-to-date, annual, year-to-date, 2-year, 3-year, 5-year, 10-year, 20-year, and "since inception" (since the fund or other investment vehicle was begun).

Examples 8 and 9 illustrate some of the many complications that arise when one tries to evaluate and compare different investments.

EXAMPLE 8 Analyzing a Stock's Annual Rate of Return

Wendell Stewart owns 50 shares of stock. His brokerage statement for the end of August showed that the stock closed that month at a price of $40 (per share), and the statement for the end of September showed a closing price of $40.12. For purposes of illustration, disregard any possible dividends, and assume that the money gained in a given month has no opportunity to earn additional returns. (Those earnings cannot be "reinvested.") Find the following.

(a) The value of these shares at the end of August

(b) Wendell's monthly return on this stock

(c) The monthly percentage return **(d)** The annual rate of return

Solution

(a)
$$\text{Value} = \underset{\underset{\text{Price per share}}{\uparrow}}{\$40} \cdot \underset{\underset{\text{Number of shares}}{\uparrow}}{50} = \$2000$$

(b)
$$\text{Return} = \underset{\underset{\text{Change in price per share}}{\uparrow}}{(\$40.12 - \$40)} \cdot \underset{\underset{\text{Number of shares}}{\uparrow}}{50}$$
$$= \$6$$

(c)
$$\text{Percentage return} = \frac{\text{Total return}}{\text{Total value}} \cdot 100\%$$
$$= \frac{\$6}{\$2000} \cdot 100\%$$
$$= 0.3\%$$

(d) Because monthly returns do not earn more returns, we use an "arithmetic" return here. Simply add 0.3% twelve times (or multiply 0.3% by 12), to obtain 3.6%. The annual rate of return is 3.6%. ∎

EXAMPLE 9 Analyzing a Mutual Fund's Annual Rate of Return

Valerie Hoskins owns 80 shares of a mutual fund with a net asset value of $10 on October 1. Assume that the only return that the fund earns is dividends of 0.4% per month, which are automatically reinvested. Find the following.

(a) The value of Valerie's holdings in this fund on October 1

(b) The monthly return   (c) The annual rate of return

Solution

(a) Value = NAV · number of shares = $10 · 80 = $800

(b) Monthly return = 0.4% of $800 = $3.20

(c) In this case, monthly returns get reinvested. Compounding occurs. Therefore, we use a "geometric" return rather than arithmetic. To make this calculation, first find the "return relative," which is 1 plus the monthly rate of return:

$$1 + 0.4\% = 1 + 0.004 = 1.004.$$

Then multiply this return relative 12 times (raise it to the 12th power) to get an annual return relative: $1.004^{12} \approx 1.049$. Finally, subtract 1 and multiply by 100 to convert back to a percentage rate. The annual rate of return is 4.9%. ∎

A geometric return such as the one just found in **Example 9** often is called the "effective annual yield." (Notice that an arithmetic return calculation would have ignored the reinvestment compounding and would have understated the effective annual rate of return by 0.1%, because $12(0.4\%) = 4.8\%$.)

Building a Nest Egg

The most important function of investing, for most people, is to build an account for some future use, probably for retirement living. In **Example 8** of **Section 13.1**, we compared two retirement programs, Plan A and Plan B, to emphasize the importance of starting a retirement investment program early in life. The performance of investments over time is generally governed by formulas based on the summation of geometric sequences (see the margin note).

A **geometric sequence** has the property that each term is the product of the previous term and a *common ratio, r*. Thus a geometric sequence of *n* terms with first term *a* is

$$a, ar, ar^2, \ldots, ar^{n-1}.$$

The sum of these *n* terms is given by the formula

$$S = \frac{a(1 - r^n)}{1 - r}.$$

There are two major barriers to building a retirement nest egg (or whatever the goal is for accumulating wealth)—*inflation* and *(income) taxes*.

Table 4 (page 714) shows the actual historical inflation rates over 33 years (1980–2013). This reflects an average of about 3.1% per year. If we had invested for retirement over that time span, the inflation proportion (see **page 715**) indicates that we would have ended up, in 2013, paying about $\frac{233.0}{82.4} \approx 2.8$ times as much for goods and services as when we started, in 1980.

This inflationary effect, if unanticipated, could make a sizable dent in retirement living, but there is a fairly painless way to make provision for it. Periodically we can simply increase contributions by enough to keep pace with inflation.

The formula below allows us to compute the future values of an account where regular deposits are systematically adjusted for inflation. If the inflation rate is i, then the amount R is deposited at the end of year 1, $R(1 + i)$ is deposited at the end of year 2, and so on, with the deposit at the end of the year n, in general, being $R(1 + i)^{n-1}$.

FUTURE VALUE OF AN INFLATION-ADJUSTED RETIREMENT ACCOUNT

Assume annual deposits into a retirement account, adjusted for inflation. If

$\qquad i$ = annual rate of inflation,

$\qquad R$ = initial deposit (at the end of year 1),

$\qquad r$ = annual rate of return on money in the account,

and $\quad n$ = number of years deposits are made,

then the value V of the account at the end of n years is calculated as follows.

$$V = R\left[\frac{(1 + r)^n - (1 + i)^n}{r - i}\right]$$

EXAMPLE 10 Adjusting a Retirement Account for Inflation

In **Example 8** of **Section 13.1,** Angie executed Plan A, depositing $2000 on each birthday from age 21 to age 30, then stopped depositing, and at age 65 had $428,378. Barry, with Plan B, waited until age 31 to make the first $2000 deposit, then deposited $2000 every year to age 65, and came out with only $344,634.

Compute the final account value, at age 65, for Barry's Plan B, assuming that the 35 annual deposits had been adjusted for a 3.1% annual inflation rate.

Solution

Use the "inflation-adjusted" formula given in the box. The final account value is

$$V = \$2000\left[\frac{(1 + 0.08)^{35} - (1 + 0.031)^{35}}{0.08 - 0.031}\right] \approx \$484,664.$$

Let R = $2000, r = 0.08, i = 0.031, and n = 35.

With inflation adjustment, Plan B, by age 65, accumulates $484,664. ■

The second barrier to building wealth—taxes—is not so easily dealt with. You will find that the more you invest and the more you earn, the higher the percentage that our progressive tax system will claim of your earnings. Probably your most effective tool in softening (though not overcoming) this effect is to take full advantage of tax-deferred accounts, such as a TSA (Tax-Sheltered Annuity, or 403(b) Plan); an IRA (Individual Retirement Account, or 401(k) plan), especially if your employer will contribute matching funds; or, if self-employed, a SEP (Simplified Employee Pension) or a SIMPLE (Savings Incentive Match Plan for Employees) plan.

With most tax-deferred retirement accounts, the accumulated money is all taxed when it is withdrawn (presumably during retirement years). To demonstrate the decided advantage of tax deferral, we will compare the two situations using the following formulas. For a fair comparison, we look at both account values *after all taxes have been paid.*

TAX-DEFERRED VERSUS TAXABLE RETIREMENT ACCOUNTS

In both cases,

R = amount withheld annually from current salary to build the retirement account,

n = number of years contributions are made,

r = annual rate of return on money in the account,

t = marginal tax rate of account holder,

V = final value of the account, **following all accumulations and payment of all taxes.**

Tax-Deferred Account

The entire amount R is contributed each year. All contributions, plus earnings, earn a return over the n years, at which time tax is paid on all money in the account. The final account value is calculated as follows.

$$V = \frac{(1 - t)R[(1 + r)^n - 1]}{r}$$

Taxable Account

Each amount R withheld from salary is taxed up front, decreasing the amount of the annual deposit to $(1 - t)R$. Furthermore, annual earnings are also taxed at the end of each year. But at the end of the accumulation period, no more tax will be due. The final account value is calculated as follows.

$$V = \frac{R[(1 + r(1 - t))^n - 1]}{r}$$

EXAMPLE 11 **Comparing Tax-Deferred and Taxable Retirement Accounts**

Leah Bell and Oliver Reed, both in a 20% marginal tax bracket, will each contribute $2000 annually for 20 years to retirement accounts that return 5% annually. Leah chooses a tax-deferred account, while Oliver chooses a taxable account. Compare their final account values at the end of the accumulation period, after payment of all taxes.

Solution

We have $R = \$2000$, $n = 20$, $r = 0.05$, and $t = 0.20 = 0.2$.

Leah: $V = \dfrac{(1 - 0.2)\$2000[(1 + 0.05)^{20} - 1]}{0.05} \approx \$52{,}906$ Substitute values in the "tax-deferred" formula, and calculate.

Oliver: $V = \dfrac{\$2000[(1 + 0.05(1 - 0.2))^{20} - 1]}{0.05} \approx \$47{,}645$ Substitute values in the "taxable" formula, and calculate.

$$\$52{,}906 - \$47{,}645 = \$5261$$

By deferring taxes, Leah ends up with $5261, or about 11%, more. ∎

Greater contributions, a longer accumulation period, a higher rate of return, and a higher tax bracket all will accentuate the positive effect of tax deferral.

EXAMPLE 12 Comparing Tax-Deferred and Taxable Retirement Accounts

Repeat **Example 11,** but this time let

$$R = \$5000, \quad n = 40, \quad r = 0.10, \quad \text{and} \quad t = 0.40 = 0.4.$$

Solution

Leah: $V = \dfrac{(1 - 0.4)\$5000[\,(1 + 0.10)^{40} - 1\,]}{0.10}$ *Substitute values in the "tax-deferred" formula.*

$\approx \$1,327,778$

Oliver: $V = \dfrac{\$5000[\,(1 + 0.10(1 - 0.4))^{40} - 1\,]}{0.10}$ *Substitute values in the "taxable" formula.*

$\approx \$464,286$

With these higher parameters, Leah's advantage is dramatic:

$$\$1,327,778 - \$464,286 = \$863,492, \quad \text{or about 186\%, more.} \qquad \blacksquare$$

Tax-deferred is not the same as "tax-free," which makes the result of **Example 12** less impressive than it may seem. This type of retirement income is taxed at the time it is withdrawn. People who sell tax-deferred plans stress that you will probably be in a lower tax bracket after retirement because of reduced income. But many find this is not so, especially compared to early career income, and especially if your retirement account investments have done very well, which could make your required withdrawals quite large. Besides, who can tell what general tax rates will be 30 years from now?

FOR FURTHER THOUGHT

Summary of Investments

The summary of various types of investments in **Table 16,** on the next page, is based on average cases and is intended only as a general comparison of investment opportunities. There are numerous exceptions to the characteristics shown. For example, although the table shows no selling fees for mutual funds, this is really true only for so-called no-load funds. The "loaded" funds do charge "early redemption fees" or other kinds of sales charges. This is not to say that a no-load fund is necessarily better. For example, the absence of sales fees may be offset by higher "management fees."

These various costs, as well as other important factors, will be clearly specified in a fund's prospectus and should be studied carefully before a purchase decision is made. Some expertise is advisable for mutual fund investing, just less so than with certain other options, like direct stock purchases, raw land, or collectibles.

To use the table, read across the columns to find the general characteristics of the investment.

"Liquidity" refers to the ability to get cash from the investment quickly.

For Group or Individual Investigation

List **(a)** some investment types you would recommend considering and **(b)** some you would recommend avoiding for a friend whose main investment objective is as follows.

1. Avoid losing money.
2. Avoid losing purchasing power.
3. Avoid having to learn about investments.
4. Avoid having to work on investments.
5. Be sure to realize a gain.
6. Be able to "cash in" at any time.
7. Get as high a return as possible.
8. Be able to "cash in" without paying fees.
9. Realize quick profits.
10. Realize growth over the long term.

Table 16 Summary of Investments

Investment	Protection of Principal	Protection against Inflation	Rate of Return (%)	Certainty of Return	Selling Fees	Liquidity	Long-term Growth	Requires Work from Investor	Expert Knowledge Required
Cash	excellent	none	0			excellent	no	no	no
Ordinary life insurance	good	poor	2–4	excellent	high	excellent	no	no	no
Savings bonds	excellent	poor	3–5	excellent	none	excellent	no	no	no
Bank savings	excellent	poor	2–4	excellent	none	excellent	no	no	no
Credit union	excellent	poor	3–4	excellent	none	excellent	no	no	no
Corporation bonds	good	poor	5–7	excellent	low	good	no	some	some
Tax-free municipal bonds	excellent	poor	2–5	excellent	medium	good	no	no	some
Corporation stock	good	good	3–12	good	medium	good	yes	some	some
Preferred stock	good	poor	4–8	excellent	medium	good	no	some	some
Mutual funds	good	good	3–12	good	none	good	yes	no	no
Your own home	good	good	4–10	good	medium	poor	yes	yes	no
Mortgages on the homes of others	fair	poor	5–8	fair	high	poor	no	yes	some
Raw land	fair	good	?	fair	medium	poor	yes	no	yes
Rental properties	good	good	4–8	fair	medium	poor	yes	yes	some
Stamps, coins, antiques	fair	good	0–10	fair	high	poor	yes	yes	yes
Your own business	fair	good	0–20	poor	high	poor	yes	yes	yes

13.5   EXERCISES

*Refer to the stock table (**Table 14** on **page 749**) for Exercises 1–34.*

Reading Stock Charts   *Find each of the following.*

1. closing price for Deere & Co (DE)

2. 52-week high for Duke Energy (DUK)

3. percent change from the previous day for General Mills (GIS)

4. sales of the day for Kimberly-Clark (KMB)

5. high price for FedEx Corp (FDX) for the day

6. change from the previous day for Hewlett-Packard (HPQ)

7. year-to-date percentage change for Pearson Plc (PSO)

8. dividend for Chevron Corp (CVX)

9. the stock with the highest price-to-earnings ratio

10. the stock with the highest dividend yield

11. the stock with the highest volume (most shares sold) for the day

12. the stock with the lowest opening price

Finding Stock Costs *Find the basic cost (ignoring any broker fees) for each stock purchase, at the day's closing price.*

13. 300 shares of Caterpillar Inc (CAT)

14. 200 shares of Walt Disney (DIS)

15. 400 shares of General Mills (GIS)

16. 700 shares of Allstate (ALL)

Finding Stock Costs *Find the cost, at the day's closing price, for each stock purchase. Include typical discount broker commissions as described in the text, using Firm B.*

	Stock Symbol	Number of Shares	Transaction Type
17.	PSO	60	broker-assisted
18.	GIS	70	broker-assisted
19.	BAC	355	phone
20.	CAT	585	phone
21.	HOG	2500	online
22.	DUK	1500	online
23.	YUM	2400	phone
24.	DIS	20,000	broker-assisted

Finding Receipts for Stock Sales *Find the amount received by the sellers of each stock (at the day's closing prices). Deduct sales expenses as described in the text, using Firm C.*

	Stock Symbol	Number of Shares	Transaction Type
25.	ALL	400	broker-assisted
26.	KMX	600	broker-assisted
27.	CVX	500	phone
28.	DE	700	phone
29.	FDX	1350	online
30.	KMB	2740	online
31.	HPQ	1480	broker-assisted
32.	PSO	1270	online

Finding Net Results of Combined Transactions *For each combined transaction (executed at the closing price of the day), find the net amount paid out or taken in. Assume typical expenses as outlined in the text, using Firm A.*

33. Paul Barnes bought 100 shares of Walt Disney and sold 120 shares of General Mills, both broker-assisted trades.

34. Jasmine Collins bought 800 shares of Bank of America and sold 200 shares of Chevron, both online trades.

Costs and Returns of Stock Investments *For each of the following stock investments, find* **(a)** *the total purchase price,* **(b)** *the total dividend amount,* **(c)** *the capital gain or loss,* **(d)** *the total return, and* **(e)** *the percentage return. Ignore broker and SEC fees.*

	Number of Shares	Purchase Price per Share	Dividend per Share	Sale Price per Share
35.	40	$20.00	$2.00	$44.00
36.	20	$25.00	$1.00	$22.00
37.	100	$12.50	$1.08	$10.15
38.	200	$8.80	$1.12	$11.30

Total Return on Bond Investments *Find the total return earned by each bond in Exercises 39–42.*

	Face Value	Annual Interest Rate	Term to Maturity
39.	$1000	5.5%	5 years
40.	$5000	6.4%	10 years
41.	$10,000	7.11%	3 months
42.	$50,000	4.88%	6 months

Net Asset Value of a Mutual Fund *For each investment, find* **(a)** *the net asset value, and* **(b)** *the number of shares purchased.*

	Amount Invested	Total Fund Assets	Total Fund Liabilities	Total Shares Outstanding
43.	$3500	$875 million	$36 million	80 million
44.	$1800	$643 million	$102 million	50 million
45.	$25,470	$2.31 billion	$135 million	263 million
46.	$83,250	$1.48 billion	$84 million	112 million

Finding Monthly and Annual Investment Returns *For each investment, assume that there is no opportunity for reinvestment of returns. In each case find* **(a)** *the monthly return,* **(b)** *the annual return, and* **(c)** *the annual percentage return.*

	Amount Invested	Monthly Percentage Return
47.	$645	1.3%
48.	$895	0.9%
49.	$2498	2.3%
50.	$4983	1.8%

Effective Annual Rate of Return of Mutual Fund Investments *Assume that each mutual fund investment earns monthly returns that are reinvested and subsequently earn at the same rate. In each case find* **(a)** *the beginning value of the investment,* **(b)** *the first monthly return, and* **(c)** *the effective annual yield.*

	Beginning NAV	Number of Shares Purchased	Monthly Percentage Return
51.	$9.63	125	1.5%
52.	$12.40	185	2.3%
53.	$11.94	350	1.83%
54.	$18.54	548	2.22%

Inflation-Adjusted Retirement Accounts *Find the future value (to the nearest dollar) of each inflation-adjusted retirement account. Deposits are made at the end of each year.*

	Annual Inflation Rate	Initial Deposit	Annual Rate of Return	Number of Years
55.	3%	$2000	5%	25
56.	2%	$1000	6%	20
57.	1%	$5000	6.5%	40
58.	4%	$2500	7%	30

The Effect of Tax Deferral on Retirement Accounts *Find the final value, after all taxes are paid, for each account if* **(a)** *taxes are deferred, or* **(b)** *taxes are not deferred. In both cases, deposits are made at the end of each year.*

	Marginal Tax Rate	Regular Deferred Contribution	Annual Rate of Return	Number of Years
59.	15%	$1000	5%	30
60.	10%	$500	7%	15
61.	35%	$1500	6%	10
62.	28%	$3000	10%	25

Decide How Much to Put Aside *Suppose that you plan to retire in 30 years and will need $1,000,000 in your retirement account to do so. In Exercises 64 and 65, determine the required annual contribution toward retirement if your retirement account is* **(a)** *tax-deferred and* **(b)** *taxable. Assume that your marginal tax rate will remain at 25%.*

63. Suppose the annual rate of return on the money in your account is 10%.

64. Suppose the annual rate of return on the money in your account is 6%.

Solve each problem.

65. *Required Annual Contributions to Reach Retirement Goal* The formulas for finding final values of tax-deferred and taxable retirement accounts can be solved for R to find the required annual contributions necessary to reach an amount V by the time you retire.

   (a) Solve the "tax-deferred" account formula for R.

   (b) Solve the "taxable" account formula for R.

66. *Monthly Deposits Adjusted for Inflation* In an inflation-adjusted retirement account, let i denote annual inflation, let r denote annual return, and let n denote number of years, as in the text, but suppose deposits are made monthly rather than yearly with initial deposit R.

   (a) What expression represents the amount that would be deposited at the end of the second month?

   (b) Modify the future value formula for an inflation-adjusted account to accommodate the more frequent deposits.

   (c) Find (to the nearest dollar) the future value for monthly deposits of $100 (initially) for 20 years if $i = 0.03$ and $r = 0.06$.

67. *Finding an Unknown Rate* Given the compound interest formula

$$A = P\left(1 + \frac{r}{n}\right)^m,$$

do the following.

   (a) Solve the formula for r.

   (b) Find the annual rate r if $50 grows to $85.49 in 10 years with interest compounded quarterly. Give r to the nearest tenth of a percent.

68. *Comparing Continuous with Annual Compounding* It was suggested in the text that inflation is more accurately reflected by the continuous compounding formula than by periodic compounding. Compute the ratio

$$\frac{Pe^{rt}}{P(1+r)^n} = \frac{e^{rt}}{(1+r)^n}$$

to find how much (to the nearest tenth of a percent) continuous compounding exceeds annual compounding. Use a rate of 3% over a period of 30 years.

Spreading Mutual Fund Investments among Asset classes Mutual funds (as well as other kinds of investments) normally are categorized into one of several "asset classes," according to the kinds of stocks or other securities they hold. Small capitalization funds are most aggressive, while cash is most conservative. Many investors, often with the help of an advisor, try to construct their portfolios in accordance with their stage of life. Basically, the idea is that a younger person can afford to be more aggressive (and assume more risk), while an older investor should be more conservative (assuming less risk). The investment diagrams here show typical percentage ranges that might be recommended by an investment advisor.

Asset Classes

a.	Aggressive Growth (small cap)
b.	Growth
c.	Growth & Income
d.	Income
e.	Cash

In Exercises 69–72, divide the given investor's money into the five categories so as to position them right at the middle of the recommended ranges.

69. Alyssa Smith, in her early investing years, with $20,000 to invest

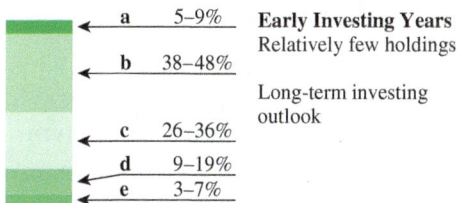

a	5–9%	**Early Investing Years**
b	38–48%	Relatively few holdings
		Long-term investing outlook
c	26–36%	
d	9–19%	
e	3–7%	

70. Justin Jones, in his good earning years, with $250,000 to invest

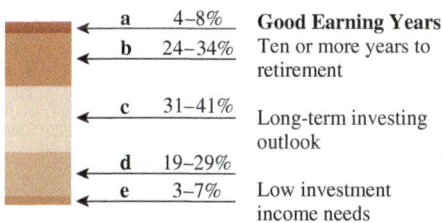

a	4–8%	**Good Earning Years**
b	24–34%	Ten or more years to retirement
c	31–41%	Long-term investing outlook
d	19–29%	
e	3–7%	Low investment income needs

71. Akira Suzuki, in his high-income/saving years, with $400,000 to invest

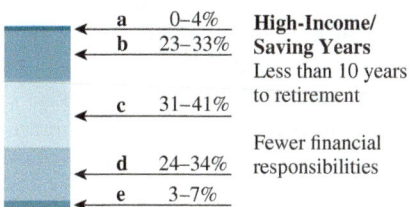

a	0–4%	**High-Income/**
b	23–33%	**Saving Years**
		Less than 10 years to retirement
c	31–41%	
		Fewer financial responsibilities
d	24–34%	
e	3–7%	

72. Eve McLean, retired, with $845,000 to invest

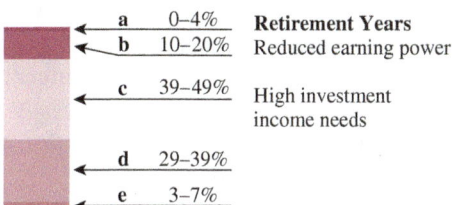

a	0–4%	**Retirement Years**
b	10–20%	Reduced earning power
c	39–49%	High investment income needs
d	29–39%	
e	3–7%	

The Effect of Taxes on Investment Returns Income from municipal bonds generally is exempt from federal, and sometimes state, income tax. For an investor with a marginal combined state and federal tax rate of 35%, a taxable return of 6% would yield only 3.9% (65% of 6%). At that tax rate, a tax-exempt return of 3.9% is equivalent to a taxable return of 6%. Find the tax-exempt rate of return that is equivalent to the given taxable rate of return for each investor.

	Investor	**Marginal Combined Tax Rate**	**Taxable Rate of Return**
73.	Joshua Sellers	25%	5%
74.	Sherry Lee	30%	7%
75.	Berna Barak	35%	8%
76.	Sean Stewart	40%	10%

Write responses to Exercises 77–84. Some research may be required.

77. Describe "dollar-cost averaging," and relate it to advantage number 1 of mutual funds as listed in the text.

78. From **Table 14** in the text, would you say that May 22, 2014, was a "good day" on Wall Street or a "bad day"?

79. Log onto www.napfa.org to research financial planners. Then report on what you would look for in a planner.

80. A great deal of research has been done in attempts to understand and predict trends in the stock market. Discuss the term *financial engineering.* In this context, what is a *rocket scientist* (or *quant*)?

81. Regarding mutual funds, discuss the difference between a *front-end load* and a *contingent deferred sales charge.*

82. What is a *growth stock?* What is an *income stock?*

83. With respect to investing in corporate America, describe the difference between being an owner and being a lender.

84. Comment on the graph shown here.

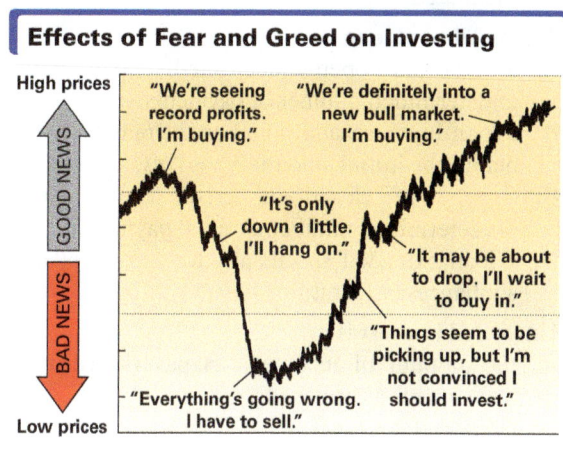

CHAPTER 13   SUMMARY

KEY TERMS

13.1

interest
principal
interest rate
simple interest
compound interest
future (maturity) value
present value
period
nominal rate
effective annual yield
price inflation
consumer price index (CPI)
deflation
recession
years to double
rule of 70

13.2

consumer credit
real estate mortgage

installment loan
closed-end credit
revolving loan
open-end credit
add-on interest
credit limit
itemized billing
finance charges
average daily balance
    method

13.3

Truth in Lending Act
annual percentage
    rate (APR)
unearned interest
actuarial method
rule of 78
early payment
    penalty
payoff amount

13.4

mortgage
term
down payment
principal amount
fixed-rate mortgage
amortizing
amortization schedule
adjustable-rate mortgage
    (ARM)
closing costs
escrow account
points
recurring costs
property taxes
homeowner's insurance
maintenance
opportunity cost

13.5

dividend

capital gain
capital loss
return on investment
stock exchanges
price-to-earnings
    ratio
commission
full-service broker
discount broker
principal
round lot
odd lot
odd-lot differential
limit order
automated trades
broker-assisted
    trades
SEC fee
face value
mutual fund
net asset value (NAV)
annual rate of return

TEST YOUR WORD POWER

See how well you have learned the vocabulary in this chapter.

1. **Compound interest** is
   A. calculated only on the original principal, and is generally more than simple interest.
   B. calculated on principal and previously earned interest.
   C. interest earned by two or more investments simultaneously.
   D. generally less than simple interest, but more easily calculated.

2. An **installment loan** is
   A. a type of open-ended credit loan with an undetermined number of payments.
   B. a loan in which additional credit may be extended before the initial amount is paid off.
   C. a type of closed-end credit that involves a predetermined number of equal payments.
   D. a loan extended specifically for the installation of appliances or furniture.

3. **Unearned interest** is
   A. the amount of income a taxpayer receives from investments as opposed to wages.

   B. the penalty paid by the borrower for being delinquent on one or more monthly payments.
   C. the amount by which the original finance charge is reduced when a loan is paid off early.
   D. the interest paid for borrowing as opposed to that received from an investment.

4. An **escrow account** (or impound account) is
   A. an account maintained by a lender and used to pay property taxes and insurance premiums.
   B. an account that is terminated when documents are signed and escrow "closes."
   C. always a requirement to obtain a mortgage.
   D. a savings account from which principal and interest payments are made.

5. A **mortgage** is
   A. an extension of unsecured credit, usually for a short period of time.
   B. a form of unsecured debt, in which wages may be garnished if payments are missed.
   C. the process of dividing principal and interest into equal periodic payments.
   D. a loan for a substantial amount, and for which property is pledged as security.

6. The **capital gain** on an investment is
  A. the overall return on the investment, including any dividends.
  B. the profit resulting from selling the investment for more than the purchase price.

C. the annual increase in the value of the investment.
D. the difference between the purchase price and the selling price.

QUICK REVIEW

Concepts	Examples

13.1 The Time Value of Money

Simple Interest
If principal P is borrowed for t years at an annual interest rate r, then (simple) interest is given by

$$I = Prt.$$

Future Value for Simple Interest

$$A = P(1 + rt)$$

If $2000 is borrowed at 3% simple interest for 5 years, find the future value (at the end of the 5-year term).

$$A = P(1 + rt)$$
$$= \$2000[1 + 0.03(5)]$$
$$= \$2300$$

Future Value for Compound Interest
If interest is compounded n times per year, then

$$A = P\left(1 + \frac{r}{n}\right)^{nt}.$$

If the $2000 is borrowed at 3% compounded monthly, find the future value (at the end of the 5-year term).

$$A = \$2000\left(1 + \frac{0.03}{12}\right)^{12(5)}$$
$$= \$2000(1.0025)^{60}$$
$$= \$2323.23$$

Thus $300 interest was paid with the simple interest loan, while $323.23 was paid with compound interest.

Present Value for Compound Interest
Solve the above equation for P.

$$P = \frac{A}{\left(1 + \frac{r}{n}\right)^{nt}}$$

A company anticipates needing $20,000 to purchase new equipment in 5 years. How much would the company need to invest now at 6% annual interest, compounded daily, in order to have the necessary amount in 5 years?

$$P = \frac{\$20,000}{\left(1 + \frac{0.06}{365}\right)^{365(5)}} = \$14,816.73$$

Effective Annual Yield
The effective annual yield Y of an account paying a nominal annual rate r, compounded n times per year, is calculated as follows.

$$Y = \left(1 + \frac{r}{n}\right)^n - 1$$

Find the effective annual yield of an account paying 2.000% annual interest, compounded quarterly.

$$Y = \left(1 + \frac{0.02}{4}\right)^4 - 1 = 2.015\%$$

Future Value for Continuous Compounding

$$A = Pe^{rt}$$

If $12,500 is invested in a money market account at 2.5% compounded continuously, find the value of the account after 7 years.

$$A = \$12,500e^{0.025(7)} = \$14,890.58$$

Concepts   *Examples*

13.2 Consumer Credit

Common Types of Consumer Credit

Installment loans involve a set amount borrowed, to be paid off by a predetermined number of equal payments **(installments).** The total of the payments will be the loan amount plus the interest added on.

A business owner purchases $5200 worth of office equipment by making a down payment of $1000 and financing the remaining balance with an installment loan with an interest rate of 5.5% over 2 years.

    Find the total cost of the purchase.

Amount financed: $P = \$5200 - \$1000 = \$4200$

Interest paid: $I = Prt$

$$= \$4200(0.055)(2)$$

$$= \$462$$

Amount to be repaid:

$$P + I = \$4200 + \$462 = \$4662$$

Monthly payment: $\dfrac{\$4662}{24} = \$194.25$

Total cost: $\$1000\ +\ \$4662\ =\ \mathbf{\$5662}$

                        ↑   ↑

                   Down   Loan

            payment repayment

Revolving loans (such as credit card accounts) involve **open-end credit** up to a predetermined **credit limit.** Charges are made and then itemized at the end of each billing period (usually a month). **Finance charges** are generally determined by the **average daily balance** for a given billing period.

A bank charges 1.4% per month on average daily balances for credit cards it issues. A billing period opens on Sept. 3 with a balance of $327.56. Transactions on the account for the period included a payment of $100 on Sept. 13 and a purchase in the amount of $158 on Sept. 25.

    Find the average daily balance, the finance charge, and the account balance for the Oct. 3 billing.

The running balance was $327.56 from Sept. 3 to Sept. 13 (for 10 days), $227.56 from Sept. 13 to Sept. 25 (for 12 days), and $385.56 from Sept. 25 to Oct. 3 (8 days).

Average daily balance for the period:

$$\frac{\$327.56(10) + \$227.56(12) + \$385.56(8)}{30} = \$303.03$$

Finance charge for Oct. 3 billing:

$$0.014(\$303.03) = \$4.24$$

The account balance for the Oct. 3 billing statement will be the sum of the last running balance and the finance charge.

$$\text{Ending balance} = \$385.56 + \$4.24$$

$$= \$389.80$$

Concepts   *Examples*

13.3 Truth in Lending

Lenders must (as required by the Truth in Lending Act) divulge the true cost of a loan by reporting the **annual percentage rate (APR)** to the borrower. **Table 5** on **page 725** gives the APR for a loan with a given number of monthly payments and a given finance charge (per $100 financed).

Makoto purchases a $10,200 car. He makes a $1500 down payment and agrees to pay 36 monthly payments of $270.62. The dealer claims a 7% annual interest rate.

    Is this dealer being truthful in lending?

Amount financed: $10,200 − $1500 = $8700

Cost of 36 monthly payments: 36 · $270.62 = $9742.32

Total finance charge: $9742.32 − $8700 = $1042.32

Finance charge per $100 financed:

$$\frac{\$1042.32}{\$8700} \cdot \$100 = \$11.98$$

Table 5 shows that for 36 monthly payments, this finance charge per $100 financed corresponds to an APR of 7.5%. *The dealer is not being truthful in lending.*

Calculating Unearned Interest

When a borrower pays off a loan early, the lender has not "earned" the full amount of interest originally disclosed when the loan was initiated. The amount of **unearned interest u** is calculated in one of the following two ways.

Makoto straightened things out with the dealer and financed the $8700 at 7% APR, which resulted in 36 monthly payments of $268.63. But he has decided to pay the loan off 12 months early.

    Which method of calculating the payoff amount would be better for Makoto?

Actuarial Method

$$u = kR\left(\frac{h}{\$100 + h}\right)$$

Here R is the regular monthly payment, k is the number of scheduled payments remaining, and h is the finance charge per $100 for a loan with the same APR and k monthly payments.

Actuarial method: We need to find h, the finance charge per $100 financed at 7% APR for 12 months. **Table 5** gives a value of $h = \$3.83$.

Unearned interest:

$$u = 12(\$268.63)\left(\frac{\$3.83}{\$100 + \$3.83}\right) = \$118.91$$

Payoff amount:

$$13 \cdot \$268.63 - \$118.91 = \textbf{\$3373.28}$$

Rule of 78

$$u = \frac{k(k + 1)}{n(n + 1)} \cdot F$$

Here F is the original finance charge, n is the number of payments originally scheduled, and k is the number of scheduled payments remaining.

Rule of 78: We need the corrected finance charge (using the smaller monthly payments).

$$F = 36 \cdot \$268.63 - \$8700 = \$970.68$$

Unearned interest:

$$u = \frac{12(13)}{36(37)} \cdot \$970.68 = \$113.68$$

Payoff Amount

To pay the loan off with k payments remaining (after the current payment), the borrower would need to pay the following.

$$\textbf{Payoff amount} = (k + 1)R - u$$

Payoff amount:

$$13 \cdot \$268.63 - \$113.68 = \textbf{\$3378.51}$$

It is to Makoto's advantage if the *actuarial method* is used, since the payoff amount would be

$$\$5.23 \text{ less.}$$

Concepts

Examples

Finance Charge Per $100 Financed
If a loan is structured in such a way that its APR or number of remaining payments (or both) are not found in **Table 5,** then h, the finance charge per $100 financed, may be calculated by the following formula.

$$h = \frac{n \cdot \frac{APR}{12} \cdot \$100}{1 - \left(1 + \frac{APR}{12}\right)^{-n}} - \$100$$

Here n is the number of monthly payments required to pay off the loan.

Makoto financed the $8700 at 7.8% interest and agreed to 42 monthly payments of $237.37. He decides to pay off the loan at the time of his 34th payment.
What is the finance charge per $100 financed?

$$h = \frac{8 \cdot \frac{0.078}{12} \cdot \$100}{1 - \left(1 + \frac{0.078}{12}\right)^{-8}} - \$100 = \$2.95 \quad \text{Here } n = 42 - 34 = 8.$$

This figure can be used to calculate the unearned interest and the payoff amount by the actuarial method.

13.4 The Costs and Advantages of Home Ownership

When a buyer borrows money to make a home purchase, the lender holds the property as security for the loan. Such an arrangement is a **mortgage. A fixed-rate mortgage** has regular periodic payments over the life **(term)** of the loan, in the amount of

$$R = \frac{P\left(\frac{r}{12}\right)}{1 - \left(\frac{12}{12 + r}\right)^{12t}}.$$

An **amortization schedule** shows all payments for the life of the loan, broken down into allocations for principal and interest for each payment.

Find the monthly payment on an $80,000 fixed-rate loan at 4.0% over 30 years.

$$R = \frac{\$80,000\left(\frac{0.04}{12}\right)}{1 - \left(\frac{12}{12 + 0.04}\right)^{12(30)}} = \$381.93$$

Amortization schedule for this loan (in part):

Payment Number	Interest Payment	Principal Payment	Balance of Principal
			$80,000.00
1	$266.67	$115.26	$79,884.74
2	$266.28	$115.65	$79,769.09
358	$3.79	$378.14	$760.06
359	$2.53	$379.40	$380.66
360	$1.27	$380.66	$0.00

An **adjustable-rate mortgage (ARM)** has a rate based on a fluctuating **index** and a **margin** determined by the lender. The introductory rate will stay fixed for a period of time, and then it will adjust according to the current value of the index.

Molly takes out a 5/1 ARM for $120,000, with a 30-year term. The introductory rate is determined by an index at 0.7% and a margin of 2.5% (a 3.2% rate). The initial payment amount for this loan is $518.96, and the amortization schedule for the first five years shows a principal balance of $107,072.98 after the 60th payment.

Payment Number	Interest Payment	Principal Payment	Balance of Principal
			$120,000.00
1	$320.00	$198.96	$119,801.04
2	$319.47	$199.49	$119,601.55
60	$286.15	$232.81	$107,072.98

Concepts

When an adjustment occurs, the new payment amount is found using the above formula, replacing P with the principal balance at the end of the last adjustment period, replacing r with the new adjusted rate, and replacing t with the number of years remaining in the original term.

In addition to the **down payment** made on a mortgage, there are other one-time expenses involved in purchasing a home. These are the **closing costs.**

Recurring Costs of Homeownership
These include

- **property taxes,**
- **homeowner's insurance,** and
- **maintenance** of the property.

Examples

At the end of five years, the index has increased to 1.9%. Adding the 2.5% margin gives an adjusted rate of 4.4% for the sixth year.

Find the new amount for payments 61–72.

$$R = \frac{\$107{,}072.98\left(\frac{0.044}{12}\right)}{1 - \left(\frac{12}{12 + 0.044}\right)^{12(25)}} = \$589.09 \quad \text{Monthly payments jump \$70.13.}$$

Mary takes out a 20-year fixed rate mortgage at 4.5% for $116,000 to purchase a $145,000 home, and uses a "direct lender" to avoid broker fees. The loan closes eight days before the end of the month. Closing costs are shown.

Loan origination fee (2%)	$?
Document fees	525
Transfer taxes	140
Title services	350
Title insurance	600
Recording fees	90
Escrow deposit	1305
Daily interest charges	?

Compute the total closing costs for this mortgage.

Loan origination fee:

$$0.02(\$116{,}000) = \$2320$$

Daily interest charges (for 8 days on the beginning balance):

$$8 \cdot \$116{,}000 \cdot \frac{0.045}{365} \approx \$114$$

Adding these amounts to those listed above gives the following.

$$\$5444 \quad \leftarrow \text{Total closing costs}$$

Mary's combined tax bracket is 30%, and property taxes in Mary's county amount to 1.2% of property value. Her homeowner's insurance rate is 0.3% of property value, and annual maintenance costs are estimated to be 0.4% of the purchase price.

Find the recurring costs (for the first year).

$$\text{Property taxes} = 0.012(\$145{,}000)(0.7)$$
$$= \$1218$$
$$\text{Insurance} = 0.003(\$145{,}000)$$
$$= \$435$$
$$\text{Maintenance} = 0.004(\$145{,}000)$$
$$= \$580$$

Concepts	Examples

13.5 Financial Investments

Classes of Investments
Investing puts capital to work for the investor.

Stocks
Purchasing company stock makes the investor a part owner of the company, so the investor shares in the profits (in the form of **dividends**) and in the growth (or decline) of the company's value over time.

An investor purchased 200 shares of FedEx Corp (FDX) stock at the closing price for May 22, 2014. He kept the stock for a year (during which one dividend was paid out) and then sold at a price of $150.02 per share. Assume the brokerage firm charged $8.95 for each transaction.

Find the return on investment for the FDX shares.

Table 14 shows that the closing price of FDX stock on May 22, 2014 was $139.33.

$$\text{Basic cost:} \quad 200 \cdot \$139.33 = \$27,866.00$$

Table 14 shows the dividend to be $0.60 per share.

$$\text{Dividend for 200 shares:} \quad 200 \cdot \$0.60 = \$120.00$$

Proceeds from sale of the shares a year later:

$$200 \cdot \$150.02 = \$30,004.00$$

$$\text{SEC fee:} \quad \$30,004.00 \cdot \left(\frac{\$22.10}{\$1,000,000} \right) \approx \$0.663 \rightarrow \$0.67$$

Expenses	
Cost of purchase	$27,866.00
Purchase commission	$8.95
Sale commission	$8.95
SEC fee	$0.67
Total:	**$27,884.57**

Income	
Proceeds from sale	$30,004.00
Dividend	$120.00
Total:	**$30,124.00**

Total income	Total expenses	Return on investment
↓	↓	↓

$$\$30,124.00 - \$27,884.57 = \mathbf{\$2239.43}$$

Bonds
When an investor purchases a bond, the issuing company agrees to pay a fixed rate of interest over the term of the bond, and then to pay back the original purchase price when the bond matures (at the end of the term).

Find the return on a $2000, 2-year corporate bond paying 4% annual interest.

$$I = Prt = \$2000(0.04)(2) = \$160$$

Mutual Funds
A mutual fund is a pool of money collected by an investment company to buy stocks, bonds, and other investments.

Net Asset Value (NAV) of a Mutual Fund

$$\text{NAV} = \frac{A - L}{N}$$

Here A = total fund assets, L = total fund liabilities, and N = number of shares outstanding.

Find the net asset value of a mutual fund having $1.5 billion in assets, $280 million in liabilities, and 40 million shares outstanding.

$$\text{NAV} = \frac{\$1,500,000,000 - \$280,000,000}{40,000,000} = \$30.50$$

An investment of $10,000 would purchase

$$\frac{\$10,000}{\$30.50} \approx 328 \text{ shares.}$$

Concepts	Examples

Annual Rate of Return

This measure of the performance of an investment may change depending on the time period for which this rate is calculated.

Joe owns 1000 shares of a mutual fund with a net asset value of $15.00 on July 1. The fund pays a 0.2% dividend per month (reinvested in the fund), and on August 1, the NAV had increased to $15.12.

Find the annual rate of return on his investment.

Holdings July 1: $1000 \cdot \$15.00 = \$15,000.00$

Dividend for July: $0.002(\$15,000.00) = \$30.00$

New value August 1: $1000 \cdot \$15.12 = \$15,120.00$

Holdings August 1: $\$15,120.00 + \$30.00 = \$15,150.00$

Rate of return for one month:

$$\frac{\$15,150 - \$15,000}{\$15,000} \cdot 100\% = 1.0\%$$

Annual rate of return for that month:

$$(1 + 0.01)^{12} - 1 \approx 0.127 = 12.7\%$$

Future Value of an Inflation-Adjusted Retirement Account

$$V = R\left[\frac{(1 + r)^n - (1 + i)^n}{r - i}\right]$$

where

i = annual rate of inflation,

R = initial deposit (end of year 1),

r = annual rate of return on account,

n = number of years deposits are made.

The inflation rate is 3% per year, and $5000 is invested in a retirement account (with a 6% annual rate of return) at the end of year 1. Deposits are increased by 3% each year.

What will the account be worth after 20 years?

$$V = \$5000\left[\frac{(1 + 0.06)^{20} - (1 + 0.03)^{20}}{0.06 - 0.03}\right]$$

$$= \$233,504.04$$

Retirement accounts can be tax-deferred (meaning that monies invested and earnings are not taxed until they are withdrawn in retirement), or taxable. If

R = amount withheld annually from salary to build the retirement account,

n = number of years contributions are made,

r = annual rate of return on account,

t = marginal tax rate of account holder,

then the final value V of the account (including all earnings and payment of taxes) is calculated as follows.

$$V = \frac{(1 - t)R[(1 + r)^n - 1]}{r} \qquad \text{Tax-deferred account}$$

$$V = \frac{R[(1 + r(1 - t))^n - 1]}{r} \qquad \text{Taxable account}$$

Monique wants to contribute $10,000 annually to a retirement account and wishes to retire in 25 years. Her marginal tax rate is 30%. Assume an inflation rate of 2.5% over the next 25 years and an annual rate of return of 8% on investments.

Should Monique invest in a tax-deferred account or a taxable account?

$$V = \frac{(1 - 0.30)(\$10,000)(1.08^{25} - 1)}{0.08} \qquad \text{Tax-deferred account}$$

$$= \$511,741.58 \quad \leftarrow \text{Better option}$$

$$V = \frac{\$10,000[(1 + 0.08(0.7))^{25} - 1]}{0.08} \qquad \text{Taxable account}$$

$$= \$363,099.06$$

The tax-deferred account is better by
$\$511,741.58 - \$363,099.06 = \$148,642.52$ or about 41%.

CHAPTER 13 | TEST

Find all monetary answers to the nearest cent. Use tables and formulas from the chapter as necessary.

Finding the Future Value of a Deposit   *Find the future value of each deposit.*

1. $1000 for 5 years at 4% simple interest

2. $500 for 2 years at 6% compounded quarterly

Solve each problem.

3. ***Effective Annual Yield of an Account***   Find the effective annual yield to the nearest hundredth of a percent for an account paying 2% compounded monthly.

4. ***Years to Double by the Rule of 70***   Use the rule of 70 to estimate the years to double at an inflation rate of 4%.

5. ***Finding the Present Value of a Deposit***   What amount deposited today in an account paying 4% compounded semiannually would grow to $100,000 in 10 years?

6. ***Finding Bank Card Interest by the Average Daily Balance Method***   Jolene's MasterCard statement shows an average daily balance of $680. Find the interest due if the rate is 1.6% per month.

Analyzing a Consumer Loan   *Julio buys a koi fish pond (and fish to put in it) for his wife on their anniversary. He pays $8000 for the pond with $2000 down. The dealer charges add-on interest of 3.5% per year, and Julio agrees to pay the loan with 36 equal monthly payments. Use this information for Exercises 7–10.*

7. Find the total amount of interest he will pay.

8. Find the monthly payment.

9. Find the APR value (to the nearest half percent).

10. Find **(a)** the unearned interest and **(b)** the payoff amount if he repays the loan in full with 12 months remaining. Use the most accurate method available.

11. ***True Annual Interest Rate in Consumer Financing***   Hot Tub Heaven wants to include financing terms in its advertising. If the price of a hot tub is $3000 and the finance charge with no down payment is $115 over a 12-month period (12 equal monthly payments), find the true annual interest rate (APR).

12. Explain what a mutual fund is, and discuss several of its advantages and disadvantages.

Finding the Monthly Payment on a Home Mortgage   *Find the monthly payment required for each home loan.*

13. The mortgage is for $150,000 at a fixed rate of 4.5% for 25 years. Amortize principal and interest only, disregarding taxes, insurance, and other costs.

14. The purchase price of the home is $218,000. The down payment is 20%. Interest is at 5.0% fixed for a term of 30 years. Annual taxes are $2500, and annual insurance is $750.

Solve each problem.

15. ***Cost of Points in a Home Loan***   If the lender in **Exercise 14** charges two *points,* how much does that add to the cost of the loan?

16. Explain in general what *closing costs* are. Are they different from *settlement charges?*

17. ***Adjusting the Rate in an Adjustable-Rate Mortgage***   To buy your home, you obtain a 1-year ARM with a 2.25% margin and a 2% periodic rate cap. The index starts at 1.85% but has increased to 4.05% by the first adjustment date. What interest rate will you pay during the second year?

18. According to **Table 14** (on **page 749**), how many shares of Pearson Plc were traded on May 22, 2014?

Finding the Return on a Stock Investment   *Barbara bought 1000 shares of stock at $12.75 per share. She received a dividend of $1.38 per share shortly after the purchase and another dividend of $1.02 per share one year later. Eighteen months after buying the stock, she sold the 1000 shares for $10.36 per share. (Disregard commissions and fees.)*

19. Find Barbara's total return on this stock.

20. Find her percentage return. Is this the *annual rate of return* on this stock transaction? Why or why not?

21. ***The Final Value of a Retirement Account*** $1800 is deposited at the end of each year in a tax-deferred retirement account. The account earns a 6% annual return, the account owner's marginal tax rate is 25%, and taxes are paid at the end of 30 years. Find the final value of the account.

22. What is meant by saying that a retirement account is "adjusted for inflation"?

Graph Theory

14

As a child, Ann Porter Wailes was often told that her name sounded like that of a writer. And as Ann grew up, she realized that her parents had aptly named her, because she absolutely loved writing. Even before entering college, Ann wanted to learn everything she could about England and its language. She was fortunate enough to be able to visit the United Kingdom several times before she graduated from high school. England was the land of her favorite literature, the study of British royalty fascinated her, and she visited many castles and courts with rich histories of the monarchy. John Keats became her favorite romantic poet.

Ann was enrolled in a mathematics survey course, and her interest was a bonus when she was assigned her first paper in an Introduction to English Literature class. The assignment was to analyze one of Keats's poems, and she chose his *Ode to Autumn*. The first stanza is shown on the next page, with the rhyme scheme illustrated by the letters A, B, C, D, and E.

Season of mists and mellow fruitfulness,	A
Close bosom-friend of the maturing sun;	B
Conspiring with him how to load and bless	A
With fruit the vines that round the thatch-eves run;	B
To bend with apples the moss'd cottage-trees	C
And fill all fruit with ripeness to the core;	D
To swell the gourd, and plump the hazel shells	E
With a sweet kernel; to set budding more,	D
And still more, later flowers for the bees,	C
Until they think that warm days will never cease,	C
For Summer has o'er brimm'd their clammy cells.	E

She learned in her Survey of Mathematics class that rhyme schemes can be analyzed by diagrams such as the following. This diagram is an example of a graph, the topic of this chapter.

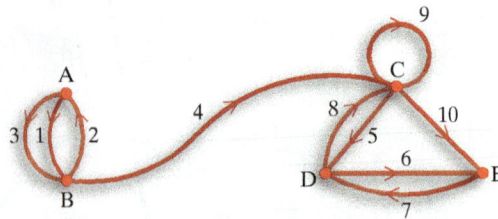

On one of Ann's visits to England, she toured Hampton Court and walked through its hedge maze, designed in 1690. In that same math class, she learned that the structure of mazes can also be analyzed by graphs. The figure on the left below is a diagram of the maze, and the one on the right offers a graphical interpretation, showing that the shortest path to the center is

$$\text{In} \rightarrow A \rightarrow B \rightarrow C \rightarrow D \rightarrow F \rightarrow G \rightarrow H \rightarrow \text{Center.}$$

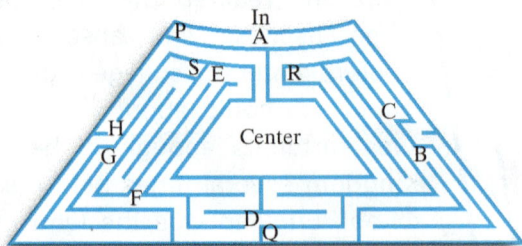

Ann soon realized that she had found her topic for a paper in yet *another* course (History of England). She had discovered mathematics in places she never dreamed it existed.

OBJECTIVES

1 Know the meanings of the terms *graph, vertex, edge, simple graph,* and *degree of a vertex.*

2 Determine whether a graph is connected or disconnected, and identify its components.

3 Determine whether two graphs are isomorphic.

4 Find the sum of the degrees of a graph.

5 Know the meanings of the terms *walk, path, circuit, weighted graph,* and *complete graph.*

6 Understand the coloring and chromatic number of a graph and how these concepts are related to the four-color theorem.

7 Find the Bacon number of an actor by using a Web site.

Graphs

Suppose a preschool teacher has ten children in her class:

Andy, Claire, Dave, Erin, Glen, Katy, Joe, Mike, Sam, and Tim.

The teacher wants to analyze the social interactions among these children. She observes with whom the children play during recess over a 2-week period and records her observations.

Child	Played with
Andy	no one
Claire	Dave, Erin, Glen, Katy, Sam
Dave	Claire, Erin, Glen, Katy
Erin	Claire, Dave, Katy
Glen	Claire, Dave, Sam
Katy	Claire, Dave, Erin
Joe	Mike, Tim
Mike	Joe
Sam	Claire, Glen
Tim	Joe

It is not easy to see patterns in the children's friendships from this list. Instead, the information may be shown in a diagram, as in **Figure 1,** where each letter is the first letter of a child's name. Such a diagram is called a *graph.** Each dot is a **vertex** (plural **vertices**) of the graph. The lines between vertices are **edges.**

> **GRAPH**
>
> A **graph** is a collection of vertices (at least one) and edges. Each edge goes from a vertex to a vertex.

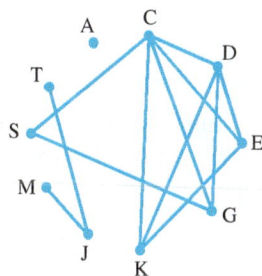

Figure 1

In **Figure 1,** each vertex represents one of the preschool children, and the edges show the relationship "play with each other." An edge must always begin and end at a vertex. We may have a vertex with no edges joined to it, such as A in **Figure 1.**

The graph in **Figure 1** has ten vertices and twelve edges. The only vertices of the graph are the dots. There is *no* vertex where edges TJ and SG intersect, because the intersection point does not represent one of the children. The positions of the vertices and the lengths of the edges have no significance. Also, the edges do not have to be drawn as straight lines. *Only the relationship between the vertices has significance, indicated by the presence or absence of an edge between them.*

Why not draw two edges for each friendship pair? For example, why not draw one edge showing that Tim plays with Joe and another edge showing that Joe plays with Tim? Since Tim and Joe play with each other, the extra edge would yield no additional information. Graphs in which there is no more than one edge between any two vertices and in which no edge goes from a vertex to the same vertex are called **simple graphs.** See **Figure 2** on the next page.

---

*Here *graph* will mean finite graph. A finite graph is a graph with finitely many edges and vertices.

These are simple graphs.

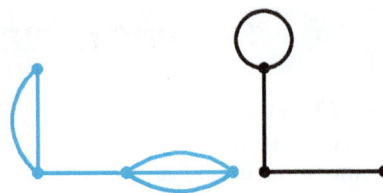

These are not simple graphs.

Figure 2

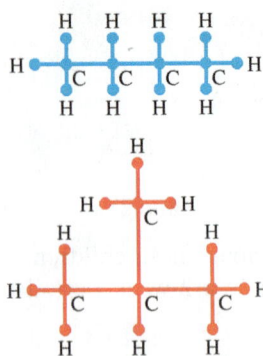

Why are they called graphs? The name was first used by chemist **A. Crum Brown.** He made a major breakthrough in chemistry in 1864 by inventing these now-familiar diagrams to show different molecules with the same chemical composition.

For example, both molecules illustrated above have chemical formula C_4H_{10}. Brown called his sketches **graphic formulae.** Mathematicians got involved in helping to classify all possible structures with a particular chemical composition and adopted the term "graph."

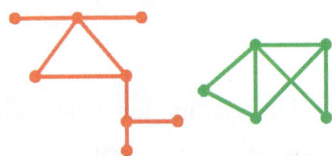

In this chapter, the word **graph** *means simple graph, unless indicated otherwise.* (The meaning of "simple" here is different from its meaning in the phrase "simple curves" used in geometry.)

From the graph in **Figure 1,** we can determine the size of a child's friendship circle by counting the number of edges coming from that child's vertex. There are five edges coming from the vertex labeled C, indicating that Claire plays with five different children. The number of edges joined to a vertex is the **degree of the vertex.** In **Figure 1,** the degree of vertex C is 5, the degree of vertex M is 1, and the degree of vertex A is 0, as no edges are joined to A. (Here, the use of *degree* has nothing to do with its use in geometry in measuring the size of an angle.)

While **Figure 1** clarifies the friendship patterns of the preschoolers, the graph can be drawn in a different, more informative way, as in **Figure 3.** The graph still has edges between the same pairs of vertices. We think of **Figure 3** as one graph, even though it has three pieces, since it represents a single situation. By drawing the graph in this way, we make the friendship patterns more apparent. Although the graph in **Figure 3** looks different from that in **Figure 1,** it shows the same relationships between the vertices. The two graphs are said to be *isomorphic.*

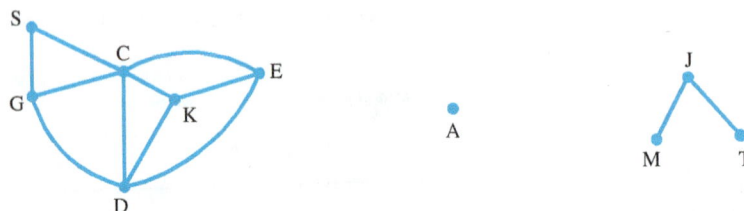

Figure 3

The word **isomorphism** comes from the Greek words *isos,* meaning "same," and *morphe,* meaning "form." Other words that come from the Greek root *isos* include the following.

isobars: lines drawn on a map connecting places with the same barometric pressure;

isosceles: a triangle with two legs (*scelos*) of equal length; and

isomers: chemical compounds with the same chemical composition but different chemical structures.

ISOMORPHIC GRAPHS

Two graphs are **isomorphic** if there is a one-to-one matching between vertices of the two graphs, with the property that whenever there is an edge between two vertices of either one of the graphs, there is an edge between the corresponding vertices of the other graph.

A useful way to think about isomorphic graphs is to imagine a graph drawn with computer software that allows you to drag vertices without detaching any of the edges from their vertices. Any graph we can obtain by simply dragging vertices in this way will be isomorphic to the original graph.

Replacing a graph with an isomorphic graph sometimes can convey more about the relationships being examined. For example, it is clear from **Figure 3** that there are no friendship links between the group consisting of Joe, Tim, and Mike, and the other children. We say that the graph is *disconnected.*

CONNECTED AND DISCONNECTED GRAPHS

A graph is **connected** if we can move from each vertex of the graph to every other vertex of the graph *along edges of the graph*. If not, the graph is **disconnected**. The connected pieces of a graph are called the **components** of the graph.

Figure 3 shows clearly that the friendship relationships among the children in the preschool class have three components. Although the graph in **Figure 1** also is disconnected and includes three components, the way in which the graph is drawn does not make this as clear.

Using the graph in **Figure 3** to analyze the relationships in her class, the teacher might make the following observations and decisions.

- Tim, Mike, and Joe do not interact with the other children.

- Andy's isolation is concerning.

- Claire seems to get along well with other children. As a result, encourage interaction between Claire and Tim, Mike, and Joe.

- Make a special attempt to encourage the children to include Andy in their games.

Showing the friendship relationships as in **Figure 3** made it much easier to analyze them.

Problem-Solving Strategy

Using different colors for different components can help determine how many components a graph has. We choose a color, begin at any vertex, and color all edges and vertices that we can get to from our starting vertex along edges of the graph. If the whole graph is not colored, we then choose a different color and an uncolored vertex, and repeat the process, continuing until the whole graph is colored.

The **World Wide Web** is sometimes studied as a very large graph. The vertices are Web pages, and the edges are links between pages. This graph is unimaginably large.

Even so, because of the way the Web is connected, one can get from any randomly chosen Web page to any other, using hyperlinks, in an average of about 19 clicks. Even if the Web becomes 10 to 100 times larger, this will still be possible in only 20 or 21 clicks. Because of this, search engines such as Google can find keywords very quickly.

(*Sources:* http://jmm.aaa.net.au/ articles/15080.htm; Antonio Gulli Universit di Pisa, Informatica; Alessio Signorini, University of Iowa, Computer Science; R. Albert, H. Jeong, and A.-L. Barabsi, *Nature* 401, 130–131, 1999.)

EXAMPLE 1 Determining Number of Components

Is the graph in **Figure 4** connected or disconnected? How many components does the graph have?

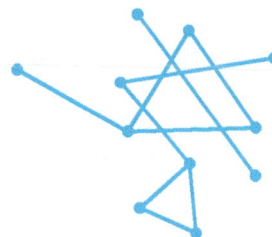

Figure 4

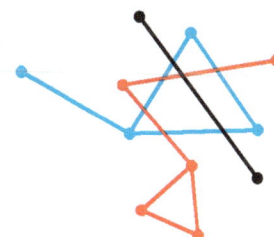

Figure 5

Solution

Coloring the graph helps to answer the question. The original graph in **Figure 4** is disconnected. The graph with color (**Figure 5**) shows three components. ∎

James Joseph Sylvester (1814–1897) was the first mathematician to use the word *graph* for these diagrams in the context of mathematics. Sylvester spent most of his life in Britain, but he had an important influence on American mathematics from 1876 to 1884 as a professor at the newly founded Johns Hopkins University in Baltimore, Maryland. As a young man he worked as an actuary and lawyer in London, tutoring math on the side. One of his students was **Florence Nightingale.** His only published book, *The Laws of Verse,* was about poetry.

EXAMPLE 2 **Deciding Whether Graphs Are Isomorphic**

Are the two graphs in **Figure 6** isomorphic? Justify your answer.

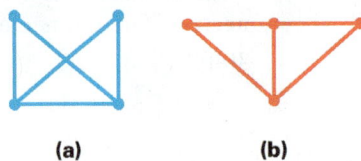

(a)   (b)

Figure 6

Solution

These graphs are isomorphic. To show this, we label corresponding vertices with the same letters and color-code the matching edges in graphs (a) and (b) of **Figure 7.**

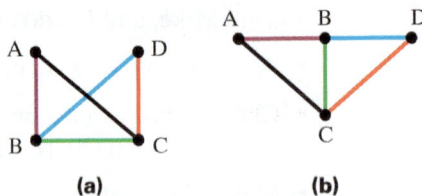

(a)   (b)

Figure 7

EXAMPLE 3 **Deciding Whether Graphs Are Isomorphic**

Are the two graphs in **Figure 8** isomorphic?

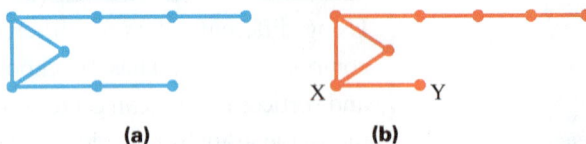

(a)   (b)

Figure 8

Solution

These graphs are not isomorphic. If we imagine graph (a) drawn with our special computer software, we can see that no matter how we drag the vertices, we can not get this graph to look exactly like graph (b). For example, graph (b) has vertex X of degree 3 joined to vertex Y of degree 1. Neither of the vertices of degree 3 in graph (a) is joined to a vertex of degree 1.

EXAMPLE 4 **Summing Degrees**

For each graph in **Figure 9,** determine the number of edges and the sum of the degrees of the vertices. Note that graph (b) is not a simple graph.

(a)   (b)

Figure 9

Solution

Start by writing the degree of each vertex next to the vertex, as in **Figure 10.**

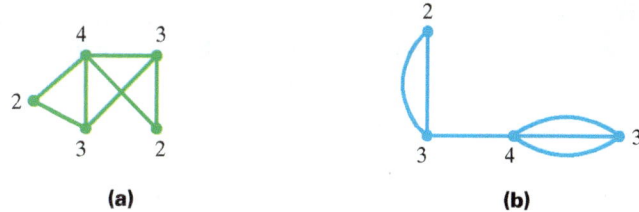

(a)   (b)

Figure 10

For graph (a):	For graph (b):
Number of edges = **7**	Number of edges = **6**
Sum of degrees of vertices	Sum of degrees of vertices
$= 2 + 3 + 2 + 3 + 4 = \mathbf{14}$	$= 2 + 3 + 4 + 3 = \mathbf{12}$

For each graph in **Example 4,** the sum of the degrees of the vertices is twice the number of edges. This is true for all graphs. Check that it is true for graphs in previous examples.

> **SUM-OF-THE-DEGREES THEOREM**
>
> In any graph, the sum of the degrees of the vertices equals twice the number of edges.

To understand why this theorem is true, we think of any graph and imagine cutting each edge at its midpoint (without adding any vertices to the graph, however). Now we have precisely twice as many half edges as we had edges in our original graph. For each vertex, we count the number of half edges joined to the vertex. This is, of course, the degree of the vertex. We add these answers to determine the sum of the degrees of the vertices. We know that the number of half edges is twice the number of edges in our original graph. So, the sum of the degrees of the vertices in the graph is twice the number of edges in the graph.

EXAMPLE 5 **Using the Sum-of-the-Degrees Theorem**

A graph has precisely six vertices, each of degree 3. How many edges does this graph have?

Solution

The sum of the degrees of the vertices of the graph is

$$3 + 3 + 3 + 3 + 3 + 3 = 18.$$

By the theorem above, this number is twice the number of edges, so the number of edges in the graph is

$$\frac{18}{2}, \quad \text{or} \quad 9.$$

Massive graphs are of great current interest in graph theory. One example is the **World Wide Web**, with Web sites as vertices, and edges indicating a link from one page to another. Social networks such as **Facebook** are another example. These networks become enormously large if one creates links from each individual to all individuals with whom that person has contact.

If we consider "people with whom an individual has contact over the past two weeks," a massive graph can describe the spread of a highly infectious disease. Massive graphs often have giant components, and just one link can cause two of them to become an even more huge component. Several years ago, the spread of SARS (sudden acute respiratory syndrome) occurred when a few individual travelers from China caused additional networks of infection in Toronto. More recently, a similar issue has arisen with the Ebola virus. In this case, West Africa was the center of infection.

The 2011 movie *Contagion* provides a fictional—but actually possible— example of the spread of a deadly disease, initiated by a single cough.

EXAMPLE 6 Planning a Tour

Suppose we decide to explore the upper Midwest next summer. We plan to travel by Greyhound bus and would like to visit Grand Forks, Fargo, Bemidji, St. Cloud, Duluth, Minneapolis, Escanaba, and Green Bay. A check of the Greyhound Web site shows direct bus links between destinations as indicated.

City	Direct bus links with
Grand Forks	Bemidji, Fargo
Fargo	Grand Forks, St. Cloud, Minneapolis
Bemidji	Grand Forks, St. Cloud
St. Cloud	Bemidji, Fargo, Minneapolis
Duluth	Escanaba, Minneapolis
Minneapolis	Fargo, St. Cloud, Duluth, Green Bay
Escanaba	Duluth, Green Bay
Green Bay	Escanaba, Minneapolis

Draw a graph with vertices representing the destinations, and with edges representing the relation "there is a direct bus link." Is this a connected graph? Explain.

Solution

Begin with eight vertices representing the eight destinations, as in **Figure 11.** Use the bus link information to fill in the edges of the graph.

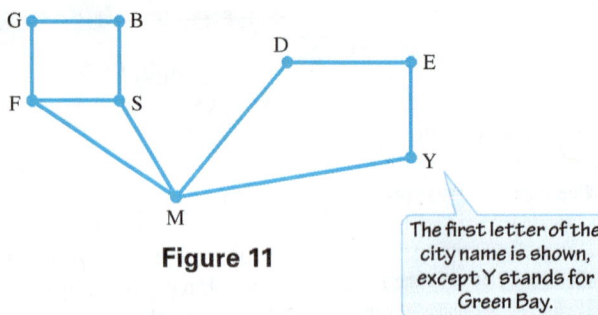

The first letter of the city name is shown, except Y stands for Green Bay.

Figure 11

Figure 11 is a connected graph. It shows that we can travel by bus from any of our chosen destinations to any other. ◼

Walks, Paths, and Circuits

Example 6 suggests several important ideas. For example, what trips could we take among the eight destinations, using only direct bus links? If we do not mind riding the same bus route more than once, we could take the following route.

$$M \rightarrow S \rightarrow F \rightarrow S \rightarrow B$$

Note that we used the "St. Cloud to Fargo" link twice. We could not take the $M \rightarrow D \rightarrow S$ route, however, since there is no direct bus link from Duluth to St. Cloud. Each possible route on the graph is called a *walk*.

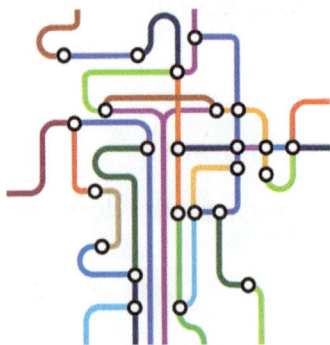

Maps of subway systems in large cities usually do not faithfully represent the positions of stations or the distances between them. They are essentially graphs with vertices representing stations, and edges showing subway links.

WALK

A **walk** in a graph is a sequence of vertices, each linked to the next vertex by a specified edge of the graph.

We can think of a walk as a route that we can trace with a pencil without lifting the pencil from the graph.

$M \rightarrow S \rightarrow F \rightarrow S \rightarrow B$ and $F \rightarrow M \rightarrow D \rightarrow E$ Walks in **Figure 11**

$M \rightarrow D \rightarrow S$ Not a walk (There is no edge between D and S in **Figure 11**.)

Suppose that the travel time for our trip is restricted, and we do not want to use any bus link more than once. A *path* is a special kind of walk.

PATH

A **path** in a graph is a walk that uses no edge more than once.*

$M \rightarrow S \rightarrow F \rightarrow M$, $F \rightarrow M \rightarrow D \rightarrow E$, $M \rightarrow S \rightarrow F \rightarrow M \rightarrow D$ Paths

On the graph in **Figure 11,** trace these paths with a finger. Note that a path may use a *vertex* more than once.

$B \rightarrow S \rightarrow F \rightarrow M \rightarrow S \rightarrow B$ Not a path (It uses edge BS more than once.)

$M \rightarrow D \rightarrow S$ Not a path (It is not even a walk.)

Perhaps we want to begin and end our tour in the same city (and still not use any bus route more than once). A path such as this is known as a *circuit*. Two circuits that we could take are shown here.

$M \rightarrow S \rightarrow F \rightarrow M$, $M \rightarrow S \rightarrow F \rightarrow M \rightarrow D \rightarrow E \rightarrow Y \rightarrow M$ Circuits

---

CIRCUIT

A **circuit** in a graph is a path that begins and ends at the same vertex.

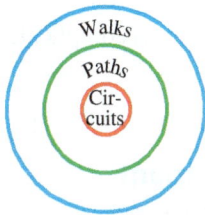

This diagram shows a **Venn diagram** relationship among walks, paths, and circuits.

Notice that a circuit is a kind of path and, therefore, is also a kind of walk.

EXAMPLE 7   Classifying Walks

Using the graph in **Figure 12,** classify each sequence as a walk, a path, or a circuit.

(a) $E \rightarrow C \rightarrow D \rightarrow E$ (b) $A \rightarrow C \rightarrow D \rightarrow E \rightarrow B \rightarrow A$

(c) $B \rightarrow D \rightarrow E \rightarrow B \rightarrow C$ (d) $A \rightarrow B \rightarrow C \rightarrow D \rightarrow B \rightarrow A$

Solution

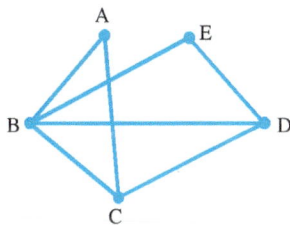

Figure 12

Walk	Path	Circuit
(a) No (no edge E to C)	No	No
(b) Yes	Yes	Yes
(c) Yes	Yes	No
(d) Yes	No (edge AB is used twice)	No

If a sequence of vertices is not a walk, it cannot possibly be either a path or a circuit.

If a sequence of vertices is not a path, it cannot possibly be a circuit, since a circuit is defined as a special kind of path. ∎

---

*Some texts call such a walk a *trail* and use the term *path* for a walk that visits no vertex more than once.

Mazes have always fascinated people. Mazes are found on coins from ancient Knossos (Greece), in the sand drawings at Nazca in Peru, on Roman pavements, on the floors of Renaissance cathedrals in Europe, as the popular cornfield mazes accompanying Halloween pumpkin patches, and as hedge mazes in gardens. Graphs can be used to clarify the structure of mazes.

When planning our tour in **Example 6,** it would be useful to know the length of time the bus rides take. We write the time for each bus ride on the graph, as shown in **Figure 13.**

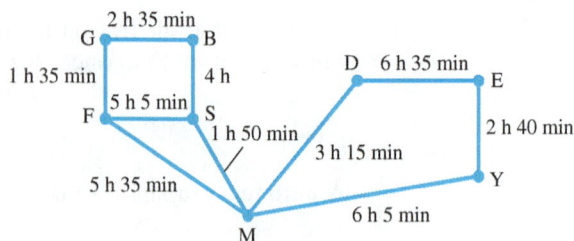

Figure 13

A graph with numbers on the edges, as in **Figure 13,** is a **weighted graph.** The numbers on the edges are **weights.**

Problem-Solving Strategy

Drawing a sketch is a useful problem-solving strategy. **Example 8** illustrates how drawing a sketch can simplify a problem that at first seems complicated.

EXAMPLE 8 Analyzing a Round-Robin Tournament

In a round-robin tournament, every contestant plays every other contestant. The winner is the contestant who wins the most games. Suppose six tennis players compete in a round-robin tournament. How many matches will be played?

Solution

Draw a graph with six vertices to represent the six players. Then, as in **Figure 14,** draw an edge for each match between a pair of contestants.

Figure 14

We can count the edges to find out how many matches must be played. Or we can reason as follows: There are 6 vertices each of degree 5, so the sum of the degrees of the vertices in this graph is 6 · 5, or 30. We know that the sum of the degrees is twice the number of edges. Therefore, there are 15 edges in this graph and, thus, 15 matches in the tournament.

Another way of solving this problem is to use combinations from **Chapter 10.** We must find $_6C_2$, the number of combinations of 6 things taken 2 at a time.

$$_6C_2 = \frac{6!}{2!(6-2)!} = \frac{6!}{2! \cdot 4!} = 15$$

Complete Graphs and Subgraphs

The graph for **Example 8** in **Figure 14** is an example of a *complete graph*.

> **COMPLETE GRAPH**
>
> A **complete graph** is a graph in which there is exactly one edge going from each vertex to each other vertex in the graph.

EXAMPLE 9 Deciding Whether a Graph Is Complete

Decide whether each of the graphs in **Figure 15** is complete. If a graph is not complete, explain why it is not.

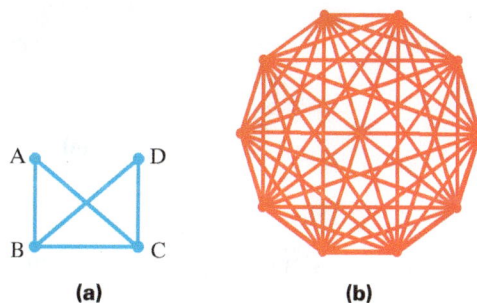

(a)   (b)

Figure 15

Solution

Graph (a) is not complete. There is no edge from vertex A to vertex D. Graph (b) is complete. (Check that there is an edge from each of the ten vertices to each of the other nine vertices in the graph.) ■

EXAMPLE 10 Finding a Complete Graph

Find a complete graph with four vertices in the preschoolers' friendship pattern depicted in **Figure 3**, repeated here.

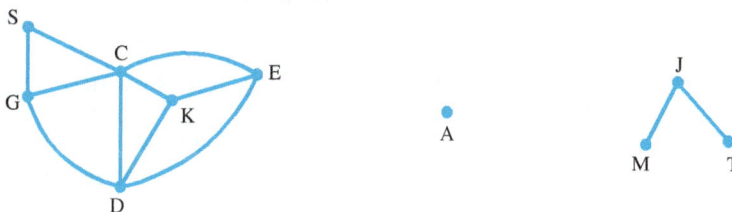

Figure 3 (repeated)

Solution

Figure 16 shows a portion of the friendship graph from **Figure 3**.

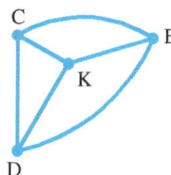

This is a complete graph with four vertices.

Figure 16

A good source for **real-life examples of graphs** is

Networks, Crowds and Markets,

by David Easley and Jon Kleinberg. It is available in book form from Cambridge University Press or as a pdf at

http://www.cs.cornell.edu/home/kleinber/networks-book/

Many of the real-life graphs shown are from scholarly research. Section 2.2 deals with graphs describing romantic relationships at a large American high school during an 18-month period. It turns out that many students who believe they are in monogamous relationships are in fact connected to much larger networks via their partner. This has obvious relevance to the transmission of sexual diseases. Section 3.4 includes examples related to Facebook. Search engines such as Google use measures like page rank to order the results of a search. Section 14.3 of the book gives a fairly simple account of page rank.

Figure 16 includes some complete graphs with three vertices. In general, a graph consisting of some of the vertices of the original graph and some of the original edges between those vertices is called a **subgraph.** Vertices or edges not included in the original graph cannot be in the subgraph.

A subgraph may include anywhere from one to all of the vertices of the original graph and anywhere from none to all of the edges of the original graph. Notice that a subgraph is a graph, and recall that in a graph, every edge goes from a vertex to a vertex. Therefore, no edge can be included in a subgraph without the vertices at both of its ends also being included.

- The subgraph shown in **Figure 16** is a complete graph, but a subgraph does not have to be complete. For example, the graphs in **Figures 17(a) and (b)** are both subgraphs of the graph in **Figure 3.**

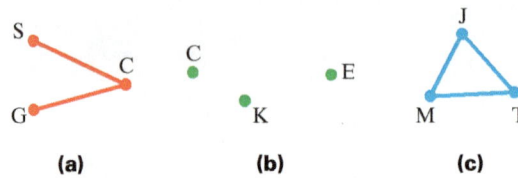

Figure 17

- The graph in **Figure 17(b)** is a subgraph even though it has no edges. (However, we could not form a subgraph by taking edges without vertices. In a graph, every edge goes from a vertex to a vertex.)

- The graph in **Figure 17(c)** is *not* a subgraph of the graph in **Figure 3,** since there was no edge between M and T in the original graph.

Graph Coloring

Tabatha schedules nurse trainees for the hospital departments listed on the left below. Trainee groups must be scheduled at different times if there are trainees working in both departments. The column on the right shows, for each department, which of the other departments have trainees in common with it. For example, the first row shows that each of the departments Radiology, Surgery, Obstetrics, and Pediatrics has one trainee, or more, also working in the ER.

Department	Other departments in which trainees in this department also work
Emergency (ER)	Radiology, Surgery, Obstetrics, Pediatrics
Obstetrics	Radiology, Surgery, Oncology, ER
Surgery	Pediatrics, Oncology, Geriatrics, ER, Obstetrics
Oncology	Pediatrics, Radiology, Geriatrics, Obstetrics, Surgery
Pediatrics	Radiology, ER, Surgery, Oncology
Radiology	Geriatrics, ER, Obstetrics, Oncology, Pediatrics
Geriatrics	Surgery, Oncology, Radiology

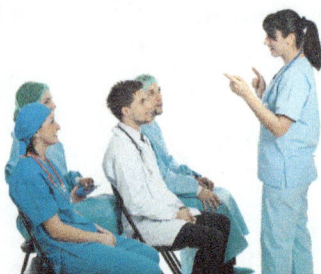

Tabatha's task is to figure out the least number of time slots needed for the seven trainee groups and to decide which (if any) can be scheduled at the same time. We can draw a graph, with a vertex for each department, and with an edge between two vertices *if there are trainees working in both departments.* See **Figure 18** on the next page.

The Mathematics of Time Scheduling

Time is money.

This short sentence says a lot when it comes to productivity in the workplace. A branch of graph theory called **mathematical theory of scheduling** investigates how tasks can be arranged for a project to maximize efficiency. The terms *processors, tasks, processing time, precedence relations, project, critical time, finishing time, schedule, scheduling algorithms, priority list,* and *optimal schedule* are used in the theory of scheduling.

Tasks are represented by vertices, and precedence relations are represented by directed arrows from one vertex to another (similar to edges in a graph). The resulting diagram is called a directed graph (or digraph).

The Web site

http://www.ctl.ua.edu/math103/scheduling/schedmnu.htm#Digraphs

gives a concise explanation of this branch of graph theory and uses the question

"How long does it take to build a house?"

to illustrate some of the problems associated with time scheduling.

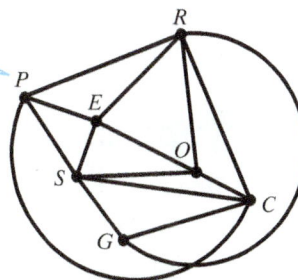

Figure 18

If there is an edge between two vertices, then the corresponding departments have trainees in common, and the trainee groups must *not* be scheduled at the same time. To show this, we use a different color for each time slot for trainee groups. We will color the vertices with the different colors to show when the trainee groups could be scheduled.

To make sure that trainee groups are scheduled at different times if the departments have trainees in common, we must make sure that any two vertices joined by an edge have different colors. To ensure the least number of time slots, we use as few colors as possible. In **Figure 19,** we show one possible coloring of the vertices using three colors. We cannot color the graph as required with fewer than three colors.

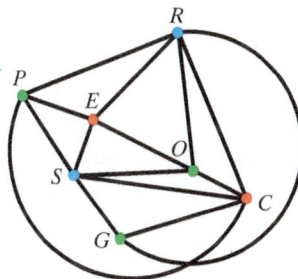

Figure 19

Tabatha can schedule the required trainee groups in no fewer than three time slots. She can schedule Radiology and Surgery at one time, Pediatrics, Obstetrics, and Geriatrics at a second time, and Oncology and ER at a third time.

The result we achieved is called a *coloring** for the graph.

COLORING AND CHROMATIC NUMBER

A **coloring** for a graph is an assignment of a color to each vertex in such a way that vertices joined by an edge have different colors. The **chromatic number** of a graph is the least number of colors needed to achieve a coloring.

Using this terminology, Tabatha had to determine the chromatic number for the graph in **Figure 18.** The chromatic number was 3, which equaled the least number of time slots needed for the trainee groups.

The idea of using vertex coloring to solve this particular problem was straightforward. However, determining the chromatic number for a graph with many vertices or producing a coloring using the least possible number of colors can be difficult. No one has yet found an efficient method for finding the exact chromatic number for arbitrarily large graphs. (The term *efficient algorithm* has a specific, technical meaning for computer scientists. See **Section 14.3.**)

---

*This method is sometimes called a *vertex* coloring for the graph.

Networks are at the forefront of research in biology. Biologists have made enormous strides in understanding the structure of molecules in living organisms. It is now important to understand how these molecules interact with each other. This can be viewed as a network, or graph.

For example, current research aims to identify how proteins in cells interact with other proteins. Insights into such networks may help explain the growth of cancerous tumors. This may lead to better treatments for cancer. It was discovered that as graphs, these biological networks are surprisingly similar to the graphs representing the World Wide Web.

The following method for coloring a graph may not give a coloring with the least number of colors, but its result still may be helpful.

COLORING A GRAPH

Step 1 Choose a vertex with greatest degree, and color it. Use the same color to color as many vertices as you can without coloring two vertices the same color if they are joined by an edge.

Step 2 Choose a new color, and repeat what you did in Step 1 for vertices not already colored.

Step 3 Repeat Step 1 until all vertices are colored.

Graph coloring is used to solve many practical problems. It is useful in management science for solving scheduling problems, such as the "trainee groups" example. Graph coloring is also related to allocating transmission frequencies to TV and radio stations and cell phone companies.

The coloring of graphs has an interesting history involving maps. Maps need to be colored so that territories with common boundaries have different colors. How many colors are needed? For map makers, four colors have always sufficed. In about 1850, mathematicians started pondering this. For over a hundred years, they were able neither to prove that four colors are always enough nor to find a map that needed more than four colors. This was called the **four-color problem.**

In 1976 two mathematicians, Kenneth Appel and Wolfgang Haken, provided a proof that four colors are always enough. This proof created a controversy in mathematics, because it relied on computer calculations so lengthy (fifty 24-hour days of computer time) that no person could check the entire proof.

The postmark shown above was used by the Urbana, Illinois, post office to commemorate the proof of the four-color theorem. (Appel and Haken were colleagues at the University of Illinois at Urbana-Champaign when they completed their proof.)

Bacon Numbers

The popular game *The Six Degrees of Kevin Bacon* was developed by three Albright College friends a number of years ago. The object of the game is to connect actor Kevin Bacon to another actor or actress using a sequence of movies, with six or fewer titles in the sequence. Thanks to the International Movie Data Base (www.imdb.com), this goal can be reached almost instantaneously. The minimum number of links necessary is called the **Bacon number** for the other performer. For example, Hayley Mills has a Bacon number of 2, because she can be connected to Kevin Bacon in 2 steps as follows:

- Hayley Mills was in *That Darn Cat!* (1965) with Roddy McDowall.

- Roddy McDowall was in *The Big Picture* (1989) with Kevin Bacon.

The process used in *The Six Degrees of Kevin Bacon* can be applied to other situations (Web pages, for example).

EXAMPLE 11 Determining a Bacon Number

Use the Web site

oracleofbacon.org

to determine the Bacon number for Mary Kaye Hebert, a lifelong friend of one of the authors of this text. She had a small speaking role in *Easy Rider* (1969). Use the default option of theatrical movies only.

Solution

According to the site, Mary Kaye Hebert has a Bacon number of 2.

- Mary Kaye Hebert was in *Easy Rider* (1969) with Jack Nicholson.
- Jack Nicholson was in *A Few Good Men* (1992) with Kevin Bacon.

It takes only 2 titles to connect Mary Kaye and Kevin. Her Bacon number is 2. ■

14.1 EXERCISES

Vertices and Edges In Exercises 1–6, determine how many vertices and how many edges each graph has.

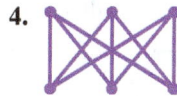

1.
2.
3.
4.
5.
6.

Sum of Degrees For each graph indicated, find the degree of each vertex in the graph. Then add the degrees to get the sum of the degrees of the vertices of the graph. What relationship do you notice between the sum of degrees and the number of edges?

7. Exercise 1
8. Exercise 2
9. Exercise 3
10. Exercise 4

Isomorphic or Not Isomorphic? In Exercises 11–16, determine whether the two graphs are isomorphic. If so, label corresponding vertices of the two graphs with the same letters and color-code corresponding edges, as in **Example 2**. (Note that there is more than one correct answer for many of these exercises.)

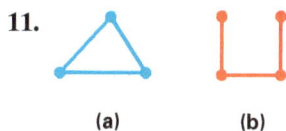

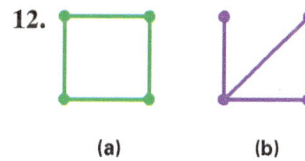

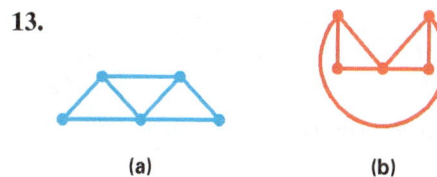

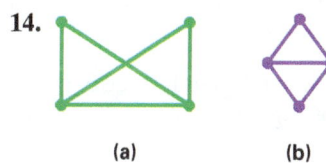

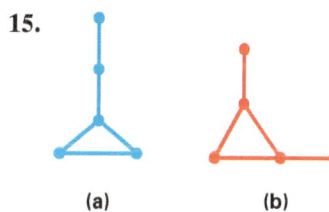

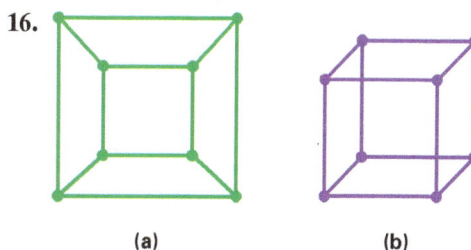

11. (a) (b)
12. (a) (b)
13. (a) (b)
14. (a) (b)
15. (a) (b)
16. (a) (b)

Connected or Disconnected?   *In Exercises 17–22, determine whether the graph is connected or disconnected. Then determine how many components the graph has.*

17.

18.

19.

20.

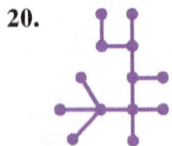

21.

22.

Number of Edges   *In Exercises 23–26, use the theorem that relates the sum of degrees to the number of edges to determine the number of edges in the graph (without drawing).*

23. A graph with 5 vertices, each of degree 4

24. A graph with 7 vertices, each of degree 4

25. A graph with 5 vertices, three of degree 1, one of degree 2, and one of degree 3

26. A graph with 8 vertices, two of degree 1, three of degree 2, one of degree 3, one of degree 5, and one of degree 6

Walks, Paths, Circuits   *In Exercises 27–29, refer to the following graph.*

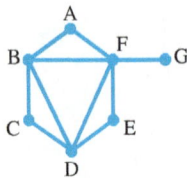

27. Which are walks in the graph? If not, why not?

(a) A → B → C

(b) B → A → D

(c) E → F → A → E

(d) B → D → F → B → D

(e) D → E

(f) C → B → C → B

28. Which are paths in the graph? If not, why not?

(a) B → D → E → F

(b) D → F → B → D

(c) B → D → F → B → D

(d) D → E → F → G → F → D

(e) B → C → D → B → A

(f) A → B → E → F → A

29. Which are circuits in the graph? If not, why not?

(a) A → B → C → D → E → F

(b) A → B → D → E → F → A

(c) C → F → E → D → C

(d) G → F → D → E → F

(e) F → D → F → E → D → F

Walks and Paths   *In Exercises 30 and 31, refer to the following graph.*

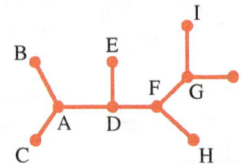

30. Which are walks in the graph? If not, why not?

(a) F → G → J → H → F

(b) D → F

(c) B → A → D → F → H

(d) B → A → D → E → D → F → H

(e) I → G → J

(f) I → G → J → I

31. Which are paths in the graph? If not, why not?

(a) A → B → C

(b) J → G → I → G → F

(c) D → E → I → G → F

(d) C → A

(e) C → A → D → E

(f) C → A → D → E → D → A → B

Walks, Paths, Circuits   *In Exercises 32–37, refer to the following graph. In each case, determine whether the sequence of vertices is* **(i)** *a walk,* **(ii)** *a path,* **(iii)** *a circuit in the graph.*

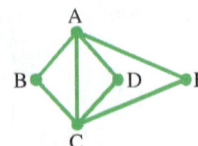

32. A → B → C → D → E

33. $A \rightarrow B \rightarrow C$

34. $A \rightarrow B \rightarrow C \rightarrow D \rightarrow A$

35. $A \rightarrow B \rightarrow A \rightarrow C \rightarrow D \rightarrow A$

36. $A \rightarrow B \rightarrow C \rightarrow A \rightarrow D \rightarrow C \rightarrow E \rightarrow A$

37. $C \rightarrow A \rightarrow B \rightarrow C \rightarrow D \rightarrow A \rightarrow E$

Complete or Not Complete? *In Exercises 38–43, determine whether the graph is a complete graph. If not, explain why it is not complete.*

38.

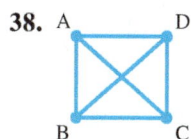

39.

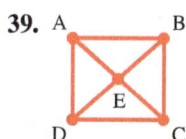

40.

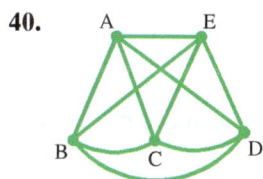

41.

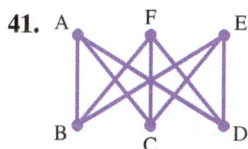

42.

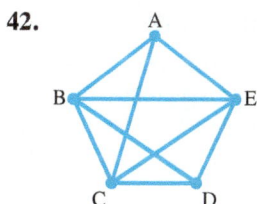

43.

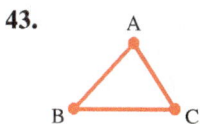

Solve each problem.

44. **Chess Competition** Students from two schools compete in chess. Each school has a team of four students. Each student must play one game against each student on the opposing team. Draw a graph with vertices representing the students, and edges representing the chess games. How many games must be played in the competition?

45. **Chess Competition** A chess master plays seven simultaneous games with seven other players. Draw a graph with vertices representing the players, and edges representing the chess games. How many games are being played?

46. **Dancing Partners** At a party there were four males and five females. During the party each male danced with each female (and no female pair or male pair danced together). Draw a graph with vertices representing the people at the party, and edges of the graph showing the relationship

"danced with."

How many edges are there in the graph?

47. **Number of Handshakes** There are six members on a hockey team (including the goalie). At the end of a hockey game, each member of the team shakes hands with each member of the opposing team. How many handshakes occur?

48. **Number of Handshakes** There are seven people at a business meeting. One of these shakes hands with four people, four shake hands with two people, and two shake hands with three people. How many handshakes occur? (*Hint:* Use the theorem that the sum of the degrees of the vertices in a graph is twice the number of edges.)

49. **Spread of a Rumor** A lawyer is preparing his argument in a libel case. He has evidence that a libelous rumor about his client was discussed in various telephone conversations among eight people. Two of the people involved had four telephone conversations in which the rumor was discussed, one person had three, four had two, and one had one such telephone conversation. How many telephone conversations were there in which the rumor was discussed among the eight people?

50. **Vertices and Edges of a Cube** Draw a graph with vertices representing the vertices (the corners) of a cube, and edges representing the edges of the cube. In your graph, find a circuit that visits four different vertices. What figure does your circuit form on the actual cube?

51. **Vertices and Edges of a Tetrahedron** Draw a graph with vertices representing the vertices (or corners) of a tetrahedron, and edges representing the edges of the tetrahedron. In the graph, identify a circuit that visits three different vertices. What figure does the circuit form on the actual tetrahedron?

52. *Students in the Same Class* Mary, Erin, Sue, Jane, Katy, and Brenda are friends at college. Mary, Erin, Sue, and Jane are in the same math class. Sue, Jane, and Katy take the same English composition class.

(a) Draw a graph with vertices representing the six students, and edges representing the relation "take a common class."

(b) Is the graph connected or disconnected? How many components does the graph have?

(c) In the graph, identify a subgraph that is a complete graph with four vertices.

(d) In the graph, identify three different subgraphs that are complete graphs with three vertices. (There are several correct answers.)

53. Here is another theorem about graphs: *In any graph, the number of vertices with odd degree must be even.* Explain why this theorem is true. (*Hint:* Use the theorem about the relationship between the number of edges and the sum of degrees.)

54. Draw two nonisomorphic (simple) graphs with 6 vertices, with each vertex having degree 3.

55. Explain why the two graphs drawn in **Exercise 54** are not isomorphic.

56. *Analyzing a Cube with a Graph* Draw a graph whose vertices represent the *faces* of a cube and in which an edge between two vertices shows that the corresponding faces of the actual cube share a common boundary. What is the degree of each vertex in the graph? What does the degree of any vertex in the graph reveal about the actual cube?

Coloring and Chromatic Numbers In Exercises 57–62, color the graph using as few colors as possible. Determine the chromatic number of the graph. (*Hint:* The chromatic number is fixed, but there may be more than one correct coloring.)

57. 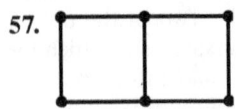 **58.**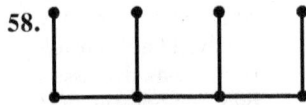

59. 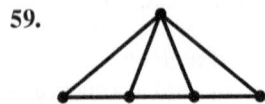 **60.**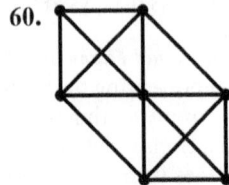

61. 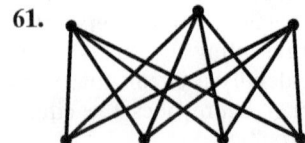 **62.**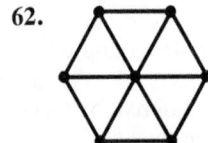

63. Color each graph using as few colors as possible. Use this to determine the chromatic number of the graph. (Graphs like these are called **cycles.**)

(a) 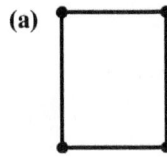 (b)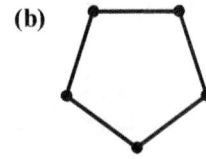

(c) 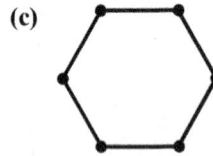 (d)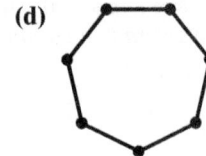

(e) Use the results from parts (a)–(d) to make a prediction about the chromatic number of a cycle. (*Hint:* Consider two cases.)

64. Sketch a complete graph with the specified number of vertices, color the graph with as few colors as possible, and use this to determine the chromatic number of the graph.

(a) 3 vertices

(b) 4 vertices

(c) 5 vertices

(d) Write a general principle by completing this statement: A complete graph with n vertices has chromatic number ___.

(e) Why is the statement in part (d) true?

(f) Generalize further by completing this statement: If a graph has a *subgraph* that is a complete graph with n vertices, then the chromatic number of the graph must be at least ___.

Coloring and Chromatic Numbers In Exercises 65–68, color the vertices using as few colors as possible. Then state the chromatic number of the graph. (*Hint:* It might help to first identify the largest subgraph that is a complete graph.)

65. 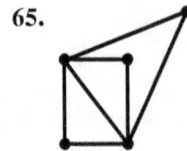 **66.**

67. 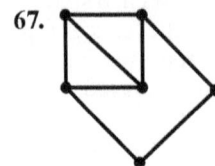 **68.**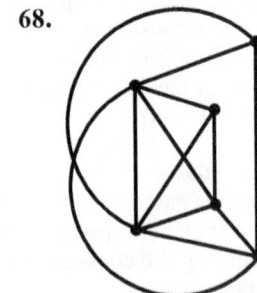

69. *Scheduling Meeting Times* Campus Life must schedule weekly meeting times for the six organizations listed below so that organizations with members in common meet at different times.

Use graph coloring to determine the least number of different meeting times needed, and which organizations should meet at the same time.

Organization	Members also belong to
Choir	Caribbean Club, Dance Club, Theater, Service Club
Caribbean Club	Dance Club, Service Club, Choir
Service Club	Forensics, Choir, Theater, Caribbean Club
Forensics	Dance Club, Service Club
Theater	Choir, Service Club
Dance Club	Choir, Caribbean Club, Forensics

70. *Assigning Frequencies to Transmitters* Interference can occur between radio stations. To avoid this, transmitters that are less than 60 miles apart must be assigned different broadcast frequencies. The transmitters for 7 radio stations are labeled A through G. The information below shows, for each transmitter, which of the others are within 60 miles.

Use graph coloring to determine the least number of different broadcast frequencies that must be assigned to the 7 transmitters. Include a plan for which transmitters could broadcast on the same frequency to use this least number of frequencies.

Transmitters	Transmitter within 60 miles of this
A	B, E, F
B	A, C, E, F, G
C	B, D, G
D	C, F, G
E	A, B, F
F	A, B, D, E, G
G	B, C, D, F

71. *Inviting Colleagues to a Gathering* Several of Tiffany's colleagues do not get along with each other, as summarized below. Tiffany plans to organize several gatherings, so that colleagues who do not get along are not there at the same time.

Use graph coloring to determine the least number of gatherings needed, and which colleagues should be invited each time.

- Joe does not get along with Brad and Phil.
- Caitlin does not get along with Lindsay and Brad.
- Lindsay does not get along with Phil.
- Mary does not get along with Lindsay.
- Eva gets along with everyone.

Graph Coloring In Exercises 72–74, each graph shows the vertices and edges of the figure specified. Color the vertices of each graph in such a way that vertices with an edge between them have different colors. Use graph coloring to find a way to do this, and specify the least number of colors needed.

72. Cube

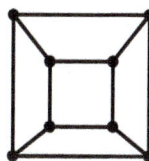

73. Octahedron

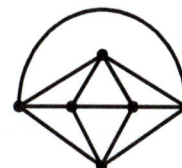

74. Dodecahedron

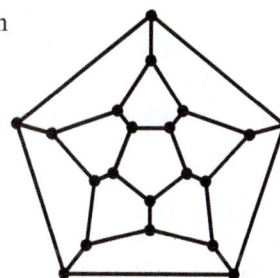

Graph Coloring In Exercises 75 and 76, draw a vertex for each region shown in the map. Draw an edge between two vertices if the two regions have a common boundary. Then find a coloring for the graph, using as few colors as possible.

75.

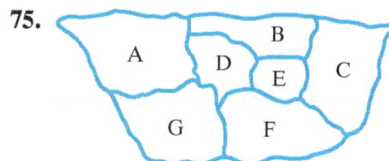

76.

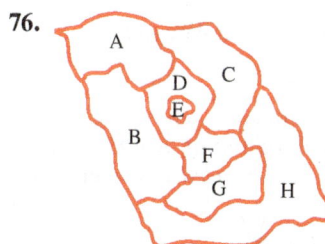

77. *Map Coloring* Use graph coloring to determine the least number of colors that can be used to color the map so that areas with common boundaries have different colors.

Upstate New York

78. Map Coloring Use graph coloring to determine the least number of colors that can be used to color the map so that countries with common boundaries have different colors.

Countries of Southern Africa

Map Drawing *In Exercises 79–81, draw a map with the specified number of countries that can be colored using the stated number of colors, and no fewer. (Countries may not consist of disconnected pieces.) If it is not possible to draw such a map, explain why not.*

79. 6 countries, 3 colors

80. 6 countries, 4 colors

81. 6 countries, 5 colors

82. Write a paper on the four-color problem. Include a discussion of the proof by Appel and Haken.

The Six Degrees of Kevin Bacon *Use the Web site*

oracleofbacon.org

to determine the Bacon number for each actor or actress in Exercises 83–92. Use the default option of theatrical movies only, and use the first performer listed in the event of duplicate names.

83. Cameron Diaz

84. Drew Barrymore

85. Mae West

86. John Wayne

87. Theda Bara

88. Jackie Robinson (baseball player)

89. Erich Bergen

90. Joe Mantegna

91. Brad Pitt

92. Jack Nicholson

Poetry Analysis *Graphs may be used to clarify the rhyme schemes in poetry. Vertices represent words at the end of a line, edges are drawn, and a path is used to indicate the rhyme scheme. See the chapter opener, where the first stanza of* Ode to Autumn *by John Keats has its rhyme scheme analyzed using a graph. In Exercises 93–96, use a graph to analyze the rhyme scheme.*

93. Lines from *She Walks in Beauty,* by Lord Byron

She walks in beauty, like the night	A
Of cloudless climes and starry skies;	B
And all that's best of dark and bright	A
Meet in her aspect and her eyes;	B
Thus mellowed to that tender light	A
Which heaven to gaudy day denies.	B

94. Lines from *Annabel Lee,* by Edgar Allan Poe

It was many and many a year ago,	A
In a kingdom by the sea,	B
That a maiden there lived whom you may know	A
By the name of Annabel Lee;	B
And this maiden she lived with no other thought	C
Than to love and be loved by me.	B

95. Sonnet 3 of William Shakespeare

Look in thy glass, and tell the face thou viewest	A
Now is the time that face should form another;	B
Whose fresh repair if now thou not renewest,	A
Thou dost beguile the world, unless some mother,	B
For where is she so fair whose unear'd womb	C
Disdains the tillage of thy husbandry?	D
Or who is he so fond will be the tomb	C
Of his self-love, to stop posterity?	D
Thou art thy mother's glass, and she in thee	D
Calls back the lovely April of her prime:	E
So thou through windows of thine age shall see	D
Despite of wrinkles this thy golden time.	E
But if thou live, remember'd not to be,	D
Die single, and thine image dies with thee.	D

96. Lines from *Sailing to Byzantium,* by William Butler Yeats

An aged man is but a paltry thing,	A
A tattered coat upon a stick, unless	B
Soul clap its hands and sing, and louder sing	A
For every tatter in its mortal dress,	B
Nor is there singing school but studying	A
Monuments of its own magnificence;	B
And therefore I have sailed the seas and come	C
To the holy city of Byzantium.	C

14.2 EULER CIRCUITS AND ROUTE PLANNING

OBJECTIVES

1 State the Königsberg bridge problem and explain its historical relevance to graph theory.

2 State and apply Euler's theorem for Euler circuits.

3 Identify a cut edge of a graph.

4 Apply Fleury's algorithm to find an Euler circuit.

5 Minimize deadheading on a street grid by converting it to a graph with an Euler circuit.

Königsberg Bridge Problem

In the early 1700s, the city of Königsberg was the capital of East Prussia. (Königsberg is now called Kaliningrad and is in Russia.) The river Pregel ran through the city in two branches with an island between the branches, as shown in **Figure 20.**

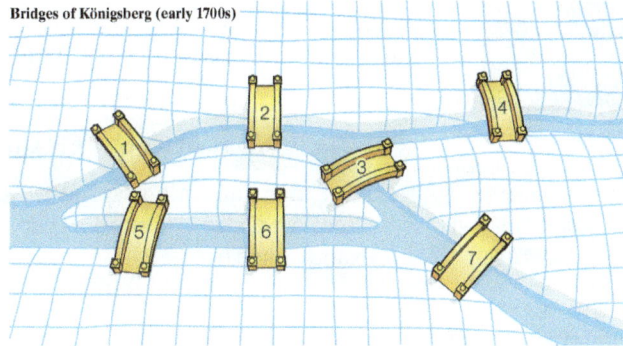

Figure 20

There were seven bridges joining various parts of the city as shown in **Figure 21.**

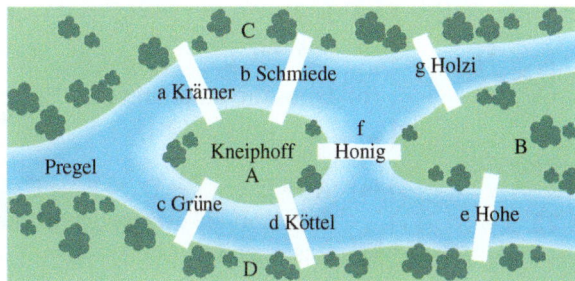

Figure 21

Leonhard Euler (1707–1783) was a devoted father and grandfather. He had thirteen children and frequently worked on mathematics with children playing around him. He lost the sight in his right eye at age 31, a couple of years after writing the Königsberg bridge paper. He became completely blind at age 58 but produced more than half his mathematical work after that. Euler wrote nearly a thousand books and papers.

According to Leonhard Euler (pronounced "oiler"), the following problem (the **Königsberg bridge problem**) was well known in his time (around 1730).

Is it possible for a citizen of Königsberg to take a stroll through the city, crossing each bridge exactly once, and beginning and ending at the same place?

Try to find such a route on the map in **Figure 21.** (Do not cross the river anywhere except at the bridges shown on the map.)

We can simplify the problem by drawing the map as a graph. See **Figure 22.** The graph focuses on the relations between the vertices. This is the relation "there is a bridge."

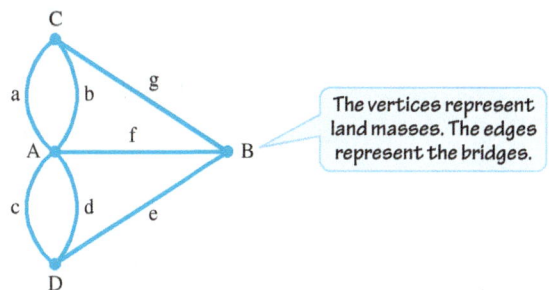

The vertices represent land masses. The edges represent the bridges.

Figure 22

In the graph in **Figure 22,** we have *two* edges between vertices A and C and *two* edges between vertices A and D. Thus, this graph is not a simple graph. We have labeled the edges with lowercase letters. Now try to find a route as required in the problem.

Euler Circuits

The Königsberg bridge problem requires more than a path and more than a circuit. (Recall that a path is a walk that uses no edge more than once, and a circuit is a path that begins and ends at the same vertex.) The problem requires a circuit that uses *every* edge, *exactly* once. This is called an *Euler circuit.*

A university was founded in Königsberg in 1544. It was established as a Lutheran center of learning. Its most famous professor was the philosopher **Immanual Kant,** who was born in Königsberg in 1724. The university was completely destroyed during World War II.

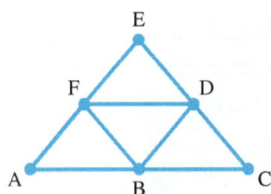

EULER PATH AND EULER CIRCUIT

An **Euler path** in a graph is a path that uses every edge of the graph exactly once. An **Euler circuit** in a graph is a circuit that uses every edge of the graph exactly once.

EXAMPLE 1 **Recognizing Euler Circuits**

Consider the graph in **Figure 23.**

(a) Is $A \rightarrow B \rightarrow C \rightarrow D \rightarrow E \rightarrow F \rightarrow A$ an Euler circuit for this graph? Justify your answer.

(b) Does the graph have an Euler circuit?

Solution

(a) $A \rightarrow B \rightarrow C \rightarrow D \rightarrow E \rightarrow F \rightarrow A$ is a circuit, but not an Euler circuit, because it does not use every edge of the graph. For example, the edge BD is not used in this circuit.

(b) In the graph in **Figure 23,** the circuit

$$A \rightarrow B \rightarrow C \rightarrow D \rightarrow E \rightarrow F \rightarrow D \rightarrow B \rightarrow F \rightarrow A$$

is an Euler circuit. Trace this path on the graph to check that it does use every edge of the graph exactly once. ∎

Figure 23

Returning to the Königsberg bridge problem, it seemed that no route of the type required could be found. Why was this so? Euler published a paper in 1736 that explained why it is impossible to find a walk of the required kind, and he provided a simple way of deciding whether a given graph has an Euler circuit. Euler did not use the terms *graph* and *circuit,* but his paper was, in fact, the first paper on graph theory. In modern terms, Euler proved the first part of the theorem below. Part 2 was not proved until 1873.

EULER'S THEOREM

Suppose we have a connected graph.

1. If the graph has an Euler circuit, then each vertex of the graph has even degree.

2. If each vertex of the graph has even degree, then the graph has an Euler circuit.

What does Euler's theorem predict about the connected graph in **Figure 23?** The degrees of the vertices are as follows.

A: 2   B: 4   C: 2   D: 4   E: 2   F: 4

Each vertex has even degree, so it follows from part 2 of the theorem that the graph has an Euler circuit. (We already knew this, since we found an Euler circuit for this graph in **Example 1.**)

What does Euler's theorem suggest about the Königsberg bridge problem? Note that the degree of vertex C in **Figure 22** is 3, which is an odd number, and the graph is connected. So, by part 1 of the theorem, the graph cannot have an Euler circuit. (In fact, the graph has many vertices with odd degree, but the presence of *any* vertex of odd degree indicates that the graph does not have an Euler circuit.)

Why is Euler's theorem true? Consider the first part of the theorem:

If a connected graph has an Euler circuit, then each vertex has even degree.

Suppose we have a connected graph with an Euler circuit in the graph. Suppose the circuit begins and ends at a vertex A. Imagine traveling along the Euler circuit, from A all the way back to A, putting an arrow on each edge in our direction of travel. (Of course, we travel along each edge exactly once.)

Now consider any vertex B in the graph. It has edges joined to it, each with an arrow. Some of the arrows point toward B, some point away from B. The number of arrows pointing toward B is the same as the number of arrows pointing away from B, for every time we visit B, we come in on one unused edge and leave via a different unused edge. In other words, we can pair off the edges joined to B, with each pair having one arrow pointing toward B and one arrow pointing away from B. Thus, the total number of edges joined to B must be an even number, and therefore, the degree of B is even. A similar argument shows that our starting vertex, A, also has even degree.

The second part of Euler's theorem states the following:

**If each vertex of a connected graph has even degree,
then the graph has an Euler circuit.**

To show that this is true, we will provide a recipe (Fleury's algorithm) for finding an Euler circuit in any such graph. First, however, we consider more examples of how Euler's theorem can be applied.

EXAMPLE 2 Using Euler's Theorem

Decide which of the graphs in **Figure 24** has an Euler circuit. Justify your answers.

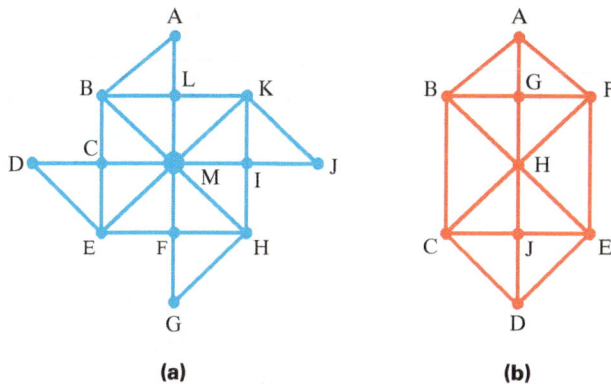

(a)

(b)

Figure 24

Solution

(a) The graph in **Figure 24(a)** is connected. If we write the degree next to each vertex on the graph, as in **Figure 25,** we see that each vertex has even degree. Since the graph is connected, and each vertex has even degree, it follows from Euler's theorem that the graph has an Euler circuit.

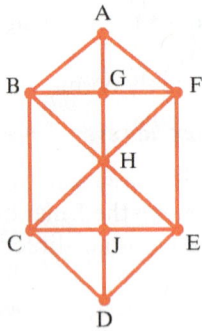

Figure 24(b) (repeated)

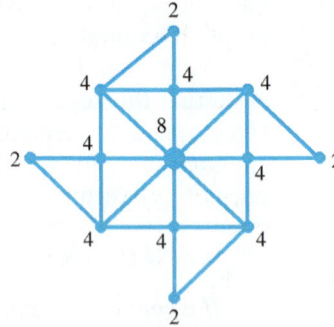

Figure 25

(b) In **Figure 24(b),** check the degrees of the vertices. Note that vertex A has degree 3, an odd degree. We need look no further. It follows from Euler's theorem that this connected graph does not have an Euler circuit. (All the other vertices in this graph except for D have even degree. However, if just one vertex has odd degree, then the graph does not have an Euler circuit.) ■

Problem-Solving Strategy

When solving a challenging problem, it is useful to ask whether the problem is related to another problem that you already know how to solve. **Example 3** illustrates this strategy.

EXAMPLE 3 **Tracing Patterns**

Beginning and ending at the same place, is it possible to trace the pattern shown in **Figure 26** below without lifting the pencil off the page and without tracing over any part of the pattern (except isolated points) more than once?

Solution

The pattern in **Figure 26** is not a graph. (It has no vertices.) But what we are being asked to do is very much like being asked to find an Euler circuit. So we imagine that there are vertices at points where lines or curves of the pattern meet, obtaining the graph in **Figure 27.**

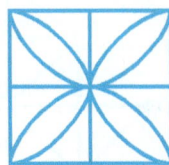

Figure 26

Figure 27

Now we can use Euler's theorem to decide whether we can trace the pattern. Vertex A has odd degree. Thus, by Euler's theorem, this graph does not have an Euler circuit. It follows that we cannot trace the original pattern in the manner required. ■

The word **algorithm** comes from the name of the Persian mathematician, **Abu Ja'far Muhammad ibn Musa Al-Khowârizmî,** who lived from about A.D. 780 to A.D. 850. One of the books he wrote was on the Hindu-Arabic system of numerals (the system we use today) and calculations using them. The Latin translation of this book was called *Algoritmi de numero Indorum,* meaning "Al-Khowârizmî on the Hindu Art of Reckoning."

The tracing problem in **Example 3** illustrates a practical problem that occurs when designing robotic arms to trace patterns. Automated machines that engrave identification tags for pets require such a mechanical arm.

Fleury's Algorithm

Fleury's algorithm can be used to find an Euler circuit in any connected graph in which each vertex has even degree. An algorithm is like a recipe—follow the steps and we achieve what we need. Before introducing Fleury's algorithm, we need a definition.

CUT EDGE

A **cut edge** in a graph is an edge whose removal disconnects a component of the graph.

We call such an edge a cut edge, since removing the edge *cuts* a connected piece of a graph into two pieces.*

EXAMPLE 4 Identifying Cut Edges

Identify the cut edges in the graph in **Figure 28**.

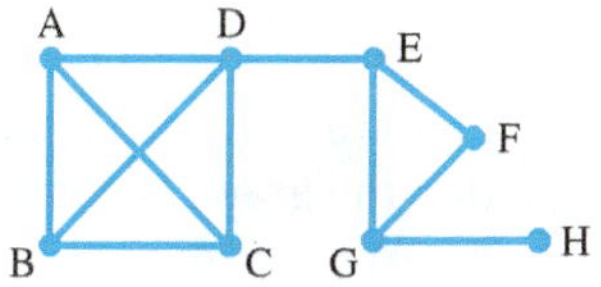

Figure 28

Solution

This graph has only one component, so we must look for edges whose removal would disconnect the graph.

DE is a cut edge. If we removed this edge, we would disconnect the graph into two components, obtaining the graph of **Figure 29(a).** HG is also a cut edge. Its removal would disconnect the graph as shown in **Figure 29(b).**

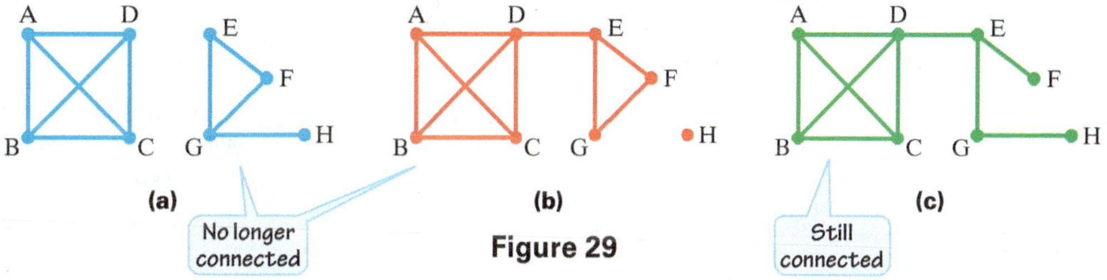

(a)   No longer connected   **(b)**   **Figure 29**   **(c)**   Still connected

None of the other edges is a cut edge. We could remove any *one* of the other edges and still have a connected graph (that is, we would still have a path from each vertex of the graph to each other vertex). For example, if we remove edge GF, we end up with a graph that still is connected, as shown in **Figure 29(c).** ∎

Fleury's algorithm is used for finding an Euler circuit in a connected graph in which each vertex has even degree.

---

*Some mathematicians use "bridge" instead of "cut edge" to describe such an edge.

FLEURY'S ALGORITHM

Step 1 Start at any vertex. Go along any edge from this vertex to another vertex. *Remove this edge from the graph.*

Step 2 You are now on a vertex of the revised graph. Choose any edge from this vertex, subject to only one condition: Do not use a cut edge (*of the revised graph*) unless you have no other option. Go along your chosen edge. *Remove this edge from the graph.*

Step 3 Repeat Step 2 until you have used all the edges and returned to the vertex at which you started.

EXAMPLE 5 Using Fleury's Algorithm

Find an Euler circuit for the graph in **Figure 30.**

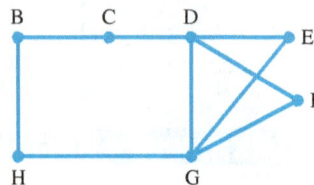

Figure 30

Solution

First we must check that this graph does indeed have an Euler circuit. A check shows that the graph is connected and that each vertex has even degree.

We can start at any vertex. We choose C. We could go from C to B or from C to D. We go to D, removing the edge CD. (We show it scratched out, but we must think of this edge as gone.) Our path begins: C → D. The revised graph is shown in **Figure 31.**

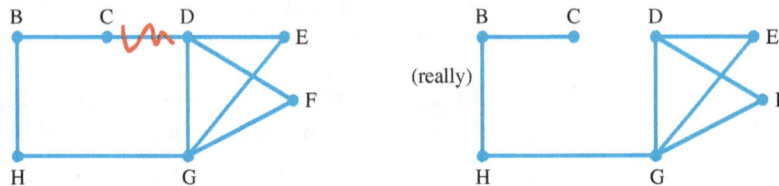

Figure 31

None of the edges DE, DF, and DG is a cut edge for our current graph, so we can choose to go along any one of these. We go from D to G, removing edge DG. So far, our path is C → D → G. (Keep a record of this path.) See **Figure 32.**

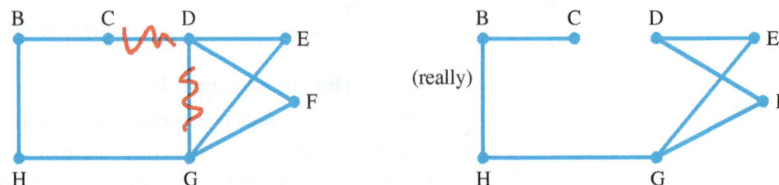

Figure 32

The remaining edges at G are GE, GF, and GH. Note, however, that GH is a cut edge for our revised graph, and we have other options. Thus, according to Fleury's algorithm, *we must not use GH at this stage.* (If we used GH now, we would never be able to get back to use the edges GF, GE, and so on.) We go from G to F.

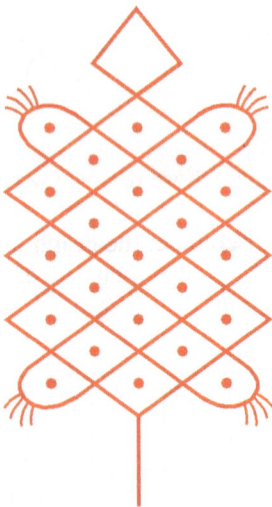

The **Tshokwe people of northeastern Angola** have a tradition involving continuous patterns in sand. The tracings are called *sona,* and they have ritual significance. They are used as mnemonics for stories about the gods and ancestors. Elders trace the drawings while telling the stories. The example shown here is called *skin of a leopard.*

(*Source:* Gerdes, Paul. *Geometry from Africa–Mathematical and Educational Explorations.* Mathematical Association of America, 1999, p. 171.)

Our path is C → D → G → F, and we show our revised graph in **Figure 33.**

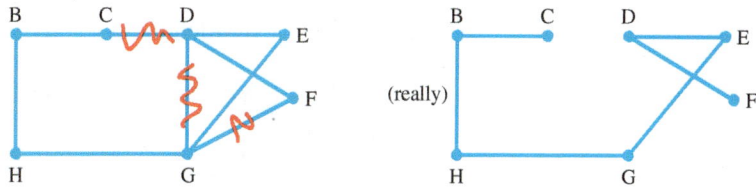

Figure 33

Now we have only one edge from F, namely, FD. This is a cut edge of the revised graph, but *since we have no other option,* we may use this edge. Our path is C → D → G → F → D, and our revised graph is shown in **Figure 34.**

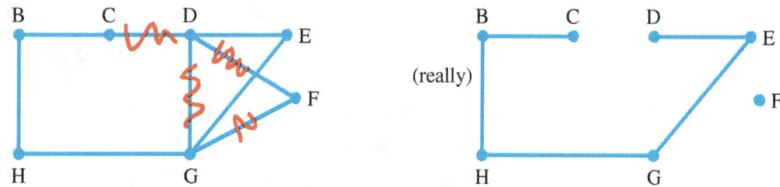

Figure 34

We have only one edge left at D, namely DE. It is a cut edge of the remaining graph, but since we have no other option, we may use it.

It is now clear how to complete the Euler circuit. (Be sure to return to the starting vertex.) The complete Euler circuit is as follows.

$$C \rightarrow D \rightarrow G \rightarrow F \rightarrow D \rightarrow E \rightarrow G \rightarrow H \rightarrow B \rightarrow C.$$ ■

If we start with a connected graph with all vertices having even degree, then Fleury's algorithm will always produce an Euler circuit for the graph. ***Note that a graph that has an Euler circuit always has more than one Euler circuit.***

Even for a large graph, a computer can successfully apply Fleury's algorithm. This is not the case for all algorithms. (See **For Further Thought** on The Speed of Algorithms in **Section 14.3.**)

Route Planning

When planning routes for mail delivery, street sweeping, or snow plowing, the roads in the region to be covered usually do not have an Euler circuit. Unavoidably, certain stretches of road must be traveled more than once. This is referred to as **deadheading.** Planners then try to find a route on which the driver will spend as little time as possible deadheading.

In **Example 6,** for simplicity we use street grids for which the distances from corner to corner in any direction are the same.

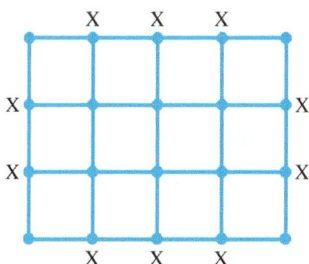

Figure 35

EXAMPLE 6 **Finding an Euler Circuit on a Street Grid**

The street grid in **Figure 35** has every vertex of odd degree marked with an X. Insert edges coinciding with existing roads to change it to a graph with an Euler circuit.

Solution

A certain amount of trial and error is needed to find the least number of edges that must be inserted to ensure that the street grid graph has an Euler circuit.

If we insert edges as shown in **Figure 36(a),** we obtain a graph having an Euler circuit. (Check that all vertices now have even degree.) This solution introduces 11 additional edges, which turns out to be more deadheading than is necessary.

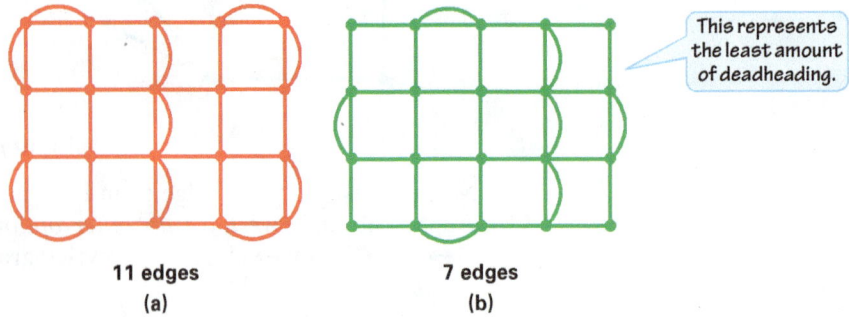

This represents the least amount of deadheading.

11 edges
(a)

7 edges
(b)

Figure 36

If we insert edges as shown in **Figure 36(b),** with just 7 additional edges we obtain a graph with an Euler circuit. This is, in fact, the least number of edges that must be inserted (keeping to the street grid) to change the original street grid graph into a graph that has an Euler circuit. (There are different ways to insert the seven edges.) ■

WHEN Will I Ever USE This ?

In the midst of the technology revolution, there is no replacement for the old-fashioned method of product delivery:

Load the product onto a vehicle, drive to the destination, and physically drop off the product there.

This is the job of courier services such as UPS, FedEx, and the United States Postal Service.

Imagine that you are in the management division of a small city post office. You have been assigned to redesign the rural mail carrier route, which is long out of date because new subdivisions have been built since the last design. Euler circuits and Fleury's algorithm can be used to create the most efficient route.

The figure shows a map of your rural district. In rural areas, mailboxes are placed along the same side of a road, so the mail delivery vehicle needs to travel along the road in one direction only. Suppose the roads off the main road in the figure show a mail delivery region.

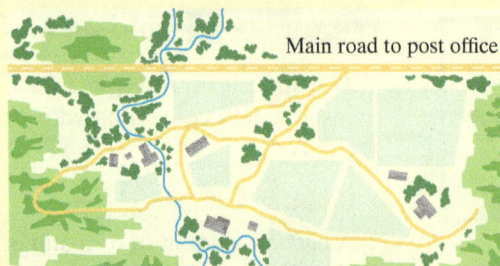

Main road to post office

As seen in Chapter 9, there are only five **regular polyhedra**: the cube, the tetrahedron, the octahedron, the dodecahedron, and the icosahedron. There is only one which has the property that you can trace your finger along each edge exactly once, beginning and ending at the same vertex. It is the octahedron, because it is the only one with an even number of edges meeting each vertex.

It would be ideal if, in your new route, you could find an Euler circuit covering the entire route. If such a circuit existed, it would minimize the distance traveled for the delivery. A glance at the map (thinking of road intersections as vertices) reveals that there is no Euler circuit for this route.

As the next best thing, the delivery vehicle should retrace its path as little as possible. For simplicity, you assume that the time to travel any of the stretches of road when not delivering mail is roughly the same. In following the figure, you added dotted edges to indicate stretches of road to be covered twice. For this delivery route, it is easy to see that you effectively inserted as few edges as possible to obtain a graph that has an Euler circuit.

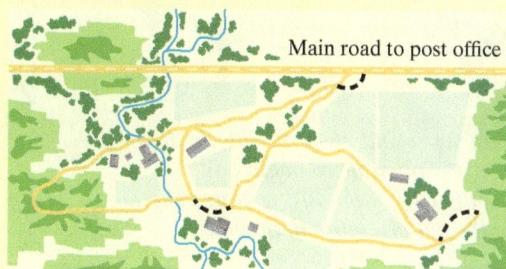
Main road to post office

Now you can use Fleury's algorithm to find an Euler circuit on the graph in the new figure, and this will give the most efficient route for mail delivery.

14.2   EXERCISES

Euler Circuits   *In Exercises 1–3, a graph is shown and some sequences of vertices are specified. Determine which of these sequences show Euler circuits. If not, explain why not.*

1.
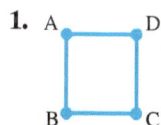

(a) $A \to B \to C \to D \to A \to B \to C \to D \to A$

(b) $C \to B \to A \to D \to C$

(c) $A \to C \to D \to B \to A$

(d) $A \to B \to C \to D$

2.
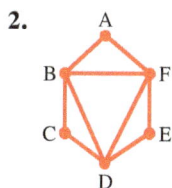

(a) $A \to B \to C \to D \to E \to F \to A$

(b) $F \to B \to D \to B$

(c) $A \to B \to C \to D \to E \to F \to B \to D \to F \to A$

(d) $A \to B \to F \to D \to B \to C \to D \to E \to F \to A$

3.
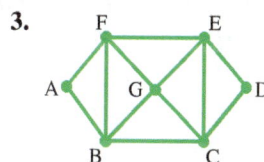

(a) $A \to B \to C \to D \to E \to F \to A$

(b) $A \to B \to C \to D \to E \to G \to C \to E \to F \to G \to B \to F \to A$

(c) $A \to B \to C \to D \to E \to C \to G \to E \to F \to G \to E \to F \to A$

(d) $A \to B \to G \to E \to D \to C \to G \to F \to B \to C \to E \to F \to A$

Euler's Theorem   *In Exercises 4–8, use Euler's theorem to decide whether the graph has an Euler circuit. (Do not actually find an Euler circuit.) Justify each answer briefly.*

4.
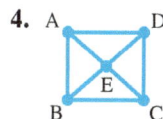

5.
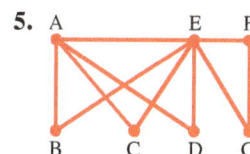

6.

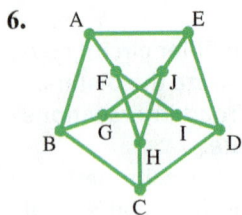

7.

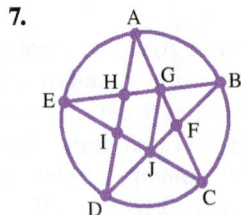

8.

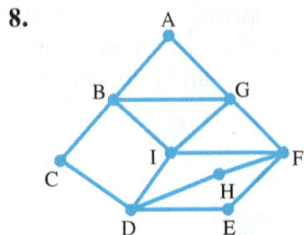

Euler's Theorem In Exercises 9 and 10, use Euler's theorem to determine whether it is possible to begin and end at the same place, trace the pattern without lifting your pencil, and trace over no line in the pattern more than once.

9.

10.

Euler's Theorem In Exercises 11–14, use Euler's theorem to determine whether the graph has an Euler circuit, justifying each answer. Then determine whether the graph has a circuit that visits each vertex exactly once, except that it returns to its starting vertex. If so, write down the circuit. (There may be more than one correct answer.)

11.

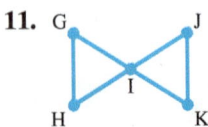

12.

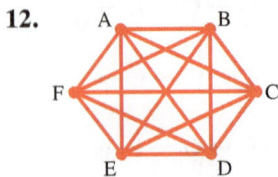

13.

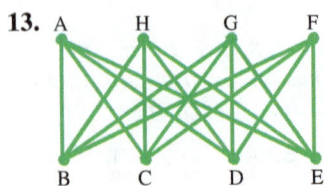

14.

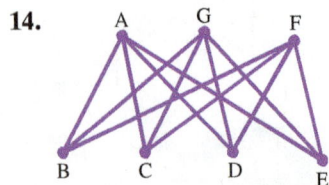

Floor Tilings In Exercises 15–18, different floor tilings are shown. The material applied between tiles is called grout. For which of these floor tilings could the grout be applied beginning and ending at the same place, without going over any joint twice, and without lifting the tool? Justify each answer.

15.

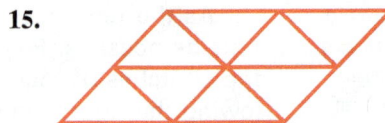

16.

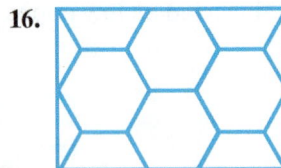

17.

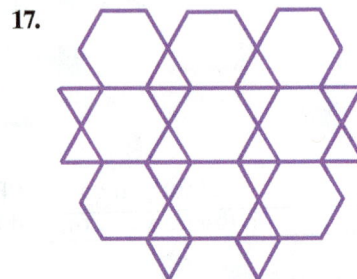

18.

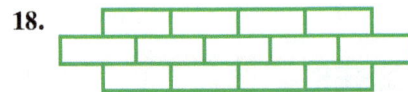

Cut Edges In Exercises 19–22, identify all cut edges in the graph. If there are none, say so.

19.

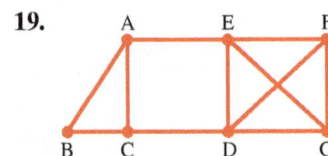

20.

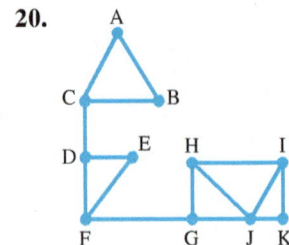

21.

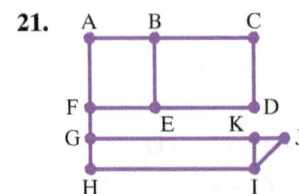

22.

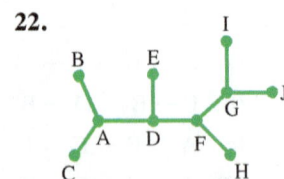

Fleury's Algorithm *In Exercises 23–25, a graph is shown for which a student has been asked to find an Euler circuit starting at A. The student's revisions of the graph after the first few steps of Fleury's algorithm are shown, and in each case the student is now at B. For each graph, determine all edges that Fleury's algorithm permits the student to use for the next step.*

23.

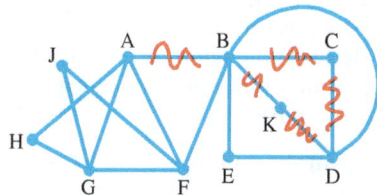

24.

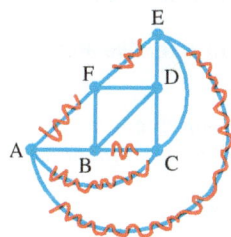

25.

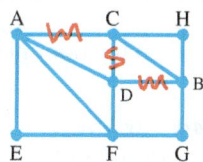

Fleury's Algorithm *In Exercises 26–28, use Fleury's algorithm to find an Euler circuit for the graph, beginning and ending at A. (There are many different correct answers.)*

26.

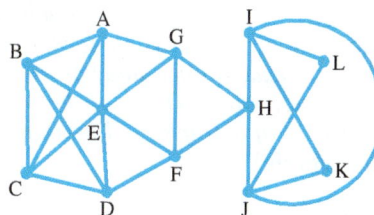

27.

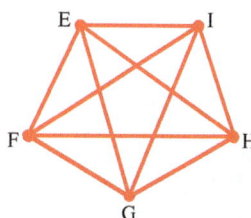

28.

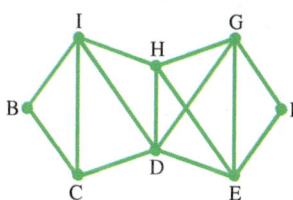

Euler's Theorem and Fleury's Algorithm *In Exercises 29–31 in the next column, use Euler's theorem to determine whether the graph has an Euler circuit. If not, explain why not. If the graph does have an Euler circuit, use Fleury's algorithm to find an Euler circuit for the graph. (There are many different correct answers.)*

29.

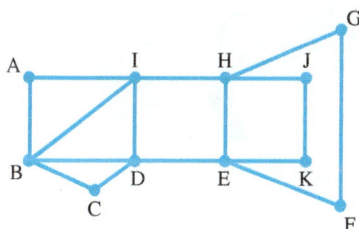

30.

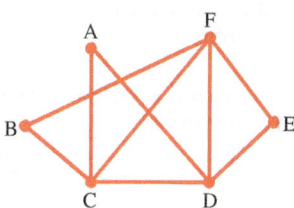

31.

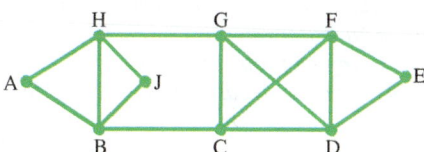

32. Garden Design The graph below shows the layout of the paths in a botanical garden. The edges represent the paths. Has the garden been designed in such a way that it is possible for a visitor to find a route that begins and ends at the entrance to the garden (represented by the vertex A) and that goes along each path exactly once? If so, use Fleury's algorithm to find such a route.

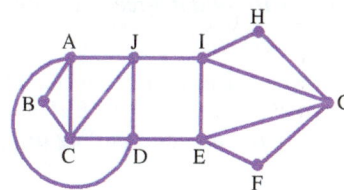

33. Parking Pattern The map shows the roads on which parking is permitted at a national monument. This is a pay-and-display facility. A security guard has the task of periodically checking that all parked vehicles have a valid parking ticket displayed. He is based at the central complex, labeled A. Is there a route that he can take to walk along each of the roads exactly once, beginning and ending at A? If so, use Fleury's algorithm to find such a route.

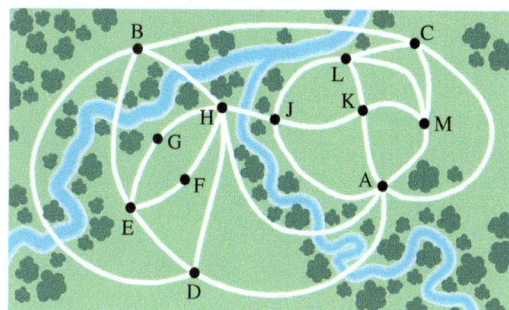

Floor Plans In Exercises 34–36, the floor plan of a building is shown. For which of these is it possible to start outside, walk through each door exactly once, and end up back outside? Justify each answer. (Hint: Think of the rooms and "outside" as the vertices of a graph, and think of the doors as the edges of the graph.)

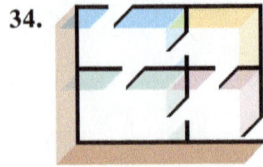

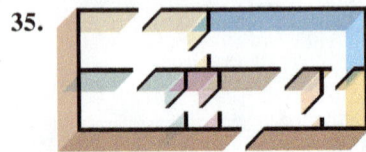

34.

35.

36.

Exercises 37–44 are based on the following theorem:

1. **If a graph has an Euler path that begins and ends at different vertices, then these two vertices are the only vertices with odd degree. (All the rest have even degree.)**

2. **If exactly two vertices in a connected graph have odd degree, then the graph has an Euler path beginning at one of these vertices and ending at the other.**

In Exercises 37–40, determine whether the graph has an Euler path that begins and ends at different vertices. Justify your answer. If the graph has such a path, say at which vertices the path must begin and end.

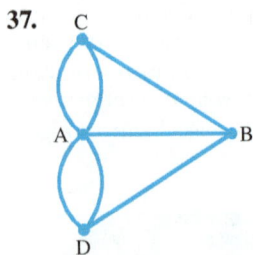

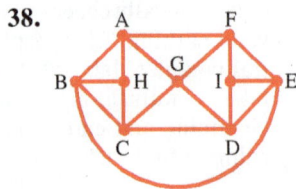

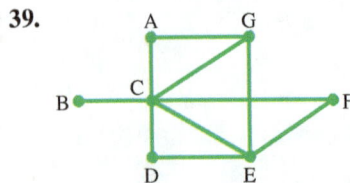

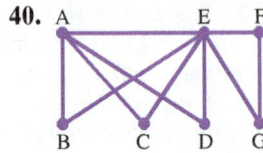

37.

38.

39.

40.

Floor Plans In Exercises 41–43, refer to the floor plan indicated, and determine whether it is possible to start in one of the rooms of the building, walk through each door exactly once, and end up in a different room from the one you started in. Justify each answer.

41. Refer to the floor plan shown in **Exercise 34.**

42. Refer to the floor plan shown in **Exercise 35.**

43. Refer to the floor plan shown in **Exercise 36.**

New York City Bridges The accompanying schematic map shows a portion of the New York City area, including tunnels and bridges.

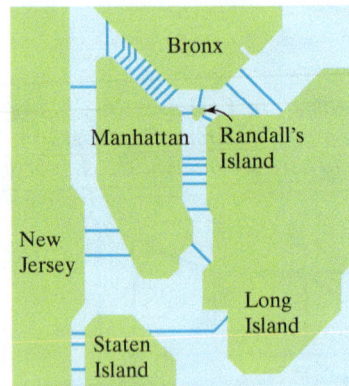

44. Is it possible to take a drive around the New York City area using each tunnel and bridge exactly once, beginning and ending in the same place?

45. Is it possible to take a drive around the New York City area using each tunnel and bridge exactly once, beginning and ending in different places? If so, where must the drive begin and end?

Circuits with Maximum Edges In Exercises 46–48, the graph does not have an Euler circuit. For each graph find a circuit that uses as many edges as possible. (There is more than one correct answer in each case.) How many edges did you use in the circuit?

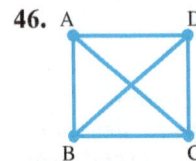

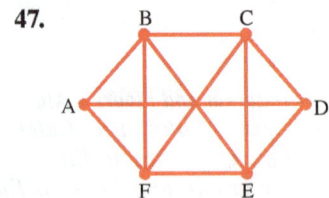

46.

47.

48.

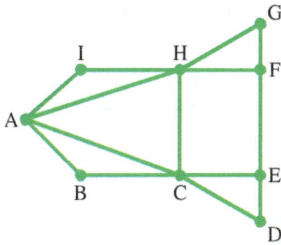

49. ***Route Planning*** In the graph shown here, we have inserted just six edges in the original street grid from **Example 6.** Why is this not a better solution than the one in **Figure 36(b)**?

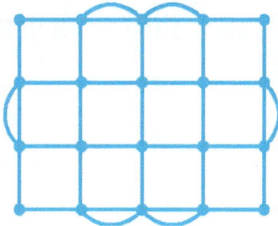

Route Planning For each street grid in Exercise 50–52, insert edges to obtain a street grid that has an Euler circuit. Try to insert as few edges as possible.

50.

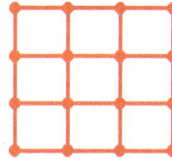

51.

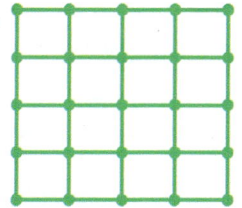

52.

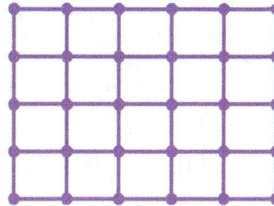

14.3 HAMILTON CIRCUITS AND ALGORITHMS

OBJECTIVES

1 Define and identify a Hamilton circuit in a graph.

2 Determine the number of Hamilton circuits in a complete graph with *n* vertices.

3 Know the meanings of the terms *weighted graph* and *minimum Hamilton circuit.*

4 Apply the brute force algorithm for a complete, weighted graph.

5 Know the meanings of the terms *efficient algorithm* and *approximate algorithm.*

6 Apply the nearest neighbor algorithm.

Hamilton Circuits

In this section we examine *Hamilton circuits* in graphs. The story of Hamilton circuits is a story of very large numbers and of unsolved problems in both mathematics and computer science. We start with a game, called the **Icosian game,** invented by Irish mathematician William Hamilton in the mid-nineteenth century. It uses a wooden board marked with the graph shown in **Figure 37.**

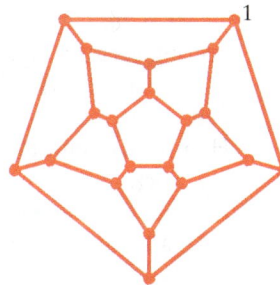

The Icosian Game

Figure 37

The game includes 20 pegs numbered 1 through 20. (The word *Icosian* comes from the Greek word for 20.) At each vertex of the graph there is a hole for a peg.

The simplest version of the game is as follows:

> *Put the pegs into the holes in order, following along the edges of the graph, in such a way that peg 20 ends up in a hole that is joined by an edge to the hole of peg 1.*

Try this now, numbering the vertices in the graph in **Figure 37,** starting at vertex 1.

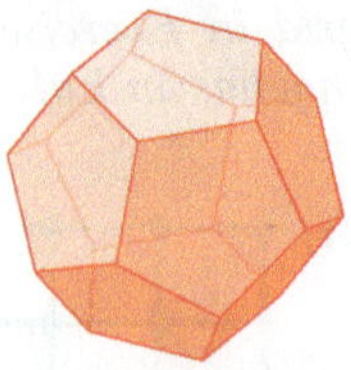

Dodecahedron

Hamilton sold the **Icosian game** idea to a games dealer for 25 British pounds. It went on the market in 1859, but it was not a huge success. A later version of the game, *A Voyage Around the World,* consisted of a regular dodecahedron with pegs at each of the 20 vertices. The vertices had names such as Brussels, Delhi, and Zanzibar. The aim was to travel to all vertices along the edges of the dodecahedron, visiting each vertex exactly once. We can show the vertices and edges of a dodecahedron precisely as in **Figure 37.**

The Icosian game asks that we find a circuit in the graph, but the circuit need not be an Euler circuit. The circuit must visit each *vertex* exactly once, except for returning to the starting vertex to complete the circuit. (The circuit may or may not travel all edges of the graph.) Circuits such as this are called *Hamilton circuits.*

HAMILTON CIRCUIT

A **Hamilton circuit** in a graph is a circuit that visits each vertex exactly once (returning to the starting vertex to complete the circuit).

EXAMPLE 1   Identifying Hamilton Circuits

Which of the following are Hamilton circuits for the graph in **Figure 38?** Justify your answers briefly.

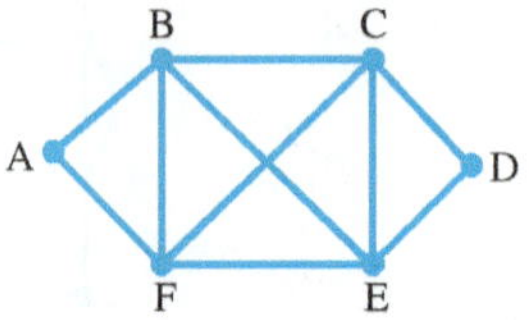

(a) $A \to B \to E \to D \to C \to F \to A$

(b) $A \to B \to C \to D \to E \to F \to C \to$
$E \to B \to F \to A$

(c) $B \to C \to D \to E \to F \to B$

Figure 38

Solution

(a) $A \to B \to E \to D \to C \to F \to A$ is a Hamilton circuit for the graph. It visits each vertex of the graph exactly once and then returns to the starting vertex. (Trace this circuit on the graph to check it.)

(b) $A \to B \to C \to D \to E \to F \to C \to E \to B \to F \to A$ is not a Hamilton circuit, since it visits vertex B (and vertices C, E, and F) more than once. (This is, however, an Euler circuit for the graph.)

(c) $B \to C \to D \to E \to F \to B$ is not a Hamilton circuit since it does not visit all vertices in the graph. ∎

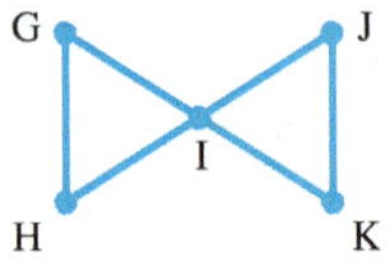

Figure 39

Some graphs, such as the graph in **Figure 39,** do not have a Hamilton circuit. There is no way to visit all the vertices and return to the starting vertex without visiting vertex I more times than allowed.

It can be difficult to determine whether a particular graph has a Hamilton circuit, since there is no theorem that gives necessary and sufficient conditions for a Hamilton circuit to exist. This remains an unsolved problem in Hamilton circuits.

Fortunately, in many real-world applications of Hamilton circuits we are dealing with complete graphs, and any complete graph with three or more vertices does have a Hamilton circuit. For example,

$$A \to B \to C \to D \to E \to F \to A$$

is a Hamilton circuit for the complete graph shown in **Figure 40.** We could form a Hamilton circuit for any complete graph with three or more vertices in a similar way.

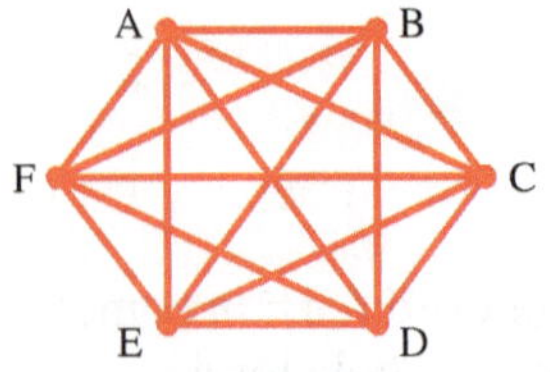

Figure 40

HAMILTON CIRCUITS FOR COMPLETE GRAPHS

Any complete graph with three or more vertices has a Hamilton circuit.

The graph in **Figure 40** has many Hamilton circuits. Try finding some that are different from the one given above.

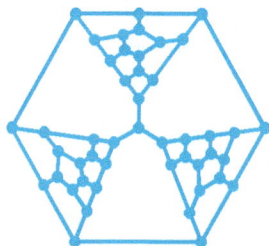

The Tutte graph, shown here, has no Hamilton circuit. This graph has a history connected to the four-color problem for map coloring (**Section 14.1**). In 1880 Peter Tate provided a "proof" for the four-color theorem. It was based on the assumption that every connected graph with all vertices of degree 3, and which can be drawn with edges crossing nowhere except at vertices, has a Hamilton circuit. But in 1946, Tutte produced the graph shown here, showing that Tate's basic assumption was wrong, and his proof was, therefore, invalid.

How many Hamilton circuits does a complete graph have? Before we count, we need an agreement about when two sequences of vertices will be considered to be *different* Hamilton circuits. For example, the Hamilton circuits

$$B \to C \to D \to E \to F \to A \to B \quad \text{and} \quad A \to B \to C \to D \to E \to F \to A$$

visit the vertices in essentially the same order in the graph in **Figure 40,** although the sequences start with different vertices. If we mark the circuit on the graph, we can describe it starting at any vertex. For our purposes, it is convenient to consider the two sequences above as representing the same Hamilton circuit.

WHEN HAMILTON CIRCUITS ARE THE SAME

Hamilton circuits that differ *only* in their starting points will be considered to be the same circuit.

Problem-Solving Strategy

Tree diagrams are often useful for counting.

We begin by counting the number of Hamilton circuits in a complete graph with four vertices. See **Figure 41.** We can use the same starting point for all the circuits we count. We choose A as the starting point. From A we can go to any of the three remaining vertices (B, C, or D). No matter which vertex we choose, we then have two unvisited vertices for our next choice. Then there is only one way to complete the Hamilton circuit. We illustrate this counting procedure in **Figure 42,** where the tree diagram shows that we have $3 \cdot 2 \cdot 1$ Hamilton circuits in the graph. We write this as 3! (read "3 factorial"). See **Chapter 10.**

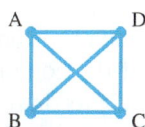

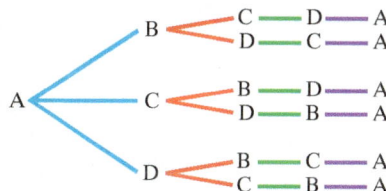

Figure 41   **Figure 42**

Returning to **Figure 40,** we can count the number of Hamilton circuits in the complete graph with 6 vertices. Start at vertex A. From A we can proceed to any of the five remaining vertices. No matter which we choose, we then have four choices for the next vertex along the circuit, three choices for the next, and two for the next. This means that we have

$$5 \cdot 4 \cdot 3 \cdot 2 \cdot 1 = 5! \quad \text{different Hamilton circuits in all.}$$

We would find that a complete graph with 10 vertices has 9! Hamilton circuits. The number of Hamilton circuits in a complete graph can be obtained by calculating the factorial of the number that is one less than the number of vertices.

NUMBER OF HAMILTON CIRCUITS IN A COMPLETE GRAPH

A complete graph with *n* vertices has the following number of Hamilton circuits.

$$(n - 1)!$$

As the number of vertices in a complete graph increases, the number of Hamilton circuits for that graph increases very quickly. Previously we considered how quickly exponential functions increase. Factorials increase quickly as well. For example,

$$25! = 15{,}511{,}210{,}043{,}330{,}985{,}984{,}000{,}000,$$

or approximately 1.6×10^{25}—more than a trillion trillion.

Minimum Hamilton Circuits

Consider that on a typical day, a UPS van might have to make deliveries to 100 different locations. For simplicity, assume that none are priority deliveries, so they can be made in any order. UPS wants to minimize the time to make these deliveries. Think of the 100 locations and the UPS distribution center as the vertices of a complete graph with 101 vertices.

In principle, the van can go from any of these locations directly to any other, shown by having an edge between each pair of vertices. Suppose we estimate the travel time between each pair of locations (of course, this would depend on distance and traffic conditions along the route). These estimates provide weights on the edges of the graph. The objective is to visit each location exactly once, to begin and end at the same place, and to take as little time as possible.

In graph theory terms, we need a Hamilton circuit for the graph with the *least possible total weight*. The **total weight** of a circuit is the sum of the weights on the edges in the circuit. We call such a circuit a *minimum Hamilton circuit* for the graph.

MINIMUM HAMILTON CIRCUIT

In a weighted graph, a **minimum Hamilton circuit** is a Hamilton circuit with the least possible total weight.

A problem whose solution requires us to find a minimum Hamilton circuit for a complete, weighted graph is often called a **traveling salesman problem** (or TSP). Think of the vertices of the complete graph as the cities that a salesperson must visit, and think of the weights on the edges as the cost of traveling directly between the cities. To minimize costs, the salesperson needs a minimum Hamilton circuit for the graph.

Traveling salesman problems are relevant to efficient routing of telephone calls and Internet connections. As another example, to manufacture integrated circuit silicon chips, many lines have to be etched on a silicon wafer. Minimizing production time involves deciding the order in which to etch the lines: a traveling salesman problem. Likewise, to manufacture circuit boards for integrated circuits, laser-drilled holes must be made for connections. Again, the order for drilling the holes is a critical factor in production time.

Brute Force Algorithm

Suppose we are given a complete, weighted graph. How can we find a minimum Hamilton circuit for the graph? One way is to systematically list all the Hamilton circuits in the graph, find the total weight of each, and choose a circuit with least total weight. (In fact, it is sufficient to add up the weights on the edges for just half of the Hamilton circuits, since the total weight of a circuit is the same as the total weight of the circuit that uses the same edges in reverse order.)

The method just described is sometimes called the **brute force algorithm** since we find the solution by checking *all* the Hamilton circuits.

Traveling Salesman Problems and Archaeology Archaeologists often excavate sites in which there is no clear evidence to show which deposits were made earlier, and which later. Some archaeologists have used minimum Hamilton circuits to solve the problem. For example, if a site consists of a number of burials, they consider a complete graph with vertices representing the various burials, and weights on the edges corresponding roughly to how dissimilar the burials are. A minimum Hamilton circuit in this graph gives the best guess for the order in which the deposits were made.

Dr. Edward C. Harris, in his *Principles of Archaeological Stratigraphy*, provided a system known today as **Harris matrices** (which are actually graphs) to assist archaeologists in determining how stratification occurred at a dig site.

(*Source:* http://enwikipedia.org/wiki/Harris_matrix)

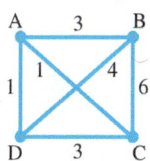

Figure 43

William Rowan Hamilton (1805–1865) spent most of his life in Dublin, Ireland. He was good friends with the poet William Wordsworth, whom he met while touring England and Scotland. Hamilton tried writing poetry, but Wordsworth tactfully suggested that his talents lay in mathematics rather than verse. Catherine Disney was the first great love of his life, but under pressure from her parents, she married another man, much wealthier than Hamilton. Hamilton never seemed to quite get over this.

> **BRUTE FORCE ALGORITHM**
>
> *Step 1* Choose a starting point.
>
> *Step 2* List all the Hamilton circuits with that starting point.
>
> *Step 3* Find the total weight of each circuit.
>
> *Step 4* Choose a Hamilton circuit with the least total weight.

EXAMPLE 2 **Using the Brute Force Algorithm**

Find a minimum Hamilton circuit for the complete, weighted graph shown in **Figure 43**.

Solution

Choose a starting point. For this example, we choose A. Now list all the Hamilton circuits starting at A. Because this is a complete graph with 4 vertices, there are

$$3! = 3 \cdot 2 \cdot 1 = 6 \text{ circuits.}$$

Thus, we must find 6 Hamilton circuits.

We need a systematic way of writing down all the Hamilton circuits. The "counting tree" of **Figure 42** on **page 807** provides a guide. We start by finding all the Hamilton circuits that begin A → B, then include those that begin A → C, and finally include those that begin A → D, as shown below. Once all the Hamilton circuits are listed, pair those that visit the vertices in precisely opposite orders. (Circuits in these pairs have the same total weight. This will save some adding.) Finally, determine the sum of the weights in each circuit.

Circuit	Total weight of the circuit
1. A → B → C → D → A	3 + 6 + 3 + 1 = 13
2. A → B → D → C → A	3 + 4 + 3 + 1 = **11** Minimum
3. A → C → B → D → A	1 + 6 + 4 + 1 = 12
4. A → C → D → B → A (opposite of 2)	11
5. A → D → B → C → A (opposite of 3)	12
6. A → D → C → B → A (opposite of 1)	13

We can now see that A → B → D → C → A is a minimum Hamilton circuit for the graph. The weight of this circuit is 11. ∎

In principle, the brute force algorithm provides a way to find a minimum Hamilton circuit in any complete, weighted graph. In practice, it takes far too long, even for a complete, weighted graph with only 7 vertices. There are 6!, or 720, Hamilton circuits in such a graph, and we would have to calculate the total weight for half of these. That means we would have to do 360 separate calculations, in addition to listing the Hamilton circuits.

Most real-world problems involve a great deal more than 7 vertices. Manufacturing a large integrated circuit can involve a graph with almost one million vertices. Likewise, telecommunications companies routinely deal with graphs with millions of vertices. As the number of vertices in the graph increases, the task of finding a minimum Hamilton circuit using the brute force algorithm soon becomes too time-consuming for even our fastest computers. For example, if one of today's supercomputers had started using the brute force algorithm on a 100-vertex traveling salesman problem when the universe was created, it would not yet be done.

The P = NP Problem The traveling salesman problem can be solved using brute force as long as the number of cities is relatively small. A computer program is required to solve problems involving more cities. One such program is the *Concorde TPS Solver*, or Concorde. Developed by a team of researchers led by William J. Cook, Concorde broke the world record for finding an optimal tour through the most locations. The problem, which involved 85,900 cities, was developed at Bell Labs in the late 1980s and took 15 years for Concorde to solve.

Probably the most important unsolved problem in computer science is either to find an efficient algorithm for the traveling salesman problem or to explain why no one could create such an algorithm. In mathematical circles, this is known as the **P = NP problem.** It was once mentioned in an episode of *The Simpsons.*

Cook has written a book on the traveling salesman problem in lay terms. It is titled *In Pursuit of the Traveling Salesman*. His Web site (see the source note below) contains links to the following related articles.

- $500 prize for best tour through 115,475 U.S. cities
- Short piece on Yogi Berra and the TSP
- Thriller movie centered around a solution of the TSP
- 49,512-city tour of Colbert Nation via the *iPhone App*
- $1000 Prize for a 100,000-city challenge problem
- Plot an optimal TSP tour with a Google interface
- Solution of an 85,900-city TSP
- Optimal route for a 99-county campaign tour

(*Sources:* www.math.uwaterloo.ca/tsp; http://jhered.oxfordjournals.org)

In contrast, Fleury's algorithm for finding an Euler circuit in a graph does not take too long for our computers, even for rather large graphs. Algorithms that do not take too much computer time are called **efficient algorithms.** Computer scientists have so far been unable to find an efficient algorithm for the traveling salesman problem, and they suspect that it is simply impossible to create such an algorithm.

For the traveling salesman problem, there are some algorithms that do not take too much computer time and that give *reasonably good* solutions *most of the time.* Such algorithms are called **approximate algorithms,** because they give an approximate solution to the problem. The underlying idea is that from each vertex, we proceed to a "nearest" available vertex.

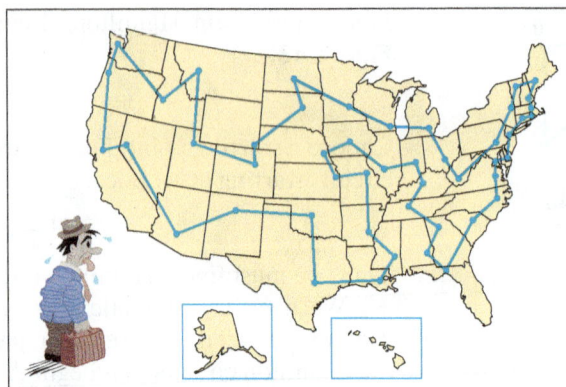

Figure 44

Nearest Neighbor Algorithm

One approximate algorithm for problems such as the traveling salesman problem is the **nearest neighbor algorithm.**

NEAREST NEIGHBOR ALGORITHM

Step 1 Choose a starting point for the circuit. Call this vertex A.

Step 2 Check all the edges joined to A, and choose one that has the least weight. Proceed along this edge to the next vertex.

Step 3 At each vertex you reach, check the edges from there *to vertices not yet visited.* Choose one with the least weight. Proceed along this edge to the next vertex.

Step 4 Repeat Step 3 until you have visited all the vertices.

Step 5 Return to the starting vertex.

EXAMPLE 3 **Using the Nearest Neighbor Algorithm**

A courier is based at the head office (A) and must deliver documents to four other offices (B, C, D, and E). The estimated time of travel (in minutes) between each of these offices is shown on the graph in **Figure 45** on the next page. The courier wants to visit the locations in an order that takes the least time. Use the nearest neighbor algorithm to find an approximate solution to this problem. Calculate the total time required to cover the chosen route.

In 1985, researcher **Shen Lin** came up with the route shown in **Figure 44** for a salesman who wants to visit all capital cities in the forty-eight contiguous states, starting and ending at the same capital and traveling the shortest possible total distance. He could not prove that his 10,628-mile route was the shortest possible, but he offered $100 to anyone who could find a shorter one.

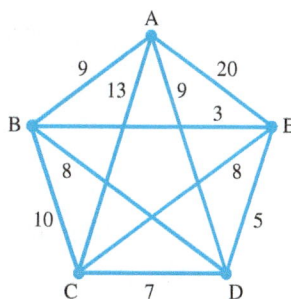

Figure 45

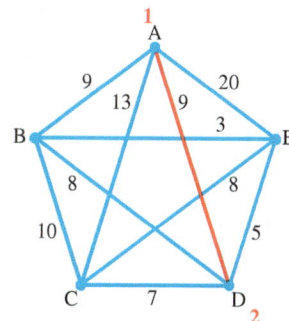

Figure 46

Solution

Step 1   Choose a starting point. Let's start at A.

Step 2   Choose an edge with least weight joined to A. Both AB and AD have weight 9, which is less than the weights of the other two edges joined to A. We choose AD* and keep a record of the circuit as we form it.

$$A \xrightarrow{9} D$$

To help ensure that we do not visit a vertex twice, we number the vertices as we visit them and color the edges used, as in **Figure 46.**

Continue to number the vertices as we work through the rest of the solution.

Step 3   Now check edges joined to D, excluding DA (since we have already been to A). DC has weight 7, DB has weight 8, and DE has weight 5. DE has the least weight, so proceed along DE to E. Our circuit begins this way.

$$A \xrightarrow{9} D \xrightarrow{5} E$$

Step 4   Now repeat the process at E. Check all edges from E that go to a vertex not yet visited. EC has weight 8 and EB has weight 3. Proceed along EB to B.

$$A \xrightarrow{9} D \xrightarrow{5} E \xrightarrow{3} B$$

Step 5   We have one vertex not yet visited, C. We go next to C and now have visited all the vertices. We return to our starting point, A. Our Hamilton circuit is

$$A \xrightarrow{9} D \xrightarrow{5} E \xrightarrow{3} B \xrightarrow{10} C \xrightarrow{13} A.$$

Its total weight is $9 + 5 + 3 + 10 + 13 = 40$.

Our advice to the courier would thus be to visit the offices in the order shown in this circuit. This route will take about 40 minutes. ■

The circuit we found in **Example 3** is not the minimum Hamilton circuit for the graph in **Figure 45**. (We shall investigate this.) However, the point of an approximate algorithm is that a computer can implement it without taking too much time, and most of the time it will give a reasonably good solution to the problem.

If we had a computer performing the nearest neighbor algorithm for us, we could make a small adjustment that would give better results without taking too much longer. We could repeat the nearest neighbor algorithm for all possible starting points, and then choose from these the Hamilton circuit with the least weight. If we do this for the example above, we obtain the results in **Table 1** on the next page.

Large Numbers 70! is the first factorial greater than 10^{100}. Most calculators will calculate numbers up to only about 10^{99}, so if you try calculating 70! you probably will get an error message. (Try it.) 10^{100} is called one googol, a name invented by a nine-year-old child who was asked to make up a name for a very, very large number.

These numbers are unimaginably large. Consider:

$$25! \approx 2 \times 10^{25}$$
$$50! \approx 3 \times 10^{64}$$
$$100! \approx 9 \times 10^{157}$$

Comparison:

Approximate number of *meters* from Saturn to the sun: 10^{12}

Estimate of the number of *inches* to the farthest object in the universe: 10^{27}

Approximate number of molecules we would have if we filled a sphere the size of Earth with water: 10^{49}

---

*If using a computer to do this, we would instruct the computer to make a random choice between AB and AD.

Table 1

Starting Vertex	Circuit Using Nearest Neighbor	Total Weight
A	A → D → E → B → C → A	9 + 5 + 3 + 10 + 13 = 40
B	B → E → D → C → A → B	3 + 5 + 7 + 13 + 9 = 37
C	C → D → E → B → A → C	7 + 5 + 3 + 9 + 13 = 37
D	D → E → B → A → C → D	5 + 3 + 9 + 13 + 7 = 37
E	E → B → D → C → A → E	3 + 8 + 7 + 13 + 20 = 51

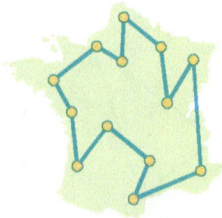

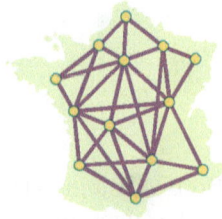

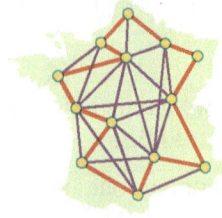

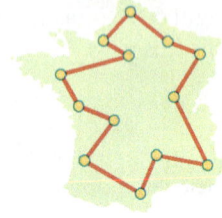

We would recommend that the courier use a circuit with total weight 37—for example,

$$D \to E \to B \to A \to C \to D.$$

This does not force him to start his journey at D. If he followed the route

$$A \to C \to D \to E \to B \to A,$$

he would be using the same edges as in $D \to E \to B \to A \to C \to D$, and his traveling time still would be 37 minutes.

Can we now be sure that we have found the minimum Hamilton circuit for the graph in **Figure 45?** No. This is still an *approximate* solution. If we check the total weights of all possible Hamilton circuits in the graph, we find the Hamilton circuit

$$A \xrightarrow{9} B \xrightarrow{3} E \xrightarrow{8} C \xrightarrow{7} D \xrightarrow{9} A$$

with total weight just 36 minutes. The nearest neighbor algorithm simply will not find this minimum Hamilton circuit for the graph. However, all we expect of an approximate algorithm is that it give a reasonably good solution for the problem in a reasonable amount of time.

One method for generating solutions for the traveling salesman problem uses ant colony simulation. It models by creating paths observed in actual ant behavior to find the shortest paths between the ant nest and food sources. See **Figure 47.** (*Source:* wikipedia.org/wiki/Traveling_salesman_problem)

Figure 47

FOR FURTHER THOUGHT

The Speed of Algorithms

How do computer scientists classify the speed of algorithms? First, they write the number of steps a computer using the algorithm takes to solve a problem as a function of the "size" of the problem. We can think of the size of the problem as the number of vertices in a graph.

Let's suppose that our graph has n vertices. For some algorithms the number of steps is a **polynomial function** of n—for example, $n^4 + 2n$. For other algorithms, the number of steps is an **exponential function** of n—for example, 2^n. (See **Chapters 7 and 8** for more on polynomial and exponential functions.) As n increases, the number of steps increases much faster for exponential functions than for polynomial functions. Functions that involve factorials increase even faster.

In the table below we show approximate values for functions of the three types for different values of n. (We use scientific notation to make it easier to compare the sizes of the numbers.)

n	n^4	2^n	$n!$
5	6.3×10^2	3.2×10	1.2×10^2
10	1.0×10^4	1.0×10^3	3.6×10^6
15	5.1×10^4	3.3×10^4	1.3×10^{12}
20	1.6×10^5	1.0×10^6	2.4×10^{18}
25	3.9×10^5	3.4×10^7	1.6×10^{25}
30	8.1×10^5	1.1×10^9	2.7×10^{32}
40	2.6×10^6	1.1×10^{12}	8.2×10^{47}
50	6.3×10^6	1.1×10^{15}	3.0×10^{64}

In the first line of the table (where $n = 5$), n^4 gives the largest value. But as n gets larger, 2^n grows much faster than n^4, while $n!$ grows extremely fast. For example, if we double the size of n from 15 to 30, n^4 becomes a little more than 10 times as large, 2^n becomes about 3×10^4 (or 30,000) times as large, while $n!$ becomes approximately 2×10^{20} times as large. 10^{20} is more than a billion billion.

Algorithms whose time functions grow no faster than a polynomial function are called **polynomial time algorithms.** These are *efficient* algorithms— they do not take too much computer time.

Algorithms whose time functions grow faster than any polynomial function are called **exponential time algorithms.** These are *not* efficient algorithms. They are by nature too time-consuming for our computers. Our silicon chip computers are getting faster each year, but even this increase in speed hardly puts a dent in the time required for a computer to implement an exponential time algorithm for a very large graph.

Because the brute force algorithm for the traveling salesman problem has a time function involving $n!$, which grows faster than an exponential function of n, this algorithm is an exponential time algorithm.

For Group or Individual Investigation

Use the table to help answer these questions:

1. By approximately what factor does n^4 grow if we double n from 10 to 20?

2. By approximately what factor does 2^n grow if we double n from 10 to 20?

3. By approximately what factor does $n!$ grow if we double n from 10 to 20?

4. Repeat Exercises 1–3 if we double n from 25 to 50.

5. Suppose it would take 2^{30} years using computers at their present speeds to solve a certain problem. (2^{30} is a little over one billion.) If we assume that computers double in speed each year, how long would we have to wait before we could solve the problem in 1 year?

14.3   EXERCISES

Hamilton Circuits   In Exercises 1 and 2, a graph is shown, and some paths in the graph are specified. Determine which paths are Hamilton circuits for the graph. If not, say why not.

1.

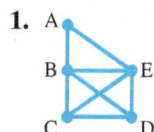

(a) $A \to E \to C \to D \to E \to B \to A$

(b) $A \to E \to C \to D \to B \to A$

(c) $D \to B \to E \to A \to B$

(d) $E \to D \to C \to B \to E$

2.

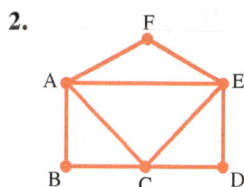

(a) $A \to B \to C \to D \to E \to C \to A \to E \to F \to A$

(b) $A \to C \to D \to E \to F \to A$

(c) $F \to A \to C \to E \to F$

(d) $C \to D \to E \to F \to A \to B$

Euler and Hamilton Circuits   In Exercises 3 and 4, determine whether each sequence of vertices is a circuit, whether it is an Euler circuit, and whether it is a Hamilton circuit. Justify your answers.

3.

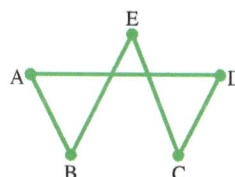

(a) $A \to B \to C \to D \to E \to A$

(b) $B \to E \to C \to D \to A \to B$

(c) $E \to B \to A \to D \to A \to D \to C \to E$

4.

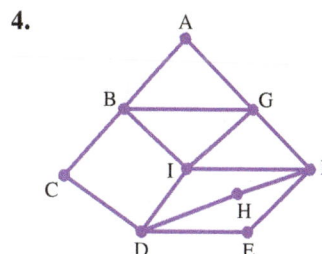

(a) $A \to B \to C \to D \to E \to F \to G \to A$

(b) $B \to I \to G \to F \to E \to D \to H \to F \to I \to D \to C \to B \to G \to A \to B$

(c) $A \to B \to C \to D \to E \to F \to G \to H \to I \to A$

Hamilton Circuits In Exercises 5–10, determine whether the graph has a Hamilton circuit. If so, find one. (There may be many different correct answers.)

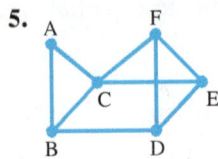

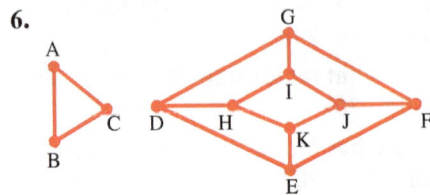

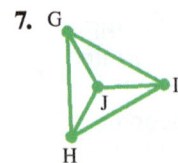

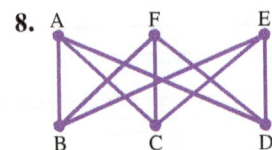

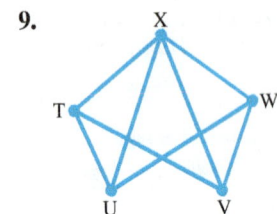

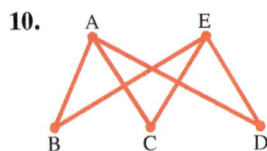

5.

6.

7.

8.

9.

10.

Hamilton and Euler Circuits Work each problem.

11. Draw a graph that has a Hamilton circuit but no Euler circuit. Specify the Hamilton circuit, and explain why the graph has no Euler circuit. (There are many different correct answers.)

12. Draw a graph that has an Euler circuit but no Hamilton circuit. Specify an Euler circuit in your graph. (There are many different correct answers.)

13. Draw a graph that has both an Euler circuit and a Hamilton circuit. Specify these circuits. (There are many different correct answers.)

14. Decide whether each statement is true or false. If the statement is false, give an example to show that it is false.

  (a) A Hamilton circuit for a graph must visit each vertex in the graph.

  (b) An Euler circuit for a graph must visit each vertex in the graph.

  (c) A Hamilton circuit for a graph must use each edge in the graph.

  (d) An Euler circuit for a graph must use each edge in the graph.

  (e) A circuit cannot be both a Hamilton circuit and an Euler circuit.

  (f) An Euler circuit must visit no vertex more than once, except the vertex where the circuit begins and ends.

Hamilton and Euler Circuits In Exercises 15–20, determine whether an Euler circuit, a Hamilton circuit, or neither would solve the problem.

15. *Bandstands at a Festival* The vertices of a graph represent bandstands at a festival, and the edges represent paths between the bandstands. A visitor wants to visit each bandstand exactly once, returning to her starting point when she is finished.

16. *Relay Team Running Order* The vertices of a complete graph represent the members of a five-person relay team. The team manager wants a circuit that will show the order in which the team members will run. (He will decide later who will start.)

17. *Paths in a Botanical Garden* The vertices of a graph represent places where paths in a botanical garden cross, and the edges represent the paths. A visitor wants to walk along each path in the garden exactly once, returning to his starting point when finished.

18. *Western Europe* The vertices of a graph represent the countries on the continent (Western Europe), and the edges represent border crossings between the countries. A traveler wants to travel over each border crossing exactly once, returning to the first country visited for his flight home to the United States.

19. *Africa* Vertices represent countries in sub-Saharan Africa, with an edge between two vertices if those countries have a common border. A traveler wants to visit each country exactly once, returning to the first country visited for her flight home to the United States.

20. *X-Rays* In using X-rays to analyze the structure of crystals, an X-ray diffractometer measures the intensity of reflected radiation from the crystal in thousands of different positions. Consider the complete graph with vertices representing the positions where measurements must be taken. The researcher must decide the order in which to take these readings, with the diffractometer returning to its starting point when finished.

Factorials *In Exercises 21–24, use a calculator, if necessary, to find the value.*

21. 4!   **22.** 6!

23. 9!   **24.** 14!

Hamilton Circuits *In Exercises 25–28, determine how many Hamilton circuits there are in a complete graph with this number of vertices. (Leave answers in factorial notation.)*

25. 10 vertices   **26.** 15 vertices

27. 18 vertices   **28.** 60 vertices

29. List all Hamilton circuits in the graph that start at P.

Hamilton Circuits *In Exercises 30–36, refer to the following graph. List all Hamilton circuits in the graph that start with the indicated vertices.*

30. Starting E → F → G

31. Starting E → H → I

32. Starting E → I → H

33. Starting E → F

34. Starting E → I

35. Starting E → G

36. Starting at E

37. **Hamilton Circuits** List all Hamilton circuits in the graph that start at A.

Brute Force Algorithm *In Exercises 38–41, use the brute force algorithm to find a minimum Hamilton circuit for the graph. In each case determine the total weight of the minimum Hamilton circuit.*

38.

39.

40.

41.

Nearest Neighbor Algorithm *In Exercises 42–44, use the nearest neighbor algorithm starting at each of the indicated vertices to determine an approximate solution to the problem of finding a minimum Hamilton circuit for the graph. In each case, find the total weight of the circuit found.*

42.

   (a) Starting at F

   (b) Starting at G

   (c) Starting at H

   (d) Starting at I

43.

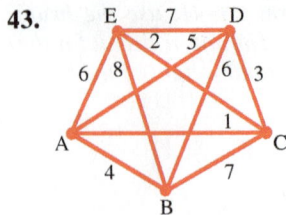

(a) Starting at A

(b) Starting at C

(c) Starting at D

(d) Starting at E

44.

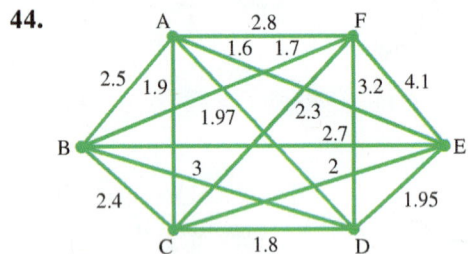

(a) Starting at A

(b) Starting at B

(c) Starting at C

(d) Starting at D

(e) Starting at E

(f) Starting at F

45. Nearest Neighbor Algorithm Refer to the accompanying graph. Complete parts (a)–(c) in order.

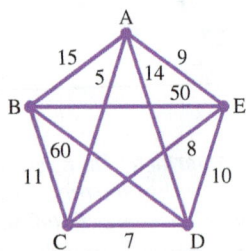

(a) Use the nearest neighbor algorithm starting at each of the vertices in turn to determine an approximate solution to the problem of finding a minimum Hamilton circuit for the graph. In each case, find the total weight of the circuit.

(b) Which of the circuits found in part (a) gives the best solution to the problem of finding a minimum Hamilton circuit for the graph?

(c) Just by looking carefully at the graph, find a Hamilton circuit in the graph that has lower total weight than any of the circuits found in part (a).

46. Complete Bipartite Graph A graph is called a complete bipartite graph if the vertices can be separated into two groups in such a way that there are no edges between vertices in the same group, and there is an edge between each vertex in the first group and each vertex in the second group. An example is shown below.

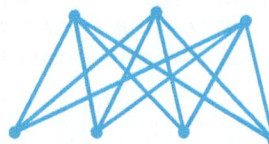

The notation we use for these graphs is $K_{m,n}$, where m and n are the numbers of vertices in the two groups. For example, the graph shown in this exercise is $K_{4,3}$ (or $K_{3,4}$). Complete parts (a)–(d) in order.

(a) Draw $K_{2,2}$, $K_{2,3}$, $K_{2,4}$, $K_{3,3}$, and $K_{4,4}$.

(b) For each of the graphs you have drawn in part (a), find a Hamilton circuit or say that the graph has no Hamilton circuit.

(c) Make a conjecture: What must be true about m and n for $K_{m,n}$ to have a Hamilton circuit?

(d) What must be true about m and n for $K_{m,n}$ to have an Euler circuit? Justify your answer.

Hamilton Circuits *In Exercises 47–50, find all Hamilton circuits in the graph that start at A. (Hint: Use counting trees.)*

47.

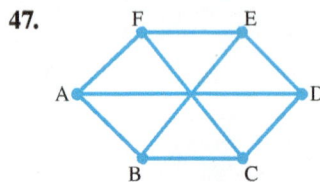

48.

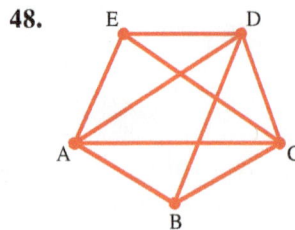

49.

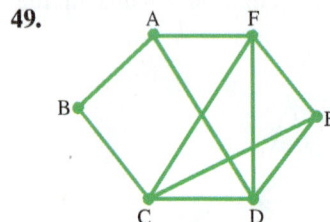

50.

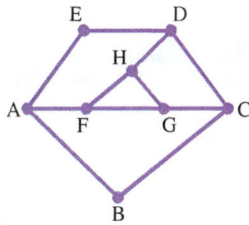

51. *Traveling Salesman Problem* The diagram represents seven cities, A through G, along with the distances in miles between many of the pairs.

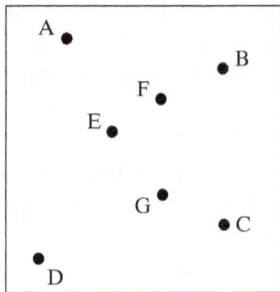

AB = 51	AF = 36	BF = 22	CG = 22	EF = 22
AD = 71	BC = 50	BG = 45	DE = 45	EG = 28
AE = 32	BE = 45	CD = 61	DG = 45	FG = 30

Determine what you believe to be the shortest possible distance that can be traveled so that the starting point and ending point are the same, and all cities have been visited.

The Icosian Game *The graph below shows the Icosian game (described in the text) with the vertices labeled. In Exercises 52–54, find a Hamilton circuit for the graph in the specified version of the game. (We suggest that you write numbers on the graph in the order in which you visit the vertices. There are different correct answers for these exercises.)*

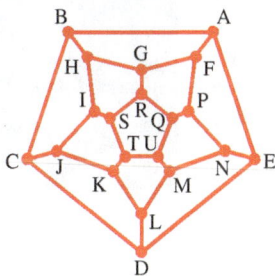

52. The circuit begins at A.

53. The circuit begins A → F.

54. The circuit begins A → B → H.

55. *Dirac's Theorem* Paul A. M. Dirac proved the following theorem in 1952.

Suppose G is a (simple) graph with n vertices, n ≥ 3. If the degree of each vertex is greater than or equal to $\frac{n}{2}$, then the graph has a Hamilton circuit.

Refer to graphs (1)–(5) and answer parts (a)–(e) in order.

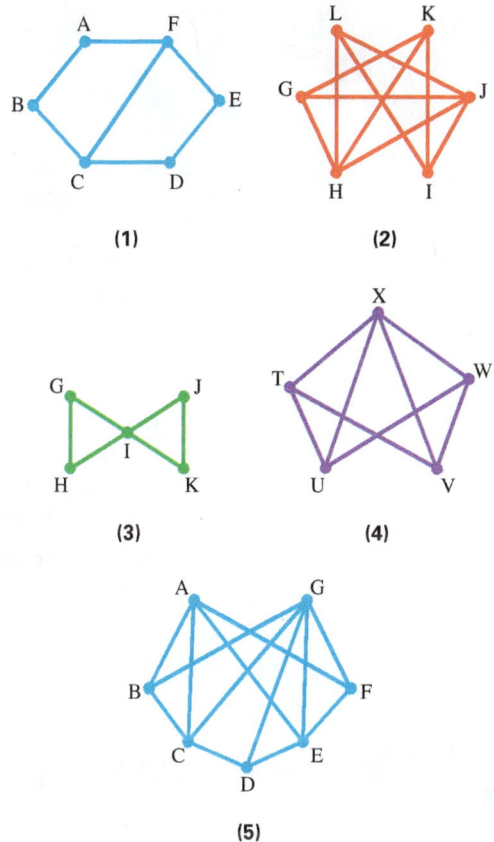

(a) Which graphs satisfy the condition that the degree of each vertex is greater than or equal to $\frac{n}{2}$?

(b) For which of the graphs can we conclude from Dirac's theorem that the graph has a Hamilton circuit?

(c) If a graph does *not* satisfy the condition that the degree of each vertex is greater than or equal to $\frac{n}{2}$, can we be sure that the graph does *not* have a Hamilton circuit? Justify your answer. (*Hint*: Study the accompanying graphs.)

(d) Is Dirac's theorem still true if $n < 3$? Justify your answer.

(e) Use Dirac's theorem to write a convincing argument that any complete graph with 3 or more vertices has a Hamilton circuit.

WHEN Will I Ever USE This ?

Do you hate waiting in lines?

TouringPlans.com is a Web site that helps its subscribers minimize their wait times for attractions at Walt Disney World and other popular theme parks. The site offers personalized, step-by-step itineraries that list the most efficient way to tour the parks.

Len Testa is the founder of TouringPlans.com, as well as the co-author of the *Unofficial Guide to Walt Disney World*. While on a vacation to Walt Disney World during the summer before he started graduate school, he waited in line almost two hours for the Great Movie Ride at the Disney-MGM Studios. Testa decided to make minimizing wait times the subject of his master's thesis in computer programming.

Using a variant of the traveling salesman problem that accounts for time dependency, Testa developed a computer program that would find the most efficient order in which to visit attractions in a theme park. The program uses a genetic algorithm that accounts for distance between attractions, duration of the attractions themselves, scheduled breaks, and average wait times. For a small fee, subscribers can create their own customized touring plans for each theme park by choosing which attractions they wish to visit and optimizing the plan for a given day. TouringPlans.com claims that it can save its users up to four hours in line per day at Walt Disney World, Disneyland, and Universal Studios Florida.

Source: Interview with Len Testa on episode 42 of the Betamouse Podcast, accessed from http://blog.touringplans.com

---

14.4 TREES AND MINIMUM SPANNING TREES

OBJECTIVES

1 Determine whether a graph is a *tree*.

2 Understand the unique path property of trees.

3 Know the meanings of the terms *spanning tree* and *minimum spanning tree*.

4 Apply Kruskal's algorithm to find a minimum spanning tree.

5 Understand the relationship between the number of vertices and the number of edges in a tree.

Connected Graphs and Trees

Peggy wants to install an underground irrigation system in her garden. The system must connect the faucet (F) to outlets at various points, specifically the rose bed (R), perennial bed (P), daffodil bed (D), annuals bed (A), berry patch (B), vegetable patch (V), cut flower bed (C), and shade bed (S). These are the vertices of the graph in **Figure 48**.

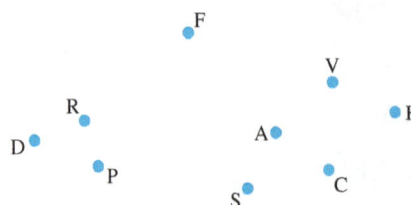

Figure 48

The irrigation system will have pipes connecting these points. For the irrigation system to work well, there should be as few pipes as possible connecting all the outlets.

In graph theory, the botanical words go beyond trees. A **forest** is a graph with all components trees, as shown here. Later in this section you will find the words **root** and **leaf** applied to graphs. The terms **pruning** and **separating** arise in applications of trees to the computer analysis of images.

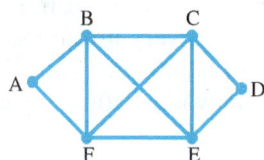

Figure 49

Think about this problem in terms of a graph with edges representing pipes. Peggy wants water to flow from the faucet (F) to each of the outlets. This means that we need to create a *connected* graph that includes all the vertices in **Figure 48.**

CONNECTED GRAPH (ALTERNATIVE DEFINITION)

A **connected graph** is one in which there is *at least one path* between each pair of vertices.

To see that this definition agrees with our definition in **Section 14.1,** observe that if we select any pair of vertices in the connected graph in **Figure 49** (for example, A and C), then there is at least one path between them. (In fact, there are many paths between A and C.) Two examples are

$$A \rightarrow B \rightarrow E \rightarrow C \quad \text{and} \quad A \rightarrow F \rightarrow C.$$

Peggy needs a *connected* graph. Also, she must use as few pipes as possible to connect all the outlets to the system. This means that the graph *must not contain any circuits*. If the graph had a circuit, we could remove one of the pipes in the circuit and still have a connected system of pipes.

Figure 50 shows one possible solution to Peggy's problem.

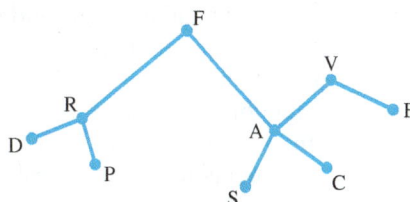

Figure 50

There is a name for graphs that are connected and contain no circuits.

TREE

A graph is a **tree** if the graph is *connected* and contains *no circuits*.

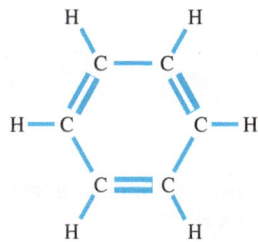

Benzene

It seems mathematician **Arthur Cayley** invented the term "tree" in graph theory in about 1857 in connection with problems related to counting all trees of certain types. Later he realized the relevance of his work to nineteenth-century organic chemistry and, in the 1870s, published a note on this. Of course, not all molecules have treelike structures. Friedrich Kekulé's realization that the structure of benzene is *not* a tree is considered one of the most brilliant breakthroughs in organic chemistry.

All five of the graphs shown in **Figure 51** are trees.

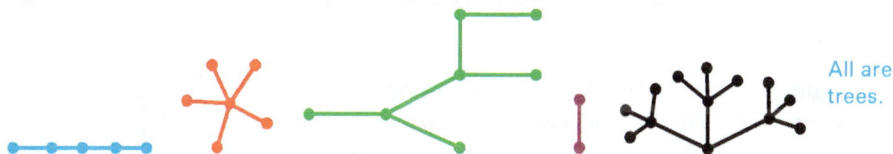

All are trees.

Figure 51

In contrast, none of the graphs shown in **Figure 52** is a tree. The graph in **Figure 52(a)** is not a tree, because it is not connected. Those in **Figures 52(b) and (c)** are not trees, because each contains at least one circuit.

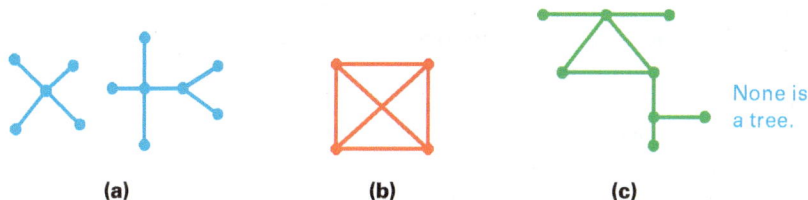

(a)   (b)   (c)

None is a tree.

Figure 52

Just how fast are the fastest computers? Computer speeds are sometimes measured in **flops** (floating point operations per second). A speed of one gigaflop means that the computer can perform about one billion (10^9) calculations per second.

UNIQUE PATH PROPERTY OF TREES

In a tree, there is always **exactly one path** from each vertex in the graph to any other vertex in the graph.

For example, consider the vertices S and B in the tree in **Figure 50**. There is exactly one path from vertex S to vertex B, namely $S \rightarrow A \rightarrow V \rightarrow B$.

This property follows from the definition of a tree. Because a tree is connected, there is always at least one path between each pair of vertices. And if the graph is a tree, there cannot be two different paths between a pair of vertices. If there were, these together would contain a circuit.

It is the unique path property of trees that makes them so important in real-world applications. This also is the reason why trees are not a very useful model for telephone networks. Each edge of a tree is a cut edge, so failure along one link would disrupt the service. Telephone companies rely on networks with many circuits to provide alternative routes for calls.

EXAMPLE 1   Renovating an Irrigation System

Joe has an old underground irrigation system in his garden, with connections at the faucet (F), outlets A through E in the garden, and existing underground pipes as shown in the graph in **Figure 53**. He wants to renovate this system, keeping only some of the pipes. Water must still be able to flow to each of the original outlets, and the graph of the irrigation system must be a tree. Design an irrigation system that meets Joe's objectives.

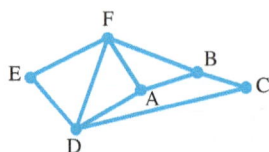

Figure 53

Solution

In graph theory terms, we need a *subgraph* of the graph in **Figure 53**. Because water still must be able to flow to each of the original outlets, we need a connected subgraph that *includes all the vertices of the original graph.* Finally, this subgraph must be a *tree*. Thus, we must remove edges (but no vertices) from the graph, without disconnecting the graph, to obtain a subgraph that is a tree. One way to achieve this is shown in **Figure 54**. ■

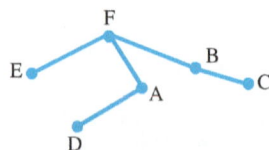

Figure 54

FOR FURTHER THOUGHT

Binary Coding

The unique path property of a tree can be used to make a code. We illustrate with a **binary code,** that is, one that uses 0s and 1s. (Recall from **Chapter 4** that when we represent numbers in binary form, we use only the symbols 0 and 1.)

To set up the code we use a special kind of tree like that shown in the figure at right. This tree is a **directed graph** (there are arrows on the edges). The vertex at the top (with no arrows pointing toward it) is called the **root** of the directed tree. The vertices with no arrows pointing away from them are called **leaves** of the directed tree. We label each leaf with a letter we want to encode. The diagram shown encodes only 8 letters, but we could easily draw a bigger tree with more leaves to represent more letters.

We now write a 0 on each branch that goes left and a 1 on each branch that goes right.

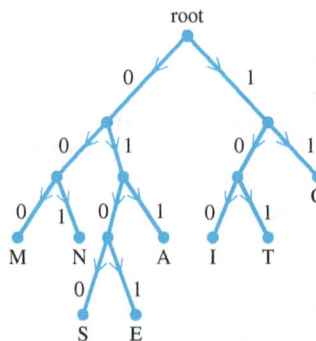

To show how the encoding works, we write the word MAT using the code. Follow the unique path from the root of the tree down to the appropriate leaf, noting in order the labels on the edges.

M is written **000**.

A is written **011**.

T is written **101**.

So MAT is written **000011101**.

We can easily translate this code using our tree. Let us see how we could decode 000011101. Referring to the tree, there is only one path from the root to a leaf that can give rise to those first three 0s, and that is the path leading to M. So we can begin to separate the code word into letters: 000–011101. Again following down from the root, the path 011 leads us unambiguously to A. So we have 000–011–101. The path 101 leads unambiguously to T.

The reason why we can translate the string of 0s and 1s back to letters without ambiguity is that no letter has a code the same as the first part of the code for a different letter.

For Group or Individual Investigation

1. Use the binary encoding tree to write the binary code for each of the following words:

ANT, SEAT

2. Use the encoding tree to find the word represented by each of the following codes:

0001111001, 0100011101

3. Decode the following messages (commas are inserted to show separation of words).

(a) 0110011010100, 100001, 1010101001101

(b) 0101011101, 0001000011010100, 100001, 101100001

(c) 011, 0000101011001, 1010101011000

(d) 00001010101101, 0000101, 011101, 1010101011

The answers are on **page 822**.

Spanning Trees

A subgraph of the type shown in **Figure 54** from **Example 1** is called a *spanning tree* for the graph. This kind of subgraph "spans" the graph in the sense that it connects all the vertices of the original graph into a subgraph.

> **SPANNING TREE**
>
> A **spanning tree** for a graph is a subgraph that includes every vertex of the original graph and is a tree.

We can find a spanning tree for any connected graph (think about that), and if the original graph has at least one circuit, it will have several different spanning trees.

Arthur Cayley (1821–1895) was a close friend of James Sylvester. They worked together as lawyers at the courts of Lincolns Inn in London and discussed mathematics with each other during work hours. Cayley considered his work as a lawyer simply a way to support himself so that he could do mathematics. In the fourteen years he worked as a lawyer, he published 250 mathematical papers. At age 42, he took a huge paycut to become a mathematics professor at Cambridge University.

EXAMPLE 2 Finding Spanning Trees

Consider the graph shown in **Figure 55.** Find two different spanning trees for the graph.

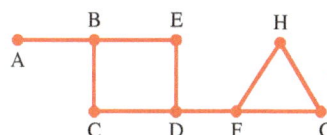

> To obtain a tree, we must break the two circuits by removing a single edge from each.

Figure 55

Solution

There are several choices for which edges to remove. We could remove edges CD and FH, leaving the subgraph in **Figure 56(a)**. Alternatively, we could remove edges ED and FG, leaving the subgraph in **Figure 56(b)**. There are other correct solutions.

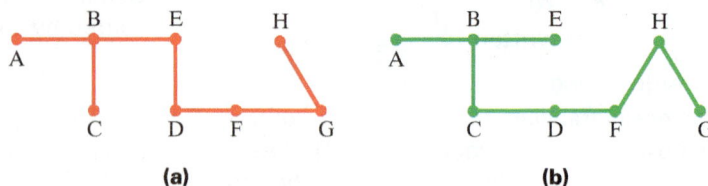

(a)   (b)

Figure 56

If two spanning trees use different edges from the original graph, we consider them to be different spanning trees even when they are isomorphic graphs. The subgraphs shown in **Figure 56** are *different* spanning trees, even though they happen to be isomorphic graphs. To decide whether they are isomorphic, change the labels in graph (a) as follows: Interchange C and E, and interchange G and H. By dragging vertices without detaching edges, we can obtain graphs that look alike.

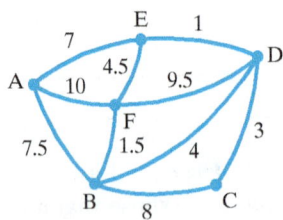

Figure 57

Minimum Spanning Trees

Suppose there are six villages in a rural district. At present all the roads in the district are gravel roads. There is not enough funding to pave all the roads, but there is a pressing need to have good paved roads so that emergency vehicles can travel easily between the villages. The vertices in the graph in **Figure 57** represent the villages in the district, and the edges represent the existing gravel roads. Estimated costs for paving the different roads are shown, in millions of dollars. (These estimates take into account distances, as well as features of the terrain.)

The district council wants to pave just enough roads so that emergency vehicles will be able to travel from any village to any other along paved roads, though possibly by a roundabout route. The council would like to achieve this at the *lowest possible cost*.

The problem calls for exactly one path from each vertex in the graph to any other, so it requires a spanning tree. This spanning tree should have minimum total weight. The total weight of a spanning tree is the sum of the weights of the edges in the tree.

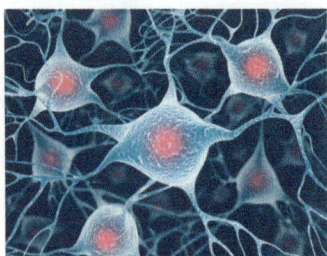

Computer analysis of images is a field of active research, particularly in medicine. (Computer analysis of images consists of having a computer detect relevant features of an image.) Some of this research uses minimum spanning trees. They have been used to identify protein fibers in photographs of cells. The computer detects some points lying on the fibers and then determines a minimum spanning tree to deduce the overall structure of the fiber. Some studies have used minimum spanning trees to analyze cancerous tissue.

MINIMUM SPANNING TREE

A spanning tree that has minimum total weight is called a **minimum spanning tree** for the graph.

Kruskal's Algorithm

Fortunately for the district planners, there is a good algorithm for finding a minimum spanning tree for any connected, weighted graph. It is called **Kruskal's algorithm.** The algorithm is an efficient algorithm, because it does not take too much computer time for even very large graphs.

Minimum Spanning Trees in Astronomy The structure of the universe may hold clues about its formation. The image shows the large-scale structure of the Milky Way against a background of many other galaxies. Galaxies are not uniformly distributed throughout the universe. Instead they seem to be arranged in clusters, which are grouped together into superclusters, which themselves seem to be arranged in vast chains called filaments. Some astronomers are using spanning trees to help map the filamentary structure of the universe.

KRUSKAL'S ALGORITHM FOR FINDING A MINIMUM SPANNING TREE FOR ANY CONNECTED, WEIGHTED GRAPH

Choose edges for the spanning tree as follows.

Step 1 First edge: Choose any edge with minimum weight.

Step 2 Next edge: Choose any edge with minimum weight from *those not yet selected*. (At this stage, the subgraph may look disconnected.)

Step 3 Continue to choose edges of minimum weight from those not yet selected, except *do not select any edge that creates a circuit* in the subgraph.

Step 4 Repeat Step 3 until the subgraph connects all vertices of the original graph.

Because Kruskal's algorithm gives a subgraph that is connected, includes all vertices of the original graph, and has no circuits, it certainly provides a spanning tree for the graph. Also, the algorithm always will give us a *minimum* spanning tree for the graph. The explanation for this fact is beyond the scope of this text. (Algorithms such as Kruskal's often are called *greedy* algorithms.)

EXAMPLE 3 Applying Kruskal's Algorithm

Use Kruskal's algorithm to find a minimum spanning tree for the graph in **Figure 57.**

Solution

Note that the graph is connected, so Kruskal's algorithm applies. Choose an edge with minimum weight. ED has the least weight of all the edges, so it is selected first. We color the edge a different color to show that it has been selected. See **Figure 58.**

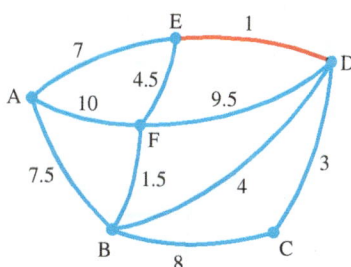

Figure 58

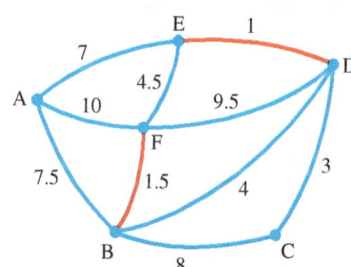

Figure 59

Of the remaining edges, BF has the least weight, so it is the second edge chosen, as shown in **Figure 59.** Note that it does not matter that at this stage the subgraph is not connected.

Continuing, we choose CD and then BD, as shown in **Figure 60** on the next page. Of the edges remaining, EF has the least weight. However, EF *would create a circuit* in the subgraph. We ignore EF and note that, of the edges remaining, AE has the least weight. Thus, AE is included in the spanning tree, as shown in **Figure 61** on the next page.

We now have a subgraph of the original that connects all vertices of the original graph into a tree. This subgraph is the required minimum spanning tree for the original graph. Its total weight is

$$1 + 1.5 + 3 + 4 + 7 = 16.5.$$

Thus, the district council can achieve its objective at a cost of 16.5 million dollars, by paving only the roads represented by edges in the minimum spanning tree.

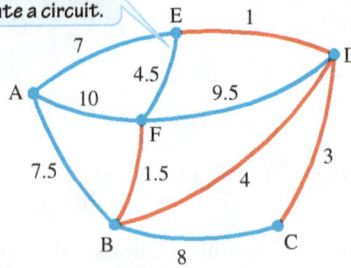

Figure 60

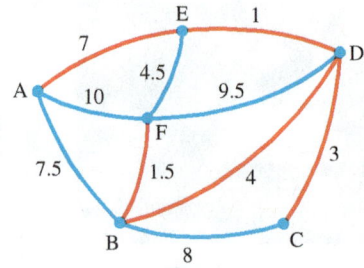

Figure 61

Number of Vertices and Edges

There is an interesting relationship between the number of vertices and the number of edges in a tree (sometimes called the **vertex/edge relationship**).

Weather data are complex, involving numerous readings of temperature, pressure, and wind velocity over huge areas. Some current research uses minimum spanning trees to help computers interpret these data. For example, minimum spanning trees are used to identify developing cold fronts.

> **NUMBER OF VERTICES AND EDGES IN A TREE**
>
> If a graph is a tree, then the number of edges in the graph is one less than the number of vertices.
>
> **A tree with n vertices has $n - 1$ edges.**

Consider the graphs shown in **Figure 62.** Information about these graphs is shown in **Table 2.** The three graphs in (a), (b), and (c) are not trees, and there is no uniform relation between the number of edges and the number of vertices for these graphs. In general, graphs may have more or fewer vertices than edges or an equal number of both. In contrast, the three graphs in (d), (e), and (f) are trees. The number of edges is indeed one less than the number of vertices for these trees.

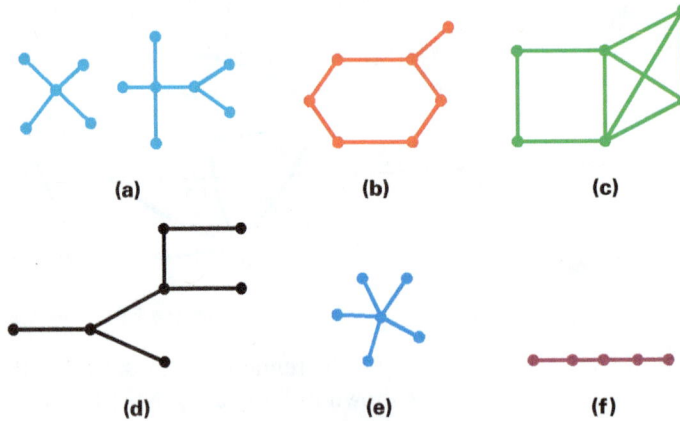

Figure 62

Table 2

Graph	Tree?	Number of Vertices	Number of Edges	
(a)	No	12	10	
(b)	No	7	7	
(c)	No	6	9	
(d)	Yes	7	6	$6 = 7 - 1$
(e)	Yes	6	5	$5 = 6 - 1$
(f)	Yes	5	4	$4 = 5 - 1$

If we consider only connected graphs, then the converse of the theorem on the previous page is also true.

For a connected graph, if the number of edges is one less than the number of vertices, then the graph is a tree.

EXAMPLE 4 **Using the Vertex/Edge Relationship**

A chemist has synthesized a new chemical compound. She knows from her analyses that a molecule of the compound contains 54 atoms and that the molecule has a tree-like structure. How many chemical bonds are there in the molecule?

Solution

Think of the atoms in the molecule as the vertices of a tree, and think of the bonds as edges of the tree. This tree has 54 vertices. Since the number of edges in a tree is one less than the number of vertices, the molecule must have 53 bonds.   ■

Problem-Solving Strategy

Trial and error is a good problem-solving strategy. Choose an arbitrary proposed solution to the problem, and check whether it works. If it does not work, we may gain insight into what to try next.

EXAMPLE 5 **Finding a Graph with Specified Properties**

Suppose a graph is a tree and has 15 vertices. What is the greatest number of vertices of degree 5 that this graph could have? Draw such a tree.

Solution

Because the graph is a tree with 15 vertices, it must have 14 edges. Recall that the sum of the degrees of the vertices of a graph is twice the number of edges. Therefore, the sum of the degrees of the vertices in the tree must be $2 \cdot 14$, which is 28.

Now we use trial and error. We want the greatest possible number of vertices of degree 5. Four vertices of degree 5 would contribute 20 to the total degree sum. This would leave $28 - 20 = 8$ as the degree sum of the remaining 11 vertices, but this would mean that some of those vertices would have no edges joined to them. The graph would not be connected and would not be a tree.

If we try 3 vertices of degree 5, this would contribute 15 to the total degree sum. It would leave $28 - 15 = 13$ as the degree sum for the remaining 12 vertices. We can draw a graph with these specifications, as in **Figure 63**. Thus, the greatest possible number of vertices of degree 5 in a tree with 15 vertices is 3.   ■

Figure 63

14.4   EXERCISES

Tree Identification   *In Exercises 1–7, determine whether the graph is a tree. If not, explain why it is not.*

1.

2.

3.

4.

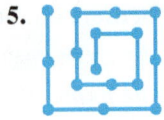

5.

6. a complete graph with 6 vertices

7. a connected graph with all vertices of even degree

Tree Identification *In Exercises 8–10, add edges (no vertices) to the graph to change the graph into a tree or explain why this is not possible. (Note: There may be more than one correct answer.)*

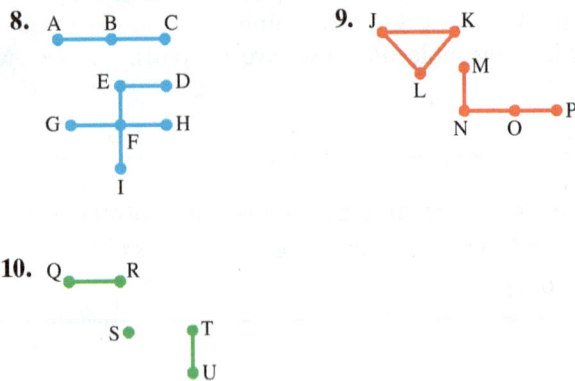

8.

9.

10.

Tree or Not a Tree? *In Exercises 11–13, determine whether the graph described must be a tree.*

11. *Spread of a Rumor* A sociologist is investigating the spread of a rumor. He finds that 18 people know of the rumor. One of these people must have started the rumor, and each must have heard the rumor for the first time from one of the others. He draws a graph with vertices representing the 18 people and edges showing from whom each person first heard the rumor.

12. *Internet Search Engine* You are using the Google search engine to search the Web for information on a topic for a paper. You start at the Google page and follow links you find, sometimes going back to a site you've already visited. To keep a record of the sites you visit, you show each site as a vertex of a graph. You draw an edge each time you connect for the *first* time to a *new* site.

13. *Tracking Ebola* A patient has contracted Ebola. An employee of the Centers for Disease Control is trying to quarantine all people who have had contact, over the past week, either with the patient or with someone already included in the contact network. The employee draws a graph with vertices representing the patient and all people who have had contact as described. The edges of the graph represent the relation "the two people have had contact."

Trees and Cut Edges *In Exercises 14–17, determine whether the statement is* true *or* false. *If the statement is false, draw an example to show that it is false.*

14. Every graph with no circuits is a tree.

15. Every connected graph in which each edge is a cut edge is a tree.

16. Every graph in which there is a path between each pair of vertices is a tree.

17. Every graph in which each edge is a cut edge is a tree.

Spanning Trees *In Exercises 18–20, find three different spanning trees for each graph. (There are many different correct answers.)*

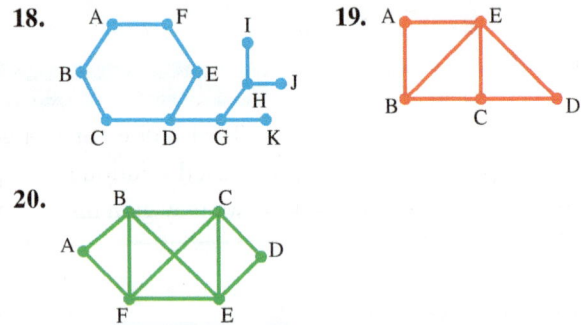

18.

19.

20.

Spanning Trees *In Exercises 21–23, find all spanning trees for the graph.*

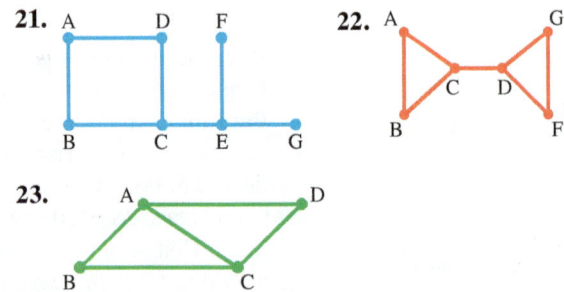

21.

22.

23.

Spanning Trees *In Exercises 24–26, determine how many spanning trees the graph has.*

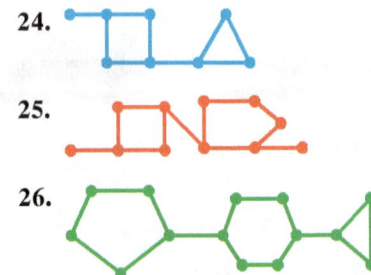

24.

25.

26.

27. What is a general principle about the number of spanning trees in graphs such as those in **Exercises 24–26?**

28. *Spanning Trees* Complete the parts of this exercise in order.

(a) Find all the spanning trees of each of the following graphs.

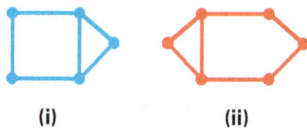

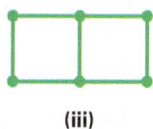

(i)   **(ii)**

(iii)

(b) For each of the following graphs, determine how many spanning trees the graph has.

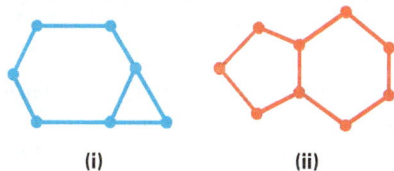

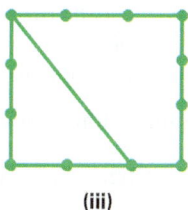

(i)   **(ii)**

(iii)

(c) What is a general principle about the number of spanning trees in graphs of the kind shown in (a) and (b)?

Minimum Spanning Trees *In Exercises 29–32, use Kruskal's algorithm to find a minimum spanning tree for the graph. Find the total weight of this minimum spanning tree.*

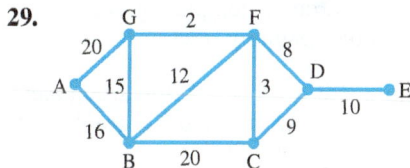

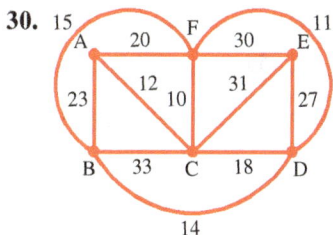

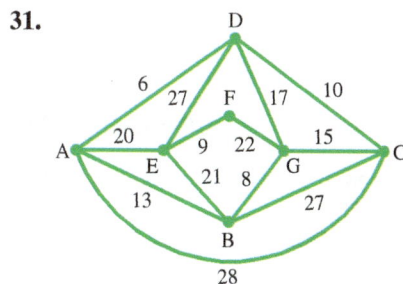

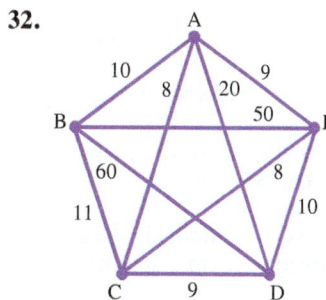

29.

30.

31.

32.

33. *School Building Layout* A school consists of 6 separate buildings, represented by the vertices in the graph.

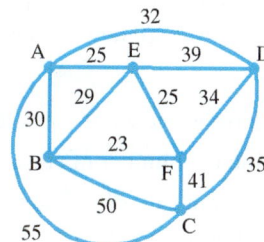

There are paths between some of the buildings as shown. The graph also shows the length in feet of each path. School administrators want to cover some of these paths with roofs so that students will be able to walk between buildings without getting wet when it rains. To minimize cost, they must select paths to be covered such that the total length to be covered is as small as possible.

Use Kruskal's algorithm to determine which paths to cover. Also determine the total length of pathways to be covered.

34. *Town Water Distribution* A town council is planning to provide town water to an area that previously relied on private wells. Water will be fed into the area at the point represented by the vertex labeled A in the graph on the next page. Water must be piped to each of five main distribution points, represented by the vertices labeled B through F in the graph. Town engineers have estimated the cost of laying the pipes to carry the water between each pair of points in millions of dollars, as indicated in the graph. They must now select which pipes should be laid, so that there is exactly one route for the water to be pumped from A to any one of the five distribution points (possibly via another distribution point), and they want to achieve this at minimum cost.

Use Kruskal's algorithm to decide which pipes should be laid. Find the total cost of laying the pipes.

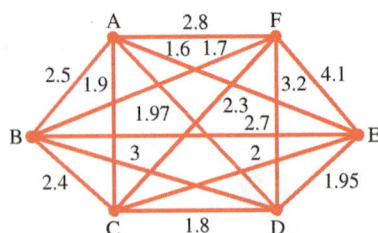

For Exercise 34

Trees, Edges, and Vertices *Work Exercises 35–42.*

35. How many edges are there in a tree with 34 vertices?

36. How many vertices are there in a tree with 40 edges?

37. How many edges are there in a spanning tree for a complete graph with 63 vertices?

38. A connected graph has 27 vertices and 43 edges. How many edges must be removed to form a spanning tree for the graph?

39. We have said that we can always find at least one spanning tree for a connected graph—and usually more than one. Could it happen that different spanning trees for the same graph have different numbers of edges? Justify your answer.

40. Suppose we have a tree with 9 vertices.

(a) Determine the number of edges in the graph.

(b) Determine the sum of the degrees of the vertices in the graph.

(c) Determine the least possible number of vertices of degree 1 in this graph.

(d) Determine the greatest possible number of vertices of degree 1 in this graph.

(e) Answer parts (a) through (d) for a tree with n vertices.

41. Suppose we have a tree with 10 vertices.

(a) Determine the number of edges in the graph.

(b) Determine the sum of the degrees of the vertices in the graph.

(c) Determine the least possible number of vertices of degree 4 in this graph.

(d) Determine the greatest possible number of vertices of degree 4 in this graph.

(e) Draw a graph to illustrate your answer in part (d).

42. Suppose we have a tree with 17 vertices.

(a) Determine the number of edges in the graph.

(b) Determine the sum of the degrees of the vertices in the graph.

(c) Determine the greatest possible number of vertices of degree 5 in this graph.

(d) Draw a graph to illustrate your answer in part (c).

43. *Computer Network Layout* A business has 23 employees, each with his or her own desk computer, all working in the same office. The managers want to network the computers. They need to install cables between individual computers so that every computer is linked into the network. Because of the way the office is laid out, it is not convenient to simply connect all the computers in one long line. Determine the least number of cables the managers need to install to achieve their objective.

44. *Design of a Garden* Maria has 12 vegetable and flower beds in her garden and wants to build flagstone paths between the beds so that she can get from each bed along flagstone paths to every other bed. She also wants a path linking her front door to one of the beds. Determine the minimum number of flagstone paths she must build to achieve this.

Tree or Not a Tree? *Exercises 45–47 refer to the following situation: We start with a tree and then draw in extra edges and vertices. For each, say whether it is possible to draw in the number of edges and vertices specified to end up with a connected graph. If it is possible, determine whether the resulting graph must be a tree, may be a tree, or cannot possibly be a tree. Justify each answer briefly.*

45. Draw in the same number of vertices as edges.

46. Draw in more vertices than edges.

47. Draw in more edges than vertices.

48. Starting with a tree, is it possible to draw in the same number of vertices as edges and end up with a disconnected graph? If so, give an example. If not, justify briefly.

Cayley's Theorem *Exercises 49–51 require the following theorem, which was proved by Cayley in 1889:* **A complete graph with n vertices has n^{n-2} spanning trees.**

49. How many spanning trees are there for a complete graph with 3 vertices? Draw a complete graph with 3 vertices, and find all the spanning trees.

50. How many spanning trees are there for a complete graph with 4 vertices? Draw a complete graph with 4 vertices, and find all the spanning trees.

51. How many spanning trees are there for a complete graph with 5 vertices?

Isomorphism *Work Exercises 52–55.*

52. Find all nonisomorphic trees with 4 vertices. How many are there?

53. Find all nonisomorphic trees with 5 vertices. How many are there?

54. Find all nonisomorphic trees with 6 vertices. How many are there?

55. Find all nonisomorphic trees with 7 vertices. How many are there?

56. *Vertex/Edge Relationship* In this exercise, we explore why the number of edges in a tree is one less than the number of vertices. Because the statement is clearly true for a tree with only one vertex, we will consider a tree *with more than one vertex*. Answer parts (a)–(h) in order.

(a) How many components does the tree have?

(b) Why must the tree have at least one edge?

(c) Remove one edge from the tree. How many components does the resulting graph have?

(d) You have not created any new circuits by removing the edge, so each of the components of the

resulting graph is a tree. If the remaining graph still has edges, choose any edge and remove it. (You have now removed 2 edges from the original tree.) Altogether, how many components remain?

(e) Repeat the procedure described in (d). If you remove 3 edges from the original tree, how many components remain? If you remove 4 edges from your original tree, how many components remain?

(f) Repeat the procedure in (d) until you have removed all the edges from the tree. If you have to remove *n* edges to achieve this, determine an expression involving *n* for the number of components remaining.

(g) What *are* the components that remain when you have removed all the edges from the tree?

(h) What can you conclude about the number of vertices in a tree with *n* edges?

CHAPTER 14 SUMMARY

KEY TERMS

14.1
vertex
edge
simple graph
degree of a vertex
isomorphic graphs
connected graph
disconnected graph
components of a graph
walk
path
circuit

weighted graph
weights
complete graph
subgraph
coloring
chromatic number
four-color problem
Bacon number

14.2
Königsberg bridge problem
Euler path

Euler circuit
cut edge
Fleury's algorithm
deadheading

14.3
Icosian game
Hamilton circuit
total weight
minimum Hamilton circuit
traveling salesman problem
brute force algorithm

efficient algorithms
approximate algorithms
nearest neighbor algorithm

14.4
connected graph
tree
spanning tree
minimum spanning tree
Kruskal's algorithm
vertex/edge relationship

NEW SYMBOLS

n! *n* factorial

A → B a walk from vertex A to vertex B

TEST YOUR WORD POWER

See how well you have learned the vocabulary in this chapter.

1. Two graphs are **isomorphic** if
  A. they have the same number of edges.
  B. they have the same number of vertices.
  C. there is a one-to-one matching between vertices, with the property that whenever there is an edge between two vertices of either one of the graphs, there is an edge between the corresponding vertices of the other graph.
  D. each graph has the same chromatic number.

2. A **circuit** in a graph is
  A. a path that begins and ends with the same vertex.
  B. a path that begins and ends with different vertices.
  C. a path whose edges are assigned weights.
  D. a path that is not a walk.

3. An **Euler circuit** in a graph
  A. is a path that uses every edge of the graph exactly once.
  B. is a circuit that uses every edge of the graph exactly once.
  C. is a circuit that is composed of cut edges.
  D. has at least one vertex of odd degree.

4. A **Hamilton circuit** in a graph

    A. is a circuit that uses every edge exactly once (returning to the starting edge to complete the circuit).

    B. is a circuit that visits each vertex exactly once (returning to the starting vertex to complete the circuit).

    C. cannot exist if there are an odd number of vertices.

    D. cannot exist if there are an even number of vertices.

5. A graph is a **tree** if

    A. it is disconnected and contains no circuits.

    B. it has more than one Euler circuit.

    C. it has more than one Hamilton circuit.

    D. it is connected and contains no circuits.

ANSWERS

1. C **2.** A **3.** B **4.** B **5.** D

QUICK REVIEW

Concepts	Examples

14.1 Basic Concepts

Graph

A **graph** is a collection of vertices (at least one) and edges. Each edge goes from a vertex to a vertex.

Isomorphic Graphs

Two graphs are **isomorphic** if there is a one-to-one matching between vertices of the two graphs, with the property that whenever there is an edge between two vertices of either one of the graphs, there is an edge between the corresponding vertices of the other graph.

These two graphs are isomorphic.

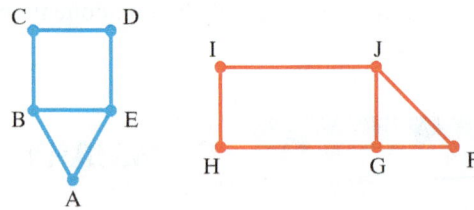

Connected and Disconnected Graphs

A graph is **connected** if we can move from each vertex of the graph to every other vertex of the graph *along edges of the graph*. If not, the graph is **disconnected.**

The connected pieces of a graph are the **components** of the graph.

The graph on the left is disconnected, as illustrated by the colors shown on the right.

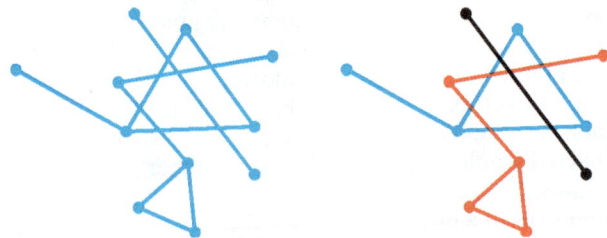

Sum-of-the-Degrees Theorem

In any graph, the sum of the degrees of the vertices equals twice the number of edges.

Consider each graph.

Number of edges = 7
Sum of degrees of vertices = 14

Number of edges = 6
Sum of degrees of vertices = 12

Walks, Paths, and Circuits

A **walk** in a graph is a sequence of vertices, each linked to the next vertex by a specified edge of the graph.

A **path** is a walk that uses no edge more than once.

A **circuit** is a path that begins and ends at the same vertex.

Consider the graph at the right.
$E \rightarrow C \rightarrow D \rightarrow E$ is not a walk (and thus is neither a path nor a circuit).

$B \rightarrow D \rightarrow E \rightarrow B \rightarrow C$ is a walk and a path, but not a circuit.

$A \rightarrow C \rightarrow D \rightarrow E \rightarrow B \rightarrow A$ is all three: walk, path, and circuit.

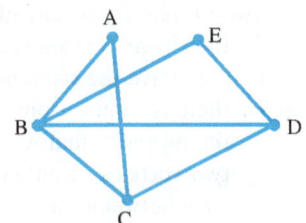

Complete Graph

A **complete graph** is a graph in which there is exactly one edge going from each vertex to each other vertex.

Concepts	*Examples*

Coloring and Chromatic Number

A **coloring** for a graph is an assignment of a color to each vertex in such a way that vertices joined by an edge have different colors.

The **chromatic number** of a graph is the least number of colors needed to achieve a coloring.

This graph has chromatic number 3.

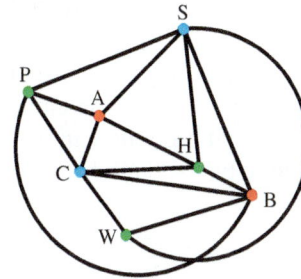

Coloring a Graph

Step 1 Choose a vertex with greatest degree, and color it. Use the same color to color as many vertices as you can without coloring two vertices the same color if they are joined by an edge.

Step 2 Choose a new color, and repeat what you did in Step 1 for vertices not already colored.

Step 3 Repeat Step 1 until all vertices are colored.

14.2 Euler Circuits and Route Planning

Euler Path and Euler Circuit

An **Euler path** in a graph is a path that uses every edge of the graph exactly once.

An **Euler circuit** in a graph is a circuit that uses every edge of the graph exactly once.

This graph has an Euler circuit, because all vertices have even degree.

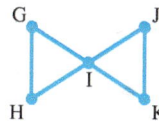

This graph has no Euler circuit, because some vertices have odd degree.

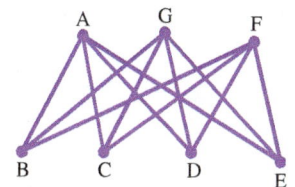

Euler's Theorem

Suppose we have a connected graph.

1. If the graph has an Euler circuit, then each vertex of the graph has even degree.

2. If each vertex of the graph has even degree, then the graph has an Euler circuit.

Cut Edge

A **cut edge** in a graph is an edge whose removal disconnects a component of the graph.

DE and HG are cut edges in this graph.

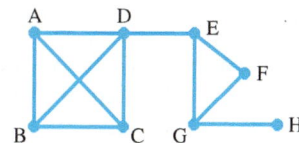

Fleury's Algorithm

Step 1 Start at any vertex. Go along any edge from this vertex to another vertex. *Remove this edge from the graph.*

Step 2 You are now on a vertex of the revised graph. Choose any edge from this vertex, subject to only one condition: Do not use a cut edge (*of the revised graph*) unless you have no other option. Go along your chosen edge. *Remove this edge from the graph.*

Step 3 Repeat Step 2 until you have used all the edges and returned to the vertex at which you started.

Fleury's algorithm can be used to show that

$$A \rightarrow C \rightarrow B \rightarrow F \rightarrow E \rightarrow D \rightarrow C \rightarrow F \rightarrow D \rightarrow A$$

is one of many Euler circuits for this graph.

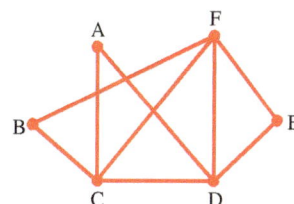

Concepts   **Examples**

14.3 Hamilton Circuits and Algorithms

Hamilton Circuit

A **Hamilton circuit** in a graph is a circuit that visits each vertex exactly once (returning to the starting vertex to complete the circuit). Any complete graph with three or more vertices has a Hamilton circuit. In this text, Hamilton circuits that differ *only* in their starting points are considered to be the same circuit.

Consider the following graph.

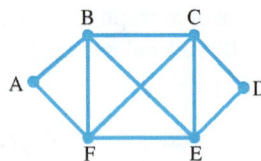

$$A \rightarrow B \rightarrow E \rightarrow D \rightarrow C \rightarrow F \rightarrow A$$

is a Hamilton circuit.

$$A \rightarrow B \rightarrow C \rightarrow D \rightarrow E \rightarrow F \rightarrow C \rightarrow E \rightarrow B \rightarrow F \rightarrow A$$

and

$$B \rightarrow C \rightarrow D \rightarrow E \rightarrow F \rightarrow B$$

are not Hamilton circuits.

Number of Hamilton Circuits in a Complete Graph

A complete graph with n vertices has the following number of Hamilton circuits.

$$(n - 1)!$$

Consider the following graph.

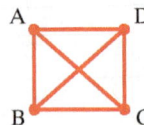

This complete graph has 4 vertices, and thus has

$$(4 - 1)! = 3! = 3 \cdot 2 \cdot 1 = 6 \text{ Hamilton circuits.}$$

Minimum Hamilton Circuit

In a weighted graph, a **minimum Hamilton circuit** is a Hamilton circuit with the least possible total weight.

Nearest Neighbor Algorithm

Step 1 Choose a starting point for the circuit. Call this vertex A.

Step 2 Check all the edges joined to A, and choose one that has the least weight. Proceed along this edge to the next vertex.

Step 3 At each vertex you reach, check the edges from there *to vertices not yet visited*. Choose one with the least weight. Proceed along this edge to the next vertex.

Step 4 Repeat Step 3 until you have visited all the vertices.

Step 5 Return to the starting vertex.

Consider the following graph.

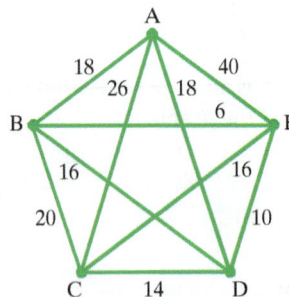

The nearest neighbor algorithm shows that an approximate minimum Hamilton circuit is

$$A \rightarrow D \rightarrow E \rightarrow B \rightarrow C \rightarrow A$$

and has weight 80.

Concepts	Examples

14.4 Trees and Minimum Spanning Trees

Connected Graph (Alternative Definition)
A **connected graph** is one in which there is *at least one path* between each pair of vertices.

Tree
A graph is a **tree** if the graph is *connected* and contains *no circuits.*

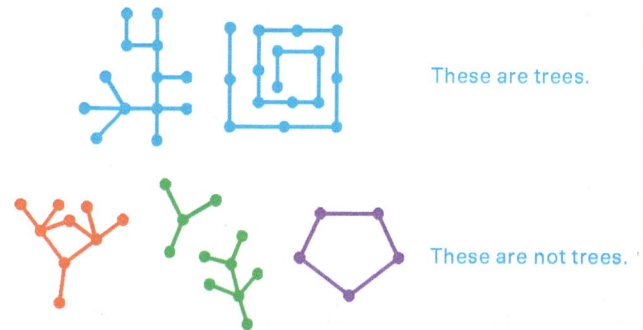

These are trees.

These are not trees.

Unique Path Property of Trees
In a tree, there is always **exactly one path** from each vertex in the graph to any other vertex in the graph.

If we choose any two vertices in the following graph, there is only one path from one to the other. See the red path as an illustration.

Spanning Tree and Minimum Spanning Tree
A **spanning tree** for a graph is a subgraph that includes every vertex of the original graph and is a tree.

A spanning tree that has minimum total weight is a **minimum spanning tree** for the graph.

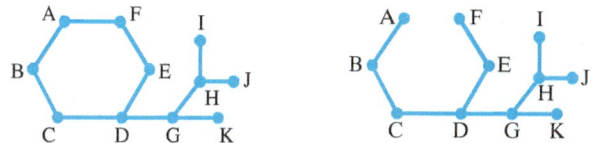

Given graph

Spanning tree
(There are others.)

Kruskal's Algorithm for Finding a Minimum Spanning Tree for Any Connected, Weighted Graph
Choose edges for the spanning tree as follows.

Step 1   First edge: Choose any edge with minimum weight.

Step 2   Next edge: Choose any edge with minimum weight from *those not yet selected.* (At this stage, the subgraph may look disconnected.)

Step 3   Continue to choose edges of minimum weight from those not yet selected, except *do not select any edge that creates a circuit in the subgraph.*

Step 4   Repeat Step 3 until the subgraph connects all vertices of the original graph.

Consider the following graph.

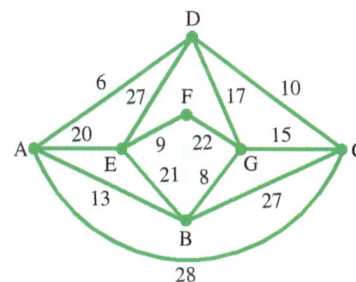

Kruskal's algorithm shows that this graph has the minimum spanning tree below, with total weight 66.

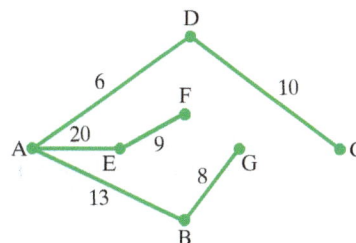

Number of Vertices and Edges in a Tree
If a graph is a tree, then the number of edges in the graph is one less than the number of vertices.

A tree with n vertices has $n - 1$ edges.

In the tree above, there are 7 vertices and

$$7 - 1 = 6 \text{ edges.}$$

CHAPTER 14 | TEST

Basic Concepts *In Exercises 1–5, refer to the following graph.*

1. Determine how many vertices the graph has.

2. Determine the sum of the degrees of the vertices.

3. Determine how many edges the graph has.

4. Which of the following are paths in the graph? If not, why not?
   (a) $B \to A \to C \to E \to B \to A$
   (b) $A \to B \to E \to A$
   (c) $A \to C \to D \to E$

5. Which of the following are circuits in the graph? If not, why not?
   (a) $A \to B \to E \to D \to A$
   (b) $A \to B \to C \to D \to E \to F \to G \to A$
   (c) $A \to B \to E \to F \to G \to E \to D \to A \to E \to C \to A$

6. **Components** Draw a graph that has 2 components.

7. **Sum of Degrees** A graph has 10 vertices, 3 of degree 4, and the rest of degree 2. Use the theorem that relates the sum of degrees to the number of edges to determine the number of edges in the graph (without drawing the graph).

8. **Isomorphism** Determine whether graphs (a) and (b) are isomorphic. If they are, justify this by labeling corresponding vertices of the two graphs with the same letters and color-coding the corresponding edges.

(a)   (b)

9. **Planning for Dinner** Julia is planning to invite some friends for dinner. She plans to invite John, Adam, Bill, Tina, Nicole, and Rita. John, Nicole, and Tina all know each other. Adam knows John and Tina. Bill also knows Tina, and he knows Rita. Draw a graph with vertices representing the six friends whom Julia plans to invite to dinner and edges representing the relationship "know each other." Is your graph connected or disconnected? Which of the guests knows the greatest number of other guests?

10. **Chess Competition** There are 8 contestants in a chess competition. Each contestant plays one chess game against every other contestant. The winner is the contestant who wins the most games. How many chess games will be played in the competition?

11. **Complete or Not Complete?** Is the accompanying graph a complete graph? Justify your answer.

Coloring and Chromatic Number *Color each graph using as few colors as possible. Use the result to determine the chromatic number of the graph.*

12.

13.

14. **Scheduling Exams** A teacher at a high school must schedule exams for the senior class. Subjects in which there are exams are listed on the left. Several students must take more than one of the exams, as shown on the right. Exams in subjects with students in common must not be scheduled at the same time. Use graph coloring to determine the least number of exam times needed and to identify exams that can be given at the same times.

Exam	Students taking this must also take
History	English, Mathematics, Biology, Geography, Psychology
English	Mathematics, Psychology, History
Mathematics	Chemistry, Biology, History, English
Chemistry	Biology, Mathematics
Biology	Mathematics, Chemistry, History
Psychology	English, History, Geography
Geography	Psychology, History

15. **Euler Circuits** Refer to the graph for **Exercises 1–5.** Which of the following are Euler circuits for the graph? If not, why not?
   (a) $A \to B \to E \to D \to A$
   (b) $A \to B \to C \to D \to E \to F \to G \to A$
   (c) $A \to B \to E \to F \to G \to E \to D \to A \to E \to C \to A$

Euler's Theorem *In Exercises 16 and 17, use Euler's theorem to decide whether the graph has an Euler circuit. (Do not actually find an Euler circuit.) Justify your answer briefly.*

16.

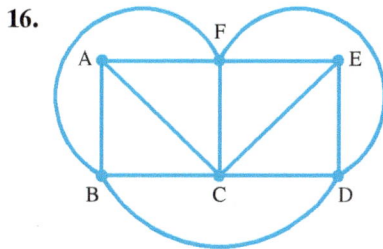

17.

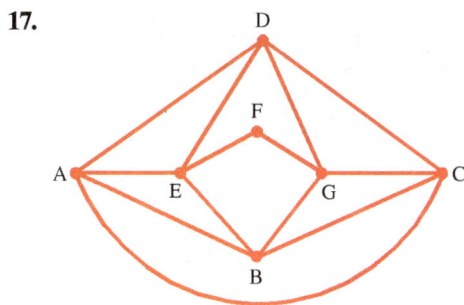

18. Floor Plan The floor plan of a building is shown. Is it possible to start outside, walk through each door exactly once, and end up back outside? Justify your answer.

19. Fleury's Algorithm Use Fleury's algorithm to find an Euler circuit for the accompanying graph, beginning

$$F \rightarrow B.$$

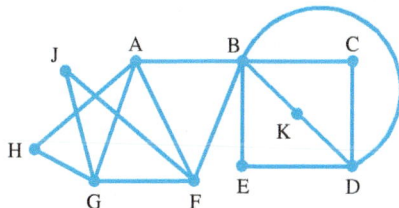

20. Hamilton Circuits Refer to the graph for **Exercises 1–5.** Which of the following are Hamilton circuits for the graph? If not, why not?

(a) $A \rightarrow B \rightarrow E \rightarrow D \rightarrow A$

(b) $A \rightarrow B \rightarrow C \rightarrow D \rightarrow E \rightarrow F \rightarrow G \rightarrow A$

(c) $A \rightarrow B \rightarrow E \rightarrow F \rightarrow G \rightarrow E \rightarrow D \rightarrow A \rightarrow$
$E \rightarrow C \rightarrow A$

21. Hamilton Circuits List all Hamilton circuits in the graph that start $F \rightarrow G$. How many are there?

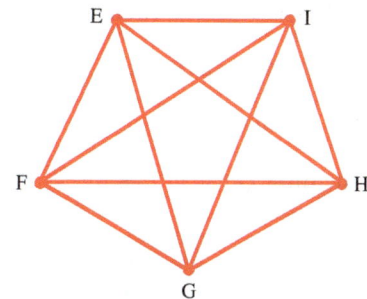

22. Brute Force Algorithm Use the brute force algorithm to find a minimum Hamilton circuit for the accompanying graph. Determine the total weight of the minimum Hamilton circuit.

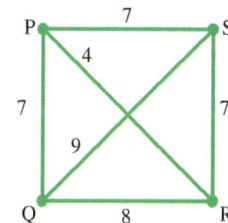

23. Nearest Neighbor Algorithm Use the nearest neighbor algorithm starting at A to find an approximate solution to the problem of finding a minimum Hamilton circuit for the graph below. Find the total weight of this circuit.

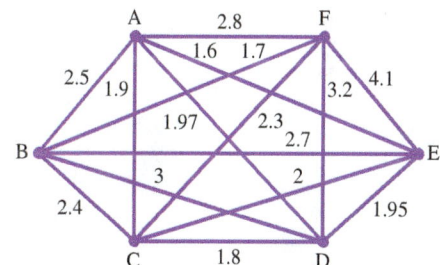

24. Hamilton Circuits How many Hamilton circuits are there in a complete graph with 25 vertices? (Leave your answer as a factorial.)

25. Rock Band Tour Plan The agent for a rock band based in Milwaukee is planning a tour for the band. He plans for the band to visit Minneapolis, Santa Barbara, Orlando, Phoenix, and St. Louis. Since the band members want to minimize the time they are away from home, they do not want to go to any of these cities more than once on the tour. Consider the complete graph with vertices representing the 6 cities mentioned. Does the problem of planning the tour require an Euler circuit, a Hamilton circuit, or neither for its solution?

26. Nonisomorphic Trees Draw three nonisomorphic trees with 7 vertices.

Trees *In Exercises 27–29, decide whether the statement is true or false.*

27. Every tree has a Hamilton circuit.

28. In a tree, each edge is a cut edge.

29. Every tree is connected.

30. Find all the different spanning trees for the following graph.

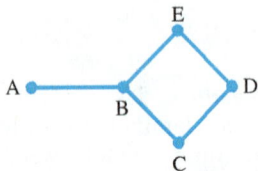

31. *Kruskal's Algorithm* Use Kruskal's algorithm to find a minimum spanning tree for the following graph.

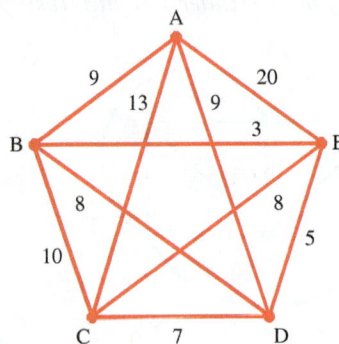

32. *Edges and Vertices* Determine the number of edges in a tree with 50 vertices.

ANSWERS TO SELECTED EXERCISES

CHAPTER 1 THE ART OF PROBLEM SOLVING

1.1 Exercises (pages 6–8)

1. inductive   **3.** deductive   **5.** deductive   **7.** inductive
9. deductive   **11.** deductive   **13.** inductive
15. inductive   **17.** Answers will vary.   **19.** 21   **21.** 3072
23. 63   **25.** $\frac{11}{12}$ **27.** 216   **29.** 52   **31.** 5
33. One such list is 10, 20, 30, 40, 50, ...
35. $(98,765 \times 9) + 3 = 888,888$
37. $3367 \times 15 = 50,505$
39. $33,334 \times 33,334 = 1,111,155,556$
41. $3 + 6 + 9 + 12 + 15 = \frac{15(6)}{2}$
43. $5(6) + 5(36) + 5(216) + 5(1296) + 5(7776)$
$= 6(7776 - 1)$
45. $\frac{1}{2} + \frac{1}{4} + \frac{1}{8} + \frac{1}{16} + \frac{1}{32} = 1 - \frac{1}{32}$ **47.** 20,100   **49.** 320,400
51. 15,400   **53.** 2550   **55.** 1 (These are the numbers of
chimes a clock rings, starting with 12 o'clock, if it rings the
number of hours on the hour, and 1 chime on the half-hour.)
57. (a) Answers will vary.   **(b)** The middle digit is
always 9, and the sum of the first and third digits is always
9 (considering 0 as the first digit if the difference has only
two digits).   **(c)** Answers will vary.
59. 142,857; 285,714; 428,571; 571,428; 714,285; 857,142.
Each result consists of the same six digits, but in a different
order. $142,857 \times 7 = 999,999$

1.2 Exercises (pages 15–19)

1. arithmetic; 56   **3.** geometric; 1215   **5.** neither
7. geometric; 8   **9.** neither   **11.** arithmetic; 22
13. 79   **15.** 450   **17.** 4032   **19.** 32,758   **21.** 57; 99
23. $(4321 \times 9) - 1 = 38,888$ **25.** $999,999 \times 4 = 3,999,996$
27. $21^2 - 15^2 = 6^3$ **29.** $5^2 - 4^2 = 5 + 4$
31. $1 + 5 + 9 + 13 = 4 \times 7$ **33.** 45,150   **35.** 228,150
37. 2601   **39.** 250,000   **41.** $S = n(n + 1)$
43. Answers will vary.   **45.** *row 1:* 28, 36; *row 2:* 36,
49, 64; *row 3:* 35, 51, 70, 92; *row 4:* 28, 45, 66, 91, 120;
row 5: 18, 34, 55, 81, 112, 148; *row 6:* 8, 21, 40, 65, 96,
133, 176
47. $8(1) + 1 = 9 = 3^2$; $8(3) + 1 = 25 = 5^2$;
$8(6) + 1 = 49 = 7^2$; $8(10) + 1 = 81 = 9^2$
49. The pattern is 1, 0, 1, 0, 1, 0, ...

51.

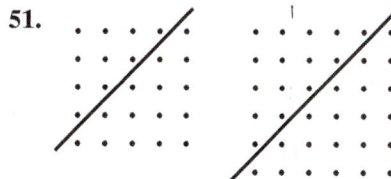

53. 256   **55.** 117   **57.** 235   **59.** $N_n = \frac{n(7n - 5)}{2}$
61. a square number   **63.** a perfect cube   **65.** 42
67. 419   **69.** $\frac{101}{2}$ **71.** 2048   **73.** $\frac{1}{2048}$ **75.** $\frac{5}{2048}$ **77.** 495
79. To find any entry within the body of the triangle, add
the two entries immediately to the left and to the right in the
row above it. For example, in the sixth row, $10 = 4 + 6$.
The next three rows of Pascal's triangle are as follows.

$$1 \quad 6 \quad 15 \quad 20 \quad 15 \quad 6 \quad 1$$
$$1 \quad 7 \quad 21 \quad 35 \quad 35 \quad 21 \quad 7 \quad 1$$
$$1 \quad 8 \quad 28 \quad 56 \quad 70 \quad 56 \quad 28 \quad 8 \quad 1$$

81. The sums along the diagonals are 1, 1, 2, 3, 5, 8. These
are the first six terms of the Fibonacci sequence.

1.3 Exercises (pages 25–30)

1. 50   **3.**

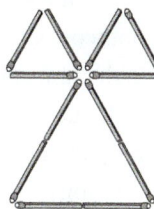

5. IN   **7.** 41   **9.** 15
11. 32   **13.** 0 (The product of the first and last digits is
the two-digit number between them.)   **15.** 11
17. infinitely many (Any line through the center of the
square will do this.)   **19.** 48.5 in.   **21.** 42   **23.** 6
25. If you multiply the two digits in the numbers in the
first row, you will get the second row of numbers. The
second row of numbers is a pattern of two numbers
(8 and 24) repeating.   **27.** 59   **29.** A   **31.** You should
choose a sock from the box labeled *red and green socks*.
Because it is mislabeled, it contains only red socks or only
green socks, determined by the sock you choose. If the
sock is green, relabel this box *green socks*. Since the other
two boxes were mislabeled, switch the remaining label to
the other box and place the label that says *red and green
socks* on the unlabeled box. No other choice guarantees
a correct relabeling because you can remove only one
sock.   **33.** D

35. One example of a solution follows.

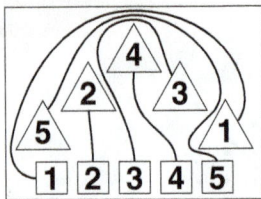

37. Here is one solution.   **39.** D   **41.** $\frac{1}{3}$

43. 90   **45.** 55 mph   **47.** 07   **49.** 437   **51.** 3
53. 21 stamps (5 five-cent stamps and 16 eight-cent stamps)   **55.** 3 socks   **57.** 6   **59.** the nineteenth day
61. 1967   **63.** Eve has $5, and Adam has $7.

65.
$$\begin{array}{r} 4\ 0\ 2 \\ \times\quad 3\ 9 \\ \hline 1\ 5,\ 6\ 7\ 8 \end{array}$$

67.

6	12	7	9
1	15	4	14
11	5	10	8
16	2	13	3

69. 25 pitches (The visiting team's pitcher retires 24 consecutive batters through the first eight innings, using only one pitch per batter. His team does not score either. Going into the bottom of the ninth tied 0–0, the first batter for the home team hits his first pitch for a home run. The pitcher threw 25 pitches and loses the game by a score of 1–0.)   **71.** Q   **73.** Here is one solution.

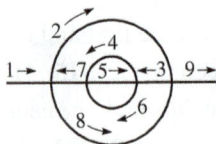

75. 6 ways   **77.** 86 cm   **79.** Dan (36) is married to Jessica (29); James (30) is married to Cathy (31).

81. 6

One of the
possibilities

1.4 Exercises (pages 39–41)

1. 43.8   **3.** 2.3589   **5.** 7.48   **7.** 7.1289   **9.** 6340.338097
11. 1   **13.** 1.061858759   **15.** 2.221441469
17. 3.141592653   **19.** 0.7782717162   **21.** yes
23. positive   **25.** 1   **27.** the same as   **29.** 0
31. negative   **33.** Answers will vary.

35. Answers will vary.   **37.** 63   **39.** 14   **41.** B   **43.** A
45. D   **47.** 6%   **49.** about 4.5 million
51. 2007, 2008, 2009, 2010   **53.** 2004: about 171 billion lb;
2010: about 193 billion lb   **55.** 2009 to 2010; about
1.5 million   **57.** The number of cars imported each year
was decreasing.

Chapter 1 Test (pages 45–46)

1. inductive   **2.** deductive   **3.** 31
4. $65,359,477,124,183 \times 68 = 4,444,444,444,444,444$
5. 351   **6.** 31,375   **7.** 65; $65 = 1 + 7 + 13 + 19 + 25$
8. 1, 8, 21, 40, 65, 96, 133, 176; The pattern is 1, 0, 1, 0, 1, 0,
1, 0, ...   **9.** The first two terms are both 1. Each term
after the second is found by adding the two previous
terms. The next term is 34.   **10.** $\frac{1}{4}$ **11.** 9   **12.** 35
13. Answers will vary. One possible solution is
$1 + 2 + 3 - 4 + 5 + 6 + 78 + 9 + 0 = 100$.
14. 108 in., or 9 ft   **15.** 3   **16.** The sum of the digits is
always 9.   **17.** 9.907572861 (Answers may vary due to the
model of calculator used.)   **18.** 34.328125   **19.** B
20. (a) between 2003 and 2004, 2004 and 2005, and 2005
and 2006   **(b)** The unemployment rate was increasing.
(c) 2008: 5.8%; 2009: 9.3%; increase: 3.5%

CHAPTER 2 THE BASIC CONCEPTS OF SET THEORY

2.1 Exercises (pages 52–54)

1. F   **3.** E   **5.** B   **7.** H   **9.** $\{1, 2, 3, 4, 5, 6\}$
11. $\{0, 1, 2, 3, 4\}$ **13.** $\{6, 7, 8, 9, 10, 11, 12, 13, 14\}$
15. $\{2, 4, 8, 16, 32, 64, 128, 256\}$ **17.** $\{0, 2, 4, 6, 8, 10\}$
19. $\{220, 240, 260, \dots\}$ **21.** $\{$Lake Erie, Lake Huron,
Lake Michigan, Lake Ontario, Lake Superior$\}$
23. $\left\{1, \frac{1}{2}, \frac{1}{3}, \frac{1}{4}, \frac{1}{5}, \dots\right\}$ **25.** $\{(0, 5), (3, 4), (4, 3), (5, 0)\}$
**In Exercises 27 and 29, there are other ways to describe
the sets.**   **27.** $\{x \mid x$ is a rational number$\}$ **29.** $\{x \mid x$ is an
odd natural number less than 76$\}$ **31.** the set of single-
digit integers   **33.** the set of states of the United States
35. finite   **37.** infinite   **39.** infinite   **41.** infinite   **43.** 8
45. 500   **47.** 26   **49.** 39   **51.** 28   **53.** Answers will vary.
55. well defined   **57.** not well defined   **59.** $\notin$ **61.** $\in$
63. $\in$ **65.** false   **67.** false   **69.** true   **71.** true
73. false   **75.** true   **77.** true   **79.** true
81. false   **83.** true   **85.** Answers will vary.
87. $\{2\}$ and $\{3, 4\}$ (Other examples are possible.)
89. $\{a, b\}$ and $\{a, c\}$ (Other examples are possible.)

91. (a) {Bernice, Heather, Marcy}, {Bernice, Heather, Natalie}, {Bernice, Susan, Marcy}, {Bernice, Susan, Natalie}, {Heather, Susan, Marcy}, {Heather, Susan, Natalie} **(b)** {Bernice, Marcy, Natalie}, {Heather, Marcy, Natalie}, {Susan, Marcy, Natalie} **(c)** {Bernice, Heather, Susan}

2.2 Exercises (pages 58–60)

1. D   **3.** B   **5.** $\not\subseteq$ **7.** $\subseteq$ **9.** $\subseteq$ **11.** $\not\subseteq$ **13.** both
15. $\subseteq$ **17.** both   **19.** neither   **21.** true   **23.** false
25. true   **27.** false   **29.** true   **31.** false   **33.** true
35. false   **37. (a)** 64   **(b)** 63   **39. (a)** 32   **(b)** 31   **41.** $\varnothing$
43. {5, 7, 9, 10}   **45.** {Higher cost, Lower cost, Educational, More time to see the sights in California, Less time to see the sights in California, Cannot visit friends along the way, Can visit friends along the way}
47. {Higher cost, More time to see the sights in California, Cannot visit friends along the way}
49. $\varnothing$ **51.** {A, B, C, D, E} (All are present.)
53. {A, B, C}, {A, B, D}, {A, B, E}, {A, C, D}, {A, C, E}, {A, D, E}, {B, C, D}, {B, C, E}, {B, D, E}, {C, D, E}   **55.** {A}, {B}, {C}, {D}, {E}   **57.** 32
59. $2^{25} - 1 = 33{,}554{,}431$ **61. (a)** 15   **(b)** 16; It is now possible to select *no* bills.
63. (a) s **(b)** s **(c)** $2s$ **(d)** Adding one more element will always double the number of subsets, so the expression 2^n is true in general.

2.3 Exercises (pages 69–71)

1. B   **3.** A   **5.** E   **7.** {a, c}   **9.** {a, b, c, d, e, f}
11. {b, d, f}   **13.** {d, f}   **15.** {a, b, c, e, g}   **17.** {e, g}
19. {a}   **21.** {e, g}   **23.** {d, f}
In Exercises 25–28, there may be other acceptable descriptions.   **25.** the set of all elements that either are in A, or are not in B and not in C **27.** the set of all elements that are in C but not in B, or are in A **29.** {e, h, c, l, b}
31. {e, h, c, l, b}   **33.** the set of all tax returns filed in 2014 without itemized deductions   **35.** the set of all tax returns with itemized deductions or showing business income, but not selected for audit   **37.** always true   **39.** not always true
41. (a) {1, 3, 5, 2}   **(b)** {1, 2, 3, 5}   **(c)** For any sets X and Y, $X \cup Y = Y \cup X$.   **43. (a)** {1, 3, 5, 2, 4}
(b) {1, 3, 5, 2, 4}   **(c)** For any sets X, Y, and Z, $X \cup (Y \cup Z) = (X \cup Y) \cup Z$.   **45.** true   **47.** true
49. $A \times B = \{(d, p), (d, i), (d, g), (o, p), (o, i), (o, g), (g, p), (g, i), (g, g)\}; B \times A = \{(p, d), (p, o), (p, g), (i, d), (i, o), (i, g), (g, d), (g, o), (g, g)\}$
51. $n(A \times B) = 210; n(B \times A) = 210$ **53.** 6

55.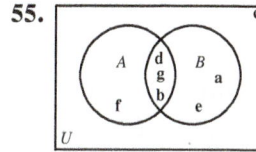
57.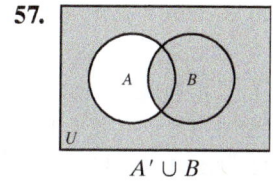
$A' \cup B$

59.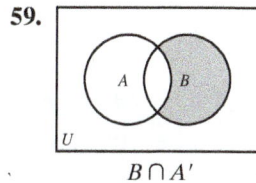
$B \cap A'$
61.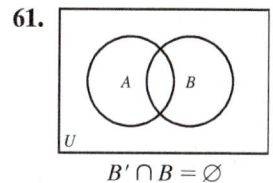
$B' \cap B = \varnothing$

63.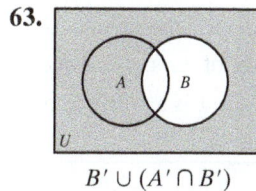
$B' \cup (A' \cap B')$
65.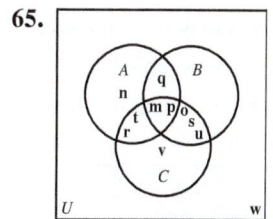

67.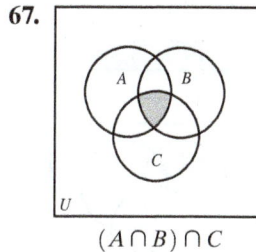
$(A \cap B) \cap C$
69.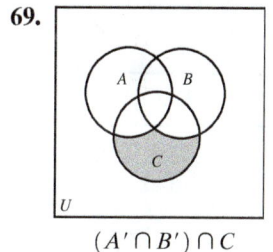
$(A' \cap B') \cap C$

71.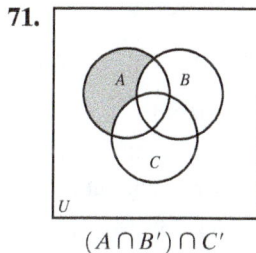
$(A \cap B') \cap C'$

73. $A' \cap B'$, or $(A \cup B)'$
75. $(A \cup B) \cap (A \cap B)'$, or $(A \cup B) - (A \cap B)$, or $(A - B) \cup (B - A)$
77. $(A \cap B) \cup (A \cap C)$, or $A \cap (B \cup C)$

79. $A \cap B = \varnothing$ **81.** This statement is true for any set A.
83. $B \subseteq A$ **85.** always true   **87.** not always true
89. Answers will vary.

2.4 Exercises (pages 75–78)

1. (a) 5   **(b)** 7   **(c)** 0   **(d)** 2   **(e)** 8   **3. (a)** 1   **(b)** 3
(c) 4   **(d)** 0   **(e)** 2   **(f)** 8   **(g)** 2   **(h)** 6   **5.** 21
7. 7   **9.** 35

11.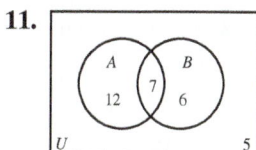
13.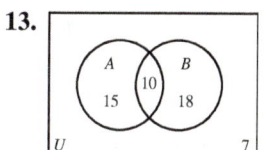

15.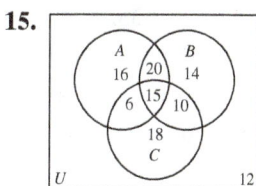
17.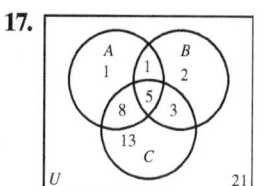

19. (a) 2 **(b)** 4 **21. (a)** 25 **(b)** 30 **(c)** 2 **(d)** 10
(e) 19 **(f)** 57 **23. (a)** 8 **(b)** 26 **(c)** 60 **25. (a)** 500
(b) 91 **27. (a)** 31 **(b)** 24 **(c)** 11 **(d)** 45 **29. (a)** 1
(b) 1, 2, 3, 4, 5, 6, 7, 8, 9, 10, 11, 12, 13, 14, 15 **(c)** 1, 2, 3, 4,
5, 9, 11 **(d)** 5, 8, 13 **31. (a)** 9 **(b)** 9 **(c)** 20 **(d)** 20
(e) 27 **(f)** 15 **33.** Answers will vary.

Chapter 2 Test *(page 82)*

1. {a, b, c, d, e} **2.** {a, b, d} **3.** {c, f, g, h} **4.** {a, c}
5. false **6.** true **7.** true **8.** true **9.** false **10.** false
11. true **12.** true **13.** 16 **14.** 15
Answers may vary in Exercises 15–18. **15.** the set of odd
integers between −4 and 10 **16.** the set of days of the week
17. $\{x \mid x \text{ is a negative integer}\}$ **18.** $\{x \mid x \text{ is a multiple of 8}$
between 20 and 90} **19.** ⊆ **20.** neither
21.

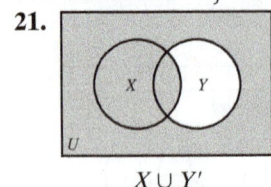

$X \cup Y'$

22.

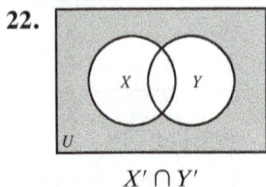

$X' \cap Y'$

23.

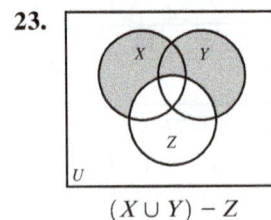

$(X \cup Y) - Z$

24.

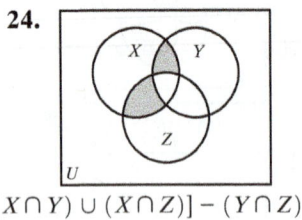

$[(X \cap Y) \cup (X \cap Z)] - (Y \cap Z)$

25. Answers will vary. **26.** {Electric razor}
27. {Adding machine, Baking powder, Pendulum clock,
Thermometer} **28.** {Telegraph, Zipper} **29. (a)** 22
(b) 12 **(c)** 28 **30. (a)** 16 **(b)** 32 **(c)** 33 **(d)** 45
(e) 14 **(f)** 26

CHAPTER 3 INTRODUCTION TO LOGIC

3.1 Exercises *(pages 88–90)*

1. statement **3.** not a statement **5.** statement
7. statement **9.** statement **11.** not a statement
13. statement **15.** compound **17.** not compound
19. not compound **21.** compound **23.** Her aunt's name
is not Hermione. **25.** No book is longer than this book.
27. At least one computer repairman can play blackjack.
29. Someone does not love somebody sometime.
31. The trash does not need to be collected. **33.** $x \leq 12$
35. $x < 5$ **37.** Answers will vary. **39.** She does not have
green eyes. **41.** She has green eyes and he is 60 years old.
43. She does not have green eyes or he is 60 years old.
45. She does not have green eyes or he is not 60 years old.

47. It is not the case that she does not have green eyes and
he is 60 years old. **49.** $p \wedge {\sim}q$ **51.** ${\sim}p \vee q$
53. ${\sim}(p \vee q)$ or, equivalently, ${\sim}p \wedge {\sim}q$
55. Answers will vary. **57.** C **59.** A, B **61.** A, C
63. B **65.** true **67.** false **69.** true **71.** true
73. false **75.** Answers will vary. **77.** Every person here
has made mistakes before. **79.** $(\forall c)({\sim}f)$

3.2 Exercises *(pages 100–101)*

1. false **3.** true **5.** true **7.** They must both be false.
9. T **11.** T **13.** F **15.** T **17.** T **19.** T **21.** It is a
disjunction, because it means "6 > 2 or 6 = 2." **23.** T
25. F **27.** T **29.** T **31.** F **33.** F **35.** T **37.** T
39. T **41.** 4 **43.** 16 **45.** 128 **47.** six **49.** FFTF
51. FTTT **53.** TTTT **55.** TFFF **57.** FFFFTFFF
59. FTFTTTTT **61.** TTTTTTTTTTTTFTTT
63. You can't pay me now and you can't pay me later.
65. It is not summer or there is snow. **67.** I did not say
yes or she did not say no. **69.** $6 - 1 \neq 5$ or $9 + 13 = 7$
71. Neither Prancer nor Vixen will lead Santa's reindeer
sleigh next Christmas. **73.** T **75.** F

77.

p	q	$p \veebar q$
T	T	F
T	F	T
F	T	T
F	F	F

79. F **81.** T

83. The lady is behind Door 2. *Reasoning:* Suppose that the
sign on Door 1 is true. Then the sign on Door 2 would also
be true, but this is impossible. So the sign on Door 2 must be
true, and the sign on Door 1 must be false. Because the sign
on Door 1 says the lady is in Room 1, and this is false, the
lady must be behind Door 2. **85.** The negation of a conjuc-
tion is equivalent to the disjunction of the negations.

3.3 Exercises *(pages 109–111)*

1. If you just believe, then you can do it. **3.** If it is an even
integer divisible by 5, then it is divisible by 10. **5.** If it is a
grizzly bear, then it does not live in California. **7.** If they
are surfers, then they can't stay away from the beach.
9. true **11.** true **13.** false **15.** true **17.** Answers will
vary. **19.** F **21.** T **23.** T **25.** If they do not collect
classics, then he fixes cars. **27.** If she sings for a living,
then they collect classics and he fixes cars. **29.** If he does
not fix cars, then they do not collect classics or she sings
for a living. **31.** $b \to p$ **33.** $p \wedge (s \to {\sim}b)$ **35.** $p \to s$
37. T **39.** T **41.** F **43.** T **45.** T **47.** Answers will
vary. **49.** TTTF **51.** TTFT **53.** TTTT; tautology
55. TTTTTTFT **57.** TTTFTTTTTTTTTTTT **59.** one

61. That is an authentic Coach bag and I am not surprised.
63. The bullfighter doesn't get going and doesn't get gored.
65. You want to be happy for the rest of your life and you make a pretty woman your wife.   **67.** You do not give your plants tender, loving care or they flourish.   **69.** She does or he will.   **71.** The person is not a resident of Pensacola or is a resident of Florida.   **73.** equivalent   **75.** equivalent
77. equivalent   **79.** equivalent   **81.** $(p \wedge q) \vee (p \wedge \sim q)$;
The statement simplifies to p.   **83.** $p \vee (\sim q \wedge r)$
85. $\sim p \vee (p \vee q)$; The statement simplifies to T.

87. The statement simplifies to $p \wedge q$.

89. The statement simplifies to F.

91. The statement simplifies to $(r \wedge \sim p) \wedge q$.

93. The statement simplifies to $p \vee q$.

95. $525.60   **97.** The circuit is equivalent to
$\sim(A \wedge A) \vee B \equiv \sim A \vee B \equiv A \rightarrow B$.

3.4 Exercises (pages 116–117)

1. (a) If you were an hour, then beauty would be a minute.
(b) If beauty were not a minute, then you would not be an hour.   **(c)** If you were not an hour, then beauty would not be a minute.
3. (a) If you don't fix it, then it ain't broke.
(b) If it's broke, then fix it.   **(c)** If you fix it, then it's broke.   **5. (a)** If it is dangerous to your health, then you walk in front of a moving car.   **(b)** If you do not walk in front of a moving car, then it is not dangerous to your health.   **(c)** If it is not dangerous to your health, then you do not walk in front of a moving car.   **7. (a)** If they flock together, then they are birds of a feather.   **(b)** If they are not birds of a feather, then they do not flock together.
(c) If they do not flock together, then they are not birds of a feather.   **9. (a)** If he comes, then you built it.   **(b)** If you don't build it, then he won't come.   **(c)** If he doesn't come, then you didn't build it.
11. (a) $\sim q \rightarrow p$ **(b)** $\sim p \rightarrow q$ **(c)** $q \rightarrow \sim p$
13. (a) $\sim q \rightarrow \sim p$ **(b)** $p \rightarrow q$ **(c)** $q \rightarrow p$
15. (a) $(q \vee r) \rightarrow p$ **(b)** $\sim p \rightarrow (\sim q \wedge \sim r)$
(c) $(\sim q \wedge \sim r) \rightarrow \sim p$ **17.** Answers will vary.
19. If the Kings go to the playoffs, then pigs will fly.
21. If it has legs of 3 and 4, then it has a hypotenuse of 5.
23. If a number is a whole number, then it is a rational number.

25. If I do logic puzzles, then I am driven crazy.
27. If the graffiti are to be covered, then two coats of paint must be used.   **29.** If employment improves, then the economy recovers.   **31.** If a number is a whole number, then it is an integer.   **33.** If their pitching improves, then the Phillies will win the pennant.   **35.** If the figure is a rectangle, then it is a parallelogram with perpendicular adjacent sides.   **37.** If a triangle has two perpendicular sides, then it is a right triangle.   **39.** If a three-digit number whose units digit is 5 is squared, then the square will end in 25.   **41.** D   **43.** Answers will vary.   **45.** true   **47.** false
49. false   **51.** contrary   **53.** contrary   **55.** consistent
57. (1) $p \rightarrow (p \rightarrow q)$ **(2)** $(p \rightarrow q) \rightarrow p$

3.5 Exercises (pages 121–122)

1. valid   **3.** invalid   **5.** valid   **7.** invalid   **9.** invalid
11. invalid   **13.** yes
15. All people with blue eyes have blond hair.
   Erin does not have blond hair.
   _____
   Erin does not have blue eyes.
17. invalid   **19.** valid   **21.** invalid   **23.** valid   **25.** invalid
27. invalid   **29.** valid

3.6 Exercises (pages 130–132)

1. valid by reasoning by transitivity   **3.** valid by modus ponens   **5.** fallacy by fallacy of the converse   **7.** valid by modus tollens   **9.** fallacy by fallacy of the inverse
11. valid by disjunctive syllogism   **13.** invalid   **15.** valid
17. invalid   **19.** valid   **21.** valid   **23.** invalid
25. invalid
27. Every time something squeaks, I use WD-40.
   Every time I use WD-40, I go to the hardware store.
   _____
   Every time something squeaks, I go to the hardware store.

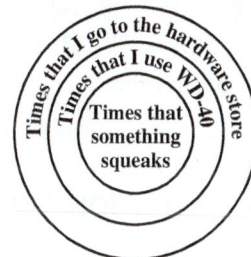

29. valid   **31.** invalid   **33.** invalid   **35.** valid   **37.** valid
39. (a) When the cable company keeps you on hold, your dad gets punched over a can of soup.   **(b)** The cable company doesn't keep you on hold.   **41.** If it is my poultry, then it is a duck.   **43.** If it is a guinea pig, then it is hopelessly ignorant of music.   **45.** If it is a teachable kitten, then it does not have green eyes.   **47.** If I can read it, then I have not filed it.   **49. (a)** $p \rightarrow \sim s$ **(b)** $r \rightarrow s$ **(c)** $q \rightarrow p$

49. (d) None of my poultry are officers. **51. (a)** $r \rightarrow \sim s$
(b) $u \rightarrow t$ **(c)** $\sim r \rightarrow p$ **(d)** $\sim u \rightarrow \sim q$ **(e)** $t \rightarrow s$
(f) All pawnbrokers are honest. **53. (a)** $r \rightarrow w$
(b) $\sim u \rightarrow \sim t$ **(c)** $v \rightarrow \sim s$ **(d)** $x \rightarrow r$ **(e)** $\sim q \rightarrow t$
(f) $y \rightarrow p$ **(g)** $w \rightarrow s$ **(h)** $\sim x \rightarrow \sim q$ **(i)** $p \rightarrow \sim u$
(j) I can't read any of Brown's letters.

Chapter 3 Test *(pages 137–138)*

1. $6 - 3 \neq 3$ **2.** Some men are not created equal.
3. No members of the class went on the field trip.
4. I fall in love and it will not be forever. **5.** She did
not apply or she got a student loan. **6.** $\sim p \rightarrow q$
7. $p \rightarrow q$ **8.** $\sim q \leftrightarrow \sim p$ **9.** You won't love me and I will
love you. **10.** It is not the case that you will love me or
I will not love you. (Equivalently: You won't love me and
I will love you.) **11.** T **12.** T **13.** T **14.** F
15. Answers will vary. **16. (a)** The antecedent must be true
and the consequent must be false. **(b)** Both component
statements must be true. **(c)** Both component statements
must be false. **(d)** Both component statements must have
the same truth value. **17.** TFFF **18.** TTTT (tautology)
19. false **20.** true
Wording may vary in the answers for Exercises 21–25.
21. If the number is an integer, then it is a rational number.
22. If a polygon is a rhombus, then it is a quadrilateral.
23. If a number is divisible by 4, then it is divisible by 2.
24. If she digs dinosaur bones, then she is a paleontologist.
25. (a) If the graph helps me understand it, then a
picture paints a thousand words. **(b)** If a picture
doesn't paint a thousand words, then the graph won't
help me understand it. **(c)** If the graph doesn't help me
understand it, then a picture doesn't paint a thousand
words. **26. (a)** $(q \wedge r) \rightarrow \sim p$ **(b)** $p \rightarrow (\sim q \vee \sim r)$
(c) $(\sim q \vee \sim r) \rightarrow p$ **27.** valid **28. (a)** A **(b)** F **(c)** C
(d) D **29.** valid **30.** invalid

CHAPTER 4 NUMERATION SYSTEMS

4.1 Exercises *(pages 146–148)*

1. 12,034 **3.** 7,610,729 **5.**

7.

9.

11. **13.**

15. 173 **17.** 14,000,000 **19.** MMDCCCLXI
21. $\overline{\text{XXV}}$DCXIX **23.** 935 **25.** 2007
27. **29.** **31.** to **33.** to

35. 206 **37.** 53,601 **39.** 113 **41.** 7598 **43.** 2021
45. 903 **47.** 1504 **49.** 8128 **51.** 622,500 shekels
53. Answers will vary. **55.** Answers will vary. **57.** 99,999
59. 3124 **61.** $10^d - 1$ **63.** $7^d - 1$ **65.** 9999

4.2 Exercises *(pages 152–153)*

1. Mayan; 12 **3.** Babylonian; 32 **5.** Greek; 234
7. Mayan; 242 **9.** Babylonian; 1282 **11.** Babylonian;
2601 **13.** Mayan; 2640 **15.** Mayan; 59,954
17. Babylonian; 80,474 **19.** Greek; 15,149
21. **23.** **25.** **27.**
29. **31.** **33.** **35.**
37. **39.** **41.** **43.** $\lambda\theta$ **45.** $\varphi\beta$ **47.** $\upsilon\iota\beta$
49. $_{\prime}\beta\psi\xi\theta$ **51.** $\overset{\epsilon}{\text{M}}_{\prime}\delta\psi\kappa\varsigma$ **53.** $\chi\pi\alpha$ **55.** $\omega\pi\eta$

4.3 Exercises *(pages 159–160)*

1. $(7 \cdot 10^1) + (3 \cdot 10^0)$
3. $(8 \cdot 10^3) + (3 \cdot 10^2) + (3 \cdot 10^1) + (5 \cdot 10^0)$
5. $(3 \cdot 10^3) + (6 \cdot 10^2) + (2 \cdot 10^1) + (4 \cdot 10^0)$
7. $(1 \cdot 10^7) + (4 \cdot 10^6) + (2 \cdot 10^5) + (0 \cdot 10^4) +$
$(6 \cdot 10^3) + (0 \cdot 10^2) + (4 \cdot 10^1) + (0 \cdot 10^0)$
9. 75 **11.** 4380 **13.** 70,401,009 **15.** 79 **17.** 53 **19.** 109
21. 733 **23.** 6 **25.** 206 **27.** 242 **29.** 49,801 **31.** 256
33. 63,259 **35.** **37.**

39. 1885 **41.** 38,325 **43.** 3,035,154 **45.** 496
47. 217,204 **49.** 460 **51.** 32,798 **53.** Answers will vary.

4.4 Exercises *(pages 170–172)*

1. 1, 2, 3, 4, 5, 6, 10, 11, 12, 13, 14, 15, 16, 20, 21, 22, 23, 24,
25, 26 **3.** 1, 2, 3, 4, 5, 6, 7, 8, 10, 11, 12, 13, 14, 15, 16, 17, 18,
20, 21, 22 **5.** 13_{five}; 20_{five} **7.** $B6E_{\text{sixteen}}$; $B70_{\text{sixteen}}$ **9.** 3
11. 11 **13.** least: $1000_{\text{three}} = 27$; greatest: $2222_{\text{three}} = 80$

15. 956   **17.** 881   **19.** 28,854   **21.** 139   **23.** 5601
25. $93_{sixteen}$ **27.** 2131101_{five} **29.** 1001001010_{two}
31. 102112101_{three} **33.** 111134_{six} **35.** 32_{seven}
37. 1031321_{four} **39.** 11110111_{two} **41.** 467_{eight}
43. 11011100_{two} **45.** $2D_{sixteen}$ **47.** 37_{eight} **49.** 1427
51. 1000011_{two} **53.** 1101011_{two} **55.** HELP
57. $100111011001011110111_{two}$ **59.** Answers will vary;
11000000.10101000.00000011.00011110   **61.** 192.172.3.64
63. Answers will vary.   **65. (a)** The binary ones digit is 1.
(b) The binary twos digit is 1.   **(c)** The binary fours digit
is 1.   **(d)** The binary eights digit is 1.   **(e)** The binary
sixteens digit is 1.   **67.** 6   **69.** yes   **71.** yes   **73.** yes
75. yes   **77.** Answers will vary.   **79.** no   **81.** yes
83. Append a 0 on the right.   **85.** Answers will vary.
87. Answers will vary.   **89.** 20120011_{three} **91.** 25657_{nine}
93. 808080   **95.** FF69B4

Chapter 4 Test *(page 176)*

1. Egyptian; 1534   **2.** Roman; 10,474   **3.** Chinese; 385
4. Babylonian; 155,540   **5.** Mayan; 50,511   **6.** Greek;
53,524   **7.** 1035   **8.** 23,862   **9.** 13,227   **10.** 73   **11.** 37
12. 48,879   **13.** 110001_{two} **14.** 43210_{five} **15.** 256_{eight}
16. $E8D_{sixteen}$ **17.** 1105   **18.** 101110101101_{two}
19. There is less repetition of symbols.   **20.** Place values
are understood by position.   **21.** There are fewer
symbols to learn.   **22.** There are fewer digits in the
numerals.   **23.** Answers will vary.
24. $7_{nine} = 21_{three}$, $6_{nine} = 20_{three}$, and $5_{nine} = 12_{three}$.
So $765_{nine} = 212012_{three}$.

CHAPTER 5 NUMBER THEORY

5.1 Exercises *(pages 182–184)*

1. true   **3.** true   **5.** false   **7.** true   **9.** 1, 2, 3, 4, 6, 12
11. 1, 2, 4, 7, 14, 28   **13. (a)** no   **(b)** yes   **(c)** no
(d) no   **(e)** no   **(f)** no   **(g)** no   **(h)** no   **(i)** no
15. (a) yes   **(b)** yes   **(c)** yes   **(d)** yes   **(e)** yes
(f) yes   **(g)** yes   **(h)** yes   **(i)** yes
17. (a) Answers will vary.   **(b)** 13   **(c)** square root;
square root; square root   **(d)** prime   **19.** 179 and 193,
181 and 197, 191 and 197, 193 and 199   **21.** 2, 3; no
23. It must be 0.
25.

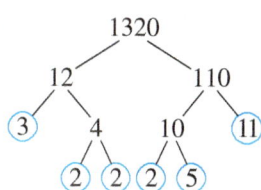

$1320 = 3 \cdot 2^3 \cdot 5 \cdot 11$
$= 2^3 \cdot 3 \cdot 5 \cdot 11$

27. $2 \cdot 3^2 \cdot 7$ **29.** $7 \cdot 13^2$ **31.** yes   **33.** no   **35.** no
37. yes   **39.** The number must be divisible by both 3 and
5. That is, the sum of the digits must be divisible by 3, and
the last digit must be 5 or 0.   **41.** 0, 2, 4, 6, 8   **43.** 0, 4, 8
45. 0, 6   **47.** 8   **49.** 27   **51.** leap year   **53.** not a leap year
55. Answers will vary.   **57.** Answers will vary.
59. 1981 and 1987   **61.** 2041

5.2 Exercises *(pages 190–191)*

1. true   **3.** false   **5.** true
7. $37 = 1^2 + 6^2$; $41 = 4^2 + 5^2$; $53 = 2^2 + 7^2$
9. Answers will vary.   **11. (a)** 1681; composite ($41 \cdot 41$)
(b) 1763; composite ($41 \cdot 43$)   **(c)** 1847; prime
13. (a) 1681; composite ($41 \cdot 41$)   **(b)** 1763; composite
($41 \cdot 43$)   **(c)** 1847; prime   **15. (a)** 65,537   **(b)** 251
17. Answers will vary.   **19.** Answers will vary.
21. composite; $30,031 = 59 \cdot 509$ **23.** 63   **25.** $2^p - 1$
27. 3 and 31   **29.** Answers will vary.

5.3 Exercises *(pages 196–198)*

1. true   **3.** true   **5.** true   **7.** false   **9.** false   **11.** The sum
of the proper divisors is 496: $1 + 2 + 4 + 8 + 16 + 31 +$
$62 + 124 + 248 = 496$.   **13.** 8191 is prime; 33,550,336
15. $1 + \frac{1}{2} + \frac{1}{4} + \frac{1}{7} + \frac{1}{14} + \frac{1}{28} = 2$ **17.** deficient
19. abundant   **21.** 12, 18, 20, 24   **23.** $1 + 3 + 5 + 7 +$
$9 + 15 + 21 + 27 + 35 + 45 + 63 + 105 + 135 + 189$
$315 = 975$, and $975 > 945$, so 945 is abundant.
25. $1 + 2 + 4 + 8 + 16 + 32 + 37 + 74 + 148 + 296 +$
$592 = 1210$ and $1 + 2 + 5 + 10 + 11 + 22 + 55 +$
$110 + 121 + 242 + 605 = 1184$ **27.** $5 + 7$ **29.** $11 + 29$
31. (a) Let $a = 5$ and $b = 3$; $11 = 5 + 2 \cdot 3$
(b) $17 = 3 + 2 \cdot 7 = 7 + 2 \cdot 5 = 11 + 2 \cdot 3 = 13 + 2 \cdot 2$
33. 59 and 61   **35. (a)** $3^4 - 1 = 80$ is divisible by 5.
(b) $2^6 - 1 = 63$ is divisible by 7.   **37.** $5^2 + 2 = 27 = 3^3$
39. Answers will vary.   **41.** False; For the first six, the
sequence is 6, 8, 6, 8, 6, 6.   **43.** one; not happy
45. both; happy   **47.** Answers will vary.   **49.** B   **51.** 7; yes
53. 15; no   **55.** 24; 23; 25; yes; no   **57.** Answers will vary.
59. B   **61.** $18 = 1 + 2 + 6 + 9 = 3 + 6 + 9$
63. $1 + 2 + 5 + 7 + 10 + 14 + 35 = 74 > 70$
65. 1,000,000,000,000,066,600,000,000,000,001   **67.** 13

5.4 Exercises *(pages 206–207)*

1. true   **3.** false   **5.** false   **7.** true   **9.** false   **11.** 28
13. 11   **15.** 17   **17.** 10   **19.** 30   **21.** 6   **23.** 6   **25.** 18
27. 45   **29.** 240   **31.** 405   **33.** 300   **35.** 108   **37.** 693
39. 2160   **41.** 180   **43.** 180   **45.** 9450   **47.** Answers
will vary.   **49. (a)** $p^a q^a r^b$ **(b)** $p^b q^b r^c$ **51.** 15   **53.** 12

55. p and q are relatively prime.   **57.** Answers will vary.
59. 144th   **61.** 48 (5 stacks of 48 pennies, 6 stacks of 48 nickels)   **63.** $600; 25 books

5.5 Exercises *(pages 212–214)*

1. 987   **3.** o, o, e, o, o, e, o, o, e, o, o, e   **5.** $\dfrac{1 + \sqrt{5}}{2}$
7. $1 + 1 + 2 + 3 + 5 = 13 - 1$; Each expression is equal to 12.   **9.** $1 + 2 + 5 + 13 + 34 = 55$; Each expression is equal to 55.   **11.** $13^2 - 5^2 = 144$; Each expression is equal to 144.   **13.** $1 - 2 + 5 - 13 + 34 = 5^2$; Each expression is equal to 25.   **15.** (There are other ways to do this.)
(a) $39 = 34 + 5$ **(b)** $59 = 55 + 3 + 1$ **(c)** $99 = 89 + 8 + 2$
17. (a) The greatest common factor of 10 and 4 is 2, and the greatest common factor of $F_{10} = 55$ and $F_4 = 3$ is $F_2 = 1$.
(b) The greatest common factor of 12 and 6 is 6, and the greatest common factor of $F_{12} = 144$ and $F_6 = 8$ is $F_6 = 8$.
(c) The greatest common factor of 14 and 6 is 2, and the greatest common factor of $F_{14} = 377$ and $F_6 = 8$ is $F_2 = 1$.
19. (a) $8^2 - 3 \cdot 21 = 1$ **(b)** $2 \cdot 34 - 8^2 = 4$
(c) $8^2 - 1 \cdot 55 = 9$ **(d)** The difference will be 25, because we are obtaining the squares of the terms of the Fibonacci sequence. $13^2 - 1 \cdot 144 = 25 = 5^2$.
21. 123   **23.** Each sum is 2 less than a Lucas number.
25. (a) $8 \cdot 18 = 144$; Each expression is equal to 144.
(b) $8 + 21 = 29$; Each expression is equal to 29.
(c) $8 + 18 = 2 \cdot 13$; Each expression is equal to 26.
(d) $18 + 47 = 5 \cdot 13$; Each expression is equal to 65.
27. 3, 4, 5   **29.** 105, 208, 233
31. The sums are 1, 1, 2, 3, 5, 8, 13. They are terms of the Fibonacci sequence.
33. $\dfrac{1 + \sqrt{5}}{2} \approx 1.618033989$ and $\dfrac{1 - \sqrt{5}}{2} \approx -0.618033989$. After the decimal point, the digits are the same.
35. 987

Chapter 5 Test *(page 218)*

1. true   **2.** false   **3.** true   **4.** true   **5.** false   **6.** true
7. (a) yes   **(b)** yes   **(c)** yes   **(d)** no   **(e)** yes   **(f)** yes
(g) no   **(h)** no   **(i)** yes   **8. (a)** composite   **(b)** prime
(c) neither   **9.** $2 \cdot 3^3 \cdot 5 \cdot 11$ **10.** Answers will vary.
11. (a) abundant   **(b)** deficient   **(c)** perfect   **12.** B
13. 71 and 73   **14.** 9   **15.** 2002   **16.** Monday
17. 28,657   **18.** $55 - (5 + 8 + 13 + 21) = 8$; Each expression is equal to 8.   **19.** B   **20.** 60   **21.** The process will yield 11 for any term chosen.   **22.** A

CHAPTER 6 THE REAL NUMBERS AND THEIR REPRESENTATIONS

6.1 Exercises *(pages 228–230)*

1. 5   **3.** 0   **5.** $\sqrt{14}$ (There are others.)   **7.** true
9. true   **11. (a)** 3, 7   **(b)** 0, 3, 7   **(c)** $-9, 0, 3, 7$
(d) $-9, -1\frac{1}{4}, -\frac{3}{5}, 0, 3, 5.9, 7$ **(e)** $-\sqrt{7}, \sqrt{5}$
(f) All are real numbers.   **13.** 2,259,000
15. -3424 **17.** 46.77   **19.** 5436   **21.** $-220°$
23. Pacific Ocean, Indian Ocean, Caribbean Sea, South China Sea, Gulf of California   **25.** true
27. (number line from -6 to 4)   **29.** (number line from -4 to 4 with marks $-3\frac{4}{5}$, $-1\frac{5}{8}$, $\frac{1}{4}$, $2\frac{1}{2}$)
31. (a) A   **(b)** A   **(c)** B   **(d)** B   **33. (a)** 2   **(b)** 2
35. (a) -6 **(b)** 6   **37. (a)** -3 **(b)** 3   **39. (a)** 0
(b) 0   **41.** -12 **43.** -8 **45.** 3   **47.** $|-3|$, or 3
49. $-|-6|$, or -6 **51.** $|5 - 3|$, or 2   **53.** true
55. true   **57.** true   **59.** false   **61.** true   **63.** false
65. (a) Las Vegas; The population increased by 45.4%.
(b) Detroit; The population decreased by 3.6%.
67. Appliances (because $|-4.5| > |-0.4|$)

6.2 Exercises *(pages 239–243)*

1. positive; $18 + 6 = 24$ **3.** greater; $-14 + 9 = -5$
5. the number with the greater absolute value is subtracted from the one with the lesser absolute value; $5 - 12 = -7$ **7.** additive inverses; $4 + (-4) = 0$
9. negative; $-5(15) = -75$ **11.** -20 **13.** -4
15. -11 **17.** 9   **19.** 20   **21.** 24   **23.** -1296 **25.** 6
27. -6 **29.** 0   **31.** -6 **33.** 27   **35.** undefined   **37.** -1
39. -4 **41.** 7   **43.** 13   **45.** A, B, C   **47.** commutative property of addition   **49.** inverse property of addition
51. identity property of multiplication   **53.** identity property of addition   **55.** associative property of multiplication
57. closure property of multiplication   **59.** associative property of addition   **61.** distributive property
63. identity   **65. (a)** messing up your room
(b) spending money   **(c)** decreasing the volume on your MP3 player   **67.** -81 **69.** 81   **71.** -81 **73.** -81
75. (a) -11; The returns decreased 11 percent.
(b) -13; The returns decreased 13 percent.   **(c)** 13; The returns increased 13 percent.   **(d)** -1; The returns decreased 1 percent.   **77.** 50,395 ft   **79.** 1345 ft
81. 136 ft   **83.** -10 **85.** $+10$ **87.** $+5$ degrees
89. -2 degrees   **91.** 0 degrees (no change)
93. $+5$ degrees   **95.** 43; 34; 29; 36; 38; 36; 37; 41
97. 27 feet   **99.** 469 B.C.   **101.** $5540
103. $-69°$F   **105.** 120°F

6.3 Exercises *(pages 254–257)*
1. rational number **3.** 1 **5.** 459; 221; 5967
7. $\frac{7}{12}; \frac{4}{3}$ **9.** 6 or 4 (Either answer is acceptable.)
11. A, C, D **13.** C **15.** $\frac{1}{3}$ **17.** $-\frac{3}{7}$ **Answers will vary
in Exercises 19 and 21.** **19.** $\frac{6}{16}, \frac{9}{24}, \frac{12}{32}$ **21.** $-\frac{10}{14}, -\frac{15}{21}, -\frac{20}{28}$
23. (a) $\frac{1}{3}$ **(b)** $\frac{1}{4}$ **(c)** $\frac{2}{5}$ **(d)** $\frac{1}{3}$ **25.** the dots in the intersection of the triangle and the rectangle as a part of the dots in the entire figure **27. (a)** Christine **(b)** Leah **(c)** Leah **(d)** Anne **(e)** Otis and Carol; $\frac{1}{2}$ **29.** $\frac{1}{2}$
31. $\frac{43}{48}$ **33.** $-\frac{5}{24}$ **35.** $\frac{23}{56}$ **37.** $\frac{27}{20}$ **39.** $\frac{5}{12}$ **41.** $\frac{1}{9}$ **43.** $\frac{3}{2}$
45. $\frac{3}{2}$ **47.** $\frac{13}{3}$ **49.** $\frac{29}{10}$ **51.** $6\frac{3}{4}$ **53.** $6\frac{1}{8}$ **55.** $-17\frac{7}{8}$
57. $1\frac{5}{16}$ **59.** $37\frac{7}{8}$ **61.** $7\frac{3}{8}$ **63.** $4\frac{1}{8}$ **65.** $\frac{9}{16}$ inch
67. $30\frac{1}{4}$ in. **69. (a)** $1\frac{1}{8}$ in. **(b)** $1\frac{7}{8}$ in. **71.** $11\frac{55}{64}$ in.
73. 8 cakes (There will be some sugar left over.)
75. $16\frac{5}{8}$ yd **77.** $6757 **79.** $\frac{5}{8}$ **81.** $\frac{19}{30}$ **83.** $-\frac{3}{4}$
85. repeating **87.** terminating **89.** terminating
91. 0.75 **93.** 0.1875 **95.** $0.\overline{27}$ **97.** $0.\overline{285714}$ **99.** $\frac{2}{5}$
101. $\frac{17}{20}$ **103.** $\frac{467}{500}$ **105.** $\frac{67}{99}$ **107.** $\frac{7}{165}$ **109.** $\frac{1}{90}$
111. (a) $0.\overline{3}$, or 0.333 . . . **(b)** $0.\overline{6}$, or 0.666 . . .
(c) $0.\overline{9}$, or 0.999 . . . **(d)** $1 = 0.\overline{9}$

6.4 Exercises *(pages 265–268)*
1. rational **3.** irrational **5.** rational **7.** rational
9. irrational **11.** rational **13.** irrational **15. (a)** $0.\overline{8}$
(b) irrational; rational **The number of digits shown
will vary among calculator models in Exercises 17–23
and 45–53.** **17.** 6.244997998 **19.** 3.885871846
21. 29.73213749 **23.** 1.060660172 **25.** 6 **27.** 2.4 m²
29. 5.4 ft **31.** 2.5 sec **33.** 1.7 amps **35.** 71 mph
37. 54 mph **39.** 392,000 mi² **41.** The area and the perimeter are both numerically equal to 36. **43.** $5\sqrt{2}$; 7.071067812 **45.** $5\sqrt{3}$; 8.660254038 **47.** $12\sqrt{2}$; 16.97056275 **49.** $\frac{5\sqrt{6}}{6}$; 2.041241452 **51.** $\frac{\sqrt{7}}{2}$; 1.322875656
53. $\frac{\sqrt{21}}{3}$; 1.527525232 **55.** $3\sqrt{17}$ **57.** $4\sqrt{7}$ **59.** $10\sqrt{2}$
61. $3\sqrt{3}$ **63.**

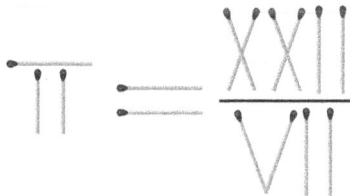

65. The result is 3.1415929, which agrees with the first seven digits in the decimal for π. **67.** 3 **69.** 4
71. The computer will forever be occupied by this task "to the exclusion of all else" and thus will no longer be able to interfere. **73.** ϕ is positive, while its conjugate is negative.

The units digit of ϕ is 1, and the units digit of its conjugate is 0. The decimal digits agree. **75.** If a 9 were under the radical, the number would equal the "nice" integer value 3. $\sqrt{3}$ is irrational, and its decimal is not so nice.
77. It is just a coincidence that 1828 appears back-to-back early in the decimal. There is no repetition indefinitely, which would be indicative of a rational number.
79. 4 **81.** 7 **83.** 6 **85.** 1 **87.** 4 **89.** 8 **The number
of decimal digits shown will vary among calculator models in Exercises 91–97.** **91.** 3.50339806
93. 5.828476683 **95.** 10.06565066 **97.** 5.578019845

6.5 Exercises *(pages 276–282)*
1. true **3.** true **5.** true **7.** false **9.** 11.315
11. −4.215 **13.** 0.8224 **15.** 47.5 **17.** 31.6 **19.** $30.13
21. (a) $466.02 **(b)** $190.68 **23.** $1229.09
25. (a) 50,000 **(b)** 55,000 **(c)** 54,800 **(d)** 54,790
27. (a) 0.89245 **(b)** 0.8925 **(c)** 0.892 **(d)** 0.89
(e) 0.9 **(f)** 1 **29. (a)** E **(b)** D **(c)** B **(d)** F **(e)** C
(f) A **31. (a)** $33\frac{1}{3}$% **(b)** 25% **(c)** 40% **(d)** $33\frac{1}{3}$%
33. 42% **35.** 36.5% **37.** 0.8% **39.** 210% **41.** 0.96
43. 0.0546 **45.** 0.003 **47.** 4.00 **49.** 0.005 **51.** 0.035
53. 20% **55.** 1% **57.** $37\frac{1}{2}$% **59.** 150% **61.** Answers
will vary. **63.** 124.8 **65.** 2.94 **67.** 150% **69.** 600
71. 1.4% **73.** 8% **75.** 9.4% **77.** 9.8% **79.** 892%
81. about 140% **83.** 21.2% **85.** $895 million **87.** A
89. C **91.** No, the price is $96.00. **93.** 0.79 m²
95. 0.44 m² **97.** 116 mg **99.** 129 mg
101. (a) 14.7 − 40 · 0.13 **(b)** 9.5 **(c)** 8.075; walking
(5 mph) **103. (a)** 0.031 **(b)** 0.035 **105.** 297
107. (a) .599 **(b)** .564 **(c)** .525 **(d)** .457 **109.** $4.50
111. $0.75 **113.** $12.00 **115.** $36.00 **117.** three (and
you would have 0.01¢ left over) **119.** $0.06, or 6¢
121. 1000 **123.** 25,000%

Chapter 6 Test *(pages 288–290)*
1. (a) 12 **(b)** 0, 12 **(c)** −4, 0, 12 **(d)** $-4, -\frac{3}{2}, -0.5,$
$0, 4.1, 12$ **(e)** $-\sqrt{5}, \sqrt{3}$ **(f)** $-4, -\sqrt{5}, -\frac{3}{2}, -0.5,$
$0, \sqrt{3}, 4.1, 12$ **2. (a)** C **(b)** B **(c)** D **(d)** A
3. (a) false **(b)** true **(c)** true **(d)** false **4.** 4
5. 10 **6.** 3 **7.** −43 **8.** 214°F **9.** 5296 ft **10. (a)** E
(b) A **(c)** B **(d)** D **(e)** F **(f)** C **11. (a)** Ben and
Jason **(b)** Ed and Charlie **(c)** Ben **(d)** Jack and
Chuck; $\frac{2}{5}$ **(e)** Jason **12.** $\frac{11}{16}$ **13.** $\frac{57}{160}$ **14.** $-\frac{2}{5}$ **15.** $\frac{4}{9}$
16. (a) $\frac{3}{50}$ **(b)** $\frac{41}{50}$ **(c)** $4\frac{4}{5}$ million **17. (a)** 0.45 **(b)** $0.41\overline{6}$
18. (a) $\frac{18}{25}$ **(b)** $\frac{58}{99}$ **19. (a)** irrational **(b)** rational
(c) rational **(d)** rational **(e)** irrational **(f)** irrational

20. (a) 12.247448714 **(b)** $5\sqrt{6}$ **21. (a)** 4.913538149
(b) $\frac{13\sqrt{7}}{7}$ **22. (a)** -45.254834 **(b)** $-32\sqrt{2}$
23. 0.08, or 8% **24. (a)** 13.81 **(b)** -0.315 **(c)** 38.7
(d) -24.3 **25. (a)** 350 **(b)** 346.04 **(c)** 346.045
26. $54.00 **27. (a)** $26\frac{2}{3}\%$ **(b)** $66\frac{2}{3}\%$ **28.** D
29. 1656; 1008; 16%; 6% **30.** 75 mg

CHAPTER 7 THE BASIC CONCEPTS OF ALGEBRA

7.1 Exercises *(pages 299–300)*

1. $\{-1\}$ **3.** $\{3\}$ **5.** $\{-7\}$ **7.** $\{0\}$ **9.** $\left\{-\frac{5}{3}\right\}$
11. $\left\{-\frac{1}{2}\right\}$ **13.** $\{2\}$ **15.** $\{-2\}$ **17.** $\{7\}$ **19.** $\{2\}$
21. $\{4\}$ **23.** $\{0\}$ **25.** $\{-2\}$ **27.** $\{2000\}$ **29.** $\{25\}$
31. $\{40\}$ **33.** contradiction; $\varnothing$ **35.** conditional; $\{-8\}$
37. conditional; $\{0\}$ **39.** identity; {all real numbers}
41. $A = \frac{150p}{w}$ **43.** $\ell = \frac{v \cdot \text{ACH}}{60}$ **45.** $t = 2\ell - 2a - 2b$
47. $\ell = \frac{r_1 x - r_2 x}{r_2}$, or $\ell = \frac{r_1 x}{r_2} - x$ **49.** $r = \frac{\ell + h^2}{2h}$
51. $t = \frac{d}{r}$ **53.** $b = \frac{sA}{h}$ **55.** $a = P - b - c$ **57.** $b = \frac{2sA}{h}$
59. $h = \frac{S - 2\pi r^2}{2\pi r}$, or $h = \frac{S}{2\pi r} - r$ **61.** $h = \frac{3V}{\pi r^2}$
63. 59 **65.** -4 **67.** 5 **69.** -25 **71.** about 40 L of air
per second **73.** minimum: 17.78%; maximum: 86.68%

7.2 Exercises *(pages 309–312)*

1. expression **3.** equation **5.** expression
7. $x - 12$ **9.** $(x - 6)(x + 4)$ **11.** $\frac{25}{x} (x \neq 0)$ **13.** 3
15. 6 **17.** -3 **19.** Madonna: $228 million; Bruce
Springsteen: $199 million **21.** wins: 66; losses: 16
23. Democrats: 52; Republicans: 46 **25.** shortest piece:
15 inches; middle piece: 20 inches; longest piece: 24 inches
27. gold: 38; silver: 27; bronze: 23 **29.** 70 milliliters
31. $250 **33.** $24.85 **35.** 4 liters **37.** 5 liters
39. 1 gallon **41.** $4000 at 3%; $8000 at 4%
43. $10,000 at 4.5%; $19,000 at 3% **45.** 17 pennies,
17 dimes, 10 quarters **47.** 305 students, 105 nonstudents
49. 54 seats on Row 1; 51 seats on Row 2 **51.** 328 miles
53. No. The distance is $55\left(\frac{1}{2}\right) = 27.5$ miles.
55. 18 miles **57.** $1\frac{3}{4}$ hours **59.** 11:00 A.M. **61.** 8 hours
63. 2.668 hours **65.** 1.715 hours **67.** 9.14 meters per
second **69.** 10.35 meters per second

7.3 Exercises *(pages 321–324)*

1. $\frac{5}{8}$ **3.** $\frac{1}{4}$ **5.** $\frac{2}{1}$ **7.** $\frac{3}{1}$ **9.** 10-lb size; $0.429
11. 32-oz size; $0.093 **13.** 128-oz size; $0.051
15. 36-oz size; $0.049 **17.** $\{35\}$ **19.** $\{-1\}$

21. $\left\{-\frac{27}{4}\right\}$ **23.** 2 tablets **25.** 0.8 milliliter
27. 910 milligrams **29.** 7.5 milliliters **31.** 12 milligrams
33. $44.55 **35.** 12,500 fish **37.** $25\frac{2}{3}$ inches
39. 2.0 inches **41.** $2\frac{5}{8}$ cups **43.** 20 pounds
45. about 302 pounds **47.** $428.82
49. 100 pounds per square inch **51.** 144 feet
53. 448.1 pounds **55.** 24 **57.** 42 **59.** 6.2 pounds
61. $239 **63.** $280 **65.** about 267 square miles

7.4 Exercises *(pages 332–334)*

1. D **3.** B **5.** F
7. $[7, \infty)$

9. $[5, \infty)$

11. $(7, \infty)$

13. $(-4, \infty)$

15. $(-\infty, -40]$

17. $(-\infty, 4]$

19. $\left(-\infty, -\frac{15}{2}\right)$

21. $\left[\frac{1}{2}, \infty\right)$

23. $(3, \infty)$

25. $(-\infty, 4)$

27. $\left(-\infty, \frac{23}{6}\right]$

29. $(1, 11)$

31. $[-14, 10]$

33. $[-5, 6]$

35. $\left[-\frac{14}{3}, 2\right]$

37. $\left[-\frac{1}{2}, \frac{35}{5}\right]$

39. $\left(-\frac{1}{3}, \frac{1}{9}\right]$

41. after 40 miles
43. 2 miles
45. 80 or greater
47. 90 or greater **49.** 71 or greater
51. $-0.15 \leq x - 18 \leq 0.15$; between and inclusive of 17.85
and 18.15 ounces **53.** $-0.75 \leq x - 52.5 \leq 0.75$; between
and inclusive of 51.75 and 53.25 milliliters
55. (a) 140 to 184 pounds **(b)** Answers will vary.
57. 26 DVDs

7.5 Exercises *(pages 342–344)*

1. 10^3 **3.** 10^6 **5.** 10^9 **7.** 625 **9.** -32 **11.** -8
13. -81 **15.** $\frac{1}{8}$ **17.** $\frac{1}{49}$ **19.** $-\frac{1}{49}$ **21.** $\frac{16}{5}$ **23.** 125

25. $\frac{25}{16}$ **27.** $\frac{9}{20}$ **29.** 1 **31.** 1 **33.** 0 **35.** 10^{16} **37.** $\frac{1}{10}$
39. $\frac{1}{5}$ **41.** 10^2 **43.** 10^3 **45.** 10^1, or 10 **47.** 5^2 **49.** 10^{11}
51. $\frac{1}{10^6}$ **53.** $\frac{4}{10^{14}}$ **55.** 2.3×10^2 **57.** 8.1×10^9
59. 1.4×10^{12} **61.** 2×10^{-2} **63.** 5.623×10^{-1}
65. 4.4×10^{-6} **67.** 6500 **69.** 28,000,000
71. 15,000,000,000 **73.** 0.42 **75.** 0.0152
77. 0.00000833 **79.** 6×10^5 **81.** 2×10^5 **83.** 2×10^5
85. (a) 3.174×10^8 (b) $\$1 \times 10^{12}$ (or $\$10^{12}$) (c) $3151
87. $40,045 **89.** approximately 9.5×10^{-7} parsec
91. 300 seconds **93.** approximately 5.87×10^{12} miles
95. (a) 5.16×10^2 people per square mile
(b) 998 square miles **97.** When Kirk said "one to the fourth power," he indicated that the sound would *not* be changed at all, because $1^4 = 1$.

7.6 Exercises *(page 350)*
1. $x^2 - x + 3$ **3.** $9x^2 - 4x + 4$ **5.** $6x^4 - 2x^3 - 7x^2 - 4x$
7. $-2x^2 - 13x + 11$ **9.** $x^2 - 5x - 24$ **11.** $28x^2 + x - 2$
13. $12x^5 - 20x^3 + 4x^2$ **15.** $4x^2 - 9$
17. $x^2 + 14x + 49$ **19.** $25x^2 - 30x + 9$
21. $10x^4 + 34x^3 + 7x^2 - 13x + 6$ **23.** $12(x + 3)$
25. $2x(4x^2 - 2x + 3)$ **27.** $2x^2(4x^2 + 3x - 6)$
29. $(x - 5)(x + 3)$ **31.** $(x + 7)(x - 5)$
33. $6(x - 10)(x + 2)$ **35.** $3x(x + 1)(x + 3)$
37. $(3x - 2)(2x + 3)$ **39.** $(5x + 3)(x - 2)$
41. $(7x + 2)(3x - 1)$ **43.** $2x^2(4x - 1)(3x + 2)$
45. $(x - 4)^2$ **47.** $(x + 7)^2$ **49.** $(3x - 2)^2$
51. $2(4x - 3)^2$ **53.** $(2x + 7)^2$ **55.** $(x + 6)(x - 6)$
57. $(10 + x)(10 - x)$ **59.** $(3x + 4)(3x - 4)$
61. $(5x^2 + 3)(5x^2 - 3)$ **63.** $(x^2 + 25)(x + 5)(x - 5)$
65. $13x(x^2 + 3x + 4)$ **67.** $(6x - 7)(2x + 5)$
69. $(2x + 7)^2$ **71.** $(x^4 + 9)(x^2 + 3)(x^2 - 3)$
73. $-1(x + 7)(x - 3)$ **75.** $-1(3x + 4)(x - 1)$
77. $-1(x + 2)(2x + 1)$

7.7 Exercises *(pages 355–357)*
1. $\{-3, 9\}$ **3.** $\left\{\frac{7}{2}, -\frac{1}{5}\right\}$ **5.** $\{-7, 0\}$ **7.** $\{-3, 4\}$
9. $\{-7, -2\}$ **11.** $\left\{-\frac{5}{2}, 1\right\}$ **13.** $\left\{-\frac{1}{2}, \frac{1}{6}\right\}$ **15.** $\{-2, 4\}$
17. $\{\pm 8\}$ **19.** $\left\{\pm 2\sqrt{6}\right\}$ **21.** $\varnothing$ **23.** $\{1, 7\}$
25. $\left\{4 \pm \sqrt{3}\right\}$ **27.** $\left\{\frac{5 \pm \sqrt{13}}{2}\right\}$ **29.** $\left\{\frac{-3 \pm \sqrt{37}}{2}\right\}$
31. $\left\{\frac{1 \pm \sqrt{3}}{2}\right\}$ **33.** $\left\{\frac{1 \pm \sqrt{5}}{2}\right\}$ **35.** $\left\{\frac{-1 \pm \sqrt{2}}{2}\right\}$
37. $\left\{\frac{1 \pm \sqrt{29}}{2}\right\}$ **39.** $\varnothing$ **41.** eastbound ship: 80 miles; southbound ship: 150 miles **43.** 120 feet **45.** 412.3 feet

47. 8, 15, 17 **49.** 5 centimeters, 12 centimeters, 13 centimeters **51.** (a) 1 second and 8 seconds (b) 9 seconds after it is projected **53.** 0.7 second and 4.0 seconds
55. 1 foot **57.** length: 26 meters; width: 16 meters

Chapter 7 Test *(pages 363–364)*
1. $\{2\}$ **2.** $\{4\}$ **3.** identity; {all real numbers}
4. $v = \frac{S + 16t^2}{t}$, or $v = \frac{S}{t} + 16t$ **5.** Hawaii: 4021 square miles; Maui: 728 square miles; Kauai: 551 square miles
6. 5 liters **7.** 2.2 hours **8.** 16 slices for $4.38
9. 3 tablets **10.** 2300 miles **11.** 200 amps
12. $(-\infty, 4]$ ⟵|++|+++⟶ **13.** $(-2, 6]$ ++⟨+++++|++⟶
        0   4   −2 0   6
14. 82 or greater **15.** $\frac{16}{9}$ **16.** -64 **17.** $\frac{64}{27}$ **18.** 0
19. 1 **20.** 10^1, or 10 **21.** (a) 693,000,000
(b) 0.000000125 **22.** 3×10^{-4} **23.** about 15,300 seconds
24. $4x^2 + 6x + 10$ **25.** $15x^2 - 14x - 8$ **26.** $16x^4 - 9$
27. $x^2 + 12x + 36$ **28.** $(x + 7)(x - 4)$
29. $(2x + 9)(x + 5)$ **30.** $(5x + 7)(5x - 7)$
31. $(x - 9)^2$ **32.** $2x(x - 5)(x - 4)$ **33.** $\left\{-\frac{3}{2}, \frac{1}{3}\right\}$
34. $\left\{\pm\sqrt{13}\right\}$ **35.** $\left\{\frac{1 \pm \sqrt{29}}{2}\right\}$ **36.** 0.87 second

CHAPTER 8 GRAPHS, FUNCTIONS, AND SYSTEMS OF EQUATIONS AND INEQUALITIES

8.1 Exercises *(pages 373–375)*
1. x **3.** $(0, 0)$ **5.** (a) I (b) III
(c) II (d) IV (e) none

7.–15.

17. (a) $\sqrt{34}$ (b) $\left(\frac{1}{2}, \frac{5}{2}\right)$
19. (a) $\sqrt{61}$ (b) $\left(\frac{1}{2}, 1\right)$
21. (a) $\sqrt{146}$ (b) $\left(-\frac{1}{2}, \frac{3}{2}\right)$
23. (a) $\sqrt{40} = 2\sqrt{10}$
(b) $(-1, 4)$

25. B **27.** D **29.** $x^2 + y^2 = 36$
31. $(x + 1)^2 + (y - 3)^2 = 16$ **33.** $x^2 + (y - 4)^2 = 3$
35. $(-2, -3); 2$ **37.** $(-5, 7); 9$ **39.** $(2, 4); 4$

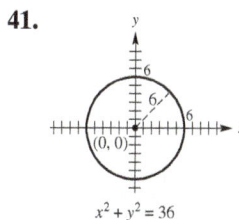

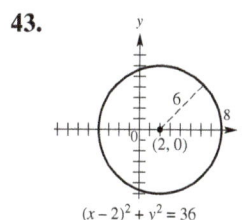

41.

$x^2 + y^2 = 36$

43.

$(x - 2)^2 + y^2 = 36$

45.

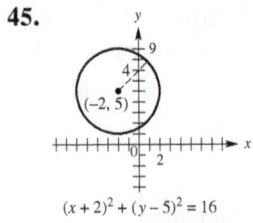

$(x+2)^2 + (y-5)^2 = 16$

47.

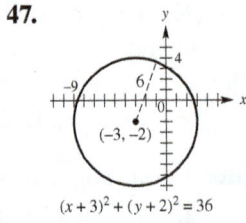

$(x+3)^2 + (y+2)^2 = 36$

49. (a) 25.35%; **(b)** This is very close to the actual figure.

51. $20,665   **53.** Answers will vary.

55. The epicenter is $(-2, -2)$.

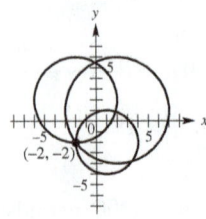

8.2 Exercises *(pages 382–385)*

1. $(0, -4), (4, 0), (2, -2),$ $(3, -1)$

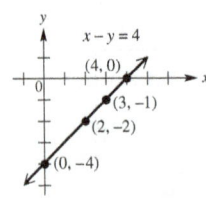

3. $(0, 5), \left(\frac{5}{2}, 0\right), (1, 3),$ $(2, 1)$

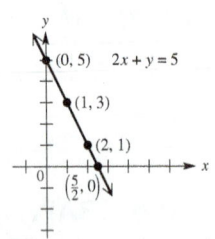

5. $(0, 4), (5, 0), \left(3, \frac{8}{5}\right),$ $\left(\frac{5}{2}, 2\right)$

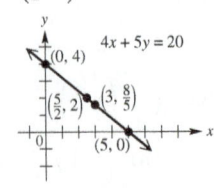

7. $(0, 4), \left(\frac{8}{3}, 0\right), (2, 1),$ $(4, -2)$

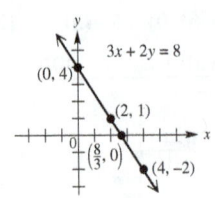

9. $(4, 0); \quad (0, 6)$

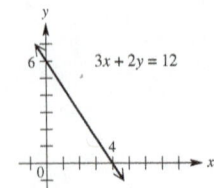

11. $(2, 0); \quad \left(0, \frac{5}{3}\right)$

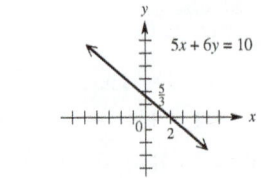

13. $\left(\frac{5}{2}, 0\right); \quad (0, -5)$

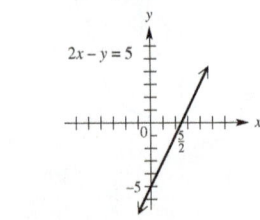

15. $(2, 0); \quad \left(0, -\frac{2}{3}\right)$

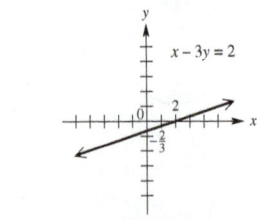

17. $(0, 0); \quad (0, 0)$

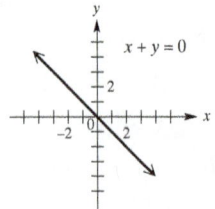

19. $(0, 0); \quad (0, 0)$

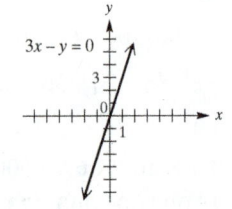

21. $(2, 0); \quad$ none

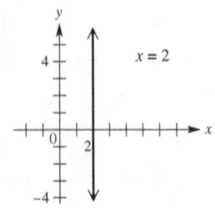

23. none; $(0, 4)$

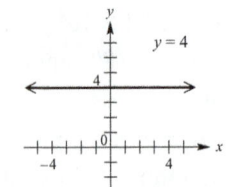

25. C **27.** A **29.** D **31.** B **33.** $\frac{3}{10}$ **35. (a)** $\frac{3}{2}$

(b) $-\frac{7}{4}$ **37.** 8 **39.** $-\frac{5}{6}$ **41.** 0 **43.** undefined

45.

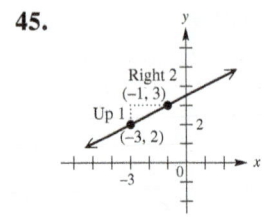

47.

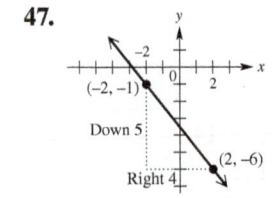

49.

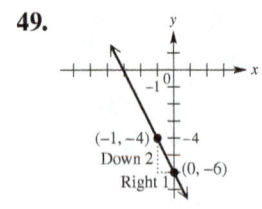

51.

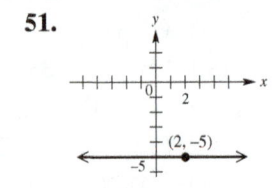

53.

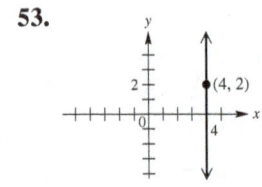

55. parallel **57.** perpendicular **59.** neither parallel nor perpendicular **61.** $\frac{7}{10}$ **63.** $-$$4000 per year; The value of the machine is decreasing by an average of $4000 per year during those years.

65. (a) 0.42 **(b)** positive; increased **(c)** 0.42%

67. (a) 14.2 **(b)** The number of subscribers increased by an average of 14.2 million per year from 2007 through 2012. **69.** $-$1859 thousand; The number of digital cameras sold decreased by an average of 1859 thousand per year from 2010 through 2013. **71.** $0.08; The price of a gallon of gasoline increased by an average of $0.08 per year from 1980 through 2012.

8.3 Exercises *(pages 389–391)*

1. D **3.** B **5.** A **7.** C **9.** H **11.** B **13.** $y = 3x - 3$

15. $y = -x + 3$ **17.** $y = -\frac{3}{4}x + \frac{5}{2}$ **19.** $y = -2x + 18$

21. $y = \frac{1}{2}x + \frac{13}{2}$ **23.** $y = 4x - 12$ **25.** $y = 5$ **27.** $x = 9$

29. $x = 0.5$ **31.** $y = 8$ **33.** $y = 2x - 2$

35. $y = -\frac{1}{2}x + 4$ **37.** $y = 5$ **39.** $x = 7$ **41.** $y = -3$

43. $y = 5x + 15$ **45.** $y = -\frac{2}{3}x + \frac{4}{5}$ **47.** $y = \frac{2}{5}x + 5$

49. (a) $y = -x + 12$ **(b)** -1 **(c)** $(0, 12)$

51. (a) $y = -\frac{5}{2}x + 10$ **(b)** $-\frac{5}{2}$ **(c)** $(0, 10)$

53. (a) $y = \frac{2}{3}x - \frac{10}{3}$ **(b)** $\frac{2}{3}$ **(c)** $\left(0, -\frac{10}{3}\right)$

55. $y = 3x - 19$ **57.** $y = \frac{1}{2}x - 1$ **59.** $y = -\frac{1}{2}x + 9$

61. $y = 7$ **63.** $C = \frac{5}{9}F - \frac{160}{9}$ **65.** -40 degrees

8.4 Exercises *(pages 397–399)*

1. set of ordered pairs   **3.** range; dependent   **5.** linear;
line   **7.** 5   **9.** 2   **11.** -1 **13.** -13 **15.** 3   **17.** 0

19.

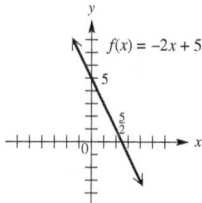

domain and range:
$(-\infty, \infty)$

21.

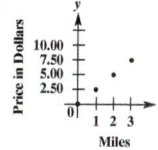

domain and range:
$(-\infty, \infty)$

23.

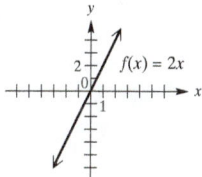

domain and range:
$(-\infty, \infty)$

25.

domain: $(-\infty, \infty)$;
range: $\{5\}$

27. 4   **29.** 16   **31. (a)** $C(x) = 0.02x + 200$
(b) $R(x) = 0.04x$ **(c)** 10,000
(d)

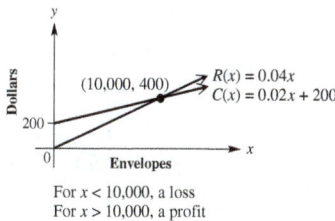

For $x < 10,000$, a loss
For $x > 10,000$, a profit

33. (a) $C(x) = 3.00x + 2300$ **(b)** $R(x) = 5.50x$ **(c)** 920
(d)

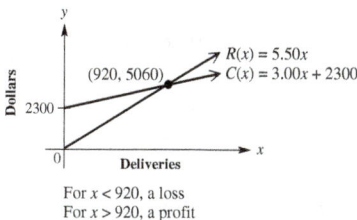

For $x < 920$, a loss
For $x > 920$, a profit

35. (a) $f(x) = 422.6x + 5492$ **(b)** 8450
37. (a) $f(x) = 0.996x + 34.3$ **(b)** 79.1%
39. (a) $f(x) = 76.9x$ **(b)** 30,760 kilometers per second

41. (a) \$0; \$2.50; \$5.00; \$7.50   **43. (a)** \$160
(b) $2.50x$ **(c)**

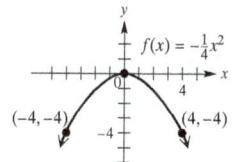

(b) 70 mph   **(c)** 66 mph
(d) for speeds more than
80 mph

8.5 Exercises *(pages 407–408)*

1. F   **3.** C   **5.** E   **7.** $(0,0)$ **9.** $(0,4)$ **11.** $(1,0)$
13. $(-3, -4)$ **15.** downward; narrower
17. upward; wider

19.

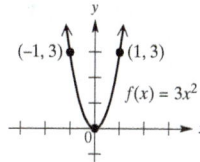

21.

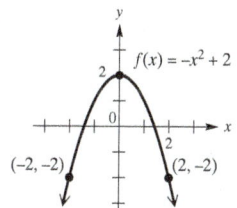

23.

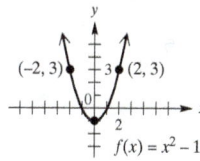

25.

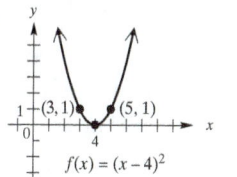

27.

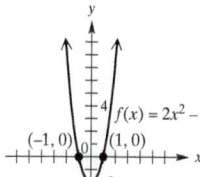

29.

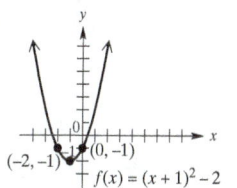

31.

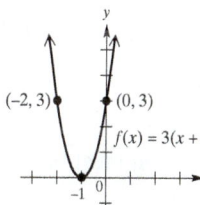

33.

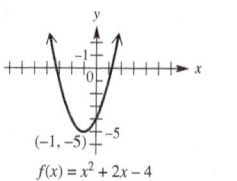

35.

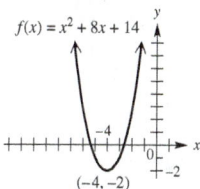

37.

39.

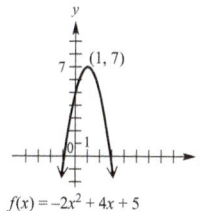

41. 25 meters; 625 square
meters   **43.** 16 feet;
2 seconds   **45.** 4.1 seconds;
81.6 meters
47. $f(45) = 161.5$; This
means that when the speed
is 45 mph, the stopping
distance is 161.5 feet.

8.6 Exercises (pages 417–419)

1. rises; falls **3.** does not **5.** rises; falls **7.** does not
9. 2.56425419972 **11.** 1.25056505582
13. 7.41309466897 **15.** 0.0000210965628481

17.

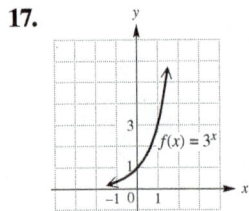

19.

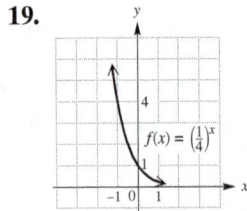

21. 20.0855369232 **23.** 0.018315638889 **25.** $2 = \log_4 16$
27. $-3 = \log_{2/3}\left(\frac{27}{8}\right)$ **29.** $2^5 = 32$ **31.** $3^1 = 3$
33. 1.38629436112 **35.** −1.0498221245

37.

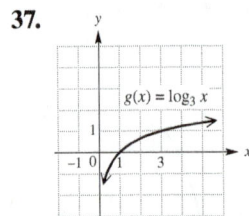

39.

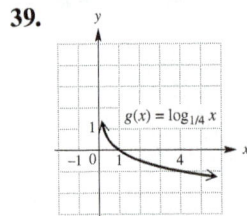

41. (a) $22,510.18 (b) $22,529.85 **43.** (a) $31,870.48
(b) $31,870.67 **45.** Plan A is better by $121.81.
47. (a) 7% compounded quarterly (b) $800.32
49. 27.73 years **51.** 13.52 years **53.** (a) 7.374 billion
(b) 27.1 years after 2010, or about 2037
55. about 9000 years **57.** (a) 440 grams (b) 387 grams
(c) 264 grams (d) 21.66 years **59.** 1611.97 years
61. 4 hours **63.** (a) 0.6°C (b) 0.3°C **65.** (a) 1.4°C
(b) 0.5°C

8.7 Exercises (pages 423–425)

1. 3; −6 **3.** D; The ordered-pair solution must be in
quadrant IV. **5.** yes **7.** no **9.** B **11.** A
13. $\{(2,2)\}$ **15.** $\{(3,-1)\}$ **17.** $\{(2,-3)\}$

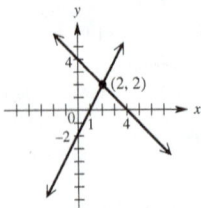

19. $\left\{\left(\frac{3}{2}, -\frac{3}{2}\right)\right\}$ **21.** $\left\{\left(\frac{6-2y}{7}, y\right)\right\}$
23. ∅ **25.** $\{(2,-4)\}$
27. $\{(1,2)\}$ **29.** $\left\{\left(\frac{22}{9}, \frac{22}{3}\right)\right\}$
31. $\{(2,3)\}$ **33.** $\{(5,4)\}$

35. $\left\{\left(-5, -\frac{10}{3}\right)\right\}$ **37.** $\{(2,6)\}$ **39.** hiking
41. (8.7, 35.9) **43.** $80x - 7y = -420$ **45.** $\{(4.375, 110)\}$

8.8 Exercises (pages 430–432)

1. wins: 97; losses: 65 **3.** length: 94 feet; width: 50 feet
5. dark clay: $5 per kilogram; light clay: $4 per kilogram
7. square: 12 cm; triangle: 8 cm **9.** cappuccino: $3.19;
cafe mocha: $3.59 **11.** Madonna: $140; Springsteen: $92

13. New York: $622; Washington, D.C: $564
15. Wal-Mart: $422 billion; ExxonMobil: $355 billion
17. NHL: $313.68; NBA: $287.85
19. (a) 12 ounces (b) 30 ounces (c) 48 ounces
(d) 60 ounces **21.** $4.29x **23.** 15% solution: $26\frac{2}{3}$ liters;
33% solution: $13\frac{1}{3}$ liters **25.** 3 liters **27.** 50% juice:
150 liters; 30% juice: 50 liters **29.** $1.20 candy:
100 pounds; $2.40 candy; 60 pounds **31.** 4%: $10,000;
3%: $5000 **33.** (a) $(10 - x)$ mph (b) $(10 + x)$ mph
35. train: 60 mph; plane: 160 mph **37.** boat: 21 mph;
current: 3 mph

8.9 Exercises (pages 437–438)

1. C **3.** B

5.

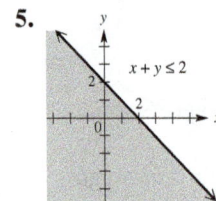

7.

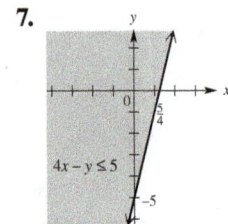

9.

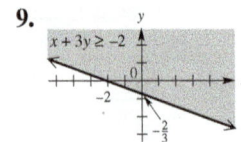

11.

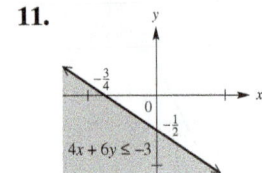

13.

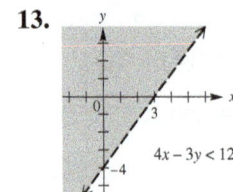

15.

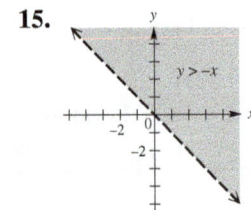

17.

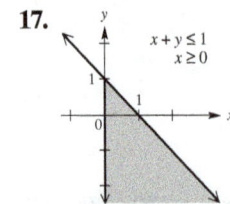

19.

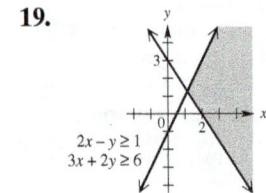

21.

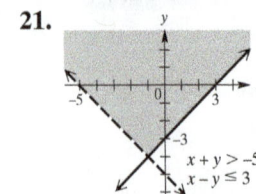

23. maximum of 65 at
(5, 10); minimum of 8 at
(1, 1)
25. $\left(\frac{6}{5}, \frac{6}{5}\right); \frac{42}{5}$ **27.** $\left(\frac{17}{3}, 5\right); \frac{49}{3}$

29. Ship 20 to A and 80 to B, for a minimum cost of $1040.
31. Take 3 red pills and 2 blue pills, for a minimum cost of
$0.70 per day. **33.** Make 3 batches of cakes and 6 batches
of cookies, for a maximum profit of $210. **35.** Ship 0
medical kits and 4000 containers of water.

Chapter 8 Test *(pages 447–448)*

1. $\sqrt{41}$ **2.** $(x+1)^2 + (y-2)^2 = 9$

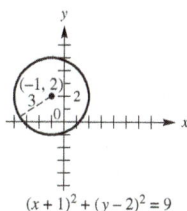

$(x+1)^2 + (y-2)^2 = 9$

3. \$151.0 billion; This amount is \$5.9 billion less than the actual figure.

4. *x*-intercept: $\left(\frac{8}{3}, 0\right)$; *y*-intercept: $(0, -4)$

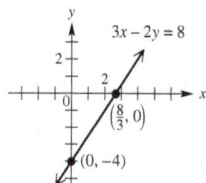

5. $\frac{2}{7}$ **6.** **(a)** $y = -\frac{2}{5}x + \frac{13}{5}$ **(b)** $y = -\frac{1}{2}x - \frac{3}{2}$

(c) $y = -\frac{1}{2}x + 2$ **7.** B   **8.** **(a)** $y = 0.238x + 63.3$

(b) 80.0 years   **9.** 17.5 million; The number of subscribers *increased* by an average of 17.5 million per year.

10. (a) $y = 0.05x + 0.50$ **(b)** $(1, 0.55), (5, 0.75), (10, 1.00)$

11. $y = \frac{2}{3}x + 1$ **12. (a)** -5 **(b)** 2   **13.** 500 units; \$30,000

14. axis: $x = -3$; vertex: $(-3, 4)$

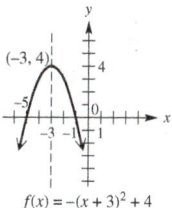

$f(x) = -(x+3)^2 + 4$

15. 80 feet by 160 feet; 12,800 square feet

16. (a) 2116.31264888

(b) 0.157237166314   **(c)** 3.15955035878

17. (a) \$12,740.13   **(b)** \$12,742.04

18. (a) 1.62 grams   **(b)** 1.18 grams

(c) 0.69 gram   **(d)** 2.00 grams

19. $\{(4, -2)\}$ **20.** $\varnothing$

21. Red Sox: \$53.38; Yankees: \$51.83   **22.** \$6-per-lb nuts: 30 lb; \$3-per-lb candy: 70 lb

23.

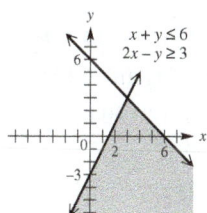

$x + y \le 6$
$2x - y \ge 3$

24. Raise 4 pigs and 12 geese, for a maximum profit of \$1120.

CHAPTER 9 GEOMETRY

9.1 Exercises *(pages 456–457)*
(The art here is not to scale with the exercise art.)

1. 90   **3.** equal   **5.** true   **7.** false   **9.** true
11. false   **13.** true   **15.** false

17. (a) $\overleftrightarrow{AB}$ **(b)**

19. (a) $\overleftrightarrow{CB}$ **(b)**

21. (a) $\overrightarrow{BC}$ **(b)**

23. (a) $\overrightarrow{BA}$ **(b)**

25. F   **27.** D   **29.** B   **31.** E

There may be other correct forms of the answers in Exercises 33–39.   **33.** $\overleftrightarrow{MO}$ **35.** $\overleftrightarrow{NO}$ **37.** $\varnothing$

39. $\overleftrightarrow{OP}$ **41.** $62°$ **43.** $1°$ **45.** $(90 - x)°$

47. $48°$ **49.** $154°$ **51.** $(180 - y)°$ **53.** $40°$

55. $52°$ **57.** $\angle CBD$ and $\angle ABE$; $\angle CBE$ and $\angle DBA$

59. (a) $52°$ **(b)** $128°$ **61.** $107°$ and $73°$ **63.** $75°$ and $75°$

65. $139°$ and $139°$ **67.** $35°$ and $55°$ **69.** $49°$ and $49°$

71. $48°$ and $132°$ **73. (a)** 3   **(b)** 6   **(c)** 7   **(d)** 7; exterior

75. (a) 180   **(b)** 180   **(c)** 180; 180   **(d)** 0   **(e)** 0   **(f)** 3

9.2 Exercises *(pages 463–465)*

1. chord   **3.** equilateral (or equiangular)   **5.** false

7. false   **9.** true   **11.** Answers will vary.   **13.** both

15. closed   **17.** closed   **19.** neither   **21.** convex

23. convex   **25.** not convex   **27.** right, scalene

29. acute, equilateral   **31.** right, scalene   **33.** right, isosceles   **35.** obtuse, scalene   **37.** An isosceles right triangle is a triangle having a $90°$ angle and two perpendicular sides of equal length.

39. $A = 50°$; $B = 70°$; $C = 60°$ **41.** $A = B = C = 60°$

43. $A = B = 55°$; $C = 70°$ **45.** $155°$ **47.** $360°$

49. (a) O **(b)** $\overrightarrow{OA}$, $\overrightarrow{OC}$, $\overrightarrow{OB}$, $\overrightarrow{OD}$ **(c)** $\overleftrightarrow{AC}$, $\overleftrightarrow{BD}$

(d) $\overleftrightarrow{AC}$, $\overleftrightarrow{BD}$, $\overleftrightarrow{BC}$, $\overleftrightarrow{AB}$ **(e)** $\overleftrightarrow{BC}$, $\overleftrightarrow{AB}$ **(f)** $\overleftrightarrow{AE}$

51. With the radius of the compass greater than one-half the length PQ, place the point of the compass at P and swing arcs above and below line r. Then, with the same radius and the point of the compass at Q, swing two more arcs above and below line r. Locate the two points of intersection of the arcs above and below, and call them A and B. With a straightedge, join A and B. AB is the perpendicular bisector of PQ.

53. With the radius of the compass greater than the distance from P to r, place the point of the compass at P and swing an arc intersecting line r in two points. Call these points A and B. Swing arcs of equal radius to the left of line r, with the point of the compass at A and at B, intersecting at point Q. With a straightedge, join P and Q. PQ is the perpendicular from P to line r.

55. With any radius, place the point of the compass at P and swing arcs to the left and right, intersecting line r in two points. Call these points A and B. With an arc of

sufficient length, place the point of the compass first at A and then at B, and swing arcs either both above or both below line r, intersecting at point Q. With a straightedge, join P and Q. PQ is perpendicular to line r at P.
57. With any radius, place the point of the compass at A and swing an arc intersecting the sides of angle A at two points. Call the point of intersection on the horizontal side B and call the other point of intersection C. Draw a horizontal working line, and locate any point A' on this line. With the same radius used earlier, place the point of the compass at A' and swing an arc intersecting the working line at B'. Return to angle A, and set the radius of the compass equal to BC. On the working line, place the point of the compass at B' and swing an arc intersecting the first arc at C'. Now draw line $A'C'$. Angle A' is equal to angle A.
59. Answers will vary.

9.3 Exercises *(pages 472–476)*

1.

STATEMENTS	REASONS
1. $AC = BD$	1. Given
2. $AD = BC$	2. Given
3. $AB = AB$	3. Reflexive property
4. $\triangle ABD \cong \triangle BAC$	4. SSS congruence property

3.

STATEMENTS	REASONS
1. $\angle BAC = \angle DAC$	1. Given
2. $\angle BCA = \angle DCA$	2. Given
3. $AC = AC$	3. Reflexive property
4. $\triangle ABC \cong \triangle ADC$	4. ASA congruence property

5.

STATEMENTS	REASONS
1. $\overleftrightarrow{DB}$ is perpendicular to $\overleftrightarrow{AC}$.	1. Given
2. $AB = BC$	2. Given
3. $\angle ABD = \angle CBD$	3. Both are right angles by definition of perpendicularity.
4. $DB = DB$	4. Reflexive property
5. $\triangle ABD \cong \triangle CBD$	5. SAS congruence property

7. $110°$ **9.** $67°, 67°$ **11.** Answers will vary.
13. $\angle H$ and $\angle F$; $\angle K$ and $\angle E$; $\angle HGK$ and $\angle FGE$; $\overleftrightarrow{HK}$ and $\overleftrightarrow{FE}$; $\overleftrightarrow{GK}$ and $\overleftrightarrow{GE}$; $\overleftrightarrow{HG}$ and $\overleftrightarrow{FG}$
15. $\angle A$ and $\angle P$; $\angle C$ and $\angle R$; $\angle B$ and $\angle Q$; $\overleftrightarrow{AC}$ and $\overleftrightarrow{PR}$; $\overleftrightarrow{CB}$ and $\overleftrightarrow{RQ}$; $\overleftrightarrow{AB}$ and $\overleftrightarrow{PQ}$
17. $\angle P = 76°$; $\angle M = 48°$; $\angle A = \angle N = 56°$

19. $\angle T = 20°$; $\angle V = 64°$; $\angle R = \angle U = 96°$
21. $\angle T = 74°$; $\angle Y = 28°$; $\angle Z = \angle W = 78°$
23. $a = 12$; $b = 9$ **25.** $x = 6$ **27.** $a = 6$; $b = \frac{15}{2}$
29. $x = 165$ **31.** $c = 111\frac{1}{9}$ **33.** $r = \frac{108}{7}$ **35.** 60 m
37. 250 m, 350 m   **39.** 112.5 ft   **41.** 10 **43.** $c = 17$
45. $a = 13$ **47.** $c = 50$ m   **49.** $a = 20$ in.
51. The sum of the squares of the two shorter sides of a right triangle is equal to the square of the longest side.
53. $(3, 4, 5)$ **55.** $(7, 24, 25)$ **57.** Answers will vary.
59. $(3, 4, 5)$ **61.** $(7, 24, 25)$ **63.** Answers will vary.
65. $(4, 3, 5)$ **67.** $(8, 15, 17)$ **69.** Answers will vary.
71. 24 m   **73.** 18 ft   **75.** 4.55 ft   **77.** 19 ft, 3 in.
79. 28 ft, 10 in.   **81. (a)** b **(b)** k **(c)** cj **(d)** ck
(e) $j + k$; $j + k$; $a^2 + b^2 = c^2$ **83.** 9 **85.** 16 **87.** $55°$
89. $360°$ **91.** Answers will vary.   **93.** Answers will vary.

9.4 Exercises *(pages 484–488)*
1. 24 **3.** 10 **5.** perimeter   **7.** 48 cm^2 **9.** 10 cm^2
11. 8 in.^{2} **13.** 418 mm^2 **15.** 8 cm^2 **17.** 3.14 cm^2
19. 1017 m^2 **21.** 4 m   **23.** 300 ft, 400 ft, 500 ft
25. 50 ft   **27.** $23{,}800.10$ ft^2 **29.** $14{,}600$ mi^2
31. 12 in., 12π in., 36π in.^{2} **33.** 5 ft, 10π ft, 25π ft^2
35. 6 cm, 12 cm, 36π cm^2 **37.** 10 in., 20 in., 20π in.
39. $\frac{20}{\pi}$ yd, $\frac{40}{\pi}$ yd, 40 yd   **41.** 14.5 **43.** 7 **45.** 5.7
47. 6 **49.** 5 **51.** 1.5 **53. (a)** 20 cm^2 **(b)** 80 cm^2
(c) 180 cm^2 **(d)** 320 cm^2 **(e)** 4 **(f)** $3; 9$ **(g)** $4; 16$
(h) n^2 **55.** n^2 **57.** $\$800$ **59.** 80 **61.** 76.26
63. 132 ft^2 **65.** 5376 cm^2 **67.** 145.34 m^2
69. 16-in. pizza   **71.** 16-in. pizza   **73.** $\frac{1}{2}(a + b)(a + b)$
75. $\frac{1}{2}(a + b)(a + b) = \frac{1}{2}ab + \frac{1}{2}ab + \frac{1}{2}c^2$
77. 26 in.   **79.** 625 ft^2 **81.** 6 cm^2 **83.** $\frac{(4 - \pi)r^2}{4}$
85. $24 + 4\sqrt{6}$ **87.** 5 in.

9.5 Exercises *(pages 493–496)*
1. true   **3.** true   **5.** false   **7. (a)** $22\frac{1}{2}$ m^3 **(b)** $49\frac{1}{4}$ m^2
9. (a) $267{,}946.67$ ft^3 **(b)** $20{,}096$ ft^2 **11. (a)** 549.5 cm^3
(b) 376.8 cm^2 **13. (a)** 65.94 m^3 **(b)** 100.00 m^2
15. 168 in.^{3} **17.** 1969.10 cm^3 **19.** 427.29 cm^3
21. 508.68 cm^3 **23.** $1{,}694{,}000$ m^3 **25.** 0.52 m^3
27. 288π in.^{3}, 144π in.^{2} **29.** 2 cm, 16π cm^2
31. 1 m, $\frac{4}{3}\pi$ m^3 **33.** volume   **35.** 270 ft^3
37. $\sqrt[3]{2}\,x$ **39.** $\$8100$ **41.** $\$37{,}500$ **43.** 65.7%
45. 2.5 **47.** 6 **49.** 210 in.^{3} **51.** 300 mi   **53.** 2 to 1
55. 288 **57.** Answers will vary.   **59.** $4, 4, 6, 2$
61. $8, 6, 12, 2$ **63.** $20, 12, 30, 2$

9.6 Exercises *(pages 505–507)*
(The answers are given in blue for this section.)

1.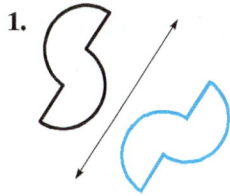

3.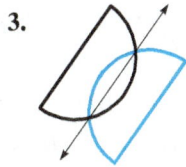

5.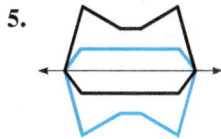

7. The figure is its own reflection image.

9.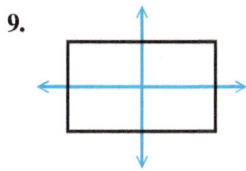

11.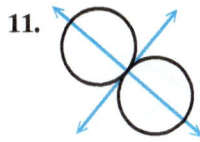

13.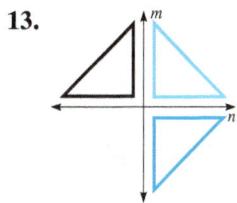

15.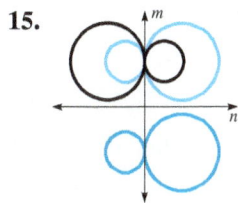

17.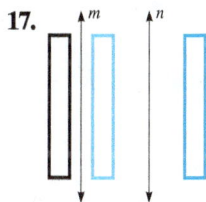

19.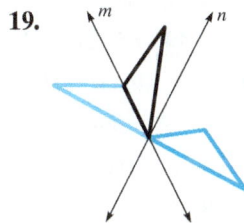

21.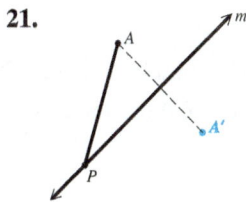

23.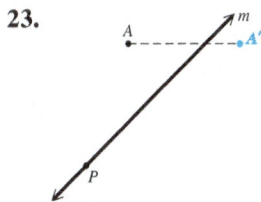

25.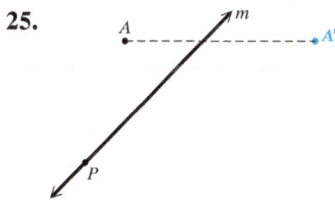

27.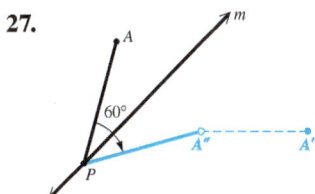

29.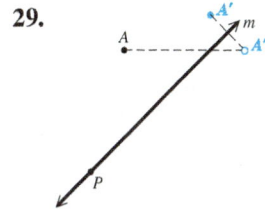

31.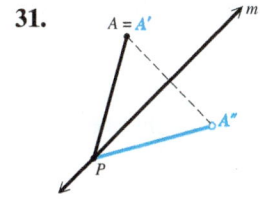

33. no

35.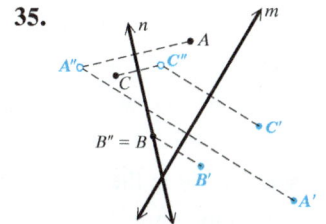

37.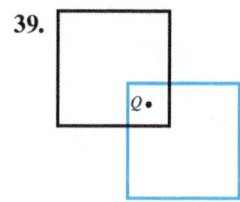

39.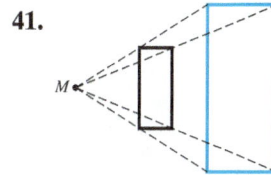

41.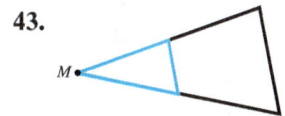

43.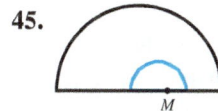

45.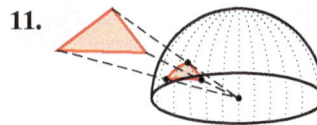

47. 6   **49.** $R_P^2 \cdot C_P^{6n+2}$

9.7 Exercises *(pages 514–516)*
1. Euclidean   **3.** Lobachevskian   **5.** greater than
7. Riemannian   **9.** Euclidean

11.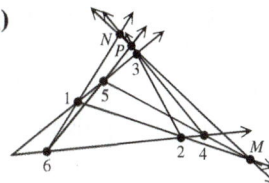

13. Yes; any two distinct lines have at least one point in common.   **15.** Yes; any two points in a plane have at least one line of the plane in common.   **17.** Yes; every point is contained by at least three lines of the plane.

19. no   **21.** yes   **23.** C   **25.** A, E   **27.** B, D
29. A, E   **31.** 1   **33.** 3   **35.** 1

37. (a)–(g)

37. (h) Suppose that a hexagon is inscribed in an angle. Let each pair of opposite sides be extended so as to intersect. Then the three points of intersection thus obtained will lie in a straight line.

9.8 Exercises *(pages 520–521)*

1. 4 **2.** 4 **3.** 2 **4.** $\frac{2}{1} = 2$ **5.** $\frac{4}{1} = 4$ **6.** $\frac{3}{1} = 3; \frac{9}{1} = 9$
7. $\frac{4}{1} = 4; \frac{16}{1} = 16$ **8.** 4, 9, 16, 25, 36, 100
9. Each ratio in the bottom row is the square of the scale factor in the top row. **10.** 4 **11.** 4, 9, 16, 25, 36, 100
12. Each ratio in the bottom row is again the square of the scale factor in the top row. **13.** Answers will vary. Some examples are: $3^d = 9$, thus $d = 2$; $5^d = 25$, thus $d = 2$; $4^d = 16$, thus $d = 2$. **14.** 8 **15.** $\frac{2}{1} = 2; \frac{8}{1} = 8$
16. 8, 27, 64, 125, 216, 1000 **17.** Each ratio in the bottom row is the cube of the scale factor in the top row.
18. Since $2^3 = 8$, the value of d in $2^d = 8$ must be 3.
19. $\frac{3}{1} = 3$ **20.** 4 **21.** 1.262, or $\frac{\ln 4}{\ln 3}$ **22.** $\frac{2}{1} = 2$ **23.** 3
24. It is between 1 and 2. **25.** 1.585, or $\frac{\ln 3}{\ln 2}$
27. 0.842, 0.452, 0.842, 0.452, ... The two attractors are approximately 0.842 and 0.452.

Chapter 9 Test *(pages 529–530)*

1. (a) 48° **(b)** 138° **(c)** acute **2.** 40°, 140°
3. 45°, 45° **4.** 30°, 60° **5.** 130°, 50° **6.** 117°, 117°
7. Answers will vary. **8.** C **9.** both **10.** neither
11. 30°, 45°, 105° **12.** 72 cm² **13.** 60 in.² **14.** 68 m²
15. 180 m² **16.** 57 cm² **17.** 24π in. **18.** 1978 ft

19.

STATEMENTS	REASONS
1. $\angle CAB = \angle DBA$	1. Given
2. $DB = CA$	2. Given
3. $AB = AB$	3. Reflexive property
4. $\triangle ABD \cong \triangle BAC$	4. SAS congruence property

20. 64 ft **21.** 29 m **22.**

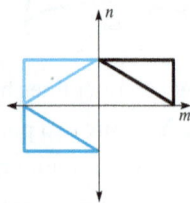

23.

24. (a) 904.32 in.³ **(b)** 452.16 in.²
25. (a) 864 ft³ **(b)** 552 ft²
26. (a) 1582.56 m³ **(b)** 753.60 m²
27. Answers will vary. **28. (a)** yes
(b) no **29.** no **30.** The only attractor is approximately 0.524.

CHAPTER 10 COUNTING METHODS

10.1 Exercises *(pages 539–541)*

1. *AB, AC, AD, AE, BA, BC, BD, BE, CA, CB, CD, CE, DA, DB, DC, DE, EA, EB, EC, ED*; 20 ways
3. *AB, AD, BA, BD, CE, DA, DB, EC*; 8 ways
5. *ACE, AEC, BCE, BEC, DCE, DEC*; 6 ways
7. *ABC, ABD, ABE, ACD, ACE, ADE, BCD, BCE, BDE, CDE*; 10 ways **9.** 1 **11.** 3 **13.** 5 **15.** 5
17. 3 **19.** 1 **21.** 18 **23.** 15 **25.** 27 **27.** 15 **29.** 30
31. 22, 33, 55, 77 **33.** 23, 37, 53, 73
35.

First coin	Second coin	Third coin	Result
			hhh
			hht
			hth
			htt
			thh
			tht
			tth
			ttt

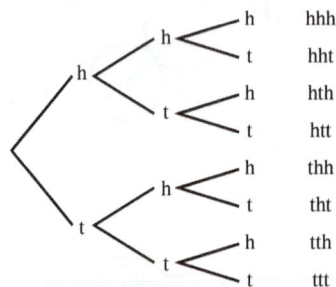

(a) hhh, hht, hth, thh **(b)** hhh **(c)** hht, hth, htt, thh, tht, tth, ttt **(d)** htt, tht, tth, ttt
37. 16 **39.** 36 **41.** 17 **43.** 72 **45.** 12 **47.** 10 **49.** 6
51. 9 **53.** 49 **55.** 21 **57.** 15 **59.** 16 **61.** 13 **63.** 4
65. (a) 1600 **(b)** $4k + 1$ for all positive integers k
(c) $4k^2$ **67.** 3

Wording may vary in the answers for Exercise 69.
69. (a) Determine the number of two-digit numbers that can be formed using only the digits 1, 2, and 3 if repetition of digits is allowed.
(b) Determine the number of two-digit numbers that can be formed using only the digits 1, 2, and 3 if the selection is done with replacement.

10.2 Exercises *(pages 549–552)*

1. 24 **3.** 72 **5.** 20 **7.** 28 **9.** 10 **11.** 1225
13. (a) 720 **(b)** 120 **15. (a)** 306 **(b)** 153
17. 3,628,800 **19.** 3,991,680 **21.** 4,151,347,200
23. 184,756 **25.** 980,179,200 **27.** 134,596 **29.** 60
31. 2,162,160 **33.** $2^3 = 8$ **35.** Answers will vary.
37. $6^3 = 216$ **39.** $2^{10} = 1024$ **41.** $5! = 120$
43. $3 \cdot 2 = 6$ **45.** $3 \cdot 3 = 9$ **47.** $3 \cdot 2 \cdot 1 = 6$
49. $5 \cdot 2 \cdot 4 = 40$ **51.** $2^6 = 64$ **53.** $2 \cdot 3 \cdot 4 \cdot 5 = 120$
55. $2 \cdot 3 \cdot 4 \cdot 3 = 72$ **57.** $2 \cdot 3 \cdot 1 \cdot 3 = 18$
59. $2 \cdot 4 \cdot 6 = 48$ **61.** $5! = 120$ **63.** 800 **65. (a)** 6
(b) 5 **(c)** 4 **(d)** 3 **(e)** 2 **(f)** 1; 720 **67. (a)** 3 **(b)** 3
(c) 2 **(d)** 2 **(e)** 1 **(f)** 1; 36 **69.** 516,243 **71.** 48

10.3 Exercises (pages 561–565)

1. 504   **3.** 95,040   **5.** 330   **7.** 45   **9.** 116,280
11. 43,680   **13.** 126   **15.** 792   **17.** $1.805037696 \times 10^{11}$
19. 225,792,840   **21.** permutation   **23.** combination
25. permutation   **27.** permutation   **29.** $_8P_5 = 6720$
31. $_{12}P_2 = 132$ **33.** $_{25}P_5 = 6{,}375{,}600$ **35.** $_6P_3 = 120$
37. (a) $_6C_3 = 20$ (b) $_6C_2 = 15$ **39.** $_{18}C_5 = 8568$
41. (a) $_{13}C_5 = 1287$ (b) $_{26}C_5 = 65{,}780$
(c) 0 (impossible)   **43.** $_9C_3 = 84$
45. $_{26}P_3 \cdot {}_{10}P_3 \cdot {}_{26}P_3 = 175{,}219{,}200{,}000$
47. $2 \cdot {}_{25}P_3 = 27{,}600$ **49.** $7 \cdot {}_{12}P_8 = 139{,}708{,}800$
51. (a) $6! = 720$ (b) $2 \cdot 4! = 48$ (c) $4! = 24$
53. $_{15}C_1 \cdot {}_{14}C_2 \cdot {}_{12}C_3 \cdot {}_9C_4 \cdot {}_5C_5 = 37{,}837{,}800$
55. $\frac{{}_8C_3 \cdot {}_5C_3 \cdot {}_2C_2}{2!} = 280$ **57.** (a) $_{13}C_4 \cdot 39 = 27{,}885$
(b) $_{12}C_2 \cdot {}_{40}C_3 = 652{,}080$ (c) $_{26}C_2 \cdot {}_{13}C_2 \cdot 13 = 329{,}550$
59. $4 \cdot {}_{13}C_5 = 5148$ **61.** (a) $_7P_2 = 42$ (b) $3 \cdot 6 = 18$
(c) $_7P_2 \cdot 5 = 210$ **63.** $_{20}C_3 = 1140$ **65.** $_8P_3 = 336$
67. $_9C_2 \cdot {}_7C_3 \cdot {}_4C_4 \cdot 2 \cdot 3 \cdot 4 = 30{,}240$
69. (a) $_6C_2 \cdot {}_6C_3 \cdot {}_6C_4 = 4500$
(b) $3! \cdot {}_6C_2 \cdot {}_6C_3 \cdot {}_6C_4 = 27{,}000$
71. (a) $6! = 720$ (b) 745,896
73. Each equals 220.

10.4 Exercises (pages 569–570)

1. 6   **3.** 20   **5.** 56   **7.** 36   **9.** $_7C_1 \cdot {}_3C_3 = 7$
11. $_7C_3 \cdot {}_3C_1 = 105$ **13.** $_8C_3 = 56$ **15.** $_8C_5 = 56$
17. $_9C_4 = 126$ **19.** $1 \cdot {}_8C_3 = 56$ **21.** 1   **23.** 10   **25.** 5
27. 32   **29.** the even-numbered rows
31. (a) All are multiples of the row number.
(b) The same pattern holds.   (c) The same pattern
holds. Each entry is a multiple of 11.   **33.** . . . 8, 13, 21,
34, . . . ; A number in this sequence is the sum of the two
preceding terms. This is the Fibonacci sequence.
35. row 8   **37.** The sum of the squares of the entries
across the top row equals the entry at the bottom vertex.

Wording may vary in the answers for Exercises 39 and 41.
39. sum $= N$; Any entry in the array equals the sum of
the column of entries from its immediate left upward to
the top of the array.   **41.** sum $= N - 1$; Any entry in the
array equals 1 more than the sum of all entries whose
cells make up the largest rectangle entirely to the left and
above that entry.

10.5 Exercises (pages 575–578)

1. $2^4 - 1 = 15$ **3.** $2^7 - 1 = 127$ **5.** $2^7 - 1 = 127$
7. 120   **9.** $36 - 6 = 30$ **11.** $6 + 6 - 1 = 11$ **13.** 51
15. 48   **17.** $90 - 9 = 81$ **19.** $29 + 9 - 2 = 36$

21. (a) $_{10}C_3 = 120$ (b) $_9C_3 = 84$ (c) $120 - 84 = 36$
23. $_7C_3 - {}_5C_3 = 25$ **25.** $_8P_4 - {}_5P_4 = 1560$
27. $_{10}P_3 - {}_7P_3 = 510$ **29.** $2 \cdot 26^2 + 2 \cdot 26^3 = 36{,}504$
31. $_{12}C_4 - {}_8C_4 = 425$ **33.** $13 + 4 - 1 = 16$
35. $12 + 26 - 6 = 32$ **37.** $9 + 8 - 5 = 12$
39. $11 + 5 = 16$ **41.** $2{,}598{,}960 - {}_{13}C_5 = 2{,}597{,}673$
43. $2{,}598{,}960 - {}_{40}C_5 = 1{,}940{,}952$ **45.** 56
47. $_{10}C_0 + {}_{10}C_1 + {}_{10}C_2 = 56$ **49.** $2^{10} - 56 = 968$
51. $26^2 \cdot 10^3 - {}_{26}P_2 \cdot {}_{10}P_3 = 208{,}000$
53. Answers will vary.   **55.** $_4C_3 + {}_3C_3 + {}_5C_3 = 15$
57. $_{12}C_3 - {}_4C_1 \cdot {}_3C_1 \cdot {}_5C_1 = 160$ **59.** Answers will vary.
61. Answers will vary.

Chapter 10 Test (pages 581–582)

1. $6 \cdot 7 \cdot 7 = 294$ **2.** $6 \cdot 7 \cdot 3 = 126$ **3.** $6 \cdot 6 \cdot 5 = 180$
4. $6 \cdot 5 \cdot 1 = 30$ end in 0; $5 \cdot 5 \cdot 1 = 25$ end in 5;
$30 + 25 = 55$ **5.** 13
6.

First toss	Second toss	Third toss	Fourth toss	Result

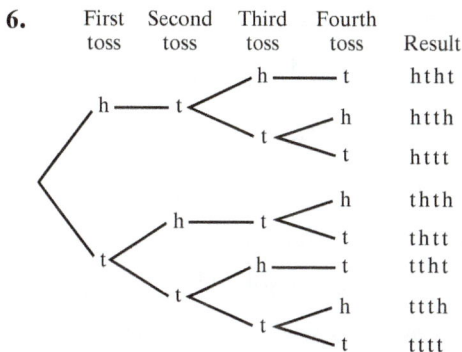

7. $4! = 24$ **8.** $2 \cdot 3 \cdot 4! = 144$ **9.** 720   **10.** 56
11. 1320   **12.** 56   **13.** $_{26}P_5 = 7{,}893{,}600$
14. $32^5 = 33{,}554{,}432$ **15.** $_7P_2 = 42$ **16.** $5! = 120$
17. $\frac{6!}{2! \cdot 3!} = 60$ **18.** $_{10}C_4 = 210$ **19.** $_{10}C_2 \cdot {}_8C_2 = 1260$
20. $_{10}C_5 \cdot {}_5C_5 = 252$ **21.** $\frac{{}_{10}C_4 \cdot {}_6C_4}{2!} = 1575$
22. $2^{10} - ({}_{10}C_0 + {}_{10}C_1 + {}_{10}C_2) = 968$ **23.** $2^5 = 32$
24. $2^3 = 8$ **25.** $2 \cdot 2^3 = 16$ **26.** 13   **27.** 2
28. $32 - (1 + 5) = 26$ **29.** $1 \cdot {}_6C_2 = 15$
30. $1 \cdot 1 \cdot {}_5C_1 = 5$ **31.** $2 \cdot {}_5C_2 = 20$ **32.** $1 \cdot {}_5C_1 = 5$
33. $_5C_3 + {}_5C_2 \cdot {}_2C_1 = 30$ **34.** $_{60}C_3 \cdot {}_{40}C_2 = 26{,}691{,}600$
35. $_9C_4 = 126$ **36.** the counting numbers

CHAPTER 11 PROBABILITY

11.1 Exercises (pages 592–597)

1. (a) $\frac{1}{3}$ (b) $\frac{1}{3}$ (c) $\frac{1}{3}$ **3.** (a) $\frac{1}{2}$ (b) $\frac{1}{3}$ (c) $\frac{1}{6}$
5. (a) $\{1, 2, 3\}$ (b) 2   (c) 1   (d) 3   (e) $\frac{2}{3}$ (f) 2 to 1
7. (a) $\{11, 12, 13, 21, 22, 23, 31, 32, 33\}$ (b) $\frac{2}{3}$ (c) $\frac{1}{3}$
(d) $\frac{1}{3}$ (e) $\frac{4}{9}$ **9.** (a) $\frac{1}{50}$ (b) $\frac{2}{50} = \frac{1}{25}$ (c) $\frac{3}{50}$ (d) $\frac{4}{50} = \frac{2}{25}$

9. (e) $\frac{5}{50} = \frac{1}{10}$ **11. (a)** $\frac{1}{36}$ **(b)** $\frac{2}{36} = \frac{1}{18}$ **(c)** $\frac{3}{36} = \frac{1}{12}$

(d) $\frac{4}{36} = \frac{1}{9}$ **(e)** $\frac{5}{36}$ **(f)** $\frac{6}{36} = \frac{1}{6}$ **(g)** $\frac{5}{36}$ **(h)** $\frac{4}{36} = \frac{1}{9}$

(i) $\frac{3}{36} = \frac{1}{12}$ **(j)** $\frac{2}{36} = \frac{1}{18}$ **(k)** $\frac{1}{36}$ **13.** $\frac{36}{2,598,960} \approx 0.00001385$

15. $\frac{624}{2,598,960} \approx 0.00024010$ **17.** $\frac{1}{4} \cdot \frac{5108}{2,598,960} \approx 0.00049135$

19. $\frac{175}{200} = 0.875$ **21.** $\frac{1}{_{26}P_3} \approx 0.000064$ **23. (a)** $\frac{5}{9}$

(b) $\frac{49}{144}$ **(c)** $\frac{5}{48}$ **25.** $\frac{1}{4}$ **27.** $\frac{1}{4}$ **29. (a)** $\frac{3}{4}$ **(b)** $\frac{1}{4}$

31. $\frac{1}{250,000} = 0.000004$ **33.** $\frac{1}{4}$ **35.** $\frac{1}{4}$ **37.** $\frac{2}{4} = \frac{1}{2}$

39. $\frac{1}{500} = 0.002$ **41.** 160   **43.** $\frac{2}{4} = \frac{1}{2}$ **45.** 7 to 4

47. 9 to 2   **49.** 4 to 7   **51.** 4 to 1   **53.** 2 to 3

55. 37 to 63   **57.** $\frac{12}{31}$ **59. (a)** 0   **(b)** no   **(c)** yes

61. (a) $3 \cdot 1 \cdot 2 \cdot 1 \cdot 1 \cdot 1 = 6$ **(b)** $\frac{6}{720} = \frac{1}{120} \approx 0.0083$

63. (a) $4 \cdot 3! \cdot 3! = 144$ **(b)** $\frac{144}{720} = \frac{1}{5} = 0.2$ **65.** $\frac{1}{9}$

67. $\frac{2}{_{7}C_2} = \frac{2}{21} \approx 0.095$ **69.** $\frac{_{5}C_3}{_{12}C_3} = \frac{10}{220} \approx 0.045$

71. $\frac{1}{_{5}P_3} = \frac{1}{60} \approx 0.017$ **73. (a)** $\frac{8}{9^2} = \frac{8}{81} \approx 0.099$

(b) $\frac{4}{_{9}C_2} = \frac{1}{9} \approx 0.111$ **75.** 1   **77.** $\frac{5}{31}$ **79.** $\frac{9}{9 \cdot 10} = \frac{1}{10}$ **81.** $\frac{1}{15}$

11.2 Exercises (pages 603–605)

1. $\frac{1}{2}$ **3.** $\frac{5}{6}$ **5.** $\frac{2}{3}$ **7. (a)** $\frac{2}{13}$ **(b)** 2 to 11   **9. (a)** $\frac{11}{26}$

(b) 11 to 15   **11. (a)** $\frac{9}{13}$ **(b)** 9 to 4   **13.** $\frac{2}{3}$ **15.** $\frac{7}{36}$

17. $\frac{5}{12}$ **19.** $\frac{2}{3}$ **21.** yes   **23.** 0.005365   **25.** 0.971285

27. 0.76   **29.** 0.92   **31.** 0.04

33.

x	$P(x)$
3	0.1
4	0.1
5	0.2
6	0.2
7	0.2
8	0.1
9	0.1

35. Answers will vary.
37. Answers will vary.
39. $n(A') = s - a$
41. $P(A) + P(A') = 1$
43. 180   **45.** 60
47. 1   **49.** $\frac{1}{4}$

11.3 Exercises (pages 613–617)

1. $\frac{7}{15}$ **3.** $\frac{2}{5}$ **5.** $\frac{1}{15}$ **7.** 1   **9.** $\frac{1}{7}$ **11.** $\frac{1}{8}$ **13.** $\frac{2}{4} = \frac{1}{2}$

15. $\frac{2}{6} = \frac{1}{3}$ **17.** $\frac{2}{4} = \frac{1}{2}$ **19.** $\frac{1}{2}$ **21.** independent

23. not independent   **25.** independent   **27.** $\frac{52}{100} = \frac{13}{25}$

29. $\frac{69}{100}$ **31.** $\frac{14}{31}$ **33.** $\frac{4}{7} \cdot \frac{4}{7} = \frac{16}{49}$ **35.** $\frac{2}{7} \cdot \frac{1}{7} = \frac{2}{49}$ **37.** $\frac{4}{7} \cdot \frac{3}{6} = \frac{2}{7}$

39. $\frac{1}{6}$ **41.** 0   **43.** $\frac{12}{51} = \frac{4}{17}$ **45.** $\frac{12}{52} \cdot \frac{11}{51} = \frac{11}{221}$ **47.** $\frac{4}{52} \cdot \frac{11}{51} = \frac{11}{663}$

49. $\frac{26}{52} \cdot \frac{26}{51} = \frac{13}{51}$ **51.** $\frac{1}{3}$ **53.** 1   **55.** $\frac{3}{10}$ (the same)

57. $\frac{1}{2} \cdot \frac{1}{2} \cdot \frac{1}{2} \cdot \frac{1}{2} \cdot \frac{1}{2} \cdot \frac{1}{2} = \frac{1}{64}$ **59.** 0.490   **61.** 0.027

63. 0.95   **65.** 0.23   **67.** $\frac{1}{20}$ **69.** $\frac{1}{10}$ **71.** $2^6 = 64$

73. 0.1479   **75.** 0.400   **77.** 0.080   **79.** $(0.90)^4 = 0.6561$

81. $_{4}C_2 \cdot (0.10) \cdot (0.20) \cdot (0.70)^2 = 0.0588$ **83.** 0.2704

85. 0.2496   **87.** Answers will vary.   **89.** 0.21   **91. (a)** $\frac{3}{4}$

(b) $\frac{1}{2}$ **(c)** $\frac{5}{16}$ **93.** $\frac{1}{5}$ **95.** $\frac{8}{33}$ **97. (a)** Answers will

vary.   **(b)** $\frac{25}{36} \approx 0.69$

11.4 Exercises (pages 621–623)

1. $\frac{1}{8}$ **3.** $\frac{3}{8}$ **5.** $\frac{3}{4}$ **7.** $\frac{1}{2}$ **9.** Answers will vary.   **11.** $\frac{1}{128}$

13. $\frac{21}{128}$ **15.** $\frac{35}{128}$ **17.** $\frac{7}{128}$ **19.** $\frac{125}{216}$ **21.** $\frac{5}{72}$ **23.** 0.041

25. 0.268   **27.** .302   **29.** Answers will vary.   **31.** 0.228

33. 0.299   **35.** 0.032   **37.** 0.259   **39.** 0.167   **41.** 0.010

43. $6p^2(1 - p)^2$ **45.** 0.883   **47.** 0.073   **49.** $\frac{1}{1024} \approx 0.001$

51. $\frac{45}{1024} \approx 0.044$ **53.** $\frac{210}{1024} = \frac{105}{512} \approx 0.205$ **55.** $\frac{772}{1024} \approx 0.754$

11.5 Exercises (pages 633–637)

1. $\frac{5}{2}$ **3.** \$1   **5.** \$0.50   **7.** no $\left(\text{expected net winnings: } -\frac{3}{4}\text{¢}\right)$

9. 1.69   **11. (a)** –\$60   **(b)** \$36,000   **(c)** \$72,000

13. \$0.46   **15.** \$2700   **17.** 2.7   **19.** a decrease of 50

21. Project B **23.** Project A **25.** \$2200   **27.** \$81,000

29. Do not purchase the insurance (because

\$86,000 > \$81,000).   **31.** \$1500; \$3000; \$17,500; \$27,000

33. \$56,000   **35.** 48.7%   **37.** $\frac{15}{47} \approx 0.319$ **39.** $\frac{18}{50} = 0.36$;

This is quite close to 0.375, the theoretical value.   **41.** no

43. $\frac{6}{50} = 0.12$ **45.** $\frac{48}{199} \approx 0.241$ **47.** $\frac{48}{198} \approx 0.242$

49. Answers will vary.   **51.** The walk ends 12 blocks

north of the starting point.   **53. (a)** Answers will vary.

(c) The value of p will vary; empirical   **(d)** Answers will

vary.   **(e)** Answers will vary.

Chapter 11 Test (pages 641–642)

1. $\frac{_{2}C_2}{_{5}C_2} = \frac{1}{10}$ **2.** $\frac{_{3}C_2}{_{5}C_2} = \frac{3}{10}$ **3.** $\frac{6}{10} = \frac{3}{5}$ **4.** $\frac{3}{10}$

5. $\frac{7}{7} \cdot \frac{6}{7} \cdot \frac{5}{7} = \frac{30}{49}$ **6.** $\frac{7}{19}$ **7.** $1 - \left(\frac{30}{49} + \frac{1}{49}\right) = \frac{18}{49}$

8. row 1: CC; row 2: cC, cc   **9.** $\frac{1}{2}$ **10.** 1 to 3   **11.** 3 to 1

12. 25 to 1   **13.** 11 to 2   **14.** $\frac{3}{10}; \frac{6}{10}; \frac{1}{10}$ **15.** $\frac{9}{10}$

16. $\frac{18}{10} = \frac{9}{5}$ **17.** $\frac{6}{36} = \frac{1}{6}$ **18.** 35 to 1   **19.** 7 to 2   **20.** $\frac{4}{36} = \frac{1}{9}$

21. $(0.78)^3 \approx 0.475$ **22.** $_{3}C_2 \cdot (0.78)^2 \cdot (0.22) \approx 0.402$

23. $1 - (0.22)^3 \approx 0.989$

24. $(0.78) \cdot (0.22) \cdot (0.78) \approx 0.134$

25. $\frac{25}{102}$ **26.** $\frac{25}{51}$ **27.** $\frac{4}{51}$ **28.** $\frac{3}{26}$ **29.** $\frac{3}{8}$ **30.** $\frac{1}{24} \approx 0.042$

31. $\frac{29}{99} \approx 0.293$ **32.** $\frac{18}{98} \approx 0.184$

CHAPTER 12 STATISTICS

12.1 Exercises *(pages 651–655)*

1. (a)

x	f	$\frac{f}{n}$
0	10	$\frac{10}{30} \approx 33\%$
1	7	$\frac{7}{30} \approx 23\%$
2	6	$\frac{6}{30} = 20\%$
3	4	$\frac{4}{30} \approx 13\%$
4	2	$\frac{2}{30} \approx 7\%$
5	1	$\frac{1}{30} \approx 3\%$

(b)

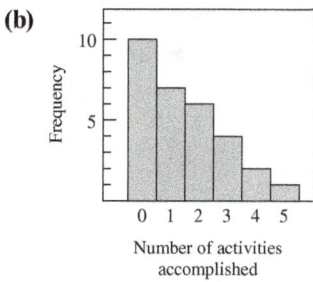

(c)

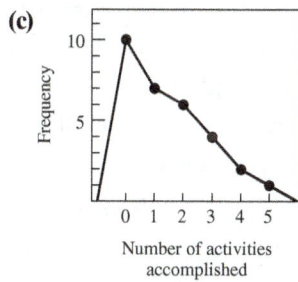

3. (a)

Class Limits	Tally	Frequency f	Relative Frequency $\frac{f}{n}$
21–25	I	1	$\frac{1}{48} \approx 2.1\%$
26–30	II	2	$\frac{2}{48} \approx 4.2\%$
31–35	IIII	5	$\frac{5}{48} \approx 10.4\%$
36–40	IIII IIII II	12	$\frac{12}{48} = 25.0\%$
41–45	IIII IIII IIII I	16	$\frac{16}{48} \approx 33.3\%$
46–50	IIII IIII II	12	$\frac{12}{48} = 25.0\%$

Total: $n = 48$

(b)

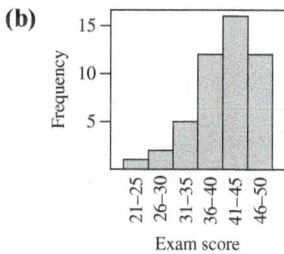

(c)

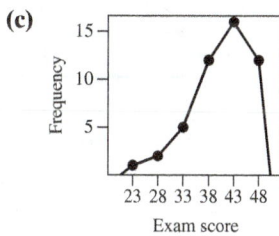

5. (a)

Class Limits	Tally	Frequency f	Relative Frequency $\frac{f}{n}$
70–74	II	2	$\frac{2}{30} \approx 6.7\%$
75–79	I	1	$\frac{1}{30} \approx 3.3\%$
80–84	III	3	$\frac{3}{30} = 10.0\%$
85–89	II	2	$\frac{2}{30} \approx 6.7\%$
90–94	IIII	5	$\frac{5}{30} \approx 16.7\%$
95–99	IIII II	7	$\frac{7}{30} \approx 23.3\%$
100–104	IIII I	6	$\frac{6}{30} = 20.0\%$
105–109	IIII	4	$\frac{4}{30} \approx 13.3\%$

Total: $n = 30$

(b)

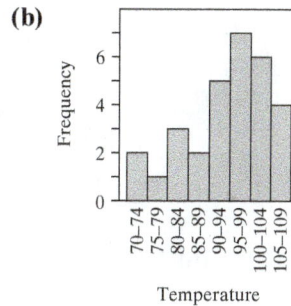

(c)

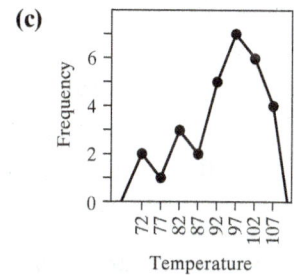

7.
```
0 | 7 9 8
1 | 1 1 2 8 9 4 3 1 0 5 0 5 5
2 | 7 0 9 6 6 2 2 5 2 3 4 4
3 | 8 1
```

9.
```
0 | 8 5 4 9 6 9 4 8
1 | 6 0 1 8 8 2 4 0 2 8 6 3
2 | 6 1 2 5 1 3
3 | 0 4 6
4 | 4
```

11. 2001 **13.** 2005, 2006, 2007, 2010, and 2012

15.

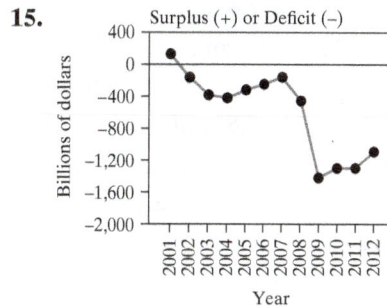

17. 3.8% in 2008   **19.** Answers will vary.
21. school; 120°
23.

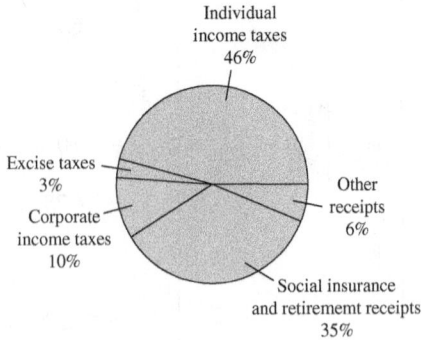

25. about 82 years   **27. (a)** about 7 years
(b) Answers will vary.   **29.** Answers will vary.
31. Answers will vary.   **33.** Answers will vary.
35. (a)

Letter	Probability
A	0.208
E	0.338
I	0.169
O	0.208
U	0.078

(b)

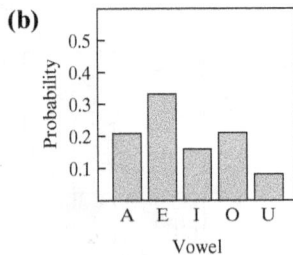

37. Answers will vary.   **39. (a)** 0.225   **(b)** 0.150
(c) 0.675   **(d)** 0.075
41. (a)

Sport	Probability
Sailing	0.175
Archery	0.100
Snowboarding	0.250
Bicycling	0.200
Rock climbing	0.200
Rafting	0.075

(b) empirical   **(c)** Answers will vary.

12.2 Exercises *(pages 667–670)*

1. (a) 12.4   **(b)** 12   **(c)** none   **3. (a)** 216.2   **(b)** 221
(c) 196   **5. (a)** 5.2   **(b)** 5.35   **(c)** 4.5 and 6.2
7. (a) 129.9   **(b)** 128   **(c)** 125 and 128   **9.** more than
14.525   **11. (a)** 10.7 million   **(b)** 10.65 million
(c) 10.6 million   **13. (a)** 73.9   **(b)** 17.5   **(c)** 0
15. mean = 47.4; median and mode remain the same
17. $297.0 billion   **19.** 5.27 seconds   **21.** 2.42 seconds
23. the mean   **25.** mean = 77; median = 80; mode = 79
27. 92   **29. (a)** 589.6   **(b)** 586   **(c)** 579   **31.** 3.19
33. (a) $2,532.4 billion   **(b)** $2,450.2 billion
35. about 511,200   **37. (a)** 8.25   **(b)** 8   **39. (a)** 66.25
(b) 64.5   **41.** 6.5   **43. (a)** 21.3   **(b)** 21.5   **(c)** 15
45. (a) 74.8   **(b)** 77.5   **(c)** 78   **47.** Answers will vary.

49. (a) mean = 3; median = 3; mode = 4
(b) mode   **(c)** Answers will vary.   **51. (a)** 4   **(b)** 4.25
53. (a) 6   **(b)** 9.33   **55.** three choices: 1, 6, 16   **57.** no
59. Answers will vary.

12.3 Exercises *(pages 677–679)*

1. the sample standard deviation   **3. (a)** 14   **(b)** 4.74
5. (a) 26   **(b)** 8.38   **7. (a)** 1.14   **(b)** 0.37   **9. (a)** 11
(b) 3.89   **11.** $\frac{3}{4}$ **13.** $\frac{21}{25}$ **15.** $\frac{3}{4}$ **17.** $\frac{15}{16}$ **19.** $\frac{1}{4}$ **21.** $\frac{4}{49}$
23. $154.58   **25.** nine   **27.** There are at least nine.
29. (a) $s_A = 2.35$; $s_B = 2.58$ **(b)** $V_A = 46.9\%$; $V_B = 36.9\%$
(c) sample B ($s_B = 2.58 > 2.35$)   **(d)** sample A
($V_A = 46.9\% > 36.9\%$)   **31.** 18.71; 4.35
33. 8.71; 4.35   **35.** 56.14; 13.04
37. (a) $\bar{x}_A = 68.8$; $\bar{x}_B = 66.6$ **(b)** $s_A = 4.21$; $s_B = 5.27$
(c) brand A ($\bar{x}_A > \bar{x}_B$)   **(d)** brand A ($s_A < s_B$)
39. Brand A ($s_B = 3539 > 2116$)   **41.** Answers will
vary.   **43.** −3.0   **45.** 4.2   **47.** 7.55 and 23.45   **49.** no
51. Answers will vary.

12.4 Exercises *(pages 683–686)*

1. 58   **3.** 59   **5.** Janet (since $z = 0.48 > 0.19$)
7. Yvette (since $z = -1.44 > -1.78$)   **9.** −0.7   **11.** 1.4
13. Russia   **15.** United States   **17.** India in consumption
(India's consumption z-score was −0.6, Canada's production z-score was −0.7, and −0.6 > −0.7.)
19. (a) The median is 28.2 quadrillion Btu.
(b) The range is 109.6 − 11.7 = 97.9 quadrillion Btu.
(c) The middle half of the items extend from 13.5 to 97.5
quadrillion Btu.   **21.** Answers will vary.   **23.** Answers
will vary.   **25.** Answers will vary.   **27.** the overall distribution   **29.** Both are skewed to the right, consumption
nearly twice as much as production.   **31. (a)** no
(b) Answers will vary.   **33.** no; Answers will vary.
35. yes; Answers will vary.   **37.** $Q_1 = 96.7$, $Q_2 = 104.8$,
$Q_3 = 109.0$ **39.** $P_{95} = 119.2$ **41.** 102.9
43.

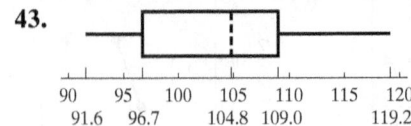

45. No; 69.4 + 20.4 = 89.8, still less than 91.6   **47.** 6
49. (a) 2.5, 4.5, 6.5   **(b)** none   **51. (a)** 3, 5.5, 8
(b) two   **53. (a)** none   **(b)** one   **(c)** two   **(d)** three

12.5 Exercises *(pages 693–695)*

1. discrete   **3.** continuous   **5.** discrete   **7.** 50   **9.** 95
11. 50%   **13.** 95%   **15.** 44.5%   **17.** 45.4%   **19.** 5.3%
21. 97.8%   **23.** 0.84   **25.** 0.77   **27.** 500   **29.** 61
31. 1000   **33.** 11.5%   **35.** 54.0%   **37.** 0.212   **39.** 0.092

41. 0.994, or 99.4%   **43.** 166 units   **45.** 0.888

47. about 2 eggs   **49.** 24.2%   **51.** Answers will vary.

53. 76   **55.** 65   **57.** 90.4   **59.** 40.3

61. (a) at least 36%   **(b)** 78.8%

Chapter 12 Test (pages 702–704)

1. (a) 59.6 million   **(b)** 116.8 million   **(c)** 48.2 million

(d) 80.9%   **2. (a)** those who copied less than 10%

(b) those with copy rates from 10% to 30% or greater

than 50%   **(c)** those with a copy rate from 30% to 50%

(d) Answers will vary.   **3. (a)** 281 feet   **(b)** 595 inches

(c) 1223   **4. (a)** 762   **(b)** 8th   **(c)** 22 inches   **(d)** 17 feet

5.

Top Six U.S. College Endowment Funds

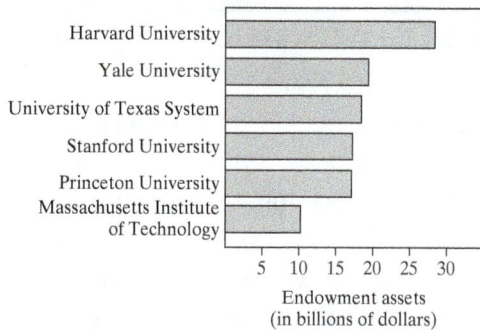

Endowment assets
(in billions of dollars)

6. 17.2%   **7.** $17.65 billion

8.

Class Limits	Frequency f	Relative Frequency $\frac{f}{n}$
6–10	3	$\frac{3}{22} \approx 0.14$
11–15	6	$\frac{6}{22} \approx 0.27$
16–20	7	$\frac{7}{22} \approx 0.32$
21–25	4	$\frac{4}{22} \approx 0.18$
26–30	2	$\frac{2}{22} \approx 0.09$

9. (a)

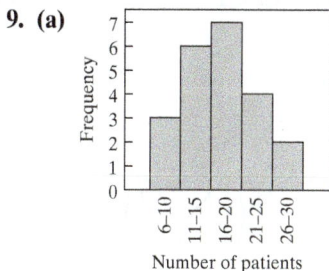

Number of patients

(b)

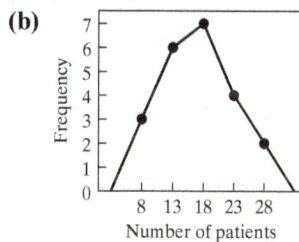

Number of patients

10. 8   **11.** 12.4   **12.** 12   **13.** 12   **14.** 10

15.
```
3 | 3  8
4 | 3  5  8  9
5 | 0  2  5
6 | 1  1  4  5  6  7  7  8
7 | 0  1  2  3  7  7  8  9  9
8 | 0  4  4
9 | 1
```

16. 35   **17.** 33   **18.** 37   **19.** 31   **20.** 49

21.

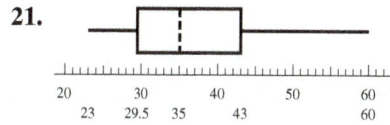

22. (a) about 95%   **(b)** about 0.3%   **(c)** about 16%

(d) about 13.5%   **23.** 0.249   **24.** 0.530

25. East   **26.** Central   **27.** 82.4

28. Red Sox (since $z = 1.20 > 0.95$)

29.

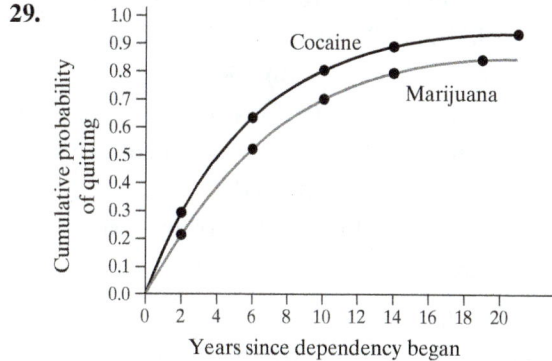

Years since dependency began

(a) about 4 years   **(b)** about 6 years

30. Answers will vary.

CHAPTER 13 PERSONAL FINANCIAL MANAGEMENT

13.1 Exercises (pages 716–719)

1. $48   **3.** $48.30   **5.** $361.13   **7. (a)** $784   **(b)** $788.31

9. (a) $2725   **(b)** $2731.82   **11.** $125   **13.** $6879.17

15. $168.26   **17.** $817.17   **19.** $17,724.34; $10,224.34

21. (a) $2339.72   **(b)** $2343.32   **(c)** $2345.16

23. (a) $18,545.42   **(b)** $18,546.80   **(c)** $18,547.49

25. (a) $62.53   **(b)** $62.66   **(c)** $62.74   **(d)** $62.79

(e) $62.79   **27.** Answers will vary.   **29.** $1461.04

31. $747.26   **33.** $3846.60   **35.** $168,910.81

37. $167,364.81   **39.** 2.000%   **41.** 2.015%

43. 2.020%   **45.** 2.020%   **47.** $112,041.31

49. 17 years, 152 days   **51.** $r = n[(Y+1)^{1/n} - 1]$

53. (a) 1.804%   **(b)** Bank A; $2.63   **(c)** no difference;

$90.81   **55.** 35 years   **57.** 8 years   **59.** 10.0%   **61.** 3.2%

63. $7.20; $8.80; $11.60; $31.60   **65.** $34,400; $42,000;

$55,600; $151,000   **67.** $237   **69.** $1632

71. $81,102,828.75

13.2 Exercises (pages 722–724)

1. $2795   **3.** $3130.40   **5.** $4130.40   **7.** $2640

9. $398.75   **11.** $607.50; $141.88   **13.** $41.63; $43.98

15. $36.38; $35.71   **17.** $229.17   **19.** $292.95

21. $12,628   **23.** 5 years   **25.** $3.50   **27.** $5.80

29. $15.30 **31. (a)** $607.33 **(b)** $7.90 **(c)** $646.87
33. (a) $473.96 **(b)** $6.16 **(c)** $505.54 **35.** $9.27
37. (a) $38.18 **(b)** $38.69 **(c)** $39.20 **39.** 16.0%
41. $32.00 **43.–47.** Answers will vary. **49. (a)** $127.44
(b) $139.08

13.3 Exercises *(pages 729–731)*

1. 7.0% **3.** 8.5% **5.** $136.46 **7.** $90.63 **9.** 6.5%
11. 9.5% **13. (a)** $8.93 **(b)** $511.60 **(c)** $6075.70
15. (a) $1.90 **(b)** $66.65 **(c)** $4103.95 **17. (a)** $206
(b) 9.5% **19. (a)** $138 **(b)** 8.5% **21. (a)** $41.29
(b) $39.70 **23. (a)** $420.68 **(b)** $368.47
25. (a) $1.80 **(b)** $14.99 **(c)** $1045.01
27. finance company APR: 12.0%; credit union APR:
11.5%; choose credit union **29.** $32.27 **31.** 13
33. 13.5% **35.** 8 **37.** 15 **39.–41.** Answers will vary.
43. $\frac{5}{39}$ **45.** Answers will vary.

13.4 Exercises *(pages 744–748)*

1. $836.44 **3.** $845.22 **5.** $2150.37 **7.** $794.26
9. (a) $513.38 **(b)** $487.50 **(c)** $25.88 **(d)** $58,474.12
11. (a) $1247.43 **(b)** $775.67 **(c)** $471.76
(d) $142,728.24 **(e)** $1247.43 **(f)** $773.11 **(g)** $474.32
(h) $142,253.92 **13. (a)** $1390.93 **(b)** $776.61
(c) $614.32 **(d)** $113,035.68 **(e)** $1390.93
(f) $772.41 **(g)** $618.52 **(h)** $112,417.16 **15.** $480.31
17. $995.95 **19.** 360 **21.** $247,532.80 **23.** $1180.66;
$425,037.60; $185,037.60 **25.** $1686.41; $303,553.80;
$63,553.80 **27.** Answers will vary. **29. (a)** $572.90
(b) $1280.71 **31. (a)** $117,886.76 **(b)** $119,609.63
33. (a) payment 153 **(b)** payment 295 **35.** $25,465.20
37. $91,030.80 **39. (a)** $674.79 **(b)** $746.36
(c) an increase of $71.57 **41. (a)** $229.17 **(b)** $582.89
43. $107.66 **45.** $140,000 **47.** $4640 **49.** $302
51. (a) $1634 **(b)** $1754 **(c)** $917 **(d)** $997 **(e)** $1555
53. (a) $2271 **(b)** $2407 **(c)** $1083 **(d)** $1174
(e) $2055 **55. (a)** $8370 **(b)** $39,199 **(c)** $23,387
57.–63. Answers will vary.

13.5 Exercises *(pages 760–763)*

1. $89.99 **3.** 0.49% higher **5.** $139.69 **7.** 13.93% lower
9. YUM **11.** BAC **13.** $30,924.00 **15.** $21,496.00
17. $1201.79 **19.** $5257.04 **21.** $179,109.99
23. $179,722.99 **25.** $23,306.53 **27.** $61,800.68
29. $188,083.39 **31.** $47,000.41 **33.** $1854.25 net paid out
35. (a) $800 **(b)** $80 **(c)** $960 **(d)** $1040 **(e)** 130%
37. (a) $1250 **(b)** $108 **(c)** −$235 **(d)** −$127
(e) −10.16% **39.** $275.00 **41.** $177.75
43. (a) $10.49 **(b)** 334 **45. (a)** $8.27 **(b)** 3080

47. (a) $8.39 **(b)** $100.68 **(c)** 15.6% **49. (a)** $57.45
(b) $689.40 **(c)** 27.6% **51. (a)** $1203.75 **(b)** $18.06
(c) 19.56% **53. (a)** $4179 **(b)** $76.48 **(c)** 24.31%
55. $129,258 **57.** $993,383 **59. (a)** $56,473.02
(b) $49,712.70 **61. (a)** $12,851.28 **(b)** $11,651.81
63. (a) $8105.66 **(b)** $12,894.98
65. (a) $R = \dfrac{rV}{(1-t)[(1+r)^n - 1]}$ **(b)** $R = \dfrac{rV}{(1 + r(1-t))^n - 1}$
67. (a) $r = n\left[\left(\dfrac{A}{P}\right)^{1/m} - 1\right]$ **(b)** 5.4%

Aggressive Growth	Growth	Growth & Income	Income	Cash
69. $1400	$8600	$6200	$2800	$1000
71. $8000	$112,000	$144,000	$116,000	$20,000

73. 3.75% **75.** 5.2% **77.–83.** Answers will vary.

Chapter 13 Test *(page 772)*

1. $1200 **2.** $563.25 **3.** 2.02% **4.** 18 years
5. $67,297.13 **6.** $10.88 **7.** $630 **8.** $184.17
9. $6.5% **10. (a)** $75.97 **(b)** $2318.24 **11.** 7.0%
12. Answers will vary. **13.** $833.75 **14.** $1207.05
15. $3488 **16.** Answers will vary. **17.** 6.1%
18. 252,300 **19.** $10.00 **20.** 0.078%; no; The
percentage return was not for a 1-year period.
21. $106,728.55 **22.** Answers will vary.

CHAPTER 14 GRAPH THEORY

14.1 Exercises *(pages 787–792)*

1. 7 vertices, 7 edges **3.** 10 vertices, 9 edges
5. 6 vertices, 9 edges **7.** Two have degree 3. Three have
degree 2. Two have degree 1. Sum of degrees is 14. This is
twice the number of edges. **9.** Six have degree 1. Four
have degree 3. Sum of degrees is 18. This is twice the number of edges. **11.** not isomorphic
13. isomorphic; Corresponding edges should be the same
color. AB should match AB, etc.

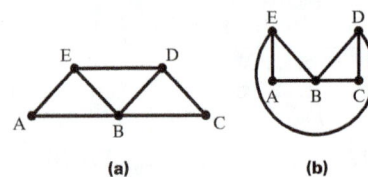

(a)   (b)

15. not isomorphic **17.** connected, 1 component
19. disconnected, 3 components **21.** disconnected,
2 components **23.** 10 **25.** 4 **27. (a)** yes **(b)** No,
because there is no edge from A to D. **(c)** No, because
there is no edge from A to E. **(d)** yes **(e)** yes **(f)** yes

29. (a) No, because it does not return to the starting vertex.
(b) yes **(c)** No, because there is no edge from C to F.
(d) No, because it does not return to starting vertex.
(e) No, because the edge from F to D is used more than
once. **31. (a)** No, because there is no edge from B to C.
(b) No, because the edge from I to G is used more than
once. **(c)** No, because there is no edge from E to I.
(d) yes **(e)** yes **(f)** No, because the edge from A to D
is used more than once. **33.** It is a walk and a path, not a
circuit. **35.** It is a walk, not a path, not a circuit.
37. It is a walk and a path, not a circuit.
39. No. For example, there is no edge from A to C.
41. No. For example, there is no edge from A to F.
43. yes
45. 7 games **47.** 36 handshakes

49. 10 telephone conversations
51. A → B → D → A corresponds to
tracing around the edges of a single
face. (The circuit is a triangle.)
53. Answers will vary. **55.** Answers will vary.
57. 2 **59.** 3
61. 2
63. (a) 2 **(b)** 3 **(c)** 2
(d) 3 **(e)** A cycle with an odd number of vertices
has chromatic number 3. A cycle
with an even number of vertices has
chromatic number 2.
65. 3 **67.** 3
69. Three times: Choir and Forensics,
Service Club and Dance Club, and
Theater and Caribbean Club.
(There are other possible
groupings.)

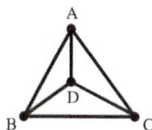

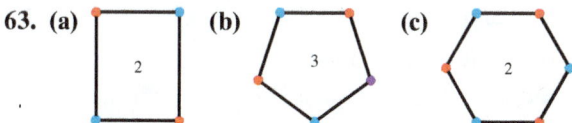

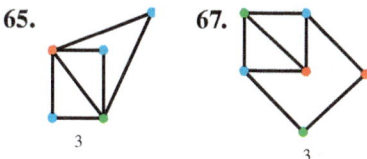

71. Three gatherings: Brad and Phil
and Mary, Joe and Lindsay, and
Caitlin and Eva. (There are other
possible groupings.)

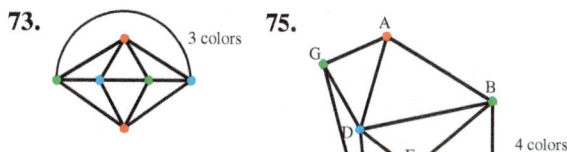

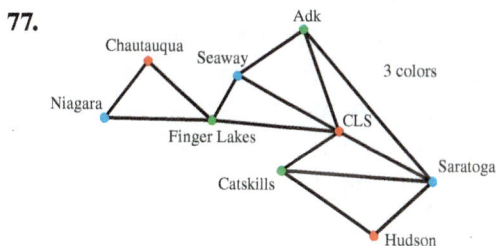

73. 3 colors **75.** 4 colors

77. 3 colors

**In Exercise 79, there are many different ways to draw the
map.**
79.

81. By the four-color theorem, this is not possible.
83. 2 **85.** 2 **87.** 3 **89.** 2 **91.** 1
93.

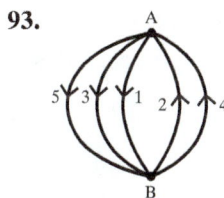

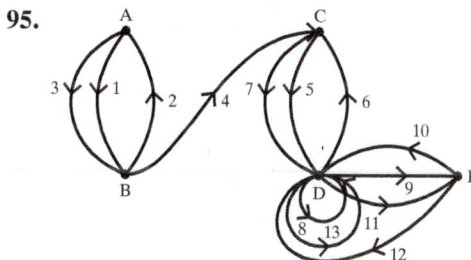

95.

14.2 Exercises *(pages 801–805)*
1. (a) No, it is not a path. **(b)** yes **(c)** No, it is not a
walk. **(d)** No, it is not a circuit. **3. (a)** No, some edges
are not used (e.g., B → F). **(b)** yes **(c)** No, it is not
a path. **(d)** yes **5.** Yes, all vertices have even degree.
7. No, some vertices (e.g., G) have odd degree.

9. yes **11.** It has an Euler circuit. All vertices have even degree. No circuit visits each vertex exactly once.

13. It has an Euler circuit. All vertices have even degree. A → B → H → C → G → D → F → E → A visits each vertex exactly once.

15. No, some vertices have odd degree. **17.** Yes, all vertices have even degree. **19.** none **21.** FG

23. B → E or B → D **25.** B → C or B → H

27. A → C → B → F → E → D → C → F → D → A

29. The graph has an Euler circuit.
A → G → H → J → I → L → J → K → I →
H → F → G → E → F → D → E → C → D →
B → C → A → E → B → A

31. The graph does not have an Euler circuit. Some vertices (e.g., C) have odd degree. **33.** There is such a route: A → D → B → C → A → H → D → E → B →
H → G → E → F → H → J → L → C → M → A → K →
M → L → K → J → A **35.** It is not possible. Some vertices (e.g., room at upper left) have odd degree.

37. No. There are more than two vertices with odd degree. **39.** yes; B and G **41.** It is possible. Exactly two of the rooms have odd numbers of doors. **43.** It is not possible. All rooms have even numbers of doors.

45. yes; Manhattan and Randall's Island.

There are other correct answers in Exercise 47.

47. A → B → C → D → E → C → F → B → E → F → A; There are 10 edges in any such circuit.

49. In the graph shown, not all vertices have even degree. For example, the vertex marked X is of degree 5. So this graph still does not have an Euler circuit.

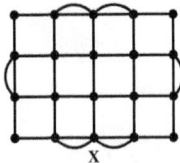

51. (There are other ways of inserting the additional edges.)

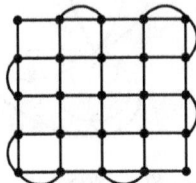

8 edges added

14.3 Exercises *(pages 813–817)*

1. (a) No, because it visits E twice. **(b)** yes
(c) No, because it does not visit C. **(d)** No, because it does not visit A. **3. (a)** None, because there is no edge from B to C. **(b)** all three **(c)** none; Edge AD is used twice. **5.** A → B → D → E → F → C → A

7. G → H → J → I → G **9.** X → T → U → W → V → X

11. Hamilton circuit: A → B → C → D → A. The graph has no Euler circuit, because at least one of the vertices has odd degree. (In fact, all have odd degree.)

13. A → B → C → D → E → F → A
is both a Hamilton and an Euler circuit.

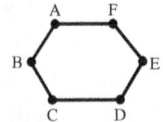

15. Hamilton circuit **17.** Euler circuit **19.** Hamilton circuit **21.** 24 **23.** 362,880 **25.** 9! **27.** 17!

29. P → Q → R → S → P   P → Q → S → R → P
P → R → Q → S → P   P → R → S → Q → P
P → S → Q → R → P   P → S → R → Q → P

31. E → H → I → F → G → E   E → H → I → G → F → E

33. E → F → G → H → I → E   E → F → G → I → H → E
E → F → H → G → I → E   E → F → H → I → G → E
E → F → I → G → H → E   E → F → I → H → G → E

35. E → G → F → H → I → E   E → G → F → I → H → E
E → G → H → F → I → E   E → G → H → I → F → E
E → G → I → F → H → E   E → G → I → H → F → E

37. A → B → C → D → E → A
A → B → C → E → D → A
A → B → D → C → E → A
A → B → D → E → C → A
A → B → E → C → D → A
A → B → E → D → C → A
A → C → B → D → E → A
A → C → B → E → D → A
A → C → D → B → E → A
A → C → D → E → B → A
A → C → E → B → D → A
A → C → E → D → B → A
A → D → B → C → E → A
A → D → B → E → C → A
A → D → C → B → E → A
A → D → C → E → B → A
A → D → E → B → C → A
A → D → E → C → B → A
A → E → B → C → D → A
A → E → B → D → C → A
A → E → C → B → D → A
A → E → C → D → B → A
A → E → D → B → C → A
A → E → D → C → B → A

39. Minimum Hamilton circuit is $P \rightarrow Q \rightarrow R \rightarrow S \rightarrow P$; Weight is 2200. **41.** Minimum Hamilton circuit is $C \rightarrow D \rightarrow E \rightarrow F \rightarrow G \rightarrow C$; Weight is 64.

43. (a) $A \rightarrow C \rightarrow E \rightarrow D \rightarrow B \rightarrow A$; Total weight is 20.
(b) $C \rightarrow A \rightarrow B \rightarrow D \rightarrow E \rightarrow C$; Total weight is 20.
(c) $D \rightarrow C \rightarrow A \rightarrow B \rightarrow E \rightarrow D$; Total weight is 23.
(d) $E \rightarrow C \rightarrow A \rightarrow B \rightarrow D \rightarrow E$; Total weight is 20.

45. (a) $A \rightarrow C \rightarrow D \rightarrow E \rightarrow B \rightarrow A$; Total weight is 87.
$B \rightarrow C \rightarrow A \rightarrow E \rightarrow D \rightarrow B$; Total weight is 95.
$C \rightarrow A \rightarrow E \rightarrow D \rightarrow B \rightarrow C$; Total weight is 95.
$D \rightarrow C \rightarrow A \rightarrow E \rightarrow B \rightarrow D$; Total weight is 131.
$E \rightarrow C \rightarrow A \rightarrow D \rightarrow B \rightarrow E$; Total weight is 137.
(b) $A \rightarrow C \rightarrow D \rightarrow E \rightarrow B \rightarrow A$ with total weight 87
(c) For example, $A \rightarrow B \rightarrow C \rightarrow D \rightarrow E \rightarrow A$ has total weight 52.

47. $A \rightarrow B \rightarrow C \rightarrow D \rightarrow E \rightarrow F \rightarrow A$
$A \rightarrow B \rightarrow C \rightarrow F \rightarrow E \rightarrow D \rightarrow A$
$A \rightarrow B \rightarrow E \rightarrow D \rightarrow C \rightarrow F \rightarrow A$
$A \rightarrow B \rightarrow E \rightarrow F \rightarrow C \rightarrow D \rightarrow A$
$A \rightarrow D \rightarrow E \rightarrow F \rightarrow C \rightarrow B \rightarrow A$
$A \rightarrow D \rightarrow E \rightarrow B \rightarrow C \rightarrow F \rightarrow A$
$A \rightarrow D \rightarrow C \rightarrow B \rightarrow E \rightarrow F \rightarrow A$
$A \rightarrow D \rightarrow C \rightarrow F \rightarrow E \rightarrow B \rightarrow A$
$A \rightarrow F \rightarrow E \rightarrow B \rightarrow C \rightarrow D \rightarrow A$
$A \rightarrow F \rightarrow E \rightarrow D \rightarrow C \rightarrow B \rightarrow A$
$A \rightarrow F \rightarrow C \rightarrow D \rightarrow E \rightarrow B \rightarrow A$
$A \rightarrow F \rightarrow C \rightarrow B \rightarrow E \rightarrow D \rightarrow A$

49. $A \rightarrow B \rightarrow C \rightarrow D \rightarrow E \rightarrow F \rightarrow A$
$A \rightarrow B \rightarrow C \rightarrow E \rightarrow D \rightarrow F \rightarrow A$
$A \rightarrow B \rightarrow C \rightarrow E \rightarrow F \rightarrow D \rightarrow A$
$A \rightarrow B \rightarrow C \rightarrow F \rightarrow E \rightarrow D \rightarrow A$
$A \rightarrow D \rightarrow E \rightarrow F \rightarrow C \rightarrow B \rightarrow A$
$A \rightarrow D \rightarrow F \rightarrow E \rightarrow C \rightarrow B \rightarrow A$
$A \rightarrow F \rightarrow D \rightarrow E \rightarrow C \rightarrow B \rightarrow A$
$A \rightarrow F \rightarrow E \rightarrow D \rightarrow C \rightarrow B \rightarrow A$

51. 252 miles

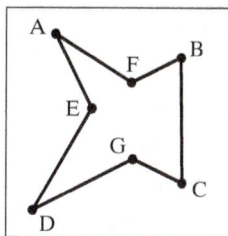

53. $A \rightarrow F \rightarrow G \rightarrow R \rightarrow S \rightarrow T \rightarrow U \rightarrow Q \rightarrow P \rightarrow N \rightarrow$
$M \rightarrow L \rightarrow K \rightarrow J \rightarrow I \rightarrow H \rightarrow B \rightarrow C \rightarrow D \rightarrow E \rightarrow A$

55. (a) graphs (2) and (4)   **(b)** graphs (2) and (4)
(c) No. Graph (1) provides a counterexample.
(d) No. If $n < 3$, the graph will have no circuits at all.
(e) The degree of each vertex in a complete graph with n vertices is $(n - 1)$. If $n \geq 3$, then $(n - 1) > \frac{n}{2}$. So we can conclude from Dirac's theorem that the graph has a Hamilton circuit.

14.4 Exercises *(pages 825–829)*
1. tree   **3.** No, because it is not connected.   **5.** tree
7. No, because it has a circuit.   **9.** It is not possible, because the graph has a circuit.   **11.** tree
13. not necessarily a tree   **15.** true
17. false

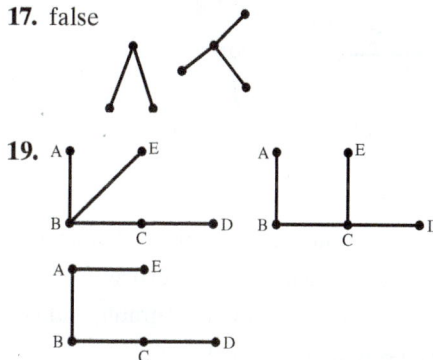

19.

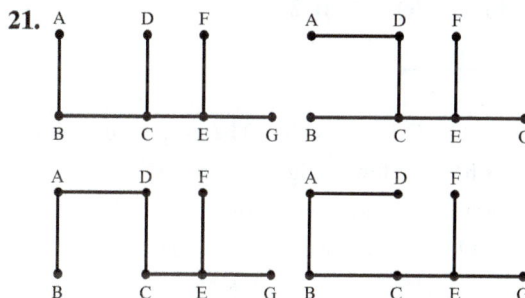

21.

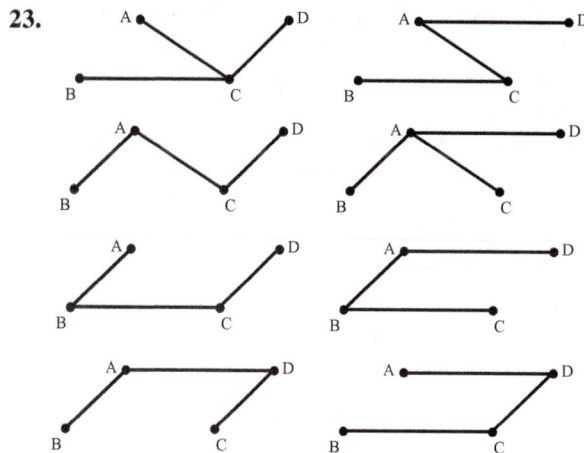

23.

25. 20   **27.** If a connected graph has circuits, none of which have common edges, then the number of spanning trees for the graph is the product of the numbers of edges in all the circuits.

29. Total weight is 51.

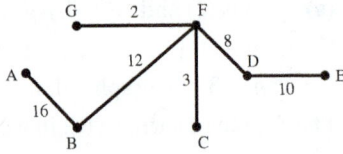

31. Total weight is 66.

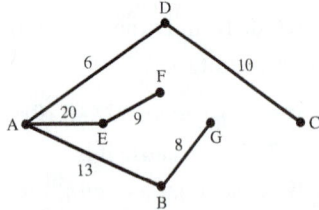

33. Total length to be covered is 140 ft.

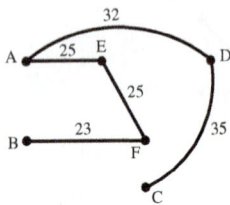

35. 33   **37.** 62   **39.** Different spanning trees must have the same number of edges. The number of vertices in the tree is the number of vertices in the original graph, and the number of edges has to be one less than this.

41. (a) 9   **(b)** 18   **(c)** 0   **(d)** 2

(e)

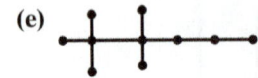

43. 22 cables   **45.** This is possible. The graph must be a tree because it has one fewer edge(s) than vertices.

47. This is possible. The graph cannot be a tree, because it would have at least as many edges as vertices.

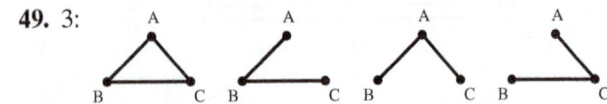

49. 3:

51. 125

53. 3 nonisomorphic trees:

55. 11 nonisomorphic trees:

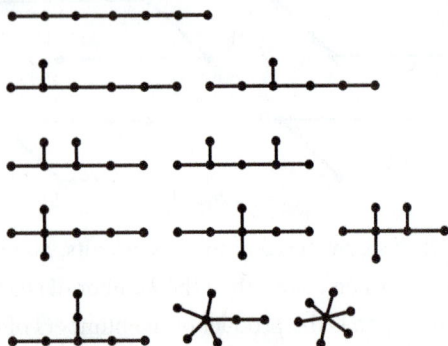

Chapter 14 Test *(pages 834–836)*

1. 7   **2.** 20   **3.** 10   **4. (a)** No, because edge AB is used twice.   **(b)** yes   **(c)** No, because there is no edge from C to D.   **5. (a)** yes   **(b)** No, because (for example) there is no edge from B to C.   **(c)** yes

6. For example:

7. 13 edges

8. The graphs are isomorphic. Corresponding edges should be the same color. AB should match AB, etc.

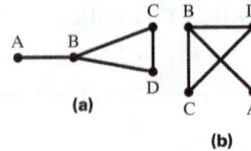

9.

The graph is connected. Tina knows the greatest number of other guests.

10. 28 games   **11.** Yes, because there is an edge from each vertex to each of the remaining 6 vertices.

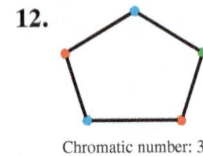

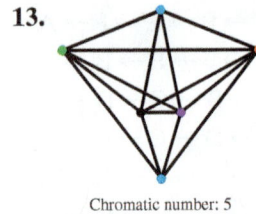

12.

Chromatic number: 3

13.

Chromatic number: 5

14.

The chromatic number is 3, so three separate exam times are needed. Exams could be at the same times as follows: Geography, Biology, and English; History and Chemistry; Mathematics and Psychology.

15. (a) No, because it does not use all the edges.
(b) No, because it is not a circuit (for example, there is no edge from B to C).   **(c)** yes   **16.** No, because some vertices have odd degree.   **17.** Yes, because all vertices have even degree.   **18.** No, because two of the rooms have odd numbers of doors.

19. F → B → E → D → B → C → D → K → B → A → H → G → F → A → G → J → F

20. (a) No, because it does not visit all vertices.
(b) No, because it is not a circuit (for example, there is no edge from B to C).   **(c)** No, because it visits some vertices twice before returning to the starting vertex.

21. $F \to G \to H \to I \to E \to F$ $F \to G \to H \to E \to I \to F$

$F \to G \to I \to H \to E \to F$ $F \to G \to I \to E \to H \to F$

$F \to G \to E \to H \to I \to F$ $F \to G \to E \to I \to H \to F$

There are 6 such Hamilton circuits.

22. $P \to Q \to S \to R \to P$; Total weight is 27.

23. $A \to E \to D \to C \to F \to B \to A$; Total weight is 11.85.   **24.** 24!   **25.** Hamilton circuit

26. Any three of these:

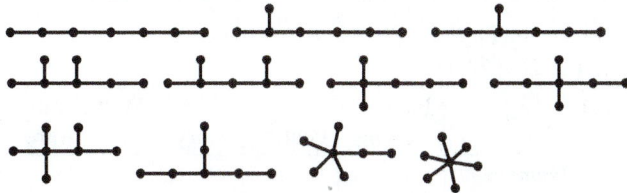

27. false   **28.** true   **29.** true

30. There are 4 spanning trees.

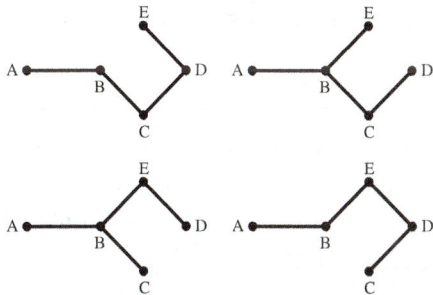

31. Weight is 24.   **32.** 49

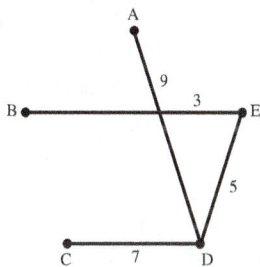

CHAPTER 15 VOTING AND APPORTIONMENT

15.1 Exercises *(pages 848–850)*

1. (a) $4! = 24$

(b)

Number of Voters	Ranking
3	$b > c > a > d$
2	$a > d > c > b$
1	$c > d > b > a$
1	$d > c > b > a$
1	$d > a > b > c$
2	$c > a > d > b$
1	$a > c > d > b$
2	$a > b > c > d$

(c) Australian shepherd

3. (a) c, with 3 (pairwise) points   **(b)** a, with 24 (Borda) points   **(c)** c   **5.** $5! = 120$; $7! = 5040$ **7.** For any value of $n \geq 5$, there are at least 120 possible rankings of the candidates. This means a voter must select one ranking from a huge number of possibilities. It also means the mechanics of any election method, except the plurality method, are difficult to manage.   **9.** $_6C_2 = 15$; $_8C_2 = 28$

11. (a) b, with 6 first-place votes   **(b)** c, with 2 (pairwise) points   **(c)** b, with 16 (Borda) points   **(d)** c

13. Logo a is selected by all the methods.

15. (a) e, with 8 first-place votes   **(b)** j, with 3 (pairwise) points   **(c)** j, with 40 (Borda) points   **(d)** h

17. (a) t, with 18 first-place votes   **(b)** h, with 4 (pairwise) points   **(c)** m, with 136 (Borda) points   **(d)** k

19. h beats e, 13 to 8.   **21.** c beats t, 37 to 18.

23. In a runoff between k and c, activity k is selected. Activity k faces activity t, and k is selected.   **25. (a)** 2   **(b)** c, with 7 pairwise points   **27. (a)** 7   **(b)** e, with 7 pairwise points   **29. (a)** 16   **(b)** c, with 16 Borda points

31. (a) 55   **(b)** c, with 55 Borda points   **33.** j

35. Answers will vary.   **37.** Answers will vary.

39. Answers will vary. One possible arrangement of the voters is 2, 4, 5, 7, 3.

15.2 Exercises *(pages 860–864)*

1. (a) Alternative a (6 of 11 first-place votes)   **(b)** b   **(c)** yes   **3. (a)** Alternative a (16 of 30 first-place votes)   **(b)** b   **(c)** yes   **5. (a)** c   **(b)** b   **(c)** b   **(d)** c   **(e)** plurality and Borda methods   **7. (a)** j   **(b)** e   **(c)** j   **(d)** h   **(e)** plurality and Hare methods   **9. (a)** h   **(b)** t   **(c)** m   **(d)** k   **(e)** all three methods   **11. (a)** a has 2 pairwise points, y and d have $1\frac{1}{2}$ pairwise points each, and n has 1 pairwise point.   **(b)** y has $2\frac{1}{2}$ pairwise points, a has 2 pairwise points, d has 1 pairwise point, and n has $\frac{1}{2}$ pairwise point, so y is selected.   **(c)** Yes; the rearranging voters moved a, the winner of the nonbinding election, to the top of their ranking, but y wins the official selection process.   **13. (a)** d drops out after one round of votes; b drops out after the second round; in the third vote, a is preferred to c by a margin of 11 to 6.   **(b)** d drops out after one round of votes; c drops out after the second round; in the third vote, b is preferred to a by a margin of 9 to 8.   **(c)** Yes; the rearranging voters moved a, the winner of the nonbinding election, to the top of their ranking, but b wins the official selection process.   **15. (a)** Candidate a has 75 first-place votes.   **(b)** Candidate c has 80 first-place votes.   **(c)** yes   **17.** No; however, the second pairwise comparison would result in a tie. **19. (a)** Round one eliminates b; a is preferred to c in the second round, by a margin of 22 to 12. **(b)** b **(c)** yes **21.** Answers will vary. **23.** Answers will vary. One possible voter profile is given.

Number of Voters	Ranking
10	$a > b > c > d > e > f$
9	$b > f > e > c > d > a$

Candidate a is a majority candidate and wins all of her or his pairwise comparisons by a margin of 10 to 9, earning 5 pairwise points. Candidate b wins all of her or his comparisons except the one with a, earning 4 pairwise points. **25.** Answers will vary. One possible profile is the voter profile for the animal shelter poster dog contest in **Exercise 2** of **Section 15.1. 27. (a)** Candidate a has 2 pairwise points. **(b)** Answers will vary. The 3 voters in the bottom row all switch to the ranking $a > z > x > y$. **29.** Delete Candidate c. **31.** Answers will vary. One possible profile is given.

Number of Voters	Ranking
21	$g > j > e$
12	$j > e > g$
8	$j > g > e$

33. Answers will vary.

15.3 Exercises (pages 876–879)

1. (a) 34,437 **(b)** 7 **3.** Virginia received 19 seats rather than 18. Delaware received only 1 seat rather than 2.
5. (a) 29,493; 123.40

(b)

State park	a	b	c	d	e
Number of trees	11	70	62	54	42

(c) $md = 122$

State park	a	b	c	d	e
Number of trees	11	70	62	54	42

(d) The traditionally rounded values of Q sum to 240, which is greater than the number of trees to be apportioned.

State park	a	b	c	d	e
Traditionally rounded Q	12	70	62	54	42

(e) The value of md for the Webster apportionment should be greater than $d = 123.40$, because greater divisors make lesser modified quotas with a lesser total sum.
(f) $md = 124$

State park	a	b	c	d	e
Number of trees	12	70	61	54	42

(g) The Huntington-Hill rounded values of Q sum to 240, which is greater than the number of trees to be apportioned.

(h) The value of md for the Huntington-Hill apportionment should be greater than $d = 123.40$, because greater divisors make lesser modified quotas with a lesser total sum.

(i)

State park	a	b	c	d	e
Number of trees	12	70	61	54	42

(j) The Hamilton and Jefferson apportionments are the same. The Webster and Huntington-Hill apportionments are the same, but they differ from the Hamilton and Jefferson apportionments.

7. (a) 269; 24.45

(b)

Course	Fiction Writing	Poetry	Short Story	Multicultural Literature
Number of sections	2	2	3	4

(c) $md = 20$

Course	Fiction Writing	Poetry	Short Story	Multicultural Literature
Number of sections	2	1	3	5

(d) The traditionally rounded values of Q sum to 10, which is less than the number of sections to be apportioned.

Course	Fiction Writing	Poetry	Short Story	Multicultural Literature
Traditionally rounded Q	2	1	3	4

(e) The value of md for the Webster apportionment should be less than $d = 24.45$ because lesser divisors make greater modified quotas with a greater total sum.
(f) $md = 23$

Course	Fiction Writing	Poetry	Short Story	Multicultural Literature
Number of sections	2	2	3	4

(g) The Huntington-Hill rounded values of Q sum to 11, which is equal to the number of sections to be apportioned.
(h) Let $md = d$, because the correct number of sections is already apportioned. **(i)** $md = d = 24.45$

Course	Fiction Writing	Poetry	Short Story	Multicultural Literature
Enrollment	2	2	3	4

(j) The Hamilton, Webster, and Huntington-Hill apportionments all agree. The Jefferson apportionment is different from the other three.
(k) The Jefferson method apportions 5 sections to Multicultural Literature, compared to only 4 sections apportioned by each of the other methods. With 5 sections, average class size would be 20 rather than 25.

(l) The Jefferson method apportions only 1 section to Poetry, for a class size of 35, whereas any of the other methods apportions 2 sections, allowing for two classes of, say, 17 and 18 students.

9. (a)

State	Abo	Boa	Cio	Dao	Effo	Foti
Number of seats	15	22	7	20	14	53

(b)

State	Abo	Boa	Cio	Dao	Effo	Foti
Number of seats	15	22	6	20	14	54

(c) The Huntington-Hill apportionment is the same as the Hamilton apportionment in part (a).

(d) The Hamilton, Webster, and Huntington-Hill methods all agree. The Jefferson method apportions one more seat to the largest state, Foti, and one fewer to the smallest state, Cio.

11. (a) 1721; 43.025

(b)

Hospital	A	B	C	D	E
Number of nurses	3	5	8	11	13

(c) $md = 40$

Hospital	A	B	C	D	E
Number of nurses	3	5	8	11	13

(d) The traditionally rounded values of Q sum to 41, which is greater than the number of nurses to be apportioned.

Hospital	A	B	C	D	E
Traditionally rounded Q	3	6	8	11	13

(e) The value of md for the Webster apportionment should be greater than $d = 43.025$, because greater divisors make lesser modified quotas with a lesser total sum.

(f) $md = 43.1$

Hospital	A	B	C	D	E
Number of nurses	3	5	8	11	13

(g) The Huntington-Hill method also agrees with the Hamilton method in part (b).

(h) In this case, all four methods agree.

13. $\sqrt{5 \cdot 6} = 5.477$ **15.** $\sqrt{32 \cdot 33} = 32.496$

17. Answers will vary. One possible ridership profile is given.

Bus route	a	b	c	d	e	Total
Number of riders	131	140	303	178	197	949

19. Answers will vary. One possible population profile is given.

State	a	b	c	d	e	Total
Population	50	230	280	320	120	1000

21. (a)

Hospital	A	B	C	D	E
Number of nurses	3	6	8	10	13

(b) The Hamilton, Jefferson, Webster, and Huntington-Hill methods all agreed, but the Adams method gives a different apportionment.

23. Answers will vary.

15.4 Exercises *(pages 888–890)*

1. $md = 595$

State	a	b	c	d
Q rounded down/up	28/29	12/13	**81/82**	9/10
Number of seats	28	12	**83**	9

3. $md = 48.4$

State	a	b	c	d	e
Q rounded down/up	52/53	30/31	**164/165**	19/20	22/23
Number of seats	53	30	**166**	19	22

5.

State	a	b	c	d
Number of seats if $n = 204$	**35**	74	42	53
Number of seats if $n = 205$	**34**	75	43	53

7.

State	a	b	c	d	e
Number of seats if $n = 126$	**10**	29	34	28	25
Number of seats if $n = 127$	**9**	30	34	29	25

9.

State	a	b	c
Initial number of seats	**1**	4	6
Percent growth	**10.91%**	18.40%	13.16%
Revised number of seats	**2**	4	5

11.

State	a	b	c
Initial number of seats	6	**5**	**2**
Percent growth	4.84%	**1.63%**	**1.45%**
Revised number of seats	6	**4**	**3**

13. Two additional seats are added for the second apportionment.

State	Original State a	Original State b	New State c
Initial number of seats	20	55	*****
Revised number of seats	21	54	2

15. Seven additional seats are added for the second apportionment.

State	Original State a	Original State b	New State c
Initial number of seats	36	47	*****
Revised number of seats	37	46	7

17. (a) The Huntington-Hill method works with the standard divisor $d = 1000$, and no violation occurs.
(b) The Adams method works with $md = 1025$ but violates the quota rule. State a, with standard quota $Q = 86.875$, should receive 86 or 87 seats but instead receives 85, which is too few.   19. (a) The Huntington-Hill method works with $md = 204$, and no violation occurs.
(b) The Adams method works with $md = 208$ but violates the quota rule. State d, with standard quota $Q = 118.250$, should receive 118 or 119 seats but instead receives 117, which is too few.   21. The new states paradox does not occur if the new population is 531, but it does occur if the new population is 532.   23. Answers will vary.
25. Answers will vary.

Chapter 15 Test (pages 895–896)
1. $7! = 5040$ 2. $_{10}C_2 = 45$ 3. Answers will vary.
4. Answers will vary.   5. Answers will vary.
6. Answers will vary.   7. Candidate a

8. Candidate a   9. 93   10. Candidate b
11. Yes, a is a majority candidate, but b wins the election.
12. Yes, a is a Condorcet candidate, but b wins the election.
13. No; b won initially, and b still wins.
14. No; b won initially, and b still wins.
15. Candidate c   16. Candidate s
17. The pairwise comparison method can violate the monotonicity criterion.   18. Candidate a
19. Candidate b   20. The plurality method can violate the irrelevant alternatives criterion.   21. The Balinski and Young Impossibility Theorem says that any apportionment method devised will either violate the quota rule or permit the possibility of a paradoxical apportionment (the Alabama paradox, the population paradox, or the new states paradox).
22. 354.52   23. a: 6.64; b: 12.69; c: 15.79; d: 64.88
24. (a) a: 6; b: 12; c: 15; d: 64
(b) a: 6; b: 12; c: 15; d: 64
(c) a: 7; b: 13; c: 16; d: 65   (d) a: 7; b: 13; c: 16; d: 65
25. a: 6; b: 13; c: 16; d: 65
26. (a) less than   (b) a: 6; b: 12; c: 16; d: 66
27. (a) greater than   (b) a: 7; b: 13; c: 16; d: 64
28. (a) greater than   (b) a: 7; b: 13; c: 16; d: 64
29. Answers will vary.   30. Answers will vary.
31. Answers will vary.   32. Answers will vary.

CREDITS

CHAPTER 11:   596 Exercises 65, 66, 75, 76, 77, and 78: From the monthly calendar of *Mathematics Teacher* by John Grant McLoughlin. Copyright © 2002 by National Council of Teachers of Mathematics. Used by permission of National Council of Teachers of Mathematics.   **604** Exercise 47: From the monthly calendar of *Mathematics Teacher* by John Grant McLoughlin. Copyright © 2002 by National Council of Teachers of Mathematics. Used by permission of National Council of Teachers of Mathematics.   **605** Exercises 49 and 50: From the monthly calendar of *Mathematics Teacher* by John Grant McLoughlin. Copyright © 2002 by National Council of Teachers of Mathematics. Used by permission of National Council of Teachers of Mathematics.   **615** Exercise 58: From the monthly calendar of *Mathematics Teacher*. Copyright © by National Council of Teachers of Mathematics. Used by permission of National Council of Teachers of Mathematics.   **617** Exercises 93 and 94: From the monthly calendar of *Mathematics Teacher*. Copyright © by National Council of Teachers of Mathematics. Used by permission of National Council of Teachers of Mathematics.   **622** Exercise 43: From the monthly calendar of *Mathematics Teacher*. Copyright © by National Council of Teachers of Mathematics. Used by permission of National Council of Teachers of Mathematics.   **627** Table 15: From *The Mathematics Teacher*, March 1967. Copyright © by National Council of Teachers of Mathematics. Used by permission of National Council of Teachers of Mathematics.   **637** Buffon's Needle Problem: Excerpt from *The History of Mathematics: An Introduction* by David Burton. Published by William C. Brown Publishers, © 1994.

CHAPTER 12:   667 Exercises 9 and 10: From the monthly calendar of *Mathematics Teacher* by John Grant McLoughlin. Copyright © 2002 by National Council of Teachers of Mathematics. Used by permission of National Council of Teachers of Mathematics; Exercise 16 (table): From "100 U.S. Corporations with Largest Revenues, 2012" by Sarah Janssen from *The World Almanac and Book of Facts*. Published by Simon and Schuster, © 2014.   **668** Exercises 33–34 (table): From "Federal Receipts, Outlays, and Surpluses or Deficits, 1901–2014" by Sarah Janssen from *The World Almanac and Book of Facts*. Published by Simon and Schuster, © 2014; Exercises 35–36 (table): From "World Cell Phone Use by Nation" by Sarah Janssen from *The World Almanac and Book of Facts*. Published by Simon and Schuster, © 2014.

CHAPTER 13:   710 Financial calculator screenshot: Used by permission of Bishinew Inc.   **725** Financial calculator screenshot: Used by permission of Bishinew Inc.   **736** Financial calculator screenshot: Used by permission of Bishinew Inc.   **750** Stock Quote screenshot: From Google Inc. Used by permission of Google Inc.

CHAPTER 14:   795 Margin note: Excerpt from "Readings from Scientific American: Mathematics in the Modern World" from *Scientific American*. Published by Graylock Press, © 1968.   **798** Margin note: Tshokwe sand tracing (sona) of leopard skin from *Geometry from Africa—Mathematical and Educational Explorations* by Paul Gerdes. Copyright © 1999 by Mathematical Association of America.

Photo Credits
COVER: Sergey Nivens/Shutterstock

FRONT MATTER: **xix** Sturti/Getty Images **xxii** (top) Courtesy of Vern E. Heeren, (middle) Callie J. Daniels, (bottom) Courtesy of Christopher Heeren

CHAPTER 1:   1 Courtesy of Terry Krieger **4** RTimages/Fotolia **6** Lakov Kalinin/Shutterstock **7** Altocumulus/Fotolia **13** Miket/Fotolia **15** Pearson Education **16** Fotofermer/Fotolia **20** (top) AP Images, (bottom) Pearson Education **21** Book shot of "We All Use Math Everyday" used by permission of National Council of Teachers of Mathematics **22** Pearson Education **23** Hunta/Fotolia **26** Bikeriderlondon/Shutterstock **28** Andrii Muzyka/Shutterstock **29** (left) Jupiterimages/Photos.com/Thinkstock, (right) Studio306fotolia/Fotolia **31** Stephen Coburn/Fotolia **32** Adrian825/iStock/360/Getty Images **33** (top) Beth Anderson/Pearson Education, (bottom) Everett Collection **34** (top) Zuma Press, Inc./Alamy, (bottom) April Saul/KRT/Newscom **35** Poznyakov/Shutterstock **36** Bertys30/Fotolia **37** Elena Schweitzer/Fotolia **40** (left) R. Mackay Photography, LLC/Shutterstock, (right) Photos.com/Thinkstock

CHAPTER 2:   47 Wavebreakmedia/Shutterstock **48** Pearson Education **51** Beth Anderson/Pearson Education **59** Beth Anderson/Pearson Education **60** Beth Anderson/Pearson Education **69** OtnaYdur/Shutterstock **73** Viki2win/Shutterstock **75** Everett Collection **76** Pictorial Press Ltd/Alamy **77** Christopher Black/Everett Collection Inc./Alamy **78** MC2 Eric C. Tretter/Defenseimagery.mil

CHAPTER 3:   83 Milanmarkovic78/Fotolia **84** Pearson Education **85** Pearson Education **86** Erich Lessing/Art Resource **89** Monkey Business Images/Shutterstock **94** (top) Pearson Education, (bottom) Dover Publications **95** Pearson Education **97** Pearson Education **101** Rich Legg/Getty Images **102** (top) Pearson Education, (bottom) Courtesy of John Hornsby **108** Denys Prykhodov/Fotolia **109** (top) urosr/Fotolia, (bottom) Courtesy of Christopher Heeren **110** (top) Olena Zaskochenko/Shutterstock, (bottom) Demarfa/Fotolia **112** Pearson Education **113** Pearson Education **114** Pictorial Press Ltd/Alamy **115** Pearson Education **117** Pearson Education **118** Pearson Education **121** Steve Liss/The LIFE Images Collection/Getty Images **126** Mary Evans/Python Pictures/Ronald Grant/Everett Collection **128** (top) Wonderstock/Alamy, (bottom) Monkey Business/Fotolia **131** (left) Michael Ciranni/Shutterstock, (right) Aliagreen/Fotolia

CHAPTER 4:   139 Dotshock/Shutterstock **140** DK Images **141** (both) DK Images **142** Kmit/Fotolia **143** Mediagram/Fotolia **146** Gary Ombler/DK Images **153** Canada [Addison Wesley Canada]/Pearson Education **154** The Art Gallery Collection/Alamy **155** Trinity College, Cambridge **156** (top) Maxx-Studio/Shutterstock, (bottom) Payless Images/Shutterstock **157** (top) Pearson Education, (bottom) Photo Researchers/Science Source **162** Photosani/Shutterstock **163** Photos.com/Thinkstock **167** Courtesy of John Hornsby **168** Wavebreakmedia/Shutterstock

CHAPTER 5:   177 Belinda Images/SuperStock **180** Hugh C. Williams, University of Manitoba **186** (top) Pearson Education, (bottom) Courtesy of Charles D. Miller **189** Michael Sohn/AP Images **194** (top) Pearson Education, (bottom) TriStar Pictures/Courtesy of Everett Collection **195** (top) Pearson Education, (bottom) AF archive/Alamy **201** Pearson Education **204** ZealPhotography/Alamy **205** Tyler Olson/Fotolia **210** (left) Bloom images/Getty Images, (right) Photodisc/Getty Images **211** (top) Vvr/Fotolia, (middle left) Scala/Art Resource, New York, (middle right) PhotoDisc, (bottom) Everett Collection

CHAPTER 6:   219 NOAA **220** Courtesy of John Hornsby **227** Lisa S./Shutterstock **229** Tusharkoley/Shutterstock **232** Pearson Education **237** Diseñador/Fotolia **240** Bizoo_n/Fotolia **247** U.S. Mint **248** Diego Cervo/Shutterstock **249** EMG Network/Pearson Education **250** (top) Dbvirago/Fotolia, (bottom) Pearson Education **253** Jubal Harshaw/Shutterstock **254** KimPinPhotography/Shutterstock **255** (left) U.S. Mint, (right) VIPDesign/Fotolia **260** (top) WavebreakmediaMicro/Fotolia, (bottom) Courtesy of Charles D. Miller **262** Pearson Education **266** Bikeriderlondon//Shutterstock **272** (top) Gravicapa/Shutterstock, (bottom) Darko Zeljkovic/Shutterstock **273** Monkey Business/Fotolia **274** Bikeriderlondon/Shutterstock **275** (all) Courtesy of John Hornsby **276** AVAVA/Fotolia **277** Neil Speers/Shutterstock **279** (left) U.S. Mint, (right) Uwimages/Fotolia **280** Lisa F. Young/Fotolia **281** (top) Nakamasa/Shutterstock, (bottom) Courtesy of John Hornsby **282** (both) Courtesy of John Hornsby **290** Cheryl Casey/Shutterstock

CHAPTER 7:   291 Andresr/Shutterstock **292** Courtesy of John Hornsby **294** Darryl Broks/Shutterstock **295** cobalt88/Shutterstock **296** Pearson Education **297** Pearson Education **298** lfong/Shutterstock **299** Myotis/Shutterstock **300** chris32m/Fotolia **302** iofoto/Shutterstock **303** Thieury/Fotolia **306** Sergey Ivanov/Fotolia **307** Bikeriderlondon/Shutterstock **309** Dusan Jankovic/Shutterstock **313** Igor Dutina/Shutterstock **316** JackF/Fotolia **317** (top) Aaron DeLonay/USGS, (bottom) defpicture/Shutterstock **319** (top) Joggie Botma/Shutterstock, (bottom) David Lee/Shutterstock **321** Kokhanchikov/Fotolia **322** (left) Mega Pixel/Shutterstock, (right) chrupka/Shutterstock **323** (left) Steve Bower/Shutterstock, (right) Ssuaphotos/Shutterstock **324** omicron/Fotolia **325** Pearson Education **326** Courtesy of Charles D. Miller **327** Pearson Education **330** Lisa F. Young/Shutterstock **331** Courtesy of John Hornsby **332** ekkasit888/Fotolia **342** StockLite/Shutterstock **344** Artalis/Shutterstock **354** Pearson Education **355** Pearson Education **363** NASA **364** Nenov Brothers/Fotolia

CHAPTER 8:   365 Mark Bowden/Getty Images **368** Courtesy of Charles D. Miller **369** Courtesy of Charles D. Miller **370** johnfoto18/Shutterstock **372** Mila Supinskaya/Shutterstock **373** Courtesy of John Hornby **379** Courtesy of John Hornsby **386** Pearson Education **389** Just Keep Drawing/Shutterstock **392** Courtesy of John Hornsby **393** Amidala/Fotolia **396** Wavebreakmedia/Shutterstock **398** Creativemarc/Shutterstock **399** Sergey Nivens/Shutterstock **400** pwrmc/Shutterstock **403** NASA **404** PPL/Shutterstock **406** Pearson Education **408** Tristan3D/Fotolia **414** Rafael Ramirez Lee/Shutterstock **415** 501room/Shutterstock **416** Yankane/Shutterstock **418** (top) Tatty/Fotolia, (bottom) Ungor/Shutterstock **419** Bruce Rolff/Shutterstock **421** The Trustees of the British Museum/Art Resource, New York **426** (left) Ollirg/Shutterstock, (right) Tudor Photography/Pearson Education Ltd. **427** Louis Lopez/Alamy **430** Layland Masuda/Shutterstock **431** Nadia Zagainova/Shutterstock **436** NOAA **448** Duckeesue/Shutterstock

CHAPTER 9:   449 Bloomberg/Getty Images **450** Bettmann/Corbis **451** Oksana Perkins/Fotolia **452** Henryk Sadura/iStockphoto **455** Georgy Shafeev/Shutterstock **462** (top) Pearson Education, (bottom) Prentice Hall/Pearson Education **463** Everett Collection **469** DK Images **470** oksana.perkins/Shutterstock **471** Courtesy of John Hornsby **472** National Council of Teachers of Mathematics **475** Courtesy of Charles D. Miller **478** GeniusMinus/Fotolia **482** Kokyat Choong/Alamy **483** Pearson Education **484** Tbradford/iStock/360/Getty Images **485** (left) Courtesy of John Hornsby, (right) Pearson Education **487** Pearson Education **491** Courtesy of Christopher Heeren **496** Peter Arnold, Inc./Alamy **497** Bloomberg/Getty Images **498** Diseñador/Fotolia **499** (top) M.C. Escher, (bottom) Hayri Er/Getty Images **500** M.C. Escher **508** Pearson Education **509** Courtesy of Tom Lehrer **510** Khoo Si Lin/Shutterstock **511** Pearson Education **513** Courtesy of Charles D. Miller **518** (top) Sabina Louise Pierce/AP Images, (bottom) Archives of the Mathematisches Forschungsinstitut Oberwolfach **519** (top left, center, right) NASA

CHAPTER 10:   531 Richard Levine/Alamy **534** (both) Courtesy of John Hornsby **535** Zev Radovan/BibleLandPictures/Alamy **538** Courtesy of Terry Krieger **542** Pearson Education **544** Paul Whitfield/Rough Guides/Dorling Kindersley **545** John Foxx Collection/Imagestate/Assetlibrary **549** StillFX/Shutterstock **550** (left) Monkey Business/Fotolia, (right) Natali/Fotolia **551** Anthony Ladd/iStockphoto **552** Philip Date/Shutterstock/Assetlibrary **554** Paul Vinten/Fotolia **557** Beth Anderson/Pearson Education **558** OtnaYdur/Shutterstock/Assetlibrary **559** Blend Images/Shutterstock/Assetlibrary **562** (left) Alisonhancock/Fotolia, (right) Scott J. Ferrell/Congressional Quarterly/Alamy **563** (left) WavebreakmediaMicro/Fotolia, (right) Gennadiy Poznyakov/Fotolia **564** (left) Courtesy of John Hornsby, (top right) Naretev/Fotolia, (bottom right) Photos.com/Thinkstock **566** Courtesy of Charles D. Miller **567** (top) Courtesy of Charles D. Miller, (bottom) Ronstik/Fotolia **572** Alvin Teo/Fotolia **574** (both) Matt Trommer/Shutterstock/Assetlibrary **576** (left) Andreykr/Fotolia, (right) Monkey Business/Fotolia **578** Steve Lovegrove/Shutterstock/Assetlibrary

CHAPTER 11:   583 Iofoto/Shutterstock **584** Courtesy of John Hornsby **585** kornienko/Fotolia **587** Library of Congress Prints and Photographs Division [LC-USZ62-75479] **590** (top) Pearson Education, (bottom) jpbadger/Fotolia **592** Pearson Education **593** Courtesy of John Hornsby **594** (top left) LianeM/Shutterstock, (bottom left) Mushy/Fotolia, (right) apops/Fotolia **595** Janice Haney Carr/CDC **596** Monkey Business/Fotolia **598** Pearson Education **599** Pearson Education **602** Wavebreakingmedia/Shutterstock **603** iChip/Fotolia **604** Courtesy of Christopher Heeren **606** Zuma Press, Inc./Alamy **607** Walter G. Arce/Shutterstock **609** Beth Anderson/Pearson Education **610** Christos Georghiou/Shutterstock **611** (left) Giovanni Benintende/Shutterstock, (right) Susan Walsh/AP Images **613** (left) NOAA, (top right) Marques/Shutterstock, (bottom right) spotmatikphoto/Fotolia **614** Denis Kadackii/Fotolia **615** Eisenhans/Fotolia **616** (left) SeanPavonePhoto/Fotolia, (right) icholakov/Fotolia **617** 2xSamara.com/Shutterstock **618** Pearson Education **621** (top) stefanolunardi/Shutterstock, (bottom) Courtesy of John Hornsby **622** Image 100/Corbis/Glow Images/Asset library **624** Christopher Bradshaw/Fotolia

INDEX OF APPLICATIONS

GENERAL INTEREST

GEOMETRY

INDEX